# Reactions at Supports and Connections for a Three-Dimensional Structure

Ball

Frictionless surface

Force with known line of action (one unknown)

Cable

Force with known line of action (one unknown)

Roller on rough surface

Wheel on rail

Two force components

Rough surface

Ball and socket

Three force components

Universal joint

Three force components and one couple

Fixed support

Three force components and three couples

Hinge and bearing supporting radial load only

Two force components (and two couples)

Pin and bracket

Hinge and bearing supporting axial thrust and radial load

Three force components (and two couples)

# VECTOR MECHANICS FOR ENGINEERS

## Dynamics
### (SI Units)

**FERDINAND P. BEER**
Late of Lehigh University

**E. RUSSELL JOHNSTON, JR.**
University of Connecticut

**WILLIAM E. CLAUSEN**
The Ohio State University

With the collaboration of
**Phillip J. Cornwell**
Rose-Hulman Institute of Technology

Adapted by
**Nilanjan Malik**
Dept. of Mechanical Engineering
Institute of Technology
Banaras Hindu University

McGraw Hill

Singapore • Boston • Burr Ridge, IL • Dubuque, IA • Madison, WI • New York • San Francisco
St. Louis • Bangkok • Kuala Lumpur • Lisbon • London • Madrid • Mexico City
Milan • Montreal • New Delhi • Seoul • Sydney • Taipei • Toronto

The McGraw·Hill Companies

**VECTOR MECHANICS FOR ENGINEERS: DYNAMICS**
**Eighth Edition in SI Units**

Exclusive rights by McGraw-Hill Education (Asia), for manufacture and export. This book cannot be re-exported from the country to which it is sold by McGraw-Hill. The International Edition is not available in North America.

The credits section for this book begins on page *pc-1* and is considered an extension of the copyright page.

10  09  08  07  06  05  04  03  02  01
20  09
CTP   MPM

When ordering this title, use ISBN: 978-007-126872-1 or MHID: 007-126872-3

Printed in Singapore

# About the Authors

As publishers of the books by Ferd Beer and Russ Johnston we are often asked how they happened to write their books together with one of them at Lehigh and the other at the University of Connecticut.

The answer to this question is simple. Russ Johnston's first teaching appointment was in the Department of Civil Engineering and Mechanics at Lehigh University. There he met Ferd Beer, who had joined that department two years earlier and was in charge of the courses in mechanics.

Ferd was delighted to discover that the young man who had been hired chiefly to teach graduate structural engineering courses was not only willing but eager to help him reorganize the mechanics courses. Both believed that these courses should be taught from a few basic principles and that the various concepts involved would be best understood and remembered by the students if they were presented to them in a graphic way. Together they wrote lecture notes in statics and dynamics, to which they later added problems they felt would appeal to future engineers, and soon they produced the manuscript of the first edition of *Mechanics for Engineers* that was published in June 1956.

The second edition of *Mechanics for Engineers* and the first edition of *Vector Mechanics for Engineers* found Russ Johnston at Worcester Polytechnic Institute and the next editions at the University of Connecticut. In the meantime, both Ferd and Russ assumed administrative responsibilities in their departments, and both were involved in research, consulting, and supervising graduate students—Ferd in the area of stochastic processes and random vibrations and Russ in the area of elastic stability and structural analysis and design. However, their interest in improving the teaching of the basic mechanics courses had not subsided, and they both taught sections of these courses as they kept revising their texts and began writing the manuscript of the first edition of their *Mechanics of Materials* text.

Their collaboration spanned more than half a century and many successful revisions of all of their textbooks, and Ferd's and Russ's contributions to engineering education have earned them a number of honors and awards. They were presented with the Western Electric Fund Award for excellence in the instruction of engineering students by their respective regional sections of the American Society for Engineering Education, and they both received the Distinguished Educator Award from the Mechanics Division of the same society. Starting in 2001, the New Mechanics Educator Award of the Mechanics Division has been named in honor of the Beer and Johnston author team.

**Ferdinand P. Beer.** Born in France and educated in France and Switzerland, Ferd received an M.S. degree from the Sorbonne and an Sc.D. degree in theoretical mechanics from the University of Geneva. He came to the United States after serving in the French army during the early part of World War II and taught for four years at Williams College in the Williams-MIT joint arts and engineering program. Following his service at Williams College, Ferd joined the faculty of Lehigh University where he taught for thirty-seven years. He held several positions, including University Distinguished Professor and chairman of the Department of Mechanical Engineering and Mechanics, and in 1995 Ferd was awarded an honorary Doctor of Engineering degree by Lehigh University.

**E. Russell Johnston, Jr.** Born in Philadelphia, Russ holds a B.S. degree in civil engineering from the University of Delaware and an Sc.D. degree in the field of structural engineering from the Massachusetts Institute of Technology. He taught at Lehigh University and Worcester Polytechnic Institute before joining the faculty of the University of Connecticut where he held the position of Chairman of the Civil Engineering Department and taught for twenty-six years. In 1991 Russ received the Outstanding Civil Engineer Award from the Connecticut Section of the American Society of Civil Engineers.

**William E. Clausen.** Bill holds a B.S. degree in engineering mechanics from Lehigh University and M.S. and Ph.D. degrees in engineering mechanics from the Ohio State University. Bill is a registered professional engineer specializing in structural dynamics and vibration measurements. He taught for thirty years and served as vice chairman in the Department of Engineering Mechanics at the Ohio State University and has also taught in the Department of Mechanical Engineering.

**Phillip J. Cornwell.** Phil holds a B.S. degree in mechanical engineering from Texas Tech University and M.A. and Ph.D. degrees in mechanical and aerospace engineering from Princeton University. He is currently a professor of mechanical engineering at Rose-Hulman Institute of Technology where he has taught since 1989. His present interests include structural dynamics, structural health monitoring, and undergraduate engineering education. Since 1995, Phil has spent his summers working at Los Alamos National Laboratory where he is a mentor in the Los Alamos Dynamics Summer School and does research in the area of structural health monitoring. Phil received an SAE Ralph R. Teetor Educational Award in 1992, the Dean's Outstanding Scholar Award at Rose-Hulman in 2000, and the Board of Trustees Outstanding Scholar Award at Rose-Hulman in 2001.

# Contents

**Preface** xiv

**List of Symbols** xxi

## 11
### KINEMATICS OF PARTICLES
601

11.1 Introduction to Dynamics 602

**Rectilinear Motion of Particles** 603
11.2 Position, Velocity, and Acceleration 603
11.3 Determination of the Motion of a Particle 607
11.4 Uniform Rectilinear Motion 616
11.5 Uniformly Accelerated Rectilinear Motion 617
11.6 Motion of Several Particles 618
*11.7 Graphical Solution of Rectilinear-Motion Problems 630
*11.8 Other Graphical Methods 631

**Curvilinear Motion of Particles** 641
11.9 Position Vector, Velocity, and Acceleration 641
11.10 Derivatives of Vector Functions 643
11.11 Rectangular Components of Velocity and Acceleration 645
11.12 Motion Relative to a Frame in Translation 646
11.13 Tangential and Normal Components 663
11.14 Radial and Transverse Components 666

**Review and Summary for Chapter 11** 680
**Review Problems** 684
**Computer Problems** 687

## 12
### KINETICS OF PARTICLES: NEWTON'S SECOND LAW
691

12.1 Introduction 692
12.2 Newton's Second Law of Motion 693
12.3 Linear Momentum of a Particle. Rate of Change of Linear Momentum 694

12.4   Systems of Units   695
12.5   Equations of Motion   697
12.6   Dynamic Equilibrium   699
12.7   Angular Momentum of a Particle. Rate of Change of Angular Momentum   718
12.8   Equations of Motion in Terms of Radial and Transverse Components   719
12.9   Motion under a Central Force. Conservation of Angular Momentum   720
12.10  Newton's Law of Gravitation   721
*12.11 Trajectory of a Particle under a Central Force   731
*12.12 Application to Space Mechanics   732
*12.13 Kepler's Laws of Planetary Motion   735

   Review and Summary for Chapter 12   **744**
   Review Problems   **784**
   Computer Problems   **751**

# 13
## KINETICS OF PARTICLES:
## ENERGY AND MOMENTUM METHODS
### 755

13.1   Introduction   756
13.2   Work of a Force   756
13.3   Kinetic Energy of a Particle. Principle of Work and Energy   760
13.4   Applications of the Principle of Work and Energy   762
13.5   Power and Efficiency   763
13.6   Potential Energy   781
*13.7  Conservative Forces   783
13.8   Conservation of Energy   784
13.9   Motion under a Conservative Central Force. Application to Space Mechanics   786
13.10  Principle of Impulse and Momentum   805
13.11  Impulsive Motion   808
13.12  Impact   820
13.13  Direct Central Impact   820
13.14  Oblique Central Impact   823
13.15  Problems Involving Energy and Momentum   826

   Review and Summary for Chapter 13   **842**
   Review Problems   **848**
   Computer Problems   **851**

# 14
## SYSTEMS OF PARTICLES
### 855

14.1   Introduction   856
14.2   Application of Newton's Laws to the Motion of a System of Particles. Effective Forces   856
14.3   Linear and Angular Momentum of a System of Particles   859
14.4   Motion of the Mass Center of a System of Particles   860
14.5   Angular Momentum of a System of Particles about Its Mass Center   862

14.6  Conservation of Momentum for a System of Particles  864
14.7  Kinetic Energy of a System of Particles  873
14.8  Work-Energy Principle. Conservation of Energy for a System
      of Particles  875
14.9  Principle of Impulse and Momentum for a System of Particles
      875
*14.10 Variable Systems of Particles  886
*14.11 Steady Stream of Particles  886
*14.12 Systems Gaining or Losing Mass  889

**Review and Summary for Chapter 14  904**
**Review Problems  908**
**Computer Problems  911**

# 15
## KINEMATICS OF RIGID BODIES
### 915

15.1  Introduction  916
15.2  Translation  918
15.3  Rotation about a Fixed Axis  919
15.4  Equations Defining the Rotation of a Rigid Body
      about a Fixed Axis  922
15.5  General Plane Motion  932
15.6  Absolute and Relative Velocity in Plane Motion  934
15.7  Instantaneous Center of Rotation in Plane Motion  945
15.8  Absolute and Relative Acceleration in Plane Motion  956
*15.9  Analysis of Plane Motion in Terms of a Parameter  958
15.10 Rate of Change of a Vector with Respect to a Rotating
      Frame  971
15.11 Plane Motion of a Particle Relative to a Rotating Frame. Coriolis
      Acceleration  973
*15.12 Motion about a Fixed Point  984
*15.13 General Motion  987
*15.14 Three-Dimensional Motion of a Particle Relative to a Rotating
      Frame. Coriolis Acceleration  998
*15.15 Frame of Reference in General Motion  999

**Review and Summary for Chapter 15  1011**
**Review Problems  1018**
**Computer Problems  1021**

# 16
## PLANE MOTION OF RIGID BODIES: FORCES AND
## ACCELERATIONS
### 1025

16.1  Introduction  1026
16.2  Equations of Motion for a Rigid Body  1027
16.3  Angular Momentum of a Rigid Body in Plane Motion  1028
16.4  Plane Motion of a Rigid Body. d'Alembert's Principle  1029
*16.5  A Remark on the Axioms of the Mechanics of Rigid
      Bodies  1030
16.6  Solution of Problems Involving the Motion of a Rigid Body  1031
16.7  Systems of Rigid Bodies  1032

16.8   Constrained Plane Motion   1051

**Review and Summary for Chapter 16   1073**
**Review Problems   1075**
**Computer Problems   1078**

# 17
## PLANE MOTION OF RIGID BODIES: ENERGY AND MOMENTUM METHODS
### 1081

17.1   Introduction   1082
17.2   Principle of Work and Energy for a Rigid Body   1082
17.3   Work of Forces Acting on a Rigid Body   1083
17.4   Kinetic Energy of a Rigid Body in Plane Motion   1084
17.5   Systems of Rigid Bodies   1085
17.6   Conservation of Energy   1086
17.7   Power   1087
17.8   Principle of Impulse and Momentum for the Plane Motion of a Rigid Body   1104
17.9   Systems of Rigid Bodies   1107
17.10  Conservation of Angular Momentum   1107
17.11  Impulsive Motion   1120
17.12  Eccentric Impact   1120

**Review and Summary for Chapter 17   1134**
**Review Problems   1138**
**Computer Problems   1141**

# 18
## KINETICS OF RIGID BODIES IN THREE DIMENSIONS
### 1145

*18.1   Introduction   1146
*18.2   Angular Momentum of a Rigid Body in Three Dimensions   1147
*18.3   Application of the Principle of Impulse and Momentum to the Three-Dimensional Motion of a Rigid Body   1151
*18.4   Kinetic Energy of a Rigid Body in Three Dimensions   1152
*18.5   Motion of a Rigid Body in Three Dimensions   1165
*18.6   Euler's Equations of Motion. Extension of d'Alembert's Principle to the Motion of a Rigid Body in Three Dimensions   1166
*18.7   Motion of a Rigid Body about a Fixed Point   1167
*18.8   Rotation of a Rigid Body about a Fixed Axis   1168
*18.9   Motion of a Gyroscope. Eulerian Angles   1183
*18.10  Steady Precession of a Gyroscope   1185
*18.11  Motion of an Axisymmetrical Body under No Force   1186

**Review and Summary for Chapter 18   1199**
**Review Problems   1204**
**Computer Problems   1208**

# 19
## MECHANICAL VIBRATIONS
### 1213

19.1  Introduction  1214

**Vibrations without Damping  1214**
19.2  Free Vibrations of Particles. Simple Harmonic Motion  1214
19.3  Simple Pendulum (Approximate Solution)  1218
*19.4  Simple Pendulum (Exact Solution)  1219
19.5  Free Vibrations of Rigid Bodies  1228
19.6  Application of the Principle of Conservation of Energy  1240
19.7  Forced Vibrations  1251

**Damped Vibrations  1261**
*19.8  Damped Free Vibrations  1261
*19.9  Damped Forced Vibrations  1264
*19.10 Electrical Analogues  1265

**Review and Summary for Chapter 19  1277**
**Review Problems  1282**
**Computer Problems  1285**

# Appendix
## FUNDAMENTALS OF ENGINEERING EXAMINATION
### app-1

**Photo Credits**  *pc-1*

**Index**  *ind-1*

**Answers to Problems**  *ans-1*

# Preface

## OBJECTIVES

The main objective of a first course in mechanics should be to develop in the engineering student the ability to analyze any problem in a simple and logical manner and to apply to its solution a few, well-understood, basic principles. This text is designed for the first courses in statics and dynamics offered in the sophomore or junior year, and it is hoped that this text, as well as the preceeding volume, *Vector Mechanics for Engineers: Statics*, will help the instructor achieve this goal.†

## GENERAL APPROACH

Vector analysis is introduced early in the text and is used throughout the presentation of statics and dynamics. This approach leads to more concise derivations of the fundamental principles of mechanics. It also results in simpler solutions of three-dimensional problems in statics and makes it possible to analyze many advanced problems in kinematics and kinetics, which could not be solved by scalar methods. The emphasis in this text, however, remains on the correct understanding of the principles of mechanics and on their application to the solution of engineering problems, and vector analysis is presented chiefly as a convenient tool.‡

**Practical Applications Are Introduced Early.** One of the characteristics of the approach used in this book is that mechanics of *particles* is clearly separated from the mechanics of *rigid bodies*. This approach makes it possible to consider simple practical applications at an early stage and to postpone the introduction of the more difficult concepts. In *Dynamics*, the basic concepts of force, mass, and

---

†This text and its companion title, *Vector Mechanics for Engineers: Statics*, is available in a single volume, *Vector Mechanics for Engineers: Statics and Dynamics*, eighth edition.

‡In a parallel text, *Mechanics for Engineers*, fourth edition, the use of vector algebra is limited to the addition and subtraction of vectors.

acceleration, of work and energy, and of impulse and momentum are introduced and first applied to problems involving only particles. Thus, students can familiarize themselves with the three basic methods used in dynamics and learn their respective advantages before facing the difficulties associated with the motion of rigid bodies.

**New Concepts Are Introduced in Simple Terms.** Since this text is designed for the first course in statics and dynamics, new concepts are presented in simple terms and every step is explained in detail. On the other hand, by discussing the broader aspects of the problems considered, and by stressing methods of general applicability, a definite maturity of approach is achieved. For example, the concept of potential energy is discussed in the general case of a conservative force. Also, the study of the plane motion of rigid bodies is designed to lead naturally to the study of their general motion in space. This is true in kinematics as well as in kinetics, where the principle of equivalence of external and effective forces is applied directly to the analysis of plane motion, thus facilitating the transition to the study of three-dimensional motion.

**Fundamental Principles Are Placed in the Context of Simple Applications.** The fact that mechanics is essentially a *deductive* science based on a few fundamental principles is stressed. Derivations have been presented in their logical sequence and with all the rigor warranted at this level. However, the learning process being largely *inductive*, simple applications are considered first. For example:

- The kinematics of particles (Chap. 11) precedes the kinematics of rigid bodies (Chap. 15).
- The fundamental principles of the kinetics rigid bodies are first applied to the solution of two-dimensional problems (Chaps. 16 and 17), which can be more easily visualized by the student, while three-dimensional problems are postponed until Chap. 18.

**The Presentation of the Principles of Kinetics is Unified.** The eighth edition of *Vector Mechanics for Engineers* retains the unified presentation of the principles of kinetics which characterized the previous seven editions. The concepts of linear and angular momentum are introduced in Chap. 12, so that Newton's second law of motion can be presented not only in its conventional form, $\mathbf{F} = m\mathbf{a}$, but also as a law relating, respectively, the sum of the forces acting on a particle and the sum of their moments to the rates of change of the linear and angular momentum of the particle. This makes possible an earlier introduction of the principle of conservation of angular momentum and a more meaningful discussion of the motion of particle under a central force (Sec. 12.9). More importantly, this approach can be readily extended to the study of the motion of a system of particles (Chap. 14) and leads to a more concise and unified treatment of the kinetics of rigid bodies in two and three dimensions (Chaps. 16 through 18).

**Free-Body Diagrams Are Used Both to Solve Equilibrium Problems and to Express the Equivalence of Force Systems.** Free-body diagrams are introduced early, and their importance is emphasized throughout the text. They are used not only to solve equilibrium problems but also to express the equivalence of two systems of forces or, more generally, of two systems of vectors. The advantage of this approach becomes apparent in the study of the dynamics of rigid bodies, where it is used to solve three-dimensional as well as two-dimensional problems. By placing the emphasis on "free-body-diagram equations" rather than on the standard algebraic equations of motion, a more intuitive and more complete understanding of the fundamental principles of dynamics can be achieved. This approach, which was first introduced in 1962 in the first edition of *Vector Mechanics for Engineers*, has now gained wide acceptance among mechanics teachers in this country. It is, therefore, used in preference to the method of dynamic equilibrium and to the equations of motion in the solution of all sample problems in this book.

**A Four-Color Presentation Uses Color to Distinguish Vectors.** Color has been used, not only to enhance the quality of the illustrations, but also to help students distinguish among the various types of vectors they will encounter. While there is no intention to "color code" this text, the same color is used in any given chapter to represent vectors of the same type. In *Dynamics*, for the chapters on kinetics, red is used for forces and couples, as well as for effective forces. Red is also used to represent impulses and momenta in free-body-diagram equations, while green is used for velocities, and blue for accelerations. In the two chapters on kinematics, which do not involve any forces, blue, green, and red are used, respectively, for displacements, velocities, and accelerations.

**Optional Sections Offer Advanced or Speciality Topics.** A large number of optional sections have been included. These sections are indicated by asterisks and thus are easily distinguished from those which form the core of the basic mechanics course. They may be omitted without prejudice to the understanding of the rest of the text.

The topics covered in the optional sections in *Dynamics* include graphical methods for the solution of rectilinear-motion problems, the trajectory of a particle under a central force, the deflection of fluid streams, problems involving jet and rocket propulsion, the kinematics and kinetics of rigid bodies in three dimensions, damped mechanical vibrations, and electrical analogues. These topics will be found of particular interest when dynamics is taught in the junior year.

The material presented in the text and most of the problems require no previous mathematical knowledge beyond algebra, trigonometry, and elementary calculus. However, special problems are included, which make use of a more advanced knowledge of calculus, and certain sections, such as Secs. 19.8 and 19.9 on damped vibrations, should be assigned only if students possess the proper mathematical background. In portions of the text using elementary calculus, a greater emphasis is placed on the correct understanding and application of the concepts of differentiation and integration than on the nimble manipulation of mathematical formulas. In this connection, it should be mentioned that the determination of the centroids of composite areas precedes the calculation of centroids by integration, thus making it possible to establish the concept of moment of area firmly before introducing the use of integration.

## CHAPTER ORGANIZATION AND PEDAGOGICAL FEATURES

**Chapter Introduction.** Each chapter begins with an introductory section setting the purpose and goals of the chapter and describing in simple terms the material to be covered and its application to the solution of engineering problems. Chapter outlines provide students with a preview of chapter topics.

**Chapter Lessons.** The body of the text is divided into units, each consisting of one or several theory sections, one or several sample problems, and a large number of problems to be assigned. Each unit corresponds to a well-defined topic and generally can be covered in one lesson. In a number of cases, however, the instructor will find it desirable to devote more than one lesson to a given topic. The *Instructor's and Solutions Manual* contains suggestions on the coverage of each lesson.

**Sample Problems.** The sample problems are set up in much the same form that students will use when solving the assigned problems. They thus serve the double purpose of amplifying the text and demonstrating the type of neat, orderly work that students should cultivate in their own solutions.

**Solving Problems on Your Own.** A section entitled *Solving Problems on Your Own* is included for each lesson, between the sample problems and the problems to be assigned. The purpose of these sections is to help students organize in their own minds the preceding theory of the text and the solution methods of the sample problems so that they can more successfully solve the homework problems. Also included in these sections are specific suggestions and strategies which will enable students to more efficiently attack any assigned problems.

**Homework Problem Sets.** Most of the problems are of a practical nature and should appeal to engineering students. They are primarily designed, however, to illustrate the material presented in the text and to help students understand the principles of mechanics. The problems are grouped according to the portions of material they illustrate and are arranged in order of increasing difficulty. Problems requiring special attention are indicated by asterisks. Answers to 70 percent of the problems are given at the end of the book. Problems for which the answers are given are set in straight type in the text, while problems for which no answer is given are set in italic.

**Chapter Review and Summary.** Each chapter ends with a review and summary of the material covered in that chapter. Marginal notes are used to help students organize their review work, and cross-references have been included to help them find the portions of material requiring their special attention.

**Review Problems.** A set of review problems is included at the end of each chapter. These problems provide students further opportunity to apply the most important concepts introduced in the chapter.

**Computer Problems.** Each chapter includes a set of problems designed to be solved with computational software. Many of these problems provide an introduction to the design process. In *Dynamics*, they may involve the determination of the motion of a particle under initial conditions, the kinematic or kinetic analysis of mechanisms in successive positions, or the numerical integration of various equations of motion. Developing the algorithm required to solve a given mechanics problem will benefit the students in two different ways: (1) it will help them gain a better understanding of the mechanics principles involved; (2) it will provide them with an opportunity to apply their computer skills to the solution of a meaningful engineering problem.

## SUPPLEMENTS

An extensive supplements package for both instructors and students is available with the text.

**Instructor's Solutions Manual.** An *Instructor's Solutions Manual* is available to instructors who are adopting this book for the course they are teaching. Instructors are encouraged to contact the local sales representatives in their respective countries.

**McGraw-Hill's ARIS—Assessment, Review and Instruction System.** ARIS is a complete homework and course management system for *Vector Mechanics for Engineers: Statics and Dynamics*. Instructors can create and share course materials and assignments with other instructors, edit questions and algorithms, import their own content, and create announcements and due dates for assignments. ARIS has automatic grading and reporting of easy-to-assign algorithmically generated homework, quizzes, and tests. Other resources available on ARIS include S.M.A.R.T. tutorials, a homework problem bank, Lecture PowerPoints, and images from the text. Visit www.mhhe.com/ beerjohnston8 for more information on the supplements available with this text.

**Hands-on Mechanics.** Hands-on Mechanics is a website designed for instructors who are interested in incorporating three-dimensional, hands-on teaching aids into their lectures. Developed through a partnership between the McGraw-Hill Engineering Team and the Department of Civil and Mechanical Engineering at the United States Military Academy at West Point, this website not only provides detailed instructions for how to build 3-D teaching tools using materials found in any lab or local hardware store but also provides a community where educators can share ideas, trade best practices, and submit their own demonstrations for posting on the site. Visit www.handsonmechanics.com.

## ACKNOWLEDGEMENTS

The authors wish to acknowledge the collaboration of David Mazurek and Phillip Cornwell to this eighth edition of *Vector Mechanics for Engineers* and thank them especially for their crucial role in making the extensive problem set revision possible.

A special thanks go to our colleagues who thoroughly checked the solutions and answers of all problems in this edition and then prepared the solutions for the accompanying *Instructor's Solution Manual:* Yohannes Ketema of University of Minnesota; Amy Mazurek of Williams Memorial Institute; David Oglesby of University of Missouri-Rolla; Gerald Rehkugler of Cornell University; Dean Updike of Lehigh University; and Daniel W. Yannitell of Louisiana State University.

We are pleased to recognize Dennis Ormond and Michael Haughey of Fine Line Illustrations of Farmingdale, New York, for the artful illustrations which contribute so much to the effectiveness of the text.

The authors thank the many companies that provided photographs for this edition. We also wish to recognize the determined efforts and patience of our photo researcher Sabina Dowell.

The authors also thank the members of the staff at McGraw-Hill for their support and dedication during the preparation of this new edition. We particularly wish to acknowledge the contributions of Sponsoring Editor Michael Hackett, Developmental Editor Katie White, and Senior Project Manager Kay Brimeyer.

The authors gratefully acknowledge the many helpful comments and suggestions offered by users of the previous editions of *Vector Mechanics for Engineers.*

E. Russell Johnston, Jr.
Elliot R. Eisenberg
William E. Clausen

# List of Symbols

| | |
|---|---|
| $\mathbf{a}$, $a$ | Acceleration |
| $a$ | Constant; radius; distance; semimajor axis of ellipse |
| $\bar{\mathbf{a}}$, $\bar{a}$ | Acceleration of mass center |
| $\mathbf{a}_{B/A}$ | Acceleration of $B$ relative to frame in translation with $A$ |
| $\mathbf{a}_{P/\mathcal{F}}$ | Acceleration of $P$ relative to rotating frame $\mathcal{F}$ |
| $\mathbf{a}_c$ | Coriolis acceleration |
| $\mathbf{A}, \mathbf{B}, \mathbf{C}, \ldots$ | Reactions at supports and connections |
| $A, B, C, \ldots$ | Points |
| $A$ | Area |
| $b$ | Width; distance; semiminor axis of ellipse |
| $c$ | Constant; coefficient of viscous damping |
| $C$ | Centroid; instantaneous center of rotation; capacitance |
| $d$ | Distance |
| $\mathbf{e}_n$, $\mathbf{e}_t$ | Unit vectors along normal and tangent |
| $\mathbf{e}_r$, $\mathbf{e}_\theta$ | Unit vectors in radial and transverse directions |
| $e$ | Coefficient of restitution; base of natural logarithms |
| $E$ | Total mechanical energy; voltage |
| $f$ | Scalar function |
| $f_f$ | Frequency of forced vibration |
| $f_n$ | Natural frequency |
| $\mathbf{F}$ | Force; friction force |
| $g$ | Acceleration of gravity |
| $G$ | Center of gravity; mass center; constant of gravitation |
| $h$ | Height; sag of cable; angular momenntum per unit mass |
| $\mathbf{H}_O$ | Angular momentum about point $O$ |
| $\dot{\mathbf{H}}_G$ | Rate of change of angular momentum $\mathbf{H}_G$ with respect to frame of fixed orientation |
| $(\dot{\mathbf{H}}_G)_{Gxyz}$ | Rate of change of angular momentum $\mathbf{H}_G$ with respect to rotating frame $Gxyz$ |
| $\mathbf{i}$, $\mathbf{j}$, $\mathbf{k}$ | Unit vectors along coordinate axes |
| $i$ | Current |
| $I, I_x, \ldots$ | Moments of inertia |

| | |
|---|---|
| $\overline{I}$ | Centroidal moment of inertia |
| $I_{xy}, \ldots$ | Products of inertia |
| $J$ | Polar moment of inertia |
| $k$ | Spring constant |
| $k_x, k_y, k_O$ | Radii of gyration |
| $\overline{k}$ | Centroidal radius of gyration |
| $l$ | Length |
| $\mathbf{L}$ | Linear momentum |
| $L$ | Length; span; inductance |
| $m$ | Mass |
| $m'$ | Mass per unit length |
| $\mathbf{M}$ | Couple; moment |
| $\mathbf{M}_O$ | Moment about point $O$ |
| $\mathbf{M}_O^R$ | Moment resultant about point $O$ |
| $M$ | Magnitude of couple or moment; mass of earth |
| $M_{OL}$ | Moment about axis $OL$ |
| $n$ | Normal direction |
| $\mathbf{N}$ | Normal component of reaction |
| $O$ | Origin of coordinates |
| $\mathbf{P}$ | Force; vector |
| $\dot{\mathbf{P}}$ | Rate of change of vector $\mathbf{P}$ with respect to frame of fixed orientation |
| $q$ | Mass rate of flow; electric charge |
| $\mathbf{Q}$ | Force; vector |
| $\dot{\mathbf{Q}}$ | Rate of change of vector $\mathbf{Q}$ with respect to frame of fixed orientation |
| $(\dot{\mathbf{Q}})_{Oxyz}$ | Rate of change of vector $\mathbf{Q}$ with respect to frame $Oxyz$ |
| $\mathbf{r}$ | Position vector |
| $\mathbf{r}_{B/A}$ | Position vector of $B$ relative to $A$ |
| $r$ | Radius; distance; polar coordinate |
| $\mathbf{R}$ | Resultant force; resultant vector; reaction |
| $R$ | Radius of earth; resistance |
| $\mathbf{s}$ | Position vector |
| $s$ | Length of arc; length of cable |
| $\mathbf{S}$ | Force; vector |
| $t$ | Time; thickness; tangential direction |
| $\mathbf{T}$ | Force |
| $T$ | Tension; kinetic energy |
| $\mathbf{u}$ | Velocity |
| $u$ | Variable |
| $U$ | Work |
| $\mathbf{v}, v$ | Velocity |
| $v$ | Speed |
| $\overline{\mathbf{v}}, \overline{v}$ | Velocity of mass center |
| $\mathbf{v}_{B/A}$ | Velocity of $B$ relative to frame in translation with $A$ |
| $\mathbf{v}_{P/\mathscr{F}}$ | Velocity of $P$ relative to rotating frame $\mathscr{F}$ |
| $\mathbf{V}$ | Vector product; shearing force |
| $V$ | Volume; potential energy; shear |
| $w$ | Load per unit length |

| | |
|---|---|
| $\mathbf{W}, W$ | Weight; load |
| $x, y, z$ | Rectangular coordinates; distances |
| $\dot{x}, \dot{y}, \dot{z}$ | Time derivatives of coordinates $x, y, z$ |
| $\bar{x}, \bar{y}, \bar{z}$ | Rectangular coordinates of centroid, center of gravity, or mass center |
| $\boldsymbol{\alpha}, \alpha$ | Angular acceleration |
| $\alpha, \beta, \gamma$ | Angles |
| $\gamma$ | Specific weight |
| $\delta$ | Elongation |
| $\delta\mathbf{r}$ | Virtual displacement |
| $\delta U$ | Virtual work |
| $\varepsilon$ | Eccentricity of conic section or of orbit |
| $\boldsymbol{\lambda}$ | Unit vector along a line |
| $\eta$ | Efficiency |
| $\theta$ | Angular coordinate; Eulerian angle; angle; polar coordinate |
| $\mu$ | Coefficient of friction |
| $\rho$ | Density; radius of curvature |
| $\tau$ | Periodic time |
| $\tau_n$ | Period of free vibration |
| $\phi$ | Angle of friction; Eulerian angle; phase angle; angle |
| $\varphi$ | Phase difference |
| $\psi$ | Eulerian angle |
| $\boldsymbol{\omega}, \omega$ | Angular velocity |
| $\omega_f$ | Circular frequency of forced vibration |
| $\omega_n$ | Natural circular frequency |
| $\boldsymbol{\Omega}$ | Angular velocity of frame of reference |

## CHAPTER 11

# Kinematics of Particles

The motion of the aircraft shown is characterized at any instant by its *position, velocity,* and *acceleration*. Also of interest is the motion of the aircraft relative to the moving air. During the approach to a crosswind landing, the velocity of the aircraft relative to the air is a vector in the direction of the longitudinal axis of the aircraft, and the velocity of the aircraft relative to the ground is a vector in the direction of the runway.

## KINEMATICS OF PARTICLES

11.1 Introduction to Dynamics
   Rectilinear Motion of Particles
11.2 Position, Velocity, and Acceleration
11.3 Determination of the Motion of a Particle
11.4 Uniform Rectilinear Motion
11.5 Uniformly Accelerated Rectilinear Motion
11.6 Motion of Several Particles
11.7 Graphical Solution of Rectilinear-Motion Problems
11.8 Other Graphical Methods
   Curvilinear Motion of Particles
11.9 Position Vector, Velocity, and Acceleration
11.10 Derivatives of Vector Functions
11.11 Rectangular Components of Velocity and Acceleration
11.12 Motion Relative to a Frame in Translation
11.13 Tangential and Normal Components
11.14 Radial and Transverse Components

## 11.1. INTRODUCTION TO DYNAMICS

Chapters 1 to 10 were devoted to *statics,* that is, to the analysis of bodies at rest. We now begin the study of *dynamics,* the part of mechanics that deals with the analysis of bodies in motion.

While the study of statics goes back to the time of the Greek philosophers, the first significant contribution to dynamics was made by Galileo (1564–1642). Galileo's experiments on uniformly accelerated bodies led Newton (1642–1727) to formulate his fundamental laws of motion.

Dynamics includes:

1. *Kinematics,* which is the study of the geometry of motion. Kinematics is used to relate displacement, velocity, acceleration, and time, without reference to the cause of the motion.
2. *Kinetics,* which is the study of the relation existing between the forces acting on a body, the mass of the body, and the motion of the body. Kinetics is used to predict the motion caused by given forces or to determine the forces required to produce a given motion.

Chapters 11 to 14 are devoted to the *dynamics of particles;* in Chap. 11 the *kinematics of particles* will be considered. The use of the word *particles* does not mean that our study will be restricted to small corpuscles; rather, it indicates that in these first chapters the motion of bodies—possibly as large as cars, rockets, or airplanes—will be considered without regard to their size. By saying that the bodies are analyzed as particles, we mean that only their motion as an entire unit will be considered; any rotation about their own mass center will be neglected. There are cases, however, when such a rotation is not negligible; the bodies cannot then be considered as particles. Such motions will be analyzed in later chapters, dealing with the *dynamics of rigid bodies.*

In the first part of Chap. 11, the rectilinear motion of a particle will be analyzed; that is, the position, velocity, and acceleration of a particle will be determined at every instant as it moves along a straight line. First, general methods of analysis will be used to study the motion of a particle; then two important particular cases will be considered, namely, the uniform motion and the uniformly accelerated motion of a particle (Secs. 11.4 and 11.5). In Sec. 11.6 the simultaneous motion of several particles will be considered, and the concept of the relative motion of one particle with respect to another will be introduced. The first part of this chapter concludes with a study of graphical methods of analysis and their application to the solution of various problems involving the rectilinear motion of particles (Secs. 11.7 and 11.8).

In the second part of this chapter, the motion of a particle as it moves along a curved path will be analyzed. Since the position, velocity, and acceleration of a particle will be defined as vector quantities, the concept of the derivative of a vector function will be introduced in Sec. 11.10 and added to our mathematical tools. Applications in which the motion of a particle is defined by the rectangular components of its velocity and acceleration will then be considered; at this point, the motion of a projectile will be analyzed (Sec. 11.11). In Sec. 11.12, the motion of a particle relative to a reference frame in translation will be considered. Finally, the curvilinear motion of a particle will be analyzed in terms of components other than rectangular. The tangential and normal components of the velocity and acceleration of a particle will be introduced in Sec. 11.13 and the radial and transverse components of its velocity and acceleration in Sec. 11.14.

# RECTILINEAR MOTION OF PARTICLES

## 11.2. POSITION, VELOCITY, AND ACCELERATION

A particle moving along a straight line is said to be in *rectilinear motion*. At any given instant $t$, the particle will occupy a certain position on the straight line. To define the position $P$ of the particle, we choose a fixed origin $O$ on the straight line and a positive direction along the line. We measure the distance $x$ from $O$ to $P$ and record it with a plus or minus sign, according to whether $P$ is reached from $O$ by moving along the line in the positive or the negative direction. The distance $x$, with the appropriate sign, completely defines the position of the particle; it is called the *position coordinate* of the particle considered. For example, the position coordinate corresponding to $P$ in Fig. 11.1$a$ is $x = +5$ m; the coordinate corresponding to $P'$ in Fig. 11.1$b$ is $x' = -2$ m.

**Fig. 11.1**

When the position coordinate $x$ of a particle is known for every value of time $t$, we say that the motion of the particle is known. The "timetable" of the motion can be given in the form of an equation in $x$ and $t$, such as $x = 6t^2 - t^3$, or in the form of a graph of $x$ versus $t$ as shown in Fig. 11.6 on page 606. The units most often used to measure the position coordinate $x$ are the meter (m) in the SI system of units† and the foot (ft) in the U.S. customary system of units. Time $t$ is usually measured in seconds (s).

Consider the position $P$ occupied by the particle at time $t$ and the corresponding coordinate $x$ (Fig. 11.2). Consider also the position $P'$ occupied by the particle at a later time $t + \Delta t$; the position coordinate of $P'$ can be obtained by adding to the coordinate $x$ of $P$ the small displacement $\Delta x$, which will be positive or negative according to whether $P'$ is to the right or to the left of $P$. The *average velocity* of the particle over the time interval $\Delta t$ is defined as the quotient of the displacement $\Delta x$ and the time interval $\Delta t$:

**Fig. 11.2**

$$\text{Average velocity} = \frac{\Delta x}{\Delta t}$$

†Cf. Sec. 1.3.

If SI units are used, $\Delta x$ is expressed in meters and $\Delta t$ in seconds; the average velocity will thus be expressed in meters per second (m/s). If U.S. customary units are used, $\Delta x$ is expressed in feet and $\Delta t$ in seconds; the average velocity will then be expressed in feet per second (ft/s).

The *instantaneous velocity* $v$ of the particle at the instant $t$ is obtained from the average velocity by choosing shorter and shorter time intervals $\Delta t$ and displacements $\Delta x$:

$$\text{Instantaneous velocity} = v = \lim_{\Delta t \to 0} \frac{\Delta x}{\Delta t}$$

The instantaneous velocity will also be expressed in m/s or ft/s. Observing that the limit of the quotient is equal, by definition, to the derivative of $x$ with respect to $t$, we write

$$v = \frac{dx}{dt} \tag{11.1}$$

The velocity $v$ is represented by an algebraic number which can be positive or negative.† A positive value of $v$ indicates that $x$ increases, that is, that the particle moves in the positive direction (Fig. 11.3$a$); a negative value of $v$ indicates that $x$ decreases, that is, that the particle moves in the negative direction (Fig. 11.3$b$). The magnitude of $v$ is known as the *speed* of the particle.

Consider the velocity $v$ of the particle at time $t$ and also its velocity $v + \Delta v$ at a later time $t + \Delta t$ (Fig. 11.4). The *average acceleration* of the particle over the time interval $\Delta t$ is defined as the quotient of $\Delta v$ and $\Delta t$:

$$\text{Average acceleration} = \frac{\Delta v}{\Delta t}$$

If SI units are used, $\Delta v$ is expressed in m/s and $\Delta t$ in seconds; the average acceleration will thus be expressed in m/s$^2$. If U.S. customary units are used, $\Delta v$ is expressed in ft/s and $\Delta t$ in seconds; the average acceleration will then be expressed in ft/s$^2$.

The *instantaneous acceleration* $a$ of the particle at the instant $t$ is obtained from the average acceleration by choosing smaller and smaller values for $\Delta t$ and $\Delta v$:

$$\text{Instantaneous acceleration} = a = \lim_{\Delta t \to 0} \frac{\Delta v}{\Delta t}$$

The instantaneous acceleration will also be expressed in m/s$^2$ or ft/s$^2$. The limit of the quotient, which is by definition the derivative of $v$

**Fig. 11.3**

**Fig. 11.4**

---

†As you will see in Sec. 11.9, the velocity is actually a vector quantity. However, since we are considering here the rectilinear motion of a particle, where the velocity of the particle has a known and fixed direction, we need only specify the sense and magnitude of the velocity; this can be conveniently done by using a scalar quantity with a plus or minus sign. The same is true of the acceleration of a particle in rectilinear motion.

with respect to $t$, measures the rate of change of the velocity. We write

$$a = \frac{dv}{dt} \qquad (11.2)$$

or, substituting for $v$ from (11.1),

$$a = \frac{d^2x}{dt^2} \qquad (11.3)$$

The acceleration $a$ is represented by an algebraic number which can be positive or negative.† A positive value of $a$ indicates that the velocity (that is, the algebraic number $v$) increases. This may mean that the particle is moving faster in the positive direction (Fig. 11.5$a$) or that it is moving more slowly in the negative direction (Fig. 11.5$b$); in both cases, $\Delta v$ is positive. A negative value of $a$ indicates that the velocity decreases; either the particle is moving more slowly in the positive direction (Fig. 11.5$c$) or it is moving faster in the negative direction (Fig. 11.5$d$).

**Photo 11.1** As the truck moves along a straight path, its motion is characterized at any instant by its position coordinate $x$, its velocity $v$, and its acceleration $a$.

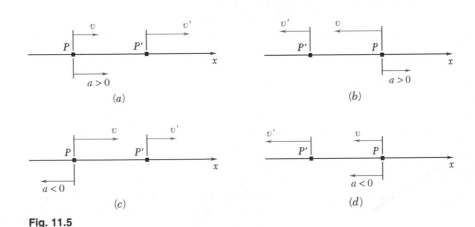

**Fig. 11.5**

The term *deceleration* is sometimes used to refer to $a$ when the speed of the particle (that is, the magnitude of $v$) decreases; the particle is then moving more slowly. For example, the particle of Fig. 11.5 is decelerated in parts $b$ and $c$; it is truly accelerated (that is, moves faster) in parts $a$ and $d$.

Another expression for the acceleration can be obtained by eliminating the differential $dt$ in Eqs. (11.1) and (11.2). Solving (11.1) for $dt$, we obtain $dt = dx/v$; substituting into (11.2), we write

$$a = v\frac{dv}{dx} \qquad (11.4)$$

†See footnote, page 604.

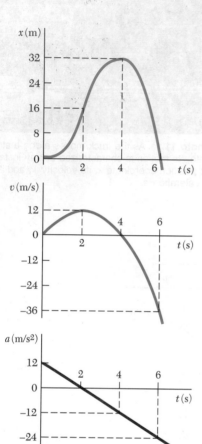

**Fig. 11.6**

*Example.* Consider a particle moving in a straight line, and assume that its position is defined by the equation

$$x = 6t^2 - t^3$$

where $t$ is expressed in seconds and $x$ in meters. The velocity $v$ at any time $t$ is obtained by differentiating $x$ with respect to $t$:

$$v = \frac{dx}{dt} = 12t - 3t^2$$

The acceleration $a$ is obtained by differentiating again with respect to $t$:

$$a = \frac{dv}{dt} = 12 - 6t$$

The position coordinate, the velocity, and the acceleration have been plotted against $t$ in Fig. 11.6. The curves obtained are known as *motion curves*. Keep in mind, however, that the particle does not move along any of these curves; the particle moves in a straight line. Since the derivative of a function measures the slope of the corresponding curve, the slope of the $x$–$t$ curve at any given time is equal to the value of $v$ at that time and the slope of the $v$–$t$ curve is equal to the value of $a$. Since $a = 0$ at $t = 2$ s, the slope of the $v$–$t$ curve must be zero at $t = 2$ s; the velocity reaches a maximum at this instant. Also, since $v = 0$ at $t = 0$ and at $t = 4$ s, the tangent to the $x$–$t$ curve must be horizontal for both of these values of $t$.

A study of the three motion curves of Fig. 11.6 shows that the motion of the particle from $t = 0$ to $t = \infty$ can be divided into four phases:

1. The particle starts from the origin, $x = 0$, with no velocity but with a positive acceleration. Under this acceleration, the particle gains a positive velocity and moves in the positive direction. From $t = 0$ to $t = 2$ s, $x$, $v$, and $a$ are all positive.

2. At $t = 2$ s, the acceleration is zero; the velocity has reached its maximum value. From $t = 2$ s to $t = 4$ s, $v$ is positive, but $a$ is negative; the particle still moves in the positive direction but more and more slowly; the particle is decelerating.

3. At $t = 4$ s, the velocity is zero; the position coordinate $x$ has reached its maximum value. From then on, both $v$ and $a$ are negative; the particle is accelerating and moves in the negative direction with increasing speed.

4. At $t = 6$ s, the particle passes through the origin; its coordinate $x$ is then zero, while the total distance traveled since the beginning of the motion is 64 m. For values of $t$ larger than 6 s, $x$, $v$, and $a$ will all be negative. The particle keeps moving in the negative direction, away from $O$, faster and faster.

We saw in the preceding section that the motion of a particle is said to be known if the position of the particle is known for every value of the time $t$. In practice, however, a motion is seldom defined by a relation between $x$ and $t$. More often, the conditions of the motion will be specified by the type of acceleration that the particle possesses. For example, a freely falling body will have a constant acceleration, directed downward and equal to 9.81 m/s$^2$, or 32.2 ft/s$^2$, a mass attached to a spring which has been stretched will have an acceleration proportional to the instantaneous elongation of the spring measured from the equilibrium position; etc. In general, the acceleration of the particle can be expressed as a function of one or more of the variables $x$, $v$, and $t$. In order to determine the position coordinate $x$ in terms of $t$, it will thus be necessary to perform two successive integrations.

Let us consider three common classes of motion:

1.  $a = f(t)$. *The Acceleration Is a Given Function of $t$.* Solving (11.2) for $dv$ and substituting $f(t)$ for $a$, we write

$$dv = a \, dt$$
$$dv = f(t) \, dt$$

Integrating both members, we obtain the equation

$$\int dv = \int f(t) \, dt$$

which defines $v$ in terms of $t$. It should be noted, however, that an arbitrary constant will be introduced as a result of the integration. This is due to the fact that there are many motions which correspond to the given acceleration $a = f(t)$. In order to uniquely define the motion of the particle, it is necessary to specify the *initial conditions* of the motion, that is, the value $v_0$ of the velocity and the value $x_0$ of the position coordinate at $t = 0$. Replacing the indefinite integrals by *definite integrals* with lower limits corresponding to the initial conditions $t = 0$ and $v = v_0$ and upper limits corresponding to $t = t$ and $v = v$, we write

$$\int_{v_0}^{v} dv = \int_{0}^{t} f(t) \, dt$$
$$v - v_0 = \int_{0}^{t} f(t) \, dt$$

which yields $v$ in terms of $t$.

Equation (11.1) can now be solved for $dx$,

$$dx = v \, dt$$

and the expression just obtained substituted for $v$. Both members are then integrated, the left-hand member with respect to $x$ from $x = x_0$ to $x = x$, and the right-hand member with respect to $t$ from $t = 0$ to $t = t$. The position coordinate $x$ is thus obtained in terms of $t$; the motion is completely determined.

Two important particular cases will be studied in greater detail in Secs. 11.4 and 11.5: the case when $a = 0$, corresponding to a *uniform motion*, and the case when $a =$ constant, corresponding to a *uniformly accelerated motion*.

2. $a = f(x)$. *The Acceleration Is a Given Function of $x$.* Rearranging Eq. (11.4) and substituting $f(x)$ for $a$, we write

$$v \, dv = a \, dx$$
$$v \, dv = f(x) \, dx$$

Since each member contains only one variable, we can integrate the equation. Denoting again by $v_0$ and $x_0$, respectively, the initial values of the velocity and of the position coordinate, we obtain

$$\int_{v_0}^{v} v \, dv = \int_{x_0}^{x} f(x) \, dx$$

$$\tfrac{1}{2}v^2 - \tfrac{1}{2}v_0^2 = \int_{x_0}^{x} f(x) \, dx$$

which yields $v$ in terms of $x$. We now solve (11.1) for $dt$,

$$dt = \frac{dx}{v}$$

and substitute for $v$ the expression just obtained. Both members can then be integrated to obtain the desired relation between $x$ and $t$. However, in most cases this last integration cannot be performed analytically and one must resort to a numerical method of integration.

3. $a = f(v)$. *The Acceleration Is a Given Function of $v$.* We can now substitute $f(v)$ for $a$ in either (11.2) or (11.4) to obtain either of the following relations:

$$f(v) = \frac{dv}{dt} \qquad f(v) = v\frac{dv}{dx}$$

$$dt = \frac{dv}{f(v)} \qquad dx = \frac{v \, dv}{f(v)}$$

Integration of the first equation will yield a relation between $v$ and $t$; integration of the second equation will yield a relation between $v$ and $x$. Either of these relations can be used in conjunction with Eq. (11.1) to obtain the relation between $x$ and $t$ which characterizes the motion of the particle.

## SAMPLE PROBLEM 11.1

The position of a particle which moves along a straight line is defined by the relation $x = t^3 - 6t^2 - 15t + 40$, where $x$ is expressed in meter and $t$ in seconds. Determine (a) the time at which the velocity will be zero, (b) the position and distance traveled by the particle at that time, (c) the acceleration of the particle at that time, (d) the distance traveled by the particle from $t = 4$ s to $t = 6$ s.

## SOLUTION

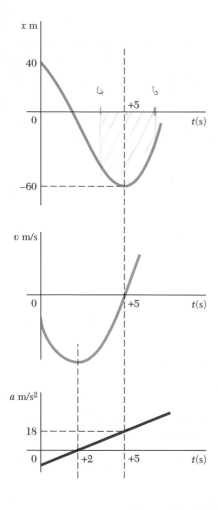

The equations of motion are

$$x = t^3 - 6t^2 - 15t + 40 \qquad (1)$$

$$v = \frac{dx}{dt} = 3t^2 - 12t - 15 \qquad (2)$$

$$a = \frac{dv}{dt} = 6t - 12 \qquad (3)$$

*a. Time at Which $v = 0$.* We set $v = 0$ in (2):

$$3t^2 - 12t - 15 = 0 \qquad t = -1 \text{ s} \qquad \text{and} \qquad t = +5 \text{ s} \quad \blacktriangleleft$$

Only the root $t = +5$ s corresponds to a time after the motion has begun: for $t < 5$ s, $v < 0$, the particle moves in the negative direction; for $t > 5$ s, $v > 0$, the particle moves in the positive direction.

*b. Position and Distance Traveled When $v = 0$.* Carrying $t = +5$ s into (1), we have

$$x_5 = (5)^3 - 6(5)^2 - 15(5) + 40 \qquad x_5 = -60 \text{ m} \quad \blacktriangleleft$$

The initial position at $t = 0$ was $x_0 = +40$ m. Since $v \neq 0$ during the interval $t = 0$ to $t = 5$ s, we have

$$\text{Distance traveled} = x_5 - x_0 = -60 \text{ m} - 40 \text{ m} = -100 \text{ m}$$

$$\text{Distance traveled} = 100 \text{ ft in the negative direction} \quad \blacktriangleleft$$

*c. Acceleration When $v = 0$.* We substitute $t = +5$ s into (3):

$$a_5 = 6(5) - 12 \qquad a_5 = +18 \text{ m/s}^2 \quad \blacktriangleleft$$

*d. Distance Traveled from $t = 4$ s to $t = 6$ s.* The particle moves in the negative direction from $t = 4$ s to $t = 5$ s and in the positive direction from $t = 5$ s to $t = 6$ s; therefore, the distance traveled during each of these time intervals will be computed separately.

From $t = 4$ s to $t = 5$ s: $\qquad x_5 = -60$ m

$$x_4 = (4)^3 - 6(4)^2 - 15(4) + 40 = -52 \text{ m}$$
$$\text{Distance traveled} = x_5 - x_4 = -60 \text{ m} - (-52 \text{ m}) = -8 \text{ m}$$
$$= 8 \text{ m in the negative direction}$$

From $t = 5$ s to $t = 6$ s: $\qquad x_5 = -60$ m

$$x_6 = (6)^3 - 6(6)^2 - 15(6) + 40 = -50 \text{ m}$$
$$\text{Distance traveled} = x_6 - x_5 = -50 \text{ m} - (-60 \text{ m}) = +10 \text{ m}$$
$$= 10 \text{ ft in the positive direction}$$

*Total distance traveled* from $t = 4$ s to $t = 6$ s is 8 m + 10 m $\quad = 18$ m $\quad \blacktriangleleft$

## SAMPLE PROBLEM 11.2

A ball is tossed with a velocity of 10 m/s directed vertically upward from a window located 20 m above the ground. Knowing that the acceleration of the ball is constant and equal to 9.81 m/s² downward, determine (*a*) the velocity *v* and elevation *y* of the ball above the ground at any time *t*, (*b*) the highest elevation reached by the ball and the corresponding value of *t*, (*c*) the time when the ball will hit the ground and the corresponding velocity. Draw the *v–t* and *y–t* curves.

## SOLUTION

*a*. **Velocity and Elevation.** The *y* axis measuring the position coordinate (or elevation) is chosen with its origin *O* on the ground and its positive sense upward. The value of the acceleration and the initial values of *v* and *y* are as indicated. Substituting for *a* in $a = dv/dt$ and noting that at $t = 0$, $v_0 = +10$ m/s, we have

$$\frac{dv}{dt} = a = -9.81 \text{ m/s}^2$$

$$\int_{v_0=10}^{v} dv = -\int_0^t 9.81 \, dt$$

$$[v]_{10}^v = -[9.81t]_0^t$$

$$v - 10 = -9.81t$$

$$v = 10 - 9.81t \quad (1) \blacktriangleleft$$

Substituting for *v* in $v = dy/dt$ and noting that at $t = 0$, $y_0 = 20$ m, we have

$$\frac{dy}{dt} = v = 10 - 9.81t$$

$$\int_{y_0=20}^{y} dy = \int_0^t (10 - 9.81t) \, dt$$

$$[y]_{20}^y = [10t - 4.905t^2]_0^t$$

$$y - 20 = 10t - 4.905t^2$$

$$y = 20 + 10t - 4.905t^2 \quad (2) \blacktriangleleft$$

*b*. **Highest Elevation.** When the ball reaches its highest elevation, we have $v = 0$. Substituting into (1), we obtain

$$10 - 9.81t = 0 \qquad t = 1.019 \text{ s} \blacktriangleleft$$

Carrying $t = 1.019$ s into (2), we have

$$y = 20 + 10(1.019) - 4.905(1.019)^2 \qquad y = 25.1 \text{ m} \blacktriangleleft$$

*c*. **Ball Hits the Ground.** When the ball hits the ground, we have $y = 0$. Substituting into (2), we obtain

$$20 + 10t - 4.905t^2 = 0 \qquad t = -1.243 \text{ s} \quad \text{and} \quad t = +3.28 \text{ s} \blacktriangleleft$$

Only the root $t = +3.28$ s corresponds to a time after the motion has begun. Carrying this value of *t* into (1), we have

$$v = 10 - 9.81(3.28) = -22.2 \text{ m/s} \qquad v = 22.2 \text{ m/s} \downarrow \blacktriangleleft$$

## SAMPLE PROBLEM 11.3

Piston

Oil

The brake mechanism used to reduce recoil in certain types of guns consists essentially of a piston attached to the barrel and moving in a fixed cylinder filled with oil. As the barrel recoils with an initial velocity $v_0$, the piston moves and oil is forced through orifices in the piston, causing the piston and the barrel to decelerate at a rate proportional to their velocity; that is, $a = -kv$. Express (a) $v$ in terms of $t$, (b) $x$ in terms of $t$, (c) $v$ in terms of $x$. Draw the corresponding motion curves.

## SOLUTION

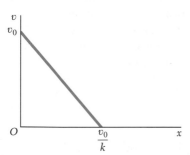

**a. $v$ in Terms of $t$.** Substituting $-kv$ for $a$ in the fundamental formula defining acceleration, $a = dv/dt$, we write

$$-kv = \frac{dv}{dt} \qquad \frac{dv}{v} = -k\,dt \qquad \int_{v_0}^{v} \frac{dv}{v} = -k \int_0^t dt$$

$$\ln \frac{v}{v_0} = -kt \qquad\qquad v = v_0 e^{-kt} \blacktriangleleft$$

**b. $x$ in Terms of $t$.** Substituting the expression just obtained for $v$ into $v = dx/dt$, we write

$$v_0 e^{-kt} = \frac{dx}{dt}$$

$$\int_0^x dx = v_0 \int_0^t e^{-kt}\,dt$$

$$x = -\frac{v_0}{k}[e^{-kt}]_0^t = -\frac{v_0}{k}(e^{-kt} - 1)$$

$$x = \frac{v_0}{k}(1 - e^{-kt}) \blacktriangleleft$$

**c. $v$ in Terms of $x$.** Substituting $-kv$ for $a$ in $a = v\,dv/dx$, we write

$$-kv = v\frac{dv}{dx}$$

$$dv = -k\,dx$$

$$\int_{v_0}^{v} dv = -k \int_0^x dx$$

$$v - v_0 = -kx \qquad\qquad v = v_0 - kx \blacktriangleleft$$

*Check.* Part $c$ could have been solved by eliminating $t$ from the answers obtained for parts $a$ and $b$. This alternative method can be used as a check. From part $a$ we obtain $e^{-kt} = v/v_0$; substituting into the answer of part $b$, we obtain

$$x = \frac{v_0}{k}(1 - e^{-kt}) = \frac{v_0}{k}\left(1 - \frac{v}{v_0}\right) \qquad v = v_0 - kx \qquad \text{(checks)}$$

# SOLVING PROBLEMS
## ON YOUR OWN

In the problems for this lesson, you will be asked to determine the *position*, the *velocity*, or the *acceleration* of a particle in *rectilinear motion*. As you read each problem, it is important that you identify both the independent variable (typically $t$ or $x$) and what is required (for example, the need to express $v$ as a function of $x$). You may find it helpful to start each problem by writing down both the given information and a simple statement of what is to be determined.

1. ***Determining $v(t)$ and $a(t)$ for a given $x(t)$.*** As explained in Sec. 11.2, the first and the second derivatives of $x$ with respect to $t$ are respectively equal to the velocity and the acceleration of the particle [Eqs. (11.1) and (11.2)]. If the velocity and the acceleration have opposite signs, the particle can come to rest and then move in the opposite direction [Sample Prob. 11.1]. Thus, when computing the total distance traveled by a particle, you should first determine if the particle will come to rest during the specified interval of time. Constructing a diagram similar to that of Sample Prob. 11.1 that shows the position and the velocity of the particle at each critical instant ($v = v_{max}$, $v = 0$, etc.) will help you to visualize the motion.

2. ***Determining $v(t)$ and $x(t)$ for a given $a(t)$.*** The solution of problems of this type was discussed in the first part of Sec. 11.3. We used the initial conditions, $t = 0$ and $v = v_0$, for the lower limits of the integrals in $t$ and $v$, but any other known state (for example, $t = t_1$, $v = v_1$) could have been used instead. Also, if the given function $a(t)$ contains an unknown constant (for example, the constant $k$ if $a = kt$), you will first have to determine that constant by substituting a set of known values of $t$ and $a$ in the equation defining $a(t)$.

3. ***Determining $v(x)$ and $x(t)$ for a given $a(x)$.*** This is the second case considered in Sec. 11.3. We again note that the lower limits of integration can be any known state (for example, $x = x_1$, $v = v_1$). In addition, since $v = v_{max}$ when $a = 0$, the positions where the maximum values of the velocity occur are easily determined by writing $a(x) = 0$ and solving for $x$.

4. ***Determining $v(x)$, $v(t)$, and $x(t)$ for a given $a(v)$.*** This is the last case treated in Sec. 11.3; the appropriate solution techniques for problems of this type are illustrated in Sample Prob. 11.3. All of the general comments for the preceding cases once again apply. Note that Sample Prob. 11.3 provides a summary of how and when to use the equations $v = dx/dt$, $a = dv/dt$, and $a = v \, dv/dx$.

# Problems†

**11.1** The motion of a particle is defined by the relation $x = t^2 - (t - 3)^3$ where $x$ and $t$ are expressed in meters and seconds, respectively. Determine (a) when the acceleration is zero, (b) the position and velocity of the particle at that time.

**11.2** The motion of a particle is defined by the relation $x = t^3 - (t - 2)^2$ where $x$ and $t$ are expressed in meters and seconds, respectively. Determine (a) when the acceleration is zero, (b) the position and velocity of the particle at that time.

**11.3** The motion of a particle is defined by the relation $x = 5t^4 - 4t^3 + 3t - 2$, where $x$ and $t$ are expressed in meter and seconds, respectively. Determine the position, the velocity, and the acceleration of the particle when $t = 2$ s.

**11.4** The motion of a particle is defined by the relation $x = 6t^4 + 8t^3 - 14t^2 - 10t + 16$, where $x$ and $t$ are expressed in meter and seconds, respectively. Determine the position, the velocity, and the acceleration of the particle when $t = 3$ s.

**11.5** The motion of the slider $A$ is defined by the relation $x = 500$ $\sin kt$, where $x$ and $t$ are expressed in millimeters and seconds, respectively, and $k$ is a constant. Knowing that $k = 10$ rad/s, determine the position, the velocity, and the acceleration of slider $A$ when $t = 0.05$ s.

**11.6** The motion of the slider $A$ is defined by the relation $x = 50$ $\sin(k_1 t - k_2 t^2)$, where $x$ and $t$ are expressed in millimeters and seconds, respectively. The constants $k_1$ and $k_2$ are known to be 1 rad/s and 0.5 rad/s², respectively. Consider the range $0 < t < 2$ s and determine the position and acceleration of slider $A$ when $v = 0$.

**Fig. P11.5 and P11.6**

**11.7** The motion of a particle is defined by the relation $x = t^3 - 6t^2 + 9t + 5$, where $x$ is expressed in meter and $t$ in seconds. Determine (a) when the velocity is zero, (b) the position, acceleration, and total distance traveled when $t = 5$ s.

**11.8** The motion of a particle is defined by the relation $x = t^2 - (t - 2)^3$, where $x$ and $t$ are expressed in meter and seconds, respectively. Determine (a) the two positions at which the velocity is zero, (b) the total distance traveled by the particle from $t = 0$ to $t = 4$ s.

**11.9** The acceleration of a particle is defined by the relation $a = 3e^{-0.2t}$, where $a$ and $t$ are expressed in m/s² and seconds, respectively. Knowing that $x = 0$ and $v = 0$ at $t = 0$, determine the velocity and position of the particle when $t = 0.5$ s.

†Answers to all problems set in straight type (such as **11.1**) are given at the end of the book. Answers to problems with a number set in italic type (such as **11.7**) are not given.

**11.10** The acceleration of point $A$ is defined by the relation $a = -5.4 \sin kt$, where $a$ and $t$ are expressed in m/s$^2$ and seconds, respectively, and $k = 3$ rad/s. Knowing that $x = 0$ and $v = 1.8$ m/s when $t = 0$, determine the velocity and position of point $A$ when $t = 0.5$ s.

**11.11** The acceleration of point $A$ is defined by the relation $a = -3.24 \sin kt - 4.32 \cos kt$, where $a$ and $t$ are expressed in m/s$^2$ and seconds, respectively, and $k = 3$ rad/s. Knowing that $x = 0.48$ m and $v = 1.08$ ft/s when $t = 0$, determine the velocity and position of point $A$ when $t = 0.5$ s.

**Fig. P11.10 and P11.11**

**11.12** The acceleration of a particle is directly proportional to the time $t$. At $t = 0$, the velocity of the particle is 400 mm/s. Knowing that $v = 370$ mm/s and $x = 500$ mm when $t = 1$ s, determine the velocity, the position, and the total distance traveled when $t = 7$ s.

**11.13** The acceleration of a particle is defined by the relation $a = 0.15$ m/s$^2$. Knowing that $x = -10$ m when $t = 0$ and $v = -0.15$ m/s when $t = 2$ s, determine the velocity, the position, and the total distance traveled when $t = 5$ s.

**11.14** The acceleration of a particle is defined by the relation $a = 9 - 3t^2$. The particle starts at $t = 0$ with $v = 0$ and $x = 5$ m. Determine (a) the time when the velocity is again zero, (b) the position and velocity when $t = 4$ s, (c) the total distance traveled by the particle from $t = 0$ to $t = 4$ s.

**11.15** The acceleration of a particle is defined by the relation $a = kt^2$. (a) Knowing that $v = -10$ m/s when $t = 0$ and that $v = 10$ m/s when $t = 2$ s, determine the constant $k$. (b) Write the equations of motion, knowing also that $x = 0$ when $t = 2$ s.

**Fig. P11.16 and P11.17**

**11.16** Point $A$ oscillates with an acceleration $a = 40 - 160x$, where $a$ and $x$ are expressed in m/s$^2$ and meters, respectively. The magnitude of the velocity is 0.3 m/s when $x = 0.4$ m. Determine (a) the maximum velocity of $A$, (b) the two positions at which the velocity of $A$ is zero.

**11.17** Point $A$ oscillates with an acceleration $a = 100(0.25 - x)$, where $a$ and $x$ are expressed in m/s$^2$ and meters, respectively. Knowing that the system starts at time $t = 0$ with $v = 0$ and $x = 0.2$ m, determine the position and the velocity of $A$ when $t = 0.2$ s.

**11.18** The acceleration of point $A$ is defined by the relation $a = 600x(1 + kx^2)$, where $a$ and $x$ are expressed in m/s$^2$ and meters, respectively, and $k$ is a constant. Knowing that the velocity of $A$ is 7.5 m/s when $x = 0$ and 15 ft/s when $x = 0.45$ m, determine the value of $k$.

**Fig. P11.18 and P11.19**

**11.19** The acceleration of point $A$ is defined by the relation $a = 800x + 3200x^3$, where $a$ and $x$ are expressed in m/s$^2$ and meters, respectively. Knowing that the velocity of $A$ is 10 m/s and $x = 0$ when $t = 0$, determine the velocity and position of $A$ when $t = 0.05$ s.

**11.20** The acceleration of a particle is defined by the relation $a = 12x - 28$, where $a$ and $x$ are expressed in m/s$^2$ and meters, respectively. Knowing that $v = 8$ m/s when $x = 0$, determine (a) the maximum value of $x$, (b) the velocity when the particle has traveled a total distance of 3 m.

**11.21** The acceleration of a particle is defined by the relation $a = k(1 - e^{-x})$, where $k$ is a constant. Knowing that the velocity of the particle is $v = +9$ m/s when $x = -3$ m and that the particle comes to rest at the origin, determine (a) the value of $k$, (b) the velocity of the particle when $x = -2$ m.

**11.22** Starting from $x = 0$ with no initial velocity, the acceleration of a race car is defined by the relation $a = 6.8e^{-0.00057x}$, where $a$ and $x$ are expressed in m/s² and meters, respectively. Determine the position of the race car when $v = 30$ m/s.

**11.23** The acceleration of a particle is defined by the relation $a = -0.4v$, where $a$ is expressed in mm/s² and $v$ in mm/s. Knowing that at $t = 0$ the velocity is 75 mm/s, determine (a) the distance the particle will travel before coming to rest, (b) the time required for the velocity of the particle to be reduced to 1 percent of its initial value.

**11.24** The acceleration of a particle is defined by the relation $a = -kv^2$, where $a$ is expressed in m/s² and $v$ in m/s. The particle starts at $x = 0$ with a velocity of 9 m/s and when $x = 13$ m the velocity is found to be 7 m/s. Determine the distance the particle will travel (a) before its velocity drops to 3 m/s, (b) before it comes to rest.

**11.25** The acceleration of a particle is defined by the relation $a = -k\sqrt{v}$, where $k$ is a constant. Knowing that $x = 0$ and $v = 25$ m/s at $t = 0$, and that $v = 12$ m/s when $x = 6$ m, determine (a) the velocity of the particle when $x = 8$ m, (b) the time required for the particle to come to rest.

**11.26** Starting from $x = 0$ with no initial velocity, a particle is given an acceleration $a = 0.8\sqrt{v^2 + 49}$, where $a$ and $v$ are expressed in m/s² and m/s, respectively. Determine (a) the position of the particle when $v = 24$ m/s, (b) the speed of the particle when $x = 40$ m.

**11.27** The acceleration of slider $A$ is defined by the relation $a = -2k\sqrt{k^2 - v^2}$, where $a$ and $v$ are expressed in m/s² and m/s, respectively, and $k$ is a constant. The system starts at time $t = 0$ with $x = 1.5$ m and $v = 0$. Knowing that $x = 1.2$ m when $t = 0.2$ s, determine the value of $k$.

**11.28** The acceleration of slider $A$ is defined by the relation $a = -2\sqrt{1 - v^2}$, where $a$ and $v$ are expressed in m/s² and m/s, respectively. The system starts at time $t = 0$ with $x = 1.5$ m and $v = 0$. Determine (a) the position of $A$ when $v = -0.6$ m/s, (b) the position of $A$ when $t = 0.3$ s.

**11.29** Starting from $x = 0$ with no initial velocity, the velocity of a race car is defined by the relation $v = 154\sqrt{1 - e^{-0.00057x}}$, where $v$ and $x$ are expressed in m/s and meters, respectively. Determine the position and acceleration of the race car when (a) $v = 20$ m/s, (b) $v = 40$ m/s.

**11.30** Based on observations, the speed of a jogger can be approximated by the relation $v = 7.5(1 - 0.04x)^{0.3}$, where $v$ and $x$ are expressed in km/h and kilometers, respectively. Knowing that $x = 0$ at $t = 0$, determine (a) the distance the jogger has run when $t = 1$ h, (b) the jogger's acceleration in m/s² at $t = 0$, (c) the time required for the jogger to run 6 km.

**Fig. P11.27 and P11.28**

**Fig. P11.29**

**Fig. P11.30**

**Fig. *P11.31***

**Fig. *P11.32***

**11.31** The acceleration due to gravity of a particle falling toward the earth is $a = -gR^2/r^2$, where $r$ is the distance from the *center* of the earth to the particle, $R$ is the radius of the earth, and $g$ is the acceleration due to gravity at the surface of the earth. If $R = 6370$ km, calculate the *escape velocity*, that is, the minimum velocity with which a particle must be projected upward from the surface of the earth if it is not to return to earth. (*Hint:* $v = 0$ for $r = \infty$.)

**11.32** The acceleration due to gravity at an altitude $y$ above the surface of the earth can be expressed as

$$a = \frac{-32.2}{[1 + (y/20.9 \times 10^6)]^2}$$

where $a$ and $y$ are expressed in m/s$^2$ and meters, respectively. Using this expression, compute the height reached by a projectile fired vertically upward from the surface of the earth if its initial velocity is (*a*) $v = 720$ m/s, (*b*) $v = 1200$ m/s, (*c*) $v = 12,000$ m/s.

**11.33** The velocity of a slider is defined by the relation $v = v' \sin(\omega_n t + \phi)$. Denoting the velocity and the position of the slider at $t = 0$ by $v_0$ and $x_0$, respectively, and knowing that the maximum displacement of the slider is $2x_0$, show that (*a*) $v' = (v_0^2 + x_0^2 \omega_n^2)/2x_0\omega_n$, (*b*) the maximum value of the velocity occurs when $x = x_0[3 - (v_0/x_0\omega_n)^2]/2$.

**11.34** The velocity of a particle is $v = v_0[1 - \sin(\pi t/T)]$. Knowing that the particle starts from the origin with an initial velocity $v_0$, determine (*a*) its position and its acceleration at $t = 3T$, (*b*) its average velocity during the interval $t = 0$ to $t = T$.

## 11.4. UNIFORM RECTILINEAR MOTION

Uniform rectilinear motion is a type of straight-line motion which is frequently encountered in practical applications. In this motion, the acceleration $a$ of the particle is zero for every value of $t$. The velocity $v$ is therefore constant, and Eq. (11.1) becomes

$$\frac{dx}{dt} = v = \text{constant}$$

The position coordinate $x$ is obtained by integrating this equation. Denoting by $x_0$ the initial value of $x$, we write

$$\int_{x_0}^{x} dx = v \int_{0}^{t} dt$$

$$x - x_0 = vt$$

$$x = x_0 + vt \tag{11.5}$$

This equation can be used *only if the velocity of the particle is known to be constant.*

Uniformly accelerated rectilinear motion is another common type of motion. In this motion, the acceleration $a$ of the particle is constant, and Eq. (11.2) becomes

$$\frac{dv}{dt} = a = \text{constant}$$

The velocity $v$ of the particle is obtained by integrating this equation:

$$\int_{v_0}^{v} dv = a \int_{0}^{t} dt$$

$$v - v_0 = at$$

$$v = v_0 + at \tag{11.6}$$

where $v_0$ is the initial velocity. Substituting for $v$ in (11.1), we write

$$\frac{dx}{dt} = v_0 + at$$

Denoting by $x_0$ the initial value of $x$ and integrating, we have

$$\int_{x_0}^{x} dx = \int_{0}^{t} (v_0 + at)\, dt$$

$$x - x_0 = v_0 t + \tfrac{1}{2}at^2$$

$$x = x_0 + v_0 t + \tfrac{1}{2}at^2 \tag{11.7}$$

We can also use Eq. (11.4) and write

$$v\frac{dv}{dx} = a = \text{constant}$$

$$v\, dv = a\, dx$$

Integrating both sides, we obtain

$$\int_{v_0}^{v} v\, dv = a \int_{x_0}^{x} dx$$

$$\tfrac{1}{2}(v^2 - v_0^2) = a(x - x_0)$$

$$v^2 = v_0^2 + 2a(x - x_0) \tag{11.8}$$

The three equations we have derived provide useful relations among position coordinate, velocity, and time in the case of a uniformly accelerated motion, as soon as appropriate values have been substituted for $a$, $v_0$, and $x_0$. The origin $O$ of the $x$ axis should first be defined and a positive direction chosen along the axis; this direction will be used to determine the signs of $a$, $v_0$, and $x_0$. Equation (11.6) relates $v$ and $t$ and should be used when the value of $v$ corresponding to a given value of $t$ is desired, or inversely. Equation (11.7) relates $x$ and $t$; Eq. (11.8) relates $v$ and $x$. An important application of uniformly accelerated motion is the motion of a *freely falling body*. The acceleration of a freely falling body (usually denoted by *g*) is equal to 9.81 m/s² or 32.2 ft/s².

It is important to keep in mind that the three equations above can be used *only when the acceleration of the particle is known to be constant*. If the acceleration of the particle is variable, its motion should be determined from the fundamental Eqs. (11.1) to (11.4) according to the methods outlined in Sec. 11.3.

## 11.6. MOTION OF SEVERAL PARTICLES

When several particles move independently along the same line, independent equations of motion can be written for each particle. Whenever possible, time should be recorded from the same initial instant for all particles, and displacements should be measured from the same origin and in the same direction. In other words, a single clock and a single measuring tape should be used.

**Fig. 11.7**

**Relative Motion of Two Particles.** Consider two particles $A$ and $B$ moving along the same straight line (Fig. 11.7). If the position coordinates $x_A$ and $x_B$ are measured from the same origin, the difference $x_B - x_A$ defines the *relative position coordinate of B with respect to A* and is denoted by $x_{B/A}$. We write

$$x_{B/A} = x_B - x_A \qquad \text{or} \qquad x_B = x_A + x_{B/A} \qquad (11.9)$$

Regardless of the positions of $A$ and $B$ with respect to the origin, a positive sign for $x_{B/A}$ means that $B$ is to the right of $A$, and a negative sign means that $B$ is to the left of $A$.

The rate of change of $x_{B/A}$ is known as the *relative velocity of B with respect to A* and is denoted by $v_{B/A}$. Differentiating (11.9), we write

$$v_{B/A} = v_B - v_A \qquad \text{or} \qquad v_B = v_A + v_{B/A} \qquad (11.10)$$

A positive sign for $v_{B/A}$ means that $B$ is *observed from A* to move in the positive direction; a negative sign means that it is observed to move in the negative direction.

The rate of change of $v_{B/A}$ is known as the *relative acceleration of B with respect to A* and is denoted by $a_{B/A}$. Differentiating (11.10), we obtain†

$$a_{B/A} = a_B - a_A \qquad \text{or} \qquad a_B = a_A + a_{B/A} \qquad (11.11)$$

†Note that the product of the subscripts $A$ and $B/A$ used in the right-hand member of Eqs. (11.9), (11.10), and (11.11) is equal to the subscript $B$ used in their left-hand member.

**Dependent Motions.** Sometimes, the position of a particle will depend upon the position of another particle or of several other particles. The motions are then said to be *dependent*. For example, the position of block *B* in Fig. 11.8 depends upon the position of block *A*. Since the rope *ACDEFG* is of constant length, and since the lengths of the portions of rope *CD* and *EF* wrapped around the pulleys remain constant, it follows that the sum of the lengths of the segments *AC*, *DE*, and *FG* is constant. Observing that the length of the segment *AC* differs from $x_A$ only by a constant and that, similarly, the lengths of the segments *DE* and *FG* differ from $x_B$ only by a constant, we write

$$x_A + 2x_B = \text{constant}$$

Since only one of the two coordinates $x_A$ and $x_B$ can be chosen arbitrarily, we say that the system shown in Fig. 11.8 has *one degree of freedom*. From the relation between the position coordinates $x_A$ and $x_B$, it follows that if $x_A$ is given an increment $\Delta x_A$, that is, if block *A* is lowered by an amount $\Delta x_A$, the coordinate $x_B$ will receive an increment $\Delta x_B = -\frac{1}{2}\Delta x_A$. In other words, block *B* will rise by half the same amount; this can easily be checked directly from Fig. 11.8.

**Fig. 11.8**

**Fig. 11.9**

In the case of the three blocks of Fig. 11.9, we can again observe that the length of the rope which passes over the pulleys is constant, and thus the following relation must be satisfied by the position coordinates of the three blocks:

$$2x_A + 2x_B + x_C = \text{constant}$$

Since two of the coordinates can be chosen arbitrarily, we say that the system shown in Fig. 11.9 has *two degrees of freedom*.

When the relation existing between the position coordinates of several particles is *linear*, a similar relation holds between the velocities and between the accelerations of the particles. In the case of the blocks of Fig. 11.9, for instance, we differentiate twice the equation obtained and write

$$2\frac{dx_A}{dt} + 2\frac{dx_B}{dt} + \frac{dx_C}{dt} = 0 \quad \text{or} \quad 2v_A + 2v_B + v_C = 0$$

$$2\frac{dv_A}{dt} + 2\frac{dv_B}{dt} + \frac{dv_C}{dt} = 0 \quad \text{or} \quad 2a_A + 2a_B + a_C = 0$$

## SAMPLE PROBLEM 11.4

A ball is thrown vertically upward from the 12-m level in an elevator shaft with an initial velocity of 18 m/s. At the same instant an open-platform elevator passes the 5-m level, moving upward with a constant velocity of 2 m/s. Determine (a) when and where the ball will hit the elevator, (b) the relative velocity of the ball with respect to the elevator when the ball hits the elevator.

## SOLUTION

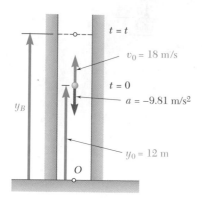

**Motion of Ball.** Since the ball has a constant acceleration, its motion is *uniformly accelerated*. Placing the origin $O$ of the $y$ axis at ground level and choosing its positive direction upward, we find that the initial position is $y_0 = +12$ m, the initial velocity is $v_0 = +18$ m/s, and the acceleration is $a = -9.81$ m/s². Substituting these values in the equations for uniformly accelerated motion, we write

$$v_B = v_0 + at \qquad v_B = 18 - 9.81t \qquad (1)$$
$$y_B = y_0 + v_0 t + \tfrac{1}{2}at^2 \qquad y_B = 12 + 18t - 4.905t^2 \qquad (2)$$

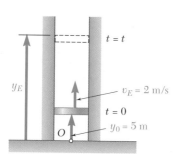

**Motion of Elevator.** Since the elevator has a constant velocity, its motion is *uniform*. Again placing the origin $O$ at the ground level and choosing the positive direction upward, we note that $y_0 = +5$ m and write

$$v_E = +2 \text{ m/s} \qquad (3)$$
$$y_E = y_0 + v_E t \qquad y_E = 5 + 2t \qquad (4)$$

**Ball Hits Elevator.** We first note that the same time $t$ and the same origin $O$ were used in writing the equations of motion of both the ball and the elevator. We see from the figure that when the ball hits the elevator,

$$y_E = y_B \qquad (5)$$

Substituting for $y_E$ and $y_B$ from (2) and (4) into (5), we have

$$5 + 2t = 12 + 18t - 4.905t^2$$
$$t = -0.39 \text{ s} \qquad \text{and} \qquad t = 3.65 \text{ s} \blacktriangleleft$$

Only the root $t = 3.65$ s corresponds to a time after the motion has begun. Substituting this value into (4), we have

$$y_E = 5 + 2(3.65) = 12.30 \text{ m}$$
$$\text{Elevation from ground} = 12.30 \text{ m} \blacktriangleleft$$

The relative velocity of the ball with respect to the elevator is

$$v_{B/E} = v_B - v_E = (18 - 9.81t) - 2 = 16 - 9.81t$$

When the ball hits the elevator at time $t = 3.65$ s, we have

$$v_{B/E} = 16 - 9.81(3.65) \qquad v_{B/E} = -19.81 \text{ m/s} \blacktriangleleft$$

The negative sign means that the ball is observed from the elevator to be moving in the negative sense (downward).

## SAMPLE PROBLEM 11.5

Collar $A$ and block $B$ are connected by a cable passing over three pulleys $C$, $D$, and $E$ as shown. Pulleys $C$ and $E$ are fixed, while $D$ is attached to a collar which is pulled downward with a constant velocity of 75 mm/s. At $t = 0$, collar $A$ starts moving downward from position $K$ with a constant acceleration and no initial velocity. Knowing that the velocity of collar $A$ is 300 mm/s as it passes through point $L$, determine the change in elevation, the velocity, and the acceleration of block $B$ when collar $A$ passes through $L$.

## SOLUTION

**Motion of Collar $A$.** We place the origin $O$ at the upper horizontal surface and choose the positive direction downward. We observe that when $t = 0$, collar $A$ is at the position $K$ and $(v_A)_0 = 0$. Since $v_A = 300$ mm/s and $x_A - (x_A)_0 = 200$ mm when the collar passes through $L$, we write

$$v_A^2 = (v_A)_0^2 + 2a_A[x_A - (x_A)_0] \qquad (300)^2 = 0 + 2a_A(200)$$
$$a_A = 225 \text{ mm/s}^2$$

The time at which collar $A$ reaches point $L$ is obtained by writing

$$v_A = (v_A)_0 + a_A t \qquad 12 = 0 + 9t \qquad t = 1.333 \text{ s}$$

**Motion of Pulley $D$.** Recalling that the positive direction is downward, we write

$$a_D = 0 \qquad v_D = 75 \text{ mm/s} \qquad x_D = (x_D)_0 + v_D t = (x_D)_0 + 75t$$

When collar $A$ reaches $L$, at $t = 1.333$ s, we have

$$x_D = (x_D)_0 + 3(1.333) = (x_D)_0 + 100$$

Thus, $$x_D - (x_D)_0 = 100 \text{ mm}$$

**Motion of Block $B$.** We note that the total length of cable $ACDEB$ differs from the quantity $(x_A + 2x_D + x_B)$ only by a constant. Since the cable length is constant during the motion, this quantity must also remain constant. Thus, considering the times $t = 0$ and $t = 1.333$ s, we write

$$x_A + 2x_D + x_B = (x_A)_0 + 2(x_D)_0 + (x_B)_0 \qquad (1)$$
$$[x_A - (x_A)_0] + 2[x_D - (x_D)_0] + [x_B - (x_B)_0] = 0 \qquad (2)$$

But we know that $x_A - (x_A)_0 = 200$ mm and $x_D - (x_D)_0 = 100$ mm; substituting these values in (2), we find

$$200 + 2(100) + [x_B - (x_B)_0] = 0 \qquad x_B - (x_B)_0 = -400 \text{ mm}$$

Thus: Change in elevation of $B$ = 400 mm ↑ ◀

Differentiating (1) twice, we obtain equations relating the velocities and the accelerations of $A$, $B$, and $D$. Substituting for the velocities and accelerations of $A$ and $D$ at $t = 1.333$ s, we have

$$v_A + 2v_D + v_B = 0: \qquad 300 + 2(75) + v_B = 0$$
$$v_B = -450 \text{ mm/s} \qquad v_B = 450 \text{ mm/s} ↑ \quad ◀$$

$$a_A + 2a_D + a_B = 0: \qquad 225 + 2(0) + a_B = 0$$
$$a_B = -225 \text{ mm/s}^2 \qquad a_B = 225 \text{ mm/s}^2 ↑ \quad ◀$$

In this lesson we derived the equations that describe *uniform rectilinear motion* (constant velocity) and *uniformly accelerated rectilinear motion* (constant acceleration). We also introduced the concept of *relative motion*. The equations for relative motion [Eqs. (11.9) to (11.11)] can be applied to the independent or dependent motions of any two particles moving along the same straight line.

**A. Independent motion of one or more particles.** The solution of problems of this type should be organized as follows:

**1. Begin your solution** by listing the given information, sketching the system, and selecting the origin and the positive direction of the coordinate axis [Sample Prob. 11.4]. It is always advantageous to have a visual representation of problems of this type.

**2. Write the equations** that describe the motions of the various particles as well as those that describe how these motions are related [Eq. (5) of Sample Prob. 11.4].

**3. Define the initial conditions,** that is, specify the state of the system corresponding to $t = 0$. This is especially important if the motions of the particles begin at different times. In such cases, either of two approaches can be used.

   *a.* Let $t = 0$ be the time when the last particle begins to move. You must then determine the initial position $x_0$ and the initial velocity $v_0$ of each of the other particles.

   *b.* Let $t = 0$ be the time when the first particle begins to move. You must then, in each of the equations describing the motion of another particle, replace $t$ with $t - t_0$, where $t_0$ is the time at which that specific particle begins to move. It is important to recognize that the equations obtained in this way are valid only for $t \geq t_0$.

**B.** *Dependent motion of two or more particles.* In problems of this type the particles of the system are connected to each other, typically by ropes or by cables. The method of solution of these problems is similar to that of the preceding group of problems, except that it will now be necessary to describe the *physical connections* between the particles. In the following problems, the connection is provided by one or more cables. For each cable, you will have to write equations similar to the last three equations of Sec. 11.6. We suggest that you use the following procedure:

1. *Draw a sketch of the system* and select a coordinate system, indicating clearly a positive sense for each of the coordinate axes. For example, in Sample Prob. 11.5 lengths are measured downward from the upper horizontal support. It thus follows that those displacements, velocities, and accelerations which have positive values are directed downward.

2. *Write the equation describing the constraint* imposed by each cable on the motion of the particles involved. Differentiating this equation twice, you will obtain the corresponding relations among velocities and accelerations.

3. *If several directions of motion are involved,* you must select a coordinate axis and a positive sense for each of these directions. You should also try to locate the origins of your coordinate axes so that the equations of constraints will be as simple as possible. For example, in Sample Prob. 11.5 it is easier to define the various coordinates by measuring them downward from the upper support than by measuring them upward from the bottom support.

*Finally, keep in mind* that the method of analysis described in this lesson and the corresponding equations can be used only for particles moving with *uniform or uniformly accelerated rectilinear motion.*

**11.35** A minivan is tested for acceleration and braking. In the street-start acceleration test, the elapsed time is 8.2 s for a velocity increase from 10 km/h to 100 km/h. In the braking test, the distance traveled is 44 m during braking to a stop from 100 km/h. Assuming constant values of acceleration and deceleration, determine (*a*) the acceleration during the street-start test, (*b*) the deceleration during the braking test.

**Fig. P11.35 and P11.36**

**11.36** In Prob. 11.35, determine (*a*) the distance traveled during the street-start acceleration test, (*b*) the elapsed time for the braking test.

**11.37** An airplane begins its take-off run at *A* with zero velocity and a constant acceleration *a*. Knowing that it becomes airborne 30 s later at *B* and that the distance *AB* is 823 m, determine (*a*) the acceleration *a*, (*b*) the take-off velocity $v_B$.

**Fig. P11.37**

**11.38** Steep safety ramps are built beside mountain highways to enable vehicles with defective brakes to stop safely. A truck enters a 229 m ramp at a high speed $v_0$ and travels 165 m in 6 s at constant deceleration before its speed is reduced to $v_0/2$. Assuming the same constant deceleration, determine (*a*) the additional time required for the truck to stop, (*b*) the additional distance traveled by the truck.

**Fig. P11.38**

**11.39** A sprinter in a 400-m race accelerates uniformly for the first 130 m and then runs with constant velocity. If the sprinter's time for the first 130 m is 25 s, determine (a) his acceleration, (b) his final velocity, (c) his time for the race.

**11.40** A group of students launches a model rocket in the vertical direction. Based on tracking data, they determine that the altitude of the rocket was 27.5 m at the end of the powered portion of the flight and that the rocket landed 16 s later. Knowing that the descent parachute failed to deploy so that the rocket fell freely to the ground after reaching its maximum altitude and assuming that $g = 9.81$ m/s$^2$, determine (a) the speed $v_1$ of the rocket at the end of powered flight, (b) the maximum altitude reached by the rocket.

Fig. *P11.40*

**11.41** Automobile A starts from O and accelerates at the constant rate of 0.75 m/s$^2$. A short time later it is passed by bus B which is traveling in the opposite direction at a constant speed of 6 m/s. Knowing that bus B passes point O 20 s after automobile A started from there, determine when and where the vehicles passed each other.

**Fig. P11.41**

**11.42** Automobiles A and B are traveling in adjacent highway lanes and at $t = 0$ have the positions and speeds shown. Knowing that automobile A has a constant acceleration of 0.6 m/s$^2$ and that B has a constant deceleration of 0.4 m/s$^2$, determine (a) when and where A will overtake B, (b) the speed of each automobile at that time.

**Fig. P11.42**

**11.43** In a close harness race, horse 2 passes horse 1 at point A, where the two velocities are $v_2 = 6.4$ m/s and $v_1 = 6.2$ m/s. Horse 1 later passes horse 2 at point B and goes on to win the race at point C, 366 m from A. The elapsed times from A to C for horse 1 and horse 2 are $t_1 = 61.5$ s and $t_2 = 62.0$ s, respectively. Assuming uniform accelerations for both horses between A and C, determine (a) the distance from A to B, (b) the position of horse 1 relative to horse 2 when horse 1 reaches the finish line C.

**Fig. P11.43**

**Fig. P11.44**

**11.44** Two rockets are launched at a fireworks performance. Rocket A is launched with an initial velocity $v_0$ and rocket B is launched 4 s later with the same initial velocity. The two rockets are timed to explode simultaneously at a height of 73 m, as A is falling and B is rising. Assuming a constant acceleration $g = 9.81$ m/s$^2$, determine (a) the initial velocity $v_0$, (b) the velocity of B relative to A at the time of the explosion.

**11.45** In a boat race, boat A is leading boat B by 38 m and both boats are traveling at a constant speed of 168 km/h. At $t = 0$, the boats accelerate at constant rates. Knowing that when B passes A, $t = 8$ s and $v_A = 228$ km/h, determine (a) the acceleration of A, (b) the acceleration of B.

**11.46** Car A is parked along the northbound lane of a highway, and car B is traveling in the southbound lane at a constant speed of 96 km/h. At $t = 0$, A starts and accelerates at a constant rate $a_A$, while at $t = 5$ s, B begins to slow down with a constant deceleration of magnitude $a_A/6$. Knowing that when the cars pass each other $x = 90$ m and $v_A = v_B$, determine (a) the acceleration $a_A$, (b) when the vehicles pass each other, (c) the distance between the vehicles at $t = 0$.

**Fig. P11.46**

**11.47** Two automobiles A and B traveling in the same direction in adjacent lanes are stopped at a traffic signal. As the signal turns green, automobile A accelerates at a constant rate of 1.98 m/s$^2$. Two seconds later, automobile B starts and accelerates at a constant rate of 3.6 m/s$^2$. Determine (a) when and where B will overtake A, (b) the speed of each automobile at that time.

**11.48** Two automobiles A and B are approaching each other in adjacent highway lanes. At $t = 0$, A and B are 1 km apart, their speeds are $v_A = 109$ km/h and $v_B = 62.5$ km/h, and they are at points P and Q, respectively. Knowing that A passes point Q 40 s after B was there and that B passes point P 42 s after A was there, determine (a) the uniform accelerations of A and B, (b) when the vehicles pass each other, (c) the speed of B at that time.

**Fig. P11.48**

**11.49** Block $A$ moves down with a constant velocity of 1 m/s. Determine (a) the velocity of block $C$, (b) the velocity of collar $B$ relative to block $A$, (c) the relative velocity of portion $D$ of the cable with respect to block $A$.

**Fig. P11.49 and P11.50**

**11.50** Block $C$ starts from rest and moves down with a constant acceleration. Knowing that after block $A$ has moved 0.5 m its velocity is 0.2 m/s, determine (a) the accelerations of $A$ and $C$, (b) the velocity and the change in position of block $B$ after 2 s.

**11.51** Block $C$ moves downward with a constant velocity of 0.6 m/s. Determine (a) the velocity of block $A$, (b) the velocity of block $D$.

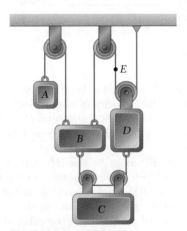

**Fig. P11.51 and P11.52**

**11.52** Block $C$ starts from rest and moves downward with a constant acceleration. Knowing that after 5 s the velocity of block $A$ relative to block $D$ is 2.4 m/s, determine (a) the acceleration of block $C$, (b) the acceleration of portion $E$ of the cable.

**11.53** In the position shown, collar $B$ moves to the left with a constant velocity of 300 mm/s. Determine ($a$) the velocity of collar $A$, ($b$) the velocity of portion $C$ of the cable, ($c$) the relative velocity of portion $C$ of the cable with respect to collar $B$.

**Fig. P11.53 and P11.54**

**11.54** Collar $A$ starts from rest and moves to the right with a constant acceleration. Knowing that after 8 s the relative velocity of collar $B$ with respect to collar $A$ is 610 mm/s, determine ($a$) the accelerations of $A$ and $B$, ($b$) the velocity and the change in position of $B$ after 6 s.

**11.55** At the instant shown, slider block $B$ is moving to the right with a constant acceleration, and its speed is 15.2 cm/s. Knowing that after slider block $A$ has moved 25.4 cm to the right its velocity is 6 cm/s, determine ($a$) the accelerations of $A$ and $B$, ($b$) the acceleration of portion $D$ of the cable, ($c$) the velocity and the change in position of slider block $B$ after 4 s.

**Fig. *P11.55* and *P11.56***

**11.56** Slider block $B$ moves to the right with a constant velocity of 30.4 cm/s. Determine ($a$) the velocity of slider block $A$, ($b$) the velocity of portion $C$ of the cable, ($c$) the velocity of portion $D$ of the cable, ($d$) the relative velocity of portion $C$ of the cable with respect to slider block $A$.

**11.57** Slider block $B$ moves to the left with a constant velocity of 50 mm/s. At $t = 0$, slider block $A$ is moving to the right with a constant acceleration and a velocity of 100 mm/s. Knowing that at $t = 2$ s slider block $C$ has moved 40 mm to the right, determine ($a$) the velocity of slider block $C$ at $t = 0$, ($b$) the velocity of portion $D$ of the cable at $t = 0$, ($c$) the accelerations of $A$ and $C$.

**11.58** Slider block $A$ starts with an initial velocity at $t = 0$ and a constant acceleration of 270 mm/s to the right. Slider block $C$ starts from rest at $t = 0$ and moves to the right with constant acceleration. Knowing that at $t = 2$ s, the velocities of $A$ and $B$ are 420 mm/s to the right and 30 mm/s to the left, respectively, determine ($a$) the accelerations of $B$ and $C$, ($b$) the initial velocities of $A$ and $B$, ($c$) the initial velocity of portion $E$ of the cable.

**Fig. P11.57 and P11.58**

**11.59**  Collar $A$ starts from rest at $t = 0$ and moves upward with a constant acceleration of 9 cm/s$^2$. Knowing that collar $B$ moves downward with a constant velocity of 45.7 cm/s, determine (a) the time at which the velocity of block $C$ is zero, (b) the corresponding position of block $C$.

**Fig. P11.59 and P11.60**

**11.60**  Collars $A$ and $B$ start from rest and move with the following accelerations: $a_A = 6.35t$ cm/s$^2$ upward and $a_B = 38$ cm/s$^2$ downward. Determine (a) the time at which the velocity of block $C$ is again zero, (b) the distance through which block $C$ will have moved at that time.

**11.61**  The system shown starts from rest, and each component moves with a constant acceleration. If the relative acceleration of block $C$ with respect to collar $B$ is 120 mm/s$^2$ upward and the relative acceleration of block $D$ with respect to block $A$ is 220 mm/s$^2$ downward, determine (a) the velocity of block $C$ after 6 s, (b) the change in position of block $D$ after 10 s.

**Fig. *P11.61* and *P11.62***

***11.62**  The system shown starts from rest, and the length of the upper cord is adjusted so that $A$, $B$, and $C$ are initially at the same level. Each component moves with a constant acceleration. Knowing that when the relative velocity of collar $B$ with respect to block $A$ is 40 mm/s downward, the displacements of $A$ and $B$ are 80 mm downward and 160 mm downward, respectively, determine (a) the accelerations of $A$ and $B$, (b) the change in position of block $D$ when the velocity of block $C$ is 300 mm/s upward.

## *11.7. GRAPHICAL SOLUTION OF RECTILINEAR-MOTION PROBLEMS

It was observed in Sec. 11.2 that the fundamental formulas

$$v = \frac{dx}{dt} \qquad \text{and} \qquad a = \frac{dv}{dt}$$

have a geometrical significance. The first formula expresses that the velocity at any instant is equal to the slope of the $x$–$t$ curve at the same instant (Fig. 11.10). The second formula expresses that the

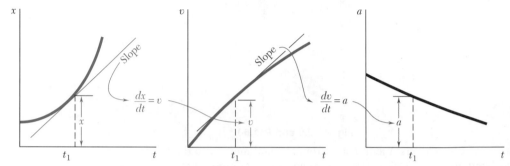

**Fig. 11.10**

acceleration is equal to the slope of the $v$–$t$ curve. These two properties can be used to determine graphically the $v$–$t$ and $a$–$t$ curves of a motion when the $x$–$t$ curve is known.

Integrating the two fundamental formulas from a time $t_1$ to a time $t_2$, we write

$$x_2 - x_1 = \int_{t_1}^{t_2} v\,dt \qquad \text{and} \qquad v_2 - v_1 = \int_{t_1}^{t_2} a\,dt \qquad (11.12)$$

The first formula expresses that the area measured under the $v$–$t$ curve from $t_1$ to $t_2$ is equal to the change in $x$ during that time interval (Fig. 11.11). Similarly, the second formula expresses that the area measured under the $a$–$t$ curve from $t_1$ to $t_2$ is equal to the change in $v$ during that time interval. These two properties can be used to determine graphically the $x$–$t$ curve of a motion when its $v$–$t$ curve or its $a$–$t$ curve is known (see Sample Prob. 11.6).

Graphical solutions are particularly useful when the motion considered is defined from experimental data and when $x$, $v$, and $a$ are not analytical functions of $t$. They can also be used to advantage when the motion consists of distinct parts and when its analysis requires writing a different equation for each of its parts. When using a graphical solution, however, one should be careful to note that (1) the area under the $v$–$t$ curve measures the *change in $x$*, not $x$ itself, and similarly, that the area under the $a$–$t$ curve measures the change in $v$; (2) an area above the $t$ axis corresponds to an *increase* in $x$ or $v$, while an area located below the $t$ axis measures a *decrease* in $x$ or $v$.

It will be useful to remember in drawing motion curves that if the velocity is constant, it will be represented by a horizontal straight line; the position coordinate $x$ will then be a linear function of $t$ and will be represented by an oblique straight line. If the acceleration is

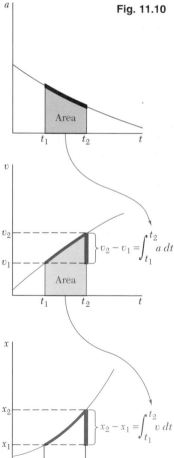

**Fig. 11.11**

constant and different from zero, it will be represented by a horizontal straight line; $v$ will then be a linear function of $t$, represented by an oblique straight line, and $x$ will be expressed as a second-degree polynomial in $t$, represented by a parabola. If the acceleration is a linear function of $t$, the velocity and the position coordinate will be equal, respectively, to second-degree and third-degree polynomials; $a$ will then be represented by an oblique straight line, $v$ by a parabola, and $x$ by a cubic. In general, if the acceleration is a polynomial of degree $n$ in $t$, the velocity will be a polynomial of degree $n + 1$ and the position coordinate a polynomial of degree $n + 2$; these polynomials are represented by motion curves of a corresponding degree.

## *11.8. OTHER GRAPHICAL METHODS

An alternative graphical solution can be used to determine the position of a particle at a given instant directly from the $a$–$t$ curve. Denoting the values of $x$ and $v$ at $t = 0$ as $x_0$ and $v_0$ and their values at $t = t_1$ as $x_1$ and $v_1$, and observing that the area under the $v$–$t$ curve can be divided into a rectangle of area $v_0 t_1$ and horizontal differential elements of area $(t_1 - t)\,dv$ (Fig. 11.12$a$), we write

$$x_1 - x_0 = \text{area under } v\text{–}t \text{ curve} = v_0 t_1 + \int_{v_0}^{v_1} (t_1 - t)\,dv$$

Substituting $dv = a\,dt$ in the integral, we obtain

$$x_1 - x_0 = v_0 t_1 + \int_0^{t_1} (t_1 - t)\,a\,dt$$

Referring to Fig. 11.12$b$, we note that the integral represents the first moment of the area under the $a$–$t$ curve with respect to the line $t = t_1$ bounding the area on the right. This method of solution is known, therefore, as the *moment-area method*. If the abscissa $\bar{t}$ of the centroid $C$ of the area is known, the position coordinate $x_1$ can be obtained by writing

$$x_1 = x_0 + v_0 t_1 + (\text{area under } a\text{–}t \text{ curve})(t_1 - \bar{t}) \quad (11.13)$$

If the area under the $a$–$t$ curve is a composite area, the last term in (11.13) can be obtained by multiplying each component area by the distance from its centroid to the line $t = t_1$. Areas above the $t$ axis should be considered as positive and areas below the $t$ axis as negative.

Another type of motion curve, the $v$–$x$ curve, is sometimes used. If such a curve has been plotted (Fig. 11.13), the acceleration $a$ can be obtained at any time by drawing the normal $AC$ to the curve and *measuring the subnormal $BC$*. Indeed, observing that the angle between $AC$ and $AB$ is equal to the angle $\theta$ between the horizontal and the tangent at $A$ (the slope of which is $\tan \theta = dv/dx$), we write

$$BC = AB \tan \theta = v \frac{dv}{dx}$$

and thus, recalling formula (11.4),

$$BC = a$$

(a)

(b)

**Fig. 11.12**

**Fig. 11.13**

## SAMPLE PROBLEM 11.6

A subway car leaves station $A$; it gains speed at the rate of 1.2 m/s² for 6 s and then at the rate of 1.8 m/s² until it has reached the speed of 14.6 m/s. The car maintains the same speed until it approaches station $B$; brakes are then applied, giving the car a constant deceleration and bringing it to a stop in 6 s. The total running time from $A$ to $B$ is 40 s. Draw the $a$–$t$, $v$–$t$, and $x$–$t$ curves, and determine the distance between stations $A$ and $B$.

## SOLUTION

**Acceleration-Time Curve.** Since the acceleration is either constant or zero, the $a$–$t$ curve is made of horizontal straight-line segments. The values of $t_2$ and $a_4$ are determined as follows:

$0 < t < 6$:  Change in $v$ = area under $a$–$t$ curve
$$v_6 - 0 = (6 \text{ s})(1.2 \text{ m/s}^2) = 7.2 \text{ m/s}$$

$6 < t < t_2$:  Since the velocity increases from 7.2 to 14.6 m/s,
Change in $v$ = area under $a$–$t$ curve
$$14.6 \text{ m/s} - 7.2 \text{ m/s} = (t_2 - 6)(1.8 \text{ m/s}^2) \qquad t_2 = 10 \text{ s}$$

$t_2 < t < 34$:  Since the velocity is constant, the acceleration is zero.

$34 < t < 40$:  Change in $v$ = area under $a$–$t$ curve
$$0 - 14.6 \text{ m/s} = (6 \text{ s})a_4 \qquad a_4 = -2.4 \text{ m/s}^2$$

The acceleration being negative, the corresponding area is below the $t$ axis; this area represents a decrease in velocity.

**Velocity-Time Curve.** Since the acceleration is either constant or zero, the $v$–$t$ curve is made of straight-line segments connecting the points determined above.

Change in $x$ = area under $v$–$t$ curve

| | |
|---|---|
| $0 < t < 6$: | $x_6 - 0 = \frac{1}{2}(6)(7.2) = 21.6$ m |
| $6 < t < 10$: | $x_{10} - x_6 = \frac{1}{2}(4)(7.2 + 14.6) = 43.6$ m |
| $10 < t < 34$: | $x_{34} - x_{10} = (24)(14.6) = 350.4$ m |
| $34 < t < 40$: | $x_{40} - x_{34} = \frac{1}{2}(14.6)(48) = 43.8$ m |

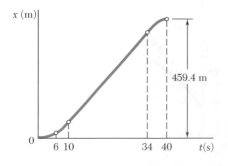

Adding the changes in $x$, we obtain the distance from $A$ to $B$:

$$d = x_{40} - 0 = 459.4 \text{ m}$$

$$d = 459.4 \text{ m} \blacktriangleleft$$

**Position-Time Curve.** The points determined above should be joined by three arcs of parabola and one straight-line segment. In constructing the $x$–$t$ curve, keep in mind that for any value of $t$ the slope of the tangent to the $x$–$t$ curve is equal to the value of $v$ at that instant.

# SOLVING PROBLEMS
## ON YOUR OWN

In this lesson (Secs. 11.7 and 11.8), we reviewed and developed several *graphical techniques for the solution of problems involving rectilinear motion*. These techniques can be used to solve problems directly or to complement analytical methods of solution by providing a visual description, and thus a better understanding, of the motion of a given body. We suggest that you sketch one or more motion curves for several of the problems in this lesson, even if these problems are not part of your homework assignment.

1. *Drawing x–t, v–t, and a–t curves and applying graphical methods.* The following properties were indicated in Sec. 11.7 and should be kept in mind as you use a graphical method of solution.

   *a. The slopes of the x–t and v–t curves* at a time $t_1$ are respectively equal to the *velocity* and the *acceleration* at time $t_1$.

   *b. The areas under the a–t and v–t curves* between the times $t_1$ and $t_2$ are respectively equal to the change $\Delta v$ in the velocity and to the change $\Delta x$ in the position coordinate during that time interval.

   *c. If one of the motion curves is known,* the fundamental properties we have summarized in paragraphs *a* and *b* will enable you to construct the other two curves. However, when using the properties of paragraph *b*, the velocity and the position coordinate at time $t_1$ must be known in order to determine the velocity and the position coordinate at time $t_2$. Thus, in Sample Prob. 11.6, knowing that the initial value of the velocity was zero allowed us to find the velocity at $t = 6$ s: $v_6 = v_0 + \Delta v = 0 + 7.2$ m/s $= 7.2$ m/s.

If you have previously studied the shear and bending-moment diagrams for a beam, you should recognize the analogy that exists between the three motion curves and the three diagrams representing respectively the distributed load, the shear, and the bending moment in the beam. Thus, any techniques that you learned regarding the construction of these diagrams can be applied when drawing the motion curves.

2. *Using approximate methods.* When the *a–t* and *v–t* curves are not represented by analytical functions or when they are based on experimental data, it is often necessary to use approximate methods to calculate the areas under these curves. In those cases, the given area is approximated by a series of rectangles of width $\Delta t$. The smaller the value of $\Delta t$, the smaller the error introduced by the approximation. The velocity and the position coordinate are obtained by writing

$$v = v_0 + \Sigma a_{\text{ave}}\, \Delta t \qquad x = x_0 + \Sigma v_{\text{ave}}\, \Delta t$$

where $a_{\text{ave}}$ and $v_{\text{ave}}$ are the heights of an acceleration rectangle and a velocity rectangle, respectively.

*(continued)*

3. **Applying the moment-area method.** This graphical technique is used when the $a$–$t$ curve is given and the change in the position coordinate is to be determined. We found in Sec. 11.8 that the position coordinate $x_1$ can be expressed as

$$x_1 = x_0 + v_0 t_1 + (\text{area under } a\text{–}t \text{ curve})(t_1 - \bar{t}) \qquad (11.13)$$

Keep in mind that when the area under the $a$–$t$ curve is a composite area, the same value of $t_1$ should be used for computing the contribution of each of the component areas.

4. **Determining the acceleration from a $v$–$x$ curve.** You saw in Sec. 11.8 that it is possible to determine the acceleration from a $v$–$x$ curve by direct measurement. It is important to note, however, that this method is applicable only if the same linear scale is used for the $v$ and $x$ axes (for example, 1 in. = 10 ft and 1 in. = 10 ft/s). When this condition is not satisfied, the acceleration can still be determined from the equation

$$a = v \frac{dv}{dx}$$

where the slope $dv/dx$ is obtained as follows: First, draw the tangent to the curve at the point of interest. Next, using appropriate scales, measure along that tangent corresponding increments $\Delta x$ and $\Delta v$. The desired slope is equal to the ratio $\Delta v / \Delta x$.

# Problems

**11.63**   A particle moves in a straight line with a constant acceleration of $-2$ m/s$^2$ for 6 s, zero acceleration for the next 4 s, and a constant acceleration of $+2$ m/s$^2$ for the next 4 s. Knowing that the particle starts from the origin and that its velocity is $-4$ m/s during the zero acceleration time interval, (a) construct the $v-t$ and $x-t$ curves for $0 \leq t \leq 14$ s, (b) determine the position and the velocity of the particle and the total distance traveled when $t = 14$ s.

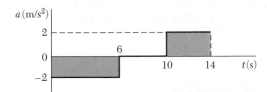

**Fig. P11.63 and P11.64**

**11.64**   A particle moves in a straight line with a constant acceleration of $-2$ m/s$^2$ for 6 s, zero acceleration for the next 4 s, and a constant acceleration of $+2$ m/s$^2$ for the next 4 s. Knowing that the particle starts from the origin with $v_0 = 8$ m/s, (a) construct the $v-t$ and $x-t$ curves for $0 \leq t \leq 14$ s, (b) determine the amount of time during which the particle is farther than 8 m from the origin.

**11.65**   A particle moves in a straight line with the velocity shown in the figure. Knowing that $x = -14.6$ m at $t = 0$, draw the $a-t$ and $x-t$ curves for $0 < t < 40$ s and determine (a) the maximum value of the position coordinate of the particle, (b) the values of $t$ for which the particle is at a distance of 33 m from the origin.

**Fig. P11.65**

**11.66**   For the particle and motion of Prob. 11.65, plot the $a-t$ and $x-t$ curves for $0 < t < 40$ s and determine (a) the total distance traveled by the particle during the period $t = 0$ to $t = 30$ s, (b) the two values of $t$ for which the particle passes through the origin.

**Fig. P11.68**

**11.67** A machine component is spray-painted while it is mounted on a pallet that travels 3.6 m in 20 s. The pallet has an initial velocity of 7.6 cm/s and can be accelerated at a maximum rate of 5 cm/s². Knowing that the painting process requires 15 s to complete and is performed as the pallet moves with a constant speed, determine the smallest possible value of the maximum speed of the pallet.

**11.68** A parachutist is in free fall at a rate of 54.8 m/s when he opens his parachute at an altitude of 579 m. Following a rapid and constant deceleration, he then descends at a constant rate of 13.4 m/s from 549 m to 30.5 m, where he maneuvers the parachute into the wind to further slow his descent. Knowing that the parachutist lands with a negligible downward velocity, determine (a) the time required for the parachutist to land after opening his parachute, (b) the initial deceleration.

**11.69** A commuter train traveling at 64 km/h is 4.8 km from a station. The train then decelerates so that its speed is 32 km/h when it is 800 m from the station. Knowing that the train arrives at the station 7.5 min after beginning to decelerate and assuming constant decelerations, determine (a) the time required to travel the first 4 km, (b) the speed of the train as it arrives at the station, (c) the final constant deceleration of the train.

**Fig. P11.69**

**11.70** Two road rally checkpoints A and B are located on the same highway and are 12.8 km apart. The speed limits for the first 8 km and the last 4.8 km are 96 km/h and 56 km/h, respectively. Drivers must stop at each checkpoint, and the specified time between points A and B is 10 min 20 s. Knowing that a driver accelerates and decelerates at the same constant rate, determine the magnitude of her acceleration if she travels at the speed limit as much as possible.

**Fig. P11.70**

**11.71** In a water-tank test involving the launching of a small model boat, the model's initial horizontal velocity is 6 m/s and its horizontal acceleration varies linearly from $-12$ m/s² at $t = 0$ to $-1.8$ m/s² at $t = t_1$ and then remains equal to $-1.8$ m/s² until $t = 1.4$ s. Knowing that $v = 1.8$ m/s² when $t = t_1$, determine (a) the value of $t_1$, (b) the velocity and position of the model at $t = 1.4$ s.

**11.72** In a 0.4 km race, runner A reaches her maximum velocity $v_A$ in 4 s with constant acceleration and maintains that velocity until she reaches the halfway point with a split time of 25 s. Runner B reaches her maximum velocity $v_B$ in 5 s with constant acceleration and maintains that velocity until she reaches the halfway point with a split time of 25.2 s. Both runners then run the second half of the race with the same constant deceleration of 0.09 m/s². Determine (a) the race times for both runners, (b) the position of the winner relative to the loser when the winner reaches the finish line.

**Fig. P11.72**

**11.73** A bus is parked along the side of a highway when it is passed by a truck traveling at a constant speed of 70 km/h. Two minutes later, the bus starts and accelerates until it reaches a speed of 100 km/h, which it then maintains. Knowing that 12 min after the truck passed the bus, the bus is 1.2 km ahead of the truck, determine (*a*) when and where the bus passed the truck, (*b*) the uniform acceleration of the bus.

**11.74** Cars A and B are *d* = 60 m apart and are traveling respectively at the constant speeds of $(v_A)_0$ = 32 km/h and $(v_B)_0$ = 24 km/h on an ice-covered road. Knowing that 45 s after driver A applies his brakes to avoid overtaking car B the two cars collide, determine (*a*) the uniform deceleration of car A, (*b*) the relative velocity of car A with respect to car B when they collide.

**Fig. *P11.74* and P11.75**

**11.75** Cars A and B are traveling respectively at the constant speeds of $(v_A)_0$ = 35 km/h and $(v_B)_0$ = 20.8 km/h on an ice-covered road. To avoid overtaking car B, the driver of car A applies his brakes so that his car decelerates at a constant rate of .04 m/s$^2$. Determine the distance *d* between the cars at which the driver of car A must apply his brakes to just avoid colliding with car B.

**11.76** An elevator starts from rest and moves upward, accelerating at a rate of 1.2 m/s$^2$ until it reaches a speed of 7.3 m/s, which it then maintains. Two seconds after the elevator begins to move, a man standing 12.2 m above the initial position of the top of the elevator throws a ball upward with an initial velocity of 19.5 m/s. Determine when the ball will hit the elevator.

12.2 m

**Fig. P11.76**

**11.77** A car and a truck are both traveling at the constant speed of 54 km/h; the car is 30 m behind the truck. The driver of the car wants to pass the truck, that is, he wishes to place his car at $B$, 30 m in front of the truck, and then resume the speed of 54 km/h. The maximum acceleration of the car is 2 m/s$^2$ and the maximum deceleration obtained by applying the brakes is 8 m/s$^2$. What is the shortest time in which the driver of the car can complete the passing operation if he does not at any time exceed a speed of 90 km/h? Draw the $v$–$t$ curve.

**Fig. P11.77**

**11.78** Solve Prob. 11.77, assuming that the driver of the car does not pay any attention to the speed limit while passing and concentrates on reaching position $B$ and resuming a speed of 54 km/h in the shortest possible time. What is the maximum speed reached? Draw the $v$–$t$ curve.

**11.79** During a manufacturing process, a conveyor belt starts from rest and travels a total of 0.36 m before temporarily coming to rest. Knowing that the jerk, or rate of change of acceleration, is limited to $\pm 1.5$ m/s$^2$ per second, determine (a) the shortest time required for the belt to move 0.36 m, (b) the maximum and average values of the velocity of the belt during that time.

**11.80** An airport shuttle train travels between two terminals that are 5 km apart. To maintain passenger comfort, the acceleration of the train is limited to $\pm 1.25$ m/s$^2$, and the jerk, or rate of change of acceleration, is limited to $\pm 0.25$ m/s$^2$ per second. If the shuttle has a maximum speed of 32 km/h, determine (a) the shortest time for the shuttle to travel between the two terminals, (b) the corresponding average velocity of the shuttle.

**11.81** An elevator starts from rest and rises 40 m to its maximum velocity in $T$ s with the acceleration record shown in the figure. Determine (a) the required time $T$, (b) the maximum velocity, (c) the velocity and position of the elevator at $t = T/2$.

**Fig. P11.81**

**11.82** An accelerometer record for the motion of a given part of a mechanism is approximated by an arc of a parabola for 0.2 s and a straight line for the next 0.2 s as shown in the figure. Knowing that $v = 0$ when $t = 0$ and $x = 0.3$ m when $t = 0.4$ s, (a) construct the $v$–$t$ curve for $0 \leq t \leq 0.4$ s. (b) determine the position of the part at $t = 0.3$ s and $t = 0.2$ s.

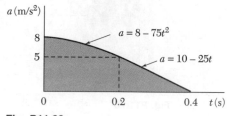

**Fig. P11.82**

**11.83** Two seconds are required to bring the piston rod of an air cylinder to rest; the acceleration record of the piston rod during the 2 s is as shown. Determine by approximate means (*a*) the initial velocity of the piston rod, (*b*) the distance traveled by the piston rod as it is brought to rest.

**Fig. P11.83**

**11.84** The acceleration record shown was obtained during the speed trials of a sports car. Knowing that the car starts from rest, determine by approximate means (*a*) the velocity of the car at *t* = 8 s, (*b*) the distance the car traveled at *t* = 20 s.

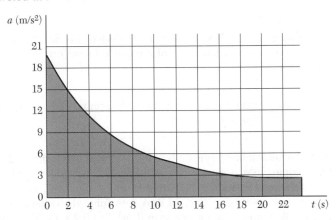

**Fig. P11.84**

**11.85** A training airplane has a velocity of 32 m/s when it lands on an aircraft carrier. As the arresting gear of the carrier brings the airplane to rest, the velocity and the acceleration of the airplane are recorded; the results are shown (solid curve) in the figure. Determine by approximate means (*a*) the time required for the airplane to come to rest, (*b*) the distance traveled in that time.

**Fig. P11.85**

**11.86**  Shown in the figure is a portion of the experimentally determined $v$–$x$ curve for a shuttle cart. Determine by approximate means the acceleration of the cart ($a$) when $x = 0.25$ m, ($b$) when $v = 2$ m/s.

**Fig. P11.86**

**11.87**  Using the method of Sec. 11.8, derive the formula $x = x_0 + v_0 t + \frac{1}{2} a t^2$ for the position coordinate of a particle in uniformly accelerated rectilinear motion.

**11.88**  Using the method of Sec. 11.8, determine the position of the particle of Prob. 11.63 when $t = 12$ s.

**11.89**  An automobile begins a braking test with a velocity of 27.4 m/s at $t = 0$ and comes to a stop at $t = t_1$ with the acceleration record shown. Knowing that the area under the $a$–$t$ curve from $t = 0$ to $t = T$ is a semiparabolic area, use the method of Sec. 11.8 to determine the distance traveled by the automobile before coming to a stop if ($a$) $T = 0.2$ s, ($b$) $T = 0.8$ s.

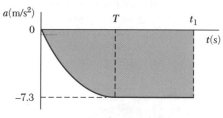

**Fig. P11.89**

**11.90**  For the particle of Prob. 11.65, draw the $a$–$t$ curve and, using the method of Sec. 11.8, determine ($a$) the position of the particle when $t = 20$ s, ($b$) the maximum value of its position coordinate.

## 11.9. POSITION VECTOR, VELOCITY, AND ACCELERATION

When a particle moves along a curve other than a straight line, we say that the particle is in *curvilinear motion*. To define the position $P$ occupied by the particle at a given time $t$, we select a fixed reference system, such as the $x$, $y$, $z$ axes shown in Fig. 11.14$a$, and draw the vector $\mathbf{r}$ joining the origin $O$ and point $P$. Since the vector $\mathbf{r}$ is characterized by its magnitude $r$ and its direction with respect to the reference axes, it completely defines the position of the particle with respect to those axes; the vector $\mathbf{r}$ is referred to as the *position vector* of the particle at time $t$.

Consider now the vector $\mathbf{r}'$ defining the position $P'$ occupied by the same particle at a later time $t + \Delta t$. The vector $\Delta\mathbf{r}$ joining $P$ and $P'$ represents the change in the position vector during the time interval $\Delta t$ since, as we can easily check from Fig. 11.14$a$, the vector $\mathbf{r}'$ is obtained by adding the vectors $\mathbf{r}$ and $\Delta\mathbf{r}$ according to the triangle rule. We note that $\Delta\mathbf{r}$ represents a change in *direction* as well as a change in *magnitude* of the position vector $\mathbf{r}$. The *average velocity* of the particle over the time interval $\Delta t$ is defined as the quotient of $\Delta\mathbf{r}$ and $\Delta t$. Since $\Delta\mathbf{r}$ is a vector and $\Delta t$ is a scalar, the quotient $\Delta\mathbf{r}/\Delta t$ is a vector attached at $P$, of the same direction as $\Delta\mathbf{r}$ and of magnitude equal to the magnitude of $\Delta\mathbf{r}$ divided by $\Delta t$ (Fig. 11.14$b$).

The *instantaneous velocity* of the particle at time $t$ is obtained by choosing shorter and shorter time intervals $\Delta t$ and, correspondingly, shorter and shorter vector increments $\Delta\mathbf{r}$. The instantaneous velocity is thus represented by the vector

$$\mathbf{v} = \lim_{\Delta t \to 0} \frac{\Delta\mathbf{r}}{\Delta t} \tag{11.14}$$

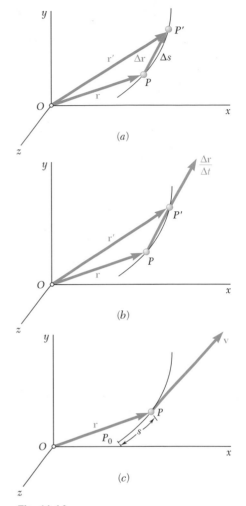

**Fig. 11.14**

As $\Delta t$ and $\Delta\mathbf{r}$ become shorter, the points $P$ and $P'$ get closer; the vector $\mathbf{v}$ obtained in the limit must therefore be tangent to the path of the particle (Fig. 11.14$c$).

Since the position vector $\mathbf{r}$ depends upon the time $t$, we can refer to it as a *vector function* of the scalar variable $t$ and denote it by $\mathbf{r}(t)$. Extending the concept of derivative of a scalar function introduced in elementary calculus, we will refer to the limit of the quotient $\Delta\mathbf{r}/\Delta t$ as the *derivative* of the vector function $\mathbf{r}(t)$. We write

$$\mathbf{v} = \frac{d\mathbf{r}}{dt} \tag{11.15}$$

The magnitude $v$ of the vector $\mathbf{v}$ is called the *speed* of the particle. It can be obtained by substituting for the vector $\Delta\mathbf{r}$ in formula (11.14) the magnitude of this vector represented by the straight-line segment $PP'$. But the length of the segment $PP'$ approaches the length $\Delta s$ of the arc $PP'$ as $\Delta t$ decreases (Fig. 11.14$a$), and we can write

$$v = \lim_{\Delta t \to 0} \frac{PP'}{\Delta t} = \lim_{\Delta t \to 0} \frac{\Delta s}{\Delta t} \qquad v = \frac{ds}{dt} \tag{11.16}$$

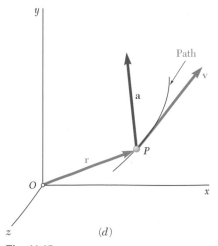

**Fig. 11.15**

The speed $v$ can thus be obtained by differentiating with respect to $t$ the length $s$ of the arc described by the particle.

Consider the velocity $\mathbf{v}$ of the particle at time $t$ and its velocity $\mathbf{v}'$ at a later time $t + \Delta t$ (Fig. 11.15$a$). Let us draw both vectors $\mathbf{v}$ and $\mathbf{v}'$ from the same origin $O'$ (Fig. 11.15$b$). The vector $\Delta\mathbf{v}$ joining $Q$ and $Q'$ represents the change in the velocity of the particle during the time interval $\Delta t$, since the vector $\mathbf{v}'$ can be obtained by adding the vectors $\mathbf{v}$ and $\Delta\mathbf{v}$. We should note that $\Delta\mathbf{v}$ represents a change in the *direction* of the velocity as well as a change in *speed*. The *average acceleration* of the particle over the time interval $\Delta t$ is defined as the quotient of $\Delta\mathbf{v}$ and $\Delta t$. Since $\Delta\mathbf{v}$ is a vector and $\Delta t$ a scalar, the quotient $\Delta\mathbf{v}/\Delta t$ is a vector of the same direction as $\Delta\mathbf{v}$.

The *instantaneous acceleration* of the particle at time $t$ is obtained by choosing smaller and smaller values for $\Delta t$ and $\Delta\mathbf{v}$. The instantaneous acceleration is thus represented by the vector

$$\mathbf{a} = \lim_{\Delta t \to 0} \frac{\Delta\mathbf{v}}{\Delta t} \tag{11.17}$$

Noting that the velocity $\mathbf{v}$ is a vector function $\mathbf{v}(t)$ of the time $t$, we can refer to the limit of the quotient $\Delta\mathbf{v}/\Delta t$ as the derivative of $\mathbf{v}$ with respect to $t$. We write

$$\mathbf{a} = \frac{d\mathbf{v}}{dt} \tag{11.18}$$

We observe that the acceleration $\mathbf{a}$ is tangent to the curve described by the tip $Q$ of the vector $\mathbf{v}$ when the latter is drawn from a fixed origin $O'$ (Fig. 11.15$c$) and that, in general, the acceleration is *not* tangent to the path of the particle (Fig. 11.15$d$). The curve described by the tip of $\mathbf{v}$ and shown in Fig. 11.15$c$ is called the *hodograph* of the motion.

# 11.10. DERIVATIVES OF VECTOR FUNCTIONS

We saw in the preceding section that the velocity **v** of a particle in curvilinear motion can be represented by the derivative of the vector function **r**(t) characterizing the position of the particle. Similarly, the acceleration **a** of the particle can be represented by the derivative of the vector function **v**(t). In this section, we will give a formal definition of the derivative of a vector function and establish a few rules governing the differentiation of sums and products of vector functions.

Let **P**(u) be a vector function of the scalar variable u. By that we mean that the scalar u completely defines the magnitude and direction of the vector **P**. If the vector **P** is drawn from a fixed origin O and the scalar u is allowed to vary, the tip of **P** will describe a given curve in space. Consider the vectors **P** corresponding, respectively, to the values u and u + Δu of the scalar variable (Fig. 11.16a). Let Δ**P** be the vector joining the tips of the two given vectors; we write

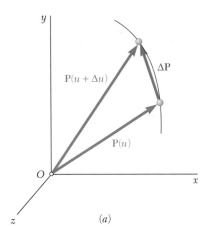

$$\Delta \mathbf{P} = \mathbf{P}(u + \Delta u) - \mathbf{P}(u)$$

Dividing through by Δu and letting Δu approach zero, *we define the derivative of the vector function* **P**(u):

$$\frac{d\mathbf{P}}{du} = \lim_{\Delta u \to 0} \frac{\Delta \mathbf{P}}{\Delta u} = \lim_{\Delta u \to 0} \frac{\mathbf{P}(u + \Delta u) - \mathbf{P}(u)}{\Delta u} \qquad (11.19)$$

As Δu approaches zero, the line of action of Δ**P** becomes tangent to the curve of Fig. 11.16a. Thus, the derivative d**P**/du of the vector function **P**(u) *is tangent to the curve described by the tip of* **P**(u) (Fig. 11.16b).

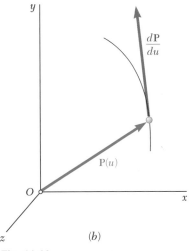

The standard rules for the differentiation of the sums and products of scalar functions can be extended to vector functions. Consider first the *sum of two vector functions* **P**(u) *and* **Q**(u) of the same scalar variable u. According to the definition given in (11.19), the derivative of the vector **P** + **Q** is

**Fig. 11.16**

$$\frac{d(\mathbf{P} + \mathbf{Q})}{du} = \lim_{\Delta u \to 0} \frac{\Delta(\mathbf{P} + \mathbf{Q})}{\Delta u} = \lim_{\Delta u \to 0} \left( \frac{\Delta \mathbf{P}}{\Delta u} + \frac{\Delta \mathbf{Q}}{\Delta u} \right)$$

or since the limit of a sum is equal to the sum of the limits of its terms,

$$\frac{d(\mathbf{P} + \mathbf{Q})}{du} = \lim_{\Delta u \to 0} \frac{\Delta \mathbf{P}}{\Delta u} + \lim_{\Delta u \to 0} \frac{\Delta \mathbf{Q}}{\Delta u}$$

$$\frac{d(\mathbf{P} + \mathbf{Q})}{du} = \frac{d\mathbf{P}}{du} + \frac{d\mathbf{Q}}{du} \qquad (11.20)$$

The *product of a scalar function* f(u) *and a vector function* **P**(u) of the same scalar variable u will now be considered. The derivative of the vector f**P** is

$$\frac{d(f\mathbf{P})}{du} = \lim_{\Delta u \to 0} \frac{(f + \Delta f)(\mathbf{P} + \Delta \mathbf{P}) - f\mathbf{P}}{\Delta u} = \lim_{\Delta u \to 0} \left( \frac{\Delta f}{\Delta u} \mathbf{P} + f \frac{\Delta \mathbf{P}}{\Delta u} \right)$$

or recalling the properties of the limits of sums and products,

$$\frac{d(f\mathbf{P})}{du} = \frac{df}{du}\mathbf{P} + f\frac{d\mathbf{P}}{du} \tag{11.21}$$

The derivatives of the *scalar product* and the *vector product* of two vector functions $\mathbf{P}(u)$ and $\mathbf{Q}(u)$ can be obtained in a similar way. We have

$$\frac{d(\mathbf{P} \cdot \mathbf{Q})}{du} = \frac{d\mathbf{P}}{du} \cdot \mathbf{Q} + \mathbf{P} \cdot \frac{d\mathbf{Q}}{du} \tag{11.22}$$

$$\frac{d(\mathbf{P} \times \mathbf{Q})}{du} = \frac{d\mathbf{P}}{du} \times \mathbf{Q} + \mathbf{P} \times \frac{d\mathbf{Q}}{du} \tag{11.23}†$$

The properties established above can be used to determine the *rectangular components of the derivative of a vector function* $\mathbf{P}(u)$. Resolving $\mathbf{P}$ into components along fixed rectangular axes $x$, $y$, $z$, we write

$$\mathbf{P} = P_x\mathbf{i} + P_y\mathbf{j} + P_z\mathbf{k} \tag{11.24}$$

where $P_x$, $P_y$, $P_z$ are the rectangular scalar components of the vector $\mathbf{P}$, and $\mathbf{i}$, $\mathbf{j}$, $\mathbf{k}$ the unit vectors corresponding, respectively, to the $x$, $y$, and $z$ axes (Sec. 2.12). By (11.20), the derivative of $\mathbf{P}$ is equal to the sum of the derivatives of the terms in the right-hand member. Since each of these terms is the product of a scalar and a vector function, we should use (11.21). But the unit vectors $\mathbf{i}$, $\mathbf{j}$, $\mathbf{k}$ have a constant magnitude (equal to 1) and fixed directions. Their derivatives are therefore zero, and we write

$$\frac{d\mathbf{P}}{du} = \frac{dP_x}{du}\mathbf{i} + \frac{dP_y}{du}\mathbf{j} + \frac{dP_z}{du}\mathbf{k} \tag{11.25}$$

Noting that the coefficients of the unit vectors are, by definition, the scalar components of the vector $d\mathbf{P}/du$, we conclude that *the rectangular scalar components of the derivative* $d\mathbf{P}/du$ *of the vector function* $\mathbf{P}(u)$ *are obtained by differentiating the corresponding scalar components of* $\mathbf{P}$.

**Rate of Change of a Vector.**  When the vector $\mathbf{P}$ is a function of the time $t$, its derivative $d\mathbf{P}/dt$ represents the *rate of change* of $\mathbf{P}$ with respect to the frame $Oxyz$. Resolving $\mathbf{P}$ into rectangular components, we have, by (11.25),

$$\frac{d\mathbf{P}}{dt} = \frac{dP_x}{dt}\mathbf{i} + \frac{dP_y}{dt}\mathbf{j} + \frac{dP_z}{dt}\mathbf{k}$$

or, using dots to indicate differentiation with respect to $t$,

$$\dot{\mathbf{P}} = \dot{P}_x\mathbf{i} + \dot{P}_y\mathbf{j} + \dot{P}_z\mathbf{k} \tag{11.25'}$$

---

†Since the vector product is not commutative (Sec. 3.4), the order of the factors in Eq. (11.23) must be maintained.

As you will see in Sec. 15.10, the rate of change of a vector as observed from a *moving frame of reference* is, in general, different from its rate of change as observed from a fixed frame of reference. However, if the moving frame $O'x'y'z'$ is in *translation*, that is, if its axes remain parallel to the corresponding axes of the fixed frame $Oxyz$ (Fig. 11.17), the same unit vectors $\mathbf{i}$, $\mathbf{j}$, $\mathbf{k}$ are used in both frames, and at any given instant the vector $\mathbf{P}$ has the same components $P_x$, $P_y$, $P_z$ in both frames. It follows from (11.25′) that the rate of change $\dot{\mathbf{P}}$ is the same with respect to the frames $Oxyz$ and $O'x'y'z'$. We state, therefore: *The rate of change of a vector is the same with respect to a fixed frame and with respect to a frame in translation.* This property will greatly simplify our work, since we will be concerned mainly with frames in translation.

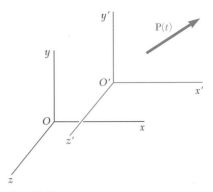

**Fig. 11.17**

## 11.11. RECTANGULAR COMPONENTS OF VELOCITY AND ACCELERATION

When the position of a particle $P$ is defined at any instant by its rectangular coordinates $x$, $y$, and $z$, it is convenient to resolve the velocity $\mathbf{v}$ and the acceleration $\mathbf{a}$ of the particle into rectangular components (Fig. 11.18).

Resolving the position vector $\mathbf{r}$ of the particle into rectangular components, we write

$$\mathbf{r} = x\mathbf{i} + y\mathbf{j} + z\mathbf{k} \tag{11.26}$$

where the coordinates $x$, $y$, $z$ are functions of $t$. Differentiating twice, we obtain

$$\mathbf{v} = \frac{d\mathbf{r}}{dt} = \dot{x}\mathbf{i} + \dot{y}\mathbf{j} + \dot{z}\mathbf{k} \tag{11.27}$$

$$\mathbf{a} = \frac{d\mathbf{v}}{dt} = \ddot{x}\mathbf{i} + \ddot{y}\mathbf{j} + \ddot{z}\mathbf{k} \tag{11.28}$$

where $\dot{x}$, $\dot{y}$, $\dot{z}$ and $\ddot{x}$, $\ddot{y}$, $\ddot{z}$ represent, respectively, the first and second derivatives of $x$, $y$, and $z$ with respect to $t$. It follows from (11.27) and (11.28) that the scalar components of the velocity and acceleration are

$$v_x = \dot{x} \qquad v_y = \dot{y} \qquad v_z = \dot{z} \tag{11.29}$$
$$a_x = \ddot{x} \qquad a_y = \ddot{y} \qquad a_z = \ddot{z} \tag{11.30}$$

A positive value for $v_x$ indicates that the vector component $\mathbf{v}_x$ is directed to the right, and a negative value indicates that it is directed to the left. The sense of each of the other vector components can be determined in a similar way from the sign of the corresponding scalar component. If desired, the magnitudes and directions of the velocity and acceleration can be obtained from their scalar components by the methods of Secs. 2.7 and 2.12.

The use of rectangular components to describe the position, the velocity, and the acceleration of a particle is particularly effective when the component $a_x$ of the acceleration depends only upon $t$, $x$, and/or $v_x$, and when, similarly, $a_y$ depends only upon $t$, $y$, and/or $v_y$,

*(a)*

*(b)*

**Fig. 11.18**

and $a_z$ upon $t$, $z$, and/or $v_z$. Equations (11.30) can then be integrated independently, and so can Eqs. (11.29). In other words, the motion of the particle in the $x$ direction, its motion in the $y$ direction, and its motion in the $z$ direction can be considered separately.

In the case of the *motion of a projectile*, for example, it can be shown (see Sec. 12.5) that the components of the acceleration are

$$a_x = \ddot{x} = 0 \qquad a_y = \ddot{y} = -g \qquad a_z = \ddot{z} = 0$$

if the resistance of the air is neglected. Denoting by $x_0$, $y_0$, and $z_0$ the coordinates of a gun, and by $(v_x)_0$, $(v_y)_0$, and $(v_z)_0$ the components of the initial velocity $\mathbf{v}_0$ of the projectile (a bullet), we integrate twice in $t$ and obtain

$$v_x = \dot{x} = (v_x)_0 \qquad v_y = \dot{y} = (v_y)_0 - gt \qquad v_z = \dot{z} = (v_z)_0$$
$$x = x_0 + (v_x)_0 t \qquad y = y_0 + (v_y)_0 t - \tfrac{1}{2}gt^2 \qquad z = z_0 + (v_z)_0 t$$

If the projectile is fired in the $xy$ plane from the origin $O$, we have $x_0 = y_0 = z_0 = 0$ and $(v_z)_0 = 0$, and the equations of motion reduce to

$$v_x = (v_x)_0 \qquad v_y = (v_y)_0 - gt \qquad v_z = 0$$
$$x = (v_x)_0 t \qquad y = (v_y)_0 t - \tfrac{1}{2}gt^2 \qquad z = 0$$

These equations show that the projectile remains in the $xy$ plane, that its motion in the horizontal direction is uniform, and that its motion in the vertical direction is uniformly accelerated. The motion of a projectile can thus be replaced by two independent rectilinear motions, which are easily visualized if we assume that the projectile is fired vertically with an initial velocity $(\mathbf{v}_y)_0$ from a platform moving with a constant horizontal velocity $(\mathbf{v}_x)_0$ (Fig. 11.19). The coordinate $x$ of the projectile is equal at any instant to the distance traveled by the platform, and its coordinate $y$ can be computed as if the projectile were moving along a vertical line.

It can be observed that the equations defining the coordinates $x$ and $y$ of a projectile at any instant are the parametric equations of a parabola. Thus, the trajectory of a projectile is *parabolic*. This result, however, ceases to be valid when the resistance of the air or the variation with altitude of the acceleration of gravity is taken into account.

(a) Motion of a projectile

(b) Equivalent rectilinear motions

**Fig. 11.19**

**Photo 11.2** When air resistance is neglected, the ski jumper may be considered as a projectile having a parabolic trajectory.

## 11.12. MOTION RELATIVE TO A FRAME IN TRANSLATION

In the preceding section, a single frame of reference was used to describe the motion of a particle. In most cases this frame was attached to the earth and was considered as fixed. Situations in which it is convenient to use several frames of reference simultaneously will now be analyzed. If one of the frames is attached to the earth, it will be called a *fixed frame of reference,* and the other frames will be referred to as *moving frames of reference.* It should be understood, however, that the selection of a fixed frame of reference is purely arbitrary. Any frame can be designated as "fixed"; all other frames not rigidly attached to this frame will then be described as "moving."

Consider two particles $A$ and $B$ moving in space (Fig. 11.20); the vectors $\mathbf{r}_A$ and $\mathbf{r}_B$ define their positions at any given instant with respect to the fixed frame of reference $Oxyz$. Consider now a system of axes $x'$, $y'$, $z'$ centered at $A$ and parallel to the $x$, $y$, $z$ axes. While the origin of these axes moves, their orientation remains the same; the frame of reference $Ax'y'z'$ is in *translation* with respect to $Oxyz$. The vector $\mathbf{r}_{B/A}$ joining $A$ and $B$ defines *the position of $B$ relative to the moving frame $Ax'y'z'$* (or, for short, *the position of $B$ relative to $A$*).

We note from Fig. 11.20 that the position vector $\mathbf{r}_B$ of particle $B$ is the sum of the position vector $\mathbf{r}_A$ of particle $A$ and of the position vector $\mathbf{r}_{B/A}$ of $B$ relative to $A$; we write

$$\mathbf{r}_B = \mathbf{r}_A + \mathbf{r}_{B/A} \tag{11.31}$$

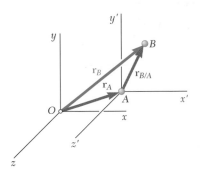

**Fig. 11.20**

Differentiating (11.31) with respect to $t$ within the fixed frame of reference, and using dots to indicate time derivatives, we have

$$\dot{\mathbf{r}}_B = \dot{\mathbf{r}}_A + \dot{\mathbf{r}}_{B/A} \tag{11.32}$$

The derivatives $\dot{\mathbf{r}}_A$ and $\dot{\mathbf{r}}_B$ represent, respectively, the velocities $\mathbf{v}_A$ and $\mathbf{v}_B$ of the particles $A$ and $B$. Since $Ax'y'z'$ is in translation, the derivative $\dot{\mathbf{r}}_{B/A}$ represents the rate of change of $\mathbf{r}_{B/A}$ with respect to the frame $Ax'y'z'$ as well as with respect to the fixed frame (Sec. 11.10). This derivative, therefore, defines *the velocity $\mathbf{v}_{B/A}$ of $B$ relative to the frame $Ax'y'z'$* (or, for short, *the velocity $\mathbf{v}_{B/A}$ of $B$ relative to $A$*). We write

$$\mathbf{v}_B = \mathbf{v}_A + \mathbf{v}_{B/A} \tag{11.33}$$

Differentiating Eq. (11.33) with respect to $t$, and using the derivative $\dot{\mathbf{v}}_{B/A}$ to define *the acceleration $\mathbf{a}_{B/A}$ of $B$ relative to the frame $Ax'y'z'$* (or, for short, *the acceleration $\mathbf{a}_{B/A}$ of $B$ relative to $A$*), we write

$$\mathbf{a}_B = \mathbf{a}_A + \mathbf{a}_{B/A} \tag{11.34}$$

The motion of $B$ with respect to the fixed frame $Oxyz$ is referred to as the *absolute motion of $B$*. The equations derived in this section show that *the absolute motion of $B$ can be obtained by combining the motion of $A$ and the relative motion of $B$ with respect to the moving frame attached to $A$*. Equation (11.33), for example, expresses that the absolute velocity $\mathbf{v}_B$ of particle $B$ can be obtained by adding vectorially the velocity of $A$ and the velocity of $B$ relative to the frame $Ax'y'z'$. Equation (11.34) expresses a similar property in terms of the accelerations.† We should keep in mind, however, that *the frame $Ax'y'z'$ is in translation;* that is, while it moves with $A$, it maintains the same orientation. As you will see later (Sec. 15.14), different relations must be used in the case of a rotating frame of reference.

**Photo 11.3**   When a frame of reference attached to the ship moves in translation, the relative motion of the helicopter with respect to the ship is characterized by Eqs. (11.31) through (11.34).

---

†Note that the product of the subscripts $A$ and $B/A$ used in the right-hand member of Eqs. (11.31) through (11.34) is equal to the subscript $B$ used in their left-hand member.

# SAMPLE PROBLEM 11.7

A projectile is fired from the edge of a 150-m cliff with an initial velocity of 180 m/s at an angle of 30° with the horizontal. Neglecting air resistance, find (*a*) the horizontal distance from the gun to the point where the projectile strikes the ground, (*b*) the greatest elevation above the ground reached by the projectile.

## SOLUTION

The vertical and the horizontal motion will be considered separately.

**Vertical Motion.**  *Uniformly Accelerated Motion.*  Choosing the positive sense of the $y$ axis upward and placing the origin $O$ at the gun, we have

$$(v_y)_0 = (180 \text{ m/s}) \sin 30° = +90 \text{ m/s}$$
$$a = -9.81 \text{ m/s}^2$$

Substituting into the equations of uniformly accelerated motion, we have

$$v_y = (v_y)_0 + at \qquad v_y = 90 - 9.81t \qquad (1)$$
$$y = (v_y)_0 t + \tfrac{1}{2}at^2 \qquad y = 90t - 4.90t^2 \qquad (2)$$
$$v_y^2 = (v_y)_0^2 + 2ay \qquad v_y^2 = 8100 - 19.62y \qquad (3)$$

**Horizontal Motion.**  *Uniform Motion.*  Choosing the positive sense of the $x$ axis to the right, we have

$$(v_x)_0 = (180 \text{ m/s}) \cos 30° = +155.9 \text{ m/s}$$

Substituting into the equation of uniform motion, we obtain

$$x = (v_x)_0 t \qquad x = 155.9t \qquad (4)$$

*a.* **Horizontal Distance.**  When the projectile strikes the ground, we have

$$y = -150 \text{ m}$$

Carrying this value into Eq. (2) for the vertical motion, we write

$$-150 = 90t - 4.90t^2 \qquad t^2 - 18.37t - 30.6 = 0 \qquad t = 19.91 \text{ s}$$

Carrying $t = 19.91$ s into Eq. (4) for the horizontal motion, we obtain

$$x = 155.9(19.91) \qquad\qquad x = 3100 \text{ m} \quad \blacktriangleleft$$

*b.* **Greatest Elevation.**  When the projectile reaches its greatest elevation, we have $v_y = 0$; carrying this value into Eq. (3) for the vertical motion, we write

$$0 = 8100 - 19.62y \qquad y = 413 \text{ m}$$

Greatest elevation above ground = 150 m + 413 m = 563 m  ◀

800 m/s

$B$

610 m

$\alpha$

$A$

3658 m

## SAMPLE PROBLEM 11.8

A projectile is fired with an initial velocity of 244 m/s at a target $B$ located 610 m above the gun $A$ and at a horizontal distance of 3658 m. Neglecting air resistance, determine the value of the firing angle $\alpha$.

## SOLUTION

$v_0 = 800$ m/s

$B$

$O$

$\alpha$

$(v_x)_0 = 800 \cos \alpha$

$x$

3658 m

The horizontal and the vertical motion will be considered separately.

**Horizontal Motion.** Placing the origin of the coordinate axes at the gun, we have

$$(v_x)_0 = 244 \cos \alpha$$

Substituting into the equation of uniform horizontal motion, we obtain

$$x = (v_x)_0 t \qquad x = (244 \cos \alpha)t$$

The time required for the projectile to move through a horizontal distance of 3658 m is obtained by setting $x$ equal to 3658 m.

$$3658 = (244 \cos \alpha)t$$

$$t = \cdot \frac{3658}{244 \cos \alpha} = \frac{15}{\cos \alpha}$$

$y$

$a = -9.81$ m/s$^2$

$B$

$O$

$\alpha$

$v_0 = 800$ m/s

610 m

$(v_y)_0 = 800 \sin \alpha$

**Vertical Motion**

$$(v_y)_0 = 244 \sin \alpha \qquad a = -9.81 \text{ m/s}^2$$

Substituting into the equation of uniformly accelerated vertical motion, we obtain

$$y = (v_y)_0 t + \tfrac{1}{2}at^2 \qquad y = (244 \sin \alpha)t - 4.9$$

**Projectile Hits Target.** When $x = 3658$ m, we must have $y = 610$ m. Substituting for $y$ and setting $t$ equal to the value found above, we write

$$610 = 244 \sin \alpha \frac{15}{\cos \alpha} - 4.9 \left( \frac{15}{\cos \alpha} \right)^2$$

Since $1/\cos^2 \alpha = \sec^2 \alpha = 1 + \tan^2 \alpha$, we have

$$610 = 244(15) \tan \alpha - 4.9(15^2)(1 + \tan^2 \alpha)$$
$$1103 \tan^2 \alpha - 3660 \tan \alpha + 1713 = 0$$

Solving this quadratic equation for $\tan \alpha$, we have

$$\tan \alpha = 0.5638 \qquad \text{and} \qquad \tan \alpha = 2.75$$

$$\alpha = 29.4° \qquad \text{and} \qquad \alpha = 70.0° \blacktriangleleft$$

The target will be hit if either of these two firing angles is used (see figure).

70.0°

$B$

$A$

29.5°

## SAMPLE PROBLEM 11.9

Automobile $A$ is traveling east at the constant speed of 36 km/h. As automobile $A$ crosses the intersection shown, automobile $B$ starts from rest 35 m north of the intersection and moves south with a constant acceleration of 1.2 m/s$^2$. Determine the position, velocity, and acceleration of $B$ relative to $A$ 5 s after $A$ crosses the intersection.

## SOLUTION

We choose $x$ and $y$ axes with origin at the intersection of the two streets and with positive senses directed respectively east and north.

**Motion of Automobile A.** First the speed is expressed in m/s:

$$v_A = \left(36 \frac{km}{h}\right)\left(\frac{1000 \text{ m}}{1 \text{ km}}\right)\left(\frac{1 \text{ h}}{3600 \text{ s}}\right) = 10 \text{ m/s}$$

Noting that the motion of $A$ is uniform, we write, for any time $t$,

$$a_A = 0$$
$$v_A = +10 \text{ m/s}$$
$$x_A = (x_A)_0 + v_A t = 0 + 10t$$

For $t = 5$ s, we have

$$a_A = 0 \qquad\qquad \mathbf{a}_A = 0$$
$$v_A = +10 \text{ m/s} \qquad\qquad \mathbf{v}_A = 10 \text{ m/s} \rightarrow$$
$$x_A = +(10 \text{ m/s})(5 \text{ s}) = +50 \text{ m} \qquad \mathbf{r}_A = 50 \text{ m} \rightarrow$$

**Motion of Automobile B.** We note that the motion of $B$ is uniformly accelerated and write

$$a_B = -1.2 \text{ m/s}^2$$
$$v_B = (v_B)_0 + at = 0 - 1.2\,t$$
$$y_B = (y_B)_0 + (v_B)_0 t + \tfrac{1}{2}a_B t^2 = 35 + 0 - \tfrac{1}{2}(1.2)t^2$$

For $t = 5$ s, we have

$$a_B = -1.2 \text{ m/s}^2 \qquad\qquad \mathbf{a}_B = 1.2 \text{ m/s}^2 \downarrow$$
$$v_B = -(1.2 \text{ m/s}^2)(5 \text{ s}) = -6 \text{ m/s} \qquad \mathbf{v}_B = 6 \text{ m/s} \downarrow$$
$$y_B = 35 - \tfrac{1}{2}(1.2 \text{ m/s}^2)(5 \text{ s})^2 = +20 \text{ m} \qquad \mathbf{r}_B = 20 \text{ m} \uparrow$$

**Motion of B Relative to A.** We draw the triangle corresponding to the vector equation $\mathbf{r}_B = \mathbf{r}_A + \mathbf{r}_{B/A}$ and obtain the magnitude and direction of the position vector of $B$ relative to $A$.

$$r_{B/A} = 53.9 \text{ m} \qquad \alpha = 21.8° \qquad \mathbf{r}_{B/A} = 53.9 \text{ m} \; \diagdown \, 21.8° \; \blacktriangleleft$$

Proceeding in a similar fashion, we find the velocity and acceleration of $B$ relative to $A$.

$$\mathbf{v}_B = \mathbf{v}_A + \mathbf{v}_{B/A}$$
$$v_{B/A} = 11.66 \text{ m/s} \qquad \beta = 31.0° \qquad \mathbf{v}_{B/A} = 11.66 \text{ m/s} \; \nearrow \, 31.0° \; \blacktriangleleft$$
$$\mathbf{a}_B = \mathbf{a}_A + \mathbf{a}_{B/A}$$
$$\mathbf{a}_{B/A} = 1.2 \text{ m/s}^2 \downarrow \; \blacktriangleleft$$

# SOLVING PROBLEMS
## ON YOUR OWN

In the problems for this lesson, you will analyze the *two- and three-dimensional motion* of a particle. While the physical interpretations of the velocity and acceleration are the same as in the first lessons of the chapter, you should remember that these quantities are vectors. In addition, you should understand from your experience with vectors in statics that it will often be advantageous to express position vectors, velocities, and accelerations in terms of their rectangular scalar components [Eqs. (11.27) and (11.28)]. Furthermore, given two vectors **A** and **B**, recall that $\mathbf{A} \cdot \mathbf{B} = 0$ if **A** and **B** are perpendicular to each other, while $\mathbf{A} \times \mathbf{B} = 0$ if **A** and **B** are parallel.

**A.  Analyzing the motion of a projectile.** Many of the following problems deal with the two-dimensional motion of a projectile, where the resistance of the air can be neglected. In Sec. 11.11, we developed the equations which describe this type of motion, and we observed that the horizontal component of the velocity remained constant (uniform motion) while the vertical component of the acceleration was constant (uniformly accelerated motion). We were able to consider separately the horizontal and the vertical motions of the particle. Assuming that the projectile is fired from the origin, we can write the two equations

$$x = (v_x)_0 t \qquad y = (v_y)_0 t - \tfrac{1}{2}gt^2$$

*1.  If the initial velocity and firing angle are known,* the value of $y$ corresponding to any given value of $x$ (or the value of $x$ for any value of $y$) can be obtained by solving one of the above equations for $t$ and substituting for $t$ into the other equation [Sample Prob. 11.7].

*2.  If the initial velocity and the coordinates of a point of the trajectory are known,* and you wish to *determine the firing angle $\alpha$*, begin your solution by expressing the components $(v_x)_0$ and $(v_y)_0$ of the initial velocity as functions of the angle $\alpha$. These expressions and the known values of $x$ and $y$ are then substituted into the above equations. Finally, solve the first equation for $t$ and substitute that value of $t$ into the second equation to obtain a trigonometric equation in $\alpha$, which you can solve for that unknown [Sample Prob. 11.8].

*(continued)*

**B. Solving translational two-dimensional relative-motion problems.** You saw in Sec. 11.12 that the absolute motion of a particle $B$ can be obtained by combining the motion of a particle $A$ and the *relative motion* of $B$ with respect to a frame attached to $A$ which is in *translation.* The velocity and acceleration of $B$ can then be expressed as shown in Eqs. (11.33) and (11.34), respectively.

*1. To visualize the relative motion of B with respect to A,* imagine that you are attached to particle $A$ as you observe the motion of particle $B$. For example, to a passenger in automobile $A$ of Sample Prob. 11.9, automobile $B$ appears to be heading in a southwesterly direction (*south* should be obvious; and *west* is due to the fact that automobile $A$ is moving to the east—automobile $B$ then appears to travel to the west). Note that this conclusion is consistent with the direction of $\mathbf{v}_{B/A}$.

*2. To solve a relative-motion problem,* first write the vector Eqs. (11.31), (11.33), and (11.34), which relate the motions of particles $A$ and $B$. You may then use either of the following methods:

*a. Construct the corresponding vector triangles* and solve them for the desired position vector, velocity, and acceleration [Sample Prob. 11.9].

*b. Express all vectors in terms of their rectangular components* and solve the two independent sets of scalar equations obtained in that way. If you choose this approach, be sure to select the same positive direction for the displacement, velocity, and acceleration of each particle.

# Problems

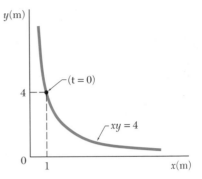

**Fig. P11.91**

**11.91** The motion of a particle is defined by the equations $x = (t + 1)^2$ and $y = 4(t + 1)^{-2}$, where $x$ and $y$ are expressed in feet and $t$ in seconds. Show that the path of the particle is part of the rectangular hyperbola shown and determine the velocity and acceleration when (a) $t = 0$; (b) $t = \frac{1}{2}$ s.

**11.92** The motion of a particle is defined by the equations $x = 6 - 0.8t(t^2 - 9t + 18)$ and $y = -4 + 0.6t(t^2 - 9t + 18)$, where $x$ and $y$ are expressed in meters and $t$ is expressed in seconds. Show that the path of the particle is a portion of a straight line, and determine the velocity and acceleration when (a) $t = 2$ s, (b) $t = 3$ s, (c) $t = 4$ s.

**11.93** The motion of a particle is defined by the equations $x = (4 \cos \pi t - 2)/(2 - \cos \pi t)$ and $y = 3 \sin \pi t/(2 - \cos \pi t)$, where $x$ and $y$ are expressed in meters and $t$ is expressed in seconds. Show that the path of the particle is part of the ellipse shown, and determine the velocity when (a) $t = 0$, (b) $t = 1/3$ s, (c) $t = 1$ s.

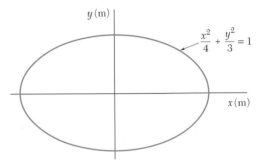

**Fig. P11.93**

**11.94** The motion of a particle is defined by the equations $x = 6t - 3 \sin t$ and $y = 6 - 3 \cos t$, where $x$ and $y$ are expressed in meters and $t$ is expressed in seconds. Sketch the path of the particle for the time interval $0 \leq t \leq 2\pi$, and determine (a) the magnitudes of the smallest and largest velocities reached by the particle, (b) the corresponding times, positions, and directions of the velocities.

**11.95** The motion of a particle is defined by the position vector $\mathbf{r} = A(\cos t + t \sin t)\mathbf{i} + A(\sin t - t \cos t)\mathbf{j}$, where $t$ is expressed in seconds. Determine the values of $t$ for which the position vector and the acceleration vector are (a) perpendicular, (b) parallel.

**11.96** The damped motion of a vibrating particle is defined by the position vector $\mathbf{r} = x_1[1 - 1/(t + 1)]\mathbf{i} + (y_1 e^{-\pi t/2} \cos 2\pi t)\mathbf{j}$, where $t$ is expressed in seconds. For $x_1 = 30$ in. and $y_1 = 20$ in., determine the position, the velocity, and the acceleration of the particle when (a) $t = 0$, (b) $t = 1.5$ s.

**Fig. P11.95**

**Fig. P11.96**

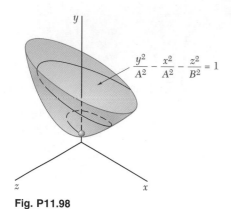

$$\frac{y^2}{A^2} - \frac{x^2}{A^2} - \frac{z^2}{B^2} = 1$$

**Fig. P11.98**

**11.97** The three-dimensional motion of a particle is defined by the position vector $\mathbf{r} = (Rt \cos \omega_n t)\mathbf{i} + ct\mathbf{j} + (Rt \sin \omega_n t)\mathbf{k}$. Determine the magnitudes of the velocity and acceleration of the particle. (The space curve described by the particle is a conic helix.)

**\*11.98** The three-dimensional motion of a particle is defined by the position vector $\mathbf{r} = (At \cos t)\mathbf{i} + (A\sqrt{t^2 + 1})\mathbf{j} + (Bt \sin t)\mathbf{k}$, where $r$ and $t$ are expressed in meter and seconds, respectively. Show that the curve described by the particle lies on the hyperboloid $(y/A)^2 - (x/A)^2 - (z/B)^2 = 1$. For $A - 3$ and $B - 1$, determine ($a$) the magnitudes of the velocity and acceleration when $t = 0$, ($b$) the smallest nonzero value of $t$ for which the position vector and the velocity vector are perpendicular to each other.

**11.99** A ski jumper starts with a horizontal take off velocity of 25 m/s and lands on a straight landing hill inclined at 30°. Determine ($a$) the time between take-off and landing, ($b$) the length $d$ of the jump, ($c$) the maximum vertical distance between the jumper and the landing hill.

**Fig. P11.99**

**11.100** A golfer aims his shot to clear the top of a tree by a distance $h$ at the peak of the trajectory and to miss the pond on the opposite side. Knowing that the magnitude of $\mathbf{v}_0$ is 30 m/s, determine the range of values of $h$ which must be avoided.

**Fig. P11.100**

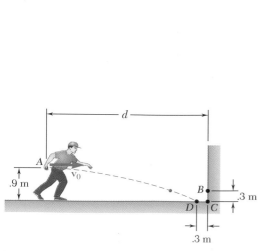

**Fig. P11.101**

**11.101** A handball player throws a ball from $A$ with a horizontal velocity $\mathbf{v}_0$. Knowing that $d = 4.57$ m, determine ($a$) the value of $v_0$ for which the ball will strike the corner $C$, ($b$) the range of values of $v_0$ for which the ball will strike the corner region $BCD$.

**11.102** A helicopter is flying with a constant horizontal velocity of 144.2 km/h and is directly above point $A$ when a loose part begins to fall. The part lands 6.5 s later at point $B$ on an inclined surface. Determine (a) the distance $d$ between points $A$ and $B$, (b) the initial height $h$.

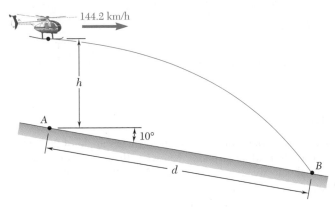

**Fig. P11.102**

**11.103** A pump is located near the edge of the horizontal platform shown. The nozzle at $A$ discharges water with an initial velocity of 7.62 m/s at an angle of 55° with the vertical. Determine the range of values of the height $h$ for which the water enters the opening $BC$.

**11.104** An oscillating water sprinkler is operated at point $A$ on an incline that forms an angle $\alpha$ with the horizontal. The sprinkler discharges water with an initial velocity $\mathbf{v}_0$ at an angle $\phi$ with the vertical which varies from $-\phi_0$ to $+\phi_0$. Knowing that $v_0 = 24$ m/s, $\phi_0 = 40°$, and $\alpha = 10°$, determine the horizontal distance between the sprinkler and points $B$ and $C$ which define the watered area.

**Fig. P11.103**

**Fig. P11.104**   **Fig. P11.105**

**11.105** In slow-pitch softball the underhand pitch must reach a maximum height of between 1.8 m and 3.7 m above the ground. A pitch is made with an initial velocity $\mathbf{v}_0$ of magnitude 13 m/s at an angle of 33° with the horizontal. Determine (a) if the pitch meets the maximum height requirement, (b) the height of the ball as it reaches the batter.

**11.106** A tennis player serves the ball at a height $h$ with an initial velocity of 40 m/s at an angle of 4° with the horizontal. Knowing that the ball clears the 0.914 m net by 152 mm, determine (a) the height $h$, (b) the distance $d$ from the net to where the ball will land.

**Fig. P11.106**

**11.107** The conveyor belt, which forms an angle of 20° with the horizontal, is used to load an airplane. Knowing that a worker tosses a package with an initial velocity $\mathbf{v}_0$ at an angle of 45° so that its velocity is parallel to the belt as it lands .9 m above the release point, determine (a) the magnitude of $v_0$, (b) the horizontal distance $d$.

**11.108** A golfer hits a ball with an initial velocity of magnitude $v_0$ at an angle $\alpha$ with the horizontal. Knowing that the ball must clear the tops of two trees and land as close as possible to the flag, determine $v_0$ and the distance $d$ when the golfer uses (a) a six-iron with $\alpha = 31°$, (b) a five-iron with $\alpha = 27°$.

**Fig. P11.107**

**Fig. P11.108**

**11.109** A homeowner uses a snowblower to clear his driveway. Knowing that the snow is discharged at an average angle of 40° with the horizontal, determine the initial speed $v_0$ of the snow.

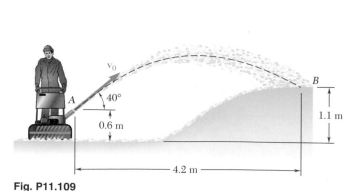

**Fig. P11.109**

**Fig. P11.110**

**11.110** A basketball player shoots when she is 5 m from the backboard. Knowing that the ball has an initial velocity $\mathbf{v}_0$ at an angle of 30° with the horizontal, determine the value of $v_0$ when $d$ is equal to (a) 228 mm, (b) 430 mm.

**11.111** A ball is projected from point $A$ with a velocity $\mathbf{v}_0$ which is perpendicular to the incline shown. Knowing that the ball strikes the incline at $B$, determine the initial speed $v_0$ in terms of the range $R$ and $\beta$.

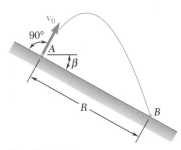

**Fig. P11.111**

**11.112** An outfielder throws a ball with an initial velocity of magnitude $v_0$ at an angle of 10° with the horizontal to the catcher 50 m away. Knowing that the ball is to arrive at a height between 0.5 m and 1.5 m, determine (a) the range of values of $v_0$, (b) the range of values of the maximum height $h$ of the trajectory.

**Fig. P11.112**

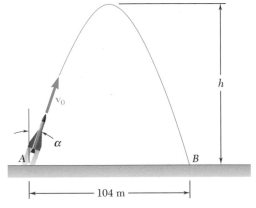

**Fig. P11.113**

**11.113** A model rocket is launched from point $A$ with an initial velocity $\mathbf{v}_0$ of 86 m/s. If the rocket's descent parachute does not deploy and the rocket lands 104 m from $A$, determine (a) the angle $\alpha$ that $\mathbf{v}_0$ forms with the vertical, (b) the maximum height $h$ reached by the rocket, (c) the duration of the flight.

**11.114** The initial velocity $\mathbf{v}_0$ of a hockey puck is 170 km/h. Determine (a) the largest value (less than 45°) of the angle $\alpha$ for which the puck will enter the net, (b) the corresponding time required for the puck to reach the net.

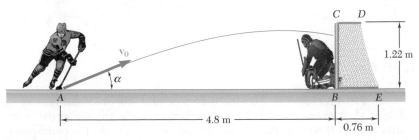

**Fig. P11.114**

**11.115** The pitcher in a softball game throws a ball with an initial velocity $\mathbf{v}_0$ of 64 km/h at an angle $\alpha$ with the horizontal. If the height of the ball at point $B$ is .67 m, determine (a) the angle $\alpha$, (b) the angle $\theta$ that the velocity of the ball at point $B$ forms with the horizontal.

**Fig. P11.115**

**11.116** A worker uses high-pressure water to clean the inside of a long drainpipe. If the water is discharged with an initial velocity $\mathbf{v}_0$ of 10.6 m/s, determine (a) the distance $d$ to the farthest point $B$ on the top of the pipe that the water can wash from his position at $A$, (b) the corresponding angle $\alpha$.

**Fig. P11.116**

**11.117** A nozzle at $A$ discharges water with an initial velocity of 11 m/s at an angle $\alpha$ with the horizontal. Determine (a) the distance $d$ to the farthest point $B$ on the roof that the water can reach, (b) the corresponding angle $\alpha$. Check that the stream will clear the edge of the roof.

**Fig. P11.117**

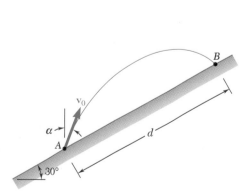

**Fig. P11.118**

**11.118** A projectile is launched from point $A$ with an initial velocity $\mathbf{v}_0$ of 36.6 m/s at an angle $\alpha$ with the vertical. Determine (a) the distance $d$ to the farthest point $B$ on the hill that the projectile can reach, (b) the corresponding angle $\alpha$, (c) the maximum height of the projectile above the surface.

**11.119**  Airplane *A* is flying due east at 700 km/h, while airplane *B* is flying at 500 km/h at the same altitude and in a direction to the west of south. Knowing that the speed of *B* with respect to *A* is 1125 km/h, determine the direction of the flight path of *B*.

**11.120**  Instruments in an airplane which is in level flight indicate that the velocity relative to the air (airspeed) is 120 km/h and the direction of the relative velocity vector (heading) is 70° east of north. Instruments on the ground indicate that the velocity of the airplane (ground speed) is 110 km/h and the direction of flight (course) is 60° east of north. Determine the wind speed and direction.

**Fig. P11.120**

**11.121**  At an intersection, car *A* is traveling south with a velocity of 40 km/h when it is struck by car *B* traveling 30° north of east with a velocity of 48 km/h. Determine the relative velocity of car *B* with respect to car *A*.

**Fig. P11.121**

**11.122**  Small wheels attached to the ends of rod *AB* roll along two surfaces. Knowing that at the instant shown the velocity $\mathbf{v}_A$ of wheel *A* is 1.4 m/s to the right and the relative velocity $\mathbf{v}_{B/A}$ of wheel *B* with respect to wheel *A* is perpendicular to rod *AB*, determine (*a*) the relative velocity $\mathbf{v}_{B/A}$, (*b*) the velocity $\mathbf{v}_B$ of wheel *B*.

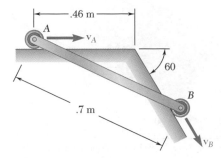

**Fig. P11.122**

*11.123* The velocities of commuter trains $A$ and $B$ are as shown. Knowing that the speed of each train is constant and that $B$ reaches the crossing 10 min after $A$ passed through the same crossing, determine (*a*) the relative velocity of $B$ with respect to $A$, (*b*) the distance between the fronts of the engines 3 min after $A$ passed through the crossing.

80 km/h

*A*

60 km/h *B*

25°

**Fig. P11.123**

*11.124* Airplanes $A$ and $B$ are flying at the same altitude and are tracking the eye of hurricane $C$. The relative velocity of $C$ with respect to $A$ is $\mathbf{v}_{C/A} = 470$ km/h ⬈ 75°, and the relative velocity of $C$ with respect to $B$ is $\mathbf{v}_{C/B} = 520$ km/h ⬊ 40°. Determine (*a*) the relative velocity of $B$ with respect to $A$, (*b*) the velocity of $A$ if ground-based radar indicates that the hurricane is moving at a speed of 48 km/h due north, (*c*) the change in position of $C$ with respect to $B$ during a 15-min interval.

*11.125* Slider block $B$ starts from rest and moves to the right with a constant acceleration of 0.3 m/s². Determine (*a*) the relative acceleration of portion $C$ of the cable with respect to slider block $A$, (*b*) the velocity of portion $C$ of the cable after 2 s.

*11.126* Pin $P$ moves at a constant speed of 20 cm/s in a counterclockwise sense along a circular slot which has been milled in the block $A$ as shown. Knowing that the block moves up the incline at a constant speed of 12 cm/s, determine the magnitude and the direction relative to the $xy$ axes of the velocity of pin $P$ when (*a*) $\theta = 30°$, (*b*) $\theta = 135°$.

*C*

*N*

*A*

*B*

**Fig. P11.124**

**Fig. P11.125**

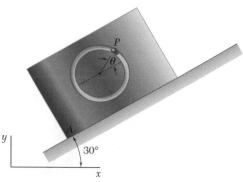

*P*

$\theta$

*y*

30°

*x*

**Fig. P11.126**

**11.127** At $t = 0$, wedge $A$ starts moving to the left with a constant acceleration of 80 mm/s$^2$ and block $B$ starts moving along the wedge toward the right with a constant acceleration of 120 mm/s$^2$ relative to the wedge. Determine ($a$) the acceleration of block $B$, ($b$) the velocity of block $B$ when $t = 3$ s.

**Fig. P11.127**

**11.128** A boat is moving to the right with a constant deceleration of 0.3 m/s$^2$ when a boy standing on the deck $D$ throws a ball with an initial velocity relative to the deck which is vertical. The ball rises to a maximum height of 8 m above the release point and the boy must step forward a distance $d$ to catch it at the same height as the release point. Determine ($a$) the distance $d$, ($b$) the relative velocity of the ball with respect to the deck when the ball is caught.

**Fig. P11.128**

**11.129** The conveyor belt $A$ moves with a constant velocity and discharges sand onto belt $B$ as shown. Knowing that the velocity of belt $B$ is 2.4 m/s, determine the velocity of the sand relative to belt $B$ as it lands on belt $B$.

**Fig. P11.129**

**11.130** A suitcase can slide down a conveyor belt on a truck that has no brakes. When $t = 0$ the suitcase is at point $A$ and the velocities of both the truck and suitcase are zero. When the suitcase reaches point $B$ its speed relative to the truck is 4.57 m/s and the truck has moved 15 cm to the right. Assuming constant accelerations, determine the velocity of the suitcase when $t = 1.2$ s.

3.5 m

20

**Fig. P11.130**

**11.131** As the driver of an automobile travels north at 20 km/h in a parking lot, he observes a truck approaching from the northwest. After he reduces his speed to 12 km/h and turns so that he is traveling in a northwest direction, the truck appears to be approaching from the west. Assuming that the velocity of the truck is constant during that period of observation, determine the magnitude and the direction of the velocity of the truck.

**11.132** Instruments in airplane $A$ indicate that with respect to the air the plane is headed 30° north of east with an airspeed of 480 km/h. At the same time, radar on ship $B$ indicates that the relative velocity of the plane with respect to the ship is 416 km/h in the direction 33° north of east. Knowing that the ship is steaming due south at 20 km/h, determine (a) the velocity of the airplane, (b) the wind speed and direction.

N

A    30°

20 km/h

B

**Fig. P11.132**

$\theta$

**Fig. P11.133**

**11.133** When a small boat travels north at 4.8 km/h, a flag mounted on its stern forms an angle $\theta = 50°$ with the centerline of the boat as shown. A short time later, when the boat travels east at 19.2 km/h, angle $\theta$ is again 50°. Determine the speed and the direction of the wind.

**11.134** As observed from a ship moving due east at 9.6 km/h, the wind appears to blow from the south. After the ship has changed course and speed, and as it is moving due north at 6.4 km/h, the wind appears to blow from the southwest. Assuming that the wind velocity is constant during the period of observation, determine the magnitude and direction of the true wind velocity.

# 11.13. TANGENTIAL AND NORMAL COMPONENTS

We saw in Sec. 11.9 that the velocity of a particle is a vector tangent to the path of the particle but that, in general, the acceleration is not tangent to the path. It is sometimes convenient to resolve the acceleration into components directed, respectively, along the tangent and the normal to the path of the particle.

**Plane Motion of a Particle.** First, let us consider a particle which moves along a curve contained in the plane of the figure. Let $P$ be the position of the particle at a given instant. We attach at $P$ a unit vector $\mathbf{e}_t$ tangent to the path of the particle and pointing in the direction of motion (Fig. 11.21a). Let $\mathbf{e}_t'$ be the unit vector corresponding to the position $P'$ of the particle at a later instant. Drawing both vectors from the same origin $O'$, we define the vector $\Delta\mathbf{e}_t = \mathbf{e}_t' - \mathbf{e}_t$ (Fig. 11.21b). Since $\mathbf{e}_t$ and $\mathbf{e}_t'$ are of unit length, their tips lie on a circle of radius 1. Denoting by $\Delta\theta$ the angle formed by $\mathbf{e}_t$ and $\mathbf{e}_t'$, we find that the magnitude of $\Delta\mathbf{e}_t$ is $2 \sin(\Delta\theta/2)$. Considering now the vector $\Delta\mathbf{e}_t/\Delta\theta$, we note that as $\Delta\theta$ approaches zero, this vector becomes tangent to the unit circle of Fig. 11.21b, that is, perpendicular to $\mathbf{e}_t$, and that its magnitude approaches

$$\lim_{\Delta\theta \to 0} \frac{2 \sin(\Delta\theta/2)}{\Delta\theta} = \lim_{\Delta\theta \to 0} \frac{\sin(\Delta\theta/2)}{\Delta\theta/2} = 1$$

**Fig. 11.21**

Thus, the vector obtained in the limit is a unit vector along the normal to the path of the particle, in the direction toward which $\mathbf{e}_t$ turns. Denoting this vector by $\mathbf{e}_n$, we write

$$\mathbf{e}_n = \lim_{\Delta\theta \to 0} \frac{\Delta\mathbf{e}_t}{\Delta\theta}$$

$$\mathbf{e}_n = \frac{d\mathbf{e}_t}{d\theta} \tag{11.35}$$

Since the velocity $\mathbf{v}$ of the particle is tangent to the path, it can be expressed as the product of the scalar $v$ and the unit vector $\mathbf{e}_t$. We have

$$\mathbf{v} = v\mathbf{e}_t \tag{11.36}$$

To obtain the acceleration of the particle, (11.36) will be differentiated with respect to $t$. Applying the rule for the differentiation of the product of a scalar and a vector function (Sec. 11.10), we write

$$\mathbf{a} = \frac{d\mathbf{v}}{dt} = \frac{dv}{dt}\mathbf{e}_t + v\frac{d\mathbf{e}_t}{dt} \tag{11.37}$$

But

$$\frac{d\mathbf{e}_t}{dt} = \frac{d\mathbf{e}_t}{d\theta}\frac{d\theta}{ds}\frac{ds}{dt}$$

Recalling from (11.16) that $ds/dt = v$, from (11.35) that $d\mathbf{e}_t/d\theta = \mathbf{e}_n$, and from elementary calculus that $d\theta/ds$ is equal to $1/\rho$, where $\rho$ is the radius of curvature of the path at $P$ (Fig. 11.22), we have

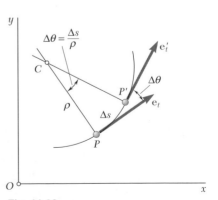

**Fig. 11.22**

$$\frac{d\mathbf{e}_t}{dt} = \frac{v}{\rho}\mathbf{e}_n \qquad (11.38)$$

Substituting into (11.37), we obtain

$$\mathbf{a} = \frac{dv}{dt}\mathbf{e}_t + \frac{v^2}{\rho}\mathbf{e}_n \qquad (11.39)$$

Thus, the scalar components of the acceleration are

$$a_t = \frac{dv}{dt} \qquad a_n = \frac{v^2}{\rho} \qquad (11.40)$$

The relations obtained express that the *tangential component* of the acceleration is equal to the *rate of change of the speed of the particle,* while the *normal component* is equal to the *square of the speed divided by the radius of curvature of the path at P.* If the speed of the particle increases, $a_t$ is positive and the vector component $\mathbf{a}_t$ points in the direction of motion. If the speed of the particle decreases, $a_t$ is negative and $\mathbf{a}_t$ points against the direction of motion. The vector component $\mathbf{a}_n$, on the other hand, *is always directed toward the center of curvature C of the path* (Fig. 11.23).

**Photo 11.4** As he travels on the curved portion of a racetrack, each cyclist is subjected to a normal component of acceleration directed toward the center of curvature of his path.

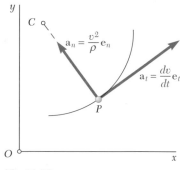

**Fig 11.23**

We conclude from the above that the tangential component of the acceleration reflects a change in the speed of the particle, while its normal component reflects a change in the direction of motion of the particle. The acceleration of a particle will be zero only if both its components are zero. Thus, the acceleration of a particle moving with constant speed along a curve will not be zero unless the particle happens to pass through a point of inflection of the curve (where the radius of curvature is infinite) or unless the curve is a straight line.

The fact that the normal component of the acceleration depends upon the radius of curvature of the path followed by the particle is taken into account in the design of structures or mechanisms as widely different as airplane wings, railroad tracks, and cams. In order to avoid sudden changes in the acceleration of the air particles flowing past a wing, wing profiles are designed without any sudden change in curvature. Similar care is taken in designing railroad curves, to avoid sudden changes in the acceleration of the cars (which would be hard on the equipment and unpleasant for the passengers). A straight section of track, for instance, is never directly followed by a circular section. Special transition sections are used to help pass smoothly from the infinite radius of curvature of the straight section to the finite radius of the circular track. Likewise, in the design of high-speed cams, abrupt changes in acceleration are avoided by using transition curves which produce a continuous change in acceleration.

**Motion of a Particle in Space.** The relations (11.39) and (11.40) still hold in the case of a particle moving along a space curve. However, since there are an infinite number of straight lines which are perpendicular to the tangent at a given point $P$ of a space curve, it is necessary to define more precisely the direction of the unit vector $\mathbf{e}_n$.

Let us consider again the unit vectors $\mathbf{e}_t$ and $\mathbf{e}_t'$ tangent to the path of the particle at two neighboring points $P$ and $P'$ (Fig. 11.24a) and the vector $\Delta\mathbf{e}_t$ representing the difference between $\mathbf{e}_t$ and $\mathbf{e}_t'$

(a)          (b)          (c)

**Fig 11.24**

(Fig. 11.24b). Let us now imagine a plane through $P$ (Fig. 11.24a) parallel to the plane defined by the vectors $\mathbf{e}_t$, $\mathbf{e}_t'$, and $\Delta\mathbf{e}_t$ (Fig. 11.24b). This plane contains the tangent to the curve at $P$ and is parallel to the tangent at $P'$. If we let $P'$ approach $P$, we obtain in the limit the plane which fits the curve most closely in the neighborhood of $P$. This plane is called the *osculating plane* at $P$.† It follows from this definition that the osculating plane contains the unit vector $\mathbf{e}_n$, since this vector represents the limit of the vector $\Delta\mathbf{e}_t/\Delta\theta$. The normal defined by $\mathbf{e}_n$ is thus contained in the osculating plane; it is called the *principal normal* at $P$. The unit vector $\mathbf{e}_b = \mathbf{e}_t \times \mathbf{e}_n$ which completes the right-handed triad $\mathbf{e}_t$, $\mathbf{e}_n$, $\mathbf{e}_b$ (Fig. 11.24c) defines the *binormal* at $P$. The binormal is thus perpendicular to the osculating plane. We conclude that the acceleration of the particle at $P$ can be resolved into two components, one along the tangent, the other along the principal normal at $P$, as indicated in Eq. (11.39). Note that the acceleration has no component along the binormal.

†From the Latin *osculari,* to kiss.

## 11.14. RADIAL AND TRANSVERSE COMPONENTS

In certain problems of plane motion, the position of the particle $P$ is defined by its polar coordinates $r$ and $\theta$ (Fig. 11.25a). It is then convenient to resolve the velocity and acceleration of the particle into components parallel and perpendicular, respectively, to the line $OP$. These components are called *radial* and *transverse components*.

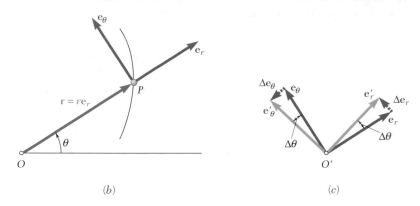

**Fig. 11.25**

We attach at $P$ two unit vectors, $\mathbf{e}_r$ and $\mathbf{e}_\theta$ (Fig. 11.25b). The vector $\mathbf{e}_r$ is directed along $OP$ and the vector $\mathbf{e}_\theta$ is obtained by rotating $\mathbf{e}_r$ through 90° counterclockwise. The unit vector $\mathbf{e}_r$ defines the *radial* direction, that is, the direction in which $P$ would move if $r$ were increased and $\theta$ were kept constant; the unit vector $\mathbf{e}_\theta$ defines the *transverse* direction, that is, the direction in which $P$ would move if $\theta$ were increased and $r$ were kept constant. A derivation similar to the one we used in Sec. 11.13 to determine the derivative of the unit vector $\mathbf{e}_t$ leads to the relations

$$\frac{d\mathbf{e}_r}{d\theta} = \mathbf{e}_\theta \qquad \frac{d\mathbf{e}_\theta}{d\theta} = -\mathbf{e}_r \qquad (11.41)$$

where $-\mathbf{e}_r$ denotes a unit vector of sense opposite to that of $\mathbf{e}_r$ (Fig. 11.25c). Using the chain rule of differentiation, we express the time derivatives of the unit vectors $\mathbf{e}_r$ and $\mathbf{e}_\theta$ as follows:

$$\frac{d\mathbf{e}_r}{dt} = \frac{d\mathbf{e}_r}{d\theta}\frac{d\theta}{dt} = \mathbf{e}_\theta\frac{d\theta}{dt} \qquad \frac{d\mathbf{e}_\theta}{dt} = \frac{d\mathbf{e}_\theta}{d\theta}\frac{d\theta}{dt} = -\mathbf{e}_r\frac{d\theta}{dt}$$

or, using dots to indicate differentiation with respect to $t$,

$$\dot{\mathbf{e}}_r = \dot{\theta}\mathbf{e}_\theta \qquad \dot{\mathbf{e}}_\theta = -\dot{\theta}\mathbf{e}_r \qquad (11.42)$$

To obtain the velocity $\mathbf{v}$ of the particle $P$, we express the position vector $\mathbf{r}$ of $P$ as the product of the scalar $r$ and the unit vector $\mathbf{e}_r$ and differentiate with respect to $t$:

$$\mathbf{v} = \frac{d}{dt}(r\mathbf{e}_r) = \dot{r}\mathbf{e}_r + r\dot{\mathbf{e}}_r$$

or, recalling the first of the relations (11.42),

$$\mathbf{v} = \dot{r}\mathbf{e}_r + r\dot{\theta}\mathbf{e}_\theta \qquad (11.43)$$

Differentiating again with respect to $t$ to obtain the acceleration, we write

$$\mathbf{a} = \frac{d\mathbf{v}}{dt} = \ddot{r}\mathbf{e}_r + \dot{r}\dot{\mathbf{e}}_r + \dot{r}\dot{\theta}\mathbf{e}_\theta + r\ddot{\theta}\mathbf{e}_\theta + r\dot{\theta}\dot{\mathbf{e}}_\theta$$

or, substituting for $\dot{\mathbf{e}}_r$ and $\dot{\mathbf{e}}_\theta$ from (11.42) and factoring $\mathbf{e}_r$ and $\mathbf{e}_\theta$,

$$\mathbf{a} = (\ddot{r} - r\dot{\theta}^2)\mathbf{e}_r + (r\ddot{\theta} + 2\dot{r}\dot{\theta})\mathbf{e}_\theta \qquad (11.44)$$

The scalar components of the velocity and the acceleration in the radial and transverse directions are, therefore,

$$v_r = \dot{r} \qquad\qquad v_\theta = r\dot{\theta} \qquad\qquad (11.45)$$
$$a_r = \ddot{r} - r\dot{\theta}^2 \qquad a_\theta = r\ddot{\theta} + 2\dot{r}\dot{\theta} \qquad (11.46)$$

It is important to note that $a_r$ is *not* equal to the time derivative of $v_r$ and that $a_\theta$ is *not* equal to the time derivative of $v_\theta$.

In the case of a particle moving along a circle of center $O$, we have $r = $ constant and $\dot{r} = \ddot{r} = 0$, and the formulas (11.43) and (11.44) reduce, respectively, to

$$\mathbf{v} = r\dot{\theta}\mathbf{e}_\theta \qquad \mathbf{a} = -r\dot{\theta}^2\mathbf{e}_r + r\ddot{\theta}\mathbf{e}_\theta \qquad (11.47)$$

**Extension to the Motion of a Particle in Space: Cylindrical Coordinates.** The position of a particle $P$ in space is sometimes defined by its cylindrical coordinates $R$, $\theta$, and $z$ (Fig. 11.26$a$). It is then convenient to use the unit vectors $\mathbf{e}_R$, $\mathbf{e}_\theta$, and $\mathbf{k}$ shown in Fig. 11.26$b$. Resolving the position vector $\mathbf{r}$ of the particle $P$ into components along the unit vectors, we write

$$\mathbf{r} = R\mathbf{e}_R + z\mathbf{k} \qquad (11.48)$$

Observing that $\mathbf{e}_R$ and $\mathbf{e}_\theta$ define, respectively, the radial and transverse directions in the horizontal $xy$ plane, and that the vector $\mathbf{k}$, which defines the *axial* direction, is constant in direction as well as in magnitude, we easily verify that

$$\mathbf{v} = \frac{d\mathbf{r}}{dt} = \dot{R}\mathbf{e}_R + R\dot{\theta}\mathbf{e}_\theta + \dot{z}\mathbf{k} \qquad (11.49)$$

$$\mathbf{a} = \frac{d\mathbf{v}}{dt} = (\ddot{R} - R\dot{\theta}^2)\mathbf{e}_R + (R\ddot{\theta} + 2\dot{R}\dot{\theta})\mathbf{e}_\theta + \ddot{z}\mathbf{k} \qquad (11.50)$$

$(a)$

$(b)$

**Fig. 11.26**

$v_A = 60$ mi/h

762 m

## SAMPLE PROBLEM 11.10

A motorist is traveling on a curved section of highway of radius 762 m at the speed of 96 km/h. The motorist suddenly applies the brakes, causing the automobile to slow down at a constant rate. Knowing that after 8 s the speed has been reduced to 72 km/h, determine the acceleration of the automobile immediately after the brakes have been applied.

## SOLUTION

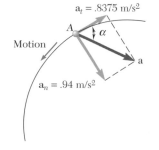

$a_t = .8375$ m/s²

A

α

Motion

a

$a_n = .94$ m/s²

**Tangential Component of Acceleration.** First the speeds are expressed in ft/s.

$$60 \text{ mi/h} = \left(60\frac{\text{mi}}{\text{h}}\right)\left(\frac{5280 \text{ ft}}{1 \text{ mi}}\right)\left(\frac{1 \text{ h}}{3600 \text{ s}}\right) = 26.7 \text{ m/s}$$

$$45 \text{ mi/h} = 20 \text{ m/s}$$

Since the automobile slows down at a constant rate, we have

$$a_t = \text{average } a_t = \frac{\Delta v}{\Delta t} = \frac{20 \text{ m/s} - 26.7 \text{ m/s}}{8 \text{ s}} = -.8375 \text{ m/s}^2$$

**Normal Component of Acceleration.** Immediately after the brakes have been applied, the speed is still 26.7 m/s, and we have

$$a_n = \frac{v^2}{\rho} = \frac{(26.7 \text{ m/s})^2}{762 \text{ m}} = .94 \text{ m/s}^2$$

**Magnitude and Direction of Acceleration.** The magnitude and direction of the resultant **a** of the components **a**$_n$ and **a**$_t$ are

$$\tan \alpha = \frac{a_n}{a_t} = \frac{.94 \text{ m/s}^2}{.8375 \text{ m/s}^2} \qquad \alpha = 48.3° \blacktriangleleft$$

$$a = \frac{a_n}{\sin \alpha} = \frac{.94 \text{ m/s}^2}{\sin 48.3°} \qquad a = 1.26 \text{ m/s}^2 \blacktriangleleft$$

## SAMPLE PROBLEM 11.11

Determine the minimum radius of curvature of the trajectory described by the projectile considered in Sample Prob. 11.7.

## SOLUTION

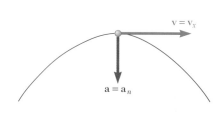

$v = v_x$

$a = a_n$

Since $a_n = v^2/\rho$, we have $\rho = v^2/a_n$. The radius will be small when $v$ is small or when $a_n$ is large. The speed $v$ is minimum at the top of the trajectory since $v_y = 0$ at that point; $a_n$ is maximum at that same point, since the direction of the vertical coincides with the direction of the normal. Therefore, the minimum radius of curvature occurs at the top of the trajectory. At this point, we have

$$v = v_x = 155.9 \text{ m/s} \qquad a_n = a = 9.81 \text{ m/s}^2$$

$$\rho = \frac{v^2}{a_n} = \frac{(155.9 \text{ m/s})^2}{9.81 \text{ m/s}^2} \qquad \rho = 2480 \text{ m} \blacktriangleleft$$

## SAMPLE PROBLEM 11.12

The rotation of the 0.9-m arm $OA$ about $O$ is defined by the relation $\theta = 0.15t^2$, where $\theta$ is expressed in radians and $t$ in seconds. Collar $B$ slides along the arm in such a way that its distance from $O$ is $r = 0.9 - 0.12t^2$, where $r$ is expressed in meters and $t$ in seconds. After the arm $OA$ has rotated through 30°, determine ($a$) the total velocity of the collar, ($b$) the total acceleration of the collar, ($c$) the relative acceleration of the collar with respect to the arm.

## SOLUTION

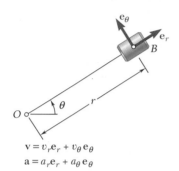

$$v = v_r e_r + v_\theta e_\theta$$
$$a = a_r e_r + a_\theta e_\theta$$

**Time $t$ at which $\theta = 30°$.**   Substituting $\theta = 30° = 0.524$ rad into the expression for $\theta$, we obtain

$$\theta = 0.15t^2 \qquad 0.524 = 0.15t^2 \qquad t = 1.869 \text{ s}$$

**Equations of Motion.**   Substituting $t = 1.869$ s in the expressions for $r$, $\theta$, and their first and second derivatives, we have

$$r = 0.9 - 0.12t^2 = 0.481 \text{ m} \qquad \theta = 0.15t^2 = 0.524 \text{ rad}$$
$$\dot{r} = -0.24t = -0.449 \text{ m/s} \qquad \dot{\theta} = 0.30t = 0.561 \text{ rad/s}$$
$$\ddot{r} = -0.24 = -0.240 \text{ m/s}^2 \qquad \ddot{\theta} = 0.30 = 0.300 \text{ rad/s}^2$$

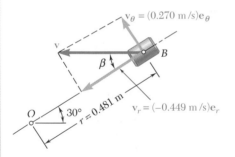

***a*. Velocity of $B$.**   Using Eqs. (11.45), we obtain the values of $v_r$ and $v_\theta$ when $t = 1.869$ s.

$$v_r = \dot{r} = -0.449 \text{ m/s}$$
$$v_\theta = r\dot{\theta} = 0.481(0.561) = 0.270 \text{ m/s}$$

Solving the right triangle shown, we obtain the magnitude and direction of the velocity,

$$v = 0.524 \text{ m/s} \qquad \beta = 31.0° \quad \blacktriangleleft$$

***b*. Acceleration of $B$.**   Using Eqs. (11.46), we obtain

$$a_r = \ddot{r} - r\dot{\theta}^2$$
$$= -0.240 - 0.481(0.561)^2 = -0.391 \text{ m/s}^2$$
$$a_\theta = r\ddot{\theta} + 2\dot{r}\dot{\theta}$$
$$= 0.481(0.300) + 2(-0.449)(0.561) = -0.359 \text{ m/s}^2$$
$$a = 0.531 \text{ m/s}^2 \qquad \gamma = 42.6° \quad \blacktriangleleft$$

***c*. Acceleration of $B$ with Respect to Arm $OA$.**   We note that the motion of the collar with respect to the arm is rectilinear and defined by the coordinate $r$. We write

$$a_{B/OA} = \ddot{r} = -0.240 \text{ m/s}^2$$
$$a_{B/OA} = 0.240 \text{ m/s}^2 \text{ toward } O. \quad \blacktriangleleft$$

# SOLVING PROBLEMS ON YOUR OWN

You will be asked in the following problems to express the velocity and the acceleration of particles in terms of either their *tangential and normal components* or their *radial and transverse components*. Although those components may not be as familiar to you as the rectangular components, you will find that they can simplify the solution of many problems and that certain types of motion are more easily described when they are used.

**1. Using tangential and normal components.** These components are most often used when the particle of interest travels along a circular path or when the radius of curvature of the path is to be determined. Remember that the unit vector $\mathbf{e}_t$ is tangent to the path of the particle (and thus aligned with the velocity) while the unit vector $\mathbf{e}_n$ is directed along the normal to the path and always points toward its center of curvature. It follows that, as the particle moves, the directions of the two unit vectors are constantly changing.

**2. Expressing the acceleration in terms of its tangential and normal components.** We derived in Sec. 11.13 the following equation, applicable to both the two-dimensional and the three-dimensional motion of a particle:

$$\mathbf{a} = \frac{dv}{dt}\mathbf{e}_t + \frac{v^2}{\rho}\mathbf{e}_n \qquad (11.39)$$

The following observations may help you in solving the problems of this lesson.

    *a. The tangential component* of the acceleration measures the rate of change of the speed: $a_t = dv/dt$. It follows that when $a_t$ is constant, the equations for uniformly accelerated motion can be used with the acceleration equal to $a_t$. Furthermore, when a particle moves at a constant speed, we have $a_t = 0$ and the acceleration of the particle reduces to its normal component.

    *b. The normal component* of the acceleration is always directed toward the center of curvature of the path of the particle, and its magnitude is $a_n = v^2/\rho$. Thus, the normal component can be easily determined if the speed of the particle and the radius of curvature $\rho$ of the path are known. Conversely, when the speed and normal acceleration of the particle are known, the radius of curvature of the path can be obtained by solving this equation for $\rho$ [Sample Prob. 11.11].

    *c. In three-dimensional motion,* a third unit vector is used, $\mathbf{e}_b = \mathbf{e}_t \times \mathbf{e}_n$, which defines the direction of the *binormal*. Since this vector is perpendicular to both the velocity and the acceleration, it can be obtained by writing

$$\mathbf{e}_b = \frac{\mathbf{v} \times \mathbf{a}}{|\mathbf{v} \times \mathbf{a}|}$$

3. *Using radial and transverse components.* These components are used to analyze the plane motion of a particle $P$, when the position of $P$ is defined by its polar coordinates $r$ and $\theta$. As shown in Fig. 11.25, the unit vector $\mathbf{e}_r$, which defines the *radial* direction, is attached to $P$ and points away from the fixed point $O$, while the unit vector $\mathbf{e}_\theta$, which defines the *transverse* direction, is obtained by rotating $\mathbf{e}_r$ *counterclockwise* through 90°. The velocity and the acceleration of a particle were expressed in terms of their radial and transverse components in Eqs. (11.43) and (11.44), respectively. You will note that the expressions obtained contain the first and second derivatives with respect to $t$ of both coordinates $r$ and $\theta$.

In the problems of this lesson, you will encounter the following types of problems involving radial and transverse components:

*a. Both $r$ and $\theta$ are known functions of $t$.* In this case, you will compute the first and second derivatives of $r$ and $\theta$ and substitute the expressions obtained into Eqs. (11.43) and (11.44).

*b. A certain relationship exists between $r$ and $\theta$.* First, you should determine this relationship from the geometry of the given system and use it to express $r$ as a function of $\theta$. Once the function $r = f(\theta)$ is known, you can apply the chain rule to determine $\dot{r}$ in terms of $\theta$ and $\dot{\theta}$, and $\ddot{r}$ in terms of $\theta$, $\dot{\theta}$ and $\ddot{\theta}$:

$$\dot{r} = f'(\theta)\dot{\theta}$$
$$\ddot{r} = f''(\theta)\dot{\theta}^2 + f'(\theta)\ddot{\theta}$$

The expressions obtained can then be substituted into Eqs. (11.43) and (11.44).

*c. The three-dimensional motion of a particle,* as indicated at the end of Sec. 11.14, can often be effectively described in terms of the *cylindrical coordinates* $R$, $\theta$, and $z$ (Fig. 11.26). The unit vectors then should consist of $\mathbf{e}_R$, $\mathbf{e}_\theta$, and $\mathbf{k}$. The corresponding components of the velocity and the acceleration are given in Eqs. (11.49) and (11.50). Please note that the radial distance $R$ is always measured in a plane parallel to the $xy$ plane, and be careful not to confuse the position vector $\mathbf{r}$ with its radial component $R\mathbf{e}_R$.

# Problems

**11.135** The diameter of the eye of a stationary hurricane is 32 km and the maximum wind speed is 160 km/h at the eye wall, $r = 16$ km. Assuming that the wind speed is constant for constant $r$ and decreases uniformly with increasing $r$ to 64 km/h at $r = 176$ km, determine the magnitude of the acceleration of the air at (*a*) $r = 16$ km, (*b*) $r = 96$ km, (*c*) $r = 176$ km.

**Fig. P11.135**

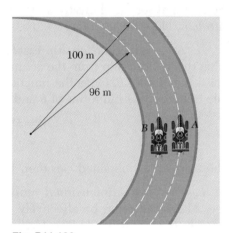

**Fig. P11.136**

**11.136** At the instant shown, race car *A* is passing race car *B* with a relative velocity of 1 m/s. Knowing that the speeds of both cars are constant and that the relative acceleration of car *A* with respect to car *B* is 0.25 m/s² directed toward the center of curvature, determine (*a*) the speed of car *A*, (*b*) the speed of car *B*.

**11.137** Determine the maximum speed that the cars of the roller-coaster can reach along the circular portion *AB* of the track if the normal component of their acceleration cannot exceed 3*g*.

**Fig. P11.137**

**11.138** As cam $A$ rotates, follower wheel $B$ rolls without slipping on the face of the cam. Knowing that the normal components of the acceleration of the points of contact at $C$ of the cam $A$ and the wheel $B$ are 0.66 m/s$^2$ and 6.8 m/s$^2$, respectively, determine the diameter of the follower wheel.

**11.139** A motorist is traveling on a curved portion of highway of radius 350 m at a speed of 72 km/h. The brakes are suddenly applied, causing the speed to decrease at a constant rate of 1.25 m/s$^2$. Determine the magnitude of the total acceleration of the automobile (a) immediately after the brakes have been applied, (b) 4 s later.

**11.140** An outdoor track is a full circle of diameter 130 m. A runner starts from rest and reaches her maximum speed in 4 s with constant tangential acceleration and then maintains that speed until she completes the circle with a total time of 54 s. Determine the magnitude of the maximum total acceleration of the runner.

**11.141** The peripheral speed of the tooth of a 25.4 cm.-diameter circular saw blade is 45.7 m/s when the power to the saw is turned off. The speed of the tooth decreases at a constant rate, and the blade comes to rest in 9 s. Determine the time at which the total acceleration of the tooth is 39.6 ms/s$^2$.

**11.142** A motorist starts from rest at point $A$ on a circular entrance ramp when $t = 0$, increases the speed of her automobile at a constant rate and enters the highway at point $B$. Knowing that her speed continues to increase at the same rate until it reaches 104 km/h at point $C$, determine (a) the speed at point $B$, (b) the magnitude of the total acceleration when $t = 15$ s.

**Fig. P11.138**

**Fig. P11.142**

**Fig. P11.143**

**11.143** At a given instant in an airplane race, airplane $A$ is flying horizontally in a straight line, and its speed is being increased at a rate of 6 m/s$^2$. Airplane $B$ is flying at the same altitude as airplane $A$ and, as it rounds a pylon, is following a circular path of 200-m radius. Knowing that at the given instant the speed of $B$ is being decreased at the rate of 2 m/s$^2$, determine, for the positions shown, (a) the velocity of $B$ relative to $A$, (b) the acceleration of $B$ relative to $A$.

**11.144** Racing cars $A$ and $B$ are traveling on circular portions of a race track. At the instant shown, the speed of $A$ is decreasing at the rate of 8 m/s$^2$, and the speed of $B$ is increasing at the rate of 3 m/s$^2$. For the positions shown, determine (a) the velocity of $B$ relative to $A$, (b) the acceleration of $B$ relative to $A$.

**Fig. P11.144**

**11.145** A nozzle discharges a stream of water in the direction shown with an initial velocity of 8 m/s. Determine the radius of curvature of the stream (a) as it leaves the nozzle, (b) at the maximum height of the stream.

**Fig. P11.145**

**11.146** A child throws a ball from point A with an initial velocity $\mathbf{v}_0$ at an angle of 3° with the horizontal. Knowing that the ball hits a wall at point B, determine (a) the magnitude of the initial velocity, (b) the minimum radius of curvature of the trajectory.

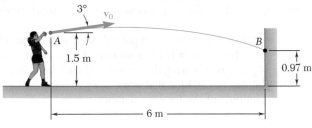

**Fig. P11.146**

**11.147** A projectile is launched from point A with an initial velocity $\mathbf{v}_0$ of 36.6 m/s at an angle of 30° with the vertical. Determine the radius of curvature of the trajectory described by the projectile (a) at point A, (b) at the point on the trajectory where the velocity is parallel to the incline.

**11.148** The motion of particle P on the elliptical path shown is defined by the equations $x = (2 \cos \pi t - 1)/(2 - \cos \pi t)$ and $y = 1.5 \sin \pi t/(2 - \cos \pi t)$, where x and y are expressed in meter and t is expressed in seconds. Determine the radius of curvature of the elliptical path when (a) t = 0, (b) t = 1/3 s, (c) t = 1 s.

**Fig. P11.147**

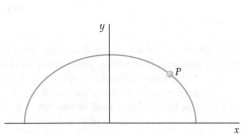

**Fig. P11.148**

**11.149** The motion of a particle is defined by the equations $x = [(t - 4)^3/6] + t^2$ and $y = (t^3/6) - (t - 1)^2/4$, where x and y are expressed in meters and t is expressed in seconds. Determine the acceleration of the particle and the radius of curvature of the path when t = 2 s.

**11.150** A horizontal pipe discharges at point $A$ a stream of water into a reservoir. Express the radius of curvature of the stream at point $B$ in terms of the magnitudes of the velocities $\mathbf{v}_A$ and $\mathbf{v}_B$.

**11.151** A projectile is fired from point $A$ with an initial velocity $\mathbf{v}_0$. (a) Show that the radius of curvature of the trajectory of the projectile reaches its minimum value at the highest point $B$ of the trajectory. (b) Denoting by $\theta$ the angle formed by the trajectory and the horizontal at a given point $C$, show that the radius of curvature of the trajectory at $C$ is $\rho = \rho_{min}/\cos^3 \theta$.

**Fig. P11.150**

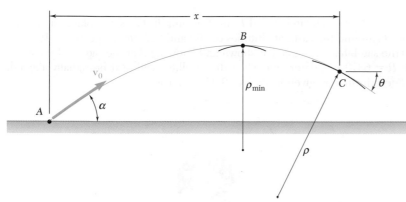

**Fig. *P11.151* and *P11.152***

**11.152** A projectile is fired from point $A$ with an initial velocity $\mathbf{v}_0$ which forms an angle $\alpha$ with the horizontal. Express the radius of curvature of the trajectory of the projectile at point $C$ in terms of $x$, $v_0$, $\alpha$, and $g$.

*11.153 Determine the radius of curvature of the path described by the particle of Prob. 11.97 when $t = 0$.

*11.154 Determine the radius of curvature of the path described by the particle of Prob. 11.98 when $t = 0$, $A = 3$, and $B = 1$.

**11.155 and 11.156** A satellite will travel indefinitely in a circular orbit around a planet if the normal component of the acceleration of the satellite is equal to $g(R/r)^2$, where $g$ is the acceleration of gravity at the surface of the planet, $R$ is the radius of the planet, and $r$ is the distance from the center of the planet to the satellite. Knowing that the diameter of the sun is 1.39 Gm and that the acceleration of gravity at its surface is 274 m/s$^2$, determine the radius of the orbit of the indicated planet around the sun assuming that the orbit is circular.

    **11.155** Earth: $(v_{mean})_{orbit} = 107$ Mm/h
    **11.156** Saturn: $(v_{mean})_{orbit} = 34.7$ Mm/h

**11.157 through 11.159** Determine the speed of a satellite relative to the indicated planet if the satellite is to travel indefinitely in a circular orbit 160 km above the surface of the planet. (See information given in Probs. 11.155–11.156.)

    **11.157** Venus: $g = 8.9$ m/s$^2$, $R = 6026.3$ km.
    **11.158** Mars: $g = 3.7$ m/s$^2$, $R = 3316.8$ km.
    *11.159* Jupiter: $g = 23$ m/s$^2$, $R = 71193.7$ km.

**11.160** A Global Positioning System (GPS) satellite is in a circular orbit 17465 km above the surface of the earth. Knowing that the radius of the earth is 6345 km, determine the time of one orbit of the satellite. (See information given in Probs. 11.155–11.156.)

**11.161** A satellite will travel indefinitely in a circular orbit around the earth if the normal component of its acceleration is equal to $g(R/r)^2$, where $g = 9.81$ m/s², $R$ = radius of the earth = 6370 km, and $r$ = distance from the center of the earth to the satellite. Assuming that the orbit of the moon is a circle of radius $384 \times 10^3$ km, determine the speed of the moon relative to the earth.

**11.162** Satellites $A$ and $B$ are traveling in the same plane in circular orbits around the earth at altitudes of 190 and 320 km, respectively. If at $t = 0$ the satellites are aligned as shown and knowing that the radius of the earth is $R = 6370$ km, determine when the satellites will next be radially aligned. (See information given in Probs. 11.155–11.156.)

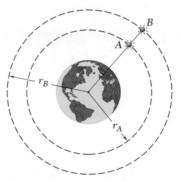

**Fig. P11.162**

**11.163** The rotation of rod $OA$ about $O$ is defined by the relation $\theta = 0.5e^{-0.8t} \sin 3\pi t$, where $\theta$ and $t$ are expressed in radians and seconds, respectively. Collar $B$ slides along the rod so that its distance from $O$ is $r = 1 + 2t - 6t^2 + 8t^3$, where $r$ and $t$ are expressed in meter and seconds, respectively. When $t = 0.5$ s, determine (a) the velocity of the collar, (b) the acceleration of the collar, (c) the acceleration of the collar relative to the rod.

**Fig. *P11.163* and P11.164**

**11.164** The oscillation of rod $OA$ about $O$ is defined by the relation $\theta = (4/\pi)(\sin \pi t)$, where $\theta$ and $t$ are expressed in radians and seconds, respectively. Collar $B$ slides along the rod so that its distance from $O$ is $r = 10/(t + 6)$, where $r$ and $t$ are expressed in mm and seconds, respectively. When $t = 1$ s, determine (a) the velocity of the collar, (b) the total acceleration of the collar, (c) the acceleration of the collar relative to the rod.

**11.165** The two-dimensional motion of a particle is defined by the relations $r = 2B \cos(At/2B)$ and $\theta = At/2B$, where $r$ is expressed in meters, $t$ in seconds, and $\theta$ in radians. Knowing that $A$ and $B$ are constants, determine (a) the magnitudes of the velocity and acceleration at any instant, (b) the radius of curvature of the path. What conclusion can you draw regarding the path of the particle?

**11.166** The path of a particle $P$ is a limaçon. The motion of the particle is defined by the relations $r = b(2 + \cos \pi t)$ and $\theta = \pi t$, where $t$ and $\theta$ are expressed in seconds and radians, respectively. Determine (a) the velocity and the acceleration of the particle when $t = 2$ s, (b) the value of $\theta$ for which the magnitude of the velocity is maximum.

**11.167** The motion of particle $P$ on the parabolic path shown is defined by the equations $r = 6t\sqrt{1 + 4t^2}$ and $\theta = \tan^{-1} 2t$, where $r$ is expressed in meter, $\theta$ in radians, and $t$ in seconds. Determine the velocity and acceleration of the particle when (a) $t = 0$, (b) $t = 0.5$ s.

**Fig. P11.166**

**Fig. P11.167**

**11.168** The two-dimensional motion of a particle is defined by the relations $r = \dfrac{3}{\sin \theta - \cos \theta}$ and $\tan \theta = 1 + \dfrac{1}{t^2}$, where $r$ and $\theta$ are expressed in meter and radians, respectively, and $t$ is expressed in seconds. Determine (a) the magnitudes of the velocity and acceleration at any instant, (b) the radius of curvature of the path. What conclusion can you draw regarding the path of the particle?

**11.169** To study the performance of a race car, a high-speed motion-picture camera is positioned at point $A$. The camera is mounted on a mechanism which permits it to record the motion of the car as the car travels on straightaway $BC$. Determine the speed of the car in terms of $b$, $\theta$, and $\dot{\theta}$.

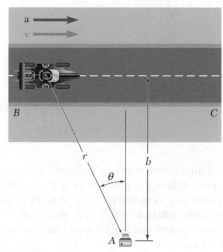

**Fig. P11.169**

**11.170** Determine the magnitude of the acceleration of the race car of Prob. 11.169 in terms of $b$, $\theta$, $\dot{\theta}$, and $\ddot{\theta}$.

**11.171** After taking off, a helicopter climbs in a straight line at a constant angle $\beta$. Its flight is tracked by radar from point $A$. Determine the speed of the helicopter in terms of $d$, $\beta$, $\theta$, and $\dot{\theta}$.

Fig. P11.171                                        Fig. P11.172

**11.172** Pin $C$ is attached to rod $BC$ and slides freely in the slot of rod $OA$ which rotates at the constant rate $\omega$. At the instant when $\beta = 60°$, determine (a) $\dot{r}$ and $\dot{\theta}$, (b) $\ddot{r}$ and $\ddot{\theta}$. Express your answers in terms of $d$ and $\omega$.

**11.173** A test rocket is fired vertically from a launching pad at $B$. When the rocket is at $P$, the angle of elevation is $\theta = 47.0°$, and 0.5 s later it is $\theta = 48.0°$. Knowing that $b = 4$ km, determine approximately the speed of the rocket during the 0.5-s interval.

Fig. P11.173                        Fig. P11.174

**11.174** An airplane passes over a radar tracking station at $A$ and continues to fly due east. When the airplane is at $P$, the distance and angle of elevation of the plane are, respectively, $r = 3840.48$ m and $\theta = 31.2°$. Two seconds later the radar station sights the plane at $r = 4145.28$ m and $\theta = 28.3°$. Determine approximately the speed and the angle of dive $\alpha$ of the plane during the 2-s interval.

**11.175 and 11.176** A particle moves along the spiral shown. Determine the magnitude of the velocity of the particle in terms of $b$, $\theta$, and $\dot{\theta}$.

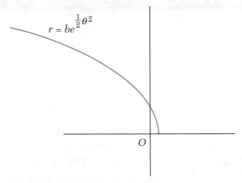

$r = be^{\frac{1}{2}\theta^2}$

$O$

**Fig. P11.175 and P11.177**

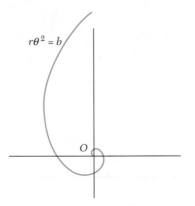

$r\theta^2 = b$

$O$

**Fig. P11.176 and P11.178**

**11.177 and 11.178** A particle moves along the spiral shown. Knowing that $\dot{\theta}$ is constant and denoting this constant by $\omega$, determine the magnitude of the acceleration of the particle in terms of $b$, $\theta$, and $\omega$.

**11.179** Show that $\dot{r} = h\dot{\phi} \sin \theta$ knowing that at the instant shown, step $AB$ of the step exerciser is rotating counterclockwise at a constant rate $\dot{\phi}$.

**11.180** The three-dimensional motion of a particle is defined by the cylindrical coordinates (see Fig. 11.26) $R = A/(t + 1)$, $\theta = Bt$, and $z = Ct/(t + 1)$. Determine the magnitudes of the velocity and acceleration when (a) $t = 0$, (b) $t = \infty$.

**11.181** The motion of a particle on the surface of a right circular cylinder is defined by the relations $R = A$, $\theta = 2\pi t$, and $z = At^2/4$, where $A$ is a constant. Determine the magnitudes of the velocity and acceleration of the particle at any time $t$.

**Fig. P11.179**

**Fig. P11.181**

*11.182 For the conic helix of Prob. 11.97, determine the angle that the oscillating plane forms with the $y$ axis.

*11.183 Determine the direction of the binormal of the path described by the particle of Prob. 11.98 when (a) $t = 0$, (b) $t = \pi/2$ s.

Position coordinate of a particle in rectilinear motion

**Fig. 11.27**

Velocity and acceleration in rectilinear motion

In the first half of the chapter, we analyzed the *rectilinear motion of a particle*, that is, the motion of a particle along a straight line. To define the position $P$ of the particle on that line, we chose a fixed origin $O$ and a positive direction (Fig. 11.27). The distance $x$ from $O$ to $P$, with the appropriate sign, completely defines the position of the particle on the line and is called the *position coordinate* of the particle [Sec. 11.2].

The *velocity* $v$ of the particle was shown to be equal to the time derivative of the position coordinate $x$,

$$v = \frac{dx}{dt} \qquad (11.1)$$

and the *acceleration* $a$ was obtained by differentiating $v$ with respect to $t$,

$$a = \frac{dv}{dt} \qquad (11.2)$$

or

$$a = \frac{d^2x}{dt^2} \qquad (11.3)$$

We also noted that $a$ could be expressed as

$$a = v\frac{dv}{dx} \qquad (11.4)$$

We observed that the velocity $v$ and the acceleration $a$ were represented by algebraic numbers which can be positive or negative. A positive value for $v$ indicates that the particle moves in the positive direction, and a negative value that it moves in the negative direction. A positive value for $a$, however, may mean that the particle is truly accelerated (that is, moves faster) in the positive direction, or that it is decelerated (that is, moves more slowly) in the negative direction. A negative value for $a$ is subject to a similar interpretation [Sample Prob. 11.1].

Determination of the velocity and acceleration by integration

In most problems, the conditions of motion of a particle are defined by the type of acceleration that the particle possesses and by the initial conditions [Sec. 11.3]. The velocity and position of the particle can then be obtained by integrating two of the equations (11.1) to (11.4). Which of these equations should be selected depends upon the type of acceleration involved [Sample Probs. 11.2 and 11.3].

Two types of motion are frequently encountered: the *uniform rectilinear motion* [Sec. 11.4], in which the velocity $v$ of the particle is constant and

$$x = x_0 + vt \qquad (11.5)$$

and the *uniformly accelerated rectilinear motion* [Sec. 11.5], in which the acceleration $a$ of the particle is constant and we have

$$v = v_0 + at \qquad (11.6)$$
$$x = x_0 + v_0 t + \tfrac{1}{2} a t^2 \qquad (11.7)$$
$$v^2 = v_0^2 + 2a(x - x_0) \qquad (11.8)$$

When two particles $A$ and $B$ move along the same straight line, we may wish to consider the *relative motion* of $B$ with respect to $A$

**Fig. 11.28**

[Sec. 11.6]. Denoting by $x_{B/A}$ the *relative position coordinate* of $B$ with respect to $A$ (Fig. 11.28), we had

$$x_B = x_A + x_{B/A} \qquad (11.9)$$

Differentiating Eq. (11.9) twice with respect to $t$, we obtained successively

$$v_B = v_A + v_{B/A} \qquad (11.10)$$
$$a_B = a_A + a_{B/A} \qquad (11.11)$$

where $v_{B/A}$ and $a_{B/A}$ represent, respectively, the *relative velocity* and the *relative acceleration* of $B$ with respect to $A$.

When several blocks are *connected by inextensible cords*, it is possible to write a *linear relation* between their position coordinates. Similar relations can then be written between their velocities and between their accelerations and can be used to analyze their motion [Sample Prob. 11.5].

It is sometimes convenient to use a *graphical solution* for problems involving the rectilinear motion of a particle [Secs. 11.7 and 11.8]. The graphical solution most commonly used involves the $x$–$t$, $v$–$t$, and $a$–$t$ curves [Sec. 11.7; Sample Prob. 11.6]. It was shown that, at any given time $t$,

$$v = \text{slope of } x\text{–}t \text{ curve}$$
$$a = \text{slope of } v\text{–}t \text{ curve}$$

while, over any given time interval from $t_1$ to $t_2$,

$$v_2 - v_1 = \text{area under } a\text{–}t \text{ curve}$$
$$x_2 - x_1 = \text{area under } v\text{–}t \text{ curve}$$

In the second half of the chapter, we analyzed the *curvilinear motion of a particle*, that is, the motion of a particle along a curved path. The position $P$ of the particle at a given time [Sec. 11.9] was

**Fig. 11.29**

Acceleration in curvilinear motion

Derivative of a vector function

Rectangular components of velocity and acceleration

Component motions

Relative motion of two particles

**Fig. 11.30**

defined by the *position vector* **r** joining the origin $O$ of the coordinates and point $P$ (Fig. 11.29). The *velocity* **v** of the particle was defined by the relation

$$\mathbf{v} = \frac{d\mathbf{r}}{dt} \tag{11.15}$$

and was found to be a *vector tangent to the path of the particle* and of magnitude $v$ (called the *speed* of the particle) equal to the time derivative of the length $s$ of the arc described by the particle:

$$v = \frac{ds}{dt} \tag{11.16}$$

The *acceleration* **a** of the particle was defined by the relation

$$\mathbf{a} = \frac{d\mathbf{v}}{dt} \tag{11.18}$$

and we noted that, in general, *the acceleration is not tangent to the path of the particle.*

Before proceeding to the consideration of the components of velocity and acceleration, we reviewed the formal definition of the derivative of a vector function and established a few rules governing the differentiation of sums and products of vector functions. We then showed that the rate of change of a vector is the same with respect to a fixed frame and with respect to a frame in translation [Sec. 11.10].

Denoting by $x$, $y$, and $z$ the rectangular coordinates of a particle $P$, we found that the rectangular components of the velocity and acceleration of $P$ equal, respectively, the first and second derivatives with respect to $t$ of the corresponding coordinates:

$$v_x = \dot{x} \qquad v_y = \dot{y} \qquad v_z = \dot{z} \tag{11.29}$$
$$a_x = \ddot{x} \qquad a_y = \ddot{y} \qquad a_z = \ddot{z} \tag{11.30}$$

When the component $a_x$ of the acceleration depends only upon $t$, $x$, and/or $v_x$, and when similarly $a_y$ depends only upon $t$, $y$, and/or $v_y$, and $a_z$ upon $t$, $z$, and/or $v_z$, Eqs. (11.30) can be integrated independently. The analysis of the given curvilinear motion can thus be reduced to the analysis of three independent rectilinear component motions [Sec. 11.11]. This approach is particularly effective in the study of the motion of projectiles [Sample Probs. 11.7 and 11.8].

For two particles $A$ and $B$ moving in space (Fig. 11.30), we considered the relative motion of $B$ with respect to $A$, or more precisely, with respect to a moving frame attached to $A$ and in translation with $A$ [Sec. 11.12]. Denoting by $\mathbf{r}_{B/A}$ the *relative position vector* of $B$ with respect to $A$ (Fig. 11.30), we had

$$\mathbf{r}_B = \mathbf{r}_A + \mathbf{r}_{B/A} \tag{11.31}$$

Denoting by $\mathbf{v}_{B/A}$ and $\mathbf{a}_{B/A}$, respectively, the *relative velocity* and the *relative acceleration* of $B$ with respect to $A$, we also showed that

$$\mathbf{v}_B = \mathbf{v}_A + \mathbf{v}_{B/A} \tag{11.33}$$

and

$$\mathbf{a}_B = \mathbf{a}_A + \mathbf{a}_{B/A} \tag{11.34}$$

It is sometimes convenient to resolve the velocity and acceleration of a particle $P$ into components other than the rectangular $x$, $y$, and $z$ components. For a particle $P$ moving along a path contained in a plane, we attached to $P$ unit vectors $\mathbf{e}_t$ tangent to the path and $\mathbf{e}_n$ normal to the path and directed toward the center of curvature of the path [Sec. 11.13]. We then expressed the velocity and acceleration of the particle in terms of tangential and normal components. We wrote

$$\mathbf{v} = v\mathbf{e}_t \tag{11.36}$$

and

$$\mathbf{a} = \frac{dv}{dt}\mathbf{e}_t + \frac{v^2}{\rho}\mathbf{e}_n \tag{11.39}$$

where $v$ is the speed of the particle and $\rho$ the radius of curvature of its path [Sample Probs. 11.10 and 11.11]. We observed that while the velocity $\mathbf{v}$ is directed along the tangent to the path, the acceleration $\mathbf{a}$ consists of a component $\mathbf{a}_t$ directed along the tangent to the path and a component $\mathbf{a}_n$ directed toward the center of curvature of the path (Fig. 11.31).

For a particle $P$ moving along a space curve, we defined the plane which most closely fits the curve in the neighborhood of $P$ as the *osculating plane*. This plane contains the unit vectors $\mathbf{e}_t$ and $\mathbf{e}_n$ which define, respectively, the tangent and principal normal to the curve. The unit vector $\mathbf{e}_b$ which is perpendicular to the osculating plane defines the *binormal*.

When the position of a particle $P$ moving in a plane is defined by its polar coordinates $r$ and $\theta$, it is convenient to use radial and transverse components directed, respectively, along the position vector $\mathbf{r}$ of the particle and in the direction obtained by rotating $\mathbf{r}$ through 90° counterclockwise [Sec. 11.14]. We attached to $P$ unit vectors $\mathbf{e}_r$ and $\mathbf{e}_\theta$ directed, respectively, in the radial and transverse directions (Fig. 11.32). We then expressed the velocity and acceleration of the particle in terms of radial and transverse components

$$\mathbf{v} = \dot{r}\mathbf{e}_r + r\dot{\theta}\mathbf{e}_\theta \tag{11.43}$$
$$\mathbf{a} = (\ddot{r} - r\dot{\theta}^2)\mathbf{e}_r + (r\ddot{\theta} + 2\dot{r}\dot{\theta})\mathbf{e}_\theta \tag{11.44}$$

where dots are used to indicate differentiation with respect to time. The scalar components of the velocity and acceleration in the radial and transverse directions are therefore

$$v_r = \dot{r} \qquad v_\theta = r\dot{\theta} \tag{11.45}$$
$$a_r = \ddot{r} - r\dot{\theta}^2 \qquad a_\theta = r\ddot{\theta} + 2\dot{r}\dot{\theta} \tag{11.46}$$

It is important to note that $a_r$ is *not* equal to the time derivative of $v_r$, and that $a_\theta$ is *not* equal to the time derivative of $v_\theta$ [Sample Prob. 11.12].

The chapter ended with a discussion of the use of cylindrical coordinates to define the position and motion of a particle in space.

Tangential and normal components

**Fig. 11.31**

Motion along a space curve

Radial and transverse components

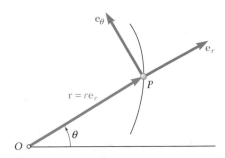

**Fig. 11.32**

**11.184** The acceleration of a particle is directly proportional to the square of the time $t$. When $t = 0$, the particle is at $x = 11$ m. Knowing that at $t = 9$ s, $x = 44$ m and $v = 8.2$ m/s, express $x$ and $v$ in terms of $t$.

**11.185** The acceleration of a particle is defined by the relation $a = 0.6(1 - kv)$, where $k$ is a constant. Knowing that at $t = 0$ the particle starts from rest at $x = 6$ m and that $v = 6$ m/s when $t = 20$ s, determine (a) the constant $k$, (b) the position of the particle when $v = 7.5$ m/s, (c) the maximum velocity of the particle.

**11.186** A motorist enters a freeway at 40 km/h and accelerates uniformly to 104 km/h. From the odometer in the car, the motorist knows that she traveled .16 km while accelerating. Determine (a) the acceleration of the car, (b) the time required to reach 104 km/h.

**Fig. P11.186**

**11.187** Block $C$ starts from rest and moves downward with a constant acceleration. Knowing that after 12 s the velocity of block $A$ is 456 mm/s, determine (a) the accelerations of $A$, $B$, and $C$, (b) the velocity and the change in position of block $B$ after 8 s.

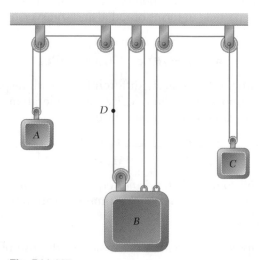

**Fig. P11.187**

**11.188** A particle moves in a straight line with the velocity shown in the figure. Knowing that $x = -540$ m at $t = 0$, (a) construct the $a$–$t$ and $x$–$t$ curves for $0 < t < 50$ s, and determine (b) the total distance traveled by the particle when $t = 50$ s, (c) the two times at which $x = 0$.

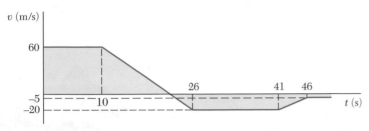

**Fig. P11.188**

**11.189** Car $A$ is traveling on a highway at a constant speed $(v_A)_0 = 100$ km/h and is 120 m from the entrance of an access ramp when car $B$ enters the acceleration lane at that point at a speed $(v_B)_0 = 25$ km/h. Car $B$ accelerates uniformly and enters the main traffic lane after traveling 70 m in 5 s. It then continues to accelerate at the same rate until it reaches a speed of 100 km/h, which it then maintains. Determine the final distance between the two cars.

**Fig. P11.189**

**11.190** A baseball pitching machine "throws" baseballs with a horizontal velocity $\mathbf{v}_0$. Knowing that height $h$ varies between 788 mm and 1068 mm, determine (a) the range of values of $v_0$, (b) the values of $\alpha$ corresponding to $h = 788$ mm and $h = 1068$ mm.

**Fig. P11.190**

**11.191** A ball is dropped onto a step at point $A$ and rebounds with a velocity $\mathbf{v}_0$ at an angle of 15° with the vertical. Determine the value of $v_0$ knowing that just before the ball bounces at point $B$ its velocity $\mathbf{v}_B$ forms an angle of 12° with the vertical.

**Fig. P11.191**                    **Fig. P11.192**

**11.192** Coal discharged from a dump truck with an initial velocity $(\mathbf{v}_C)_0 = 1.8$ m/s ⬏ 50° falls onto conveyor belt $B$. Determine the required velocity $\mathbf{v}_B$ of the belt if the relative velocity with which the coal hits the belt is to be (a) vertical, (b) as small as possible.

**11.193** Pin $A$, which is attached to link $AB$, is constrained to move in the circular slot $CD$. Knowing that at $t = 0$ the pin starts from rest and moves so that its speed increases at a constant rate of 2 cm/s², determine the magnitude of its total acceleration when (a) $t = 0$, (b) $t = 2$ s.

**Fig. P11.193**                    **Fig. P11.194**

**11.194** Coal is discharged from the tailgate of a dump truck with an initial velocity $\mathbf{v}_A = 2$ m/s ⬏ 50°. Determine the radius of curvature of the trajectory described by the coal (a) at point $A$, (b) at the point of the trajectory 1 m below point $A$.

**11.195** The two-dimensional motion of a particle is defined by the relations $r = 6(4 - 2e^{-t})$ and $\theta = 2(2t + 4e^{-2t})$, where $r$ is expressed in meter, $t$ in seconds, and $\theta$ in radians. Determine the velocity and the acceleration of the particle (a) when $t = 0$, (b) as $t$ approaches infinity. What conclusions can you draw regarding the final path of the particle?

# Computer Problems

**11.C1**  A projectile enters a resisting medium at $x = 0$ with an initial velocity $v_0 = 300$ m/s and travels a distance $d$ before coming to rest. The velocity of the projectile is defined by the relation $v = v_0 - 2700x$, where $v$ is expressed in meters/second and $x$ in meters. (*a*) Determine the distance $d$ and derive expressions for the position, velocity, and acceleration of the projectile as functions of time. (*b*) Plot the position, velocity, and acceleration of the projectile as functions of time from the time the projectile enters the medium until it penetrates a distance $0.99d$.

**11.C2**  The magnitude in m/s$^2$ of the deceleration due to air resistance of the nose cone of a small experimental rocket is known to be $0.00015v^2$, where $v$ is expressed in m/s. The nose cone is projected vertically from the ground with an initial velocity of 128 m/s. Derive expressions for the speed of the cone as a function of height as the cone moves up to its maximum height and as it returns to the ground. Plot the speed of the cone as a function of height for the upward motion and for the downward motion. [*Hint:* The deceleration of the cone as it moves up is $g + 0.00015v^2$ and the acceleration of the cone as it moves down is $g - 0.00015v^2$, where $g = 9.81$ m/s$^2$.]

**11.C3**  The motion of a particle is defined by the equations $x = 30t^2 - 120t$ and $y = 120t^2 - 40t^3$, where $x$ and $y$ are expressed in millimeters and $t$ in seconds. Derive expressions for the velocity and acceleration of the particle as functions of $t$. Consider the time interval $0 \leq t \leq 2$ s and plot (*a*) the path of the particle in the $xy$ plane, (*b*) the components of the velocity $v_x$ and $v_y$ and the magnitude of the velocity $v$, (*c*) the components of the acceleration $a_x$ and $a_y$ and the magnitude of the acceleration $a$.

**11.C4**  The rotation of rod $OA$ about $O$ is defined by the relation $\theta = t^3 + 4t$, where $\theta$ is expressed in radians and $t$ in seconds. Collar $B$ slides along the rod in such a way that its distance from $O$ is $r = t^3 + 2t^2$, where $r$ is expressed in inches and $t$ in seconds. Derive expressions for the velocity and acceleration of the collar as functions of $t$. Consider one revolution of the rod starting at $t = 0$ and plot (*a*) the path of the particle in the $xy$ plane, (*b*) the components of the velocity $v_r$ and $v_\theta$ and the magnitude of the velocity $v$, (*c*) the components of the acceleration $a_r$ and $a_\theta$ and the magnitude of the acceleration $a$.

**Fig. P11.C4**

**11.C5** The path of particle $P$ is the ellipse defined by the relations $r = 1.75/(1 - 0.75 \cos \pi t)$ and $\theta = \pi t$, where $r$ is expressed in cm, $t$ in seconds, and $\theta$ in radians. Derive expressions for the velocity and acceleration of $P$ as functions of $t$. Consider the time interval $0 \le t \le 2$ s and plot (a) the components of the velocity $v_r$ and $v_\theta$ and the magnitude of the velocity $v$, (b) the components of the acceleration $a_r$ and $a_\theta$ and the magnitude of the acceleration $a$.

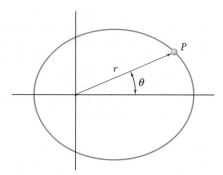

**Fig. P11.C5**

**11.C6** The mechanism shown is known as a Whitworth quick-return mechanism. The input rod $AP$ rotates at a constant rate $\dot{\phi}$, and the pin $P$ is free to slide in the slot of the output rod $BD$. Using computational software, calculate and plot $\dot{\theta}$ versus $\phi$ and $\theta$ versus $\phi$ for one revolution of rod $AP$. Assume $\dot{\phi} = 1$ rad/s, $l = 80$ mm, and (a) $b = 50$ mm, (b) $b = 60$ mm, (c) $b = 70$ mm.

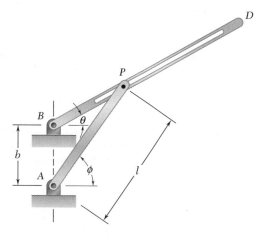

**Fig. P11.C6**

**11.C7** In an amusement park ride, "airplane" $A$ is attached to the 10.7 m long rigid member $OB$. To operate the ride, the airplane and $OB$ are rotated so that $70° \leq \theta_0 \leq 130°$ and then are allowed to swing freely about $O$. The airplane is subjected to the acceleration of gravity and to a deceleration due to air resistance, $-kv^2$, which acts in a direction opposite to that of its velocity $\mathbf{v}$. Neglecting the mass and the aerodynamic drag of $OB$ and the friction in the bearing at $O$, use computational software to determine the speed of the airplane for given values of $\theta_0$ and $\theta$ and the value of $\theta$ at which the airplane first comes to rest after being released. Use values of $\theta_0$ from $70°$ to $130°$ in $30°$ increments, and determine the maximum speed of the airplane and the first two values of $\theta$ at which $v = 0$. For each value of $\theta_0$, let (a) $k = 0$, (b) $k = 4 \times 10^{-6}$ in$^{-1}$, (c) $k = 1 \times 10^{-3}$ in$^{-1}$. (*Hint:* Express the tangential acceleration of the airplane in terms of $g$, $k$, and $\theta$. Recall that $v_\theta = r\dot{\theta}$ and use time intervals $\Delta t = 0.008$ s.)

**Fig. P11.C7**

# CHAPTER 12

# Kinetics of Particles: Newton's Second Law

As a car travels on the curved portion of a racetrack, it is subjected to a component of acceleration directed toward the center of curvature of its path. The force of gravity and the other external forces which act on the car must be considered as well as the components of acceleration. The relation existing among force, mass, and acceleration will be studied in this chapter.

**KINETICS OF PARTICLES: NEWTON'S SECOND LAW**

12.1   Introduction
12.2   Newton's Second Law of Motion
12.3   Linear Momentum of a Particle. Rate of Change of Linear Momentum
12.4   Systems of Units
12.5   Equations of Motion
12.6   Dynamic Equilibrium
12.7   Angular Momentum of a Particle. Rate of Change of Angular Momentum
12.8   Equations of Motion in Terms of Radial and Transverse Components
12.9   Motion under a Central Force. Conservation of Angular Momentum
12.10  Newton's Law of Gravitation
12.11  Trajectory of a Particle under a Central Force
12.12  Application to Space Mechanics
12.13  Kepler's Laws of Planetary Motion

## 12.1. INTRODUCTION

Newton's first and third laws of motion were used extensively in statics to study bodies at rest and the forces acting upon them. These two laws are also used in dynamics; in fact, they are sufficient for the study of the motion of bodies which have no acceleration. However, when bodies are accelerated, that is, when the magnitude or the direction of their velocity changes, it is necessary to use Newton's second law of motion to relate the motion of the body with the forces acting on it.

In this chapter we will discuss Newton's second law and apply it to the analysis of the motion of particles. As we state in Sec. 12.2, if the resultant of the forces acting on a particle is not zero, the particle will have an acceleration proportional to the magnitude of the resultant and in the direction of this resultant force. Moreover, the ratio of the magnitudes of the resultant force and of the acceleration can be used to define the *mass* of the particle.

In Sec. 12.3, the *linear momentum* of a particle is defined as the product $\mathbf{L} = m\mathbf{v}$ of the mass $m$ and velocity $\mathbf{v}$ of the particle, and it is demonstrated that Newton's second law can be expressed in an alternative form relating the rate of change of the linear momentum with the resultant of the forces acting on that particle.

Section 12.4 stresses the need for consistent units in the solution of dynamics problems and provides a review of the International System of Units (SI units) and the system of U.S. customary units.

In Secs. 12.5 and 12.6 and in the Sample Problems which follow, Newton's second law is applied to the solution of engineering problems, using either rectangular components or tangential and normal components of the forces and accelerations involved. We recall that an actual body—including bodies as large as a car, rocket, or airplane—can be considered as a particle for the purpose of analyzing its motion as long as the effect of a rotation of the body about its mass center can be ignored.

The second part of the chapter is devoted to the solution of problems in terms of radial and transverse components, with particular emphasis on the motion of a particle under a central force. In Sec. 12.7, the *angular momentum* $\mathbf{H}_O$ of a particle about a point $O$ is defined as the moment about $O$ of the linear momentum of the particle: $\mathbf{H}_O = \mathbf{r} \times m\mathbf{v}$. It then follows from Newton's second law that the rate of change of the angular momentum $\mathbf{H}_O$ of a particle is equal to the sum of the moments about $O$ of the forces acting on that particle.

Section 12.9 deals with the motion of a particle under a *central force*, that is, under a force directed toward or away from a fixed point $O$. Since such a force has zero moment about $O$, it follows that the angular momentum of the particle about $O$ is conserved. This property greatly simplifies the analysis of the motion of a particle under a central force; in Sec. 12.10 it is applied to the solution of problems involving the orbital motion of bodies under gravitational attraction.

Sections 12.11 through 12.13 are optional. They present a more extensive discussion of orbital motion and contain a number of problems related to space mechanics.

## 12.2. NEWTON'S SECOND LAW OF MOTION

Newton's second law can be stated as follows:

*If the resultant force acting on a particle is not zero, the particle will have an acceleration proportional to the magnitude of the resultant and in the direction of this resultant force.*

Newton's second law of motion is best understood by imagining the following experiment: A particle is subjected to a force $\mathbf{F}_1$ of constant direction and constant magnitude $F_1$. Under the action of that force, the particle is observed to move in a straight line and *in the direction of the force* (Fig. 12.1a). By determining the position of the particle at various instants, we find that its acceleration has a constant magnitude $a_1$. If the experiment is repeated with forces $\mathbf{F}_2$, $\mathbf{F}_3$, . . . , of different magnitude or direction (Fig. 12.1b and c), we find each time that the particle moves in the direction of the force acting on it and that the magnitudes $a_1$, $a_2$, $a_3$, . . . , of the accelerations are proportional to the magnitudes $F_1$, $F_2$, $F_3$, . . . , of the corresponding forces:

$$\frac{F_1}{a_1} = \frac{F_2}{a_2} = \frac{F_3}{a_3} = \cdots = \text{constant}$$

The constant value obtained for the ratio of the magnitudes of the forces and accelerations is a characteristic of the particle under consideration; it is called the *mass* of the particle and is denoted by $m$. When a particle of mass $m$ is acted upon by a force $\mathbf{F}$, the force $\mathbf{F}$ and the acceleration $\mathbf{a}$ of the particle must therefore satisfy the relation

$$\mathbf{F} = m\mathbf{a} \tag{12.1}$$

This relation provides a complete formulation of Newton's second law; it expresses not only that the magnitudes of $\mathbf{F}$ and $\mathbf{a}$ are proportional but also (since $m$ is a positive scalar) that the vectors $\mathbf{F}$ and $\mathbf{a}$ have the same direction (Fig. 12.2). We should note that Eq. (12.1) still holds when $\mathbf{F}$ is not constant but varies with time in magnitude or direction. The magnitudes of $\mathbf{F}$ and $\mathbf{a}$ remain proportional, and the two vectors have the same direction at any given instant. However, they will not, in general, be tangent to the path of the particle.

When a particle is subjected simultaneously to several forces, Eq. (12.1) should be replaced by

$$\Sigma\mathbf{F} = m\mathbf{a} \tag{12.2}$$

where $\Sigma\mathbf{F}$ represents the sum, or resultant, of all the forces acting on the particle.

It should be noted that the system of axes with respect to which the acceleration $\mathbf{a}$ is determined is not arbitrary. These axes must have a constant orientation with respect to the stars, and their origin must either be attached to the sun† or move with a constant velocity with

†More accurately, to the mass center of the solar system.

**Fig. 12.1**

Fig. 12.2

respect to the sun. Such a system of axes is called a *newtonian frame of reference*.† A system of axes attached to the earth does *not* constitute a newtonian frame of reference, since the earth rotates with respect to the stars and is accelerated with respect to the sun. However, in most engineering applications, the acceleration **a** can be determined with respect to axes attached to the earth and Eqs. (12.1) and (12.2) used without any appreciable error. On the other hand, these equations do not hold if **a** represents a relative acceleration measured with respect to moving axes, such as axes attached to an accelerated car or to a rotating piece of machinery.

We observe that if the resultant $\Sigma \mathbf{F}$ of the forces acting on the particle is zero, it follows from Eq. (12.2) that the acceleration **a** of the particle is also zero. If the particle is initially at rest ($\mathbf{v}_0 = 0$) with respect to the newtonian frame of reference used, it will thus remain at rest ($\mathbf{v} = 0$). If originally moving with a velocity $\mathbf{v}_0$, the particle will maintain a constant velocity $\mathbf{v} = \mathbf{v}_0$; that is, it will move with the constant speed $v_0$ in a straight line. This, we recall, is the statement of Newton's first law (Sec. 2.10). Thus, Newton's first law is a particular case of Newton's second law and can be omitted from the fundamental principles of mechanics.

## 12.3. LINEAR MOMENTUM OF A PARTICLE. RATE OF CHANGE OF LINEAR MOMENTUM

Replacing the acceleration **a** by the derivative $d\mathbf{v}/dt$ in Eq. (12.2), we write

$$\Sigma \mathbf{F} = m \frac{d\mathbf{v}}{dt}$$

or, since the mass $m$ of the particle is constant,

$$\Sigma \mathbf{F} = \frac{d}{dt}(m\mathbf{v}) \tag{12.3}$$

The vector $m\mathbf{v}$ is called the *linear momentum*, or simply the *momentum*, of the particle. It has the same direction as the velocity of the particle, and its magnitude is equal to the product of the mass $m$ and the speed $v$ of the particle (Fig. 12.3). Equation (12.3) expresses that *the resultant of the forces acting on the particle is equal to the rate of change of the linear momentum of the particle*. It is in this form that the second law of motion was originally stated by Newton. Denoting by **L** the linear momentum of the particle,

$$\mathbf{L} = m\mathbf{v} \tag{12.4}$$

and by $\dot{\mathbf{L}}$ its derivative with respect to $t$, we can write Eq. (12.3) in the alternative form

$$\Sigma \mathbf{F} = \dot{\mathbf{L}} \tag{12.5}$$

**Fig. 12.3**

---

†Since stars are not actually fixed, a more rigorous definition of a newtonian frame of reference (also called an *inertial system*) is *one with respect to which Eq. (12.2) holds*.

It should be noted that the mass $m$ of the particle is assumed to be constant in Eqs. (12.3) to (12.5). Equation (12.3) or (12.5) should therefore not be used to solve problems involving the motion of bodies, such as rockets, which gain or lose mass. Problems of that type will be considered in Sec. 14.12.[†]

It follows from Eq. (12.3) that the rate of change of the linear momentum $m\mathbf{v}$ is zero when $\Sigma\mathbf{F} = 0$. Thus, *if the resultant force acting on a particle is zero, the linear momentum of the particle remains constant, in both magnitude and direction*. This is the principle of *conservation of linear momentum* for a particle, which can be recognized as an alternative statement of Newton's first law (Sec. 2.10).

## 12.4. SYSTEMS OF UNITS

In using the fundamental equation $\mathbf{F} = m\mathbf{a}$, the units of force, mass, length, and time cannot be chosen arbitrarily. If they are, the magnitude of the force $\mathbf{F}$ required to give an acceleration $\mathbf{a}$ to the mass $m$ will *not* be numerically equal to the product $ma$; it will be only proportional to this product. Thus, we can choose three of the four units arbitrarily but must choose the fourth unit so that the equation $\mathbf{F} = m\mathbf{a}$ is satisfied. The units are then said to form a system of consistent kinetic units.

Two systems of consistent kinetic units are currently used by American engineers, the International System of Units (SI units[‡]) and the system of U.S. customary units. Both systems were discussed in detail in Sec. 1.3 and are described only briefly in this section.

**International System of Units (SI Units).** In this system, the base units are the units of length, mass, and time, and are called, respectively, the *meter* (m), the *kilogram* (kg), and the *second* (s). All three are arbitrarily defined (Sec. 1.3). The unit of force is a derived unit. It is called the *newton* (N) and is defined as the force which gives an acceleration of 1 m/s² to a mass of 1 kg (Fig. 12.4). From Eq. (12.1) we write

**Fig 12.4**

$$1\ \text{N} = (1\ \text{kg})(1\ \text{m/s}^2) = 1\ \text{kg} \cdot \text{m/s}^2$$

The SI units are said to form an *absolute* system of units. This means that the three base units chosen are independent of the location where measurements are made. The meter, the kilogram, and the second may be used anywhere on the earth; they may even be used on another planet. They will always have the same significance.

The *weight* $\mathbf{W}$ of a body, or *force of gravity* exerted on that body, should, like any other force, be expressed in newtons. Since a body subjected to its own weight acquires an acceleration equal to the acceleration of gravity $g$, it follows from Newton's second law that the magnitude $W$ of the weight of a body of mass $m$ is

$$W = mg \tag{12.6}$$

[†]On the other hand, Eqs. (12.3) and (12.5) do hold in *relativistic mechanics*, where the mass $m$ of the particle is assumed to vary with the speed of the particle.

[‡]SI stands for *Système International d'Unités* (French).

**Fig 12.5**

Recalling that $g = 9.81$ m/s$^2$, we find that the weight of a body of mass 1 kg (Fig. 12.5) is

$$W = (1 \text{ kg})(9.81 \text{ m/s}^2) = 9.81 \text{ N}$$

Multiples and submultiples of the units of length, mass, and force are frequently used in engineering practice. They are, respectively, the *kilometer* (km) and the *millimeter* (mm); the *megagram*† (Mg) and the *gram* (g); and the *kilonewton* (kN). By definition,

$$1 \text{ km} = 1000 \text{ m} \qquad 1 \text{ mm} = 0.001 \text{ m}$$
$$1 \text{ Mg} = 1000 \text{ kg} \qquad 1 \text{ g} = 0.001 \text{ kg}$$
$$1 \text{ kN} = 1000 \text{ N}$$

The conversion of these units to meters, kilograms, and newtons, respectively, can be effected simply by moving the decimal point three places to the right or to the left.

Units other than the units of mass, length, and time can all be expressed in terms of these three base units. For example, the unit of linear momentum can be obtained by recalling the definition of linear momentum and writing

$$mv = (\text{kg})(\text{m/s}) = \text{kg} \cdot \text{m/s}$$

**U.S. Customary Units.**   Most practicing American engineers still commonly use a system in which the base units are the units of length, force, and time. These units are, respectively, the *foot* (ft), the *pound* (lb), and the *second* (s). The second is the same as the corresponding SI unit. The foot is defined as 0.3048 m. The pound is defined as the *weight* of a platinum standard, called the *standard pound*, which is kept at the National Institute of Standards and Technology outside Washington and the mass of which is 0.453 592 43 kg. Since the weight of a body depends upon the gravitational attraction of the earth, which varies with location, it is specified that the standard pound should be placed at sea level and at a latitude of 45° to properly define a force of 1 lb. Clearly, the U.S. customary units do not form an absolute system of units. Because of their dependence upon the gravitational attraction of the earth, they are said to form a *gravitational* system of units.

While the standard pound also serves as the unit of mass in commercial transactions in the United States, it cannot be so used in engineering computations since such a unit would not be consistent with the base units defined in the preceding paragraph. Indeed, when acted upon by a force of 1 lb, that is, when subjected to its own weight, the standard pound receives the acceleration of gravity, $g = 32.2$ ft/s$^2$ (Fig. 12.6), and not the unit acceleration required by Eq. (12.1). The unit of mass consistent with the foot, the pound, and the second is the mass which receives an acceleration of 1 ft/s$^2$ when a force of 1 lb is applied to it (Fig. 12.7). This unit, sometimes called a *slug*, can be derived from the equation $F = ma$ after substituting 1 lb and 1 ft/s$^2$ for $F$ and $a$, respectively. We write

$$F = ma \qquad 1 \text{ lb} = (1 \text{ slug})(1 \text{ ft/s}^2)$$

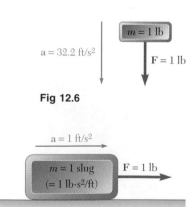

**Fig 12.6**

**Fig 12.7**

†Also known as a *metric ton*.

and obtain

$$1 \text{ slug} = \frac{1 \text{ lb}}{1 \text{ ft/s}^2} = 1 \text{ lb} \cdot \text{s}^2/\text{ft}$$

Comparing Figs. 12.6 and 12.7, we conclude that the slug is a mass 32.2 times larger than the mass of the standard pound.

The fact that bodies are characterized in the U.S. customary system of units by their weight in pounds rather than by their mass in slugs was a convenience in the study of statics, where we were dealing for the most part with weights and other forces and only seldomly with masses. However, in the study of kinetics, which involves forces, masses, and accelerations, it will be repeatedly necessary to express in slugs the mass $m$ of a body, the weight $W$ of which has been given in pounds. Recalling Eq. (12.6), we will write

$$m = \frac{W}{g} \tag{12.7}$$

where $g$ is the acceleration of gravity ($g = 32.2 \text{ ft/s}^2$).

Units other than the units of force, length, and time can all be expressed in terms of these three base units. For example, the unit of linear momentum can be obtained by using the definition of linear momentum to write

$$mv = (\text{lb} \cdot \text{s}^2/\text{ft})(\text{ft/s}) = \text{lb} \cdot \text{s}$$

**Conversion from One System of Units to Another.**   The conversion from U.S. customary units to SI units, and vice versa, was discussed in Sec. 1.4. You will recall that the conversion factors obtained for the units of length, force, and mass are, respectively,

Length:            1 ft = 0.3048 m
Force:             1 lb = 4.448 N
Mass:     1 slug = 1 lb · s²/ft = 14.59 kg

Although it cannot be used as a consistent unit of mass, the mass of the standard pound is, by definition,

1 pound-mass = 0.4536 kg

This constant can be used to determine the *mass* in SI units (kilograms) of a body which has been characterized by its *weight* in U.S. customary units (pounds).

## 12.5. EQUATIONS OF MOTION

Consider a particle of mass $m$ acted upon by several forces. We recall from Sec. 12.2 that Newton's second law can be expressed by the equation

$$\Sigma \mathbf{F} = m\mathbf{a} \tag{12.2}$$

which relates the forces acting on the particle and the vector $m\mathbf{a}$ (Fig. 12.8). In order to solve problems involving the motion of a particle, however, it will be found more convenient to replace Eq. (12.2) by equivalent equations involving scalar quantities.

**Fig. 12.8**

**Rectangular Components.** Resolving each force $\mathbf{F}$ and the acceleration $\mathbf{a}$ into rectangular components, we write

$$\Sigma(F_x\mathbf{i} + F_y\mathbf{j} + F_z\mathbf{k}) = m(a_x\mathbf{i} + a_y\mathbf{j} + a_z\mathbf{k})$$

from which it follows that

$$\Sigma F_x = ma_x \qquad \Sigma F_y = ma_y \qquad \Sigma F_z = ma_z \qquad (12.8)$$

Recalling from Sec. 11.11 that the components of the acceleration are equal to the second derivatives of the coordinates of the particle, we have

$$\Sigma F_x = m\ddot{x} \qquad \Sigma F_y = m\ddot{y} \qquad \Sigma F_z = m\ddot{z} \qquad (12.8')$$

Consider, as an example, the motion of a projectile. If the resistance of the air is neglected, the only force acting on the projectile after it has been fired is its weight $\mathbf{W} = -W\mathbf{j}$. The equations defining the motion of the projectile are therefore

$$m\ddot{x} = 0 \qquad m\ddot{y} = -W \qquad m\ddot{z} = 0$$

and the components of the acceleration of the projectile are

$$\ddot{x} = 0 \qquad \ddot{y} = -\frac{W}{m} = -g \qquad \ddot{z} = 0$$

where $g$ is 9.81 m/s$^2$ or 32.2 ft/s$^2$. The equations obtained can be integrated independently, as shown in Sec. 11.11, to obtain the velocity and displacement of the projectile at any instant.

When a problem involves two or more bodies, equations of motion should be written for each of the bodies (see Sample Probs. 12.3 and 12.4). You will recall from Sec. 12.2 that all accelerations should be measured with respect to a newtonian frame of reference. In most engineering applications, accelerations can be determined with respect to axes attached to the earth, but relative accelerations measured with respect to moving axes, such as axes attached to an accelerated body, cannot be substituted for $\mathbf{a}$ in the equations of motion.

**Tangential and Normal Components.** Resolving the forces and the acceleration of the particle into components along the tangent to the path (in the direction of motion) and the normal (toward

**Fig. 12.9**

the inside of the path) (Fig. 12.9), and substituting into Eq. (12.2), we obtain the two scalar equations

$$\Sigma F_t = ma_t \qquad \Sigma F_n = ma_n \qquad (12.9)$$

Substituting for $a_t$ and $a_n$ from Eqs. (11.40), we have

$$\Sigma F_t = m\frac{dv}{dt} \qquad \Sigma F_n = m\frac{v^2}{\rho} \qquad (12.9')$$

The equations obtained may be solved for two unknowns.

**Photo 12.1** As it travels on the curved portion of a track, the bobsled is subjected to a normal component of acceleration directed toward the center of curvature of its path.

# 12.6. DYNAMIC EQUILIBRIUM

Returning to Eq. (12.2) and transposing the right-hand member, we write Newton's second law in the alternative form

$$\Sigma\mathbf{F} - m\mathbf{a} = 0 \qquad (12.10)$$

which expresses that if we add the vector $-m\mathbf{a}$ to the forces acting on the particle, *we obtain a system of vectors equivalent to zero* (Fig. 12.10). The vector $-m\mathbf{a}$, of magnitude $ma$ and of *direction opposite* to that of the acceleration, is called an *inertia vector*. The particle may thus be considered to be in equilibrium under the given forces and the inertia vector. The particle is said to be in *dynamic equilibrium,* and the problem under consideration can be solved by the methods developed earlier in statics.

In the case of coplanar forces, all the vectors shown in Fig. 12.10, *including the inertia vector,* can be drawn tip-to-tail to form a closed-vector polygon. Or the sums of the components of all the vectors in Fig. 12.10, again including the inertia vector, can be equated to zero. Using rectangular components, we therefore write

$$\Sigma F_x = 0 \qquad \Sigma F_y = 0 \qquad \textit{\textbf{including inertia vector}} \qquad (12.11)$$

**Fig. 12.10**

When tangential and normal components are used, it is more convenient to represent the inertia vector by its two components $-m\mathbf{a}_t$ and $-m\mathbf{a}_n$ in the sketch itself (Fig. 12.11). The tangential component of the inertia vector provides a measure of the resistance the particle offers to a change in speed, while its normal component (also called *centrifugal force*) represents the tendency of the particle to leave its curved path. We should note that either of these two components may be zero under special conditions: (1) if the particle starts from rest, its initial velocity is zero and the normal component of the inertia vector is zero at $t = 0$; (2) if the particle moves at constant speed along its path, the tangential component of the inertia vector is zero and only its normal component needs to be considered.

Because they measure the resistance that particles offer when we try to set them in motion or when we try to change the conditions of their motion, inertia vectors are often called *inertia forces*. The inertia forces, however, are not forces like the forces found in statics, which are either contact forces or gravitational forces (weights). Many people, therefore, object to the use of the word "force" when referring to the vector $-m\mathbf{a}$ or even avoid altogether the concept of dynamic equilibrium. Others point out that inertia forces and actual forces, such as gravitational forces, affect our senses in the same way and cannot be distinguished by physical measurements. A man riding in an elevator which is accelerated upward will have the feeling that his weight has suddenly increased; and no measurement made within the elevator could establish whether the elevator is truly accelerated or whether the force of attraction exerted by the earth has suddenly increased.

Sample problems have been solved in this text by the direct application of Newton's second law, as illustrated in Figs. 12.8 and 12.9, rather than by the method of dynamic equilibrium.

**Fig. 12.11**

## SAMPLE PROBLEM 12.1

A 90.7 kg block rests on a horizontal plane. Find the magnitude of the force **P** required to give the block an acceleration of 3 m/s² to the right. The coefficient of kinetic friction between the block and the plane is $\mu_k = 0.25$.

## SOLUTION

The mass of the block is

$$\text{Weight of block, } W = mg = 890 \text{ N.}$$
$$m = 90.7 \text{ kg}$$

We note that $F = \mu_k R = 0.25R$ and that $a = 3$ m/s². Expressing that the forces acting on the block are equivalent to the vector *m***a**, we write

$$\xrightarrow{+} \Sigma F_x = ma: \qquad P \cos 30° - 0.25R = (90.7 \text{ kg} (3 \text{ m/s}^2)$$
$$P \cos 30° - 0.25R = 272 \text{ N} \qquad (1)$$
$$+\uparrow \Sigma F_y = 0: \qquad R - P \sin 30° - 890 \text{ N} = 0 \qquad (2)$$

Solving (2) for $N$ and substituting the result into (1), we obtain

$$R = P \sin 30° + 890 \text{ N}$$
$$P \cos 30° - 0.25(P \sin 30° + 890 \text{ N}) = 272 \text{ N} \qquad P = 667.3 \text{ N} \quad \blacktriangleleft$$

## SAMPLE PROBLEM 12.2

An 80-kg block rests on a horizontal plane. Find the magnitude of the force **P** required to give the block an acceleration of 2.5 m/s² to the right. The coefficient of kinetic friction between the block and the plane is $\mu_k = 0.25$.

## SOLUTION

The weight of the block is

$$W = mg = (80 \text{ kg})(9.81 \text{ m/s}^2) = 785 \text{ N}$$

We note that $F = \mu_k N = 0.25N$ and that $a = 2.5$ m/s². Expressing that the forces acting on the block are equivalent to the vector *m***a**, we write

$$\xrightarrow{+} \Sigma F_x = ma: \qquad P \cos 30° - 0.25N = (80 \text{ kg})(2.5 \text{ m/s}^2)$$
$$P \cos 30° - 0.25N = 200 \text{ N} \qquad (1)$$
$$+\uparrow \Sigma F_y = 0: \qquad N - P \sin 30° - 785 \text{ N} = 0 \qquad (2)$$

Solving (2) for $N$ and substituting the result into (1), we obtain

$$N = P \sin 30° + 785 \text{ N}$$
$$P \cos 30° - 0.25(P \sin 30° + 785 \text{ N}) = 200 \text{ N} \qquad P = 535 \text{ N} \quad \blacktriangleleft$$

## SAMPLE PROBLEM 12.3

The two blocks shown start from rest. The horizontal plane and the pulley are frictionless, and the pulley is assumed to be of negligible mass. Determine the acceleration of each block and the tension in each cord.

## SOLUTION

**Kinematics.** We note that if block $A$ moves through $x_A$ to the right, block $B$ moves down through

$$x_B = \tfrac{1}{2}x_A$$

Differentiating twice with respect to $t$, we have

$$a_B = \tfrac{1}{2}a_A \qquad (1)$$

**Kinetics.** We apply Newton's second law successively to block $A$, block $B$, and pulley $C$.

**Block A.** Denoting by $T_1$ the tension in cord $ACD$, we write

$$\xrightarrow{+}\Sigma F_x = m_A a_A: \qquad\qquad T_1 = 100a_A \qquad (2)$$

**Block B.** Observing that the weight of block $B$ is

$$W_B = m_B g = (300 \text{ kg})(9.81 \text{ m/s}^2) = 2940 \text{ N}$$

and denoting by $T_2$ the tension in cord $BC$, we write

$$+\downarrow\Sigma F_y = m_B a_B: \qquad\qquad 2940 - T_2 = 300a_B$$

or, substituting for $a_B$ from (1),

$$2940 - T_2 = 300(\tfrac{1}{2}a_A)$$
$$T_2 = 2940 - 150a_A \qquad (3)$$

**Pulley C.** Since $m_C$ is assumed to be zero, we have

$$+\downarrow\Sigma F_y = m_C a_C = 0: \qquad\qquad T_2 - 2T_1 = 0 \qquad (4)$$

Substituting for $T_1$ and $T_2$ from (2) and (3), respectively, into (4) we write

$$2940 - 150a_A - 2(100a_A) = 0$$
$$2940 - 350a_A = 0 \qquad\qquad a_A = 8.40 \text{ m/s}^2 \quad \blacktriangleleft$$

Substituting the value obtained for $a_A$ into (1) and (2), we have

$$a_B = \tfrac{1}{2}a_A = \tfrac{1}{2}(8.40 \text{ m/s}^2) \qquad\qquad a_B = 4.20 \text{ m/s}^2 \quad \blacktriangleleft$$
$$T_1 = 100a_A = (100 \text{ kg})(8.40 \text{ m/s}^2) \qquad T_1 = 840 \text{ N} \quad \blacktriangleleft$$

Recalling (4), we write

$$T_2 = 2T_1 \qquad T_2 = 2(840 \text{ N}) \qquad T_2 = 1680 \text{ N} \quad \blacktriangleleft$$

We note that the value obtained for $T_2$ is *not* equal to the weight of block $B$.

## SAMPLE PROBLEM 12.4

The 5.4 kg block $B$ starts from rest and slides on the 13.6 kg wedge $A$, which is supported by a horizontal surface. Neglecting friction, determine (a) the acceleration of the wedge, (b) the acceleration of the block relative to the wedge.

## SOLUTION

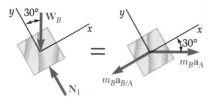

**Kinematics.** We first examine the acceleration of the wedge and the acceleration of the block.

*Wedge A.* Since the wedge is constrained to move on the horizontal surface, its acceleration $\mathbf{a}_A$ is horizontal. We will assume that it is directed to the right.

*Block B.* The acceleration $\mathbf{a}_B$ of block $B$ can be expressed as the sum of the acceleration of $A$ and the acceleration of $B$ relative to $A$. We have

$$\mathbf{a}_B = \mathbf{a}_A + \mathbf{a}_{B/A}$$

where $\mathbf{a}_{B/A}$ is directed along the inclined surface of the wedge.

**Kinetics.** We draw the free-body diagrams of the wedge and of the block and apply Newton's second law.

*Wedge A.* We denote the forces exerted by the block and the horizontal surface on wedge $A$ by $\mathbf{N}_1$ and $\mathbf{N}_2$, respectively.

$$\xrightarrow{+}\Sigma F_x = m_A a_A: \qquad N_1 \sin 30° = m_A a_A$$
$$0.5 N_1 = (W_A/g) a_A \qquad (1)$$

*Block B.* Using the coordinate axes shown and resolving $\mathbf{a}_B$ into its components $\mathbf{a}_A$ and $\mathbf{a}_{B/A}$, we write

$$+\nearrow\Sigma F_x = m_B a_x: \quad -W_B \sin 30° = m_B a_A \cos 30° - m_B a_{B/A}$$
$$-W_B \sin 30° = (W_B/g)(a_A \cos 30° - a_{B/A})$$
$$a_{B/A} = a_A \cos 30° + g \sin 30° \qquad (2)$$
$$+\nwarrow\Sigma F_y = m_B a_y: \quad N_1 - W_B \cos 30° = -m_B a_A \sin 30°$$
$$N_1 - W_B \cos 30° = -(W_B/g) a_A \sin 30° \qquad (3)$$

*a.* **Acceleration of Wedge A.** Substituting for $N_1$ from Eq. (1) into Eq. (3), we have

$$2(W_A/g) a_A - W_B \cos 30° = -(W_B/g) a_A \sin 30°$$

mass of block B $m_B$ = 5.4 kg
Weight of block B $W_B = m_B g$ = 53 N
mass of Wedge A $m_A$ = 13.6 kg
Weight of Wedge A $W_A = m_A g$ = 133.4 N

Solving for $a_A$ and substituting the numerical data, we write

$$a_A = \frac{W_B \cos 30°}{2 W_A + W_B \sin 30°} g = \frac{(53\,\text{N}) \cos 30°}{2\,(133.4\,\text{N}) + (53\,\text{N}) \sin 30°}(9.81\,\text{m/s}^2)$$

$$a_A = +1.54\,\text{m/s}^2 \qquad \mathbf{a}_A = 1.53\,\text{m/s}^2 \rightarrow \quad \blacktriangleleft$$

*b.* **Acceleration of Block B Relative to A.** Substituting the value obtained for $a_A$ into Eq. (2), we have

$$a_{B/A} = (1.54\,\text{m/s}^2) \cos 30° + (9.81\,\text{m/s}^2) \sin 30°$$
$$a_{B/A} = +6.24\,\text{m/s}^2 \qquad \mathbf{a}_{B/A} = 6.24\,\text{m/s}^2 \ \diagup\!\!\!30° \quad \blacktriangleleft$$

## SAMPLE PROBLEM 12.5

The bob of a 2-m pendulum describes an arc of a circle in a vertical plane. If the tension in the cord is 2.5 times the weight of the bob for the position shown, find the velocity and the acceleration of the bob in that position.

## SOLUTION

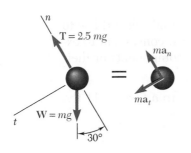

The weight of the bob is $W = mg$; the tension in the cord is thus 2.5 $mg$. Recalling that $\mathbf{a}_n$ is directed toward $O$ and assuming $\mathbf{a}_t$ as shown, we apply Newton's second law and obtain

$+\swarrow \Sigma F_t = ma_t$:         $mg \sin 30° = ma_t$
                              $a_t = g \sin 30° = +4.90 \text{ m/s}^2$    $a_t = 4.90 \text{ m/s}^2 \swarrow$ ◄

$+\nwarrow \Sigma F_n = ma_n$:         $2.5 \, mg - mg \cos 30° = ma_n$
                              $a_n = 1.634 \, g = +16.01 \text{ m/s}^2$    $a_n = 16.01 \text{ m/s}^2 \nwarrow$ ◄

Since $a_n = v^2/\rho$, we have $v^2 = \rho a_n = (2 \text{ m})(16.03 \text{ m/s}^2)$

$v = \pm 5.66 \text{ m/s}$      $v = 5.66 \text{ m/s} \swarrow$ (up or down) ◄

## SAMPLE PROBLEM 12.6

Determine the rated speed of a highway curve of radius $\rho = 122$ m banked through an angle $\theta = 18°$. The *rated speed* of a banked highway curve is the speed at which a car should travel if no lateral friction force is to be exerted on its wheels.

## SOLUTION

The car travels in a *horizontal* circular path of radius $\rho$. The normal component $\mathbf{a}_n$ of the acceleration is directed toward the center of the path; its magnitude is $a_n = v^2/\rho$, where $v$ is the speed of the car in m/s. The mass $m$ of the car is $W/g$, where $W$ is the weight of the car. Since no lateral friction force is to be exerted on the car, the reaction $\mathbf{R}$ of the road is shown perpendicular to the roadway. Applying Newton's second law, we write

$+\uparrow \Sigma F_y = 0$:         $R \cos \theta - W = 0$     $R = \dfrac{W}{\cos \theta}$     (1)

$\xrightarrow{+} \Sigma F_n = ma_n$:         $R \sin \theta = \dfrac{W}{g} a_n$     (2)

Substituting for $R$ from (1) into (2), and recalling that $a_n = v^2/\rho$,

$$\dfrac{W}{\cos \theta} \sin \theta = \dfrac{W}{g} \dfrac{v^2}{\rho}     v^2 = g \rho \tan \theta$$

Substituting $\rho = 122$ m and $\theta = 18°$ into this equation, we obtain

$v^2 = (9.81 \text{ m/s}^2)(122 \text{ m}) \tan 18°$
$v = 19.7 \text{ m/s}$         $v = 19.7 \text{ m/s}$ ◄

# SOLVING PROBLEMS
## ON YOUR OWN

In the problems for this lesson, you will apply *Newton's second law of motion*, $\Sigma \mathbf{F} = m\mathbf{a}$, to relate the forces acting on a particle to the motion of the particle.

**1. Writing the equations of motion.** When applying Newton's second law to the types of motion discussed in this lesson, you will find it most convenient to express the vectors $\mathbf{F}$ and $\mathbf{a}$ in terms of either their rectangular components or their tangential and normal components.

   ***a. When using rectangular components,*** and recalling from Sec. 11.11 the expressions found for $a_x$, $a_y$, and $a_z$, you will write

$$\Sigma F_x = m\ddot{x} \qquad \Sigma F_y = m\ddot{y} \qquad \Sigma F_z = m\ddot{z}$$

   ***b. When using tangential and normal components,*** and recalling from Sec. 11.13 the expressions found for $a_t$ and $a_n$, you will write

$$\Sigma F_t = m\frac{dv}{dt} \qquad \Sigma F_n = m\frac{v^2}{\rho}$$

**2. Drawing a free-body diagram** showing the applied forces *and an equivalent diagram* showing the vector $m\mathbf{a}$ or its components will provide you with a pictorial representation of Newton's second law [Sample Probs. 12.1 through 12.6]. These diagrams will be of great help to you when writing the equations of motion. Note that when a problem involves two or more bodies, it is usually best to consider each body separately.

**3. Applying Newton's second law.** As we observed in Sec. 12.2, the acceleration used in the equation $\Sigma \mathbf{F} = m\mathbf{a}$ should always be the *absolute acceleration* of the particle (that is, it should be measured with respect to a newtonian frame of reference). Also, *if the sense of the acceleration* $\mathbf{a}$ *is unknown* or is not easily deduced, assume an arbitrary sense for $\mathbf{a}$ (usually the positive direction of a coordinate axis) and then let the solution provide the correct sense. Finally, note how the solutions of Sample Probs. 12.3 and 12.4 were divided into a *kinematics* portion and a *kinetics* portion, and how in Sample Prob. 12.4 we used two systems of coordinate axes to simplify the equations of motion.

**4. When a problem involves dry friction,** be sure to review the relevant sections of *Statics* [Secs. 8.1 to 8.3] before attempting to solve that problem. In particular, you should know when each of the equations $F = \mu_s N$ and $F = \mu_k N$ may be used. You should also recognize that if the motion of a system is not specified, it is necessary first to assume a possible motion and then to check the validity of that assumption.

**5. Solving problems involving relative motion.** When a body $B$ moves with respect to a body $A$, as in Sample Prob. 12.4, it is often convenient to express the acceleration of $B$ as

$$\mathbf{a}_B = \mathbf{a}_A + \mathbf{a}_{B/A}$$

where $\mathbf{a}_{B/A}$ is the acceleration of $B$ relative to $A$, that is, the acceleration of $B$ as observed from a frame of reference attached to $A$ and in translation. If $B$ is observed to move in a straight line, $\mathbf{a}_{B/A}$ will be directed along that line. On the other hand, if $B$ is observed to move along a circular path, the relative acceleration $\mathbf{a}_{B/A}$ should be resolved into components tangential and normal to that path.

**6. Finally, always consider the implications of any assumption you make.** Thus, in a problem involving two cords, if you assume that the tension in one of the cords is equal to its maximum allowable value, check whether any requirements set for the other cord will then be satisfied. For instance, will the tension $T$ in that cord satisfy the relation $0 \leq T \leq T_{\text{max}}$? That is, will the cord remain taut and will its tension be less than its maximum allowable value?

Fig. P12.3

Fig. P12.5

**12.1**  The acceleration due to gravity on Mars is 3.75 m/s². Knowing that the mass of a silver bar has been officially designated as 20 kg, determine, on Mars, its weight in newtons.

**12.2**  The value of the acceleration of gravity at any latitude $\phi$ is given by $g = 9.7087(1 + 0.0053 \sin^2 \phi)$ m/s², where the effect of the rotation of the earth as well as the fact that the earth is not spherical have been taken into account. Knowing that the mass of a gold bar has been officially designated as 2 kg, determine to four significant figures its mass in kilograms and its weight in newtons at a latitude of (*a*) 0°, (*b*) 45°, (*c*) 60°.

*12.3*  A spring scale *A* and a lever scale *B* having equal lever arms are fastened to the roof on an elevator, and identical packages are attached to the scales as shown. Knowing that when the elevator moves downward with an acceleration of .6 m/s² the spring scale indicates a load of 3.2 kg, determine (*a*) the weight of the packages, (*b*) the load indicated by the spring scale and the mass needed to balance the lever scale when the elevator moves upward with an acceleration of .6 m/s².

*12.4*  A Global Positioning System (GPS) satellite is in a circular orbit 20157 km above the surface of the earth and completes one orbit every 12 h. Knowing that the magnitude of the linear momentum of the satellite is 340194 kg · s and the radius of the earth is 6370 km, determine (*a*) the mass of the satellite, (*b*) the weight of the satellite before it was launched from earth.

**12.5**  The 18 kg block starts from rest and moves upward when constant forces of 20 N and 40 N are applied to supporting ropes. Neglecting the masses of the pulleys and the effect of friction, determine the speed of the block after it has moved .457 m.

**12.6**  A motorist traveling at a speed of 108 km/h suddenly applies the brakes and comes to a stop after skidding 75 m. Determine (*a*) the time required for the car to stop, (*b*) the coefficient of friction between the tires and the pavement.

**12.7**  A 1400-kg automobile is driven down a 4° incline at a speed of 88 km/h when the brakes are applied, causing a total braking force of 7500 N to be applied to the automobile. Determine the distance traveled by the automobile before it comes to a stop.

**12.8**  In the braking test of a sports car its velocity is reduced from 112.16 km/h to zero in a distance of 51.8 m with slipping impending. Knowing that the coefficient of kinetic friction is 80 percent of the coefficient of static friction, determine (*a*) the coefficient of static friction, (*b*) the stopping distance for the same initial velocity if the car skids. Ignore air resistance and rolling resistance.

**12.9** A .09 kg model rocket is launched vertically from rest at time $t = 0$ with a constant thrust of 4 N for one second and no thrust for $t > 1$ s. Neglecting air resistance and the decrease in mass of the rocket, determine (a) the maximum height $h$ reached by the rocket, (b) the time required to reach this maximum height.

**12.10** A 40-kg package is at rest on an incline when a force **P** is applied to it. Determine the magnitude of **P** if 4 s is required for the package to travel 10 m up the incline. The static and kinetic coefficients of friction between the package and the incline are 0.30 and 0.25, respectively.

**Fig. P12.10**

**Fig. P12.9**

**12.11** If an automobile's braking distance from 100 km/h is 60 m on level pavement, determine the automobile's braking distance from 100 km/h when it is (a) going up a 6° incline, (b) going down a 2-percent incline.

**12.12** The two blocks shown are originally at rest. Neglecting the masses of the pulleys and the effect of friction in the pulleys and between the blocks and the incline, determine (a) the acceleration of each block, (b) the tension in the cable.

**Fig. P12.12 and P12.13**

**12.13** The two blocks shown are originally at rest. Neglecting the masses of the pulleys and the effect of friction in the pulleys and assuming that the coefficients of friction between both blocks and the incline are $\mu_s = 0.25$ and $\mu_k = 0.20$, determine (a) the acceleration of each block, (b) the tension in the cable.

**12.14** A light train made up of two cars is traveling at 88 km/h when the brakes are applied to both cars. Knowing that car A has a weight of 24947.56 kg and car B has a weight of 19958 kg and that the braking force is 31137.5 N on each car, determine (a) the distance traveled by the train before it comes to a stop, (b) the force in the coupling between the cars while the train is slowing down.

**Fig. P12.14**

**12.15** Solve Prob. 12.14, assuming that the brakes of car $B$ fail to operate.

**12.16** Block $A$ weighs 356 N, and block $B$ weighs 71 N. The coefficients of friction between all surfaces of contact are $\mu_s = 0.20$ and $\mu_k = 0.15$. Knowing that $P = 0$, determine ($a$) the acceleration of block $B$, ($b$) the tension in the cord.

**Fig. P12.16 and P12.17**

**12.17** Block $A$ weighs 356 N, and block $B$ weighs 71 N. The coefficients of friction between all surfaces of contact are $\mu_s = 0.20$ and $\mu_k = 0.15$. Knowing that $\mathbf{P} = 44.5$ N $\rightarrow$, determine ($a$) the acceleration of block $B$, ($b$) the tension in the cord.

**12.18** Boxes $A$ and $B$ are at rest on a conveyor belt that is initially at rest. The belt is suddenly started in an upward direction so that slipping occurs between the belt and the boxes. Knowing that the coefficients of kinetic friction between the belt and the boxes are $(\mu_k)_A = 0.30$ and $(\mu_k)_B = 0.32$, determine the initial acceleration of each box.

**12.19** The system shown is initially at rest. Neglecting the masses of the pulleys and the effect of friction in the pulleys, determine ($a$) the acceleration of each block, ($b$) the tension in each cable.

**12.20** Each of the systems shown is initially at rest. Neglecting axle friction and the masses of the pulleys, determine for each system ($a$) the acceleration of block $A$, ($b$) the velocity of block $A$ after it has moved through 1.5 m, ($c$) the time required for block $A$ to reach a velocity of 3 m/s.

**Fig. P12.18**

**Fig. P12.19**

**Fig. P12.20**

**12.21** The flatbed trailer carries two 1360.8 kg beams with the upper beam secured by a cable. The coefficients of static friction between the two beams and between the lower beam and the bed of the trailer are 0.25 and 0.30, respectively. Knowing that the load does not shift, determine (*a*) the maximum acceleration of the trailer and the corresponding tension in the cable, (*b*) the maximum deceleration of the trailer.

**Fig. P12.21**

**12.22** The 10-kg block *B* is supported by the 40-kg block *A* which is pulled up an incline by a constant 500 N force. Neglecting friction between the block and the incline and knowing that block *B* does not slip on block *A*, determine the smallest allowable value of the coefficient of static friction between the blocks.

**Fig. P12.22**

**12.23** A package is at rest on a conveyor belt which is initially at rest. The belt is started and moves to the right for 1.5 s with a constant acceleration of 3.2 m/s². The belt then moves with a constant deceleration $a_2$ and comes to a stop after a total displacement of 4.6 m. Knowing that the coefficients of friction between the package and the belt are $\mu_s = 0.35$ and $\mu_k = 0.25$, determine (*a*) the deceleration $a_2$ of the belt, (*b*) the displacement of the package relative to the belt as the belt comes to a stop.

**Fig. P12.23**

**12.24** To transport a series of bundles of shingles *A* to a roof, a contractor uses a motor-driven lift consisting of a horizontal platform *BC* which rides on rails attached to the sides of a ladder. The lift starts from rest and initially moves with a constant acceleration $a_1$ as shown. The lift then decelerates at a constant rate $a_2$ and comes to rest at *D*, near the top of the ladder. Knowing that the coefficient of static friction between the bundle of shingles and the horizontal platform is 0.30, determine the largest allowable acceleration $a_1$ and the largest allowable deceleration $a_2$ if the bundle is not to slide on the platform.

**Fig. P12.24**

**Fig. P12.25**

**Fig. P12.27**

**Fig. P12.29**

**Fig. P12.30**

**12.25** To unload a bound stack of plywood from a truck, the driver first tilts the bed of the truck and then accelerates from rest. Knowing that the coefficients of friction between the bottom sheet of plywood and the bed are $\mu_s = 0.40$ and $\mu_k = 0.30$, determine (a) the smallest acceleration of the truck which will cause the stack of plywood to slide, (b) the acceleration of the truck which causes corner A of the stack of plywood to reach the end of the bed in 0.4 s.

**12.26** The propellers of a ship of mass $m$ can produce a propulsive force $\mathbf{F}_0$; they produce a force of the same magnitude but opposite direction when the engines are reversed. Knowing that the ship was proceeding forward at its maximum speed $v_0$ when the engines were put into reverse, determine the distance the ship travels before coming to a stop. Assume that the frictional resistance of the water varies directly with the square of the velocity.

**12.27** A constant force $\mathbf{P}$ is applied to a piston and rod of total mass $m$ to make them move in a cylinder filled with oil. As the piston moves, the oil is forced through orifices in the piston and exerts on the piston a force of magnitude $kv$ in a direction opposite to the motion of the piston. Knowing that the piston starts from rest at $t = 0$ and $x = 0$, show that the equation relating $x$, $v$, and $t$, where $x$ is the distance traveled by the piston and $v$ is the speed of the piston, is linear in each of the variables.

**12.28** A 4-kg projectile is fired vertically with an initial velocity of 90 m/s, reaches a maximum height, and falls to the ground. The aerodynamic drag $\mathbf{D}$ has a magnitude $D = 0.0024\,v^2$ where $D$ and $v$ are expressed in newtons and m/s, respectively. Knowing that the direction of the drag is always opposite to the direction of the velocity, determine (a) the maximum height of the trajectory, (b) the speed of the projectile when it reaches the ground.

**12.29** A spring $AB$ of constant $k$ is attached to a support $A$ and to a collar of mass $m$. The unstretched length of the spring is $l$. Knowing that the collar is released from rest at $x = x_0$ and neglecting friction between the collar and the horizontal rod, determine the magnitude of the velocity of the collar as it passes through point $C$.

**12.30** The system of three 10-kg blocks is supported in a vertical plane and is initially at rest. Neglecting the masses of the pulleys and the effect of friction in the pulleys, determine (a) the change in position of block $A$ after 0.5 s, (b) the tension in the cable.

**Fig. P12.31**

**12.31** The coefficients of friction between block $B$ and block $A$ are $\mu_s = 0.12$ and $\mu_k = 0.10$ and the coefficients of friction between block $A$ and the incline are $\mu_s = 0.24$ and $\mu_k = 0.20$. The masses of block $A$ and block $B$ are 10 kg and 5 kg, respectively. Knowing that the system is released from rest in the position shown, determine (a) the acceleration of $A$, (b) the velocity of $B$ relative to $A$ at $t = 0.5$ s.

**12.32** The weights of blocks $A$, $B$, and $C$ are $w_A = w_C = 89$ N, and $w_B = 44.5$ N. Knowing that $P = 222.4$ N and neglecting the masses of the pulleys and the effect of friction, determine (*a*) the acceleration of each block, (*b*) the tension in the cable.

**Fig. P12.32 and P12.33**

**12.33** The coefficients of friction between the three blocks and the horizontal surfaces are $\mu_s = 0.25$ and $\mu_k = 0.20$. The weights of the blocks are $w_A = w_C = 89$ N, and $w_B = 44.5$ N. Knowing that the blocks are initially at rest and that $C$ moves to the right through .73 m in 0.4 s, determine (*a*) the acceleration of each block, (*b*) the tension in the cable, (*c*) the force **P**. Neglect axle friction and the masses of the pulleys.

**Fig. *P12.34***

**12.34** A 25-kg block $A$ rests on an inclined surface, and a 15-kg counterweight $B$ is attached to a cable as shown. Neglecting friction, determine the acceleration of $A$ and the tension in the cable immediately after the system is released from rest.

**12.35** A 250-kg crate $B$ is suspended from a cable attached to a 20-kg trolley $A$ which rides on an inclined I-beam as shown. Knowing that at the instant shown the trolley has an acceleration of 0.4 m/s² up to the right, determine (*a*) the acceleration of $B$ relative to $A$, (*b*) the tension in cable $CD$.

**12.36** A 2-kg ball revolves in a horizontal circle as shown at a constant speed of 1.5 m/s. Knowing that $L = 600$ mm, determine (*a*) the angle $\theta$ that the cord forms with the vertical, (*b*) the tension in the cord.

**Fig. *P12.35***

**12.37** A single wire $ACB$ of length 2 m passes through a ring at $C$ that is attached to a sphere which revolves at a constant speed $v$ in the horizontal circle shown. Knowing that $\theta_1 = 60°$ and $\theta_2 = 30°$ and that the tension is the same in both portions of the wire, determine the speed $v$.

**Fig. P12.36**

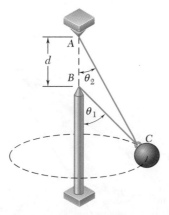

**Fig. P12.37 and *P12.38***

**12.38** Two wires $AC$ and $BC$ are tied to a 6.8 kg sphere which revolves at a constant speed $v$ in the horizontal circle shown. Knowing that $\theta_1 = 50°$ and $\theta_2 = 25°$ and that $d = 1.2$ m, determine the range of values of $v$ for which both wires are taut.

**12.39** During a hammer thrower's practice swings, the 7.3 kg head $A$ of the hammer revolves at a constant speed $v$ in a horizontal circle as shown. If $\rho = 0.9$ m and $\theta = 60°$, determine ($a$) the tension in wire $BC$, ($b$) the speed of the hammer's head.

**Fig. P12.39**

**Fig. P12.40**

**12.40** A 1-kg sphere is at rest relative to a parabolic dish which rotates at a constant rate about a vertical axis. Neglecting friction and knowing that $r = 1$ m, determine ($a$) the speed $v$ of the sphere, ($b$) the magnitude of the normal force exerted by the sphere on the inclined surface of the dish.

**12.41** A 1-kg collar $C$ slides without friction along the rod $OA$ and is attached to rod $BC$ by a frictionless pin. The rods rotate in the horizontal plane. At the instant shown, $BC$ is rotating counterclockwise and the speed of $C$ is 1 m/s, increasing at a rate of 1.3 m/s. Determine at this instant, ($a$) the tension in rod $BC$, ($b$) the force exerted by the collar on rod $OA$.

**Fig. P12.41**

*12.42 The 0.5-kg flyballs of a centrifugal governor revolve at a constant speed $v$ in the horizontal circle of 150-mm radius shown. Neglecting the mass of links $AB$, $BC$, $AD$, and $DE$ and requiring that the links support only tensile forces, determine the range of the allowable values of $v$ so that the magnitudes of the forces in the links do not exceed 75 N.

**Fig. P12.42**

**Fig. *P12.43***

*12.43 As part of an outdoor display, a 5-kg model $C$ of the earth is attached to wires $AC$ and $BC$ and revolves at a constant speed $v$ in the horizontal circle shown. Determine the range of the allowable values of $v$ if both wires are to remain taut and if the tension in either of the wires is not to exceed 116 N.

12.44 A small sphere of weight $W$ is held as shown by two wires $AB$ and $CD$. If wire $AB$ is cut, determine the tension in the other wire $(a)$ before $AB$ is cut, $(b)$ immediately after $AB$ has been cut.

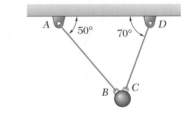

**Fig. P12.44**

12.45 A series of small packages are being moved by a thin conveyor belt that passes over a 300-mm-radius idler pulley. The belt starts from rest at time $t = 0$ and its speed increases at a constant rate of 150 mm/s². Knowing that the coefficient of static friction between the packages and the belt is 0.75, determine the time at which the first package slips.

**Fig. P12.45**

12.46 An airline pilot climbs to a new flight level along the path shown. Knowing that the speed of the airplane decreases at a constant rate from 165 m/s at point $A$ to 146 m/s at point $C$, determine the magnitude of the abrupt change in the force exerted on a 90.7 kg passenger as the airplane passes point $B$.

12.47 An airline pilot climbs to a new flight level along the path shown. The motion of the airplane between $A$ and $B$ is defined by the relation $s = 3t(180 - t)$, where $s$ is the arc length in meter, $t$ is the time in seconds, and $t = 0$ when the airplane is at point $A$. Determine the force exerted by his seat on a 74.8 kg passenger $(a)$ just after the airplane passes point $A$, $(b)$ just before the airplane reaches point $B$.

**Fig. P12.46 and P12.47**

**Fig. P12.48**

**Fig. P12.50**

**12.48** During a high-speed chase, an 1100-kg sports car traveling at a speed of 160 km/h just loses contact with the road as it reaches the crest $A$ of a hill. (*a*) Determine the radius of curvature $\rho$ of the vertical profile of the road at $A$. (*b*) Using the value of $\rho$ found in part *a*, determine the force exerted on a 70-kg driver by the seat of his 1400-kg car as the car, traveling at a constant speed of 80 km/h, passes through $A$.

**12.49** A small 0.2-kg sphere $B$ is given a downward velocity $\mathbf{v}_0$ and swings freely in the vertical plane, first about $O$ and then about the peg $A$ after the cord comes in contact with the peg. Determine the largest allowable velocity $\mathbf{v}_0$ if the tension in the cord is not to exceed 10 N.

**Fig. P12.49**

**12.50** A .23 kg block $B$ fits inside a small cavity cut in arm $OA$, which rotates in the vertical plane at a constant rate such that $v = 2.74$ m/s. Knowing that the spring exerts on block $B$ a force of magnitude $P = 1.33$ N and neglecting the effect of friction, determine the range of values of $\theta$ for which block $B$ is in contact with the face of the cavity closest to the axis of rotation.

**12.51** A 54.4 kg pilot flies a jet trainer in a half vertical loop of 1097-m radius so that the speed of the trainer decreases at a constant rate. Knowing that the pilot's apparent weights at points $A$ and $C$ are 1690 N and 356 N, respectively, determine the force exerted on her by the seat of the trainer when the trainer is at point $B$.

**Fig. P12.51**

**12.52** A car is traveling on a banked road at a constant speed $v$. Determine the range of values of $v$ for which the car does not skid. Express your answer in terms of the radius $r$ of the curve, the banking angle $\theta$, and the angle of static friction $\phi_s$ between the tires and the pavement.

**12.53** A curve in a speed track has a radius of 200 m and a rated speed of 180 km/h. (See Sample Prob. 12.6 for the definition of rated speed.) Knowing that a racing car starts skidding on the curve when traveling at a speed of 320 km/h, determine (*a*) the banking angle $\theta$, (*b*) the coefficient of static friction between the tires and the track under the prevailing conditions, (*c*) the minimum speed at which the same car could negotiate that curve.

**Fig. P12.52 and P12.53**

**12.54** Tilting trains such as the *Acela*, which runs from Washington to New York to Boston, are designed to travel safely at high speeds on curved sections of track which were built for slower, conventional trains. As it enters a curve, each car is tilted by hydraulic actuators mounted on its trucks. The tilting feature of the cars also increases passenger comfort by eliminating or greatly reducing the side force $\mathbf{F}_s$ (parallel to the floor of the car) to which passengers feel subjected. For a train traveling at 200 km/h on a curved section of track banked at an angle $\theta = 8°$ and with a rated speed of 120 km/h, determine (*a*) the magnitude of the side force felt by a passenger of weight $W$ in a standard car with no tilt ($\phi = 0$), (*b*) the required angle of tilt $\phi$ if the passenger is to feel no side force. (See Sample Prob. 12.6 for the definition of rated speed.)

**Fig. P12.54 and P12.55**

**12.55** Tests carried out with the tilting trains described in Prob. 12.54 revealed that passengers feel queasy when they see through the car windows that the train is rounding a curve at high speed, yet do not feel any side force. Designers, therefore, prefer to reduce, but not eliminate that force. For the train of Prob. 12.54, determine the required angle of tilt $\phi$ if passengers are to feel side forces equal to 12 percent of their weights.

**Fig. P12.56**

**12.56**   A small 250-g collar $C$ can slide on a semicircular rod which is made to rotate about the vertical $AB$ at a constant rate of 7.5 rad/s. Determine the three values of $\theta$ for which the collar will not slide on the rod, assuming no friction between the collar and the rod.

**12.57**   For the collar and rod of Prob. 12.56, and assuming that the coefficient of friction are $\mu_s = 0.25$ and $\mu_k = 0.20$, indicate whether the collar will slide on the rod if it is released in the position corresponding to (a) $\theta = 75°$, (b) $\theta = 40°$. Also, determine the magnitude and direction of the friction force exerted on the collar immediately after release.

**12.58**   A small block $B$ fits inside a slot cut in arm $OA$ which rotates in a vertical plane at a constant rate. The block remains in contact with the end of the slot closest to $A$ and its speed is 1.28 m/s for $0 \le \theta \le 150°$. Knowing that the block begins to slide when $\theta = 150°$, determine the coefficient of static friction between the block and the slot.

**Fig. P12.58**

**12.59**   A 2.7 kg block is at rest relative to a parabolic dish which rotates at a constant rate about a vertical axis. Knowing that the coefficient of static friction is 0.5 and that $r = 1.8$ m, determine the maximum allowable speed $v$ of the block.

**Fig. P12.59**

**12.60**   Four seconds after a polisher is started from rest, small tufts of fleece from along the circumference of the 25.4 cm-diameter polishing pad are observed to fly free of the pad. If the polisher is started so that the fleece along the circumference undergoes a constant tangential acceleration of 3.7 m/s², determine (a) the speed $v$ of a tuft as it leaves the pad, (b) the magnitude of the force required to free the tuft if the average weight of a tuft is $1.5 \times 10^{-6}$ kg.

**Fig. P12.60**

**12.61** A turntable $A$ is built into a stage for use in a theatrical production. It is observed during a rehearsal that a trunk $B$ starts to slide on the turntable 12 s after the turntable begins to rotate. Knowing that the trunk undergoes a constant tangential acceleration of 0.23 m/s², determine the coefficient of static friction between the trunk and the turntable.

**12.62** The parallel-link mechanism $ABCD$ is used to transport a component $I$ between manufacturing processes at stations $E$, $F$, and $G$ by picking it up at a station when $\theta = 0$ and depositing it at the next station when $\theta = 180°$. Knowing that member $BC$ remains horizontal throughout its motion and that links $AB$ and $CD$ rotate at a constant rate in a vertical plane in such a way that $v_B = 0.7$ m/s, determine (a) the minimum value of the coefficient of static friction between the component and $BC$ if the component is not to slide on $BC$ while being transferred, (b) the values of $\theta$ for which sliding is impending.

**Fig. P12.61**

**Fig. P12.62**

**12.63** Knowing that the coefficients of friction between the component $I$ and member $BC$ of the mechanism of Prob. 12.62 are $\mu_s = 0.35$ and $\mu_k = 0.25$, determine (a) the maximum allowable speed $v_B$ if the component is not to slide on $BC$ while being transferred, (b) the values of $\theta$ for which sliding is impending.

**12.64** In the cathode-ray tube shown, electrons emitted by the cathode and attracted by the anode pass through a small hole in the anode and then travel in a straight line with a speed $v_0$ until they strike the screen at $A$. However, if a difference of potential $V$ is established between the two parallel plates, the electrons will be subjected to a force $\mathbf{F}$ perpendicular to the plates while they travel between the plates and will strike the screen at point $B$, which is at a distance $\delta$ from $A$. The magnitude of the force $\mathbf{F}$ is $F = eV/d$, where $-e$ is the charge of an electron and $d$ is the distance between the plates. Neglecting the effects of gravity, derive an expression for the deflection $\delta$ in terms of $V$, $v_0$, the charge $-e$ and the mass $m$ of an electron, and the dimensions $d$, $l$, and $L$.

**12.65** In Prob. 12.64, determine the smallest allowable value of the ratio $d/l$ in terms of $e$, $m$, $v_0$, and $V$ if at $x = l$ the minimum permissible distance between the path of the electrons and the positive plate is $0.075d$.

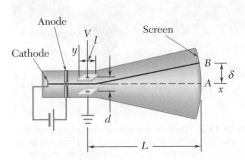

**Fig. P12.64 and P12.65**

## 12.7. ANGULAR MOMENTUM OF A PARTICLE. RATE OF CHANGE OF ANGULAR MOMENTUM

Consider a particle $P$ of mass $m$ moving with respect to a newtonian frame of reference $Oxyz$. As we saw in Sec. 12.3, the linear momentum of the particle at a given instant is defined as the vector $m\mathbf{v}$ obtained by multiplying the velocity $\mathbf{v}$ of the particle by its mass $m$. The moment about $O$ of the vector $m\mathbf{v}$ is called the *moment of momentum*, or the *angular momentum*, of the particle about $O$ at that instant and is denoted by $\mathbf{H}_O$. Recalling the definition of the moment of a vector (Sec. 3.6) and denoting by $\mathbf{r}$ the position vector of $P$, we write

$$\mathbf{H}_O = \mathbf{r} \times m\mathbf{v} \tag{12.12}$$

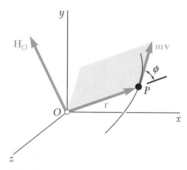

**Fig. 12.12**

and note that $\mathbf{H}_O$ is a vector perpendicular to the plane containing $\mathbf{r}$ and $m\mathbf{v}$ and of magnitude

$$H_O = rmv \sin \phi \tag{12.13}$$

where $\phi$ is the angle between $\mathbf{r}$ and $m\mathbf{v}$ (Fig. 12.12). The sense of $\mathbf{H}_O$ can be determined from the sense of $m\mathbf{v}$ by applying the right-hand rule. The unit of angular momentum is obtained by multiplying the units of length and of linear momentum (Sec. 12.4). With SI units, we have

$$(\text{m})(\text{kg} \cdot \text{m/s}) = \text{kg} \cdot \text{m}^2/\text{s}$$

With U.S. customary units, we write

$$(\text{ft})(\text{lb} \cdot \text{s}) = \text{ft} \cdot \text{lb} \cdot \text{s}$$

Resolving the vectors $\mathbf{r}$ and $m\mathbf{v}$ into components and applying formula (3.10), we write

$$\mathbf{H}_O = \begin{vmatrix} \mathbf{i} & \mathbf{j} & \mathbf{k} \\ x & y & z \\ mv_x & mv_y & mv_z \end{vmatrix} \tag{12.14}$$

The components of $\mathbf{H}_O$, which also represent the moments of the linear momentum $m\mathbf{v}$ about the coordinate axes, can be obtained by expanding the determinant in (12.14). We have

$$\begin{aligned} H_x &= m(yv_z - zv_y) \\ H_y &= m(zv_x - xv_z) \\ H_z &= m(xv_y - yv_x) \end{aligned} \tag{12.15}$$

In the case of a particle moving in the $xy$ plane, we have $z = v_z = 0$ and the components $H_x$ and $H_y$ reduce to zero. The angular momentum is thus perpendicular to the $xy$ plane; it is then completely defined by the scalar

$$H_O = H_z = m(xv_y - yv_x) \tag{12.16}$$

**Photo 12.2** During a hammer thrower's practice swings, the hammer acquires angular momentum about a vertical axis at the center of its circular path.

which will be positive or negative according to the sense in which the particle is observed to move from $O$. If polar coordinates are used, we resolve the linear momentum of the particle into radial and transverse components (Fig. 12.13) and write

$$H_O = rmv \sin \phi = rmv_\theta \qquad (12.17)$$

or, recalling from (11.45) that $v_\theta = r\dot{\theta}$,

$$H_O = mr^2\dot{\theta} \qquad (12.18)$$

**Fig. 12.13**

Let us now compute the derivative with respect to $t$ of the angular momentum $\mathbf{H}_O$ of a particle $P$ moving in space. Differentiating both members of Eq. (12.12), and recalling the rule for the differentiation of a vector product (Sec. 11.10), we write

$$\dot{\mathbf{H}}_O = \dot{\mathbf{r}} \times m\mathbf{v} + \mathbf{r} \times m\dot{\mathbf{v}} = \mathbf{v} \times m\mathbf{v} + \mathbf{r} \times m\mathbf{a}$$

Since the vectors $\mathbf{v}$ and $m\mathbf{v}$ are collinear, the first term of the expression obtained is zero; and, by Newton's second law, $m\mathbf{a}$ is equal to the sum $\Sigma\mathbf{F}$ of the forces acting on $P$. Noting that $\mathbf{r} \times \Sigma\mathbf{F}$ represents the sum $\Sigma\mathbf{M}_O$ of the moments about $O$ of these forces, we write

$$\Sigma\mathbf{M}_O = \dot{\mathbf{H}}_O \qquad (12.19)$$

Equation (12.19), which results directly from Newton's second law, states that *the sum of the moments about $O$ of the forces acting on the particle is equal to the rate of change of the moment of momentum, or angular momentum, of the particle about $O$.*

## 12.8. EQUATIONS OF MOTION IN TERMS OF RADIAL AND TRANSVERSE COMPONENTS

Consider a particle $P$, of polar coordinates $r$ and $\theta$, which moves in a plane under the action of several forces. Resolving the forces and the acceleration of the particle into radial and transverse components (Fig. 12.14) and substituting into Eq. (12.2), we obtain the two scalar equations

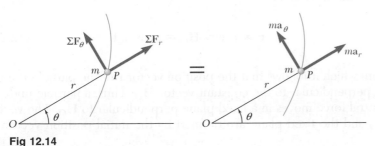

**Fig 12.14**

$$\Sigma F_r = ma_r \qquad \Sigma F_\theta = ma_\theta \qquad (12.20)$$

Substituting for $a_r$ and $a_\theta$ from Eqs. (11.46), we have

$$\Sigma F_r = m(\ddot{r} - r\dot{\theta}^2) \qquad (12.21)$$

$$\Sigma F_\theta = m(r\ddot{\theta} + 2\dot{r}\dot{\theta}) \qquad (12.22)$$

The equations obtained can be solved for two unknowns.

Equation (12.22) could have been derived from Eq. (12.19). Recalling (12.18) and noting that $\Sigma M_O = r\Sigma F_\theta$, Eq. (12.19) yields

$$r\Sigma F_\theta = \frac{d}{dt}(mr^2\dot{\theta})$$
$$= m(r^2\ddot{\theta} + 2r\dot{r}\dot{\theta})$$

and, after dividing both members by $r$,

$$\Sigma F_\theta = m(r\ddot{\theta} + 2\dot{r}\dot{\theta}) \qquad (12.22)$$

**Photo 12.3** The path of specimen being tested in a centrifuge is a horizontal circle. The forces acting on the specimen and its acceleration can be resolved into radial and transverse components with $r =$ constant.

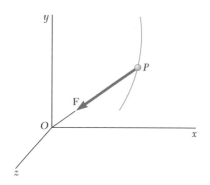

**Fig. 12.15**

## 12.9. MOTION UNDER A CENTRAL FORCE. CONSERVATION OF ANGULAR MOMENTUM

When the only force acting on a particle $P$ is a force **F** directed toward or away from a fixed point $O$, the particle is said to be moving *under a central force,* and the point $O$ is referred to as the *center of force* (Fig. 12.15). Since the line of action of **F** passes through $O$, we must have $\Sigma \mathbf{M}_O = 0$ at any given instant. Substituting into Eq. (12.19), we therefore obtain

$$\dot{\mathbf{H}}_O = 0$$

for all values of $t$ and, integrating in $t$,

$$\mathbf{H}_O = \text{constant} \qquad (12.23)$$

We thus conclude that *the angular momentum of a particle moving under a central force is constant, in both magnitude and direction.*

Recalling the definition of the angular momentum of a particle (Sec. 12.7), we write

$$\mathbf{r} \times m\mathbf{v} = \mathbf{H}_O = \text{constant} \qquad (12.24)$$

from which it follows that the position vector **r** of the particle $P$ must be perpendicular to the constant vector $\mathbf{H}_O$. Thus, a particle under a central force moves in a fixed plane perpendicular to $\mathbf{H}_O$. The vector $\mathbf{H}_O$ and the fixed plane are defined by the initial position vector $\mathbf{r}_0$

and the initial velocity $\mathbf{v}_0$ of the particle. For convenience, let us assume that the plane of the figure coincides with the fixed plane of motion (Fig. 12.16).

Since the magnitude $H_O$ of the angular momentum of the particle $P$ is constant, the right-hand member in Eq. (12.13) must be constant. We therefore write

$$rmv \sin \phi = r_0 m v_0 \sin \phi_0 \qquad (12.25)$$

This relation applies to the motion of any particle under a central force. Since the gravitational force exerted by the sun on a planet is a central force directed toward the center of the sun, Eq. (12.25) is fundamental to the study of planetary motion. For a similar reason, it is also fundamental to the study of the motion of space vehicles in orbit about the earth.

Alternatively, recalling Eq. (12.18), we can express the fact that the magnitude $H_O$ of the angular momentum of the particle $P$ is constant by writing

$$mr^2\dot{\theta} = H_O = \text{constant} \qquad (12.26)$$

or, dividing by $m$ and denoting by $h$ the angular momentum per unit mass $H_O/m$,

$$r^2\dot{\theta} = h \qquad (12.27)$$

Equation (12.27) can be given an interesting geometric interpretation. Observing from Fig. 12.17 that the radius vector $OP$ sweeps an infinitesimal area $dA = \frac{1}{2}r^2\,d\theta$ as it rotates through an angle $d\theta$, and defining the *areal velocity* of the particle as the quotient $dA/dt$, we note that the left-hand member of Eq. (12.27) represents twice the areal velocity of the particle. We thus conclude that *when a particle moves under a central force, its areal velocity is constant.*

**Fig. 12.16**

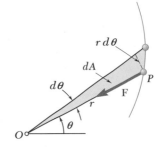

**Fig. 12.17**

## 12.10. NEWTON'S LAW OF GRAVITATION

As you saw in the preceding section, the gravitational force exerted by the sun on a planet or by the earth on an orbiting satellite is an important example of a central force. In this section, you will learn how to determine the magnitude of a gravitational force.

In his *law of universal gravitation,* Newton states that two particles of masses $M$ and $m$ at a distance $r$ from each other attract each other with equal and opposite forces $\mathbf{F}$ and $-\mathbf{F}$ directed along the line joining the particles (Fig. 12.18). The common magnitude $F$ of the two forces is

$$F = G\frac{Mm}{r^2} \qquad (12.28)$$

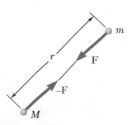

**Fig. 12.18**

where $G$ is a universal constant, called the *constant of gravitation*. Experiments show that the value of $G$ is $(66.73 \pm 0.03) \times 10^{-12}$ m$^3$/kg $\cdot$ s$^2$ in SI units or approximately $34.4 \times 10^{-9}$ ft$^4$/lb $\cdot$ s$^4$ in U.S. customary units. Gravitational forces exist between any pair of bodies, but their effect is appreciable only when one of the bodies has a very large mass. The effect of gravitational forces is apparent in the cases of the motion of a planet about the sun, of satellites orbiting about the earth, or of bodies falling on the surface of the earth.

Since the force exerted by the earth on a body of mass $m$ located on or near its surface is defined as the weight **W** of the body, we can substitute the magnitude $W = mg$ of the weight for $F$, and the radius $R$ of the earth for $r$, in Eq. (12.28). We obtain

$$W = mg = \frac{GM}{R^2} m \qquad \text{or} \qquad g = \frac{GM}{R^2} \qquad (12.29)$$

where $M$ is the mass of the earth. Since the earth is not truly spherical, the distance $R$ from the center of the earth depends upon the point selected on its surface, and the values of $W$ and $g$ will thus vary with the altitude and latitude of the point considered. Another reason for the variation of $W$ and $g$ with latitude is that a system of axes attached to the earth does not constitute a newtonian frame of reference (see Sec. 12.2). A more accurate definition of the weight of a body should therefore include a component representing the centrifugal force due to the rotation of the earth. Values of $g$ at sea level vary from 9.781 m/s$^2$, or 32.09 ft/s$^2$, at the equator to 9.833 m/s$^2$, or 32.26 ft/s$^2$, at the poles.†

The force exerted by the earth on a body of mass $m$ located in space at a distance $r$ from its center can be found from Eq. (12.28). The computations will be somewhat simplified if we note that according to Eq. (12.29), the product of the constant of gravitation $G$ and the mass $M$ of the earth can be expressed as

$$GM = gR^2 \qquad (12.30)$$

where $g$ and the radius $R$ of the earth will be given their average values $g = 9.81$ m/s$^2$ and $R = 6.37 \times 10^6$ m in SI units‡ and $g = 32.2$ ft/s$^2$ and $R = (3960$ mi$)(5280$ ft/mi$)$ in U.S. customary units.

The discovery of the law of universal gravitation has often been attributed to the belief that, after observing an apple falling from a tree, Newton had reflected that the earth must attract an apple and the moon in much the same way. While it is doubtful that this incident actually took place, it may be said that Newton would not have formulated his law if he had not first perceived that the acceleration of a falling body must have the same cause as the acceleration which keeps the moon in its orbit. This basic concept of the continuity of gravitational attraction is more easily understood today, when the gap between the apple and the moon is being filled with artificial earth satellites.

---

†A formula expressing $g$ in terms of the latitude $\phi$ was given in Prob. 12.1.

‡The value of $R$ is easily found if one recalls that the circumference of the earth is $2\pi R = 40 \times 10^6$ m.

# SAMPLE PROBLEM 12.7

A block $B$ of mass $m$ can slide freely on a frictionless arm $OA$ which rotates in a horizontal plane at a constant rate $\dot{\theta}_0$. Knowing that $B$ is released at a distance $r_0$ from $O$, express as a function of $r$, (a) the component $v_r$ of the velocity of $B$ along $OA$, (b) the magnitude of the horizontal force $\mathbf{F}$ exerted on $B$ by the arm $OA$.

## SOLUTION

Since all other forces are perpendicular to the plane of the figure, the only force shown acting on $B$ is the force $\mathbf{F}$ perpendicular to $OA$.

**Equations of Motion.** Using radial and transverse components.

$$+\nearrow\Sigma F_r = ma_r: \qquad\qquad 0 = m(\ddot{r} - r\dot{\theta}^2) \qquad\qquad (1)$$
$$+\nwarrow\Sigma F_\theta = ma_\theta: \qquad\qquad F = m(r\ddot{\theta} + 2\dot{r}\dot{\theta}) \qquad\qquad (2)$$

*a. Component $v_r$ of Velocity.* Since $v_r = \dot{r}$, we have

$$\ddot{r} = \dot{v}_r = \frac{dv_r}{dt} = \frac{dv_r}{dr}\frac{dr}{dt} = v_r\frac{dv_r}{dr}$$

Substituting for $\ddot{r}$ in (1), recalling that $\dot{\theta} = \dot{\theta}_0$, and separating the variables,

$$v_r\, dv_r = \dot{\theta}_0^2 r\, dr$$

Multiplying by 2, and integrating from 0 to $v_r$ and from $r_0$ to $r$,

$$v_r^2 = \dot{\theta}_0^2(r^2 - r_0^2) \qquad v_r = \dot{\theta}_0(r^2 - r_0^2)^{1/2} \blacktriangleleft$$

*b. Horizontal Force F.* Setting $\dot{\theta} = \dot{\theta}_0$, $\ddot{\theta} = 0$, $\dot{r} = v_r$ in Eq. (2), and substituting for $v_r$ the expression obtained in part $a$,

$$F = 2m\dot{\theta}_0(r^2 - r_0^2)^{1/2}\dot{\theta}_0 \qquad F = 2m\dot{\theta}_0^2(r^2 - r_0^2)^{1/2} \blacktriangleleft$$

# SAMPLE PROBLEM 12.8

A satellite is launched in a direction parallel to the surface of the earth with a velocity of 30155 km/h from an altitude of 385 km. Determine the velocity of the satellite as it reaches its maximum altitude of 3749 km. It is recalled that the radius of the earth is 6345 km.

## SOLUTION

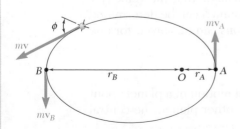

Since the satellite is moving under a central force directed toward the center $O$ of the earth, its angular momentum $\mathbf{H}_O$ is constant. From Eq. (12.13) we have

$$rmv \sin\phi = H_O = \text{constant}$$

which shows that $v$ is minimum at $B$, where both $r$ and $\sin\phi$ are maximum. Expressing conservation of angular momentum between $A$ and $B$.

$$r_A m v_A = r_B m v_B$$

$$v_B = v_A \frac{r_A}{r_B} = (30{,}155 \text{ km/h}) \cdot \frac{6345 \text{ km} + 385 \text{ km}}{6345 \text{ km} + 3749 \text{ km}}$$

$$v_B = 20{,}105 \text{ km/h} \blacktriangleleft$$

In this lesson we continued our study of Newton's second law by expressing the force and the acceleration in terms of their *radial and transverse components,* where the corresponding equations of motion are

$\Sigma F_r = ma_r:$            $\Sigma F_r = m(\ddot{r} - r\dot{\theta}^2)$
$\Sigma F_\theta = ma_\theta:$            $\Sigma F_\theta = m(r\ddot{\theta} + 2\dot{r}\dot{\theta})$

We introduced the *moment of the momentum,* or the *angular momentum,* $\mathbf{H}_O$ of a particle about $O$:

$$\mathbf{H}_O = \mathbf{r} \times m\mathbf{v} \qquad (12.12)$$

and found that $\mathbf{H}_O$ is constant when the particle moves under a *central force* with its center located at $O$.

1.   *Using radial and transverse components.*    Radial and transverse components were introduced in the last lesson of Chap. 11 [Sec. 11.14]; you should review that material before attempting to solve the following problems. Also, our comments in the preceding lesson regarding the application of Newton's second law (drawing a free-body diagram and a $m\mathbf{a}$ diagram, etc.) still apply [Sample Prob. 12.7]. Finally, note that the solution of that sample problem depends on the application of techniques developed in Chap. 11—you will need to use similar techniques to solve some of the problems of this lesson.

2.   *Solving problems involving the motion of a particle under a central force.* In problems of this type, the angular momentum $\mathbf{H}_O$ of the particle about the center of force $O$ is conserved. You will find it convenient to introduce the constant $h = H_O/m$ representing the angular momentum per unit mass. Conservation of the angular momentum of the particle $P$ about $O$ can then be expressed by either of the following equations

$$rv \sin \phi = h \qquad \text{or} \qquad r^2\dot{\theta} = h$$

where $r$ and $\theta$ are the polar coordinates of $P$, and $\phi$ is the angle that the velocity $\mathbf{v}$ of the particle forms with the line $OP$ (Fig. 12.16). The constant $h$ can be determined from the initial conditions and either of the above equations can be solved for one unknown.

3.   *In space-mechanics problems*   involving the orbital motion of a planet about the sun, or a satellite about the earth, the moon, or some other planet, the central

force **F** is the force of gravitational attraction; it is directed *toward* the center of force $O$ and has the magnitude

$$F = G\frac{Mm}{r^2} \tag{12.28}$$

Note that in the particular case of the gravitational force exerted by the earth, the product $GM$ can be replaced by $gR^2$, where $R$ is the radius of the earth [Eq. 12.30].

The following two cases of orbital motion are frequently encountered:

    *a. For a satellite in a circular orbit,* the force **F** is normal to the orbit and you can write $F = ma_n$; substituting for $F$ from Eq. (12.28) and observing that $a_n = v^2/\rho = v^2/r$, you will obtain

$$G\frac{Mm}{r^2} = m\frac{v^2}{r} \qquad \text{or} \qquad v^2 = \frac{GM}{r}$$

    *b. For a satellite in an elliptic orbit,* the radius vector **r** and the velocity **v** of the satellite are perpendicular to each other at the points $A$ and $B$ which are, respectively, farthest and closest to the center of force $O$ [Sample Prob. 12.8]. Thus, conservation of angular momentum of the satellite between these two points can be expressed as

$$r_A m v_A = r_B m v_B$$

**12.66** A 0.5-kg block $B$ slides without friction inside a slot cut in arm $OA$ which rotates in a vertical plane at a constant rate, $\dot\theta = 2$ rad/s. At the instant when $\theta = 30°$, $r = 0.6$ m and the force exerted on the block by the arm is zero. Determine, at this instant, (a) the relative velocity of the block with respect to the arm, (b) the relative acceleration of the block with respect to the arm.

**Fig. P12.66 and P12.67**

**12.67** A 0.5-kg block $B$ slides without friction inside a slot cut in arm $OA$ which rotates in a vertical plane. The motion of the rod is defined by the relation $\ddot\theta = 10$ rad/s², constant. At the instant when $\theta = 45°$, $r = 0.8$ m and the velocity of the block is zero. Determine, at this instant, (a) the force exerted on the block by the arm, (b) the relative acceleration of the block with respect to the arm.

**12.68** The motion of a 1.8 kg block $B$ in a horizontal plane is defined by the relations $r = 3t^2 - t^3$ and $\theta = 2t^2$, where $r$ is expressed in meter, $t$ in seconds, and $\theta$ in radians. Determine the radial and transverse components of the force exerted on the block when (a) $t = 0$, (b) $t = 1$ s.

**Fig. P12.68 and P12.69**

**12.69** The motion of a .45 kg block $B$ in a horizontal plane is defined by the relations $r = 6(1 + \cos 2\pi t)$ and $\theta = 2\pi t$, where $r$ is expressed in meter, $t$ in seconds, and $\theta$ in radians. Determine the radial and transverse components of the force exerted on the block when (a) $t = 0$, (b) $t = 0.75$ s.

**12.70** The 2.7 kg collar $B$ slides on the frictionless arm $AA'$. The arm is attached to drum $D$ and rotates about $O$ in a horizontal plane at the rate $\dot\theta = 0.8t$, where $\dot\theta$ and $t$ are expressed in rad/s and seconds, respectively. As the arm-drum assembly rotates, a mechanism within the drum releases cord so that the collar moves outward from $O$ with a constant speed of .457 m/s. Knowing that at $t = 0$, $r = 0$, determine the time at which the tension in the cord is equal to the magnitude of the horizontal force exerted on $B$ by arm $AA'$.

**Fig. P12.70**

**12.71** The horizontal rod $OA$ rotates about a vertical shaft according to the relation $\dot{\theta} = 10t$, where $\dot{\theta}$ and $t$ are expressed in rad/s and seconds, respectively. A .23 kg collar $B$ is held by a cord with a breaking strength of 1.8 kg. Neglecting friction, determine, immediately after the cord breaks, (a) the relative acceleration of the collar with respect to the rod, (b) the magnitude of the horizontal force exerted on the collar by the rod.

**Fig. P12.71**

**12.72** Disk $A$ rotates in a horizontal plane about a vertical axis at the constant rate of $\dot{\theta}_0 = 15$ rad/s. Slider $B$ has a mass of 230 g and moves in a frictionless slot cut in the disk. The slider is attached to a spring of constant $k = 60$ N/m, which is undeformed when $r = 0$. Knowing that at a given instant the acceleration of the slider relative to the disk is $\ddot{r} = -12$ m/s$^2$ and that the horizontal force exerted on the slider by the disk is 9 N, determine at that instant (a) the distance $r$, (b) the radial component of the velocity of the slider.

**12.73** A 1.5-kg collar is attached to a spring and slides without friction along a circular rod in a vertical plane. Knowing that the tension in the spring is 70 N and the speed of the collar is 3.8 m/s as it passes through point $A$, determine, at that instant, the radial and transverse components of acceleration of the collar.

**Fig. P12.72**

**Fig. P12.73**

**12.74** The two blocks are released from rest when $r = .73$ m and $\theta = 30°$. Neglecting the mass of the pulley and the effect of friction in the pulley and between block $A$ and the horizontal surface, determine (a) the initial tension in the cable, (b) the initial acceleration of block $A$, (c) the initial acceleration of block $B$.

**12.75** The velocity of block $A$ is 1.8 m/s to the right at the instant when $r = 2.4$ ft and $\theta = 30°$. Neglecting the mass of the pulley and the effect of friction in the pulley and between block $A$ and the horizontal surface, determine, at this instant, (a) the tension in the cable, (b) the acceleration of block $A$, (c) the acceleration of block $B$.

**Fig. P12.74 and P12.75**

**Fig. P12.76 and P12.77**

**Fig. P12.78 and *P12.79***

**12.76** A particle of mass $m$ is projected from point $A$ with an initial velocity $\mathbf{v}_0$ perpendicular to line $OA$ and moves under a central force $\mathbf{F}$ directed away from the center of force $O$. Knowing that the particle follows a path defined by the equation $r = r_0/\sqrt{\cos 2\theta}$ and using Eq. (12.27), express the radial and transverse components of the velocity $\mathbf{v}$ of the particle as functions of $\theta$.

**12.77** For the particle of Prob. 12.76, show (*a*) that the velocity of the particle and the central force $\mathbf{F}$ are proportional to the distance $r$ from the particle to the center of force $O$, (*b*) that the radius of curvature of the path is proportional to $r^3$.

**12.78** A particle of mass $m$ is projected from point $A$ with an initial velocity $\mathbf{v}_0$ perpendicular to line $OA$ and moves under a central force $\mathbf{F}$ along a semicircular path of diameter $OA$. Observing that $r = r_0 \cos \theta$ and using Eq. (12.27), show that the speed of the particle is $v = v_0/\cos^2 \theta$.

**12.79** For the particle of Prob. 12.78, determine the tangential component $F_t$ of the central force $\mathbf{F}$ along the tangent to the path of the particle for (*a*) $\theta = 0$, (*b*) $\theta = 45°$.

**12.80** The radius of the orbit of a moon of a given planet is three times as large as the radius of that planet. Denoting by $\rho$ the mean density of the planet, show that the time required by the moon to complete one full revolution about the planet is $9(\pi/G\rho)^{1/2}$, where $G$ is the constant of gravitation.

**12.81** Communication satellites are placed in a geosynchronous orbit, that is, in a circular orbit such that they complete one full revolution about the earth in one sidereal day (23.934 h), and thus appear stationary with respect to the ground. Determine (*a*) the altitude of these satellites above the surface of the earth, (*b*) the velocity with which they describe their orbit. Give the answers in both SI and U.S. customary units.

**12.82** Show that the radius $r$ of the orbit of a moon of a given planet can be determined from the radius $R$ of the planet, the acceleration of gravity at the surface of the planet, and the time $\tau$ required by the moon to complete one full revolution about the planet. Determine the acceleration of gravity at the surface of the planet Jupiter knowing that $R = 71142$ km, $\tau = 3.551$ days, and $r = 668162$ km for its moon Europa.

**12.83** The orbit of the planet Venus is nearly circular with an orbital velocity of 125460.7 km/h. Knowing that the mean distance from the center of the sun to the center of Venus is $107.7 \times 10^6$ km and that the radius of the sun is $692 \times 10^3$ km, determine (*a*) the mass of the sun, (*b*) the acceleration of gravity at the surface of the sun.

**12.84** The periodic times of the planet Jupiter's satellites, Ganymede and Callisto, have been observed to be 7.15 days and 16.69 days, respectively. Knowing that the mass of Jupiter is 319 times that of the earth and that the orbits of the two satellites are circular, determine (*a*) the radius of the orbit of Ganymede, (*b*) the velocity with which Callisto describes its orbit. Give the answers in SI units. (The *periodic time* of a satellite is the time it requires to complete one full revolution about the planet.)

**12.85** The periodic time (see Prob. 12.84) of an earth satellite in a circular polar orbit is 120 min. Determine (*a*) the altitude $h$ of the satellite, (*b*) the time during which the satellite is above the horizon for an observer located at the north pole.

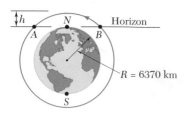

**Fig. P12.85**

**12.86**    A space vehicle is in a circular orbit 320.5 km above the surface of the moon. Knowing that the radius and mass of the moon are 1730.5 km and $2.3 \times 10^{21}$ kg respectively, determine (*a*) the acceleration of gravity at the surface of the moon, (*b*) the periodic time (see Prob. 12.84) of the space vehicle.

**12.87**    The periodic times (see Prob. 12.84) of the planet Saturn's satellites Tethys and Rhea have been observed to be 1.888 days and 4.52 days, respectively. Assuming the orbits are circular and knowing that the radius of the orbit of Tethys is $293.7 \times 10^3$ km, determine (*a*) the radius of the orbit of Rhea, (*b*) the mass of Saturn.

**12.88**    During a flyby of the earth, the velocity of a spacecraft is $10.4 \times 10^3$ m/s as it reaches its minimum altitude of 960 km above the surface at point *A*. At point *B* the spacecraft is observed to have an altitude of 8300 km. Assuming that the trajectory of the spacecraft is parabolic, determine its velocity at *B*.

**Fig. P12.88**

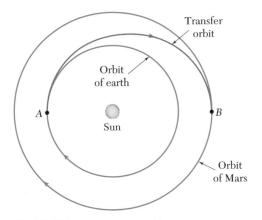

**Fig. P12.89**

**12.89**    As a first approximation to the analysis of a space flight from the earth to Mars, assume the orbits of the earth and Mars are circular and coplanar. The mean distances from the sun to the earth and to Mars are $149 \times 10^6$ km and $226.7 \times 10^6$ km respectively. To place the spacecraft into an elliptical transfer orbit at point *A*, its speed is increased over a short interval of time to $v_A$ which is 2.9 km/s faster than the earth's orbital speed. When the spacecraft reaches point *B* on the elliptical transfer orbit, its speed $v_B$ is increased to the orbital speed of Mars. Knowing that the mass of the sun is $332.8 \times 10^3$ times the mass of the earth, determine the increase in speed required at *B*.

**12.90**    A space vehicle is in a circular orbit of 2243 km radius around the moon. To transfer to a smaller orbit of 2083 km radius, the vehicle is first placed in an elliptic path *AB* by reducing its speed by 26.2 m/s as it passes through *A*. Knowing that the mass of the moon is $2.28 \times 10^{21}$ kg determine (*a*) the speed of the vehicle as it approaches *B* on the elliptic path, (*b*) the amount by which its speed should be reduced as it approaches *B* to insert it into the smaller circular orbit.

**Fig. *P12.90***

**12.91** A space shuttle $S$ and a satellite $A$ are in the circular orbits shown. In order for the shuttle to recover the satellite, the shuttle is first placed in an elliptic path $BC$ by increasing its speed by $\Delta v_B = 85$ m/s as it passes through $B$. As the shuttle approaches $C$, its speed is increased by $\Delta v_C = 79$ m/s to insert it into a second elliptic transfer orbit $CD$. Knowing that the distance from $O$ to $C$ is 6900 km, determine the amount by which the speed of the shuttle should be increased as it approaches $D$ to insert it into the circular orbit of the satellite.

610 km

290 km

**Fig. P12.91**

**12.92** Two 1.18 kg collars $A$ and $B$ can slide without friction on a frame, consisting of the horizontal rod $OE$ and the vertical rod $CD$, which is free to rotate about $CD$. The two collars are connected by a cord running over a pulley that is attached to the frame at $O$ and a stop prevents collar $B$ from moving. The frame is rotating at the rate $\dot\theta = 10$ rad/s and $r = .18$ m when the stop is removed allowing collar $A$ to move out along rod $OE$. Neglecting friction and the mass of the frame, determine $(a)$ the tension in the cord and the acceleration of collar $A$ relative to rod $OE$ immediately after the stop is removed, $(b)$ the transverse component of the velocity of collar $A$ when $r = .27$ m.

**12.93** Two 1.18 kg collars $A$ and $B$ can slide without friction on a frame, consisting of the horizontal rod $OE$ and the vertical rod $CD$, which is free to rotate about $CD$. The two collars are connected by a cord running over a pulley that is attached to the frame at $O$ and a stop prevents collar $B$ from moving. The frame is rotating at the rate $\dot\theta = 12$ rad/s and $r = .18$ m when the stop is removed allowing collar $A$ to move out along rod $OE$. Neglecting friction and the mass of the frame, determine, for the position $r = .37$ m, $(a)$ the transverse component of the velocity of collar $A$, $(b)$ the tension in the cord and the acceleration of collar $A$ relative to the rod $OE$.

**Fig. P12.92 and P12.93**

**12.94** A 300-g collar can slide on a horizontal rod which is free to rotate about a vertical shaft. The collar is initially held at $A$ by a cord attached to the shaft and compresses a spring of constant 5 N/m, which is undeformed when the collar is located 750 mm from the shaft. As the rod rotates at the rate $\dot\theta = 12$ rad/s, the cord is cut and the collar moves out along the rod. Neglecting friction and the mass of the rod, determine for position $B$ of the collar $(a)$ the transverse component of the velocity of the collar, $(b)$ the radial and transverse components of its acceleration, $(c)$ the acceleration of the collar relative to the rod.

600 mm

150 mm

**Fig. P12.94**

**12.95** In Prob. 12.94, determine for position $B$ of the collar, $(a)$ the radial component of the velocity of the collar, $(b)$ the value of $\ddot\theta$ .

# *12.11. TRAJECTORY OF A PARTICLE UNDER A CENTRAL FORCE

Consider a particle $P$ moving under a central force $\mathbf{F}$. We propose to obtain the differential equation which defines its trajectory.

Assuming that the force $\mathbf{F}$ is directed toward the center of force $O$, we note that $\Sigma F_r$ and $\Sigma F_\theta$ reduce, respectively, to $-F$ and zero in Eqs. (12.21) and (12.22). We therefore write

$$m(\ddot{r} - r\dot{\theta}^2) = -F \tag{12.31}$$
$$m(r\ddot{\theta} + 2\dot{r}\dot{\theta}) = 0 \tag{12.32}$$

These equations define the motion of $P$. We will, however, replace Eq. (12.32) by Eq. (12.27), which is equivalent to Eq. (12.32), as can easily be checked by differentiating it with respect to $t$, but which is more convenient to use. We write

$$r^2\dot{\theta} = h \qquad \text{or} \qquad r^2\frac{d\theta}{dt} = h \tag{12.33}$$

Equation (12.33) can be used to eliminate the independent variable $t$ from Eq. (12.31). Solving Eq. (12.33) for $\dot{\theta}$ or $d\theta/dt$, we have

$$\dot{\theta} = \frac{d\theta}{dt} = \frac{h}{r^2} \tag{12.34}$$

from which it follows that

$$\dot{r} = \frac{dr}{dt} = \frac{dr}{d\theta}\frac{d\theta}{dt} = \frac{h}{r^2}\frac{dr}{d\theta} = -h\frac{d}{d\theta}\left(\frac{1}{r}\right) \tag{12.35}$$

$$\ddot{r} = \frac{d\dot{r}}{dt} = \frac{d\dot{r}}{d\theta}\frac{d\theta}{dt} = \frac{h}{r^2}\frac{d\dot{r}}{d\theta}$$

or, substituting for $\dot{r}$ from (12.35),

$$\ddot{r} = \frac{h}{r^2}\frac{d}{d\theta}\left[-h\frac{d}{d\theta}\left(\frac{1}{r}\right)\right]$$

$$\ddot{r} = -\frac{h^2}{r^2}\frac{d^2}{d\theta^2}\left(\frac{1}{r}\right) \tag{12.36}$$

Substituting for $\dot{\theta}$ and $\ddot{r}$ from (12.34) and (12.36), respectively, in Eq. (12.31) and introducing the function $u = 1/r$, we obtain after reductions

$$\frac{d^2u}{d\theta^2} + u = \frac{F}{mh^2u^2} \tag{12.37}$$

In deriving Eq. (12.37), the force $\mathbf{F}$ was assumed directed toward $O$. The magnitude $F$ should therefore be positive if $\mathbf{F}$ is actually directed toward $O$ (attractive force) and negative if $\mathbf{F}$ is directed away from $O$ (repulsive force). If $F$ is a known function of $r$ and thus of $u$, Eq. (12.37) is a differential equation in $u$ and $\theta$. This differential equation defines the trajectory followed by the particle under the central force $\mathbf{F}$. The equation of the trajectory can be obtained by solving the differential Eq. (12.37) for $u$ as a function of $\theta$ and determining the constants of integration from the initial conditions.

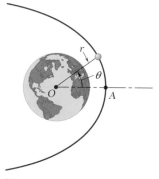

**Photo 12.4** The International Space Station and the space vehicle are subjected to the gravitational pull of the earth and if all other forces are neglected, the trajectory of each will be a circle, an ellipse, a parabola, or a hyperbola.

**Fig. 12.19**

## *12.12. APPLICATION TO SPACE MECHANICS

After the last stages of their launching rockets have burned out, earth satellites and other space vehicles are subjected to only the gravitational pull of the earth. Their motion can therefore be determined from Eqs. (12.33) and (12.37), which govern the motion of a particle under a central force, after $F$ has been replaced by the expression obtained for the force of gravitational attraction.† Setting in Eq. (12.37)

$$F = \frac{GMm}{r^2} = GMmu^2$$

where $M$ = mass of earth
$m$ = mass of space vehicle
$r$ = distance from center of earth to vehicle
$u = l/r$

we obtain the differential equation

$$\frac{d^2u}{d\theta^2} + u = \frac{GM}{h^2} \qquad (12.38)$$

where the right-hand member is observed to be a constant.

The solution of the differential equation (12.38) is obtained by adding the particular solution $u = GM/h^2$ to the general solution $u = C \cos(\theta - \theta_0)$ of the corresponding homogeneous equation (that is, the equation obtained by setting the right-hand member equal to zero). Choosing the polar axis so that $\theta_0 = 0$, we write

$$\frac{1}{r} = u = \frac{GM}{h^2} + C \cos\theta \qquad (12.39)$$

Equation (12.39) is the equation of a *conic section* (ellipse, parabola, or hyperbola) in the polar coordinates $r$ and $\theta$. The origin $O$ of the coordinates, which is located at the center of the earth, is a *focus* of this conic section, and the polar axis is one of its axes of symmetry (Fig. 12.19).

The ratio of the constants $C$ and $GM/h^2$ defines the *eccentricity* $\varepsilon$ of the conic section; letting

$$\varepsilon = \frac{C}{GM/h^2} = \frac{Ch^2}{GM} \qquad (12.40)$$

we can write Eq. (12.39) in the form

$$\frac{1}{r} = \frac{GM}{h^2}(1 + \varepsilon \cos\theta) \qquad (12.39')$$

This equation represents three possible trajectories.

1. $\varepsilon > 1$, or $C > GM/h^2$: There are two values $\theta_1$ and $-\theta_1$ of the polar angle, defined by $\cos\theta_1 = -GM/Ch^2$, for which

---

†It is assumed that the space vehicles considered here are attracted by the earth only and that their mass is negligible compared with the mass of the earth. If a vehicle moves very far from the earth, its path may be affected by the attraction of the sun, the moon, or another planet.

the right-hand member of Eq. (12.39) becomes zero. For both these values, the radius vector $r$ becomes infinite; the conic section is a *hyperbola* (Fig. 12.20).

2.  $\varepsilon = 1$, or $C = GM/h^2$: The radius vector becomes infinite for $\theta = 180°$; the conic section is a *parabola*.
3.  $\varepsilon < 1$, or $C < GM/h^2$: The radius vector remains finite for every value of $\theta$; the conic section is an *ellipse*. In the particular case when $\varepsilon = C = 0$, the length of the radius vector is constant; the conic section is a circle.

Let us now see how the constants $C$ and $GM/h^2$, which characterize the trajectory of a space vehicle, can be determined from the vehicle's position and velocity at the beginning of its free flight. We will assume that, as is generally the case, the powered phase of its flight has been programmed in such a way that as the last stage of the launching rocket burns out, the vehicle has a velocity parallel to the surface of the earth (Fig. 12.21). In other words, we will assume that the space vehicle begins its free flight at the vertex $A$ of its trajectory.[†]

Denoting the radius vector and speed of the vehicle at the beginning of its free flight by $r_0$ and $v_0$, respectively, we observe that the velocity reduces to its transverse component and, thus, that $v_0 = r_0 \dot{\theta}_0$. Recalling Eq. (12.27), we express the angular momentum per unit mass $h$ as

$$h = r_0^2 \dot{\theta}_0 = r_0 v_0 \qquad (12.41)$$

The value obtained for $h$ can be used to determine the constant $GM/h^2$. We also note that the computation of this constant will be simplified if we use the relation obtained in Sec. 12.10:

$$GM = gR^2 \qquad (12.30)$$

where $R$ is the radius of the earth ($R = 6.37 \times 10^6$ m or 3960 mi) and $g$ is the acceleration of gravity at the surface of the earth.

The constant $C$ is obtained by setting $\theta = 0$, $r = r_0$ in (12.39):

$$C = \frac{1}{r_0} - \frac{GM}{h^2} \qquad (12.42)$$

Substituting for $h$ from (12.41), we can then easily express $C$ in terms of $r_0$ and $v_0$.

Let us now determine the initial conditions corresponding to each of the three fundamental trajectories indicated above. Considering first the parabolic trajectory, we set $C$ equal to $GM/h^2$ in Eq. (12.42) and eliminate $h$ between Eqs. (12.41) and (12.42). Solving for $v_0$, we obtain

$$v_0 = \sqrt{\frac{2GM}{r_0}}$$

We can easily check that a larger value of the initial velocity corresponds to a hyperbolic trajectory and a smaller value corresponds to an elliptic orbit. Since the value of $v_0$ obtained for the parabolic

**Fig. 12.20**

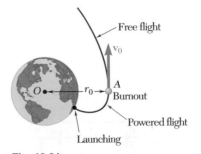

**Fig. 12.21**

[†]Problems involving oblique launchings will be considered in Sec. 13.9.

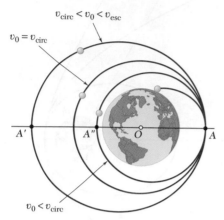

$$v_{\text{circ}} < v_0 < v_{\text{esc}}$$

$$v_0 = v_{\text{circ}}$$

$$v_0 < v_{\text{circ}}$$

**Fig. 12.22**

trajectory is the smallest value for which the space vehicle does not return to its starting point, it is called the *escape velocity*. We write therefore

$$v_{\text{esc}} = \sqrt{\frac{2GM}{r_0}} \qquad \text{or} \qquad v_{\text{esc}} = \sqrt{\frac{2gR^2}{r_0}} \qquad (12.43)$$

if we make use of Eq. (12.30). We note that the trajectory will be (1) hyperbolic if $v_0 > v_{\text{esc}}$, (2) parabolic if $v_0 = v_{\text{esc}}$, and (3) elliptic if $v_0 < v_{\text{esc}}$.

Among the various possible elliptic orbits, the one obtained when $C = 0$, the *circular orbit*, is of special interest. The value of the initial velocity corresponding to a circular orbit is easily found to be

$$v_{\text{circ}} = \sqrt{\frac{GM}{r_0}} \qquad \text{or} \qquad v_{\text{circ}} = \sqrt{\frac{gR^2}{r_0}} \qquad (12.44)$$

if Eq. (12.30) is taken into account. We note from Fig. 12.22 that for values of $v_0$ larger than $v_{\text{circ}}$ but smaller than $v_{\text{esc}}$, point A where free flight begins is the point of the orbit closest to the earth; this point is called the *perigee*, while point A', which is farthest away from the earth, is known as the *apogee*. For values of $v_0$ smaller than $v_{\text{circ}}$, point A is the apogee, while point A'', on the other side of the orbit, is the perigee. For values of $v_0$ much smaller than $v_{\text{circ}}$, the trajectory of the space vehicle intersects the surface of the earth; in such a case, the vehicle does not go into orbit.

Ballistic missiles, which were designed to hit the surface of the earth, also travel along elliptic trajectories. In fact, we should now realize that any object projected in vacuum with an initial velocity $v_0$ smaller than $v_{\text{esc}}$ will move along an elliptic path. It is only when the distances involved are small that the gravitational field of the earth can be assumed uniform and that the elliptic path can be approximated by a parabolic path, as was done earlier (Sec. 11.11) in the case of conventional projectiles.

**Periodic Time.**  An important characteristic of the motion of an earth satellite is the time required by the satellite to describe its orbit. This time, known as the *periodic time* of the satellite, is denoted by $\tau$. We first observe, in view of the definition of areal velocity (Sec. 12.9), that $\tau$ can be obtained by dividing the area inside the orbit by the areal velocity. Noting that the area of an ellipse is equal to $\pi ab$, where $a$ and $b$ denote the semimajor and semiminor axes, respectively, and that the areal velocity is equal to $h/2$, we write

$$\tau = \frac{2\pi ab}{h} \qquad (12.45)$$

While $h$ can be readily determined from $r_0$ and $v_0$ in the case of a satellite launched in a direction parallel to the surface of the earth, the semiaxes $a$ and $b$ are not directly related to the initial conditions. Since, on the other hand, the values $r_0$ and $r_1$ of $r$ corresponding to the perigee and apogee of the orbit can easily be determined from Eq. (12.39), we will express the semiaxes $a$ and $b$ in terms of $r_0$ and $r_1$.

Consider the elliptic orbit shown in Fig. 12.23. The earth's center is located at O and coincides with one of the two foci of the

**Fig. 12.23**

ellipse, while the points $A$ and $A'$ represent, respectively, the perigee and apogee of the orbit. We easily check that

$$r_0 + r_1 = 2a$$

and thus

$$a = \tfrac{1}{2}(r_0 + r_1) \tag{12.46}$$

Recalling that the sum of the distances from each of the foci to any point of the ellipse is constant, we write

$$O'B + BO = O'A + OA = 2a \qquad \text{or} \qquad BO = a$$

On the other hand, we have $CO = a - r_0$. We can therefore write

$$b^2 = (BC)^2 = (BO)^2 - (CO)^2 = a^2 - (a - r_0)^2$$
$$b^2 = r_0(2a - r_0) = r_0 r_1$$

and thus

$$b = \sqrt{r_0 r_1} \tag{12.47}$$

Formulas (12.46) and (12.47) indicate that the semimajor and semiminor axes of the orbit are equal, respectively, to the arithmetic and geometric means of the maximum and minimum values of the radius vector. Once $r_0$ and $r_1$ have been determined, the lengths of the semiaxes can be easily computed and substituted for $a$ and $b$ in formula (12.45).

## *12.13. KEPLER'S LAWS OF PLANETARY MOTION

The equations governing the motion of an earth satellite can be used to describe the motion of the moon around the earth. In that case, however, the mass of the moon is not negligible compared with the mass of the earth, and the results obtained are not entirely accurate.

The theory developed in the preceding sections can also be applied to the study of the motion of the planets around the sun. Although another error is introduced by neglecting the forces exerted by the planets on one another, the approximation obtained is excellent. Indeed, even before Newton had formulated his fundamental theory, the properties expressed by Eq. (12.39), where $M$ now represents the mass of the sun, and by Eq. (12.33) had been discovered by the German astronomer Johann Kepler (1571–1630) from astronomical observations of the motion of the planets.

Kepler's three *laws of planetary motion* can be stated as follows:

1. Each planet describes an ellipse, with the sun located at one of its foci.
2. The radius vector drawn from the sun to a planet sweeps equal areas in equal times.
3. The squares of the periodic times of the planets are proportional to the cubes of the semimajor axes of their orbits.

The first law states a particular case of the result established in Sec. 12.12, and the second law expresses that the areal velocity of each planet is constant (see Sec. 12.9). Kepler's third law can also be derived from the results obtained in Sec. 12.12.†

†See Prob. 12.120.

36,900 km/h

Earth

Maximum altitude

500 km

## SAMPLE PROBLEM 12.9

A satellite is launched in a direction parallel to the surface of the earth with a velocity of 36 900 km/h from an altitude of 500 km. Determine (a) the maximum altitude reached by the satellite, (b) the periodic time of the satellite.

## SOLUTION

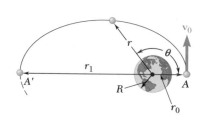

*a.* **Maximum Altitude.** After the satellite is launched, it is subjected only to the gravitational attraction of the earth; its motion is thus governed by Eq. (12.39),

$$\frac{1}{r} = \frac{GM}{h^2} + C \cos \theta \tag{1}$$

Since the radial component of the velocity is zero at the point of launching $A$, we have $h = r_0 v_0$. Recalling that for the earth $R = 6370$ km, we compute

$$r_0 = 6370 \text{ km} + 500 \text{ km} = 6870 \text{ km} = 6.87 \times 10^6 \text{ m}$$

$$v_0 = 36\,900 \text{ km/h} = \frac{36.9 \times 10^6 \text{ m}}{3.6 \times 10^3 \text{ s}} = 10.25 \times 10^3 \text{ m/s}$$

$$h = r_0 v_0 = (6.87 \times 10^6 \text{ m})(10.25 \times 10^3 \text{ m/s}) = 70.4 \times 10^9 \text{ m}^2/\text{s}$$

$$h^2 = 4.96 \times 10^{21} \text{ m}^4/\text{s}^2$$

Since $GM = gR^2$, where $R$ is the radius of the earth, we have

$$GM = gR^2 = (9.81 \text{ m/s}^2)(6.37 \times 10^6 \text{ m})^2 = 397.6 \times 10^{12} \text{ m}^3/\text{s}^2$$

$$\frac{GM}{h^2} = \frac{398 \times 10^{12} \text{ m}^3/\text{s}^2}{4.96 \times 10^{21} \text{ m}^4/\text{s}^2} = 8.02 \times 10^{-8} \text{ m}^{-1}$$

Substituting this value into (1), we obtain

$$\frac{1}{r} = 80.3 \times 10^{-9} \text{ m}^{-1} + C \cos \theta \tag{2}$$

Noting that at point $A$ we have $\theta = 0$ and $r = r_0 = 6.87 \times 10^6$ m, we compute the constant $C$:

$$\frac{1}{6.87 \times 10^6 \text{ m}} = 80.3 \times 10^{-9} \text{ m}^{-1} + C \cos 0° \qquad C = 65.3 \times 10^{-9} \text{ m}^{-1}$$

At $A'$, the point on the orbit farthest from the earth, we have $\theta = 180°$. Using (2), we compute the corresponding distance $r_1$:

$$\frac{1}{r_1} = 80.3 \times 10^{-9} \text{ m}^{-1} + (65.3 \times 10^{-9} \text{ m}^{-1}) \cos 180°$$

$$r_1 = 67.1 \times 10^6 \text{ m} = 67\,100 \text{ km}$$

$$\textit{Maximum altitude} = 66\,700 \text{ km} - 6370 \text{ km} = 60\,730 \text{ km} \quad \blacktriangleleft$$

*b.* **Periodic Time.** Since $A$ and $A'$ are the perigee and apogee, respectively, of the elliptic orbit, we use Eqs. (12.46) and (12.47) and compute the semimajor and semiminor axes of the orbit:

$$a = \tfrac{1}{2}(r_0 + r_1) = \tfrac{1}{2}(6.87 + 66.7)(10^6) \text{ m} = 37 \times 10^6 \text{ m}$$

$$b = \sqrt{r_0 r_1} = \sqrt{(6.87)(66.7)} \times 10^6 \text{ m} = 21.5 \times 10^5 \text{ m}$$

$$\tau = \frac{2\pi ab}{h} = \frac{2\pi(36.8 \times 10^6 \text{ m})(21.4 \times 10^6 \text{ m})}{70.4 \times 10^9 \text{ m}^2/\text{s}}$$

$$\tau = 70.99 \times 10^3 \text{ s} = 1183.17 \text{ min} = 19 \text{ h } 43 \text{ min} \quad \blacktriangleleft$$

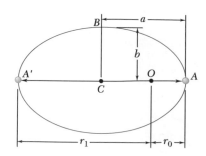

# SOLVING PROBLEMS
## ON YOUR OWN

In this lesson, we continued our study of the motion of a particle under a central force and applied the results to problems in space mechanics. We found that the trajectory of a particle under a central force is defined by the differential equation

$$\frac{d^2u}{d\theta^2} + u = \frac{F}{mh^2u^2} \tag{12.37}$$

where $u$ is the reciprocal of the distance $r$ of the particle to the center of force ($u = 1/r$), $F$ is the magnitude of the central force $\mathbf{F}$, and $h$ is a constant equal to the angular momentum per unit mass of the particle. In space-mechanics problems, $\mathbf{F}$ is the force of gravitational attraction exerted on the satellite or spacecraft by the sun, earth, or other planet about which it travels. Substituting $F = GMm/r^2 = GMmu^2$ into Eq. (12.37), we obtain for that case

$$\frac{d^2u}{d\theta^2} + u = \frac{GM}{h^2} \tag{12.38}$$

where the right-hand member is a constant.

1. *Analyzing the motion of satellites and spacecraft.* The solution of the differential equation (12.38) defines the trajectory of a satellite or spacecraft. It was obtained in Sec. 12.12 and was given in the alternative forms

$$\frac{1}{r} = \frac{GM}{h^2} + C\cos\theta \qquad \text{or} \qquad \frac{1}{r} = \frac{GM}{h^2}(1 + \varepsilon\cos\theta) \qquad (12.39, \ 12.39')$$

Remember when applying these equations that $\theta = 0$ always corresponds to the perigee (the point of closest approach) of the trajectory (Fig. 12.19) and that $h$ is a constant for a given trajectory. Depending on the value of the eccentricity $\varepsilon$, the trajectory will be a hyperbola, a parabola, or an ellipse.

   *a.* $\varepsilon > 1$: *The trajectory is a hyperbola,* so that for this case the spacecraft never returns to its starting point.

   *b.* $\varepsilon = 1$: *The trajectory is a parabola.* This is the limiting case between open (hyperbolic) and closed (elliptic) trajectories. We had observed for this case that the velocity $v_0$ at the perigee is equal to the escape velocity $v_{\text{esc}}$,

$$v_0 = v_{\text{esc}} = \sqrt{\frac{2GM}{r_0}} \tag{12.43}$$

Note that the escape velocity is the smallest velocity for which the spacecraft does not return to its starting point.

*c. $\varepsilon < 1$: The trajectory is an elliptic orbit.* For problems involving elliptic orbits, you may find that the relation derived in Prob. 12.102,

$$\frac{1}{r_0} + \frac{1}{r_1} = \frac{2GM}{h^2}$$

will be useful in the solution of subsequent problems. When you apply this equation, remember that $r_0$ and $r_1$ are the distances from the center of force to the perigee ($\theta = 0$) and apogee ($\theta = 180°$), respectively; that $h = r_0v_0 = r_1v_1$; and that, for a satellite orbiting the earth, $GM_{earth} = gR^2$, where $R$ is the radius of the earth. Also recall that the trajectory is a circle when $\varepsilon = 0$.

**2.  Determining the point of impact of a descending spacecraft.**  For problems of this type, you may assume that the trajectory is elliptic and that the initial point of the descent trajectory is the apogee of the path (Fig. 12.22). Note that at the point of impact, the distance $r$ in Eqs. (12.39) and (12.39') is equal to the radius $R$ of the body on which the spacecraft lands or crashes. In addition, we have $h = Rv_I \sin \phi_I$, where $v_I$ is the speed of the spacecraft at impact and $\phi_I$ is the angle that its path forms with the vertical at the point of impact.

**3.  Calculating the time to travel between two points on a trajectory.**  For central force motion, the time $t$ required for a particle to travel along a portion of its trajectory can be determined by recalling from Sec. 12.9 that the rate at which area is swept per unit time by the position vector $\mathbf{r}$ is equal to one-half of the angular momentum per unit mass $h$ of the particle: $dA/dt = h/2$. It follows, since $h$ is a constant for a given trajectory, that

$$t = \frac{2A}{h}$$

where $A$ is the total area swept in the time $t$.

*a. In the case of an elliptic trajectory,* the time required to complete one orbit is called the *periodic time* and is expressed as

$$\tau = \frac{2(\pi ab)}{h} \tag{12.45}$$

where $a$ and $b$ are the semimajor and semiminor axes, respectively, of the ellipse and are related to the distances $r_0$ and $r_1$ by

$$a = \frac{1}{2}(r_0 + r_1) \qquad \text{and} \qquad b = \sqrt{r_0 r_1} \tag{12.46, 12.47}$$

*b. Kepler's third law* provides a convenient relation between the periodic times of two satellites describing elliptic orbits about the same body [Sec. 12.13]. Denoting the semimajor axes of the two orbits by $a_1$ and $a_2$, respectively, and the corresponding periodic times by $\tau_1$ and $\tau_2$, we have

$$\frac{\tau_1^2}{\tau_2^2} = \frac{a_1^3}{a_2^3}$$

*c. In the case of a parabolic trajectory,* you may be able to use the expression given on the inside of the front cover of the book for a parabolic or a semiparabolic area to calculate the time required to travel between two points of the trajectory.

**12.96** A particle of mass $m$ is projected from point $A$ with an initial velocity $\mathbf{v}_0$ perpendicular to $OA$ and moves under a central force $\mathbf{F}$ along an elliptic path defined by the equation $r = r_0/(2 - \cos \theta)$. Using Eq. (12.37), show that $\mathbf{F}$ is inversely proportional to the square of the distance $r$ from the particle to the center of force $O$.

**Fig. P12.96**

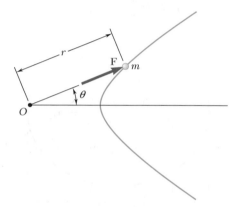

**Fig. P12.97**

**12.97** A particle of mass $m$ describes the path defined by the equation $r = r_0/(6 \cos \theta - 5)$ under a central force $\mathbf{F}$ directed away from the center of force $O$. Using Eq. (12.37), show that $\mathbf{F}$ is inversely proportional to the square of the distance $r$ from the particle to $O$.

**12.98** A particle of mass $m$ describes the parabola $y = x^2/4r_0$ under a central force $\mathbf{F}$ directed toward the center of force $C$. Using Eq. (12.37) and Eq. (12.39′) with $\varepsilon = 1$, show that $\mathbf{F}$ is inversely proportional to the square of the distance $r$ from the particle to the center of force and that the angular momentum per unit mass $h = \sqrt{2GMr_0}$.

**12.99** A particle of mass $m$ describes the logarithmic spiral $r = r_0 e^{b\theta}$ under a central force $\mathbf{F}$ directed toward the center of force $O$. Using Eq. (12.37) show that $\mathbf{F}$ is inversely proportional to the cube of the distance $r$ from the particle to $O$.

**Fig. P12.98**

**12.100** It was observed that as the spacecraft Voyager 1 reached the point on its trajectory closest to the planet Jupiter, it was at a distance of $350 \times 10^3$ km from the center of the planet and had a velocity of 26.9 km/s. Determine the mass of Jupiter, assuming that the trajectory of the spacecraft was parabolic.

280.4 km

**Fig. P12.101**

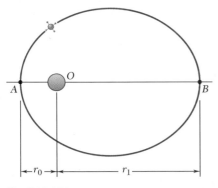

**Fig. P12.104**

**12.101** As a space probe approaching the planet Venus on a parabolic trajectory reaches point A closest to the planet, its velocity is decreased to insert it into a circular orbit. Knowing that the mass and the radius of Venus are $151.5 \times 10^{21}$ kg and 6026 km respectively, determine (a) the velocity of the probe as it approaches A, (b) the decrease in velocity required to insert it into a circular orbit.

**12.102** It was observed that as the Galileo spacecraft reached the point of its trajectory closest to Io, a moon of the planet Jupiter, it was at a distance of 2820 km from the center of Io and had a velocity of 15 km/s. Knowing that the mass of Io is 0.01496 times the mass of the earth, determine the eccentricity of the trajectory of the spacecraft as it approached Io.

**12.103** A small satellite of Jupiter, discovered in 2002, was reported to be an orbit of semimajor axis $23.6 \times 10^6$ km and eccentricity 0.45. Knowing that the mass of Jupiter is 318 times the mass of the earth, determine the maximum and minimum speeds of the satellite.

**12.104** A satellite describes an elliptic orbit about a planet of mass $M$. Denoting by $r_0$ and $r_1$, respectively, the minimum and maximum values of the distance $r$ from the satellite to the center of the planet, derive the relation

$$\frac{1}{r_0} + \frac{1}{r_1} = \frac{2GM}{h^2}$$

where $h$ is the angular momentum per unit mass of the satellite.

**12.105** A satellite describes a circular orbit at an altitude of 19019.4 km above the surface of the earth. Determine (a) the increase in speed required at point A for the satellite to achieve the escape velocity and enter a parabolic orbit, (b) the decrease in speed required at point A for the satellite to enter an elliptic orbit of minimum altitude 6345 km, (c) the eccentricity $\varepsilon$ of the elliptic orbit.

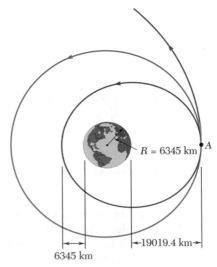

$R = 6345$ km

19019.4 km

6345 km

**Fig. P12.105**

**12.106** The Chandra X-ray observatory, launched in 1999, achieved an elliptical orbit of minimum altitude 10 000 km and maximum altitude 140 000 km above the surface of the earth. Assuming that the observatory was transferred to this orbit from a circular orbit of altitude 10 000 km at point $A$, determine (a) the increase in speed required at $A$, (b) the speed of the observatory at $B$.

**12.107** As it describes an elliptic orbit about the sun, a spacecraft reaches a maximum distance of $325 \times 10^6$ km from the center of the sun at point $A$ (called the *aphelion*) and a minimum distance of $148 \times 10^6$ km at point $B$ (called the *perihelion*). To place the spacecraft in a smaller elliptic orbit with aphelion $A'$ and perihelion $B'$, where $A'$ and $B'$ are located $264.7 \times 10^6$ km and $137.6 \times 10^6$ km, respectively, from the center of the sun, the speed of the spacecraft is first reduced as it passes through $A$ and then is further reduced as it passes through $B'$. Knowing that the mass of the sun is $332.8 \times 10^3$ times the mass of the earth, determine (a) the speed of the spacecraft at $A$, (b) the amounts by which the speed of the spacecraft should be reduced at $A$ and $B'$ to insert it into the desired elliptic orbit.

**Fig. P12.106**

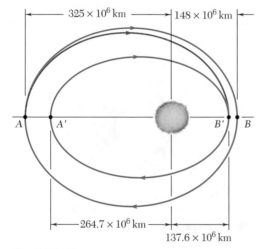

**Fig. P12.107**

**12.108** A space probe is to be placed in a circular orbit of 9000-km radius about the planet Venus in a specified plane. As the probe reaches $A$, the point of its original trajectory closest to Venus, it is inserted in a first elliptic transfer orbit by reducing its speed of $\Delta v_A$. This orbit brings it to point $B$ with a much reduced velocity. There the probe is inserted in a second transfer orbit located in the specified plane by changing the direction of its velocity and further reducing its speed by $\Delta v_B$. Finally, as the probe reaches point $C$, it is inserted in the desired circular orbit by reducing its speed by $\Delta v_C$. Knowing that the mass of Venus is 0.82 times the mass of the earth, that $r_A = 15 \times 10^3$ km and $r_B = 300 \times 10^3$ km, and that the probe approaches $A$ on a parabolic trajectory, determine by how much the velocity of the probe should be reduced (a) at $A$, (b) at $B$, (c) at $C$.

**12.109** For the space probe of Prob. 12.108, it is known that $r_A = 15 \times 10^3$ km and that the velocity of the probe is reduced to 6500 m/s as it passes through $A$. Determine (a) the distance from the center of Venus to point $B$, (b) the amounts by which the velocity of the probe should be reduced at $B$ and $C$, respectively.

**Fig. P12.108**

**Fig. P12.110**

**Fig. P12.114**

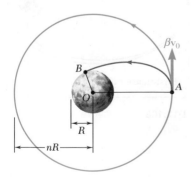

**Fig. P12.116**

**12.110** A space probe is to be placed in a circular orbit of radius 4005.8 km about the planet Mars. As the probe reaches $A$, the point of its original trajectory closest to Mars, it is inserted into a first elliptic transfer orbit by reducing its speed. This orbit brings it to point $B$ with a much reduced velocity. There the probe is inserted into a second transfer orbit by further reducing its speed. Knowing that the mass of Mars is 0.1074 times the mass of the earth, that $r_A = 9013$ km and $r_B = 180260$ km and that the probe approaches $A$ on a parabolic trajectory, determine the time needed for the space probe to travel from $A$ to $B$ on its first transfer orbit.

**12.111** Halley's comet travels in an elongated elliptic orbit for which the minimum distance from the sun in approximately $\frac{1}{2}r_E$, where $r_E = 148.9 \times 10^6$ km is the mean distance from the sun to the earth. Knowing that the periodic time of Halley's comet is about 76 years, determine the maximum distance from the sun reached by the comet.

**12.112** A spacecraft and a satellite are at diametrically opposite positions in the same circular orbit of altitude 497 km above the earth. As it passes through point $A$, the spacecraft fires its engine for a short interval of time to increase its speed and enter an elliptic orbit. Knowing that the spacecraft returns to $A$ at the same time the satellite reaches $A$ after completing one and a half orbits, determine (*a*) the increase in speed required, (*b*) the periodic time for the elliptic orbit.

**Fig. P12.112**

**12.113** Based on observations made during the 1996 sighting of the comet Hyakutake, it was concluded that the trajectory of the comet is a highly elongated ellipse for which the eccentricity is approximately $\varepsilon = 0.999887$. Knowing that for the 1996 sighting the minimum distance between the comet and the sun was $0.230R_E$, where $R_E$ is the mean distance from the sun to the earth, determine the periodic time of the comet.

**12.114** It was observed that during its first flyby of the earth, the Galileo spacecraft had a velocity of 10.4 km/s as it reached its minimum distance of 7307 km from the center of the earth. Assuming that the trajectory of the spacecraft was parabolic, determine the time needed for the spacecraft to travel from $B$ to $C$ on its trajectory.

**12.115** A space shuttle is in an elliptic orbit of eccentricity 0.0356 and a minimum altitude of 292 km above the surface of the earth. Knowing that the radius of the earth is 6345 km, determine the periodic time for the orbit.

**12.116** A space probe is describing a circular orbit of radius $nR$ with a speed $v_0$ about a planet of radius $R$ and center $O$. As the probe passes through point $A$, its speed is reduced from $v_0$ to $\beta v_0$, where $\beta < 1$, to place the probe on a crash trajectory. Express in terms of $n$ and $\beta$ the angle $AOB$, where $B$ denotes the point of impact of the probe on the planet.

**12.117**   Prior to the Apollo missions to the moon, several Lunar Orbiter spacecraft were used to photograph the lunar surface to obtain information regarding possible landing sites. At the conclusion of each mission, the trajectory of the spacecraft was adjusted so that the spacecraft would crash on the moon to further study the characteristics of the lunar surface. Shown below is the elliptic orbit of Lunar Orbiter 2. Knowing that the mass of the moon is 0.01230 times the mass of the earth, determine the amount by which the speed of the orbiter should be reduced at point $B$ so that it impacts the lunar surface at point $C$. (*Hint.* Point $B$ is the apogee of the elliptic impact trajectory.)

**Fig. P12.117**

**12.118**   A long-range ballistic trajectory between points $A$ and $B$ on the earth's surface consists of a portion of an ellipse with the apogee at point $C$. Knowing that point $C$ is 1490 km above the surface of the earth and the range $R\phi$ of the trajectory is 5928.5 km, determine (*a*) the velocity of the projectile at $C$, (*b*) the eccentricity $\varepsilon$ of the trajectory.

**Fig. P12.118**

**12.119**   A space shuttle is describing a circular orbit at an altitude of 320.5 km above the surface of the earth. As it passes through $A$ it fires its engine for a short interval of time to reduce its speed by 6 percent and begin its descent toward the earth. Determine the altitude of the shuttle at point $B$, knowing that the angle $AOB$ is equal to 50°. (*Hint.* Point $A$ is the apogee of the elliptic descent trajectory.)

**12.120**   A satellite describes an elliptic orbit about a planet. Denoting by $r_0$ and $r_1$ the distances corresponding, respectively, to the perigee and apogee of the orbit, show that the curvature of the orbit at each of these two points can be expressed as

$$\frac{1}{\rho} = \frac{1}{2}\left(\frac{1}{r_0} + \frac{1}{r_1}\right)$$

**Fig. P12.119**

**12.121**   (*a*) Express the eccentricity $\varepsilon$ of the elliptic orbit described by a satellite about a planet in terms of the distances $r_0$ and $r_1$ corresponding, respectively, to the perigee and apogee of the orbit. (*b*) Use the result obtained in part *a* and the data given in Prob. 12.113, where $R_\varepsilon = 149 \times 10^6$ km, to determine the appropriate maximum distance from the sun reached by comet Hyakutake.

**12.122**   Derive Kepler's third law of planetary motion from Eqs. (12.39) and (12.45).

**12.123**   Show that the angular momentum per unit mass $h$ of a satellite describing an elliptic orbit of semimajor axis $a$ and eccentricity $\varepsilon$ about a planet of mass $M$ can be expressed as

$$h = \sqrt{GMa(1 - \varepsilon^2)}$$

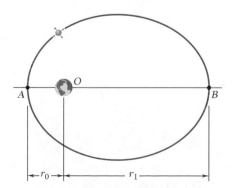

**Fig. P12.120 and P12.121**

This chapter was devoted to Newton's second law and its application to the analysis of the motion of particles.

**Newton's second law**

Denoting by $m$ the mass of a particle, by $\Sigma\mathbf{F}$ the sum, or resultant, of the forces acting on the particle, and by $\mathbf{a}$ the acceleration of the particle relative to a *newtonian frame of reference* [Sec. 12.2], we wrote

$$\Sigma\mathbf{F} = m\mathbf{a} \qquad (12.2)$$

**Linear momentum**

Introducing the *linear momentum* of a particle, $\mathbf{L} = m\mathbf{v}$ [Sec. 12.3], we saw that Newton's second law can also be written in the form

$$\Sigma\mathbf{F} = \dot{\mathbf{L}} \qquad (12.5)$$

which expresses that *the resultant of the forces acting on a particle is equal to the rate of change of the linear momentum of the particle.*

**Consistent systems of units**

Equation (12.2) holds only if a consistent system of units is used. With SI units, the forces should be expressed in newtons, the masses in kilograms, and the accelerations in m/s$^2$; with U.S. customary units, the forces should be expressed in pounds, the masses in lb · s$^2$/ft (also referred to as *slugs*), and the accelerations in ft/s$^2$ [Sec. 12.4].

**Equations of motion for a particle**

To solve a problem involving the motion of a particle, Eq. (12.2) should be replaced by equations containing scalar quantities [Sec. 12.5]. Using *rectangular components* of $\mathbf{F}$ and $\mathbf{a}$, we wrote

$$\Sigma F_x = ma_x \qquad \Sigma F_y = ma_y \qquad \Sigma F_z = ma_z \qquad (12.8)$$

Using *tangential and normal components,* we had

$$\Sigma F_t = m\frac{dv}{dt} \qquad \Sigma F_n = m\frac{v^2}{\rho} \qquad (12.9')$$

**Dynamic equilibrium**

We also noted [Sec. 12.6] that the equations of motion of a particle can be replaced by equations similar to the equilibrium equations used in statics if a vector $-m\mathbf{a}$ of magnitude $ma$ but of sense opposite to that of the acceleration is added to the forces applied to the particle; the particle is then said to be in *dynamic equilibrium.* For the sake of uniformity, however, all the Sample Problems were solved by using the equations of motion, first with rectangular components [Sample Probs. 12.1 through 12.4], then with tangential and normal components [Sample Probs. 12.5 and 12.6].

In the second part of the chapter, we defined the *angular momentum* $\mathbf{H}_O$ of a particle about a point $O$ as the moment about $O$ of the linear momentum $m\mathbf{v}$ of that particle [Sec. 12.7]. We wrote

$$\mathbf{H}_O = \mathbf{r} \times m\mathbf{v} \qquad (12.12)$$

and noted that $\mathbf{H}_O$ is a vector perpendicular to the plane containing $\mathbf{r}$ and $m\mathbf{v}$ (Fig. 12.24) and of magnitude

$$H_O = rmv \sin \phi \qquad (12.13)$$

Resolving the vectors $\mathbf{r}$ and $m\mathbf{v}$ into rectangular components, we expressed the angular momentum $\mathbf{H}_O$ in the determinant form

$$\mathbf{H}_O = \begin{vmatrix} \mathbf{i} & \mathbf{j} & \mathbf{k} \\ x & y & z \\ mv_x & mv_y & mv_z \end{vmatrix} \qquad (12.14)$$

In the case of a particle moving in the $xy$ plane, we have $z = v_z = 0$. The angular momentum is perpendicular to the $xy$ plane and is completely defined by its magnitude. We wrote

$$H_O = H_z = m(xv_y - yv_x) \qquad (12.16)$$

Computing the rate of change $\dot{\mathbf{H}}_O$ of the angular momentum $\mathbf{H}_O$, and applying Newton's second law, we wrote the equation

$$\Sigma \mathbf{M}_O = \dot{\mathbf{H}}_O \qquad (12.19)$$

which states that *the sum of the moments about $O$ of the forces acting on a particle is equal to the rate of change of the angular momentum of the particle about $O$.*

In many problems involving the plane motion of a particle, it is found convenient to use *radial and transverse components* [Sec. 12.8, Sample Prob. 12.7] and to write the equations

$$\Sigma F_r = m(\ddot{r} - r\dot{\theta}^2) \qquad (12.21)$$
$$\Sigma F_\theta = m(r\ddot{\theta} + 2\dot{r}\dot{\theta}) \qquad (12.22)$$

When the only force acting on a particle $P$ is a force $\mathbf{F}$ directed toward or away from a fixed point $O$, the particle is said to be moving *under a central force* [Sec. 12.9]. Since $\Sigma \mathbf{M}_O = 0$ at any given instant, it follows from Eq. (12.19) that $\dot{\mathbf{H}}_O = 0$ for all values of $t$ and, thus, that

$$\mathbf{H}_O = \text{constant} \qquad (12.23)$$

We concluded that *the angular momentum of a particle moving under a central force is constant, both in magnitude and direction,* and that the particle moves in a plane perpendicular to the vector $\mathbf{H}_O$.

Angular momentum

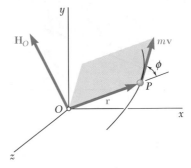

**Fig. 12.24**

Rate of change of angular momentum

Radial and transverse components

Motion under a central force

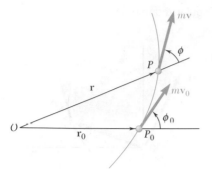

**Fig. 12.25**

Recalling Eq. (12.13), we wrote the relation

$$rmv \sin \phi = r_0 mv_0 \sin \phi_0 \qquad (12.25)$$

for the motion of any particle under a central force (Fig. 12.25). Using polar coordinates and recalling Eq. (12.18), we also had

$$r^2 \dot{\theta} = h \qquad (12.27)$$

where $h$ is a constant representing the angular momentum per unit mass, $H_O/m$, of the particle. We observed (Fig. 12.26) that the infinitesimal area $dA$ swept by the radius vector $OP$ as it rotates through $d\theta$ is equal to $\frac{1}{2}r^2 \, d\theta$ and, thus, that the left-hand member of Eq. (12.27) represents twice the *areal velocity dA/dt* of the particle. Therefore, *the areal velocity of a particle moving under a central force is constant.*

**Fig. 12.26**

Newton's law of universal gravitation

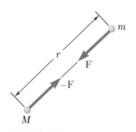

**Fig. 12.27**

Orbital motion

An important application of the motion under a central force is provided by the orbital motion of bodies under gravitational attraction [Sec. 12.10]. According to *Newton's law of universal gravitation,* two particles at a distance $r$ from each other and of masses $M$ and $m$, respectively, attract each other with equal and opposite forces **F** and $-$**F** directed along the line joining the particles (Fig. 12.27). The common magnitude $F$ of the two forces is

$$F = G\frac{Mm}{r^2} \qquad (12.28)$$

where $G$ is the *constant of gravitation.* In the case of a body of mass $m$ subjected to the gravitational attraction of the earth, the product $GM$, where $M$ is the mass of the earth, can be expressed as

$$GM = gR^2 \qquad (12.30)$$

where $g = 9.81$ m/s$^2$ $= 32.2$ ft/s$^2$ and $R$ is the radius of the earth.

It was shown in Sec. 12.11 that a particle moving under a central force describes a trajectory defined by the differential equation

$$\frac{d^2u}{d\theta^2} + u = \frac{F}{mh^2u^2} \qquad (12.37)$$

where $F > 0$ corresponds to an attractive force and $u = 1/r$. In the case of a particle moving under a force of gravitational attraction [Sec. 12.12], we substituted for $F$ the expression given in Eq. (12.28). Measuring $\theta$ from the axis $OA$ joining the focus $O$ to the point $A$ of the trajectory closest to $O$ (Fig. 12.28), we found that the solution to Eq. (12.37) was

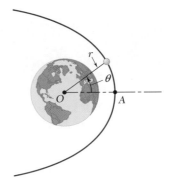

**Fig. 12.28**

$$\frac{1}{r} = u = \frac{GM}{h^2} + C \cos \theta \qquad (12.39)$$

This is the equation of a conic of eccentricity $\varepsilon = Ch^2/GM$. The conic is an *ellipse* if $\varepsilon < 1$, a *parabola* if $\varepsilon = 1$, and a *hyperbola* if $\varepsilon > 1$. The constants $C$ and $h$ can be determined from the initial conditions; if the particle is projected from point $A$ ($\theta = 0$, $r = r_0$) with an initial velocity $\mathbf{v}_0$ perpendicular to $OA$, we have $h = r_0 v_0$ [Sample Prob. 12.9].

It was also shown that the values of the initial velocity corresponding, respectively, to a parabolic and a circular trajectory were

Escape velocity

$$v_{\text{esc}} = \sqrt{\frac{2GM}{r_0}} \qquad (12.43)$$

$$v_{\text{circ}} = \sqrt{\frac{GM}{r_0}} \qquad (12.44)$$

and that the first of these values, called the *escape velocity,* is the smallest value of $v_0$ for which the particle will not return to its starting point.

The *periodic time* $\tau$ of a planet or satellite was defined as the time required by that body to describe its orbit. It was shown that

Periodic time

$$\tau = \frac{2\pi ab}{h} \qquad (12.45)$$

where $h = r_0 v_0$ and where $a$ and $b$ represent the semimajor and semiminor axes of the orbit. It was further shown that these semi-axes are respectively equal to the arithmetic and geometric means of the maximum and minimum values of the radius vector $r$.

The last section of the chapter [Sec. 12.13] presented *Kepler's laws of planetary motion* and showed that these empirical laws, obtained from early astronomical observations, confirm Newton's laws of motion as well as his law of gravitation.

Kepler's laws

**12.124** Determine the maximum theoretical speed that an automobile starting from rest can reach after traveling 402 km. Assume that the coefficient of static friction is 0.80 between the tires and the pavement and that (a) the automobile has front-wheel drive and the front wheels support 65 percent of the automobile's weight, (b) the automobile has rear-wheel drive and the rear wheels support 42 percent of the automobile's weight.

**12.125** A tractor-trailer is traveling at 90 km/h when the driver applies his brakes. Knowing that the braking forces of the tractor and the trailer are 16 kN and 60 kN, respectively, determine (a) the distance traveled by the tractor-trailer before it comes to a stop, (b) the horizontal component of the force in the hitch between the tractor and the trailer while they are slowing down.

**Fig. P12.125**

**Fig. *P12.126***

**12.126** The 13.6 kg block B is supported by the 25 kg block A and is attached to a cord to which a 222.4 N horizontal force is applied as shown. Neglecting friction, determine (a) the acceleration of block A, (b) the acceleration of block B relative to A.

**12.127** A 81.6 kg wrecking ball B is attached to a 12.2 m long steel cable AB and swings in the vertical arc shown. Determine the tension in the cable (a) at the top C of the swing, where $\theta = 30°$, (b) at the bottom D of the swing, where the speed of B is 5.7 m/s.

**Fig. P12.127**

**12.128**  A car traveling at a speed of 110 km/h approaches a curve of radius 50 m. Knowing that the coefficient of static friction between the tires and the road is 0.65, determine by how much the driver must reduce his speed to safely negotiate the curve if the banking angle is (*a*) $\theta = 10°$, (*b*) $\theta = -5°$, because of the presence of a sinkhole under the roadbed.

**12.129**  A small .24 kg collar *D* can slide on portion *AB* of a rod which is bent as shown. Knowing that the rod rotates about the vertical *AC* at a constant rate and that $\alpha = 40°$ and $r = 61$ cm, determine the range of values of the speed *v* for which the collar will not slide on the rod if the coefficient of static friction between the rod and the collar is 0.35.

**Fig. P12.129**

**12.130**  Rod *OA* rotates about *O* in a horizontal plane. The motion of the 400-g collar *B* is defined by the relations $r = 500 + 300 \sin \pi t$ and $\theta = 2\pi(t^2 - 2t)$, where *r* is expressed in millimeters, *t* in seconds, and $\theta$ in radians. Determine the radial and transverse components of the force exerted on the collar when (*a*) $t = 0$, (*b*) $t = 0.8$ s.

**Fig. P12.130**

**12.131**  The .12 kg pin *B* slides along the slot in the rotating arm *OC* and along the slot *DE* which is cut in a fixed horizontal plate. Neglecting friction and knowing that arm *OC* rotates at a constant rate $\dot\theta_0 = 10$ rad/s, determine for any given value of $\theta$ (*a*) the radial and transverse components of the resultant force **F** exerted on pin *B*, (*b*) the forces **P** and **Q** exerted on pin *B* by arm *OC* and the wall of the slot *DE*, respectively.

**Fig. P12.131**

**12.132** Determine the mass of the earth knowing that the mean radius of the moon's orbit about the earth is 384.5 Mm and that the moon requires 27.32 days to complete one full revolution about the earth.

**12.133** A 540-kg spacecraft first is placed into a circular orbit about the earth at an altitude of 4500 km and then is transferred to a circular orbit about the moon. Knowing that the mass of the moon is 0.01230 times the mass of the earth and that the radius of the moon is 1740 km, determine (*a*) the gravitational force exerted on the spacecraft as it was orbiting the earth, (*b*) the required radius of the orbit of the spacecraft about the moon if the periodic times of the two orbits are to be equal, (*c*) the acceleration of gravity at the surface of the moon. (The *periodic time* of a satellite is the time it requires to complete one full revolution about a planet.)

**12.134** A .45 kg ball *A* and a .9 kg ball *B* are mounted on a horizontal rod which rotates freely about a vertical shaft. The balls are held in the positions shown by pins. The pin holding *B* is suddenly removed and the ball moves to position *C* as the rod rotates. Neglecting friction and the mass of the rod and knowing that the initial speed of *A* is $v_A = 2.4$ m/s, determine (*a*) the radial and transverse components of the acceleration of ball *B* immediately after the pin is removed, (*b*) the acceleration of ball *B* relative to the rod at that instant, (*c*) the speed of ball *A* after ball *B* has reached the stop at *C*.

**Fig. P12.134**

**12.135** A space shuttle is describing a circular orbit at an altitude of 563 km above the surface of the earth. As it passes through point *A*, it fires its engine for a short interval of time to reduce its speed by 152 m/s and begin its descent toward the earth. Determine the angle *AOB* so that the altitude of the shuttle at point *B* is 121 km. (*Hint.* Point *A* is the apogee of the elliptic descent orbit.)

**Fig. P12.135**

# Computer Problems

**12.C1**  Block $B$ of weight 80 N is initially at rest as shown on the upper surface of a 200 N wedge $A$ which is supported by a horizontal surface. A 18 N block $C$ is connected to block $B$ by a cord, which passes over a pulley of negligible mass. Using computational software and denoting by $\mu$ the coefficient of friction at all surfaces, calculate the initial acceleration of the wedge and the initial acceleration of block $B$ relative to the wedge for values of $\mu \geq 0$. Use 0.01 increments for $\mu$ until the wedge does not move and then use 0.1 increments until no motion occurs.

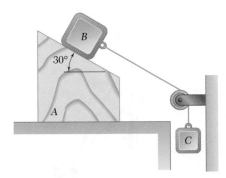

**Fig. P12.C1**

**12.C2**  A small 0.50-kg block is at rest at the top of a cylindrical surface. The block is given an initial velocity $\mathbf{v}_0$ to the right of magnitude 3 m/s, which causes it to slide on the cylindrical surface. Using computational software calculate and plot the values of $\theta$ at which the block leaves the surface for values of $\mu_k$, the coefficient of kinetic friction between the block and the surface, from 0 to 0.4 using 0.05 increments.

**Fig. P12.C2**

**Fig. P12.C3**

**12.C3**   A block of mass $m$ is attached to a spring of constant $k$. The block is released from rest when the spring is in a horizontal and undeformed position. Knowing that $r_0 = 1.2$ m use computational software to determine (*a*) for $k/m = 15$ s$^{-2}$, 20s$^{-2}$, and 25 s$^{-2}$, the length of the spring and the magnitude and direction of the velocity of the block as the block passes directly under the point of suspension of the spring, (*b*) the value of $k/m$ for which that velocity is horizontal.

**12.C4**   An airplane has a weight of 266893 N and its engines develop a constant thrust of 55603 N during take-off. The drag **D** exerted on the airplane has a magnitude $D = 0.060v^2$, where $D$ is expressed in lb and $v$ is the speed in m/s. The airplane starts from rest at the end of the runway and becomes airborne at a speed of 76.2 m/s. Determine and plot the position and speed of the airplane as functions of time and the speed as a function of position as the airplane moves down the runway.

**12.C5**   Two wires $AC$ and $BC$ are tied at point $C$ to a small sphere which revolves at a constant speed $v$ in the horizontal circle shown. Calculate and plot the tensions in the wires as functions of $v$. Determine the range of values of $v$ for which both wires remain taut.

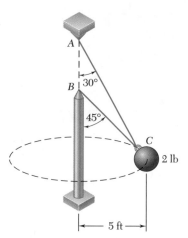

**Fig. P12.C5**

**12.C6**   A 4.5 kg bag is gently pushed off the top of a wall at point A and swings in a vertical plane at the end of a rope of length $l = 1.5$ m. Calculate and plot the speed of the bag and the magnitude of the tension in the rope as functions of the angle $\theta$ from 0 to 90°.

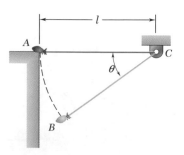

**Fig. P12.C6**

**12.C7**   The two-dimensional motion of particle $B$ is defined by the relations $r = t^3 - 2t^2$ and $\theta = t^3 - 4t$, where $r$ is expressed in m, $t$ in seconds, and $\theta$ in radians. Knowing that the particle has a mass of 0.25 kg and moves in a horizontal plane, calculate and plot the radial and transverse components and the magnitude of the force acting on the particle as functions of $t$ from 0 to 1.5 s.

**Fig. P12.C7**

## CHAPTER 13

# Kinetics of Particles: Energy and Momentum Methods

As the head of the golf club hits the ball and the sand, some of its energy and momentum are transferred to the ball and to the sand particles. Considering the energy and/or momentum of a particle is often an effective way of studying its motion.

## KINETICS OF PARTICLES: ENERGY AND MOMENTUM METHODS

13.1   Introduction
13.2   Work of a Force
13.3   Kinetic Energy of a Particle. Principle of Work and Energy
13.4   Applications of the Principle of Work and Energy
13.5   Power and Efficiency
13.6   Potential Energy
13.7   Conservative Forces
13.8   Conservation of Energy
13.9   Motion under a Conservative Central Force. Application to Space Mechanics
13.10  Principle of Impulse and Momentum
13.11  Impulsive Motion
13.12  Impact
13.13  Direct Central Impact
13.14  Oblique Central Impact
13.15  Problems Involving Energy and Momentum

## 13.1. INTRODUCTION

In the preceding chapter, most problems dealing with the motion of particles were solved through the use of the fundamental equation of motion $\mathbf{F} = m\mathbf{a}$. Given a particle acted upon by a force $\mathbf{F}$, we could solve this equation for the acceleration $\mathbf{a}$; then, by applying the principles of kinematics, we could determine from $\mathbf{a}$ the velocity and position of the particle at any time.

Using the equation $\mathbf{F} = m\mathbf{a}$ together with the principles of kinematics allows us to obtain two additional methods of analysis, the *method of work and energy* and the *method of impulse and momentum*. The advantage of these methods lies in the fact that they make the determination of the acceleration unnecessary. Indeed, the method of work and energy directly relates force, mass, velocity, and displacement, while the method of impulse and momentum relates force, mass, velocity, and time.

The method of work and energy will be considered first. In Secs. 13.2 through 13.4, the *work of a force* and the *kinetic energy of a particle* are discussed and the principle of work and energy is applied to the solution of engineering problems. The concepts of *power* and *efficiency* of a machine are introduced in Sec. 13.5.

Sections 13.6 through 13.8 are devoted to the concept of *potential energy* of a conservative force and to the application of the principle of conservation of energy to various problems of practical interest. In Sec. 13.9, the principles of conservation of energy and of conservation of angular momentum are used jointly to solve problems of space mechanics.

The second part of the chapter is devoted to the *principle of impulse and momentum* and to its application to the study of the motion of a particle. As you will see in Sec. 13.11, this principle is particularly effective in the study of the *impulsive motion* of a particle, where very large forces are applied for a very short time interval.

In Secs. 13.12 through 13.14, the *central impact* of two bodies will be considered. It will be shown that a certain relation exists between the relative velocities of the two colliding bodies before and after impact. This relation, together with the fact that the total momentum of the two bodies is conserved, can be used to solve a number of problems of practical interest.

Finally, in Sec. 13.15, you will learn to select from the three fundamental methods presented in Chaps. 12 and 13 the method best suited for the solution of a given problem. You will also see how the principle of conservation of energy and the method of impulse and momentum can be combined to solve problems involving only conservative forces, except for a short impact phase during which impulsive forces must also be taken into consideration.

## 13.2. WORK OF A FORCE

We will first define the terms *displacement* and *work* as they are used in mechanics.† Consider a particle which moves from a point $A$ to a

---

†The definition of work was given in Sec. 10.2, and the basic properties of the work of a force were outlined in Secs. 10.2 and 10.6. For convenience, we repeat here the portions of this material which relate to the kinetics of particles.

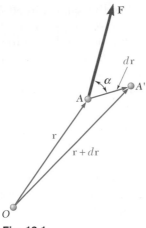

**Fig. 13.1**

neighboring point $A'$ (Fig. 13.1). If $\mathbf{r}$ denotes the position vector corresponding to point $A$, the small vector joining $A$ and $A'$ can be denoted by the differential $d\mathbf{r}$; the vector $d\mathbf{r}$ is called the *displacement* of the particle. Now, let us assume that a force $\mathbf{F}$ is acting on the particle. The *work of the force* $\mathbf{F}$ *corresponding to the displacement* $d\mathbf{r}$ is defined as the quantity

$$dU = \mathbf{F} \cdot d\mathbf{r} \tag{13.1}$$

obtained by forming the scalar product of the force $\mathbf{F}$ and the displacement $d\mathbf{r}$. Denoting by $F$ and $ds$ respectively, the magnitudes of the force and of the displacement, and by $\alpha$ the angle formed by $\mathbf{F}$ and $d\mathbf{r}$, and recalling the definition of the scalar product of two vectors (Sec. 3.9), we write

$$dU = F \, ds \cos \alpha \tag{13.1'}$$

Using formula (3.30), we can also express the work $dU$ in terms of the rectangular components of the force and of the displacement:

$$dU = F_x dx + F_y dy + F_z dz \tag{13.1''}$$

Being a *scalar quantity,* work has a magnitude and a sign but no direction. We also note that work should be expressed in units obtained by multiplying units of length by units of force. Thus, if U.S. customary units are used, work should be expressed in ft · lb or in · lb. If SI units are used, work should be expressed in N · m. The unit of work N · m is called a *joule* (J).† Recalling the conversion factors indicated in Sec. 12.4, we write

$$1 \text{ ft} \cdot \text{lb} = (1 \text{ ft})(1 \text{ lb}) = (0.3048 \text{ m})(4.448 \text{ N}) = 1.356 \text{ J}$$

It follows from (13.1′) that the work $dU$ is positive if the angle $\alpha$ is acute and negative if $\alpha$ is obtuse. Three particular cases are of special

---

†The joule (J) is the SI unit of *energy,* whether in mechanical form (work, potential energy, kinetic energy) or in chemical, electrical, or thermal form. We should note that even though N · m = J, the moment of a force must be expressed in N · m and not in joules, since the moment of a force is not a form of energy.

**Fig. 13.2**

**Fig. 13.3**

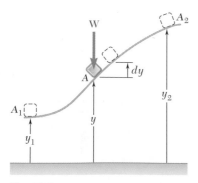

**Fig. 13.4**

interest. If the force $\mathbf{F}$ has the same direction as $d\mathbf{r}$, the work $dU$ reduces to $F\,ds$. If $\mathbf{F}$ has a direction opposite to that of $d\mathbf{r}$, the work is $dU = -F\,ds$. Finally, if $\mathbf{F}$ is perpendicular to $d\mathbf{r}$, the work $dU$ is zero.

The work of $\mathbf{F}$ during a *finite* displacement of the particle from $A_1$ to $A_2$ (Fig. 13.2$a$) is obtained by integrating Eq. (13.1) along the path described by the particle. This work, denoted by $U_{1\rightarrow2}$, is

$$U_{1\rightarrow2} = \int_{A_1}^{A_2} \mathbf{F} \cdot d\mathbf{r} \qquad (13.2)$$

Using the alternative expression (13.1$'$) for the elementary work $dU$, and observing that $F\cos\alpha$ represents the tangential component $F_t$ of the force, we can also express the work $U_{1\rightarrow2}$ as

$$U_{1\rightarrow2} = \int_{s_1}^{s_2} (F\cos\alpha)\,ds = \int_{s_1}^{s_2} F_t\,ds \qquad (13.2')$$

where the variable of integration $s$ measures the distance traveled by the particle along the path. The work $U_{1\rightarrow2}$ is represented by the area under the curve obtained by plotting $F_t = F\cos\alpha$ against $s$ (Fig. 13.2$b$).

When the force $\mathbf{F}$ is defined by its rectangular components, the expression (13.1$''$) can be used for the elementary work. We then write

$$U_{1\rightarrow2} = \int_{A_1}^{A_2} (F_x\,dx + F_y\,dy + F_z\,dz) \qquad (13.2'')$$

where the integration is to be performed along the path described by the particle.

**Work of a Constant Force in Rectilinear Motion.** When a particle moving in a straight line is acted upon by a force $\mathbf{F}$ of constant magnitude and of constant direction (Fig. 13.3), formula (13.2$'$) yields

$$U_{1\rightarrow2} = (F\cos\alpha)\,\Delta x \qquad (13.3)$$

where $\alpha$ = angle the force forms with direction of motion
$\Delta x$ = displacement from $A_1$ to $A_2$

**Work of the Force of Gravity.** The work of the weight $\mathbf{W}$ of a body, that is, of the force of gravity exerted on that body, is obtained by substituting the components of $\mathbf{W}$ into (13.1$''$) and (13.2$''$). With the $y$ axis chosen upward (Fig. 13.4), we have $F_x = 0$, $F_y = -W$, and $F_z = 0$, and we write

$$dU = -W\,dy$$

$$U_{1\rightarrow2} = -\int_{y_1}^{y_2} W\,dy = Wy_1 - Wy_2 \qquad (13.4)$$

or

$$U_{1\rightarrow2} = -W(y_2 - y_1) = -W\,\Delta y \qquad (13.4')$$

where $\Delta y$ is the vertical displacement from $A_1$ to $A_2$. The work of the weight $\mathbf{W}$ is thus equal to *the product of W and the vertical displacement of the center of gravity of the body*. The work is *positive* when $\Delta y < 0$, that is, *when the body moves down*.

**Work of the Force Exerted by a Spring.** Consider a body $A$ attached to a fixed point $B$ by a spring; it is assumed that the spring is undeformed when the body is at $A_0$ (Fig. 13.5*a*). Experimental evidence shows that the magnitude of the force $\mathbf{F}$ exerted by the spring on body $A$ is proportional to the deflection $x$ of the spring measured from the position $A_0$. We have

$$F = kx \tag{13.5}$$

where $k$ is the *spring constant,* expressed in N/m or kN/m if SI units are used and in lb/ft or lb/in. if U.S. customary units are used.†

The work of the force $\mathbf{F}$ exerted by the spring during a finite displacement of the body from $A_1(x = x_1)$ to $A_2(x = x_2)$ is obtained by writing

$$dU = -F\,dx = -kx\,dx$$

$$U_{1\rightarrow2} = -\int_{x_1}^{x_2} kx\,dx = \tfrac{1}{2}kx_1^2 - \tfrac{1}{2}kx_2^2 \tag{13.6}$$

Care should be taken to express $k$ and $x$ in consistent units. For example, if U.S. customary units are used, $k$ should be expressed in lb/ft and $x$ in feet, or $k$ in lb/in. and $x$ in inches; in the first case, the work is obtained in ft · lb, in the second case, in in · lb. We note that the work of the force $\mathbf{F}$ exerted by the spring on the body is *positive* when $x_2 < x_1$, that is, *when the spring is returning to its undeformed position.*

Since Eq. (13.5) is the equation of a straight line of slope $k$ passing through the origin, the work $U_{1\rightarrow2}$ of $\mathbf{F}$ during the displacement from $A_1$ to $A_2$ can be obtained by evaluating the area of the trapezoid shown in Fig. 13.5*b*. This is done by computing $F_1$ and $F_2$ and multiplying the base $\Delta x$ of the trapezoid by its mean height $\tfrac{1}{2}(F_1 + F_2)$. Since the work of the force $\mathbf{F}$ exerted by the spring is positive for a negative value of $\Delta x$, we write

$$U_{1\rightarrow2} = -\tfrac{1}{2}(F_1 + F_2)\,\Delta x \tag{13.6'}$$

Formula (13.6′) is usually more convenient to use than (13.6) and affords fewer chances of confusing the units involved.

**Work of a Gravitational Force.** We saw in Sec. 12.10 that two particles of mass $M$ and $m$ at a distance $r$ from each other attract each other with equal and opposite forces $\mathbf{F}$ and $-\mathbf{F}$, directed along the line joining the particles and of magnitude

$$F = G\frac{Mm}{r^2}$$

---

†The relation $F = kx$ is correct under static conditions only. Under dynamic conditions, formula (13.5) should be modified to take the inertia of the spring into account. However, the error introduced by using the relation $F = kx$ in the solution of kinetics problems is small if the mass of the spring is small compared with the other masses in motion.

Spring undeformed

(*a*)

(*b*)

**Fig. 13.5**

**Fig. 13.6**

Let us assume that the particle $M$ occupies a fixed position $O$ while the particle $m$ moves along the path shown in Fig. 13.6. The work of the force $\mathbf{F}$ exerted on the particle $m$ during an infinitesimal displacement of the particle from $A$ to $A'$ can be obtained by multiplying the magnitude $F$ of the force by the radial component $dr$ of the displacement. Since $\mathbf{F}$ is directed toward $O$, the work is negative and we write

$$dU = -F\,dr = -G\frac{Mm}{r^2}\,dr$$

The work of the gravitational force $\mathbf{F}$ during a finite displacement from $A_1(r = r_1)$ to $A_2(r = r_2)$ is therefore

$$U_{1\rightarrow2} = -\int_{r_1}^{r_2} \frac{GMm}{r^2}\,dr = \frac{GMm}{r_2} - \frac{GMm}{r_1} \qquad (13.7)$$

where $M$ is the mass of the earth. This formula can be used to determine the work of the force exerted by the earth on a body of mass $m$ at a distance $r$ from the center of the earth, when $r$ is larger than the radius $R$ of the earth. Recalling the first of the relations (12.29), we can replace the product $GMm$ in Eq. (13.7) by $WR^2$, where $R$ is the radius of the earth ($R = 6.37 \times 10^6$ m or 3960 mi) and $W$ is the weight of the body at the surface of the earth.

A number of forces frequently encountered in problems of kinetics *do no work*. They are forces applied to fixed points ($ds = 0$) or acting in a direction perpendicular to the displacement ($\cos \alpha = 0$). Among the forces which do no work are the following: the reaction at a frictionless pin when the body supported rotates about the pin, the reaction at a frictionless surface when the body in contact moves along the surface, the reaction at a roller moving along its track, and the weight of a body when its center of gravity moves horizontally.

## 13.3. KINETIC ENERGY OF A PARTICLE. PRINCIPLE OF WORK AND ENERGY

Consider a particle of mass $m$ acted upon by a force $\mathbf{F}$ and moving along a path which is either rectilinear or curved (Fig. 13.7). Expressing Newton's second law in terms of the tangential components of the force and of the acceleration (see Sec. 12.5), we write

$$F_t = ma_t \qquad \text{or} \qquad F_t = m\,\frac{dv}{dt}$$

where $v$ is the speed of the particle. Recalling from Sec. 11.9 that $v = ds/dt$, we obtain

$$F_t = m\,\frac{dv}{ds}\frac{ds}{dt} = mv\,\frac{dv}{ds}$$

$$F_t\,ds = mv\,dv$$

Integrating from $A_1$, where $s = s_1$ and $v = v_1$, to $A_2$, where $s = s_2$ and $v = v_2$, we write

$$\int_{s_1}^{s_2} F_t\,ds = m\int_{v_1}^{v_2} v\,dv = \tfrac{1}{2}mv_2^2 - \tfrac{1}{2}mv_1^2 \qquad (13.8)$$

The left-hand member of Eq. (13.8) represents the work $U_{1\rightarrow2}$ of the force $\mathbf{F}$ exerted on the particle during the displacement from $A_1$ to

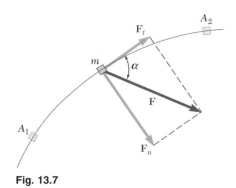

**Fig. 13.7**

$A_2$; as indicated in Sec. 13.2, the work $U_{1\to2}$ is a scalar quantity. The expression $\frac{1}{2}mv^2$ is also a scalar quantity; it is defined as the kinetic energy of the particle and is denoted by $T$. We write

$$T = \tfrac{1}{2}mv^2 \tag{13.9}$$

Substituting into (13.8), we have

$$U_{1\to2} = T_2 - T_1 \tag{13.10}$$

which expresses that, when a particle moves from $A_1$ to $A_2$ under the action of a force **F**, *the work of the force **F** is equal to the change in kinetic energy of the particle.* This is known as the *principle of work and energy.* Rearranging the terms in (13.10), we write

$$T_1 + U_{1\to2} = T_2 \tag{13.11}$$

Thus, *the kinetic energy of the particle at $A_2$ can be obtained by adding to its kinetic energy at $A_1$ the work done during the displacement from $A_1$ to $A_2$ by the force **F** exerted on the particle.* Like Newton's second law from which it is derived, the principle of work and energy applies only with respect to a newtonian frame of reference (Sec. 12.2). The speed $v$ used to determine the kinetic energy $T$ should therefore be measured with respect to a newtonian frame of reference.

Since both work and kinetic energy are scalar quantities, their sum can be computed as an ordinary algebraic sum, the work $U_{1\to2}$ being considered as positive or negative according to the direction of **F**. When several forces act on the particle, the expression $U_{1\to2}$ represents the total work of the forces acting on the particle; it is obtained by adding algebraically the work of the various forces.

As noted above, the kinetic energy of a particle is a scalar quantity. It further appears from the definition $T = \frac{1}{2}mv^2$ that regardless of the direction of motion of the particle the kinetic energy is always positive. Considering the particular case when $v_1 = 0$ and $v_2 = v$, and substituting $T_1 = 0$ and $T_2 = T$ into (13.10), we observe that the work done by the forces acting on the particle is equal to $T$. Thus, the kinetic energy of a particle moving with a speed $v$ represents the work which must be done to bring the particle from rest to the speed $v$. Substituting $T_1 = T$ and $T_2 = 0$ into (13.10), we also note that when a particle moving with a speed $v$ is brought to rest, the work done by the forces acting on the particle is $-T$. Assuming that no energy is dissipated into heat, we conclude that the work done by the forces exerted *by the particle* on the bodies which cause it to come to rest is equal to $T$. Thus, the kinetic energy of a particle also represents *the capacity to do work associated with the speed of the particle.*

The kinetic energy is measured in the same units as work, that is, in joules if SI units are used and in ft · lb if U.S. customary units are used. We check that, in SI units,

$$T = \tfrac{1}{2}mv^2 = \text{kg(m/s)}^2 = (\text{kg} \cdot \text{m/s}^2)\text{m} = \text{N} \cdot \text{m} = \text{J}$$

while, in customary units,

$$T = \tfrac{1}{2}mv^2 = (\text{lb} \cdot \text{s}^2/\text{ft})(\text{ft/s})^2 = \text{ft} \cdot \text{lb}$$

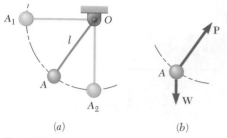

Fig. 13.8

## 13.4. APPLICATIONS OF THE PRINCIPLE OF WORK AND ENERGY

The application of the principle of work and energy greatly simplifies the solution of many problems involving forces, displacements, and velocities. Consider, for example, the pendulum $OA$ consisting of a bob $A$ of weight $W$ attached to a cord of length $l$ (Fig. 13.8a). The pendulum is released with no initial velocity from a horizontal position $OA_1$ and allowed to swing in a vertical plane. We wish to determine the speed of the bob as it passes through $A_2$, directly under $O$.

We first determine the work done during the displacement from $A_1$ to $A_2$ by the forces acting on the bob. We draw a free-body diagram of the bob, showing all the *actual* forces acting on it, that is, the weight $\mathbf{W}$ and the force $\mathbf{P}$ exerted by the cord (Fig. 13.8b). (An inertia vector is not an actual force and *should not* be included in the free-body diagram.) We note that the force $\mathbf{P}$ does no work, since it is normal to the path; the only force which does work is thus the weight $\mathbf{W}$. The work of $\mathbf{W}$ is obtained by multiplying its magnitude $W$ by the vertical displacement $l$ (Sec. 13.2); since the displacement is downward, the work is positive. We therefore write $U_{1 \rightarrow 2} = Wl$.

Now considering the kinetic energy of the bob, we find $T_1 = 0$ at $A_1$ and $T_2 = \frac{1}{2}(W/g)v_2^2$ at $A_2$. We can now apply the principle of work and energy; recalling formula (13.11), we write:

$$T_1 + U_{1 \rightarrow 2} = T_2 \qquad 0 + Wl = \frac{1}{2}\frac{W}{g}v_2^2$$

Solving for $v_2$, we find $v_2 = \sqrt{2gl}$. We note that the speed obtained is that of a body falling freely from a height $l$.

The example we have considered illustrates the following advantages of the method of work and energy:

1. In order to find the speed at $A_2$, there is no need to determine the acceleration in an intermediate position $A$ and to integrate the expression obtained from $A_1$ to $A_2$.
2. All quantities involved are scalars and can be added directly, without using $x$ and $y$ components.
3. Forces which do no work are eliminated from the solution of the problem.

What is an advantage in one problem, however, may be a disadvantage in another. It is evident, for instance, that the method of work and energy cannot be used to directly determine an acceleration. It is also evident that in determining a force which is normal to the path of the particle, a force which does no work, the method of work and energy must be supplemented by the direct application of Newton's second law. Suppose, for example, that we wish to determine the tension in the cord of the pendulum of Fig. 13.8a as the bob passes through $A_2$. We draw a free-body diagram of the bob in that position (Fig. 13.9) and express Newton's second law in terms of tangential and normal components. The equations $\Sigma F_t = ma_t$ and $\Sigma F_n = ma_n$ yield, respectively, $a_t = 0$ and

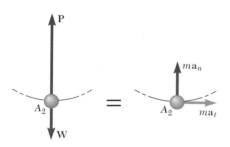

Fig. 13.9

$$P - W = ma_n = \frac{W}{g}\frac{v_2^2}{l}$$

But the speed at $A_2$ was determined earlier by the method of work and energy. Substituting $v_2^2 = 2gl$ and solving for $P$, we write

$$P = W + \frac{W}{g}\frac{2gl}{l} = 3W$$

When a problem involves two particles or more, the principle of work and energy can be applied to each particle separately. Adding the kinetic energies of the various particles, and considering the work of all the forces acting on them, we can also write a single equation of work and energy for all the particles involved. We have

$$T_1 + U_{1\to 2} = T_2 \qquad (13.11)$$

where $T$ represents the arithmetic sum of the kinetic energies of the particles involved (all terms are positive) and $U_{1\to 2}$ is the work of all the forces acting on the particles, *including the forces of action and reaction exerted by the particles on each other.* In problems involving bodies connected by *inextensible cords or links,* however, the work of the forces exerted by a given cord or link on the two bodies it connects cancels out, since the points of application of these forces move through equal distances (see Sample Prob. 13.2).†

Since friction forces have a direction opposite of that of the displacement of the body on which they act, *the work of friction forces is always negative.* This work represents energy dissipated into heat and always results in a decrease in the kinetic energy of the body involved (see Sample Prob. 13.3).

## 13.5. POWER AND EFFICIENCY

*Power* is defined as the time rate at which work is done. In the selection of a motor or engine, power is a much more important criterion than is the actual amount of work to be performed. Either a small motor or a large power plant can be used to do a given amount of work; but the small motor may require a month to do the work done by the power plant in a matter of minutes. If $\Delta U$ is the work done during the time interval $\Delta t$, then the average power during that time interval is

$$\text{Average power} = \frac{\Delta U}{\Delta t}$$

Letting $\Delta t$ approach zero, we obtain at the limit

$$\text{Power} = \frac{dU}{dt} \qquad (13.12)$$

†The application of the method of work and energy to a system of particles is discussed in detail in Chap. 14.

Substituting the scalar product $\mathbf{F} \cdot d\mathbf{r}$ for $dU$, we can also write

$$\text{Power} = \frac{dU}{dt} = \frac{\mathbf{F} \cdot d\mathbf{r}}{dt}$$

and, recalling that $d\mathbf{r}/dt$ represents the velocity $\mathbf{v}$ of the point of application of $\mathbf{F}$,

$$\text{Power} = \mathbf{F} \cdot \mathbf{v} \qquad (13.13)$$

Since power was defined as the time rate at which work is done, it should be expressed in units obtained by dividing units of work by the unit of time. Thus, if SI units are used, power should be expressed in J/s; this unit is called a *watt* (W). We have

$$1 \text{ W} = 1 \text{ J/s} = 1 \text{ N} \cdot \text{m/s}$$

If U.S. customary units are used, power should be expressed in ft · lb/s or in *horsepower* (hp), with the latter defined as

$$1 \text{ hp} = 550 \text{ ft} \cdot \text{lb/s}$$

Recalling from Sec. 13.2 that 1 ft · lb = 1.356 J, we verify that

$$1 \text{ ft} \cdot \text{lb/s} = 1.356 \text{ J/s} = 1.356 \text{ W}$$
$$1 \text{ hp} = 550(1.356 \text{ W}) = 746 \text{ W} = 0.746 \text{ kW}$$

The *mechanical efficiency* of a machine was defined in Sec. 10.5 as the ratio of the output work to the input work:

$$\eta = \frac{\text{output work}}{\text{input work}} \qquad (13.14)$$

This definition is based on the assumption that work is done at a constant rate. The ratio of the output to the input work is therefore equal to the ratio of the rates at which output and input work are done, and we have

$$\eta = \frac{\text{power output}}{\text{power input}} \qquad (13.15)$$

Because of energy losses due to friction, the output work is always smaller than the input work, and consequently the power output is always smaller than the power input. The mechanical efficiency of a machine is therefore always less than 1.

When a machine is used to transform mechanical energy into electric energy, or thermal energy into mechanical energy, its *overall efficiency* can be obtained from formula (13.15). The overall efficiency of a machine is always less than 1; it provides a measure of all the various energy losses involved (losses of electric or thermal energy as well as frictional losses). Note that it is necessary to express the power output and the power input in the same units before using formula (13.15).

## SAMPLE PROBLEM 13.1

An automobile weighing 17793 N is driven down a 5° incline at a speed of 96 km when the brakes are applied, causing a constant total braking force (applied by the road on the tires) of 6672 N. Determine the distance traveled by the automobile as it comes to a stop.

## SOLUTION

**Kinetic Energy**

**Position 1:** $v_1 = \left(96(km/h)\right)\left(\dfrac{10^3 \text{ m}}{3600 \text{ s}}\right) = 26.7$ m/s

$$T_1 = \tfrac{1}{2}mv_1^2 = \tfrac{1}{2}(17793/9.81)(26.7)^2 = 646506.2 \text{ N} \cdot \text{m}$$

**Position 2:** $\qquad v_2 = 0 \qquad T_2 = 0$

**Work** $\qquad U_{1\rightarrow2} = -6672x + (17793 \sin 5°)x = -5121.2x$

**Principle of Work and Energy**

$$T_1 + U_{1\rightarrow2} = T_2$$
$$646506.2 - 5121.2x = 0 \qquad x = 126.2 \text{ m} \quad \blacktriangleleft$$

## SAMPLE PROBLEM 13.2

Two blocks are joined by an inextensible cable as shown. If the system is released from rest, determine the velocity of block A after it has moved 2 m. Assume that the coefficient of kinetic friction between block A and the plane is $\mu_k = 0.25$ and that the pulley is weightless and frictionless.

## SOLUTION

**Work and Energy for Block A.** We denote the friction force by $\mathbf{F}_A$ and the force exerted by the cable by $\mathbf{F}_C$, and write

$$m_A = 200 \text{ kg} \qquad W_A = (200 \text{ kg})(9.81 \text{ m/s}^2) = 1962 \text{ N}$$
$$F_A = \mu_k N_A = \mu_k W_A = 0.25(1962 \text{ N}) = 490.5 \text{ N}$$
$$T_1 + U_{1\rightarrow2} = T_2: \qquad 0 + F_C(2 \text{ m}) - F_A(2 \text{ m}) = \tfrac{1}{2}m_A v^2$$
$$F_C(2 \text{ m}) - (490 \text{ N})(2 \text{ m}) = \tfrac{1}{2}(200 \text{ kg})v^2 \qquad (1)$$

**Work and Energy for Block B.** We write

$$m_B = 300 \text{ kg} \qquad W_B = (300 \text{ kg})(9.81 \text{ m/s}^2) = 2940 \text{ N}$$
$$T_1 + U_{1\rightarrow2} = T_2: \qquad 0 + W_B(2 \text{ m}) - F_C(2 \text{ m}) = \tfrac{1}{2}m_B v^2$$
$$(2940 \text{ N})(2 \text{ m}) - F_C(2 \text{ m}) = \tfrac{1}{2}(300 \text{ kg})v^2 \qquad (2)$$

Adding the left-hand and right-hand members of (1) and (2), we observe that the work of the forces exerted by the cable on A and B cancels out:

$$(2940 \text{ N})(2 \text{ m}) - (490 \text{ N})(2 \text{ m}) = \tfrac{1}{2}(200 \text{ kg} + 300 \text{ kg})v^2$$
$$4900 \text{ J} = \tfrac{1}{2}(500 \text{ kg})v^2 \qquad v = 4.43 \text{ m/s} \quad \blacktriangleleft$$

2.5 m/s
60 kg
600 mm
Cable

# SAMPLE PROBLEM 13.3

A spring is used to stop a 60-kg package which is sliding on a horizontal surface. The spring has a constant $k = 20$ kN/m and is held by cables so that it is initially compressed 120 mm. Knowing that the package has a velocity of 2.5 m/s in the position shown and that the maximum additional deflection of the spring is 40 mm, determine (a) the coefficient of kinetic friction between the package and the surface, (b) the velocity of the package as it passes again through the position shown.

## SOLUTION

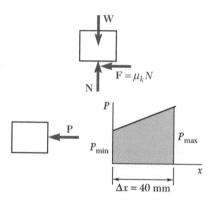

*a. Motion from Position 1 to Position 2*
*Kinetic Energy.* **Position 1:**  $v_1 = 2.5$ m/s

$$T_1 = \tfrac{1}{2}mv_1^2 = \tfrac{1}{2}(60 \text{ kg})(2.5 \text{ m/s})^2 = 187.5 \text{ N} \cdot \text{m} = 187.5 \text{ J}$$

**Position 2:** (maximum spring deflection):  $v_2 = 0$   $T_2 = 0$
*Work*
*Friction Force* **F**. We have

$$F = \mu_k N = \mu_k W = \mu_k mg = \mu_k(60 \text{ kg})(9.81 \text{ m/s}^2) = (588.6 \text{ N})\mu_k$$

The work of **F** is negative and equal to

$$(U_{1 \to 2})_f = -Fx = -(588.6 \text{ N})\mu_k(0.600 \text{ m} + 0.040 \text{ m}) = -(377 \text{ J})\mu_k$$

*Spring Force* **P**. The variable force **P** exerted by the spring does an amount of negative work equal to the area under the force-deflection curve of the spring force. We have

$$P_{min} = kx_0 = (20 \text{ kN/m})(120 \text{ mm}) = (20\,000 \text{ N/m})(0.120 \text{ m}) = 2400 \text{ N}$$
$$P_{max} = P_{min} + k\,\Delta x = 2400 \text{ N} + (20 \text{ kN/m})(40 \text{ mm}) = 3200 \text{ N}$$
$$(U_{1 \to 2})_e = -\tfrac{1}{2}(P_{min} + P_{max})\,\Delta x = -\tfrac{1}{2}(2400 \text{ N} + 3200 \text{ N})(0.040 \text{ m}) = -112.0 \text{ J}$$

The total work is thus

$$U_{1 \to 2} = (U_{1 \to 2})_f + (U_{1 \to 2})_e = -(377 \text{ J})\mu_k - 112.0 \text{ J}$$

*Principle of Work and Energy*

$T_1 + U_{1 \to 2} = T_2$:    $187.5 \text{ J} - (377 \text{ J})\mu_k - 112.0 \text{ J} = 0$    $\mu_k = 0.20$  ◄

*b. Motion from Position 2 to Position 3*
*Kinetic Energy.* **Position 2:**    $v_2 = 0$    $T_2 = 0$

**Position 3:**        $T_3 = \tfrac{1}{2}mv_3^2 = \tfrac{1}{2}(60 \text{ kg})v_3^2$

*Work.*   Since the distances involved are the same, the numerical values of the work of the friction force **F** and of the spring force **P** are the same as above. However, while the work of **F** is still negative, the work of **P** is now positive.

$$U_{2 \to 3} = -(377 \text{ J})\mu_k + 112.0 \text{ J} = -75.5 \text{ J} + 112.0 \text{ J} = +36.5 \text{ J}$$

*Principle of Work and Energy*

$T_2 + U_{2 \to 3} = T_3$:    $0 + 36.5 \text{ J} = \tfrac{1}{2}(60 \text{ kg})v_3^2$
$$v_3 = 1.103 \text{ m/s}$$        $v_3 = 1.103$ m/s ←  ◄

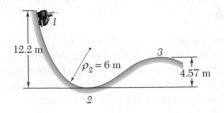

## SAMPLE PROBLEM 13.4

A 907 kg car starts from rest at point *1* and moves without friction down the track shown. (*a*) Determine the force exerted by the track on the car at point *2*, where the radius of curvature of the track is 6 m. (*b*) Determine the minimum safe value of the radius of curvature at point *3*.

## SOLUTION

*a.* **Force Exerted by the Track at Point 2.**   The principle of work and energy is used to determine the velocity of the car as it passes through point 2. mass of car m = 907 kg

*Kinetic Energy.*   $T_1 = 0$     $T_2 = \frac{1}{2}mv_2^2$

*Work.*   The only force which does work is the weight **W** (= mg). Since the vertical displacement from point *1* to point *2* is 12.2 m downward, the work of the weight is

$$U_{1\to2} = +mg \ (12.2 \text{ m})$$

*Principle of Work and Energy*

$$T_1 + U_{1\to2} = T_2 \qquad 0 + mg(12.2) = \frac{1}{2}mv_2^2$$

$$v_2^2 = 24.4 \ g = 24.4(9.81) \qquad v_2 = 15.47 \text{ m/s}$$

*Newton's Second Law at Point 2.*   The acceleration $\mathbf{a}_n$ of the car at point 2 has a magnitude $a_n = v_2^2/\rho$ and is directed upward. Since the external forces acting on the car are **W** and **N**, we write

$+\uparrow \Sigma F_n = ma_n:$      $-W + N = ma_n$

$$= \frac{W}{g}\frac{v_2^2}{\rho}$$

$$= \frac{W}{g}\frac{24.4}{6} \times g$$

$$N = 5W \qquad N = 45081.53 \text{ N} \uparrow \quad \blacktriangleleft$$
$$= 5mg$$

*b.* **Minimum Value of $\rho$ at Point 3.**   *Principle of Work and Energy.*   Applying the principle of work and energy between point *1* and point *3*, we obtain

$$T_1 + U_{1\to3} = T_3 \qquad 0 + W(12.2 - 4.57) = \frac{1}{2}\frac{W}{g}v_3^2$$

$$v_3^2 = 15.26g = 15.26(9.81) \qquad v_3 = 12.2 \text{ m/s}$$

*Newton's Second Law at Point 3.*   The minimum safe value of $\rho$ occurs when **N** = 0. In this case, the acceleration $\mathbf{a}_n$, of magnitude $a_n = v_3^2/\rho$, is directed downward, and we write

$+\downarrow \Sigma F_n = ma_n:$      $W = \frac{W}{g}\frac{v_3^2}{\rho}$

$$= \frac{W}{g}\frac{15.26 \ g}{\rho} \qquad\qquad \rho = 15.26 \text{ m} \quad \blacktriangleleft$$

## SAMPLE PROBLEM 13.5

The dumbwaiter $D$ and its load have a combined weight of 2668.9 N, while the counterweight $C$ weighs 3558.6 N. Determine the power delivered by the electric motor $M$ when the dumbwaiter ($a$) is moving up at a constant speed of 2.44 m/s, ($b$) has an instantaneous velocity of 2.44 m/s and an acceleration of 0.76 m/s$^2$, both directed upward. (1 hp = 745.7 N · m/s)

## SOLUTION

Since the force $\mathbf{F}$ exerted by the motor cable has the same direction as the velocity $\mathbf{v}_D$ of the dumbwaiter, the power is equal to $Fv_D$, where $v_D = 2.44$ m/s. To obtain the power, we must first determine $\mathbf{F}$ in each of the two given situations.

*a.* **Uniform Motion.**   We have $\mathbf{a}_C = \mathbf{a}_D = 0$; both bodies are in equilibrium.

Free Body $C$:    $+\uparrow\Sigma F_y = 0$:    $2T - 3558.6 = 0$    $T = 1779.3$ N
Free Body $D$:    $+\uparrow\Sigma F_y = 0$:                $F + T - 2668.9$ N $= 0$

$$F = 2668.9 \text{ N} - T = 2668.9 \text{ N} - 1779.3 \text{ N} = 889.6 \text{ N}$$
$$Fv_D = (889.6 \text{ N})(2.44 \text{ m/s}) = 2170.6 \text{ N} \cdot \text{m/s}$$

$$\text{Power} = (2170.6 \text{ N . m/s}) \frac{1 \text{ hp}}{745.7 \text{ N.m/s}} = 2.91 \text{ hp} \quad \blacktriangleleft$$

*b.* **Accelerated Motion.**   We have

$$\mathbf{a}_D = .76 \text{ m/s}^2 \uparrow \qquad \mathbf{a}_C = -\tfrac{1}{2}\mathbf{a}_D = 0.38 \text{ m/s}^2 \downarrow$$

The equations of motion are

Free Body $C$: $+\downarrow \Sigma F_y = m_C a_C$: $3558.6 - 2T = \dfrac{3558.6}{9.81} (.38) T = 1710.4$ N

Free Body $D$:    $+\uparrow\Sigma F_y = m_D a_D$:    $F + T - 2668.9 = \dfrac{2668.9}{9.81} (.76)$

$$F + 1710.4 - 2608.9 = 206.76 \qquad F = 1165.26 \text{ N}$$
$$Fv_D = (1165.26 \text{ N})(2.44 \text{ m/s}) = 2843.2 \text{ N} \cdot \text{m/s}$$

$$\text{Power} = (2843.2 \text{ N} \cdot \text{m/s}) \frac{1 \text{ hp}}{745.7 \text{ N . m/s}} = 3.81 \text{ hp} \quad \blacktriangleleft$$

In the preceding chapter, you solved problems dealing with the motion of a particle by using the fundamental equation $\mathbf{F} = m\mathbf{a}$ to determine the acceleration $\mathbf{a}$. By applying the principles of kinematics you were then able to determine from $\mathbf{a}$ the velocity and displacement of the particle at any time. In this lesson we combined $\mathbf{F} = m\mathbf{a}$ and the principles of kinematics to obtain an additional method of analysis called the *method of work and energy*. This eliminates the need to calculate the acceleration and will enable you to relate the velocities of the particle at two points along its path of motion. To solve a problem by the method of work and energy you will follow these steps:

1. *Computing the work of each of the forces.* The work $U_{1\rightarrow2}$ of a given force $\mathbf{F}$ during the finite displacement of the particle from $A_1$ to $A_2$ is defined as

$$U_{1\rightarrow2} = \int \mathbf{F} \cdot d\mathbf{r} \qquad \text{or} \qquad U_{1\rightarrow2} = \int (F \cos \alpha)\, ds \quad (13.2, 13.2')$$

where $\alpha$ is the angle between $\mathbf{F}$ and the displacement $d\mathbf{r}$. The work $U_{1\rightarrow2}$ is a scalar quantity and is expressed in ft · lb or in · lb in the U.S. customary system of units and in N · m or joules (J) in the SI system of units. Note that the work done is zero for a force perpendicular to the displacement ($\alpha = 90°$). Negative work is done for $90° < \alpha < 180°$ and in particular for a friction force, which is always opposite in direction to the displacement ($\alpha = 180°$).

The work $U_{1\rightarrow2}$ can be easily evaluated in the following cases that you will encounter:

   a. *Work of a constant force in rectilinear motion*

$$U_{1\rightarrow2} = (F \cos \alpha)\, \Delta x \qquad\qquad (13.3)$$

where $\alpha$ = angle the force forms with the direction of motion
   $\Delta x$ = displacement from $A_1$ to $A_2$ (Fig. 13.3)

   b. *Work of the force of gravity*

$$U_{1\rightarrow2} = -W\, \Delta y \qquad\qquad (13.4')$$

where $\Delta y$ is the vertical displacement of the center of gravity of the body of weight $W$. Note that the work is positive when $\Delta y$ is negative, that is, when the body moves down (Fig. 13.4).

   c. *Work of the force exerted by a spring*

$$U_{1\rightarrow2} = \tfrac{1}{2}kx_1^2 - \tfrac{1}{2}kx_2^2 \qquad\qquad (13.6)$$

where $k$ is the spring constant and $x_1$ and $x_2$ are the elongations of the spring corresponding to the positions $A_1$ and $A_2$ (Fig. 13.5).　　　　　　　(*continued*)

### d. Work of a gravitational force

$$U_{1\to2} = \frac{GMm}{r_2} - \frac{GMm}{r_1} \qquad (13.7)$$

for a displacement of the body from $A_1(r = r_1)$ to $A_2(r = r_2)$ (Fig. 13.6).

2. **Calculate the kinetic energy at $A_1$ and $A_2$.** The kinetic energy $T$ is

$$T = \tfrac{1}{2}mv^2 \qquad (13.9)$$

where $m$ is the mass of the particle and $v$ is the magnitude of its velocity. The units of kinetic energy are the same as the units of work, that is, ft · lb or in · lb if U.S. customary units are used and N · m or joules (J) if SI units are used.

3. **Substitute the values for the work done $U_{1\to2}$ and the kinetic energies $T_1$ and $T_2$** into the equation

$$T_1 + U_{1\to2} = T_2 \qquad (13.11)$$

You will now have *one equation* which you can solve for *one unknown.* Note that this equation does not yield the time of travel or the acceleration directly. However, if you know the radius of curvature $\rho$ of the path of the particle at a point where you have obtained the velocity $v$, you can express the normal component of the acceleration as $a_n = v^2/\rho$ and obtain the normal component of the force exerted on the particle by writing $F_n = mv^2/\rho$.

4. **Power was introduced in this lesson as the time rate at which work is done, $P = dU/dt$.** Power is measured in ft · lb/s or *horsepower* (hp) in U.S. customary units and in J/s or *watts* (W) in the SI system of units. To calculate the power, you can use the equivalent formula,

$$P = \mathbf{F} \cdot \mathbf{v} \qquad (13.13)$$

where $\mathbf{F}$ and $\mathbf{v}$ denote the force and the velocity, respectively, at a given time [Sample Prob. 13.5]. In some problems [see, e.g., Prob. 13.50], you will be asked for the *average power,* which can be obtained by dividing the total work by the time interval during which the work is done.

# Problems

**13.1** A 454 kg satellite is placed in a circular orbit 4807 km above the surface of the earth. At this elevation the acceleration of gravity is 2.45 m/s². Knowing that its orbital speed is 22,432 km/h, determine the kinetic energy of the satellite.

**13.2** A 500-kg communications satellite is in a circular geosynchronous orbit and completes one revolution about the earth in 23 h and 56 min at an altitude of 35 800 km above the surface of the earth. Knowing that the radius of the earth is 6370 km, determine the kinetic energy of the satellite.

**13.3** A 2-kg stone is dropped from a height $h$ and strikes the ground with a velocity of 24 m/s. (*a*) Find the kinetic energy of the stone as it strikes the ground and the height $h$ from which it was dropped. (*b*) Solve part *a*, assuming that the same stone is dropped on the moon. (Acceleration of gravity on the moon = 1.62 m/s².)

**13.4** A golfer hits a 48 gm ball with an initial velocity of 49 m/s at an angle of 25° with the horizontal. Determine (*a*) the initial kinetic energy of the ball, (*b*) the kinetic energy of the ball when it reaches its maximum height.

**13.5** Packages are thrown down an incline at *A* with a velocity of 1.2 m/s. The packages slide along the surface *ABC* to a conveyor belt which moves with a velocity of 2.4 m/s. Knowing that $\mu_k = 0.25$ between the packages and the surface *ABC*, determine the distance $d$ if the packages are to arrive at *C* with a velocity of 2.4 m/s.

**Fig. P13.5**

**13.6** A 22.7 kg package is projected up a 20° incline with an initial velocity of 12.2 m/s. Knowing that the coefficient of kinetic friction between the package and the incline is 0.15, determine (*a*) the maximum distance *x* that the package will move up the incline, (*b*) the velocity of the package as it returns to its original position, (*c*) the total amount of energy dissipated due to friction.

**Fig. P13.6**

**13.7**  A 2000-kg automobile starts from rest at point $A$ on a 6° incline and coasts through a distance of 150 m to point $B$. The brakes are then applied, causing the automobile to come to a stop at point $C$, 20 m from $B$. Knowing that slipping is impending during the braking period and neglecting air resistance and rolling resistance, determine (a) the speed of the automobile at point $B$, (b) the coefficient of static friction between the tires and the road.

**Fig. P13.7 and P13.8**

**13.8**  A 2000-kg automobile starts from rest at point $A$ on a 6° incline and coasts through a distance of 150 m to point $B$. The brakes are then fully applied, causing the automobile to skid to a stop at point $C$, 20 m from $B$. Knowing that the coefficient of dynamic friction between the tires and the road is 0.75, determine the work done on the automobile by the combined effects of air resistance and rolling resistance between points $A$ and $C$.

*13.9*  A 40.8 kg package is at rest on an incline when a constant force $P$ is applied to it. The coefficient of kinetic friction between the package and the incline is 0.35. Knowing that the speed of the package is 0.6 m/s after it has moved 0.9 m up the incline, determine the magnitude of the force $P$.

**Fig. *P13.9***

*13.10*  A 1.36 kg model rocket is launched vertically from rest with a constant thrust of 24.5 N until the rocket reaches an altitude of 15.2 m and the thrust ends. Neglecting air resistance, determine (a) the speed of the rocket when the thrust ends, (b) the maximum height reached by the rocket, (c) the speed of the rocket when it returns to the ground.

**13.11**  The 7-kg block $A$ is released from rest in the position shown. Neglecting the effect of friction and the masses of the pulleys, determine the velocity of the block after it has moved 0.6 m up the incline.

**13.12**  The 7-kg block $A$ is released in the position shown with a velocity of 1.5 m/s up the incline. Knowing that the velocity of the block is 3 m/s after it has moved 0.6 m up the incline, determine the work done by the friction force exerted on the block. Neglect the masses of the pulleys.

**Fig. P13.11 and P13.12**

**13.13**  Boxes are transported by a conveyor belt with a velocity $\mathbf{v}_0$ to a fixed incline at $A$ where they slide and eventually fall off at $B$. Knowing that $\mu_k = 0.40$, determine the velocity of the conveyor belt if the boxes leave the incline at $B$ with a velocity of 2 m/s.

Fig. **P13.13** and **P13.14**

**13.14**  Boxes are transported by a conveyor belt with a velocity $\mathbf{v}_0$ to a fixed incline at $A$ where they slide and eventually fall off at $B$. Knowing that $\mu_k = 0.40$, determine the velocity of the conveyor belt if the boxes are to have a zero velocity at $B$.

**13.15**  The subway train shown is traveling at a speed of 54 km/h when the brakes are fully applied on the wheels of cars $B$ and $C$, causing them to slide on the track, but are not applied on the wheels of car $A$. Knowing that the coefficient of kinetic friction is 0.35 between the wheels and the track, determine ($a$) the distance required to bring the train to a stop, ($b$) the force in each coupling.

Fig. **P13.15**

**13.16**  Solve Prob. 13.15, assuming that the brakes are applied only on the wheels of car $A$.

**13.17**  Car $B$ is towing car $A$ with a 4.6 m cable at a constant speed of 9 m/s on an uphill grade when the brakes of car $B$ are fully applied causing it to skid to a stop. Car $A$, whose driver had not observed that car $B$ was slowing down, then strikes the rear of car $B$. Neglecting air resistance and rolling resistance and assuming a coefficient of kinetic friction of 0.9, determine the speed of car $A$ just before the collision.

Fig. **P13.17** and **P13.18**

**13.18**  Car $B$ is towing car $A$ at a constant speed of 9 m/s on an uphill grade when the brakes of car $A$ are fully applied causing all four wheels to skid. The driver of car $B$ does not change the throttle setting or change gears. The weights of cars $A$ and $B$ are 13345 N and 11121 N, respectively, and the coefficient of kinetic friction is 0.8. Neglecting air resistance and rolling resistance, determine ($a$) the distance traveled by the cars before they come to a stop, ($b$) the tension in the cable.

**13.19** The two blocks shown are released from rest. Neglecting the masses of the pulleys and the effect of friction in the pulleys and between the blocks and the incline, determine (*a*) the velocity of block *A* after it has moved 0.46 m, (*b*) the tension in the cable.

**Fig. P13.19 and P13.20**

**13.20** The two blocks shown are released from rest. Neglecting the masses of the pulleys and the effect of friction in the pulleys and knowing that the coefficients of friction between both blocks and the incline are $\mu_s = 0.25$ and $\mu_k = 0.20$, determine (*a*) the velocity of block *A* after it has moved 0.46 m, (*b*) the tension in the cable.

**13.21** The system shown, consisting of a 20-kg collar *A* and a 10-kg counterweight *B*, is at rest when a constant 500-N force is applied to collar *A*. (*a*) Determine the velocity of *A* just before it hits the support at *C*. (*b*) Solve part *a* assuming that the counterweight *B* is replaced by a 98.1-N downward force. Ignore friction and the mass of the pulleys.

**Fig. *P13.21***

**Fig. *P13.22***

**13.22** The 10-kg block *A* and the 4-kg block *B* are both at a height $h = 0.5$ m above the ground when the system is released from rest. After *A* hits the ground without rebound it is observed that *B* reaches a maximum height of 1.18 m. Determine (*a*) the speed of *A* just before impact, (*b*) the amount of energy dissipated by axle friction in the pulley.

**13.23**  Two blocks $A$ and $B$, of mass 8 kg and 10 kg, respectively, are connected by a cord which passes over pulleys as shown. A 6-kg collar $C$ is placed on block $A$ and the system is released from rest. After the blocks move 1.8 m, collar $C$ is removed and blocks $A$ and $B$ continue to move. Determine the speed of block $A$ just before it strikes the ground.

**13.24**  Four 3-kg packages are held in place by friction on a conveyor which is disengaged from its drive motor. When the system is released from rest, package *1* leaves the belt at *A* just as package *4* comes onto the inclined portion of the belt at *B*. Determine (*a*) the velocity of package *2* as it leaves the belt at *A*, (*b*) the velocity of package *3* as it leaves the belt at *A*. Neglect the mass of the belt and rollers.

**Fig. P13.23**

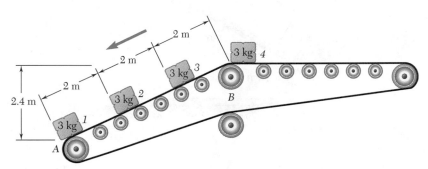

**Fig. P13.24**

**13.25**  An 8-kg plunger is released from rest in the position shown and is stopped by two nested springs; the constant of the outer spring is $k_1 = 3$ kN/m and the constant of the inner spring is $k_2 = 10$ kN/m. If the maximum deflection of the outer spring is observed to be 150 mm, determine the height $h$ from which the plunger was released.

**Fig. *P13.25* and *P13.26***

**13.26**  An 8-kg plunger is released from rest in the position shown and is stopped by two nested springs; the constant of the outer spring is $k_1 = 3$ kN/m and the constant of the inner spring is $k_2 = 10$ kN/m. If the plunger is released from a height $h = 600$ mm, determine the maximum deflection of the outer spring.

**13.27**  A 0.3 kg block rests on top of a 0.23 kg block supported by but not attached to a spring of constant 131.3 N/m. The upper block is suddenly removed. Determine (*a*) the maximum velocity reached by the .23 kg block, (*b*) the maximum height reached by the .23 kg block.

**Fig. P13.27**

**13.28** Solve Prob. 13.27, assuming that the .23 kg block is attached to the spring.

**13.29** A 3.4 kg collar is released from rest in the position shown, slides down the inclined rod, and compresses the spring. The direction of motion is reversed and the collar slides up the rod. Knowing that the maximum deflection of the spring is 12.7 cm, determine (*a*) the coefficient of kinetic friction between the collar and the rod, (*b*) the maximum speed of the collar.

**Fig. P13.29**

*13.30* A 10-kg block is attached to spring *A* and connected to spring *B* by a cord and pulley. The block is held in the position shown with both springs unstretched when the support is removed and the block is released with no initial velocity. Knowing that the constant of each spring is 2 kN/m, determine (*a*) the velocity of the block after it has moved down 50 mm, (*b*) the maximum velocity achieved by the block.

**Fig. *P13.30* and *P13.31***

*13.31* A 10-kg block is attached to spring *A* and connected to spring *B* by a cord and pulley. The block is held in the position shown with spring *A* stretched 25 mm and spring *B* unstretched when the support is removed and the block is released with no initial velocity. Knowing that the constant of each spring is 2 kN/m, determine (*a*) the velocity of the block after it has moved down 25 mm, (*b*) the maximum velocity achieved by the block.

**13.32** An uncontrolled automobile traveling at 100 km/h strikes squarely a highway crash cushion of the type shown in which the automobile is brought to rest by successively crushing steel barrels. The magnitude $F$ of the force required to crush the barrels is shown as a function of the distance $x$ the automobile has moved into the cushion. Knowing that the automobile has a mass of 1000 kg and neglecting the effect of friction, determine (a) the distance the automobile will move into the cushion before it comes to rest, (b) the maximum deceleration of the automobile.

**Fig. P13.32**

**13.33** A piston of mass $m$ and cross-sectional area $A$ is in equilibrium under the pressure $p$ at the center of a cylinder closed at both ends. Assuming that the piston is moved to the left a distance $a/2$ and released, and knowing that the pressure on each side of the piston varies inversely with the volume, determine the velocity of the piston as it again reaches the center of the cylinder. Neglect friction between the piston and the cylinder and express your answer in terms of $m$, $a$, $p$, and $A$.

**Fig. P13.33**

**13.34** Express the acceleration of gravity $g_h$ at an altitude $h$ above the surface of the earth in terms of the acceleration of gravity $g_0$ at the surface of the earth, the altitude $h$, and the radius $R$ of the earth. Determine the percent error if the weight that an object has on the surface of the earth is used as its weight at an altitude of (a) 1 km, (b) 1000 km.

**13.35** A rocket is fired vertically from the surface of the moon with a speed $v_0$. Derive a formula for the ratio $h_n/h_u$ of heights reached with a speed $v$, if Newton's law of gravitation is used to calculate $h_n$ and a uniform gravitational field is used to calculate $h_u$. Express your answer in terms of the acceleration of gravity $g_m$ on the surface of the moon, the radius $R_m$ of the moon, and the speeds $v$ and $v_0$.

**13.36** A meteor starts from rest at a very great distance from the earth. Knowing that the radius of the earth is 6345 km and neglecting all forces except the gravitational attraction of the earth, determine the speed of the meteor (a) when it enters the ionosphere at an altitude of 993 km, (b) when it enters the stratosphere at an altitude of 48 km, (c) when it strikes the earth's surface.

**13.37** During a flyby of the earth, the velocity of a spacecraft is 10.4 km/s as it reaches its minimum altitude of 993 km above the surface at point $O$. At point $B$ the spacecraft is observed to have an altitude of 8332 km. Assuming that the trajectory of the spacecraft is parabolic, determine its speed at (a) point $A$, (b) point $B$.

**Fig. P13.37**

**13.38** A bullet is fired straight up from the surface of the moon with an initial velocity of 600 m/s. Determine the maximum elevation reached by the bullet, (*a*) assuming a uniform gravitational field with $g = 1.62$ m/s$^2$, (*b*) using Newton's law of gravitation. (Radius of moon = 1740 km.)

**13.39** A 1.5-kg block *A* rests on a 1.5-kg block *B* and is attached to a spring of constant 180 N/m. The coefficients of friction between the two blocks are $\mu_s = 0.95$ and $\mu_k = 0.90$ and the coefficients of friction between block *B* and the horizontal surface are $\mu_s = 0.15$ and $\mu_k = 0.10$. Knowing that the blocks are released from rest when the spring is stretched 100 mm, determine (*a*) the velocity of block *A* as it reaches the position in which the spring is unstretched, (*b*) the maximum velocity of block *A*.

$k = 180$ N/m

**Fig. P13.39 and P13.40**

**13.40** A 1.5-kg block *A* rests on a 1.5-kg block *B* and is attached to a spring of constant 180 N/m. The coefficients of friction between the two blocks are $\mu_s = 0.35$ and $\mu_k = 0.30$ and the coefficients of friction between block *B* and the horizontal surface are $\mu_s = 0.15$ and $\mu_k = 0.10$. Knowing that the blocks are released from rest when the spring is stretched 100 mm, determine (*a*) the velocity of block *A* as it reaches the position in which the spring is unstretched, (*b*) the maximum velocity of block *A*.

**13.41** The sphere at *A* is given a downward velocity $\mathbf{v}_0$ and swings in a vertical circle of radius *l* and center *O*. Determine the smallest velocity $\mathbf{v}_0$ for which the sphere will reach point *B* as it swings about point *O* (*a*) if *AO* is a rope, (*b*) if *AO* is a slender rod of negligible mass.

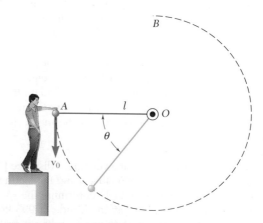

**Fig. P13.41 and P13.42**

**13.42** The sphere at *A* is given a downward velocity $\mathbf{v}_0$ of magnitude 4.9 m/s and swings in a vertical plane at the end of a rope of length $l = 1.8$ m attached to a support at *O*. Determine the angle $\theta$ at which the rope will break, knowing that it can withstand a maximum tension equal to twice the weight of the sphere.

**13.43** Sphere $C$ and block $A$ are both moving to the left with a velocity $\mathbf{v}_0$ when the block is suddenly stopped by the wall. Determine the smallest velocity $\mathbf{v}_0$ for which the sphere $C$ will swing in a full circle about the pivot $B$ (a) if $BC$ is a slender rod of negligible mass, (b) if $BC$ is a cord.

**Fig. P13.43**

**13.44** A bag is gently pushed off the top of a wall at $A$ and swings in a vertical plane at the end of a rope of length $l$. Determine the angle $\theta$ for which the rope will break, knowing that it can withstand a maximum tension equal to twice the weight of the bag.

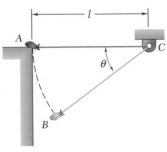

**Fig. P13.44**

**13.45** A section of track for a roller coaster consists of two circular arcs $AB$ and $CD$ joined by a straight portion $BC$. The radius of $AB$ is 27 m and the radius of $CD$ is 72 m. The car and its occupants, of total mass 250 kg, reach point $A$ with practically no velocity and then drop freely along the track. Determine the normal force exerted by the track on the car as the car reaches point $B$. Ignore air resistance and rolling resistance.

**Fig. P13.45 and P13.46**

**13.46** A section of track for a roller coaster consists of two circular arcs $AB$ and $CD$ joined by a straight portion $BC$. The radius of $AB$ is 27 m and the radius of $CD$ is 72 m. The car and its occupants, of total mass 250 kg, reach point $A$ with practically no velocity and then drop freely along the track. Determine the maximum and minimum values of the normal force exerted by the track on the car as the car travels from $A$ to $D$. Ignore air resistance and rolling resistance.

**13.47** A 68 kg sprinter starts from rest and accelerates uniformly for 5.4 s over a distance of 33.5 m. Neglecting air resistance, determine the average power developed by the sprinter.

**13.48** (a) A 60-kg woman rides a 7-kg bicycle up a 3 percent slope at a constant speed of 2 m/s. How much power must be developed by the woman? (b) A 90-kg man on a 9-kg bicycle starts down the same slope and maintains a constant speed of 6 m/s by braking. How much power is dissipated by the brakes? Ignore air resistance and rolling resistance.

2 m/s      6 m/s

3% slope

(a)                    (b)

**Fig. P13.48**

**13.49** It takes 16 s to raise a 1270 kg car and the supporting 295 kg hydraulic car-lift platform to a height of 2 m. Knowing that the overall conversion efficiency from electric to mechanical power for the system is 82 percent, determine (a) the average output power delivered by the hydraulic pump to lift the system, (b) the average electric power required.

v

**Fig. P13.49**

**13.50** A power specification formula is to be derived for electric motors which drive conveyor belts moving solid material at different rates to different heights and distances. Denoting the efficiency of the motors by $\eta$ and neglecting the power needed to drive the belt itself, derive a formula (a) in the SI system of units, for the power $P$ in kW, in terms of the mass flow rate $m$ in kg/h, the height $b$, and the horizontal distance $l$ in meters, and (b) in U.S. customary units, for the power in hp, in terms of the mass flow rate $m$ in tons/h, and the height $b$ and horizontal distance $l$ in feet.

b

l

**Fig. P13.50**

**13.51** The fluid transmission of a 15-Mg truck allows the engine to deliver an essentially constant power of 50 kW to the driving wheels. Determine the time required and the distance traveled as the speed of the truck is increased (a) from 36 km/h to 54 km/h, (b) from 54 km/h to 72 km/h.

**13.52** A 60-kg runner increases her speed from 2 m/s to 4.3 m/s in 5 s. Assuming she develops constant power during this time interval and neglecting air resistance, determine (a) the power developed, (b) the distance traveled.

**13.53** A 1361 kg automobile starts from rest and travels 366 m during a performance test. The motion of the automobile is defined by the relation $x = 12{,}000 \ln(\cosh 0.03t)$, where $x$ and $t$ are expressed in meter and seconds, respectively. The magnitude of the aerodynamic drag is $D = 0.01v^2$, where $D$ and $v$ are expressed in N and m/s, respectively. Determine the power dissipated by the aerodynamic drag when (a) $t = 10$ s, (b) $t = 15$ s.

**Fig. *P13.53* and *P13.54***

**13.54** A 1361 k automobile starts from rest and travels 366 m during a performance test. The motion of the automobile is defined by the relation $a = 11e^{-0.0005x}$, where $a$ and $x$ are expressed in m/s$^2$ and meter, respectively. The magnitude of the aerodynamic drag is $D = 0.01\,v^2$, where $D$ and $v$ are expressed in N and m/s, respectively. Determine the power dissipated by the aerodynamic drag when (a) $x = 183$ m, (b) $x = 366$ m.

## 13.6. POTENTIAL ENERGY†

Let us consider again a body of weight **W** which moves along a curved path from a point $A_1$ of elevation $y_1$ to a point $A_2$ of elevation $y_2$ (Fig. 13.4). We recall from Sec. 13.2 that the work of the force of gravity **W** during this displacement is

$$U_{1 \to 2} = Wy_1 - Wy_2 \qquad (13.4)$$

The work of **W** may thus be obtained by subtracting the value of the function $Wy$ corresponding to the second position of the body from its value corresponding to the first position. The work of **W** is independent of the actual path followed; it depends only upon the initial and final values of the function $Wy$. This function is called the *potential energy* of the body with respect to the *force of gravity* **W** and is denoted by $V_g$. We write

$$U_{1 \to 2} = (V_g)_1 - (V_g)_2 \qquad \text{with } V_g = Wy \qquad (13.16)$$

We note that if $(V_g)_2 > (V_g)_1$, that is, *if the potential energy increases during the displacement* (as in the case considered here), *the work $U_{1 \to 2}$ is negative*. If, on the other hand, the work of **W** is positive, the potential energy decreases. Therefore, the potential energy $V_g$ of the body provides a measure of the work which can be done by its weight **W**. Since only the *change* in potential energy, and not the actual value of $V_g$, is involved in formula (13.16), an arbitrary constant can be added to the expression obtained for $V_g$. In other words, the level, or datum, from which the elevation $y$ is measured can be chosen arbitrarily. Note that potential energy is expressed in the same units as work, that is, in joules if SI units are used and in ft · lb or in · lb if U.S. customary units are used.

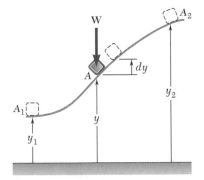

**Fig. 13.4 (*repeated*)**

†Some of the material in this section has already been considered in Sec. 10.7.

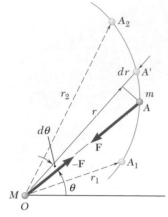

**Fig. 13.6 (repeated)**

It should be noted that the expression just obtained for the potential energy of a body with respect to gravity is valid only as long as the weight **W** of the body can be assumed to remain constant, that is, as long as the displacements of the body are small compared with the radius of the earth. In the case of a space vehicle, however, we should take into consideration the variation of the force of gravity with the distance $r$ from the center of the earth. Using the expression obtained in Sec. 13.2 for the work of a gravitational force, we write (Fig. 13.6).

$$U_{1 \to 2} = \frac{GMm}{r_2} - \frac{GMm}{r_1} \qquad (13.7)$$

The work of the force of gravity can therefore be obtained by subtracting the value of the function $-GMm/r$ corresponding to the second position of the body from its value corresponding to the first position. Thus, the expression which should be used for the potential energy $V_g$ when the variation in the force of gravity cannot be neglected is

$$V_g = -\frac{GMm}{r} \qquad (13.17)$$

Taking the first of the relations (12.29) into account, we write $V_g$ in the alternative form

$$V_g = -\frac{WR^2}{r} \qquad (13.17')$$

where $R$ is the radius of the earth and $W$ is the value of the weight of the body at the surface of the earth. When either of the relations (13.17) or (13.17') is used to express $V_g$, the distance $r$ should, of course, be measured from the center of the earth.† Note that $V_g$ is always negative and that it approaches zero for very large values of $r$.

Consider now a body attached to a spring and moving from a position $A_1$, corresponding to a deflection $x_1$ of the spring, to a position $A_2$, corresponding to a deflection $x_2$ of the spring (Fig. 13.5). We recall from Sec. 13.2 that the work of the force **F** exerted by the spring on the body is

$$U_{1 \to 2} = \tfrac{1}{2}kx_1^2 - \tfrac{1}{2}kx_2^2 \qquad (13.6)$$

The work of the elastic force is thus obtained by subtracting the value of the function $\tfrac{1}{2}kx^2$ corresponding to the second position of the body from its value corresponding to the first position. This function is denoted by $V_e$ and is called the *potential energy* of the body with respect to the *elastic force* **F**. We write

$$U_{1 \to 2} = (V_e)_1 - (V_e)_2 \qquad \text{with } V_e = \tfrac{1}{2}kx^2 \qquad (13.18)$$

and observe that during the displacement considered, the work of the force **F** exerted by the spring on the body is negative and the

Spring undeformed

**Fig. 13.5 (repeated)**

---

†The expressions given for $V_g$ in (13.17) and (13.17') are valid only when $r \geq R$, that is, when the body considered is above the surface of the earth.

potential energy $V_e$ increases. You should note that the expression obtained for $V_e$ is valid only if the deflection of the spring is measured from its undeformed position. On the other hand, formula (13.18) can be used even when the spring is rotated about its fixed end (Fig. 13.10a). The work of the elastic force depends only upon the initial and final deflections of the spring (Fig. 13.10b).

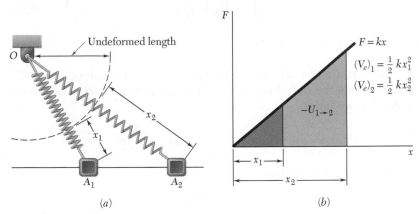

**Fig. 13.10**

The concept of potential energy can be used when forces other than gravity forces and elastic forces are involved. Indeed, it remains valid as long as the work of the force considered is independent of the path followed by its point of application as this point moves from a given position $A_1$ to a given position $A_2$. Such forces are said to be *conservative forces;* the general properties of conservative forces are studied in the following section.

## *13.7. CONSERVATIVE FORCES

As indicated in the preceding section, a force **F** acting on a particle $A$ is said to be conservative *if its work $U_{1\to2}$ is independent of the path followed by the particle $A$ as it moves from $A_1$ to $A_2$* (Fig. 13.11a). We can then write

$$U_{1\to2} = V(x_1, y_1, z_1) - V(x_2, y_2, z_2) \qquad (13.19)$$

or, for short,

$$U_{1\to2} = V_1 - V_2 \qquad (13.19')$$

The function $V(x, y, z)$ is called the potential energy, or *potential function,* of **F**.

We note that if $A_2$ is chosen to coincide with $A_1$, that is, if the particle describes a closed path (Fig. 13.11b), we have $V_1 = V_2$ and the work is zero. Thus for any conservative force **F** we can write

$$\oint \mathbf{F} \cdot d\mathbf{r} = 0 \qquad (13.20)$$

where the circle on the integral sign indicates that the path is closed.

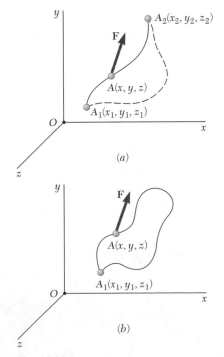

**Fig. 13.11**

Let us now apply (13.19) between two neighboring points $A(x, y, z)$ and $A'(x + dx, y + dy, z + dz)$. The elementary work $dU$ corresponding to the displacement $d\mathbf{r}$ from $A$ to $A'$ is

$$dU = V(x, y, z) - V(x + dx, y + dy, z + dz)$$

or

$$dU = -dV(x, y, z) \tag{13.21}$$

Thus, the elementary work of a conservative force is an *exact differential*.

Substituting for $dU$ in (13.21) the expression obtained in (13.1″) and recalling the definition of the differential of a function of several variables, we write

$$F_x \, dx + F_y \, dy + F_z \, dz = -\left( \frac{\partial V}{\partial x} dx + \frac{\partial V}{\partial y} dy + \frac{\partial V}{\partial z} dz \right)$$

from which it follows that

$$F_x = -\frac{\partial V}{\partial x} \qquad F_y = -\frac{\partial V}{\partial y} \qquad F_z = -\frac{\partial V}{\partial z} \tag{13.22}$$

It is clear that the components of $\mathbf{F}$ must be functions of the coordinates $x$, $y$, and $z$. Thus, a *necessary* condition for a conservative force is that it depend only upon the position of its point of application. The relations (13.22) can be expressed more concisely if we write

$$\mathbf{F} = F_x\mathbf{i} + F_y\mathbf{j} + F_z\mathbf{k} = -\left( \frac{\partial V}{\partial x}\mathbf{i} + \frac{\partial V}{\partial y}\mathbf{j} + \frac{\partial V}{\partial z}\mathbf{k} \right)$$

The vector in parentheses is known as the *gradient of the scalar function* $V$ and is denoted by **grad** $V$. We thus write for any conservative force

$$\mathbf{F} = -\mathbf{grad}\ V \tag{13.23}$$

The relations (13.19) to (13.23) were shown to be satisfied by any conservative force. It can also be shown that if a force $\mathbf{F}$ satisfies one of these relations, $\mathbf{F}$ must be a conservative force.

## 13.8. CONSERVATION OF ENERGY

We saw in the preceding two sections that the work of a conservative force, such as the weight of a particle or the force exerted by a spring, can be expressed as a change in potential energy. When a particle moves under the action of conservative forces, the principle of work and energy stated in Sec. 13.3 can be expressed in a modified form. Substituting for $U_{1\rightarrow2}$ from (13.19′) into (13.10), we write

$$V_1 - V_2 = T_2 - T_1$$

$$T_1 + V_1 = T_2 + V_2 \tag{13.24}$$

Formula (13.24) indicates that when a particle moves under the action of conservative forces, *the sum of the kinetic energy and of the potential energy of the particle remains constant*. The sum $T + V$ is called the *total mechanical energy* of the particle and is denoted by $E$.

Consider, for example, the pendulum analyzed in Sec. 13.4, which is released with no velocity from $A_1$ and allowed to swing in a vertical plane (Fig. 13.12). Measuring the potential energy from the level of $A_2$, we have, at $A_1$,

$$T_1 = 0 \qquad V_1 = Wl \qquad T_1 + V_1 = Wl$$

Recalling that at $A_2$ the speed of the pendulum is $v_2 = \sqrt{2gl}$, we have

$$T_2 = \tfrac{1}{2}mv_2^2 = \frac{1}{2}\frac{W}{g}(2gl) = Wl \qquad V_2 = 0$$

$$T_2 + V_2 = Wl$$

We thus check that the total mechanical energy $E = T + V$ of the pendulum is the same at $A_1$ and $A_2$. Whereas the energy is entirely potential at $A_1$, it becomes entirely kinetic at $A_2$, and as the pendulum keeps swinging to the right, the kinetic energy is transformed back into potential energy. At $A_3$, $T_3 = 0$ and $V_3 = Wl$.

Since the total mechanical energy of the pendulum remains constant and since its potential energy depends only upon its elevation, the kinetic energy of the pendulum will have the same value at any two points located on the same level. Thus, the speed of the pendulum is the same at $A$ and at $A'$ (Fig. 13.12). This result can be extended to the case of a particle moving along any given path, regardless of the shape of the path, as long as the only forces acting on the particle are its weight and the normal reaction of the path. The particle of Fig. 13.13, for example, which slides in a vertical plane along a frictionless track, will have the same speed at $A$, $A'$, and $A''$.

While the weight of a particle and the force exerted by a spring are conservative forces, *friction forces are nonconservative forces*. In other words, *the work of a friction force cannot be expressed as a change in potential energy*. The work of a friction force depends upon the path followed by its point of application; and while the work $U_{1\to2}$ defined by (13.19) is positive or negative according to the sense of motion, *the work of a friction force*, as we noted in Sec. 13.4, *is always negative*. It follows that when a mechanical system involves friction, its total mechanical energy does not remain constant but decreases. The energy of the system, however, is not lost; it is transformed into heat, and the sum of the *mechanical energy* and of the *thermal energy* of the system remains constant.

Other forms of energy can also be involved in a system. For instance, a generator converts mechanical energy into *electric energy;* a gasoline engine converts *chemical energy* into mechanical energy; a nuclear reactor converts *mass* into thermal energy. If all forms of energy are considered, the energy of any system can be considered as constant and the principle of conservation of energy remains valid under all conditions.

**Fig. 13.12**

**Fig. 13.13**

**Fig. 13.14**

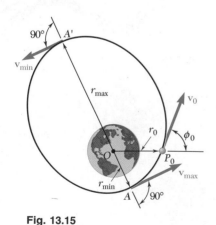

**Fig. 13.15**

## 13.9. MOTION UNDER A CONSERVATIVE CENTRAL FORCE. APPLICATION TO SPACE MECHANICS

We saw in Sec. 12.9 that when a particle $P$ moves under a central force $\mathbf{F}$, the angular momentum $\mathbf{H}_O$ of the particle about the center of force $O$ is constant. If the force $\mathbf{F}$ is also conservative, there exists a potential energy $V$ associated with $\mathbf{F}$, and the total energy $E = T + V$ of the particle is constant (Sec. 13.8). Thus, when a particle moves under a conservative central force, both the principle of conservation of angular momentum and the principle of conservation of energy can be used to study its motion.

Consider, for example, a space vehicle of mass $m$ moving under the earth's gravitational force. Let us assume that it begins its free flight at point $P_0$ at a distance $r_0$ from the center of the earth, with a velocity $\mathbf{v}_0$ forming an angle $\phi_0$ with the radius vector $OP_0$ (Fig. 13.14). Let $P$ be a point of the trajectory described by the vehicle; we denote by $r$ the distance from $O$ to $P$, by $\mathbf{v}$ the velocity of the vehicle at $P$, and by $\phi$ the angle formed by $\mathbf{v}$ and the radius vector $OP$. Applying the principle of conservation of angular momentum about $O$ between $P_0$ and $P$ (Sec. 12.9), we write

$$r_0 m v_0 \sin \phi_0 = r m v \sin \phi \tag{13.25}$$

Recalling the expression (13.17) obtained for the potential energy due to a gravitational force, we apply the principle of conservation of energy between $P_0$ and $P$ and write

$$T_0 + V_0 = T + V$$

$$\tfrac{1}{2}mv_0^2 - \frac{GMm}{r_0} = \tfrac{1}{2}mv^2 - \frac{GMm}{r} \tag{13.26}$$

where $M$ is the mass of the earth.

Equation (13.26) can be solved for the magnitude $v$ of the velocity of the vehicle at $P$ when the distance $r$ from $O$ to $P$ is known; Eq. (13.25) can then be used to determine the angle $\phi$ that the velocity forms with the radius vector $OP$.

Equations (13.25) and (13.26) can also be used to determine the maximum and minimum values of $r$ in the case of a satellite launched from $P_0$ in a direction forming an angle $\phi_0$ with the vertical $OP_0$ (Fig. 13.15). The desired values of $r$ are obtained by making $\phi = 90°$ in (13.25) and eliminating $v$ between Eqs. (13.25) and (13.26).

It should be noted that the application of the principles of conservation of energy and of conservation of angular momentum leads to a more fundamental formulation of the problems of space mechanics than does the method indicated in Sec. 12.12. In all cases involving oblique launchings, it will also result in much simpler computations. And while the method of Sec. 12.12 must be used when the actual trajectory or the periodic time of a space vehicle is to be determined, the calculations will be simplified if the conservation principles are first used to compute the maximum and minimum values of the radius vector $r$.

## SAMPLE PROBLEM 13.6

A 9 kg collar slides without friction along a vertical rod as shown. The spring attached to the collar has an undeformed length of 10 cm and a constant of 525 N/m. If the collar is released from rest in position *1*, determine its velocity after it has moved 15 cm to position *2*.

## SOLUTION

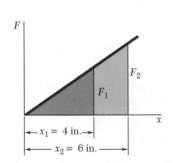

**Position *1*.** **Potential Energy.** The elongation of the spring is

$$x_1 = 20 \text{ cm} - 10 \text{ cm} = 10 \text{ cm}.$$

and we have

$$V_e = \tfrac{1}{2}kx_1^2 = \tfrac{1}{2}(525 \text{ N/m})(.1 \text{ m})^2 = 2.625 \text{ N} \cdot \text{m}$$

Choosing the datum as shown, we have $V_g = 0$. Therefore,

$$V_1 = V_e + V_g = 2.625 \text{ N} \cdot \text{m}$$

**Kinetic Energy.** Since the velocity in position *1* is zero, $T_1 = 0$.

**Position *2*.** **Potential Energy.** The elongation of the spring is

$$x_2 = 25 \text{ cm} - 10 \text{ cm} = 15 \text{ cm}.$$

and we have

$$V_e = \tfrac{1}{2}kx_2^2 = \tfrac{1}{2}(525 \text{ N/m})(.15 \text{ m})^2 = 5.9 \text{ N} \cdot \text{m}$$
$$V_g = Wy = (88.3 \text{ N})(- .15 \text{ m}) = - 13.25 \text{ N} \cdot \text{m}$$

Therefore,

$$V_2 = V_e + V_g = 5.9 - 13.25 = -7.35 \text{ N} \cdot \text{m}$$

**Kinetic Energy**

$$T_2 = \tfrac{1}{2}mv_2^2 = \frac{1}{2} \ (9 \text{ kg})v_2^2 = 4.5 \ v_2^2$$

**Conservation of Energy.** Applying the principle of conservation of energy between positions *1* and *2*, we write

$$T_1 + V_1 = T_2 + V_2$$
$$0 + 2.625 \text{ N} \cdot \text{m} = 4.5 \ v_2^2 - 7.35 \text{ N} \cdot \text{m}$$
$$v_2 = \pm \ 1.5 \text{ m/s}$$

$$\mathbf{v}_2 = 1.5 \text{ m/s} \downarrow \ \blacktriangleleft$$

## SAMPLE PROBLEM 13.7

The 0.23 kg pellet is pushed against the spring at $A$ and released from rest. Neglecting friction, determine the smallest deflection of the spring for which the pellet will travel around the loop $ABCDE$ and remain at all times in contact with the loop.

## SOLUTION

**Required Speed at Point $D$.** As the pellet passes through the highest point $D$, its potential energy with respect to gravity is maximum and thus, its kinetic energy and speed are minimum. Since the pellet must remain in contact with the loop, the force **N** exerted on the pellet by the loop must be equal to or greater than zero. Setting **N** = 0, we compute the smallest possible speed $v_D$.

$$+\downarrow \Sigma F_n = ma_n: \qquad W = ma_n \qquad mg = ma_n \qquad a_n = g$$

$$a_n = \frac{v_D^2}{r}: \qquad v_D^2 = ra_n = rg = (.6 \text{ m})(9.81 \text{ m/s}^2) = 5.886 \text{ ft}^2/\text{s}^2$$

**Position 1. Potential Energy.** Denoting by $x$ the deflection of the spring and noting that $k = 525$ N/m = 525 N/m, we write

$$V_e = \tfrac{1}{2}kx^2 = \tfrac{1}{2}(525 \text{ N/m})x^2 = 262.5 \ x^2$$

Choosing the datum at $A$, we have $V_g = 0$; therefore

$$V_1 = V_e + V_g = 262.5 \ x^2$$

**Kinetic Energy.** Since the pellet is released from rest, $v_A = 0$ and we have $T_1 = 0$.

**Position 2. Potential Energy.** The spring is now undeformed; thus $V_e = 0$. Since the pellet is 1.2 m above the datum, we have

$$V_g = Wy = (2.26 \text{ N})(1.2 \text{ m}) = 2.7 \text{ N} \cdot \text{m}$$
$$V_2 = V_e + V_g = 2.7 \text{ N} \cdot \text{m}$$

**Kinetic Energy.** Using the value of $v_D^2$ obtained above, we write

$$T_2 = \tfrac{1}{2}mv_D^2 = \frac{1}{2} \times (0.23 \text{ kg}) \times (5.886 \text{ ft}^2/\text{s}^2) = 0.68 \text{ N} \cdot \text{m}$$

**Conservation of Energy.** Applying the principle of conservation of energy between positions 1 and 2, we write

$$T_1 + V_1 = T_2 + V_2$$
$$0 + 262.5 \ x^2 = 0.68 + 2.7$$
$$x = .1135 \text{ m} \qquad\qquad x = 0.1135 \text{ m} \blacktriangleleft$$

## SAMPLE PROBLEM 13.8

A sphere of mass $m = 0.6$ kg is attached to an elastic cord of constant $k = 100$ N/m, which is undeformed when the sphere is located at the origin $O$. Knowing that the sphere may slide without friction on the horizontal surface and that in the position shown its velocity $\mathbf{v}_A$ has a magnitude of 20 m/s, determine (a) the maximum and minimum distances from the sphere to the origin $O$, (b) the corresponding values of its speed.

## SOLUTION

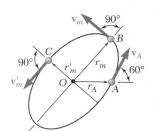

The force exerted by the cord on the sphere passes through the fixed point $O$, and its work can be expressed as a change in potential energy. It is therefore a conservative central force, and both the total energy of the sphere and its angular momentum about $O$ are conserved.

**Conservation of Angular Momentum about $O$.** At point $B$, where the distance from $O$ is maximum, the velocity of the sphere is perpendicular to $OB$ and the angular momentum is $r_m m v_m$. A similar property holds at point $C$, where the distance from $O$ is minimum. Expressing conservation of angular momentum between $A$ and $B$, we write

$$r_A m v_A \sin 60° = r_m m v_m$$
$$(0.5 \text{ m})(0.6 \text{ kg})(20 \text{ m/s}) \sin 60° = r_m(0.6 \text{ kg})v_m$$
$$v_m = \frac{8.66}{r_m} \qquad (1)$$

**Conservation of Energy**

At point $A$: $\quad T_A = \frac{1}{2}mv_A^2 = \frac{1}{2}(0.6 \text{ kg})(20 \text{ m/s})^2 = 120 \text{ J}$
$\qquad\qquad V_A = \frac{1}{2}kr_A^2 = \frac{1}{2}(100 \text{ N/m})(0.5 \text{ m})^2 = 12.5 \text{ J}$

At point $B$: $\quad T_B = \frac{1}{2}mv_m^2 = \frac{1}{2}(0.6 \text{ kg})v_m^2 = 0.3v_m^2$
$\qquad\qquad V_B = \frac{1}{2}kr_m^2 = \frac{1}{2}(100 \text{ N/m})r_m^2 = 50r_m^2$

Applying the principle of conservation of energy between points $A$ and $B$, we write

$$T_A + V_A = T_B + V_B$$
$$120 + 12.5 = 0.3v_m^2 + 50r_m^2 \qquad (2)$$

**a. Maximum and Minimum Values of Distance.** Substituting for $v_m$ from Eq. (1) into Eq. (2) and solving for $r_m^2$, we obtain

$$r_m^2 = 2.468 \text{ or } 0.1824 \qquad r_m = 1.571 \text{ m}, \; r_m' = 0.427 \text{ m} \blacktriangleleft$$

**b. Corresponding Values of Speed.** Substituting the values obtained for $r_m$ and $r_m'$ into Eq. (1), we have

$$v_m = \frac{8.66}{1.571} \qquad\qquad v_m = 5.51 \text{ m/s} \blacktriangleleft$$

$$v_m' = \frac{8.66}{0.427} \qquad\qquad v_m' = 20.3 \text{ m/s} \blacktriangleleft$$

*Note.* It can be shown that the path of the sphere is an ellipse of center $O$.

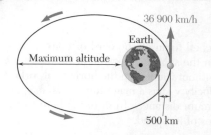

# SAMPLE PROBLEM 13.9

A satellite is launched in a direction parallel to the surface of the earth with a velocity of 36 900 km/h from an altitude of 500 km. Determine (a) the maximum altitude reached by the satellite, (b) the maximum allowable error in the direction of launching if the satellite is to go into orbit and come no closer than 200 km to the surface of the earth.

## SOLUTION

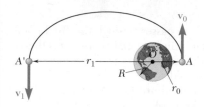

**a. Maximum Altitude.** We denote by $A'$ the point of the orbit farthest from the earth and by $r_1$ the corresponding distance from the center of the earth. Since the satellite is in free flight between $A$ and $A'$, we apply the principle of conservation of energy:

$$T_A + V_A = T_{A'} + V_{A'}$$

$$\tfrac{1}{2}mv_0^2 - \frac{GMm}{r_0} = \tfrac{1}{2}mv_1^2 - \frac{GMm}{r_1} \tag{1}$$

Since the only force acting on the satellite is the force of gravity, which is a central force, the angular momentum of the satellite about $O$ is conserved. Considering points $A$ and $A'$, we write

$$r_0 m v_0 = r_1 m v_1 \qquad v_1 = v_0 \frac{r_0}{r_1} \tag{2}$$

Substituting this expression for $v_1$ into Eq. (1), dividing each term by the mass $m$, and rearranging the terms, we obtain

$$\tfrac{1}{2}v_0^2 \left(1 - \frac{r_0^2}{r_1^2}\right) = \frac{GM}{r_0}\left(1 - \frac{r_0}{r_1}\right) \qquad 1 + \frac{r_0}{r_1} = \frac{2GM}{r_0 v_0^2} \tag{3}$$

Recalling that the radius of the earth is $R = 6370$ km, we compute

$$r_0 = 6370 \text{ km} + 500 \text{ km} = 6870 \text{ km} = 6.87 \times 10^6 \text{ m}$$
$$v_0 = 36\ 900 \text{ km/h} = (36.9 \times 10^6 \text{ m})/(3.6 \times 10^3 \text{ s}) = 10.25 \times 10^3 \text{ m/s}$$
$$GM = gR^2 = (9.81 \text{ m/s}^2)(6.37 \times 10^6 \text{ m})^2 = 398 \times 10^{12} \text{ m}^3/\text{s}^2$$

Substituting these values into (3), we obtain $r_1 = 66.8 \times 10^6$ m.

Maximum altitude $= 66.8 \times 10^6$ m $- 6.37 \times 10^6$ m $= 60.4 \times 10^6$ m $=$ 60 400 km ◄

**b. Allowable Error in Direction of Launching.** The satellite is launched from $P_0$ in a direction forming an angle $\phi_0$ with the vertical $OP_0$. The value of $\phi_0$ corresponding to $r_{min} = 6370$ km $+ 200$ km $= 6570$ km is obtained by applying the principles of conservation of energy and of conservation of angular momentum between $P_0$ and $A$:

$$\tfrac{1}{2}mv_0^2 - \frac{GMm}{r_0} = \tfrac{1}{2}mv_{max}^2 - \frac{GMm}{r_{min}} \tag{4}$$

$$r_0 m v_0 \sin \phi_0 = r_{min} m v_{max} \tag{5}$$

Solving (5) for $v_{max}$ and then substituting for $v_{max}$ into (4), we can solve (4) for $\sin \phi_0$. Using the values of $v_0$ and $GM$ computed in part $a$ and noting that $r_0/r_{min} = 6870/6570 = 1.0457$, we find

$$\sin \phi_0 = 0.9801 \qquad \phi_0 = 90° \pm 11.5° \qquad \text{Allowable error} = \pm 11.5° \blacktriangleleft$$

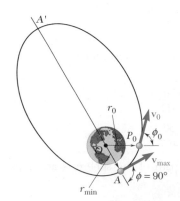

In this lesson you learned that when the work done by a force $\mathbf{F}$ acting on a particle $A$ *is independent of the path followed by the particle* as it moves from a given position $A_1$ to a given position $A_2$ (Fig. 13.11*a*), then a function $V$, called *potential energy,* can be defined for the force $\mathbf{F}$. Such forces are said to be *conservative forces,* and you can write

$$U_{1\to2} = V(x_1, y_1, z_1) - V(x_2, y_2, z_2) \qquad (13.19)$$

or, for short,

$$U_{1\to2} = V_1 - V_2 \qquad (13.19')$$

Note that the work is negative when the change in the potential energy is positive, that is, when $V_2 > V_1$.

Substituting the above expression into the equation for work and energy, you can write

$$T_1 + V_1 = T_2 + V_2 \qquad (13.24)$$

which shows that when a particle moves under the action of a conservative force *the sum of the kinetic and potential energies of the particle remains constant.*

Your solution of problems using the above formula will consist of the following steps.

**1. Determine whether all the forces involved are conservative.** If some of the forces are not conservative, for example if friction is involved, you must use the method of work and energy from the previous lesson, since the work done by such forces depends upon the path followed by the particle and a potential function does not exist. If there is no friction and if all the forces are conservative, you can proceed as follows.

**2. Determine the kinetic energy $T = \frac{1}{2}mv^2$ at each end of the path.**

*(continued)*

3. *Compute the potential energy for all the forces involved at each end of the path.* You will recall that the following expressions for the potential energy were derived in this lesson.

   *a. The potential energy of a weight W* close to the surface of the earth and at a height $y$ above a given datum,

$$V_g = Wy \tag{13.16}$$

   *b. The potential energy of a mass m located at a distance r from the center of the earth,* large enough so that the variation of the force of gravity must be taken into account,

$$V_g = -\frac{GMm}{r} \tag{13.17}$$

where the distance $r$ is measured from the center of the earth and $V_g$ is equal to zero at $r = \infty$.

   *c. The potential energy of a body with respect to an elastic force F = kx,*

$$V_e = \tfrac{1}{2}kx^2 \tag{13.18}$$

where the distance $x$ is the deflection of the elastic spring measured from its *undeformed* position and $k$ is the spring constant. Note that $V_e$ *depends only upon the deflection x* and not upon the path of the body attached to the spring. Also, $V_e$ is always positive, whether the spring is compressed or elongated.

4. *Substitute your expressions for the kinetic and potential energies* into Eq. (13.24) above. You will be able to solve this equation for one unknown, for example, for a velocity [Sample Prob. 13.6]. If more than one unknown is involved, you will have to search for another condition or equation, such as the minimum speed [Sample Prob. 13.7] or the minimum potential energy of the particle. For problems involving a central force, a second equation can be obtained by using conservation of angular momentum [Sample Prob. 13.8]. This is especially useful in applications to space mechanics [Sec. 13.9].

# Problems

**13.55** A force **P** is slowly applied to a plate that is attached to two springs and causes a deflection $x_0$. In each of the two cases shown, derive an expression for the constant $k_e$, in terms of $k_1$ and $k_2$, of the single spring equivalent to the given system, that is, of the single spring which will undergo the same deflection $x_0$ when subjected to the same force **P**.

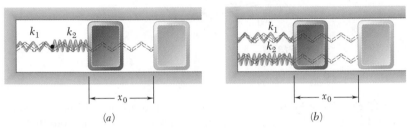

(a)              (b)

**Fig. P13.55**

**13.56** A collar $C$ of weight $m$ slides without friction on a horizontal rod between springs $A$ and $B$. If the collar is pushed to the left until spring $A$ is compressed 0.1 m and released, determine the distance through which the collar will travel and the maximum velocity it will reach (a) if $m = 1$ kg, (b) if $m = 2.5$ kg.

**Fig. P13.56**

**13.57** A 1.8 kg collar can slide without friction along a horizontal rod and is in equilibrium at $A$ when it is pushed 2.5 cm to the right and released from rest. The springs are undeformed when the collar is at $A$ and the constant of each spring is 490 kN/m. Determine the maximum velocity of the collar.

**Fig. P13.57 and P13.58**

**13.58** A 1.8 kg collar can slide without friction along a horizontal rod and is released from rest at $A$. The undeformed lengths of springs $BA$ and $CA$ are 25 cm and 23 cm, respectively, and the constant of each spring is 490 kN/m. Determine the velocity of the collar when it has moved 2.5 cm to the right.

**13.59** A 750-g collar can slide along the horizontal rod shown. It is attached to an elastic cord with an undeformed length of 300 mm and a spring constant of 150 N/m. Knowing that the collar is released from rest at $A$ and neglecting friction, determine the speed of the collar (a) at $B$, (b) at $E$.

**Fig. P13.59**

**13.60** Solve Prob. 13.59, assuming that the elastic cord has an undeformed length of 450 mm and a spring constant of 150 N/m. Note that the cord becomes slack during part of the motion.

**13.61** A 2.7 kg collar can slide without friction on a vertical rod and is resting in equilibrium on a spring. It is pushed down, compressing the spring 15 cm, and released. Knowing that the spring constant is $k = 2627$ N/m, determine (a) the maximum height $h$ reached by the collar above its equilibrium position, (b) the maximum velocity of the collar.

**Fig. P13.61 and P13.62**

**13.62** A 2.7 kg collar can slide without friction on a vertical rod and is held so it just touches an undeformed spring. Determine the maximum deflection of the spring (a) if the collar is slowly released until it reaches an equilibrium position, (b) if the collar is suddenly released.

**13.63** A thin circular rod is supported in a *vertical plane* by a bracket at A. Attached to the bracket and loosely wound around the rod is a spring of constant $k = 40$ N/m and undeformed length equal to the arc of circle AB. A 200-g collar C, not attached to the spring, can slide without friction along the rod. Knowing that the collar is released from rest when $\theta = 30°$, determine (a) the maximum height above point B reached by the collar, (b) the maximum velocity of the collar.

**13.64** A thin circular rod is supported in a *vertical plane* by a bracket at A. Attached to the bracket and loosely wound around the rod is a spring of constant $k = 40$ N/m and undeformed length equal to the arc of circle AB. A 200-g collar C, not attached to the spring, can slide without friction along the rod. Knowing that the collar is released from rest at an angle $\theta$ with respect to the vertical, determine (a) the smallest value of $\theta$ for which the collar will pass through D and reach point A, (b) the velocity of the collar as it reaches point A.

**Fig. P13.63 and P13.64**

**13.65** A spring is used to stop a 90.7 kg package which is moving down a 20° incline. The spring has a constant $k = 22$ kN/m and is held by cables so that it is initially compressed 15 cm. Knowing that the velocity of the package is 2.4 m/s when it is 7.6 m from the spring and neglecting friction, determine the maximum additional deformation of the spring in bringing the package to rest.

**Fig. P13.65**

**13.66** A 4.5 kg collar is attached to a spring and slides without friction along a fixed rod in a vertical plane. The spring has an undeformed length of 35.6 cm and a constant $k = 700$ N/m. Knowing that the collar is released from rest in the position shown, determine the speed of the collar at (a) point A, (b) point B.

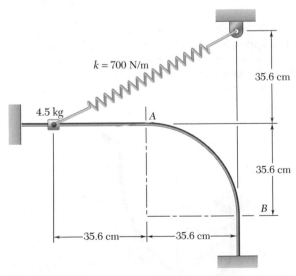

**Fig. P13.66**

**13.67** Blocks A and B have masses of 4 kg and 1.5 kg, respectively, and are connected by a cord-and-pulley system and released from rest in the position shown with the spring undeformed. Knowing that the constant of the spring is 300 N/m, determine (a) the velocity of block B after it has moved 150 mm, (b) the maximum velocity of block B, (c) the maximum displacement of block B. Ignore friction and the masses of the pulleys and spring.

**Fig. P13.67**

**13.68**  A 4-kg collar $A$ can slide without friction along a vertical rod and is released from rest in the position shown with the springs undeformed. Knowing that the constant of each spring is 300 N/m, determine the velocity of the collar after it has moved ($a$) 100 mm, ($b$) 190 mm.

**Fig. P13.68**

**13.69**  A thin circular rod is supported in a *vertical plane* by a bracket at $A$. Attached to the bracket and loosely wound around the rod is a spring of constant $k = 40$ N/m and undeformed length equal to the arc of circle $AB$. A 200-g collar $C$ is unattached to the spring and can slide without friction along the rod. Knowing that the collar is released from rest when $\theta = 30°$, determine ($a$) the velocity of the collar as it passes through point $B$, ($b$) the force exerted by the rod on the collar as it passes through $B$.

**13.70**  A 500-g collar can slide without friction along the semicircular rod $BCD$. The spring is of constant 320 N/m and its undeformed length is 200 mm. Knowing that the collar is released from rest at $B$, determine ($a$) the speed of the collar as it passes through $C$, ($b$) the force exerted by the rod on the collar at $C$.

**Fig. P13.69**

**Fig. P13.70**

**13.71**   A 1.1 kg collar is attached to a spring and slides without friction along a circular rod in a *vertical plane*. The spring has an undeformed length of 10 cm and a constant $k = 292$ N/m. Knowing that the collar is at rest at $C$ and is given a slight push to get it moving, determine the velocity of the collar and the force exerted by the rod on the collar as it passes through (*a*) point $A$, (*b*) point $B$.

**13.72**   A 2.5-kg collar is attached to a spring and slides without friction along a circular rod in a *vertical plane*. The spring has an undeformed length of 10 cm. and a constant $k$. The collar is at rest at $C$ and is given a slight push to get it moving. Knowing that the maximum velocity of the collar is achieved as it passes through point $A$, determine (*a*) the spring constant $k$, (*b*) the maximum velocity of the collar.

**13.73**   An .24 kg package is projected upward with a velocity $\mathbf{v}_0$ by a spring at $A$; it moves around a frictionless loop and is deposited at $C$. For each of the two loops shown, determine (*a*) the smallest velocity $\mathbf{v}_0$ for which the package will reach $C$, (*b*) the corresponding force exerted by the package on the loop just before the package leaves the loop at $C$.

**Fig. P13.71 and P13.72**

**Fig. P13.73**

**13.74**   If the package of Prob. 13.73 is not to hit the horizontal surface at $C$ with a speed greater than 3 m/s, (*a*) show that this requirement can be satisfied only by the second loop, (*b*) determine the largest allowable initial velocity $\mathbf{v}_0$ when the second loop is used.

**13.75**   The pendulum shown is released from rest at $A$ and swings through 90° before the cord touches the fixed peg $B$. Determine the smallest value of $a$ for which the pendulum bob will describe a circle about the peg.

**Fig. P13.75**

**Fig. P13.76**

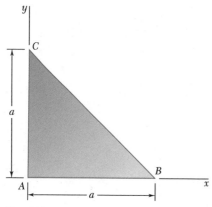

**Fig. P13.80**

**13.76** A 70-kg skier starts from rest at point $A$ and skis down the slope shown. Neglecting friction, determine the force exerted on the skier by the snow (a) just before the skier reaches $B$, (b) immediately after he has passed through $B$. (c) Does the skier become airborne before reaching $C$? Neglect the height of the skier compared to the radii of curvature.

**13.77** Prove that a force $F(x, y, z)$ is conservative if, and only if, the following relations are satisfied:

$$\frac{\partial F_x}{\partial y} = \frac{\partial F_y}{\partial x} \qquad \frac{\partial F_y}{\partial z} = \frac{\partial F_z}{\partial y} \qquad \frac{\partial F_z}{\partial x} = \frac{\partial F_x}{\partial z}$$

**13.78** The force $\mathbf{F} = (yz\mathbf{i} + zx\mathbf{j} + xy\mathbf{k})/xyz$ acts on the particle $P(x, y, z)$ which moves in space. (a) Using the relation derived in Prob. 13.77, show that this force is a conservative force. (b) Determine the potential function associated with $\mathbf{F}$.

**13.79** The force $\mathbf{F} = (x\mathbf{i} + y\mathbf{j} + z\mathbf{k})/(x^2 + y^2 + z^2)^{3/2}$ acts on the particle $P(x, y, z)$ which moves in space. (a) Using the relation derived in Prob. 13.77, prove that $\mathbf{F}$ is a conservative force. (b) Determine the potential function $V(x, y, z)$ associated with $\mathbf{F}$.

**\*13.80** A force $\mathbf{F}$ acts on a particle $P(x, y)$ which moves in the $xy$ plane. Determine whether $\mathbf{F}$ is a conservative force and compute the work of $\mathbf{F}$ when $P$ describes the path $ABCA$ knowing that (a) $\mathbf{F} = (kx + y)\mathbf{i} + (kx + y)\mathbf{j}$, (b) $\mathbf{F} = (kx + y)\mathbf{i} + (x + ky)\mathbf{j}$.

**13.81** Certain springs are characterized by increasing stiffness with increasing deformation according to the relation $F = k_1 x + k_2 x^3$, where $F$ is the force exerted by the spring, $k_1$ and $k_2$ are positive constants, and $x$ is the deflection of the spring measured from its undeformed position. Determine (a) the potential energy $V_e$ as a function of $x$, (b) the maximum velocity of a particle of mass $m$ attached to the spring and released from rest with $x = x_0$. Neglect friction.

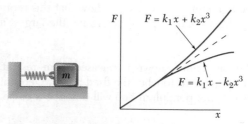

**Fig. P13.81 and P13.82**

**13.82** Certain springs are characterized by decreasing stiffness with increasing deformation according to the relation $F = k_1 x - k_2 x^3$, where $F$ is the force exerted by the spring, $k_1$ and $k_2$ are positive constants, and $x$ is the deflection of the spring measured from its undeformed position. Determine (a) the potential energy $V_e$ as a function of $x$, (b) the maximum velocity of a particle of mass $m$ attached to the spring and released from rest with $x = x_0$. Neglect friction.

**13.83** A spacecraft is describing a circular orbit at an altitude of 1490 km above the surface of the earth. As it passes through point $A$, its speed is reduced by 40 percent and it enters an elliptic crash trajectory with the apogee at point $A$. Neglecting air resistance, determine the speed of the spacecraft when it reaches the earth's surface at point $B$.

**13.84** A satellite describes an elliptic orbit of minimum altitude 602.5 km above the surface of the earth. The semimajor and semiminor axes are 17369 km and 13892 km, respectively. Knowing that the speed of the satellite at point $C$ is 4.8 km/s, determine (a) the speed at point $A$, the perigee, (b) the speed at point $B$, the apogee.

**Fig. P13.83**

**Fig. P13.84**

**13.85** Knowing that the velocity of an experimental space probe fired from the earth has a magnitude $v_A = 32.5$ Mm/h at point $A$, determine the velocity of the probe as it passes through point $B$.

**13.86** A lunar excursion module (LEM) was used in the Apollo moon-landing missions to save fuel by making it unnecessary to launch the entire Apollo spacecraft from the moon's surface on its return trip to earth. Check the effectiveness of this approach by computing the energy per kilogram required for a spacecraft to escape the moon's gravitational field if the spacecraft starts from (a) the moon's surface, (b) a circular orbit 80 km above the moon's surface. Neglect the effect of the earth's gravitational field. (The radius of the moon is 1740 km and its mass is 0.0123 times the mass of the earth.)

**13.87** A satellite is placed into an elliptic orbit about the earth. Knowing that the ratio $v_A/v_P$ of the velocity at the apogee $A$ to the velocity at the perigee $P$ is equal to the ratio $r_P/r_A$ of the distance to the center of the earth at $P$ to that at $A$, and that the distance between $A$ and $P$ is 80,115 km, determine the energy per unit weight required to place the satellite in its orbit by launching it from the surface of the earth. Exclude the additional energy needed to overcome the weight of the booster rocket, air resistance, and maneuvering.

**13.88** While describing a circular orbit 320.5 km above the earth a space vehicle launches a 2722 kg communications satellite. Determine (a) the additional energy required to place the satellite in a geosynchronous orbit at an altitude of 35250.8 km above the surface of the earth, (b) the energy required to place the satellite in the same orbit by launching it from the surface of the earth, excluding the energy needed to overcome air resistance. (A *geosynchronous orbit* is a circular orbit in which the satellite appears stationary with respect to the ground.)

**Fig. P13.85**

**Fig. P13.87**

**13.89** A space shuttle ejects two communication satellites while describing a circular orbit, 296 km above the surface of the earth. Knowing that one of these satellites had a mass of 3600 kg, determine (a) the additional energy required to place the satellite in a geosynchronous orbit (see Prob. 12.81) at an altitude of 35 780 km above the surface of the earth, (b) the energy required to place it in the same orbit by launching it from the surface of the earth.

**13.90** A satellite of mass $m$ describes a circular orbit of radius $r$ about the earth. Express as a function of $r$ (a) the potential energy of the satellite, (b) its kinetic energy, (c) its total energy. Denote the radius of the earth by $R$ and the acceleration of gravity at the surface of the earth by $g$, and assume that the potential energy of the satellite is zero on its launching pad.

**13.91** The orbit of the planet Venus is nearly circular with an orbital velocity of $125.5 \times 10^3$ km/h. Knowing that the mean distance from the center of the sun to the center of Venus is $107.7 \times 10^6$ km and that the mass of the sun is $407 \times 10^3$ times the mass of Venus, determine (a) the mass of the sun, (b) the total energy of Venus.

**13.92** (a) Show that, by setting $r = R + y$ in the right-hand member of Eq. (13.17′) and expanding that member in a power series in $y/R$, the expression in Eq. (13.16) for the potential energy $V_g$ due to gravity is a first-order approximation for the expression given in Eq. (13.17′). (b) Using the same expansion, derive a second-order approximation for $V_g$.

**13.93** Collar $B$ has a mass of 4 kg and is attached to a spring of constant 1500 N/m and of undeformed length 0.4 m. The system is set in motion with $r = 0.2$ m, $v_\theta = 6$ m/s, and $v_r = 0$. Neglecting the mass of the rod and the effect of friction, determine the radial and transverse components of the velocity of the collar when $r = 0.5$ m.

**13.94** For the motion described in Prob. 13.93, determine (a) the maximum distance between the origin and the collar, (b) the corresponding velocity. (*Hint.* Solve by trial and error the equation obtained for $r$.)

**13.95** Two identical 1.8 kg collars, $A$ and $B$, are attached to a spring of constant 102.2 N/m and can slide on a horizontal rod which is free to rotate about a vertical shaft. Collar $B$ is initially prevented from sliding by a stop as the rod rotates at a constant rate $\dot{\theta}_0 = 5$ rad/s and the spring is in compression with $r_A = .9$ m and $r_B = 2.3$ m. After the stop is removed both collars move out along the rod. At the instant when $r_B = 2.7$ m, determine (a) $r_A$, (b) $\dot{\theta}$, (c) the total kinetic energy. Neglect friction and the mass of the rod.

**13.96** Two identical 1.8 kg collars, $A$ and $B$, are attached to a spring of constant 102.2 N/m and can slide on a horizontal rod which is free to rotate about a vertical shaft. Collar $B$ is initially prevented from sliding by a stop as the rod rotates at a constant rate $\dot{\theta}_0 = 5$ rad/s and the spring is in compression with $r_A = .9$ m and $r_B = 2.3$ m. After the stop is removed both collars move along the rod. At the instant when the spring is in compression and the total kinetic energy is 135.6 N · m, determine (a) $r_A$, (b) $r_B$, (c) $\dot{\theta}$. Neglect friction and the mass of the rod.

**Fig. P13.93**

**Fig. P13.95 and P13.96**

**13.97** Using the principles of conservation of energy and conservation of angular momentum, solve part *a* of Sample Prob. 13.9.

**13.98** A 0.7-kg ball that can slide on a *horizontal* frictionless surface is attached to a fixed point *O* by means of an elastic cord of constant $k =$ 150 N/m and undeformed length 600 mm. The ball is placed at point *A*, 800 mm from *O*, and given an initial velocity $\mathbf{v}_0$ perpendicular to *OA*. Determine (*a*) the smallest allowable value of the initial speed $v_0$ if the cord is not to become slack, (*b*) the closest distance *d* that the ball will come to point *O* if it is given half the initial speed found in part *a*.

**Fig. *P13.98* and *P13.99***

**13.99** A 0.7-kg ball that can slide on a *horizontal* frictionless surface is attached to a fixed point *O* by means of an elastic cord of constant $k =$ 150 N/m and undeformed length 600 mm. The ball is placed at point *A*, 800 mm from *O*, and given an initial velocity $\mathbf{v}_0$ perpendicular to *OA*, allowing the ball to come within a distance $d = 270$ mm of point *O* after the cord has become slack. Determine (*a*) the initial speed $v_0$ of the ball, (*b*) its maximum speed.

**13.100** After completing their moon-exploration mission, the two astronauts forming the crew of an *Apollo* lunar excursion module (LEM) would prepare to rejoin the command module which was orbiting the moon at an altitude of 140 km. They would fire the LEM's engine, bring it along a curved path to a point *A*, 8 km above the moon's surface, and shut off the engine. Knowing that the LEM was moving at that time in a direction parallel to the moon's surface and that it then coasted along an elliptic path to a rendezvous at *B* with the command module, determine (*a*) the speed of the LEM at engine shutoff, (*b*) the relative velocity with which the command module approached the LEM at *B*. (The radius of the moon is 1740 km and its mass is 0.01230 times the mass of the earth.)

**Fig. P13.100**

**13.101** While describing a circular orbit, 264 km above the surface of the earth, a space shuttle ejects at point *A* an inertial upper stage (IUS) carrying a communication satellite to be placed in a geosynchronous orbit (see Prob. 12.81) at an altitude of 35 780 km above the surface of the earth. Determine (*a*) the velocity of the IUS relative to the shuttle after its engine has been fired at *A*, (*b*) the increase in velocity required at *B* to place the satellite in its final orbit.

**Fig. P13.101**

**13.102** As a first approximation to the analysis of a space flight from the earth to Mars, it is assumed that the orbits of the earth and Mars are circular and coplanar. The mean distances from the sun to the earth and to Mars are $149 \times 10^6$ km and $226.7 \times 10^6$ km, respectively. To place the spacecraft into an elliptical transfer orbit at point $A$, its speed is increased over a short interval of time to $v_A$ which is faster than the earth's orbital speed. When the spacecraft reaches point $B$ on the elliptical transfer orbit, its speed $v_B$ is increased to the orbital speed of Mars. Knowing that the mass of the sun is $332.8 \times 10^3$ times the mass of the earth, determine the increase in velocity required (*a*) at $A$, (*b*) at $B$.

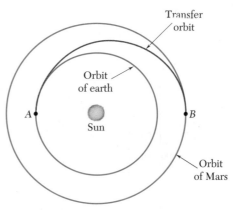

**Fig. P13.102**

**13.103** A spacecraft describing an elliptic orbit about a planet has a maximum speed $v_A = 8$ km/s at its minimum altitude $h_A = 1922.8$ km above the surface of the planet and a minimum speed $v_B = 1.9$ km/s at its maximum altitude $h_B = 26117.6$ km. Determine (*a*) the radius of the planet, (*b*) the mass of the planet.

**13.104** The optimal way of transferring a space vehicle from an inner circular orbit to an outer coplanar orbit is to fire its engines as it passes through $A$ to increase its speed and place it in an elliptic transfer orbit. Another increase in speed as it passes through $B$ will place it in the desired circular orbit. For a vehicle in a circular orbit about the earth at an altitude $h_1 = 320$ km, which is to be transferred to a circular orbit at an altitude $h_2 = 800$ km, determine (*a*) the required increase in speed at $A$ and $B$, (*b*) the total energy per unit mass required to execute the transfer.

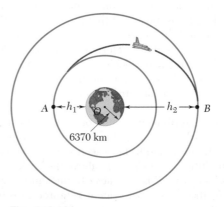

**Fig. *P13.104***

**13.105** A spacecraft approaching the planet Saturn reaches point $A$ with a velocity $\mathbf{v}_A$ of magnitude 21 km/s. It is placed in an elliptic orbit about Saturn so that it will be able to periodically examine Tethys, one of Saturn's moons. Tethys is in a circular orbit of radius $295 \times 10^3$ km about the center of Saturn, traveling at a speed of 11.3 km/s. Determine (a) the decrease in speed required by the spacecraft at $A$ to achieve the desired orbit, (b) the speed of the spacecraft when it reaches the orbit of Tethys at $B$.

**13.106** A space shuttle is to rendezvous with an orbiting laboratory which circles the earth at a constant altitude of 360 km. The shuttle has reached an altitude of 60 km when its engine is shut off, and its velocity $\mathbf{v}_0$ forms an angle $\phi_0 = 50°$ with the vertical $OB$ at that time. What magnitude should $\mathbf{v}_0$ have if the shuttle's trajectory is to be tangent at $A$ to the orbit of the laboratory?

**Fig. P13.105**

**Fig. P13.106**

**13.107** An experimental space probe is launched from the earth and enters a highly eccentric elliptic orbit as it reaches point $A$. Knowing that its velocity there is observed to be $v_A = 30\,000$ km/h at an angle of 60° with the vertical, determine the minimum altitude of the orbit.

**13.108** During a flyby of the earth, the velocity of a spacecraft is 10.4 km/s as it reaches its minimum altitude of 908.5 km above the surface at point $A$. At point $B$ the spacecraft is observed to have an altitude of 8316 km. Determine (a) the magnitude of the velocity at point $B$, (b) the angle $\phi_B$.

**Fig. P13.107**

**Fig. P13.108**

**13.109** A satellite describes an elliptic orbit of minimum altitude 604 km above the surface of the earth and semimajor axis 17369 km. Knowing that the speed of the satellite at point $C$ is 4.8 km/s, apply the principles of conservation of energy and conservation of angular momentum to determine (a) the speed at point $A$, the perigee, (b) the semiminor axis $b$.

**Fig. P13.109**

Fig. *P13.110*

Fig. *P13.111*

Fig. **P13.112**

**13.110** When the lunar excursion module (LEM) was set adrift after returning two of the Apollo astronauts to the command module, which was orbiting the moon at an altitude of 139.4 km, its speed was reduced to let it crash on the moon's surface. Determine (*a*) the smallest amount by which the speed of the LEM should have been reduced to make sure that it would crash on the moon's surface, (*b*) the amount by which its speed should have been reduced to cause it to hit the moon's surface at a 45° angle. (*Hint:* Point *A* is at the apogee of the elliptic crash trajectory. Recall also that the mass of the moon is 0.0123 times the mass of the earth.)

**13.111** A satellite is projected into space with a velocity $\mathbf{v}_0$ at a distance $r_0$ from the center of the earth by the last stage of its launching rocket. The velocity $\mathbf{v}_0$ was designed to send the satellite into a circular orbit of radius $r_0$. However, owing to a malfunction of control, the satellite is not projected horizontally but at an angle $\alpha$ with the horizontal and, as a result, is propelled into an elliptic orbit. Determine the maximum and minimum values of the distance from the center of the earth to the satellite.

**13.112** A space vehicle is in a circular orbit at an altitude of 362 km above the earth. To return to earth, it decreases its speed as it passes through *A* by firing its engine for a short interval of time in a direction opposite to the direction of its motion. Knowing that the velocity of the space vehicle should form an angle $\phi_B = 60°$ with the vertical as it reaches point *B* at an altitude of 64.4 km, determine (*a*) the required speed of the vehicle as it leaves its circular orbit at *A*, (*b*) its speed at point *B*.

**\*13.113** In Prob. 13.112, the speed of the space vehicle was decreased as it passed through *A* by firing its engine in a direction opposite to the direction of motion. An alternative strategy for taking the space vehicle out of its circular orbit would be to turn it around so that its engine would point away from the earth and then give it an incremental velocity $\Delta\mathbf{v}_A$ toward the center *O* of the earth. This would likely require a smaller expenditure of energy when firing the engine at *A*, but might result in too fast a descent at *B*. Assuming this strategy is used with only 50 percent of the energy expenditure used in Prob. 13.112, determine the resulting values of $\phi_B$ and $v_B$.

**13.114** Show that the values $v_A$ and $v_P$ of the speed of an earth satellite at the apogee *A* and perigee *P* of an elliptic orbit are defined by the relations

$$v_A^2 = \frac{2GM}{r_A + r_P}\frac{r_P}{r_A} \qquad v_P^2 = \frac{2GM}{r_A + r_P}\frac{r_A}{r_P}$$

where *M* is the mass of the earth, and $r_A$ and $r_P$ represent, respectively, the maximum and minimum distances of the orbit to the center of the earth.

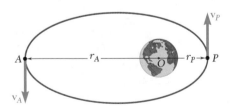

Fig. **P13.114 and P13.115**

**13.115** Show that the total energy *E* of an earth satellite of mass *m* describing an elliptic orbit is $E = -GMm/(r_A + r_P)$, where *M* is the mass of the earth, and $r_A$ and $r_P$ represent, respectively, the maximum and minimum distances of the orbit to the center of the earth. (Recall that the gravitational potential energy of a satellite was defined as being zero at an infinite distance from the earth.)

**13.116** A spacecraft of mass $m$ describes a circular orbit of radius $r_1$ around the earth. (*a*) Show that the additional energy $\Delta E$ which must be imparted to the spacecraft to transfer it to a circular orbit of larger radius $r_2$ is

$$\Delta E = \frac{GMm(r_2 - r_1)}{2r_1r_2}$$

where $M$ is the mass of the earth. (*b*) Further show that if the transfer from one circular orbit to the other circular orbit is executed by placing the spacecraft on a transitional semielliptic path $AB$, the amounts of energy $\Delta E_A$ and $\Delta E_B$ which must be imparted at $A$ and $B$ are, respectively, proportional to $r_1$ and $r_2$:

$$\Delta E_A = \frac{r_2}{r_1 + r_2}\Delta E \qquad \Delta E_B = \frac{r_1}{r_1 + r_2}\Delta E$$

**13.117** Using the answers obtained in Prob. 13.111, show that the intended circular orbit and the resulting elliptic orbit intersect at the ends of the minor axis of the elliptic orbit.

**13.118** A missile is fired from the ground with an initial velocity $\mathbf{v}_0$ forming an angle $\phi_0$ with the vertical. If the missile is to reach a maximum altitude equal to $\alpha R$, where $R$ is the radius of the earth, (*a*) show that the required angle $\phi_0$ is defined by the relation

$$\sin\phi_0 = (1 + \alpha)\sqrt{1 - \frac{\alpha}{1 + \alpha}\left(\frac{v_{esc}}{v_0}\right)^2}$$

where $v_{esc}$ is the escape velocity, (*b*) determine the range of allowable values of $v_0$.

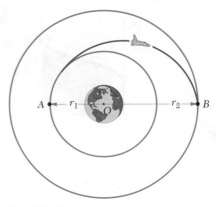

**Fig. P13.116**

## 13.10. PRINCIPLE OF IMPULSE AND MOMENTUM

A third basic method for the solution of problems dealing with the motion of particles will be considered now. This method is based on the principle of impulse and momentum and can be used to solve problems involving force, mass, velocity, and time. It is of particular interest in the solution of problems involving impulsive motion and problems involving impact (Secs. 13.11 and 13.12).

Consider a particle of mass $m$ acted upon by a force $\mathbf{F}$. As we saw in Sec. 12.3, Newton's second law can be expressed in the form

$$\mathbf{F} = \frac{d}{dt}(m\mathbf{v}) \tag{13.27}$$

where $m\mathbf{v}$ is the linear momentum of the particle. Multiplying both sides of Eq. (13.27) by $dt$ and integrating from a time $t_1$ to a time $t_2$, we write

$$\mathbf{F}\,dt = d(m\mathbf{v})$$
$$\int_{t_1}^{t_2}\mathbf{F}\,dt = m\mathbf{v}_2 - m\mathbf{v}_1$$

or, transposing the last term,

$$m\mathbf{v}_1 + \int_{t_1}^{t_2}\mathbf{F}\,dt = m\mathbf{v}_2 \tag{13.28}$$

$F_x$

$O \quad t_1 \quad t_2 \quad t$

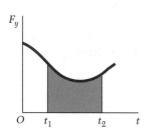

$F_y$

$O \quad t_1 \quad t_2 \quad t$

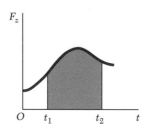

$F_z$

$O \quad t_1 \quad t_2 \quad t$

**Fig. 13.16**

The integral in Eq. (13.28) is a vector known as the *linear impulse,* or simply the *impulse,* of the force **F** during the interval of time considered. Resolving **F** into rectangular components, we write

$$\mathbf{Imp}_{1\to2} = \int_{t_1}^{t_2} \mathbf{F}\, dt$$

$$= \mathbf{i} \int_{t_1}^{t_2} F_x\, dt + \mathbf{j} \int_{t_1}^{t_2} F_y\, dt + \mathbf{k} \int_{t_1}^{t_2} F_z\, dt \quad (13.29)$$

and note that the components of the impulse of the force **F** are, respectively, equal to the areas under the curves obtained by plotting the components $F_x$, $F_y$, and $F_z$ against $t$ (Fig. 13.16). In the case of a force **F** of constant magnitude and direction, the impulse is represented by the vector $\mathbf{F}(t_2 - t_1)$, which has the same direction as **F**.

If SI units are used, the magnitude of the impulse of a force is expressed in N · s. But, recalling the definition of the newton, we have

$$\text{N} \cdot \text{s} = (\text{kg} \cdot \text{m/s}^2) \cdot \text{s} = \text{kg} \cdot \text{m/s}$$

which is the unit obtained in Sec. 12.4 for the linear momentum of a particle. We thus check that Eq. (13.28) is dimensionally correct. If U.S. customary units are used, the impulse of a force is expressed in lb · s, which is also the unit obtained in Sec. 12.4 for the linear momentum of a particle.

Equation (13.28) expresses that when a particle is acted upon by a force **F** during a given time interval, *the final momentum* $m\mathbf{v}_2$ *of the particle can be obtained by adding vectorially its initial momentum* $m\mathbf{v}_1$ *and the impulse of the force* **F** *during the time interval considered*

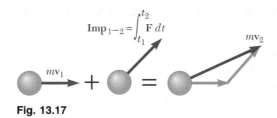

$$\mathbf{Imp}_{1\to2} = \int_{t_1}^{t_2} \mathbf{F}\, dt$$

$m\mathbf{v}_1 \quad + \quad = \quad m\mathbf{v}_2$

**Fig. 13.17**

(Fig. 13.17). We write

$$m\mathbf{v}_1 + \mathbf{Imp}_{1\to2} = m\mathbf{v}_2 \quad (13.30)$$

We note that while kinetic energy and work are scalar quantities, momentum and impulse are vector quantities. To obtain an analytic solution, it is thus necessary to replace Eq. (13.30) by the corresponding component equations

$$(mv_x)_1 + \int_{t_1}^{t_2} F_x\, dt = (mv_x)_2$$

$$(mv_y)_1 + \int_{t_1}^{t_2} F_y\, dt = (mv_y)_2 \quad (13.31)$$

$$(mv_z)_1 + \int_{t_1}^{t_2} F_z\, dt = (mv_z)_2$$

When several forces act on a particle, the impulse of each of the forces must be considered. We have

$$m\mathbf{v}_1 + \Sigma\ \mathbf{Imp}_{1\to2} = m\mathbf{v}_2 \tag{13.32}$$

Again, the equation obtained represents a relation between vector quantities; in the actual solution of a problem, it should be replaced by the corresponding component equations.

When a problem involves two particles or more, each particle can be considered separately and Eq. (13.32) can be written for each particle. We can also add vectorially the momenta of all the particles and the impulses of all the forces involved. We write then

$$\Sigma m\mathbf{v}_1 + \Sigma\ \mathbf{Imp}_{1\to2} = \Sigma m\mathbf{v}_2 \tag{13.33}$$

Since the forces of action and reaction exerted by the particles on each other form pairs of equal and opposite forces, and since the time interval from $t_1$ to $t_2$ is common to all the forces involved, the impulses of the forces of action and reaction cancel out and only the impulses of the external forces need be considered.†

If no external force is exerted on the particles or, more generally, if the sum of the external forces is zero, the second term in Eq. (13.33) vanishes and Eq. (13.33) reduces to

$$\Sigma m\mathbf{v}_1 = \Sigma m\mathbf{v}_2 \tag{13.34}$$

which expresses that *the total momentum of the particles is conserved.* Consider, for example, two boats, of mass $m_A$ and $m_B$, initially at rest, which are being pulled together (Fig. 13.18). If the resistance of the

$m_A v_A = 0$    $m_B v_B = 0$    $m_A v_A'$    $m_B v_B'$

**Fig. 13.18**

water is neglected, the only external forces acting on the boats are their weights and the buoyant forces exerted on them. Since these forces are balanced, we write

$$\Sigma m\mathbf{v}_1 = \Sigma m\mathbf{v}_2$$
$$0 = m_A\mathbf{v}_A' + m_B\mathbf{v}_B'$$

where $\mathbf{v}_A'$ and $\mathbf{v}_B'$ represent the velocities of the boats after a finite interval of time. The equation obtained indicates that the boats move in opposite directions (toward each other) with velocities inversely proportional to their masses.‡

---

†We should note the difference between this statement and the corresponding statement made in Sec. 13.4 regarding the work of the forces of action and reaction between several particles. While the sum of the impulses of these forces is always zero, the sum of their work is zero only under special circumstances, for example, when the various bodies involved are connected by inextensible cords or links and are thus constrained to move through equal distances.

‡Blue equals signs are used in Fig. 13.18 and throughout the remainder of this chapter to express that two systems of vectors are *equipollent,* that is, that they have the same resultant and moment resultant (cf. Sec. 3.19). Red equals signs will continue to be used to indicate that two systems of vectors are *equivalent,* that is, that they have the same effect. This and the concept of conservation of momentum for a system of particles will be discussed in greater detail in Chap. 14.

**Fig. 13.19**

**Photo 13.1** A heavy weight is lifted by the pile driver and then dropped. This delivers a large impulsive force to the pile which is driven into the ground.

## 13.11. IMPULSIVE MOTION

A force acting on a particle during a very short time interval that is large enough to produce a definite change in momentum is called an *impulsive force* and the resulting motion is called an *impulsive motion*. For example, when a baseball is struck, the contact between bat and ball takes place during a very short time interval $\Delta t$. But the average value of the force $\mathbf{F}$ exerted by the bat on the ball is very large, and the resulting impulse $\mathbf{F}\,\Delta t$ is large enough to change the sense of motion of the ball (Fig. 13.19).

When impulsive forces act on a particle, Eq. (13.32) becomes

$$m\mathbf{v}_1 + \Sigma\mathbf{F}\,\Delta t = m\mathbf{v}_2 \qquad (13.35)$$

Any force which is not an impulsive force may be neglected, since the corresponding impulse $\mathbf{F}\,\Delta t$ is very small. *Nonimpulsive forces* include the weight of the body, the force exerted by a spring, or any other force which is *known* to be small compared with an impulsive force. Unknown reactions may or may not be impulsive; their impulses should therefore be included in Eq. (13.35) as long as they have not been proved negligible. The impulse of the weight of the baseball considered above, for example, may be neglected. If the motion of the bat is analyzed, the impulse of the weight of the bat can also be neglected. The impulses of the reactions of the player's hands on the bat, however, should be included; these impulses will not be negligible if the ball is incorrectly hit.

We note that the method of impulse and momentum is particularly effective in the analysis of the impulsive motion of a particle, since it involves only the initial and final velocities of the particle and the impulses of the forces exerted on the particle. The direct application of Newton's second law, on the other hand, would require the determination of the forces as functions of the time and the integration of the equations of motion over the time interval $\Delta t$.

In the case of the impulsive motion of several particles, Eq. (13.33) can be used. It reduces to

$$\Sigma m\mathbf{v}_1 + \Sigma\mathbf{F}\,\Delta t = \Sigma m\mathbf{v}_2 \qquad (13.36)$$

where the second term involves only impulsive, external forces. If all the external forces acting on the various particles are nonimpulsive, the second term in Eq. (13.36) vanishes and this equation reduces to Eq. (13.34). We write

$$\Sigma m\mathbf{v}_1 = \Sigma m\mathbf{v}_2 \qquad (13.34)$$

which expresses that the total momentum of the particles is conserved. This situation occurs, for example, when two particles which are moving freely collide with one another. We should note, however, that while the total momentum of the particles is conserved, their total energy is generally *not* conserved. Problems involving the collision or *impact* of two particles will be discussed in detail in Secs. 13.12 through 13.14.

## SAMPLE PROBLEM 13.10

An automobile weighing 8818.5 kg is driven down a 5° incline at a speed of 96 km/h when the brakes are applied, causing a constant total braking force (applied by the road on the tires) of 6672 N. Determine the time required for the automobile to come to a stop.

## SOLUTION

We apply the principle of impulse and momentum. Since each force is constant in magnitude and direction, each corresponding impulse is equal to the product of the force and of the time interval $t$.

$$m\mathbf{v}_1 + \Sigma\,\mathbf{Imp}_{1\rightarrow2} = m\mathbf{v}_2$$

$+\searrow$ components:   $mv_1 + (W\sin 5°)t - Ft = 0$

$(8818.5 \times 9.81)\,(26.7\ \text{m/s}) + (8818.5 \times 9.81\sin 5°)t - 6672t = 0$  $t = 2661.68\ \text{s}$ ◀

## SAMPLE PROBLEM 13.11

A .12 kg baseball is pitched with a velocity of 24.4 m/s toward a batter. After the ball is hit by the bat $B$, it has a velocity of 36.6 m/s in the direction shown. If the bat and ball are in contact 0.015 s, determine the average impulsive force exerted on the ball during the impact.

## SOLUTION

We apply the principle of impulse and momentum to the ball. Since the weight of the ball is a nonimpulsive force, it can be neglected.

$$m\mathbf{v}_1 + \Sigma\,\mathbf{Imp}_{1\rightarrow2} = m\mathbf{v}_2$$

$\xrightarrow{+} x$ components:     $-mv_1 + F_x\,\Delta t = mv_2\cos 40°$

$-(.12 \times 9.81\ \text{N})\,(24.4\ \text{m/s}) + F_x(0.015\ \text{s}) = (.12 \times 9.81\ \text{N})(36.6\ \text{m/s})\cos 40°$

$$F_x = +4115.3\ \text{N}$$

$+\uparrow y$ components:      $0 + F_y\,\Delta t = mv_2\sin 40°$

$$F_y(0.015\ \text{s}) = (.12 \times 9.81\ \text{N})(36.6\ \text{m/s})\sin 40°$$

$$F_y = +1846.3\ \text{N}$$

From its components $F_x$ and $F_y$ we determine the magnitude and direction of the force $\mathbf{F}$:

$$\mathbf{F} = 4510.5\ \text{N}\ \measuredangle\ 24.2° \quad◀$$

## SAMPLE PROBLEM 13.12

A 10-kg package drops from a chute into a 25-kg cart with a velocity of 3 m/s. Knowing that the cart is initially at rest and can roll freely, determine (a) the final velocity of the cart, (b) the impulse exerted by the cart on the package, (c) the fraction of the initial energy lost in the impact.

## SOLUTION

We first apply the principle of impulse and momentum to the package-cart system to determine the velocity $\mathbf{v}_2$ of the cart and package. We then apply the same principle to the package alone to determine the impulse $\mathbf{F}\,\Delta t$ exerted on it.

### a. Impulse-Momentum Principle: Package and Cart

$$m_P\mathbf{v}_1 + \Sigma\,\mathbf{Imp}_{1\rightarrow 2} = (m_P + m_C)\mathbf{v}_2$$

$\xrightarrow{+}\ x$ components:
$$m_P v_1 \cos 30° + 0 = (m_P + m_C)v_2$$
$$(10\text{ kg})(3\text{ m/s}) \cos 30° = (10\text{ kg} + 25\text{ kg})v_2$$
$$\mathbf{v}_2 = 0.742\text{ m/s} \rightarrow \quad \blacktriangleleft$$

We note that the equation used expresses conservation of momentum in the $x$ direction.

### b. Impulse-Momentum Principle: Package

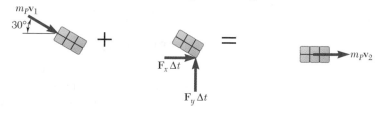

$$m_P\mathbf{v}_1 + \Sigma\,\mathbf{Imp}_{1\rightarrow 2} = m_P\mathbf{v}_2$$
$\xrightarrow{+}\ x$ components: $\quad (10\text{ kg})(3\text{ m/s}) \cos 30° + F_x\,\Delta t = (10\text{ kg})(0.742\text{ m/s})$
$$F_x\,\Delta t = -18.56\text{ N}\cdot\text{s}$$
$+\uparrow y$ components: $\quad -m_P v_1 \sin 30° + F_y\,\Delta t = 0$
$$-(10\text{ kg})(3\text{ m/s}) \sin 30° + F_y\,\Delta t = 0$$
$$F_y\,\Delta t = +15\text{ N}\cdot\text{s}$$

The impulse exerted on the package is $\qquad \mathbf{F}\,\Delta t = 23.9\text{ N}\cdot\text{s} \ \measuredangle\ 38.9° \quad \blacktriangleleft$

### c. Fraction of Energy Lost.

The initial and final energies are
$$T_1 = \tfrac{1}{2}m_P v_1^2 = \tfrac{1}{2}(10\text{ kg})(3\text{ m/s})^2 = 45\text{ J}$$
$$T_2 = \tfrac{1}{2}(m_P + m_C)v_2^2 = \tfrac{1}{2}(10\text{ kg} + 25\text{ kg})(0.742\text{ m/s})^2 = 9.63\text{ J}$$

The fraction of energy lost is $\qquad \dfrac{T_1 - T_2}{T_1} = \dfrac{45\text{ J} - 9.63\text{ J}}{45\text{ J}} = 0.786 \quad \blacktriangleleft$

# SOLVING PROBLEMS
## ON YOUR OWN

In this lesson we integrated Newton's second law to derive the *principle of impulse and momentum* for a particle. Recalling that the *linear momentum* of a particle was defined as the product of its mass $m$ and its velocity $\mathbf{v}$ [Sec. 12.3], we wrote

$$m\mathbf{v}_1 + \Sigma\,\mathbf{Imp}_{1\to2} = m\mathbf{v}_2 \qquad (13.32)$$

This equation expresses that the linear momentum $m\mathbf{v}_2$ of a particle at time $t_2$ can be obtained by adding to its linear momentum $m\mathbf{v}_1$ at time $t_1$ the *impulses* of the forces exerted on the particle during the time interval $t_1$ to $t_2$. For computing purposes, the momenta and impulses may be expressed in terms of their rectangular components, and Eq. (13.32) can be replaced by the equivalent scalar equations. The units of momentum and impulse are $N \cdot s$ in the SI system of units and $lb \cdot s$ in U.S. customary units. To solve problems using this equation you can follow these steps:

1. **Draw a diagram** showing the particle, its momentum at $t_1$ and at $t_2$, and the impulses of the forces exerted on the particle during the time interval $t_1$ to $t_2$.

2. **Calculate the impulse of each force,** expressing it in terms of its rectangular components if more than one direction is involved. You may encounter the following cases:

   a. **The time interval is finite and the force is constant.**

   $$\mathbf{Imp}_{1\to2} = \mathbf{F}(t_2 - t_1)$$

   b. **The time interval is finite and the force is a function of t.**

   $$\mathbf{Imp}_{1\to2} = \int_{t_1}^{t_2} \mathbf{F}(t)\,dt$$

   c. **The time interval is very small and the force is very large.** The force is called an *impulsive force* and its impulse over the time interval $t_2 - t_1 = \Delta t$ is

   $$\mathbf{Imp}_{1\to2} = \mathbf{F}\,\Delta t$$

Note that this impulse is *zero for a nonimpulsive force* such as the *weight* of a body, the force exerted by a *spring,* or any other force which is known to be small by comparison with the impulsive forces. Unknown reactions, however, *cannot be assumed* to be nonimpulsive and their impulses should be taken into account.

3. **Substitute the values obtained for the impulses into Eq. (13.32)** or into the equivalent scalar equations. You will find that the forces and velocities in the problems of this lesson are contained in a plane. You will, therefore, write two scalar equations and solve these equations for *two unknowns*. These unknowns may be a *time* [Sample Prob. 13.10], a *velocity* and an *impulse* [Sample Prob. 13.12], or an *average impulsive force* [Sample Prob. 13.11]. *(continued)*

**4. When several particles are involved,** a separate diagram should be drawn for each particle, showing the initial and final momentum of the particle, as well as the impulses of the forces exerted on the particle.

*a. It is usually convenient, however,* to first consider a diagram including all the particles. This diagram leads to the equation

$$\Sigma m\mathbf{v}_1 + \Sigma \, \mathbf{Imp}_{1 \to 2} = \Sigma m\mathbf{v}_2 \tag{13.33}$$

where the impulses of *only the forces external to the system* need be considered. Therefore, the two equivalent scalar equations will not contain any of the impulses of the unknown internal forces.

*b. If the sum of the impulses of the external forces is zero,* Eq. (13.33) reduces to

$$\Sigma m\mathbf{v}_1 = \Sigma m\mathbf{v}_2 \tag{13.34}$$

which expresses that *the total momentum of the particles is conserved.* This occurs either if the resultant of the external forces is zero or, when the time interval $\Delta t$ is very short (impulsive motion), if all the external forces are nonimpulsive. Keep in mind, however, that the total momentum may be conserved *in one direction,* but not in another [Sample Prob. 13.12].

# Problems

**13.119**   A 1.36 kg particle is acted upon by a force $\mathbf{F} = (20 \sin 2t)\mathbf{i} + (24 \cos 2t)\mathbf{j}$, where $F$ is expressed in pounds and $t$ in seconds. Determine the magnitude and direction of the velocity of the particle at $t = 6$ s knowing that its velocity is zero at $t = 0$.

**13.120**   A 4-kg particle is acted upon by a force $\mathbf{F} = (4t - 8)\mathbf{i} - 2\mathbf{j}$ where $F$ is expressed in newtons and $t$ in seconds. Knowing that the velocity of the particle is zero at $t = 0$, determine its maximum speed in the interval $0 \le t \le 3$ s.

**13.121**   A 440-kg sailboat with its occupants is running downwind at 12 km/h when its spinnaker is raised to increase its speed. Determine the net force provided by the spinnaker over the 10-s interval that it takes for the boat to reach a speed of 18 km/h.

**13.122**   The initial velocity of the block in position $A$ is 9 m/s. Knowing that the coefficient of kinetic friction between the block and the plane is $\mu_k = 0.30$, determine the time it takes for the block to reach $B$ with zero velocity, if (a) $\theta = 0$, (b) $\theta = 20°$.

**13.123**   Steep safety ramps are built beside mountain highways to enable vehicles with defective brakes to stop. A 10,000 kg truck enters a 15° ramp at a high speed $v_0 = 32.9$ m/s and travels for 6 s before its speed is reduced to 11 m/s. Assuming constant deceleration, determine (a) the magnitude of the braking force, (b) the additional time required for the truck to stop. Neglect air resistance and rolling resistance.

**Fig. P13.121**

**Fig. P13.122**

**Fig. P13.123 and P13.124**

**13.124**   A 10,000 kg truck enters a 15° safety ramp at a high speed $v_0$ and travels 164.6 m in 5.5 s before its speed is reduced to $v_0/2$. Assuming constant deceleration, determine (a) the initial speed $v_0$, (b) the magnitude of the braking force. Neglect air resistance and rolling resistance.

**13.125**   In anticipation of a long 6° upgrade, a bus driver accelerates at a constant rate from 80 km/h to 100 km/h in 8 s while still on a level section of the highway. Knowing that the speed of the bus is 100 km/h as it begins to climb the grade at time $t = 0$ and that the driver does not change the setting of the throttle or shift gears, determine (a) the speed of the bus when $t = 10$ s, (b) the time when the speed is 60 km/h.

**813**

**Fig. P13.126**

**Fig. P13.128**

**13.126** A 20-kg block is at rest on an incline when a constant horizontal force **P** is applied to it. The static and kinetic coefficients of friction between the block and the incline are 0.4 and 0.3, respectively. Knowing that the speed of the block is 15 m/s after 6 s, determine the magnitude of **P**.

**13.127** The coefficients of friction between the load and the flatbed trailer shown are $\mu_s = 0.40$ and $\mu_k = 0.35$. Knowing that the speed of the rig is 88 km/h, determine the shortest time in which the rig can be brought to a stop if the load is not to shift.

**Fig. P13.127**

**13.128** The 18 kg block is moving downward with a speed of .9 m/s at $t = 0$ when constant forces **P** and 2**P** are applied through the ropes. Knowing that the block is moving upward with a speed of .6 m/s when $t = 4$ s, determine (a) the magnitude of **P**, (b) the time at which the speed is zero. Neglect the effect of friction and the masses of the pulleys.

**13.129** A tractor-trailer rig with a 2000-kg tractor, a 4500-kg trailer, and a 3600-kg trailer is traveling on a level road at 90 km/h. The brakes on the rear trailer fail and the antiskid system of the tractor and front trailer provide the largest possible force which will not cause the wheels to slide. Knowing that the coefficient of static friction is 0.75, determine (a) the shortest time for the rig to come to a stop, (b) the force in the coupling between the two trailers during that time. Assume that the force exerted by the coupling on each of the two trailers is horizontal.

**Fig. P13.129**

**13.130** A light train made of two cars travels at 72 km/h. The mass of car A is 18 Mg and the mass of car B is 13 Mg. When the brakes are suddenly applied, a constant braking force of 19 kN is applied to each car. Determine (a) the time required for the train to stop after the brakes are applied, (b) the force in the coupling between the cars while the train is slowing down.

**Fig. P13.130**

**13.131** Solve Prob. 13.130, assuming that a constant braking force of 19 kN is applied to car $B$ but the brakes on car $A$ are not applied.

**13.132** The two blocks shown are released from rest at time $t = 0$. Neglecting the masses of the pulleys and the effect of friction in the pulleys and between the blocks and the incline, determine (a) the velocity of block $A$ at $t = 0.5$ s, (b) the tension in the cable.

**13.133** The coefficients of friction between the three blocks and the horizontal surfaces are $\mu_s = 0.25$ and $\mu_k = 0.20$. The weights of the blocks are $W_A = W_C = 89$ N, and $W_B = 44.5$ N. The velocities of blocks $A$ and $C$ at time $t = 0$ are $v_A = 2.7$ m/s and $v_C = 4.6$ m/s, both to the right. Determine (a) the velocity of each block at $t = 0.5$ s, (b) the tension in the cable.

**Fig. P13.132**

**Fig. P13.133**

**13.134** The system shown is released from rest. Determine the time it takes for the velocity of $A$ to reach 0.6 m/s. Neglect friction and the mass of the pulleys.

**13.135** Two packages are placed on an incline as shown. The coefficients of friction are $\mu_s = 0.30$ and $\mu_k = 0.25$ between the incline and package $A$, and $\mu_s = 0.20$ and $\mu_k = 0.15$ between the incline and package $B$. Knowing that the packages are in contact when released, determine (a) the velocity of each package after 3 s, (b) the force exerted by package $A$ on package $B$.

**Fig. *P13.134***

**Fig. *P13.135***

**13.136** A 2-kg collar which can slide on a frictionless vertical rod is acted upon by a force $\mathbf{P}$ which varies in magnitude as shown. Knowing that the collar is initially at rest, determine its velocity at (a) $t = 2$ s, (b) $t = 3$ s.

**13.137** A 2-kg collar which can slide on a frictionless vertical rod is acted upon by a force $\mathbf{P}$ which varies in magnitude as shown. Knowing that the collar is initially at rest, determine (a) the maximum velocity of the collar, (b) the time when the velocity is zero.

**Fig. P13.136 and P13.137**

**13.138**  A 5.4 kg block which can slide on a frictionless inclined surface is acted upon by a force **P** which varies in magnitude as shown. Knowing that the block is initially at rest, determine (a) the velocity of the block at $t = 5$ s, (b) the time at which the velocity of the block is zero.

**Fig. P13.138**

**13.139**  A simplified model consisting of a single straight line is to be obtained for the variation of pressure inside the 1 cm.-diameter barrel of a rifle as a .02 kg bullet is fired. Knowing that it takes 1.6 ms for the bullet to travel the length of the barrel and that the velocity of the bullet upon exit is 640 m/s, determine the value of $p_0$.

**Fig. P13.139**

**13.140**  A player hits a .06 kg tennis ball with a horizontal initial velocity of 16.5 m/s at a height of 1.4 m. The ball bounces at point A and rises to a maximum height of .9 m where the velocity is 9 m/s. Knowing that the duration of the impact is 0.004 s, determine the impulsive force exerted on the ball at point A.

**Fig. P13.140**

**13.141**  The triple jump is a track-and-field event in which an athlete gets a running start and tries to leap as far as he can with a hop, step, and jump. Shown in the figure is the initial hop of the athlete. Assuming that he approaches the takeoff line from the left with a horizontal velocity of 9 m/s, remains in contact with the ground for 0.18 s, and takes off at a 50° angle with a velocity of 11 m/s, determine the vertical component of the average impulsive force exerted by the ground on his foot. Give your answer in terms of the weight W of the athlete.

**Fig. P13.141**

**13.142**  A 28-g steel-jacketed bullet is fired with a velocity of 650 m/s toward a steel plate and ricochets along path CD with a velocity of 500 m/s. Knowing that the bullet leaves a 50-mm scratch on the surface of the plate and assuming that it has an average speed of 600 m/s while in contact with the plate, determine the magnitude and direction of the impulsive force exerted by the plate on the bullet.

**Fig. P13.142**

**13.143**  A 1.5-kg collar can slide on a horizontal rod which is free to rotate about a vertical shaft. The collar is initially held at A by a cord attached to the shaft. As the rod rotates at a rate $\dot{\theta} = 18$ rad/s, the cord is cut and the collar moves out along the rod and strikes the stop at B without rebounding. Neglecting friction and the mass of the rod, determine the magnitude of the impulse of the force exerted by the stop on the collar.

**Fig. P13.143**

**13.144**  A 45-g golf ball is hit with a golf club and leaves it with a velocity of 38 m/s. Assume that for $0 \leq t \leq t_0$, where $t_0$ is the duration of the impact, the magnitude $F$ of the force exerted on the ball can be expressed as $F = F_m \sin(\pi t/t_0)$. Knowing that $t_0 = 0.5$ ms, determine the maximum value $F_m$ of the force exerted on the ball.

**13.145**  An estimate of the expected load on over-the-shoulder seat belts is made before designing prototype belts that will be evaluated in automobile crash tests. Assuming that an automobile traveling at 72 km/h is brought to a stop in 110 ms, determine (*a*) the average impulsive force exerted by a 100-kg man on the belt, (*b*) the maximum force $F_m$ exerted on the belt if the force-time diagram has the shape shown.

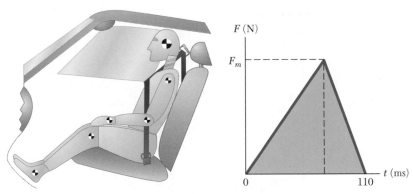

**Fig. P13.145**

**13.146**  A 120,000 kg tugboat is moving at 1.8 m/s with a slack towing cable attached to a 100,000 kg barge which is at rest. The cable is being unwound from a drum on the tugboat at a constant rate of 1.6 m/s and that rate is maintained after the cable becomes taut. Neglecting the resistance of the water, determine (*a*) the velocity of the tugboat after the cable becomes taut, (*b*) the impulse exerted on the barge as the cable becomes taut.

**Fig. P13.146**

**13.147**  A .03 kg bullet is fired into an 3.6 kg wooden block and becomes embedded in it. Knowing that the block and bullet then move up the smooth incline for 1.2 s before they come to a stop, determine (*a*) the magnitude of the initial velocity of the bullet, (*b*) the magnitude of the impulse of the force exerted by the bullet on the block.

**Fig. P13.147**

**13.148**  A 2-kg block *A* can slide without friction down a slot in a 10-kg block *B* which can slide without friction on a horizontal surface. Both blocks are at rest when block *A* is released in the position shown. Just before block *A* reaches the end of the slot its speed relative to block *B* is 3.59 m/s. Neglecting friction and knowing that the blocks stick together after block *A* strikes the end of the slot, determine (*a*) the impulse exerted by block *B* on block *A*, (*b*) the energy lost in the impact.

**Fig. P13.148**

**Fig. P13.149**

**13.149** Two swimmers $A$ and $B$, of mass 75 kg and 50 kg, respectively, dive off the end of a 200-kg boat. Each swimmer has a relative horizontal velocity of 3 m/s when leaving the boat. If the boat is initially at rest, determine its final velocity, assuming that ($a$) the two swimmers dive simultaneously, ($b$) swimmer $A$ dives first, ($c$) swimmer $B$ dives first.

**13.150** A .09 kg ball is projected from a height of 1.46 m with a horizontal velocity of 1.8 m/s and bounces from a .42 kg smooth plate supported by springs. Knowing that the height of the rebound is .55 m, determine ($a$) the velocity of the plate immediately after the impact, ($b$) the energy lost due to the impact.

**Fig. P13.150**

**13.151** In order to test the resistance of a chain to impact, the chain is suspended from a 120-kg rigid beam supported by two columns. A rod attached to the last link is then hit by a 30-kg block dropped from a 2-m height. Determine the initial impulse exerted on the chain and the energy absorbed by the chain, assuming that the block does not rebound from the rod and that the columns supporting the beam are ($a$) perfectly rigid, ($b$) equivalent to two perfectly elastic springs.

**Fig. P13.151**

**13.152** A small rivet connecting two pieces of sheet metal is being clinched by hammering. Determine the impulse exerted on the rivet and the energy absorbed by the rivet under each blow, knowing that the head of the hammer has a mass of 750 g and that it strikes the rivet with a velocity of 6 m/s. Assume that the hammer does not rebound and that the anvil is supported by springs and (a) has an infinite mass (rigid support), (b) has a mass of 4 kg.

6 m/s

**Fig. P13.152**

**13.153** A baseball player catching a ball can soften the impact by pulling his hand back. Assuming that a .15 kg ball reaches his glove at 154 km/h and that the player pulls his hand back during the impact at an average speed of 7.6 m/s over a distance of 20 cm., bringing the ball to a stop, determine the average impulsive force exerted on the player's hand.

20 cm

154 km/h

**13.154** A 1.8 kg sphere A is connected to a fixed point O by an inextensible cord of length 1.1 m. The sphere is resting on a frictionless horizontal surface at a distance of .46 m from O when it is given a velocity $\mathbf{v}_0$ in a direction perpendicular to line OA. It moves freely until it reaches position A', when the cord becomes taut. Determine the maximum allowable velocity $\mathbf{v}_0$ if the impulse of the force exerted on the cord is not to exceed 3.2 N · s.

**Fig. P13.153**

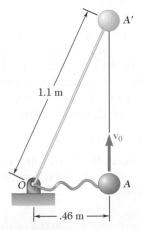

A'

1.1 m

$v_0$

O

A

.46 m

**Fig. P13.154**

## 13.12. IMPACT

A collision between two bodies which occurs in a very small interval of time and during which the two bodies exert relatively large forces on each other is called an *impact*. The common normal to the surfaces in contact during the impact is called the *line of impact*. If the mass centers on the two colliding bodies are located on this line, the impact is a *central impact*. Otherwise, the impact is said to be *eccentric*. Our present study will be limited to the central impact of two particles. The analysis of the eccentric impact of two rigid bodies will be considered later, in Sec. 17.12.

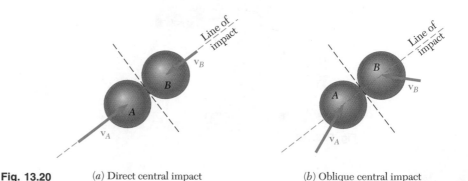

**Fig. 13.20**      (*a*) Direct central impact        (*b*) Oblique central impact

If the velocities of the two particles are directed along the line of impact, the impact is said to be a *direct impact* (Fig. 13.20*a*). If either or both particles move along a line other than the line of impact, the impact is said to be an *oblique impact* (Fig. 13.20*b*).

## 13.13. DIRECT CENTRAL IMPACT

Consider two particles $A$ and $B$, of mass $m_A$ and $m_B$, which are moving in the same straight line and to the right with known velocities $\mathbf{v}_A$ and $\mathbf{v}_B$ (Fig. 13.21*a*). If $\mathbf{v}_A$ is larger than $\mathbf{v}_B$, particle $A$ will eventually strike particle $B$. Under the impact, the two particles, will *deform* and, at the end of the period of deformation, they will have the same velocity $\mathbf{u}$ (Fig. 13.21*b*). A period of *restitution* will then take place, at the end of which, depending upon the magnitude of the impact forces and upon the materials involved, the two particles either will have regained their original shape or will stay permanently deformed. Our purpose here is to determine the velocities $\mathbf{v}'_A$ and $\mathbf{v}'_B$ of the particles at the end of the period of restitution (Fig. 13.21*c*).

Considering first the two particles as a single system, we note that there is no impulsive, external force. Thus, the total momentum of the two particles is conserved, and we write

$$m_A\mathbf{v}_A + m_B\mathbf{v}_B = m_A\mathbf{v}'_A + m_B\mathbf{v}'_B$$

Since all the velocities considered are directed along the same axis, we can replace the equation obtained by the following relation involving only scalar components:

$$m_A v_A + m_B v_B = m_A v'_A + m_B v'_B \qquad (13.37)$$

(*a*) Before impact

(*b*) At maximum deformation

(*c*) After impact

**Fig. 13.21**

A positive value for any of the scalar quantities $v_A$, $v_B$, $v'_A$, or $v'_B$ means that the corresponding vector is directed to the right; a negative value indicates that the corresponding vector is directed to the left.

To obtain the velocities $\mathbf{v}'_A$ and $\mathbf{v}'_B$, it is necessary to establish a second relation between the scalars $v'_A$ and $v'_B$. For this purpose, let us now consider the motion of particle $A$ during the period of deformation and apply the principle of impulse and momentum. Since the only impulsive force acting on $A$ during this period is the force $\mathbf{P}$ exerted by $B$ (Fig. 13.22$a$), we write, using again scalar components,

$$m_A v_A - \int P \, dt = m_A u \qquad (13.38)$$

where the integral extends over the period of deformation. Considering now the motion of $A$ during the period of restitution, and denoting by $\mathbf{R}$ the force exerted by $B$ on $A$ during this period (Fig. 13.22$b$), we write

$$m_A u - \int R \, dt = m_A v'_A \qquad (13.39)$$

where the integral extends over the period of restitution.

($a$) Period of deformation

($b$) Period of restitution

**Fig. 13.22**

In general, the force $\mathbf{R}$ exerted on $A$ during the period of restitution differs from the force $\mathbf{P}$ exerted during the period of deformation, and the magnitude $\int R \, dt$ of its impulse is smaller than the magnitude $\int P \, dt$ of the impulse of $\mathbf{P}$. The ratio of the magnitudes of the impulses corresponding, respectively, to the period of restitution and to the period of deformation is called the *coefficient of restitution* and is denoted by $e$. We write

$$e = \frac{\int R \, dt}{\int P \, dt} \qquad (13.40)$$

The value of the coefficient $e$ is always between 0 and 1. It depends to a large extent on the two materials involved, but it also varies considerably with the impact velocity and the shape and size of the two colliding bodies.

Solving Eqs. (13.38) and (13.39) for the two impulses and substituting into (13.40), we write

$$e = \frac{u - v'_A}{v_A - u} \qquad (13.41)$$

A similar analysis of particle $B$ leads to the relation

$$e = \frac{v'_B - u}{u - v_B} \tag{13.42}$$

Since the quotients in (13.41) and (13.42) are equal, they are also equal to the quotient obtained by adding, respectively, their numerators and their denominators. We have, therefore,

$$e = \frac{(u - v'_A) + (v'_B - u)}{(v_A - u) + (u - v_B)} = \frac{v'_B - v'_A}{v_A - v_B}$$

and

$$v'_B - v'_A = e(v_A - v_B) \tag{13.43}$$

Since $v'_B - v'_A$ represents the relative velocity of the two particles after impact and $v_A - v_B$ represents their relative velocity before impact, formula (13.43) expresses that *the relative velocity of the two particles after impact can be obtained by multiplying their relative velocity before impact by the coefficient of restitution.* This property is used to determine experimentally the value of the coefficient of restitution of two given materials.

The velocities of the two particles after impact can now be obtained by solving Eqs. (13.37) and (13.43) simultaneously for $v'_A$ and $v'_B$. It is recalled that the derivation of Eqs. (13.37) and (13.43) was based on the assumption that particle $B$ is located to the right of $A$, and that both particles are initially moving to the right. If particle $B$ is initially moving to the left, the scalar $v_B$ should be considered negative. The same sign convention holds for the velocities after impact: a positive sign for $v'_A$ will indicate that particle $A$ moves to the right after impact, and a negative sign will indicate that it moves to the left.

Two particular cases of impact are of special interest:

1. $e = 0$, *Perfectly Plastic Impact.* When $e = 0$, Eq. (13.43) yields $v'_B = v'_A$. There is no period of restitution, and both particles stay together after impact. Substituting $v'_B = v'_A = v'$ into Eq. (13.37), which expresses that the total momentum of the particles is conserved, we write

$$m_A v_A + m_B v_B = (m_A + m_B)v' \tag{13.44}$$

This equation can be solved for the common velocity $v'$ of the two particles after impact.

2. $e = 1$, *Perfectly Elastic Impact.* When $e = 1$, Eq. (13.43) reduces to

$$v'_B - v'_A = v_A - v_B \tag{13.45}$$

which expresses that the relative velocities before and after impact are equal. The impulses received by each particle during the period of deformation and during the period of restitution are equal. The particles move away from each other after impact with the same velocity with which they approached each other before impact. The velocities $v'_A$ and $v'_B$ can be obtained by solving Eqs. (13.37) and (13.45) simultaneously.

It is worth noting that *in the case of a perfectly elastic impact, the total energy of the two particles,* as well as their total momentum, *is conserved.* Equations (13.37) and (13.45) can be written as follows:

$$m_A(v_A - v'_A) = m_B(v'_B - v_B) \qquad (13.37')$$

$$v_A + v'_A = v_B + v'_B \qquad (13.45')$$

Multiplying (13.37') and (13.45') member by member, we have

$$m_A(v_A - v'_A)(v_A + v'_A) = m_B(v'_B - v_B)(v'_B + v_B)$$

$$m_A v_A^2 - m_A(v'_A)^2 = m_B(v'_B)^2 - m_B v_B^2$$

Rearranging the terms in the equation obtained and multiplying by $\frac{1}{2}$, we write

$$\tfrac{1}{2}m_A v_A^2 + \tfrac{1}{2}m_B v_B^2 = \tfrac{1}{2}m_A(v'_A)^2 + \tfrac{1}{2}m_B(v'_B)^2 \qquad (13.46)$$

**Photo 13.2** Because the coefficient of restitution is less than one, the tennis ball loses some energy with each bounce.

which expresses that the kinetic energy of the particles is conserved. It should be noted, however, that *in the general case of impact,* that is, when $e$ is not equal to 1, *the total energy of the particles is not conserved.* This can be shown in any given case by comparing the kinetic energies before and after impact. The lost kinetic energy is in part transformed into heat and in part spent in generating elastic waves within the two colliding bodies.

## 13.14. OBLIQUE CENTRAL IMPACT

Let us now consider the case when the velocities of the two colliding particles are *not* directed along the line of impact (Fig. 13.23). As indicated in Sec. 13.12, the impact is said to be *oblique.* Since the

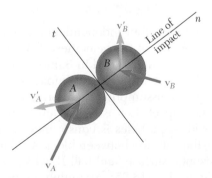

**Fig. 13.23**

velocities $\mathbf{v}'_A$ and $\mathbf{v}'_B$ of the particles after impact are unknown in direction as well as in magnitude, their determination will require the use of four independent equations.

We choose as coordinate axes the $n$ axis along the line of impact, that is, along the common normal to the surfaces in contact, and the $t$ axis along their common tangent. Assuming that the particles are perfectly *smooth and frictionless,* we observe that the only impulses

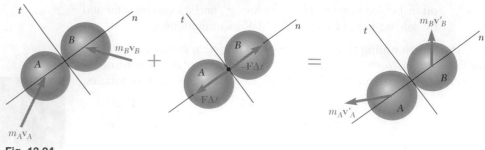

**Fig. 13.24**

exerted on the particles during the impact are due to internal forces directed along the line of impact, that is, along the $n$ axis (Fig. 13.24). It follows that

1. The component along the $t$ axis of the momentum of each particle, considered separately, is conserved; hence the $t$ component of the velocity of each particle remains unchanged. We write

$$(v_A)_t = (v_A')_t \qquad (v_B)_t = (v_B')_t \qquad (13.47)$$

2. The component along the $n$ axis of the total momentum of the two particles is conserved. We write

$$m_A(v_A)_n + m_B(v_B)_n = m_A(v_A')_n + m_B(v_B')_n \qquad (13.48)$$

3. The component along the $n$ axis of the relative velocity of the two particles after impact is obtained by multiplying the $n$ component of their relative velocity before impact by the coefficient of restitution. Indeed, a derivation similar to that given in Sec. 13.13 for direct central impact yields

$$(v_B')_n - (v_A')_n = e[(v_A)_n - (v_B)_n] \qquad (13.49)$$

We have thus obtained four independent equations which can be solved for the components of the velocities of $A$ and $B$ after impact. This method of solution is illustrated in Sample Prob. 13.15.

Our analysis of the oblique central impact of two particles has been based so far on the assumption that both particles moved freely before and after the impact. Let us now examine the case when one or both of the colliding particles is constrained in its motion. Consider, for instance, the collision between block $A$, which is constrained to move on a horizontal surface, and ball $B$, which is free to move in the plane of the figure (Fig. 13.25). Assuming no friction between the block and the ball, or between the block and the horizontal surface, we note that the impulses exerted on the system consist of the impulses of the internal forces $\mathbf{F}$ and $-\mathbf{F}$ directed along the line of impact, that is, along the $n$ axis, and of the impulse of the external force $\mathbf{F}_{\text{ext}}$ exerted by the horizontal surface on block $A$ and directed along the vertical (Fig. 13.26).

The velocities of block $A$ and ball $B$ immediately after the impact are represented by three unknowns: the magnitude of the velocity $\mathbf{v}_A'$ of block $A$, which is known to be horizontal, and the magnitude and

**Photo 13.3** In a game of pool, the cue ball has just hit another ball, and most of its energy and momentum have been transferred to that ball.

**Fig. 13.25**

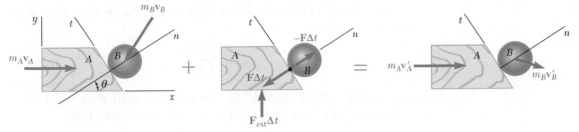

**Fig. 13.26**

direction of the velocity $\mathbf{v}'_B$ of ball $B$. We must therefore write three equations by expressing that

1.  The component along the $t$ axis of the momentum of ball $B$ is conserved; hence the $t$ component of the velocity of ball $B$ remains unchanged. We write

$$(v_B)_t = (v'_B)_t \qquad (13.50)$$

2.  The component along the horizontal $x$ axis of the total momentum of block $A$ and ball $B$ is conserved. We write

$$m_A v_A + m_B (v_B)_x = m_A v'_A + m_B (v'_B)_x \qquad (13.51)$$

3.  The component along the $n$ axis of the relative velocity of block $A$ and ball $B$ after impact is obtained by multiplying the $n$ component of their relative velocity before impact by the coefficient of restitution. We write again

$$(v'_B)_n - (v'_A)_n = e[(v_A)_n - (v_B)_n] \qquad (13.49)$$

We should note, however, that in the case considered here, the validity of Eq. (13.49) cannot be established through a mere extension of the derivation given in Sec. 13.13 for the direct central impact of two particles moving in a straight line. Indeed, these particles were not subjected to any external impulse, while block $A$ in the present analysis is subjected to the impulse exerted by the horizontal surface. To prove that Eq. (13.49) is still valid, we will first apply the principle of impulse and momentum to block $A$ over the period of deformation (Fig. 13.27). Considering only the horizontal components, we write

$$m_A v_A - (\textstyle\int P\, dt) \cos\theta = m_A u \qquad (13.52)$$

where the integral extends over the period of deformation and where $\mathbf{u}$ represents the velocity of block $A$ at the end of that period. Considering now the period of restitution, we write in a similar way

$$m_A u - (\textstyle\int R\, dt) \cos\theta = m_A v'_A \qquad (13.53)$$

where the integral extends over the period of restitution.

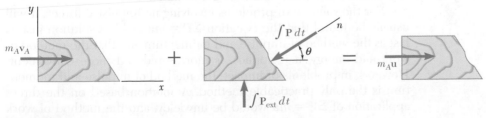

**Fig. 13.27**

Recalling from Sec. 13.13 the definition of the coefficient of restitution, we write

$$e = \frac{\int R\, dt}{\int P\, dt} \tag{13.40}$$

Solving Eqs. (13.52) and (13.53) for the integrals $\int P\, dt$ and $\int R\, dt$, and substituting into Eq. (13.40), we have, after reductions,

$$e = \frac{u - v_A'}{v_A - u}$$

or, multiplying all velocities by $\cos \theta$ to obtain their projections on the line of impact,

$$e = \frac{u_n - (v_A')_n}{(v_A)_n - u_n} \tag{13.54}$$

We note that Eq. (13.54) is identical to Eq. (13.41) of Sec. 13.13, except for the subscripts $n$ which are used here to indicate that we are considering velocity components along the line of impact. Since the motion of ball $B$ is unconstrained, the proof of Eq. (13.49) can be completed in the same manner as the derivation of Eq. (13.43) of Sec. 13.13. Thus, we conclude that the relation (13.49) between the components along the line of impact of the relative velocities of two colliding particles remains valid when one of the particles is constrained in its motion. The validity of this relation is easily extended to the case when both particles are constrained in their motion.

## 13.15. PROBLEMS INVOLVING ENERGY AND MOMENTUM

You now have at your disposal three different methods for the solution of kinetics problems: the direct application of Newton's second law, $\Sigma\mathbf{F} = m\mathbf{a}$; the method of work and energy; and the method of impulse and momentum. To derive maximum benefit from these three methods, you should be able to choose the method best suited for the solution of a given problem. You should also be prepared to use different methods for solving the various parts of a problem when such a procedure seems advisable.

You have already seen that the method of work and energy is in many cases more expeditious than the direct application of Newton's second law. As indicated in Sec. 13.4, however, the method of work and energy has limitations, and it must sometimes be supplemented by the use of $\Sigma\mathbf{F} = m\mathbf{a}$. This is the case, for example, when you wish to determine an acceleration or a normal force.

For the solution of problems involving no impulsive forces, it will usually be found that the equation $\Sigma\mathbf{F} = m\mathbf{a}$ yields a solution just as fast as the method of impulse and momentum and that the method of work and energy, if it applies, is more rapid and more convenient. However, in problems of impact, the method of impulse and momentum is the only practicable method. A solution based on the direct application of $\Sigma\mathbf{F} = m\mathbf{a}$ would be unwieldy, and the method of work

**Photo 13.4** As the automobile crashes into the barrier, considerable energy is dissipated as heat during the permanent deformation of the automobile and the barrier.

and energy cannot be used since impact (unless perfectly elastic) involves a loss of mechanical energy.

Many problems involve only conservative forces, except for a short impact phase during which impulsive forces act. The solution of such problems can be divided into several parts. The part corresponding to the impact phase calls for the use of the method of impulse and momentum and of the relation between relative velocities, and the other parts can usually be solved by the method of work and energy. If the problem involves the determination of a normal force, however, the use of $\Sigma \mathbf{F} = m\mathbf{a}$ is necessary.

Consider, for example, a pendulum $A$, of mass $m_A$ and length $l$, which is released with no velocity from a position $A_1$ (Fig. 13.28$a$). The pendulum swings freely in a vertical plane and hits a second pendulum $B$, of mass $m_B$ and same length $l$, which is initially at rest. After the impact (with coefficient of restitution $e$), pendulum $B$ swings through an angle $\theta$ that we wish to determine.

The solution of the problem can be divided into three parts:

1. *Pendulum A Swings from $A_1$ to $A_2$.* The principle of conservation of energy can be used to determine the velocity $(\mathbf{v}_A)_2$ of the pendulum at $A_2$ (Fig. 13.28$b$).
2. *Pendulum A Hits Pendulum B.* Using the fact that the total momentum of the two pendulums is conserved and the relation between their relative velocities, we determine the velocities $(\mathbf{v}_A)_3$ and $(\mathbf{v}_B)_3$ of the two pendulums after impact (Fig. 13.28$c$).
3. *Pendulum B Swings from $B_3$ to $B_4$.* Applying the principle of conservation of energy to pendulum $B$, we determine the maximum elevation $y_4$ reached by that pendulum (Fig. 13.28$d$). The angle $\theta$ can then be determined by trigonometry.

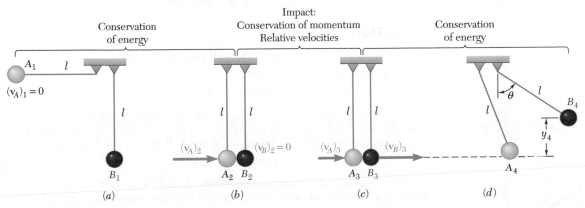

**Fig. 13.28**

We note that if the tensions in the cords holding the pendulums are to be determined, the method of solution just described should be supplemented by the use of $\Sigma \mathbf{F} = m\mathbf{a}$.

## SAMPLE PROBLEM 13.13

A 20-Mg railroad car moving at a speed of 0.5 m/s to the right collides with a 35-Mg car which is at rest. If after the collision the 35-Mg car is observed to move to the right at a speed of 0.3 m/s, determine the coefficient of restitution between the two cars.

## SOLUTION

We express that the total momentum of the two cars is conserved.

$$m_A\mathbf{v}_A + m_B\mathbf{v}_B = m_A\mathbf{v}'_A + m_B\mathbf{v}'_B$$
$$(20\ \text{Mg})(+0.5\ \text{m/s}) + (35\ \text{Mg})(0) = (20\ \text{Mg})v'_A + (35\ \text{Mg})(+0.3\ \text{m/s})$$
$$v'_A = -0.025\ \text{m/s} \qquad \mathbf{v}'_A = 0.025\ \text{m/s} \leftarrow$$

The coefficient of restitution is obtained by writing

$$e = \frac{v'_B - v'_A}{v_A - v_B} = \frac{+0.3 - (-0.025)}{+0.5 - 0} = \frac{0.325}{0.5} \qquad e = 0.65 \blacktriangleleft$$

## SAMPLE PROBLEM 13.14

A ball is thrown against a frictionless, vertical wall. Immediately before the ball strikes the wall, its velocity has a magnitude $v$ and forms an angle of 30° with the horizontal. Knowing that $e = 0.90$, determine the magnitude and direction of the velocity of the ball as it rebounds from the wall.

## SOLUTION

We resolve the initial velocity of the ball into components respectively perpendicular and parallel to the wall:

$$v_n = v\cos 30° = 0.866v \qquad v_t = v\sin 30° = 0.500v$$

**Motion Parallel to the Wall.** Since the wall is frictionless, the impulse it exerts on the ball is perpendicular to the wall. Thus, the component parallel to the wall of the momentum of the ball is conserved and we have

$$\mathbf{v}'_t = \mathbf{v}_t = 0.500v \uparrow$$

**Motion Perpendicular to the Wall.** Since the mass of the wall (and earth) is essentially infinite, expressing that the total momentum of the ball and wall is conserved would yield no useful information. Using the relation (13.49) between relative velocities, we write

$$0 - v'_n = e(v_n - 0)$$
$$v'_n = -0.90(0.866v) = -0.779v \qquad \mathbf{v}'_n = 0.779v \leftarrow$$

**Resultant Motion.** Adding vectorially the components $\mathbf{v}'_n$ and $\mathbf{v}'_t$,

$$\mathbf{v}' = 0.926v \ \text{\reflectbox{$\nwarrow$}} \ 32.7° \blacktriangleleft$$

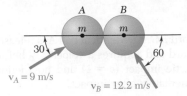

# SAMPLE PROBLEM 13.15

The magnitude and direction of the velocities of two identical frictionless balls before they strike each other are as shown. Assuming $e = 0.90$, determine the magnitude and direction of the velocity of each ball after the impact.

## SOLUTION

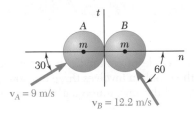

The impulsive forces that the balls exert on each other during the impact are directed along a line joining the centers of the balls called the *line of impact*. Resolving the velocities into components directed, respectively, along the line of impact and along the common tangent to the surfaces in contact, we write

$$(v_A)_n = v_A \cos 30° = +7.8 \text{ m/s}$$
$$(v_A)_t = v_A \sin 30° = +4.5 \text{ m/s}$$
$$(v_B)_n = -v_B \cos 60° = -6.1 \text{ m/s}$$
$$(v_B)_t = v_B \sin 60° = +10.6 \text{ m/s}$$

**Principle of Impulse and Momentum.** In the adjoining sketches we show in turn the initial momenta, the impulses, and the final momenta.

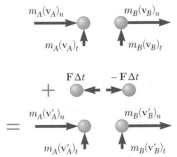

**Motion along the Common Tangent.** Considering only the $t$ components, we apply the principle of impulse and momentum to each ball *separately*. Since the impulsive forces are directed along the line of impact, the $t$ component of the momentum, and hence the $t$ component of the velocity of each ball, is unchanged. We have

$$(\mathbf{v}'_A)_t = 4.5 \text{ m/s} \uparrow \qquad (\mathbf{v}'_B)_t = 10.6 \text{ m/s} \uparrow$$

**Motion along the Line of Impact.** In the $n$ direction, we consider the two balls as a single system and note that by Newton's third law, the internal impulses are, respectively, $\mathbf{F} \Delta t$ and $-\mathbf{F} \Delta t$ and cancel. We thus write that the total momentum of the balls is conserved:

$$m_A(v_A)_n + m_B(v_B)_n = m_A(v'_A)_n + m_B(v'_B)_n$$
$$m(7.8) + m(-6.1) = m(v'_A)_n + m(v'_B)_n$$
$$(v_A)_n + (v'_B)_n = 1.7 \quad (1)$$

Using the relation (13.49) between relative velocities, we write

$$(v'_B)_n - (v'_A)_n = e[(v_A)_n - (v_B)_n]$$
$$(v'_B)_n - (v'_A)_n = (0.90)[7.8 - (-6.1)]$$
$$(v'_B)_n - (v'_A)_n = 12.5 \quad (2)$$

Solving Eqs. (1) and (2) simultaneously, we obtain

$$(v'_A)_n = -5.4 \qquad (v'_B)_n = +7.1$$
$$(\mathbf{v}'_A)_n = 5.4 \text{ m/s} \leftarrow \qquad (\mathbf{v}'_B)_n = 7.1 \text{ m/s} \rightarrow$$

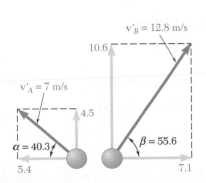

**Resultant Motion.** Adding vectorially the velocity components of each ball, we obtain

$$\mathbf{v}'_A = 7 \text{ m/s} \searrow 39.8° \qquad \mathbf{v}'_B = 12.8 \text{ m/s} \nearrow 56.2° \blacktriangleleft$$

## SAMPLE PROBLEM 13.16

Ball $B$ is hanging from an inextensible cord $BC$. An identical ball $A$ is released from rest when it is just touching the cord and acquires a velocity $\mathbf{v}_0$ before striking ball $B$. Assuming perfectly elastic impact ($e = 1$) and no friction, determine the velocity of each ball immediately after impact.

## SOLUTION

Since ball $B$ is constrained to move in a circle of center $C$, its velocity $\mathbf{v}_B$ after impact must be horizontal. Thus the problem involves three unknowns: the magnitude $v'_B$ of the velocity of $B$, and the magnitude and direction of the velocity $\mathbf{v}'_A$ of $A$ after impact.

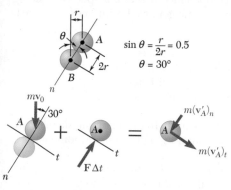

$$\sin \theta = \frac{r}{2r} = 0.5$$
$$\theta = 30°$$

**Impulse-Momentum Principle: Ball $A$**

$$m\mathbf{v}_A + \mathbf{F}\,\Delta t = m\mathbf{v}'_A$$

$+\searrow t$ components:
$$mv_0 \sin 30° + 0 = m(v'_A)_t$$
$$(v'_A)_t = 0.5v_0 \tag{1}$$

We note that the equation used expresses conservation of the momentum of ball $A$ along the common tangent to balls $A$ and $B$.

**Impulse-Momentum Principle: Balls $A$ and $B$**

$$m\mathbf{v}_A + \mathbf{T}\,\Delta t = m\mathbf{v}'_A + m\mathbf{v}'_B$$

$\xrightarrow{+} x$ components:
$$0 = m(v'_A)_t \cos 30° - m(v'_A)_n \sin 30° - mv'_B$$

We note that the equation obtained expresses conservation of the total momentum in the $x$ direction. Substituting for $(v'_A)_t$ from Eq. (1) and rearranging terms, we write

$$0.5(v'_A)_n + v'_B = 0.433v_0 \tag{2}$$

**Relative Velocities along the Line of Impact.** Since $e = 1$, Eq. (13.49) yields

$$(v'_B)_n - (v'_A)_n = (v_A)_n - (v_B)_n$$
$$v'_B \sin 30° - (v'_A)_n = v_0 \cos 30° - 0$$
$$0.5v'_B - (v'_A)_n = 0.866v_0 \tag{3}$$

Solving Eqs. (2) and (3) simultaneously, we obtain

$$(v'_A)_n = -0.520v_0 \qquad v'_B = 0.693v_0$$
$$\mathbf{v}'_B = 0.693v_0 \leftarrow \quad \blacktriangleleft$$

Recalling Eq. (1) we draw the adjoining sketch and obtain by trigonometry

$$v'_A = 0.721v_0 \qquad \beta = 46.1° \qquad \alpha = 46.1° - 30° = 16.1°$$
$$\mathbf{v}'_A = 0.721v_0 \measuredangle 16.1° \quad \blacktriangleleft$$

## SAMPLE PROBLEM 13.17

A 30-kg block is dropped from a height of 2 m onto the 10-kg pan of a spring scale. Assuming the impact to be perfectly plastic, determine the maximum deflection of the pan. The constant of the spring is $k = 20$ kN/m.

## SOLUTION

The impact between the block and the pan *must* be treated separately; therefore we divide the solution into three parts.

**Conservation of Energy.** Block: $W_A = (30$ kg$)(9.81$ m/s$^2) = 294$ N

$$T_1 = \tfrac{1}{2}m_A(v_A)_1^2 = 0 \qquad V_1 = W_A y = (294 \text{ N})(2 \text{ m}) = 588 \text{ J}$$
$$T_2 = \tfrac{1}{2}m_A(v_A)_2^2 = \tfrac{1}{2}(30 \text{ kg})(v_A)_2^2 \qquad V_2 = 0$$
$$T_1 + V_1 = T_2 + V_2: \qquad 0 + 588 \text{ J} = \tfrac{1}{2}(30 \text{ kg})(v_A)_2^2 + 0$$
$$(v_A)_2 = +6.26 \text{ m/s} \qquad (\mathbf{v}_A)_2 = 6.26 \text{ m/s} \downarrow$$

**Impact: Conservation of Momentum.** Since the impact is perfectly plastic, $e = 0$; the block and pan move together after the impact.

$$m_A(v_A)_2 + m_B(v_B)_2 = (m_A + m_B)v_3$$
$$(30 \text{ kg})(6.26 \text{ m/s}) + 0 = (30 \text{ kg} + 10 \text{ kg})v_3$$
$$v_3 = +4.70 \text{ m/s} \qquad \mathbf{v}_3 = 4.70 \text{ m/s} \downarrow$$

**Conservation of Energy.** Initially the spring supports the weight $W_B$ of the pan; thus the initial deflection of the spring is

$$x_3 = \frac{W_B}{k} = \frac{(10 \text{ kg})(9.81 \text{ m/s}^2)}{20 \times 10^3 \text{ N/m}} = \frac{98.1 \text{ N}}{20 \times 10^3 \text{ N/m}} = 4.91 \times 10^{-3} \text{ m}$$

Denoting by $x_4$ the total maximum deflection of the spring, we write

$$T_3 = \tfrac{1}{2}(m_A + m_B)v_3^2 = \tfrac{1}{2}(30 \text{ kg} + 10 \text{ kg})(4.70 \text{ m/s})^2 = 442 \text{ J}$$
$$V_3 = V_g + V_e = 0 + \tfrac{1}{2}kx_3^2 = \tfrac{1}{2}(20 \times 10^3)(4.91 \times 10^{-3})^2 = 0.241 \text{ J}$$
$$T_4 = 0$$
$$V_4 = V_g + V_e = (W_A + W_B)(-h) + \tfrac{1}{2}kx_4^2 = -(392)h + \tfrac{1}{2}(20 \times 10^3)x_4^2$$

Noting that the displacement of the pan is $h = x_4 - x_3$, we write

$$T_3 + V_3 = T_4 + V_4:$$
$$442 + 0.241 = 0 - 392(x_4 - 4.91 \times 10^{-3}) + \tfrac{1}{2}(20 \times 10^3)x_4^2$$
$$x_4 = 0.230 \text{ m} \qquad h = x_4 - x_3 = 0.230 \text{ m} - 4.91 \times 10^{-3} \text{ m}$$
$$h = 0.225 \text{ m} \qquad\qquad h = 225 \text{ mm} \blacktriangleleft$$

This lesson deals with the *impact of two bodies,* that is, with a collision occurring in a very small interval of time. You will solve a number of impact problems by expressing that the total momentum of the two bodies is conserved and noting the relationship which exists between the relative velocities of the two bodies before and after impact.

**1. As a first step in your solution** you should select and draw the following coordinate axes: the $t$ axis, which is tangent to the surfaces of contact of the two colliding bodies, and the $n$ axis, which is normal to the surfaces of contact and defines the *line of impact.* In all the problems of this lesson the line of impact passes through the mass centers of the colliding bodies, and the impact is referred to as a *central impact.*

**2. Next you will draw a diagram** showing the momenta of the bodies before impact, the impulses exerted on the bodies during impact, and the final momenta of the bodies after impact (Fig. 13.24). You will then observe whether the impact is a *direct central impact* or an *oblique central impact.*

**3. Direct central impact.** This occurs when the velocities of bodies $A$ and $B$ before impact are *both directed along the line of impact* (Fig. 13.20a).

    *a. Conservation of momentum.* Since the impulsive forces are internal to the system, you can write that the *total momentum of $A$ and $B$ is conserved,*

$$m_A v_A + m_B v_B = m_A v'_A + m_B v'_B \qquad (13.37)$$

where $v_A$ and $v_B$ denote the velocities of bodies $A$ and $B$ before impact and $v'_A$ and $v'_B$ denote their velocities after impact.

    *b. Coefficient of restitution.* You can also write the following relation between the *relative velocities* of the two bodies before and after impact,

$$v'_B - v'_A = e(v_A - v_B) \qquad (13.43)$$

where $e$ is the coefficient of restitution between the two bodies.

Note that Eqs. (13.37) and (13.43) are scalar equations which can be solved for two unknowns. Also, be careful to adopt a consistent sign convention for all velocities.

**4. Oblique central impact.** This occurs when *one or both* of the initial velocities of the two bodies is *not directed* along the line of impact (Fig. 13.20b). To solve problems of this type, you should *first resolve into components* along the $t$ axis and the $n$ axis the momenta and impulses shown in your diagram.

*a. Conservation of momentum.* Since the impulsive forces act along the line of impact, that is, along the $n$ axis, the component along the $t$ axis of the momentum *of each body* is conserved. Therefore, you can write for each body that the $t$ components of its velocity before and after impact are equal,

$$(v_A)_t = (v_A')_t \qquad (v_B)_t = (v_B')_t \qquad (13.47)$$

Also, the component along the $n$ axis of the *total momentum* of the system is conserved,

$$m_A(v_A)_n + m_B(v_B)_n = m_A(v_A')_n + m_B(v_B')_n \qquad (13.48)$$

*b. Coefficient of restitution.* The relation between the relative velocities of the two bodies before and after impact can be written in the $n$ direction only,

$$(v_B')_n - (v_A')_n = e[(v_A)_n - (v_B)_n] \qquad (13.49)$$

You now have four equations that you can solve for four unknowns. Note that after finding all the velocities, you can determine the impulse exerted by body $A$ on body $B$ by drawing an impulse-momentum diagram for $B$ alone and equating components in the $n$ direction.

*c. **When the motion of one of the colliding bodies is constrained,*** you must include the impulses of the external forces in your diagram. You will then observe that some of the above relations do not hold. However, in the example shown in Fig. 13.26 the total momentum of the system is conserved in a direction perpendicular to the external impulse. You should also note that when a body $A$ bounces off a fixed surface $B$, the only conservation of momentum equation which can be used is the first of Eqs. (13.47) [Sample Prob. 13.14].

**5. *Remember that energy is lost during most impacts.*** The only exception is for *perfectly elastic* impacts ($e = 1$), where energy is conserved. Thus, in the general case of impact, where $e < 1$, the energy is not conserved. Therefore, be careful *not to apply* the principle of conservation of energy through an impact situation. Instead, apply this principle separately to the motions preceding and following the impact [Sample Prob. 13.17].

# Problems

Fig. P13.155

Fig. *P13.156*

**13.155**  Collars $A$ and $B$, of the same mass $m$, are moving toward each other with the velocities shown. Knowing that the coefficient of restitution between the collars is 0 (plastic impact), show that after impact (*a*) the common velocity of the collars is equal to half the difference in their speeds before impact, (*b*) the loss in kinetic energy is $\frac{1}{4}m(v_A + v_B)^2$.

*13.156*  Collars $A$ and $B$, of the same mass $m$, are moving in the same direction with $v_A > v_B$. Neglecting friction and knowing that the coefficient of restitution is 0.5, determine (*a*) the velocity of each collar after impact, (*b*) the energy loss during impact.

**13.157**  Two steel blocks slide without friction on a horizontal surface; immediately before impact their velocities are as shown. Knowing that $e = 0.75$, determine (*a*) their velocities after impact, (*b*) the energy loss during impact.

Fig. **P13.157 and P13.158**

**13.158**  The velocities of two steel blocks before impact are as shown. If after impact the velocity of block $B$ is observed to be 2.5 m/s to the right, determine the coefficient of restitution between the two blocks.

**13.159**  A .54 kg ball $A$ is moving with a velocity $\mathbf{v}_A$ when it is struck by a 1 kg ball $B$ which has a velocity $\mathbf{v}_B$ of magnitude $v_B = 5.5$ m/s. Knowing that the velocity of ball $B$ is zero after impact and that the coefficient of restitution is 0.8, determine the velocity of ball $A$ (*a*) before impact, (*b*) after impact.

**13.160**  A .54 kg ball $A$ is moving with a velocity $\mathbf{v}_A$ of magnitude $v_A = 7.3$ m/s when it is struck by a 1 kg ball $B$ which has a velocity $\mathbf{v}_B$. Knowing that the velocity of ball $A$ is zero after impact and that the coefficient of restitution is 0.2, determine the velocity of ball $B$ (*a*) before impact, (*b*) after impact.

*13.161*  Two disks sliding on a frictionless horizontal plane with opposite velocities of the same magnitude $v_0$ hit each other squarely. Disk $A$ is known to have a mass of 6 kg and is observed to have zero velocity after impact. Determine (*a*) the mass of disk $B$, knowing that the coefficient of restitution between the two disks is 0.5, (*b*) the range of possible values of the mass of disk $B$ if the coefficient of restitution between the two disks is unknown.

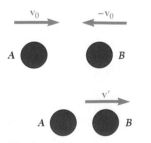

Fig. *P13.159 and P13.160*

Fig. *P13.161*

**13.162** Three steel spheres of equal mass are suspended from the ceiling by cords of equal length which are spaced at a distance slightly greater than the diameter of the spheres. After being pulled back and released, sphere A hits sphere B, which then hits sphere C. Denoting by e the coefficient of restitution between the spheres and by $\mathbf{v}_0$ the velocity of A just before it hits B, determine (a) the velocities of A and B immediately after the first collision, (b) the velocities of B and C immediately after the second collision. (c) Assuming now that n spheres are suspended from the ceiling and that the first sphere is pulled back and released as described above, determine the velocity of the last sphere after it is hit for the first time. (d) Use the result of part c to obtain the velocity of the last sphere when $n = 8$ and $e = 0.9$.

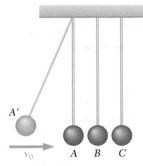

**Fig. P13.162**

**13.163** Packages in an automobile parts supply house are transported to the loading dock by pushing them along a roller track with very little friction. At the instant shown packages B and C are at rest and package A has a velocity of 1.8 m/s. Knowing that the coefficient of restitution between the packages is 0.3, determine (a) the velocity of package C after A hits B and B hits C, (b) the velocity of A after it hits B for the second time.

**Fig. P13.163**

**13.164** A 1.1-kg ball A is falling vertically with a velocity of magnitude $v_A = 2.5$ m/s when it is hit as shown by a 0.7-kg ball B which has a velocity of magnitude $v_B = 2$ m/s. Knowing that the coefficient of restitution between the two balls is $e = 0.75$ and assuming no friction, determine the velocity of each ball immediately after impact.

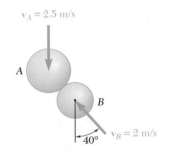

**Fig. P13.164**

**13.165** Two identical billiard balls can move freely on a horizontal table. Ball A has a velocity $\mathbf{v}_0$ as shown and hits ball B, which is at rest, at a point C defined by $\theta = 45°$. Knowing that the coefficient of restitution between the two balls is $e = 0.8$ and assuming no friction, determine the velocity of each ball after impact.

**Fig. P13.165**

**13.166** A 7.9 kg sphere $A$ of radius 11.4 cm. moving with a velocity $\mathbf{v}_0$ of magnitude $v_0 = 1.8$ cm/s strikes a .73 kg sphere $B$ of radius 5 cm. which was at rest. Both spheres are hanging from identical light flexible cords. Knowing that the coefficient of restitution is 0.8, determine the velocity of each sphere immediately after impact.

**Fig. P13.166**

**13.167** Solve Prob. 13.166, assuming that the flexible cords from which the spheres are hanging are replaced by light rigid rods.

**13.168** A 26.3 kg sphere $A$ of radius 11.4 cm. moving with a velocity of magnitude $v_0 = 1.8$ cm/s strikes a 2.4 kg sphere $B$ of radius 5 cm. which is initially at rest. All surfaces of contact are frictionless. Assuming perfectly elastic impact ($e = 1$) and no friction, determine the velocity of each sphere immediately after impact.

**Fig. P13.168**

**13.169** (a) Show that when two identical spheres $A$ and $B$ with coefficient of restitution $e = 1$ collide while moving with velocities $\mathbf{v}_A$ and $\mathbf{v}_B$ which are perpendicular to each other they will rebound with velocities $\mathbf{v}_A'$ and $\mathbf{v}_B'$ which are also perpendicular to each other. (b) To verify this property, solve Sample Prob. 13.15, assuming $e = 1$, and determine the angle formed by $\mathbf{v}_A'$ and $\mathbf{v}_B'$.

**Fig. P13.169**

**13.170** The coefficient of restitution is 0.9 between the two 60-mm-diameter billiard balls $A$ and $B$. Ball $A$ is moving in the direction shown with a velocity of 1 m/s when it strikes ball $B$, which is at rest. Knowing that after impact $B$ is moving in the $x$ direction, determine (a) the angle $\theta$, (b) the velocity of $B$ after impact.

**Fig. P13.170**

**13.171**  In problem 13.170, determine (*a*) the velocity of ball *A* after impact, (*b*) the fraction of the initial energy lost during impact.

**13.172**  A sphere rebounds as shown after striking an inclined plane with a vertical velocity $v_0$ of magnitude $v_0 = 5$ m/s. Knowing that $\alpha = 30°$ and $e = 0.8$ between the sphere and the plane, determine the height $h$ reached by the sphere.

**13.173**  A boy releases a ball with an initial horizontal velocity at a height of 0.6 m. The ball bounces off the ground at point *A*, bounces off the wall at point *B*, and hits the ground again at point *C*. Neglecting friction, determine (*a*) the coefficient of restitution at point *A*, (*b*) the coefficient of restitution at point *B*.

**Fig. P13.172**

**Fig. P13.173**

**13.174**  A girl throws a ball at an inclined wall from a height of .9 m, hitting the wall at *A* with a horizontal velocity $v_0$ of magnitude 7.6 m/s. Knowing that the coefficient of restitution between the ball and the wall is 0.9 and neglecting friction, determine the distance $d$ from the foot of the wall to the point *B* where the ball will hit the ground after bouncing off the wall.

**Fig. P13.174**

**13.175**  A ball is thrown into a 90° corner with an initial velocity **v**. Denoting the coefficient of restitution by $e$, show that the final velocity is of magnitude $ev$ and that the initial and final paths *AB* and *CD* are parallel.

**Fig. P13.175**

**13.176** A 1.1 kg block $B$ is moving with a velocity $v_0$ of magnitude $v_0 = 1.8$ m/s as it hits the .68 kg sphere $A$, which is at rest and hanging from a cord attached at $O$. Knowing that $\mu_k = 0.6$ between the block and the horizontal surface and $e = 0.8$ between the block and the sphere, determine after impact, (*a*) the maximum height $h$ reached by the sphere, (*b*) the distance $x$ traveled by the block.

**Fig. P13.176**

**Fig. P13.177**

**13.177** After having been pushed by an airline employee, an empty 36 kg luggage carrier $A$ hits with a velocity of 4.57 m/s an identical carrier $B$ containing a 13.6 kg suitcase equipped with rollers. The impact causes the suitcase to roll into the left wall of carrier $B$. Knowing that the coefficient of restitution between the two carriers is 0.80 and the coefficient of restitution between the suitcase and the wall of the carrier is 0.30, determine (*a*) the velocity of carrier $B$ after the suitcase hits the wall for the first time, (*b*) the total energy lost in the impact.

**13.178** Two cars of the same mass run head-on into each other at $C$. After the collision, the cars skid with their brakes locked and come to a stop in the position shown in the lower part of the figure. Knowing that the speed of car $A$ just before impact was 5 km/h and that the coefficient of kinetic friction between the pavement and the tires of both cars is 0.30, determine (*a*) the speed of car $B$ just before impact, (*b*) the effective coefficient of restitution between the two cars.

**Fig. P13.178**

**13.179**  Blocks $A$ and $B$ each have a mass of 0.4 kg and block $C$ has a mass of 1.2 kg. The coefficient of friction between the blocks and the plane is $\mu_k = 0.30$. Initially block $A$ is moving at a speed $v_0 = 3$ m/s and blocks $B$ and $C$ are at rest (Fig. 1). After $A$ strikes $B$ and $B$ strikes $C$, all three blocks come to a stop in the positions shown (Fig. 2). Determine (*a*) the coefficients of restitution between $A$ and $B$ and between $B$ and $C$, (*b*) the displacement $x$ of block $C$.

(1)

(2)

**Fig. P13.179**

**13.180**  An 8-kg cylinder $C$ is released from rest in the position shown and drops onto a 5-kg platform $A$ which is at rest and is supported by an inextensible cord attached to a 5-kg counterweight $B$. Knowing that the coefficient of restitution between cylinder $C$ and platform $A$ is 0.8, determine (*a*) the velocities of $C$ and $A$ immediately after the first impact, (*b*) the impulse of the force exerted on platform $A$ by the cord during the first impact.

**Fig. P13.180 and P13.181**

**13.181**  An 8-kg cylinder $C$ is released from rest in the position shown and drops onto a 5-kg platform $A$ which is at rest and is supported by an inextensible cord attached to a 5-kg counterweight $B$. Knowing that the coefficient of restitution between cylinder $C$ and platform $A$ is 0.8 and that the velocity of $C$ is zero immediately after the first impact, determine (*a*) the velocities of $C$ and $A$ immediately before the second impact, (*b*) the velocities of $C$ and $A$ immediately after the second impact.

**13.182** A .27 kg collar $A$ is released from rest, slides down a frictionless rod, and strikes a .82 kg collar $B$ which is at rest and supported by a spring of constant 496.2 N/m. Knowing that the coefficient of restitution between the two collars is 0.9, determine (*a*) the maximum distance collar $A$ moves up the rod after impact, (*b*) the maximum distance collar $B$ moves down the rod after impact.

**Fig. *P13.182* and *P13.183***

**13.183** A .27 kg collar $A$ is released from rest, slides down a frictionless rod, and strikes a .82 kg collar $B$ which is at rest and supported by a spring of constant 496.2 N/m. Knowing that the velocity of collar $A$ is zero immediately after impact, determine (*a*) the coefficient of restitution between the two collars, (*b*) the energy lost in the impact, (*c*) the maximum distance collar $B$ moves down the rod after impact.

**13.184** Ball $B$ is hanging from an inextensible cord. An identical ball $A$ is released from rest when it is just touching the cord and drops through the vertical distance $h_A = 200$ mm before striking ball $B$. Assuming perfectly elastic impact ($e = 1$) and no friction, determine the resulting maximum vertical displacement $h_B$ of ball $B$.

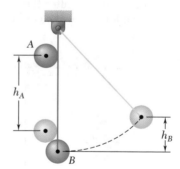

**Fig. P13.184**

**13.185** A 2-kg sphere moving to the right with a velocity of 5 m/s strikes at $A$ the surface of a 9-kg quarter cylinder which is initially at rest and in contact with a spring of constant 20 kN/m. The spring is held by cables so that it is initially compressed 50 mm. Neglecting friction and knowing that the coefficient of restitution is 0.6, determine (*a*) the velocity of the sphere immediately after impact, (*b*) the maximum compressive force in the spring.

**Fig. P13.185**

**13.186** A 22.7 kg sphere A of radius 11.4 cm. moving with a velocity of magnitude $v_0 = 1.8$ m/s strikes a 2 kg sphere B of radius 5 cm. which is hanging from an inextensible cord and is initially at rest. Knowing that sphere B swings to a maximum height $h = .23$ m, determine the coefficient of restitution between the two spheres.

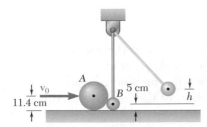

**Fig. P13.186 and P13.187**

**13.187** A 22.7 kg sphere A of radius 11.4 cm. moving with a velocity of magnitude $v_0 = 1.8$ m/s strikes a 2 kg sphere B of radius 5 cm. which is hanging from an inextensible cord and is initially at rest. Sphere B swings to a maximum height $h$ after the impact. Determine the range of values of $h$ for values of the coefficient of restitution $e$ between zero and one.

*13.188* A 340-g ball B is hanging from an inextensible cord attached to a support C. A 170-g ball A strikes B with a velocity $v_0$ of magnitude 1.5 m/s at an angle of 60° with the vertical. Assuming perfectly elastic impact $(e = 1)$ and no friction, determine the height $h$ reached by ball B.

**Fig. *P13.188***

*13.189* A 2-kg sphere A strikes the frictionless inclined surface of a 6-kg wedge B at a 90° angle with a velocity of magnitude 4 m/s. The wedge can roll freely on the ground and is initially at rest. Knowing that the coefficient of restitution between the wedge and the sphere is 0.50 and that the inclined surface of the wedge forms an angle $\theta = 40°$ with the horizontal, determine (a) the velocities of the sphere and of the wedge immediately after impact, (b) the energy lost due to the impact.

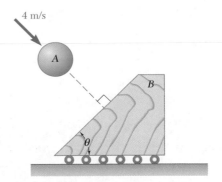

**Fig. *P13.189***

This chapter was devoted to the method of work and energy and to the method of impulse and momentum. In the first half of the chapter we studied the method of work and energy and its application to the analysis of the motion of particles.

Work of a force

We first considered a force $\mathbf{F}$ acting on a particle $A$ and defined the *work of* $\mathbf{F}$ *corresponding to the small displacement* $d\mathbf{r}$ [Sec. 13.2] as the quantity

$$dU = \mathbf{F} \cdot d\mathbf{r} \qquad (13.1)$$

or, recalling the definition of the scalar product of two vectors,

$$dU = F \, ds \cos \alpha \qquad (13.1')$$

where $\alpha$ is the angle between $\mathbf{F}$ and $d\mathbf{r}$ (Fig. 13.29). The work of $\mathbf{F}$ during a finite displacement from $A_1$ to $A_2$, denoted by $U_{1 \to 2}$, was obtained by integrating Eq. (13.1) along the path described by the particle:

$$U_{1 \to 2} = \int_{A_1}^{A_2} \mathbf{F} \cdot d\mathbf{r} \qquad (13.2)$$

For a force defined by its rectangular components, we wrote

$$U_{1 \to 2} = \int_{A_1}^{A_2} (F_x \, dx + F_y \, dy + F_z \, dz) \qquad (13.2'')$$

Work of a weight

The work of the weight $\mathbf{W}$ of a body as its center of gravity moves from the elevation $y_1$ to $y_2$ (Fig. 13.30) was obtained by substituting $F_x = F_z = 0$ and $F_y = -W$ into Eq. (13.2'') and integrating. We found

$$U_{1 \to 2} = -\int_{y_1}^{y_2} W \, dy = Wy_1 - Wy_2 \qquad (13.4)$$

**Fig. 13.29**

**Fig. 13.30**

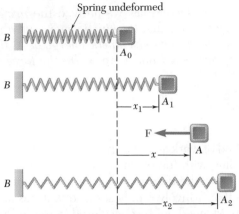

Spring undeformed

**Fig. 13.31**

The work of a force **F** exerted by a spring on a body $A$ during a finite displacement of the body (Fig. 13.31) from $A_1(x = x_1)$ to $A_2(x = x_2)$ was obtained by writing

$$dU = -F\,dx = -kx\,dx$$
$$U_{1\rightarrow 2} = -\int_{x_1}^{x_2} kx\,dx = \tfrac{1}{2}kx_1^2 - \tfrac{1}{2}kx_2^2 \qquad (13.6)$$

The work of **F** is therefore positive *when the spring is returning to its undeformed position.*

Work of the force exerted by a spring

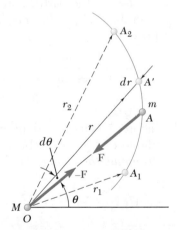

**Fig. 13.32**

The *work of the gravitational force* **F** exerted by a particle of mass $M$ located at $O$ on a particle of mass $m$ as the latter moves from $A_1$ to $A_2$ (Fig. 13.32) was obtained by recalling from Sec. 12.10 the expression for the magnitude of **F** and writing

$$U_{1\rightarrow 2} = -\int_{r_1}^{r_2} \frac{GMm}{r^2}\,dr = \frac{GMm}{r_2} - \frac{GMm}{r_1} \qquad (13.7)$$

Work of the gravitational force

The *kinetic energy of a particle* of mass $m$ moving with a velocity **v** [Sec. 13.3] was defined as the scalar quantity

$$T = \tfrac{1}{2}mv^2 \qquad (13.9)$$

Kinetic energy of a particle

**Principle of work and energy**

From Newton's second law we derived the *principle of work and energy,* which states that *the kinetic energy of a particle at $A_2$ can be obtained by adding to its kinetic energy at $A_1$ the work done during the displacement from $A_1$ to $A_2$ by the force $\mathbf{F}$ exerted on the particle:*

$$T_1 + U_{1\to2} = T_2 \qquad (13.11)$$

**Method of work and energy**

The method of work and energy simplifies the solution of many problems dealing with forces, displacements, and velocities, since it does not require the determination of accelerations [Sec. 13.4]. We also note that it involves only scalar quantities and that forces which do no work need not be considered [Sample Probs. 13.1 and 13.3]. However, this method should be supplemented by the direct application of Newton's second law to determine a force normal to the path of the particle [Sample Prob. 13.4].

**Power and mechanical efficiency**

The power developed by a machine and its mechanical efficiency were discussed in Sec. 13.5. Power was defined as the time rate at which work is done:

$$\text{Power} = \frac{dU}{dt} = \mathbf{F} \cdot \mathbf{v} \qquad (13.12, \ 13.13)$$

where $\mathbf{F}$ is the force exerted on the particle and $\mathbf{v}$ the velocity of the particle [Sample Prob. 13.5]. The *mechanical efficiency,* denoted by $\eta$, was expressed as

$$\eta = \frac{\text{power output}}{\text{power input}} \qquad (13.15)$$

**Conservative force. Potential energy**

When the work of a force $\mathbf{F}$ is independent of the path followed [Secs. 13.6 and 13.7], the force $\mathbf{F}$ is said to be a *conservative force,* and its work is equal to *minus the change in the potential energy $V$ associated with $\mathbf{F}$:*

$$U_{1\to2} = V_1 - V_2 \qquad (13.19')$$

The following expressions were obtained for the potential energy associated with each of the forces considered earlier:

Force of gravity (weight): $\qquad\qquad V_g = Wy \qquad (13.16)$

Gravitational force: $\qquad\qquad V_g = -\dfrac{GMm}{r} \qquad (13.17)$

Elastic force exerted by a spring: $\quad V_e = \tfrac{1}{2}kx^2 \qquad (13.18)$

Substituting for $U_{1\rightarrow 2}$ from Eq. (13.19′) into Eq. (13.11) and rearranging the terms [Sec. 13.8], we obtained

$$T_1 + V_1 = T_2 + V_2 \qquad (13.24)$$

This is the *principle of conservation of energy,* which states that when a particle moves under the action of conservative forces, *the sum of its kinetic and potential energies remains constant.* The application of this principle facilitates the solution of problems involving only conservative forces [Sample Probs. 13.6 and 13.7].

Principle of conservation of energy

Recalling from Sec. 12.9 that, when a particle moves under a central force **F**, its angular momentum about the center of force $O$ remains constant, we observed [Sec. 13.9] that, if the central force **F** is also conservative, the principles of conservation of angular momentum and of conservation of energy can be used jointly to analyze the motion of the particle [Sample Prob. 13.8]. Since the gravitational force exerted by the earth on a space vehicle is both central and conservative, this approach was used to study the motion of such vehicles [Sample Prob. 13.9] and was found particularly effective in the case of an *oblique launching.* Considering the initial position $P_0$ and an arbitrary position $P$ of the vehicle (Fig. 13.33), we wrote

Motion under a gravitational force

$(H_O)_0 = H_O$: $\qquad r_0 m v_0 \sin \phi_0 = rmv \sin \phi \qquad (13.25)$

$T_0 + V_0 = T + V$: $\quad \frac{1}{2}mv_0^2 - \frac{GMm}{r_0} = \frac{1}{2}mv^2 - \frac{GMm}{r} \qquad (13.26)$

where $m$ was the mass of the vehicle and $M$ the mass of the earth.

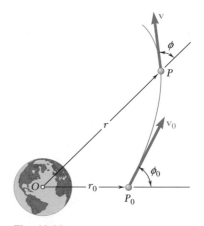

**Fig. 13.33**

The second half of the chapter was devoted to the method of impulse and momentum and to its application to the solution of various types of problems involving the motion of particles.

Principle of impulse and momentum for a particle

The *linear momentum of a particle* was defined [Sec. 13.10] as the product $m\mathbf{v}$ of the mass $m$ of the particle and its velocity $\mathbf{v}$. From Newton's second law, $\mathbf{F} = m\mathbf{a}$, we derived the relation

$$m\mathbf{v}_1 + \int_{t_1}^{t_2} \mathbf{F}\, dt = m\mathbf{v}_2 \qquad (13.28)$$

where $m\mathbf{v}_1$ and $m\mathbf{v}_2$ represent the momentum of the particle at a time $t_1$ and a time $t_2$, respectively, and where the integral defines the *linear impulse of the force* **F** during the corresponding time interval. We wrote therefore

$$m\mathbf{v}_1 + \mathbf{Imp}_{1\rightarrow 2} = m\mathbf{v}_2 \qquad (13.30)$$

which expresses the principle of impulse and momentum for a particle.

When the particle considered is subjected to several forces, the sum of the impulses of these forces should be used; we had

$$m\mathbf{v}_1 + \Sigma\,\mathbf{Imp}_{1\to2} = m\mathbf{v}_2 \tag{13.32}$$

Since Eqs. (13.30) and (13.32) involve *vector quantities*, it is necessary to consider their $x$ and $y$ components separately when applying them to the solution of a given problem [Sample Probs. 13.10 and 13.11].

*Impulsive motion*

The method of impulse and momentum is particularly effective in the study of the *impulsive motion* of a particle, when very large forces, called *impulsive forces*, are applied for a very short interval of time $\Delta t$, since this method involves the impulses $\mathbf{F}\,\Delta t$ of the forces, rather than the forces themselves [Sec. 13.11]. Neglecting the impulse of any nonimpulsive force, we wrote

$$m\mathbf{v}_1 + \Sigma\mathbf{F}\,\Delta t = m\mathbf{v}_2 \tag{13.35}$$

In the case of the impulsive motion of several particles, we had

$$\Sigma m\mathbf{v}_1 + \Sigma\mathbf{F}\,\Delta t = \Sigma m\mathbf{v}_2 \tag{13.36}$$

where the second term involves only impulsive, external forces [Sample Prob. 13.12].

In the particular case *when the sum of the impulses of the external forces is zero,* Eq. (13.36) reduces to $\Sigma m\mathbf{v}_1 = \Sigma m\mathbf{v}_2$; that is, *the total momentum of the particles is conserved.*

*Direct central impact*

In Secs. 13.12 through 13.14, we considered the *central impact* of two colliding bodies. In the case of a *direct central impact* [Sec. 13.13], the two colliding bodies $A$ and $B$ were moving along the *line of impact* with velocities $\mathbf{v}_A$ and $\mathbf{v}_B$, respectively (Fig. 13.34). Two equations could be used to determine their

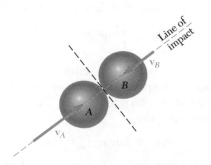

**Fig. 13.34**

velocities $\mathbf{v}_A'$ and $\mathbf{v}_B'$ after the impact. The first expressed conservation of the total momentum of the two bodies,

$$m_A v_A + m_B v_B = m_A v_A' + m_B v_B' \qquad (13.37)$$

where a positive sign indicates that the corresponding velocity is directed to the right, while the second related the *relative velocities* of the two bodies before and after the impact,

$$v_B' - v_A' = e(v_A - v_B) \qquad (13.43)$$

The constant $e$ is known as the *coefficient of restitution;* its value lies between 0 and 1 and depends in a large measure on the materials involved. When $e = 0$, the impact is said to be *perfectly plastic;* when $e = 1$, it is said to be *perfectly elastic* [Sample Prob. 13.13].

In the case of an *oblique central impact* [Sec. 13.14], the velocities of the two colliding bodies before and after the impact were resolved into $n$ components along the line of impact and $t$ components along the common tangent to the surfaces in contact (Fig. 13.35). We observed that the $t$ component of the velocity of

Oblique central impact

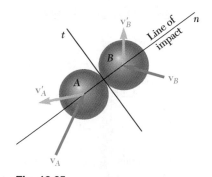

**Fig. 13.35**

each body remained unchanged, while the $n$ components satisfied equations similar to Eqs. (13.37) and (13.43) [Sample Probs. 13.14 and 13.15]. It was shown that although this method was developed for bodies moving freely before and after the impact, it could be extended to the case when one or both of the colliding bodies is constrained in its motion [Sample Prob. 13.16].

In Sec. 13.15, we discussed the relative advantages of the three fundamental methods presented in this chapter and the preceding one, namely, Newton's second law, work and energy, and impulse and momentum. We noted that the method of work and energy and the method of impulse and momentum can be combined to solve problems involving a short impact phase during which impulsive forces must be taken into consideration [Sample Prob. 13.17].

Using the three fundamental methods of kinetic analysis

**Fig. P13.190**

**13.190** Skid marks on a drag race track indicate that the rear (drive) wheels of a car skid for the first 18 m and roll with slipping impending during the remaining 382 m. The front wheels of the car are just off the ground for the first 18 m, and for the remainder of the race 75 percent of the weight of the car is on the rear wheels. Knowing that the speed of the car is 58 km/h at the end of the first 18 m and that the coefficient of kinetic friction is 80 percent of the coefficient of static friction, determine the speed of the car at the end of the 400-m track. Ignore air resistance and rolling resistance.

**13.191** A 3-kg collar $C$ slides on a frictionless vertical rod. It is pushed up into the position shown, compressing the upper spring by 50 mm and released. Determine (a) the maximum deflection of the lower spring, (b) the maximum velocity of the collar.

$k = 3$ N/mm

150 mm

$C$

300 mm

$k = 2$ N/mm

**Fig. P13.191**

**13.192** A small block slides at a speed $v = 3$ m/s on a horizontal surface at a height $h = 1$ m above the ground. Determine (a) the angle $\theta$ at which it will leave the cylindrical surface $BCD$, (b) the distance $x$ at which it will hit the ground. Neglect friction and air resistance.

v

B

C

$h$

$\theta$

D  E

$x$

**Fig. P13.192**

848

**13.193**   The system shown is in equilibrium when $\phi = 0$. Knowing that initially $\phi = 90°$ and that block $C$ is given a slight nudge when the system is in that position, determine the velocity of the block as it passes through the equilibrium position $\phi = 0$. Neglect the mass of the rod.

**13.194**   A spacecraft is describing an elliptic orbit of minimum altitude $h_A = 2403.5$ km and maximum altitude $h_B = 9613.8$ km above the surface of the earth. Determine the speed of the spacecraft at $A$.

**Fig. P13.193**

**Fig. P13.194**

**13.195**   A truck is traveling on a level road at a speed of 96 km/h when its brakes are applied to slow it down to 32 km/h. An antiskid braking system limits the braking force to a value at which the wheels of the truck are just about to slide. Knowing that the coefficient of static friction between the road and the wheels is 0.65, determine the shortest time needed for the truck to slow down.

**13.196**   The last segment of the triple jump track-and-field event is the jump, in which the athlete makes a final leap, landing in a sand-filled pit. Assuming that the velocity of an 84-kg athlete just before landing is 9.14 m/s at an angle of 35° with the horizontal and that the athlete comes to a complete stop in 0.22 s after landing, determine the horizontal component of the average impulsive force exerted on his feet during landing.

**Fig. P13.196**

**13.197**   At an intersection car $B$ was traveling south and car $A$ was traveling 30° north of east when they slammed into each other. Upon investigation it was found that after the crash the two cars got stuck and skidded off at an angle of 10° north of east. Each driver claimed that he was going at the speed limit of 48 km/h and that he tried to slow down but couldn't avoid the crash because the other driver was going a lot faster. Knowing that the weights of cars $A$ and $B$ were 16 kN and 12.5 kN, respectively, determine (a) which car was going faster, (b) the speed of the faster of the two cars if the slower car was traveling at the speed limit.

**Fig. P13.197**

**13.198** Two identical cars $A$ and $B$ are at rest on a loading dock with brakes released. Car $C$, of a slightly different style but of the same weight, has been pushed by dockworkers and hits car $B$ with a velocity of 1.37 m/s. Knowing that the coefficient of restitution is 0.8 between $B$ and $C$ and 0.5 between $A$ and $B$, determine the velocity of each car after all collisions have taken place.

**Fig. P13.198**

**13.199** A .68 kg ball $A$ is moving with a velocity of magnitude 5.5 m/s when it is hit by a 1.1 kg ball $B$ which has a velocity of magnitude 3.7 m/s. Knowing that the coefficient of restitution is 0.8 and assuming no friction, determine the velocity of each ball after impact.

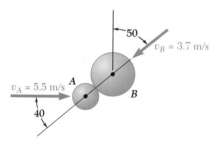

**Fig. P13.199**

**13.200** A ball hits the ground at $A$ with a velocity $\mathbf{v}_0$ of 6 m/s at an angle of 60° with the horizontal. Knowing that $e = 0.6$ between the ball and the ground and that after rebounding the ball reaches point $B$ with a horizontal velocity, determine (a) the distances $h$ and $d$, (b) the velocity of the ball as it reaches $B$.

**Fig. P13.200**

**13.201** A .68 kg sphere $A$ moving with a velocity $\mathbf{v}_0$ parallel to the ground strikes the inclined face of a 2 kg wedge $B$ which can roll freely on the ground and is initially at rest. After impact the sphere is observed from the ground to be moving straight up. Knowing that the coefficient of restitution between the sphere and the wedge is $e = 0.6$, determine (a) the angle $\theta$ that the inclined face of the wedge makes with the horizontal, (b) the energy lost due to the impact.

**Fig. P13.201**

# Computer Problems

**13.C1**   A 6-kg collar is attached to a spring anchored at point $C$ and can slide on a frictionless rod forming an angle of 30° with the vertical. The spring is of constant $k$ and is unstretched when the collar is at $A$. Knowing that the collar is released from rest at $A$, use computational software to calculate and plot the speed of the collar at point $B$ for values of $k$ from 20 N/m to 400 N/m.

**Fig. P13.C1**

**13.C2**   A 4.5 kg bag is gently pushed off the top of a wall and swings in a vertical plane at the end of a .3 m rope which can withstand a maximum tension $F_m$ before it breaks. Derive expressions for (a) the difference in elevation $h$ between point $A$ and point $B$ where the rope will break, (b) the distance $d$ from the vertical wall to the point where the bag will strike the floor. Using computational software, calculate $h$ and $d$ for values of $F_m$ from 4.5 kg to 11.3 kg, using .45 kg increments.

**Fig. P13.C2**

**Fig. P13.C3**

**13.C3** The two blocks shown are released from rest. The coefficients of static and kinetic friction between the block $A$ and the incline are $\mu_s = 0.25$ and $\mu_k = 0.20$. Neglect the masses of the pulleys and the effect of friction in the pulleys. ($a$) Denoting by $s$ the distance block $A$ moves, determine and plot the speed of block $A$ as a function of $s$ for $0 \leq s \leq 1.8$ m. ($b$) Using the expression derived in part ($a$), derive an expression for the distance $s$ as a function of time $t$ and plot this function for values of $s$ from 0 to 1.8 m.

**13.C4** A 3-kg block is attached to a cable and to a spring as shown and is initially at rest. The constant of the spring is $k = 2.5$ kN/m and the initial tension in the cable is 18 N. Knowing that the cable is cut, ($a$) derive an expression for the velocity of the block as a function of its displacement $s$, ($b$) determine the maximum displacement $s_m$ and the maximum speed $v_m$, ($c$) plot the speed of the block as a function of $s$ for $0 \leq s \leq s_m$.

**Fig. P13.C4**

**13.C5** Nonlinear springs are classified as hard or soft, depending upon the curvature of their force-deflection curve (see figure). A delicate instrument having a weight of 4.5 kg is placed on a spring of length $l$ so that its base is just touching the undeformed spring and then inadvertently released from that position. For each of the following three cases, ($a$) derive an expression for the velocity of the instrument as a function of its displacement $x$, ($b$) determine the maximum displacement $x_m$, ($c$) plot the speed of the instrument as a function of $x$ for $0 \leq x \leq x_m$.

(1) Linear spring with a constant $k = 2627$ N/m.
(2) Hard spring for which $F = (2627 \text{ N/m})x(1 + 0.064x^2)$
(3) Soft spring for which $F = (2627 \text{ N/m})x(1 - 0.064x^2)$

**Fig. P13.C5**

**13.C6** A 1-kg collar is attached to a spring and slides without friction along a circular rod which lies in a vertical plane. The spring has a constant $k = 0.60$ kN/m and is undeformed when the collar is at point $B$. Knowing that the collar is released from rest at point $C$, (a) determine and plot the speed of the collar as a function of $\theta$ for values of $\theta$ from 90° to $\theta_m$, where $\theta_m$ is the maximum value of $\theta$, (b) determine the maximum speed of the collar and the corresponding value of $\theta$.

**Fig. P13.C6**

**13.C7** A 25-kg block initially at rest is acted upon by a force **P** which varies in magnitude with time as shown. The coefficients of static and kinetic friction between the block and the horizontal surface are $\mu_s = 0.5$ and $\mu_k = 0.4$. (a) Determine and plot the speed of the block as a function of $t$ for $0 \leq t \leq 16$ s. (b) Determine the maximum velocity of the block and the corresponding value of $t$.

**Fig. P13.C7**

## CHAPTER 14

# Systems of Particles

The determination of the trust created by the rotating blades of a hovering helicopter is based on the analysis of the motion of a *variable system of particles,* i.e., the motion of a large number of air particles considered together, rather than separately.

**SYSTEMS OF PARTICLES**

14.1 Introduction
14.2 Application of Newton's Laws to the Motion of a System of Particles. Effective Forces
14.3 Linear and Angular Momentum of a System of Particles
14.4 Motion of the Mass Center of a System of Particles
14.5 Angular Momentum of a System of Particles about Its Mass Center
14.6 Conservation of Momentum for a System of Particles
14.7 Kinetic Energy of a System of Particles
14.8 Work-Energy Principle. Conservation of Energy for a System of Particles
14.9 Principle of Impulse and Momentum for a System of Particles
14.10 Variable Systems of Particles
14.11 Steady Stream of Particles
14.12 Systems Gaining or Losing Mass

## 14.1. INTRODUCTION

In this chapter you will study the motion of *systems of particles*, that is, the motion of a large number of particles considered together. The first part of the chapter is devoted to systems consisting of well-defined particles; the second part considers the motion of variable systems, that is, systems which are continually gaining or losing particles, or doing both at the same time.

In Sec. 14.2, Newton's second law will first be applied to each particle of the system. Defining the *effective force* of a particle as the product $m_i\mathbf{a}_i$ of its mass $m_i$ and its acceleration $\mathbf{a}_i$, we will show that the *external forces* acting on the various particles form a system equipollent to the system of the effective forces, that is, both systems have the same resultant and the same moment resultant about any given point. In Sec. 14.3, it will be further shown that the resultant and moment resultant of the external forces are equal, respectively, to the rate of change of the total linear momentum and of the total angular momentum of the particles of the system.

In Sec. 14.4, the *mass center* of a system of particles is defined and the motion of that point is described, while in Sec. 14.5 the motion of the particles about their mass center is analyzed. The conditions under which the linear momentum and the angular momentum of a system of particles are conserved are discussed in Sec. 14.6, and the results obtained in that section are applied to the solution of various problems.

Sections 14.7 and 14.8 deal with the application of the work-energy principle to a system of particles, and Sec. 14.9 with the application of the impulse-momentum principle. These sections also contain a number of problems of practical interest.

It should be noted that while the derivations given in the first part of this chapter are carried out for a system of independent particles, they remain valid when the particles of the system are rigidly connected, that is, when they form a rigid body. In fact, the results obtained here will form the foundation of our discussion of the kinetics of rigid bodies in Chaps. 16 through 18.

The second part of this chapter is devoted to the study of variable systems of particles. In Sec. 14.11 you will consider steady streams of particles, such as a stream of water diverted by a fixed vane, or the flow of air through a jet engine, and learn to determine the force exerted by the stream on the vane and the thrust developed by the engine. Finally, in Sec. 14.12, you will learn how to analyze systems which gain mass by continually absorbing particles or lose mass by continually expelling particles. Among the various practical applications of this analysis will be the determination of the thrust developed by a rocket engine.

## 14.2. APPLICATION OF NEWTON'S LAWS TO THE MOTION OF A SYSTEM OF PARTICLES. EFFECTIVE FORCES

In order to derive the equations of motion for a system of $n$ particles, let us begin by writing Newton's second law for each individual particle of the system. Consider the particle $P_i$, where $1 \le i \le n$. Let $m_i$ be the mass of $P_i$ and $\mathbf{a}_i$ its acceleration with respect to the newtonian frame of reference $Oxyz$. The force exerted on $P_i$ by another particle $P_j$ of the system (Fig. 14.1), called an *internal force,* will be denoted by $\mathbf{f}_{ij}$. The resultant of the internal forces exerted on $P_i$ by all the other particles of the system is thus $\sum\limits_{j=1}^{n} \mathbf{f}_{ij}$ (where $\mathbf{f}_{ii}$ has no meaning and is assumed to be equal to zero). Denoting, on the other hand, by $\mathbf{F}_i$ the resultant of all the *external forces* acting on $P_i$, we write Newton's second law for the particle $P_i$ as follows:

$$\mathbf{F}_i + \sum_{j=1}^{n} \mathbf{f}_{ij} = m_i \mathbf{a}_i \tag{14.1}$$

Denoting by $\mathbf{r}_i$ the position vector of $P_i$ and taking the moments about $O$ of the various terms in Eq. (14.1), we also write

$$\mathbf{r}_i \times \mathbf{F}_i + \sum_{j=1}^{n} (\mathbf{r}_i \times \mathbf{f}_{ij}) = \mathbf{r}_i \times m_i \mathbf{a}_i \tag{14.2}$$

Repeating this procedure for each particle $P_i$ of the system, we obtain $n$ equations of the type (14.1) and $n$ equations of the type (14.2), where $i$ takes successively the values $1, 2, \ldots, n$. The vectors $m_i \mathbf{a}_i$ are referred to as the *effective forces* of the particles.† Thus the equations obtained express the fact that the external forces $\mathbf{F}_i$ and the internal forces $\mathbf{f}_{ij}$ acting on the various particles form a system equivalent to the system of the effective forces $m_i \mathbf{a}_i$ (that is, one system may be replaced by the other) (Fig. 14.2).

**Fig. 14.1**

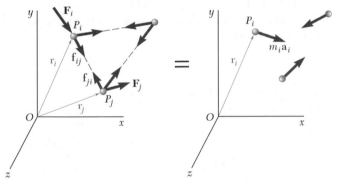

**Fig. 14.2**

Before proceeding further with our derivation, let us examine the internal forces $\mathbf{f}_{ij}$. We note that these forces occur in pairs $\mathbf{f}_{ij}$, $\mathbf{f}_{ji}$, where $\mathbf{f}_{ij}$ represents the force exerted by the particle $P_j$ on the particle $P_i$ and $\mathbf{f}_{ji}$ represents the force exerted by $P_i$ on $P_j$ (Fig. 14.2). Now, according to Newton's third law (Sec. 6.1), as extended by Newton's law of gravitation to particles acting at a distance (Sec. 12.10), the forces $\mathbf{f}_{ij}$ and $\mathbf{f}_{ji}$ are equal and opposite and have the same line of action. Their sum is therefore $\mathbf{f}_{ij} + \mathbf{f}_{ji} = 0$, and the sum of their moments about $O$ is

$$\mathbf{r}_i \times \mathbf{f}_{ij} + \mathbf{r}_j \times \mathbf{f}_{ji} = \mathbf{r}_i \times (\mathbf{f}_{ij} + \mathbf{f}_{ji}) + (\mathbf{r}_j - \mathbf{r}_i) \times \mathbf{f}_{ji} = 0$$

since the vectors $\mathbf{r}_j - \mathbf{r}_i$ and $\mathbf{f}_{ji}$ in the last term are collinear. Adding

†Since these vectors represent the resultants of the forces acting on the various particles of the system, they can truly be considered as forces.

all the internal forces of the system and summing their moments about $O$, we obtain the equations

$$\sum_{i=1}^{n}\sum_{j=1}^{n} \mathbf{f}_{ij} = 0 \qquad \sum_{i=1}^{n}\sum_{j=1}^{n} (\mathbf{r}_i \times \mathbf{f}_{ij}) = 0 \qquad (14.3)$$

which express the fact that the resultant and the moment resultant of the internal forces of the system are zero.

Returning now to the $n$ Eq. (14.1), where $i = 1, 2, \ldots, n$, we sum their left-hand members and sum their right-hand members. Taking into account the first of Eqs. (14.3), we obtain

$$\sum_{i=1}^{n} \mathbf{F}_i = \sum_{i=1}^{n} m_i\mathbf{a}_i \qquad (14.4)$$

Proceeding similarly with Eqs. (14.2) and taking into account the second of Eqs. (14.3), we have

$$\sum_{i=1}^{n} (\mathbf{r}_i \times \mathbf{F}_i) = \sum_{i=1}^{n} (\mathbf{r}_i \times m_i\mathbf{a}_i) \qquad (14.5)$$

Equations (14.4) and (14.5) express the fact that the system of the external forces $\mathbf{F}_i$ and the system of the effective forces $m_i\mathbf{a}_i$ have the same resultant and the same moment resultant. Referring to the definition given in Sec. 3.19 for two equipollent systems of vectors, we can therefore state that *the system of the external forces acting on the particles and the system of the effective forces of the particles are equipollent*† (Fig. 14.3).

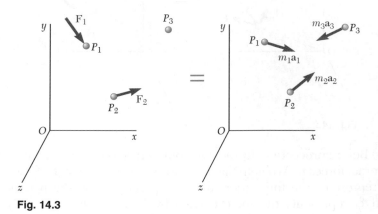

**Fig. 14.3**

---

†The result just obtained is often referred to as *d'Alembert's principle,* after the French mathematician Jean le Rond d'Alembert (1717–1783). However, d'Alembert's original statement refers to the motion of a system of connected bodies, with $\mathbf{f}_{ij}$ representing constraint forces which if applied by themselves will not cause the system to move. Since, as it will now be shown, this is in general not the case for the internal forces acting on a system of free particles, the consideration of d'Alembert's principle will be postponed until the motion of rigid bodies is considered (Chap. 16).

Equations (14.3) express the fact that the system of the internal forces $\mathbf{f}_{ij}$ is equipollent to zero. Note, however, that it does *not* follow that the internal forces have no effect on the particles under consideration. Indeed, the gravitational forces that the sun and the planets exert on one another are internal to the solar system and equipollent to zero. Yet these forces are alone responsible for the motion of the planets about the sun.

Similarly, it does not follow from Eqs. (14.4) and (14.5) that two systems of external forces which have the same resultant and the same moment resultant will have the same effect on a given system of particles. Clearly, the systems shown in Figs. 14.4*a* and 14.4*b* have

(a)

(b)

**Fig. 14.4**

the same resultant and the same moment resultant; yet the first system accelerates particle $A$ and leaves particle $B$ unaffected, while the second accelerates $B$ and does not affect $A$. It is important to recall that when we stated in Sec. 3.19 that two equipollent systems of forces acting on a rigid body are also equivalent, we specifically noted that this property could *not* be extended to a system of forces acting on a set of independent particles such as those considered in this chapter.

In order to avoid any confusion, blue equal signs are used to connect equipollent systems of vectors, such as those shown in Figs. 14.3 and 14.4. These signs indicate that the two systems of vectors have the same resultant and the same moment resultant. Red equals signs will continue to be used to indicate that two systems of vectors are equivalent, that is, that one system can actually be replaced by the other (Fig. 14.2).

## 14.3. LINEAR AND ANGULAR MOMENTUM OF A SYSTEM OF PARTICLES

Equations (14.4) and (14.5), obtained in the preceding section for the motion of a system of particles, can be expressed in a more condensed form if we introduce the linear and the angular momentum of the system of particles. Defining the linear momentum $\mathbf{L}$ of the system of particles as the sum of the linear momenta of the various particles of the system (Sec. 12.3), we write

$$\mathbf{L} = \sum_{i=1}^{n} m_i \mathbf{v}_i \qquad (14.6)$$

Defining the angular momentum $\mathbf{H}_O$ about $O$ of the system of particles in a similar way (Sec. 12.7), we have

$$\mathbf{H}_O = \sum_{i=1}^{n} (\mathbf{r}_i \times m_i\mathbf{v}_i) \tag{14.7}$$

Differentiating both members of Eqs. (14.6) and (14.7) with respect to $t$, we write

$$\dot{\mathbf{L}} = \sum_{i=1}^{n} m_i\dot{\mathbf{v}}_i = \sum_{i=1}^{n} m_i\mathbf{a}_i \tag{14.8}$$

and

$$\dot{\mathbf{H}}_O = \sum_{i=1}^{n} (\dot{\mathbf{r}}_i \times m_i\mathbf{v}_i) + \sum_{i=1}^{n} (\mathbf{r}_i \times m_i\dot{\mathbf{v}}_i)$$

$$= \sum_{i=1}^{n} (\mathbf{v}_i \times m_i\mathbf{v}_i) + \sum_{i=1}^{n} (\mathbf{r}_i \times m_i\mathbf{a}_i)$$

which reduces to

$$\dot{\mathbf{H}}_O = \sum_{i=1}^{n} (\mathbf{r}_i \times m_i\mathbf{a}_i) \tag{14.9}$$

since the vectors $\mathbf{v}_i$ and $m_i\mathbf{v}_i$ are collinear.

We observe that the right-hand members of Eqs. (14.8) and (14.9) are respectively identical with the right-hand members of Eqs. (14.4) and (14.5). It follows that the left-hand members of these equations are respectively equal. Recalling that the left-hand member of Eq. (14.5) represents the sum of the moments $\mathbf{M}_O$ about $O$ of the external forces acting on the particles of the system, and omitting the subscript $i$ from the sums, we write

$$\Sigma\mathbf{F} = \dot{\mathbf{L}} \tag{14.10}$$

$$\Sigma\mathbf{M}_O = \dot{\mathbf{H}}_O \tag{14.11}$$

These equations express that *the resultant and the moment resultant about the fixed point $O$ of the external forces are respectively equal to the rates of change of the linear momentum and of the angular momentum about $O$ of the system of particles.*

## 14.4. MOTION OF THE MASS CENTER OF A SYSTEM OF PARTICLES

Equation (14.10) may be written in an alternative form if the *mass center* of the system of particles is considered. The mass center of the system is the point $G$ defined by the position vector $\bar{\mathbf{r}}$, which satisfies

$$m\bar{\mathbf{r}} = \sum_{i=1}^{n} m_i \mathbf{r}_i \qquad (14.12)$$

where $m$ represents the total mass $\sum_{i=1}^{n} m_i$ of the particles. Resolving the position vectors $\bar{\mathbf{r}}$ and $\mathbf{r}_i$ into rectangular components, we obtain the following three scalar equations, which can be used to determine the coordinates $\bar{x}, \bar{y}, \bar{z}$ of the mass center:

$$m\bar{x} = \sum_{i=1}^{n} m_i x_i \qquad m\bar{y} = \sum_{i=1}^{n} m_i y_i \qquad m\bar{z} = \sum_{i=1}^{n} m_i z_i \qquad (14.12')$$

Since $m_i g$ represents the weight of the particle $P_i$, and $mg$ the total weight of the particles, $G$ is also the center of gravity of the system of particles. However, in order to avoid any confusion, $G$ will be referred to as the *mass center* of the system of particles when properties associated with the *mass* of the particles are being discussed, and as the *center of gravity* of the system when properties associated with the *weight* of the particles are being considered. Particles located outside the gravitational field of the earth, for example, have a mass but no weight. We can then properly refer to their mass center, but obviously not to their center of gravity.†

Differentiating both members of Eq. (14.12) with respect to $t$, we write

$$m\dot{\bar{\mathbf{r}}} = \sum_{i=1}^{n} m_i \dot{\mathbf{r}}_i$$

or

$$m\bar{\mathbf{v}} = \sum_{i=1}^{n} m_i \mathbf{v}_i \qquad (14.13)$$

where $\bar{\mathbf{v}}$ represents the velocity of the mass center $G$ of the system of particles. But the right-hand member of Eq. (14.13) is, by definition, the linear momentum $\mathbf{L}$ of the system (Sec. 14.3). We therefore have

$$\mathbf{L} = m\bar{\mathbf{v}} \qquad (14.14)$$

and, differentiating both members with respect to $t$,

$$\dot{\mathbf{L}} = m\bar{\mathbf{a}} \qquad (14.15)$$

†It may also be pointed out that the mass center and the center of gravity of a system of particles do not exactly coincide, since the weights of the particles are directed toward the center of the earth and thus do not truly form a system of parallel forces.

where $\bar{\mathbf{a}}$ represents the acceleration of the mass center $G$. Substituting for $\dot{\mathbf{L}}$ from (14.15) into (14.10), we write the equation

$$\Sigma\mathbf{F} = m\bar{\mathbf{a}} \tag{14.16}$$

which defines the motion of the mass center $G$ of the system of particles.

We note that Eq. (14.16) is identical with the equation we would obtain for a particle of mass $m$ equal to the total mass of the particles of the system, acted upon by all the external forces. We therefore state that *the mass center of a system of particles moves as if the entire mass of the system and all the external forces were concentrated at that point.*

This principle is best illustrated by the motion of an exploding shell. We know that if air resistance is neglected, it can be assumed that a shell will travel along a parabolic path. After the shell has exploded, the mass center $G$ of the fragments of shell will continue to travel along the same path. Indeed, point $G$ must move as if the mass and the weight of all fragments were concentrated at $G$; it must, therefore, move as if the shell had not exploded.

It should be noted that the preceding derivation does not involve the moments of the external forces. Therefore, *it would be wrong to assume* that the external forces are equipollent to a vector $m\bar{\mathbf{a}}$ attached at the mass center $G$. This is not in general the case since, as you will see in the next section, the sum of the moments about $G$ of the external forces is not in general equal to zero.

## 14.5. ANGULAR MOMENTUM OF A SYSTEM OF PARTICLES ABOUT ITS MASS CENTER

In some applications (for example, in the analysis of the motion of a rigid body) it is convenient to consider the motion of the particles of the system with respect to a centroidal frame of reference $Gx'y'z'$ which translates with respect to the newtonian frame of reference $Oxyz$ (Fig. 14.5). Although a centroidal frame is not, in general, a newtonian frame of reference, it will be seen that the fundamental relation (14.11) holds when the frame $Oxyz$ is replaced by $Gx'y'z'$.

Denoting, respectively, by $\mathbf{r}'_i$ and $\mathbf{v}'_i$ the position vector and the velocity of the particle $P_i$ relative to the moving frame of reference $Gx'y'z'$, we define the *angular momentum* $\mathbf{H}'_G$ of the system of particles *about the mass center $G$* as follows:

$$\mathbf{H}'_G = \sum_{i=1}^{n} (\mathbf{r}'_i \times m_i\mathbf{v}'_i) \tag{14.17}$$

We now differentiate both members of Eq. (14.17) with respect to $t$. This operation is similar to that performed in Sec. 14.3 on Eq. (14.7), and so we write immediately

$$\dot{\mathbf{H}}'_G = \sum_{i=1}^{n} (\mathbf{r}'_i \times m_i\mathbf{a}'_i) \tag{14.18}$$

**Fig. 14.5**

where $\mathbf{a}_i'$ denotes the acceleration of $P_i$ relative to the moving frame of reference. Referring to Sec. 11.12, we write

$$\mathbf{a}_i = \overline{\mathbf{a}} + \mathbf{a}_i'$$

where $\mathbf{a}_i$ and $\overline{\mathbf{a}}$ denote, respectively, the accelerations of $P_i$ and $G$ relative to the frame $Oxyz$. Solving for $\mathbf{a}_i'$ and substituting into (14.18), we have

$$\dot{\mathbf{H}}_G' = \sum_{i=1}^{n} (\mathbf{r}_i' \times m_i \mathbf{a}_i) - \left( \sum_{i=1}^{n} m_i \mathbf{r}_i' \right) \times \overline{\mathbf{a}} \qquad (14.19)$$

But, by (14.12), the second sum in Eq. (14.19) is equal to $m\overline{\mathbf{r}}'$ and thus to zero, since the position vector $\overline{\mathbf{r}}'$ of $G$ relative to the frame $Gx'y'z'$ is clearly zero. On the other hand, since $\mathbf{a}_i$ represents the acceleration of $P_i$ relative to a newtonian frame, we can use Eq. (14.1) and replace $m_i\mathbf{a}_i$ by the sum of the internal forces $\mathbf{f}_{ij}$ and of the resultant $\mathbf{F}_i$ of the external forces acting on $P_i$. But a reasoning similar to that used in Sec. 14.2 shows that the moment resultant about $G$ of the internal forces $\mathbf{f}_{ij}$ of the entire system is zero. The first sum in Eq. (14.19) therefore reduces to the moment resultant about $G$ of the external forces acting on the particles of the system, and we write

$$\Sigma\mathbf{M}_G = \dot{\mathbf{H}}_G' \qquad (14.20)$$

which expresses that *the moment resultant about $G$ of the external forces is equal to the rate of change of the angular momentum about $G$ of the system of particles.*

It should be noted that in Eq. (14.17) we defined the angular momentum $\mathbf{H}_G'$ as the sum of the moments about $G$ of the momenta of the particles $m_i\mathbf{v}_i'$ *in their motion relative to the centroidal frame of reference $Gx'y'z'$.* We may sometimes want to compute the sum $\mathbf{H}_G$ of the moments about $G$ of the momenta of the particles $m_i\mathbf{v}_i$ *in their absolute motion,* that is, in their motion as observed from the newtonian frame of reference $Oxyz$ (Fig. 14.6):

$$\mathbf{H}_G = \sum_{i=1}^{n} (\mathbf{r}_i' \times m_i \mathbf{v}_i) \qquad (14.21)$$

Remarkably, the angular momenta $\mathbf{H}_G'$ and $\mathbf{H}_G$ are identically equal. This can be verified by referring to Sec. 11.12 and writing

$$\mathbf{v}_i = \overline{\mathbf{v}} + \mathbf{v}_i' \qquad (14.22)$$

Substituting for $\mathbf{v}_i$ from (14.22) into Eq. (14.21), we have

$$\mathbf{H}_G = \left( \sum_{i=1}^{n} m_i \mathbf{r}_i' \right) \times \overline{\mathbf{v}} + \sum_{i=1}^{n} (\mathbf{r}_i' \times m_i \mathbf{v}_i')$$

But, as observed earlier, the first sum is equal to zero. Thus $\mathbf{H}_G$ reduces to the second sum, which, by definition, is equal to $\mathbf{H}_G'$.[†]

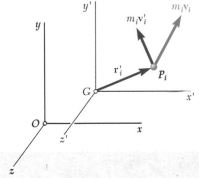

**Fig. 14.6**

---

[†]Note that this property is peculiar to the centroidal frame $Gx'y'z'$ and does not, in general, hold for other frames of reference (see Prob. 14.29).

Taking advantage of the property we have just established, we simplify our notation by dropping the prime ($'$) from Eq. (14.20) and writing

$$\Sigma \mathbf{M}_G = \dot{\mathbf{H}}_G \qquad (14.23)$$

where it is understood that the angular momentum $\mathbf{H}_G$ can be computed by forming the moments about $G$ of the momenta of the particles in their motion with respect to either the newtonian frame $Oxyz$ or the centroidal frame $Gx'y'z'$:

$$\mathbf{H}_G = \sum_{i=1}^{n} (\mathbf{r}_i' \times m_i \mathbf{v}_i) = \sum_{i=1}^{n} (\mathbf{r}_i' \times m_i \mathbf{v}_i') \qquad (14.24)$$

## 14.6. CONSERVATION OF MOMENTUM FOR A SYSTEM OF PARTICLES

If no external force acts on the particles of a system, the left-hand members of Eqs. (14.10) and (14.11) are equal to zero and these equations reduce to $\dot{\mathbf{L}} = 0$ and $\dot{\mathbf{H}}_O = 0$. We conclude that

$$\mathbf{L} = \text{constant} \qquad \mathbf{H}_O = \text{constant} \qquad (14.25)$$

The equations obtained express that the linear momentum of the system of particles and its angular momentum about the fixed point $O$ are conserved.

In some applications, such as problems involving central forces, the moment about a fixed point $O$ of each of the external forces can be zero without any of the forces being zero. In such cases, the second of Eqs. (14.25) still holds; the angular momentum of the system of particles about $O$ is conserved.

The concept of conservation of momentum can also be applied to the analysis of the motion of the mass center $G$ of a system of particles and to the analysis of the motion of the system about $G$. For example, if the sum of the external forces is zero, the first of Eqs. (14.25) applies. Recalling Eq. (14.14), we write

$$\bar{\mathbf{v}} = \text{constant} \qquad (14.26)$$

which expresses that the mass center $G$ of the system moves in a straight line and at a constant speed. On the other hand, if the sum of the moments about $G$ of the external forces is zero, it follows from Eq. (14.23) that the angular momentum of the system about its mass center is conserved:

$$\mathbf{H}_G = \text{constant} \qquad (14.27)$$

**Photo 14.1** If the system of particles is defined as the two stages of the space vehicle and no external forces act, the linear momentum of the system and its angular momentum about its mass center are conserved.

## SAMPLE PROBLEM 14.1

A 200-kg space vehicle is observed at $t = 0$ to pass through the origin of a newtonian reference frame $Oxyz$ with velocity $\mathbf{v}_0 = (150 \text{ m/s})\mathbf{i}$ relative to the frame. Following the detonation of explosive charges, the vehicle separates into three parts $A$, $B$, and $C$, of mass 100 kg, 60 kg, and 40 kg, respectively. Knowing that at $t = 2.5$ s the positions of parts $A$ and $B$ are observed to be $A(555, -180, 240)$ and $B(255, 0, -120)$, where the coordinates are expressed in meters, determine the position of part $C$ at that time.

### SOLUTION

Since there is no external force, the mass center $G$ of the system moves with the constant velocity $\mathbf{v}_0 = (150 \text{ m/s})\mathbf{i}$. At $t = 2.5$ s, its position is

$$\bar{\mathbf{r}} = \mathbf{v}_0 t = (150 \text{ m/s})\mathbf{i}(2.5 \text{ s}) = (375 \text{ m})\mathbf{i}$$

Recalling Eq. (14.12), we write

$$m\bar{\mathbf{r}} = m_A\mathbf{r}_A + m_B\mathbf{r}_B + m_C\mathbf{r}_C$$
$$(200 \text{ kg})(375 \text{ m})\mathbf{i} = (100 \text{ kg})[(555 \text{ m})\mathbf{i} - (180 \text{ m})\mathbf{j} + (240 \text{ m})\mathbf{k}]$$
$$+ (60 \text{ kg})[(255 \text{ m})\mathbf{i} - (120 \text{ m})\mathbf{k}] + (40 \text{ kg})\mathbf{r}_C$$

$$\mathbf{r}_C = (105 \text{ m})\mathbf{i} + (450 \text{ m})\mathbf{j} - (420 \text{ m})\mathbf{k} \quad \blacktriangleleft$$

## SAMPLE PROBLEM 14.2

A 9 kg projectile is moving with a velocity of 30.5 m/s when it explodes into two fragments $A$ and $B$, weighing 2.3 kg and 6.8 kg, respectively. Knowing that immediately after the explosion, fragments $A$ and $B$ travel in directions defined respectively by $\theta_A = 45°$ and $\theta_B = 30°$, determine the velocity of each fragment.

### SOLUTION

Since there is no external force, the linear momentum of the system is conserved, and we write

$$m_A\mathbf{v}_A + m_B\mathbf{v}_B = m\mathbf{v}_0$$
$$(2.3)\mathbf{v}_A + (6.8)\mathbf{v}_B = (9)\mathbf{v}_0$$

$\xrightarrow{+}$ $x$ components: $\quad 2.3v_A \cos 45° + 6.8v_B \cos 30° = 9(30.5)$
$+\uparrow$ $y$ components: $\quad 2.3v_A \sin 45° - 6.8v_B \sin 30° = 0$

Solving simultaneously the two equations for $v_A$ and $v_B$, we have

$$v_A = 61.68 \text{ m/s} \qquad v_B = 29.5 \text{ m/s}$$

$$\mathbf{v}_A = 61.68 \text{ m/s} \measuredangle 45° \qquad \mathbf{v}_B = 29.5 \text{ m/s} \searrow 30° \quad \blacktriangleleft$$

# SOLVING PROBLEMS
## ON YOUR OWN

This chapter deals with the motion of *systems of particles*, that is, with the motion of a large number of particles considered together, rather than separately. In this first lesson you learned to compute the *linear momentum* and the *angular momentum* of a system of particles. We defined the linear momentum **L** of a system of particles as the sum of the linear momenta of the particles and we defined the angular momentum $\mathbf{H}_O$ of the system as the sum of the angular momenta of the particles about $O$:

$$\mathbf{L} = \sum_{i=1}^{n} m_i\mathbf{v}_i \qquad \mathbf{H}_O = \sum_{i=1}^{n} (\mathbf{r}_i \times m_i\mathbf{v}_i) \tag{14.6, 14.7}$$

In this lesson, you will solve a number of problems of practical interest, either by observing that the linear momentum of a system of particles is conserved or by considering the motion of the mass center of a system of particles.

1. *Conservation of the linear momentum of a system of particles.* This occurs *when the resultant of the external forces acting on the particles of the system is zero.* You may encounter such a situation in the following types of problems.

   *a. Problems involving the rectilinear motion* of objects such as colliding automobiles and railroad cars. After you have checked that the resultant of the external forces is zero, equate the algebraic sums of the initial momenta and final momenta to obtain an equation which can be solved for one unknown.

   *b. Problems involving the two-dimensional or three-dimensional motion* of objects such as exploding shells, or colliding aircraft, automobiles, or billiard balls. After you have checked that the resultant of the external forces is zero, add vectorially the initial momenta of the objects, add vectorially their final momenta, and equate the two sums to obtain a vector equation expressing that the linear momentum of the system is conserved.

   In the case of a two-dimensional motion, this equation can be replaced by two scalar equations which can be solved for two unknowns, while in the case of a three-dimensional motion it can be replaced by three scalar equations which can be solved for three unknowns.

2. *Motion of the mass center of a system of particles.* You saw in Sec. 14.4 that *the mass center of a system of particles moves as if the entire mass of the system and all of the external forces were concentrated at that point.*

   *a. In the case of a body exploding while in motion,* it follows that the mass center of the resulting fragments moves as the body itself would have moved if the explosion had not occurred. Problems of this type can be solved by writing the equation of motion of the mass center of the system in vectorial form and expressing the position vector of the mass center in terms of the position vectors of the various fragments [Eq. (14.12)]. You can then rewrite the vector equation as two or three scalar equations and solve the equations for an equivalent number of unknowns.

   *b. In the case of the collision of several moving bodies,* it follows that the motion of the mass center of the various bodies is unaffected by the collision. Problems of this type can be solved by writing the equation of motion of the mass center of the system in vectorial form and expressing its position vector before and after the collision in terms of the position vectors of the relevant bodies [Eq. (14.12)]. You can then rewrite the vector equation as two or three scalar equations and solve these equations for an equivalent number of unknowns.

# Problems

**14.1**  Two identical 1350-kg automobiles $A$ and $B$ are at rest with their brakes released when $B$ is struck by a 5400-kg truck $C$ which is moving to the left at 8 km/h. A second collision then occurs when $B$ strikes $A$. Assuming the first collision is perfectly plastic and the second collision is perfectly elastic, determine the velocities of the three vehicles just after the second collision.

**Fig. P14.1**

**14.2**  A 28-g bullet is fired with a velocity of 550 m/s into a block $A$, which has a mass of 5 kg. The coefficient of kinetic friction between block $A$ and the cart $BC$ is 0.50. Knowing that the cart has a mass of 4 kg and can roll freely, determine (*a*) the final velocity of the cart and block, (*b*) the final position of the block on the cart.

**Fig. P14.2**

**14.3**  Car $A$ weighing 17.79 kN and car $B$ weighing 16.46 kN are at rest on a 22,000 kg flatcar which is also at rest. Cars $A$ and $B$ then accelerate and quickly reach constant speeds relative to the flatcar of 2.1 m/s and 1.07 m/s, respectively, before decelerating to a stop at the opposite end of the flatcar. Neglecting friction and rolling resistance, determine the velocity of the flatcar when the cars are moving at constant speeds.

**Fig. P14.3 and P14.4**

**14.4**  Car $A$ weighing 17.79 kN and car $B$ weighing 16.46 kN are at rest on a flatcar which is also at rest. Cars $A$ and $B$ then accelerate and quickly reach constant speeds relative to the flatcar of 2.33 m/s and 2.29 m/s, respectively, before decelerating to a stop at the opposite end of the flatcar. Knowing that the speed of the flatcar is 0.31 m/s when the cars are moving at constant speeds, determine the weight of the flatcar. Neglect friction and rolling resistance.

**14.5**  An 80-Mg railroad engine $A$ coasting at 6.5 km/h strikes a 20-Mg flatcar $C$ carrying a 30-Mg load $B$ which can slide along the floor of the car ($\mu_k = 0.25$). Knowing that the flatcar was at rest with its brakes released and that it automatically coupled with the engine upon impact, determine the velocity of the flatcar ($a$) immediately after impact, ($b$) after the load has slid to a stop relative to the flatcar.

**Fig. P14.5**

**14.6**  A bullet is fired with a horizontal velocity of 500 m/s through a 3-kg block $A$ and becomes embedded in a 2.5-kg block $B$. Knowing that blocks $A$ and $B$ start moving with velocities of 3 m/s and 5 m/s, respectively, determine ($a$) the mass of the bullet, ($b$) its velocity as it travels from block $A$ to block $B$.

**Fig. P14.6**

*14.7*  A 81.6 kg man and a 54.4 kg woman stand side by side at the same end of a 136 kg boat, ready to dive, each with a 4.87 m/s velocity relative to the boat. Determine the velocity of the boat after they have both dived, if ($a$) the woman dives first, ($b$) the man dives first.

**Fig. *P14.7***

*14.8*  A 81.6 kg man and a 54.4 kg woman stand at opposite ends of a 136 kg boat, ready to dive, each with a 4.87 m/s velocity relative to the boat. Determine the velocity of the boat after they have both dived, if ($a$) the woman dives first, ($b$) the man dives first.

**Fig. *P14.8***

**14.9**  A system consists of three identical 9-kg particles $A$, $B$, and $C$. The velocities of the particles are, respectively, $\mathbf{v}_A = v_A\mathbf{j}$, $\mathbf{v}_B = v_B\mathbf{i}$, and $\mathbf{v}_C = v_C\mathbf{k}$. Knowing that the angular momentum of the system about $O$, expressed in $\text{kg} \cdot \text{m}^2/\text{s}$ is $\mathbf{H}_O = -1.8\ \mathbf{k}$, determine ($a$) the velocities of the particles, ($b$) the angular momentum of the system about its mass center $G$.

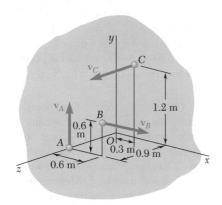

**Fig. P14.9 and P14.10**

**14.10**  A system consists of three identical 9-kg particles $A$, $B$, and $C$. The velocities of the particles are, respectively, $\mathbf{v}_A = v_A\mathbf{j}$, $\mathbf{v}_B = v_B\mathbf{i}$, and $\mathbf{v}_C = v_C\mathbf{k}$, and the magnitude of the linear momentum $\mathbf{L}$ of the system is $45\ \text{kg} \cdot \text{m}/\text{s}$. Knowing that $\mathbf{H}_G = \mathbf{H}_O$, where $\mathbf{H}_G$ is the angular momentum of the system about its mass center $G$ and $\mathbf{H}_O$ is the angular momentum of the system about $O$, determine ($a$) the velocities of the particles, ($b$) the angular momentum of the system about $O$.

**14.11**  A system consists of three particles $A$, $B$, and $C$. We know that $W_A = W_B = 17.79\ \text{N}$ and $W_C = 124.55\ \text{N}$ and that the velocities of the particles, expressed in m/s are, respectively, $\mathbf{v}_A = 42\mathbf{i} + 63\mathbf{j}$, $\mathbf{v}_B = -42\mathbf{i} + 63\mathbf{j}$, and $\mathbf{v}_C = -9\mathbf{j} - 6\mathbf{k}$. Determine the angular momentum $\mathbf{H}_O$ of the system about $O$.

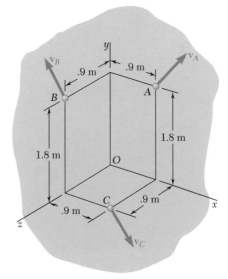

**Fig. P14.11 and P14.12**

**14.12**  For the system of particles of Prob. 14.11, determine ($a$) the position vector $\bar{\mathbf{r}}$ of the mass center $G$ of the system, ($b$) the linear momentum $m\bar{\mathbf{v}}$ of the system, ($c$) the angular momentum $\mathbf{H}_G$ of the system about $G$. Also verify that the answers to this problem and to Prob. 14.11 satisfy the equation given in Prob. 14.28.

**14.13**  A system consists of three particles $A$, $B$, and $C$. We know that $m_A = 1\ \text{kg}$, $m_B = 2\ \text{kg}$, and $m_C = 3\ \text{kg}$ and that the velocities of the particles expressed in m/s are, respectively, $\mathbf{v}_A = 3\mathbf{i} - 2\mathbf{j} + 4\mathbf{k}$, $\mathbf{v}_B = 4\mathbf{i} + 3\mathbf{j}$, and $\mathbf{v}_C = 2\mathbf{i} + 5\mathbf{j} - 3\mathbf{k}$. Determine the angular momentum $\mathbf{H}_O$ of the system about $O$.

**14.14**  For the system of particles of Prob. 14.13, determine ($a$) the position vector $\bar{\mathbf{r}}$ of the mass center $G$ of the system, ($b$) the linear momentum $m\bar{\mathbf{v}}$ of the system, ($c$) the angular momentum $\mathbf{H}_G$ of the system about $G$. Also verify that the answers to this problem and to Prob. 14.13 satisfy the equations given in Prob. 14.28.

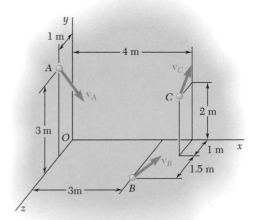

**Fig. P14.13**

**14.15** A 20-kg projectile is passing through the origin $O$ with a velocity $\mathbf{v}_0 = (60 \text{ m/s})\mathbf{i}$ when it explodes into two fragments $A$ and $B$, of 8-kg and 12-kg, respectively. Knowing that 2 s later the position of fragment $A$ is (120 m, −10 m, −20 m), determine the position of fragment $B$ at the same instant. Assume $a_y = -g = -9.81 \text{ m/s}^2$ and neglect air resistance.

**14.16** A 500-kg space vehicle traveling with a velocity $\mathbf{v}_0 = (450 \text{ m/s})\mathbf{i}$ passes through the origin $O$ at $t = 0$. Explosive charges then separate the vehicle into three parts $A$, $B$, and $C$ of masses 300-kg, 150-kg, and 50-kg, respectively. Knowing that at $t = 4$ s, the positions of parts $A$ and $B$ are observed to be $A$ (1200 m, −350 m, −600 m) and $B$ (2500 m, 450 m, 900 m), determine the corresponding position of part $C$. Neglect the effect of gravity.

**14.17 and 14.18** Car $A$ was at rest 8.47 m south of the point $O$ when it was struck in the rear by car $B$, which was traveling north at a speed $v_B$. Car $C$, which was traveling west at a speed $v_C$, was 36.6 m east of point $O$ at the time of the collision. Cars $A$ and $B$ stuck together and, because the pavement was covered with ice, they slid into the intersection and were struck by car $C$ which had not changed its speed. Measurements based on a photograph taken from a traffic helicopter shortly after the second collision indicated that the positions of the cars, expressed in meter, were $\mathbf{r}_A = -30.3\mathbf{i} + 50.7\mathbf{j}$, $\mathbf{r}_B = -30.3\mathbf{i} + 61.2\mathbf{j}$, and $\mathbf{r}_C = -59.4\mathbf{i} - 45.6\mathbf{j}$. Knowing that the weights of cars $A$, $B$, and $C$ are, respectively, 12.46 kN, 16 kN, and 14.23 kN, solve the problems indicated.

**Fig. P14.17 and P14.18**

**14.17** Knowing that $v_C = 72.1$ km/h, determine the initial speed of car $B$ and the time elapsed between the first collision and the time the photograph was taken.

**14.18** Knowing that the time elapsed between the first collision and the time the photograph was taken was 3.4 s, determine the initial speeds of cars $B$ and $C$.

**14.19**  Two 15-kg cannonballs are chained together and fired horizontally with a velocity of 165 m/s from the top of a 15-m wall. The chain breaks during the flight of the cannonballs and one of them strikes the ground at $t = 1.5$ s, at a distance of 240 m from the foot of the wall, and 7 m to the right of the line of fire. Determine the position of the other cannon ball at that instant. Neglect the resistance of the air.

**Fig. P14.19**

**14.20**  A 3-kg model rocket is launched vertically and reaches an altitude of 60 m with an speed of 28 m/s at the end of powered flight, time $t = 0$. As the rocket approaches its maximum altitude it explodes into two parts of masses $m_A = 1$ kg and $m_B = 2$ kg. Part A is observed to strike the ground 74.4 m west of the launch point at $t = 5.85$ s. Determine the position of part B at that time.

**Fig. P14.20**

**14.21**  Two spheres, each of mass $m$, can slide freely on a frictionless, horizontal surface. Sphere A is moving at a speed $v_0 = 4.8$ m/s when it strikes sphere B which is at rest and the impact causes sphere B to break into two pieces, each of mass $m/2$. Knowing that 0.7 s after the collision one piece reaches point C and 0.9 s after the collision the other piece reaches point D, determine (a) the velocity of sphere A after the collision, (b) the angle $\theta$ and the speeds of the two pieces after the collision.

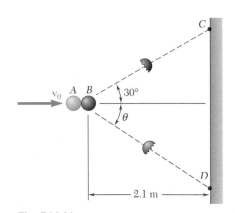

**Fig. P14.21**

**14.22**  In a game of pool, ball A is traveling with a velocity $v_0$ when it strikes balls B and C which are at rest and aligned as shown. Knowing that after the collision the three balls move in the directions indicated and that $v_0 = 4$ m/s and $v_C = 2.1$ m/s, determine the magnitude of the velocity of (a) ball A, (b) ball B.

**14.23**  A 4536 kg helicopter A was traveling due east at a speed of 304.4 km/h and an altitude of 731.5 m when it was hit by a 5897 kg helicopter B. As a result of the collision, both helicopters lost their lift, and their entangled wreckage fell to the ground in 12 s at a point located 488 m east and 122 m south of the point of impact. Neglecting air resistance, determine the velocity components of helicopter B just before the collision.

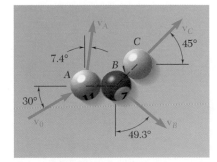

**Fig. P14.22**

**14.24**  A 2.7 kg game bird flying due east 13.7 m above the ground with a velocity $\mathbf{v}_B = (9$ m/s$)\mathbf{i}$ is hit by a .06 kg arrow with a velocity $\mathbf{v}_A = (54.86$ m/s$)\mathbf{j} + (73.15$ m/s$)\mathbf{k}$, where $\mathbf{j}$ is directed upward. Determine the position of the point P where the bird will hit the ground, relative to the point O located directly under the point of impact.

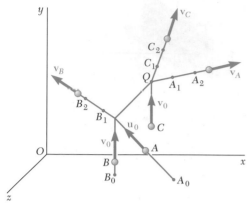

**Fig. P14.25**

**14.25**   In a scattering experiment, an alpha particle $A$ is projected with the velocity $\mathbf{u}_0 = -(600 \text{ m/s})\mathbf{i} + (750 \text{ m/s})\mathbf{j} - (800 \text{ m/s})\mathbf{k}$ into a stream of oxygen nuclei moving with a common velocity $\mathbf{v}_0 = (600 \text{ m/s})\mathbf{j}$. After colliding successively with nuclei $B$ and $C$, particle $A$ is observed to move along the path defined by the points $A_1$ (280, 240, 120) and $A_2$ (360, 320, 160), while nuclei $B$ and $C$ are observed to move along paths defined, respectively, by $B_1$ (147, 220, 130) and $B_2$ (114, 290, 120), and by $C_1$ (240, 232, 90) and $C_2$ (240, 280, 75). All paths are along straight lines and all coordinates are expressed in millimeters. Knowing that the mass of an oxygen nucleus is four times that of an alpha particle, determine the speed of each of the three particles after the collisions.

**14.26**   An 8.16 kg shell moving with a velocity $\mathbf{v}_0 = (18.29 \text{ m/s})\mathbf{i} - (13.7 \text{ m/s})\mathbf{j} - (548.6 \text{ m/s})\mathbf{k}$ explodes at point $D$ into three fragments $A$, $B$, and $C$ weighing, respectively, 35.58 N, 26.69 N, and 17.79 N. Knowing that the fragments hit the vertical wall at the points indicated, determine the speed of each fragment immediately after the explosion.

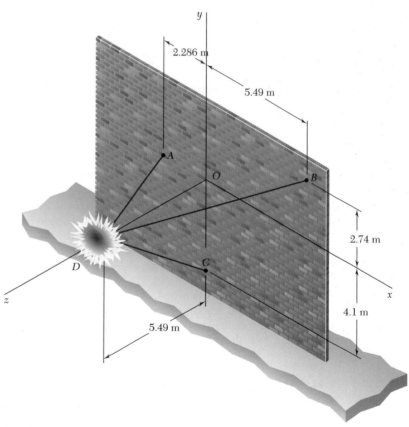

**Fig. *P14.26* and *P14.27***

**14.27**   A 2.7 kg shell moving with a velocity $-v_0\mathbf{k}$ explodes at point $D$ into three fragments which hit the vertical wall at the points indicated. Fragments $A$, $B$, and $C$ hit the wall 0.010 s, 0.018 s, and 0.012 s, respectively, after the explosion. Determine (*a*) the weights of the three fragments, (*b*) the initial speed $v_0$.

**14.28**  Derive the relation

$$\mathbf{H}_O = \bar{\mathbf{r}} \times m\bar{\mathbf{v}} + \mathbf{H}_G$$

between the angular momenta $\mathbf{H}_O$ and $\mathbf{H}_G$ defined in Eqs. (14.7) and (14.24), respectively. The vectors $\bar{\mathbf{r}}$ and $\bar{\mathbf{v}}$ define, respectively, the position and velocity of the mass center $G$ of the system of particles relative to the newtonian frame of reference $Oxyz$, and $m$ represents the total mass of the system.

**14.29**  Consider the frame of reference $Ax'y'z'$ in translation with respect to the newtonian frame of reference $Oxyz$. We define the angular momentum $\mathbf{H}'_A$ of a system of $n$ particles about $A$ as the sum

$$\mathbf{H}'_A = \sum_{i=1}^{n} \mathbf{r}'_i \times m_i\mathbf{v}'_i \qquad (1)$$

of the moments about $A$ of the momenta $m_i\mathbf{v}'_i$ of the particles in their motion relative to the frame $Ax'y'z'$. Denoting by $\mathbf{H}_A$ the sum

$$\mathbf{H}_A = \sum_{i=1}^{n} \mathbf{r}'_i \times m_i\mathbf{v}_i \qquad (2)$$

of the moments about $A$ of the momenta $m_i\mathbf{v}_i$ of the particles in their motion relative to the newtonian frame $Oxyz$, show that $\mathbf{H}_A = \mathbf{H}'_A$ at a given instant if, and only if, one of the following conditions is satisfied at that instant: ($a$) $A$ has zero velocity with respect to the frame $Oxyz$, ($b$) $A$ coincides with the mass center $G$ of the system, ($c$) the velocity $\mathbf{v}_A$ relative to $Oxyz$ is directed along the line $AG$.

**14.30**  Show that the relation $\Sigma\mathbf{M}_A = \dot{\mathbf{H}}'_A$, where $\mathbf{H}'_A$ is defined by Eq. (1) of Prob. 14.29 and where $\Sigma\mathbf{M}_A$ represents the sum of the moments about $A$ of the external forces acting on the system of particles, is valid if, and only if, one of the following conditions is satisfied: ($a$) the frame $Ax'y'z'$ is itself a newtonian frame of reference, ($b$) $A$ coincides with the mass center $G$, ($c$) the acceleration $\mathbf{a}_A$ of $A$ relative to $Oxyz$ is directed along the line $AG$.

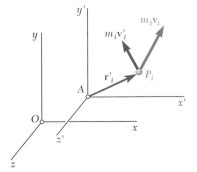

**Fig. P14.29**

## 14.7. KINETIC ENERGY OF A SYSTEM OF PARTICLES

The kinetic energy $T$ of a system of particles is defined as the sum of the kinetic energies of the various particles of the system. Referring to Sec. 13.3, we therefore write

$$T = \frac{1}{2} \sum_{i=1}^{n} m_i v_i^2 \qquad (14.28)$$

**Using a Centroidal Frame of Reference.**  It is often convenient when computing the kinetic energy of a system comprising a large number of particles (as in the case of a rigid body) to consider separately the motion of the mass center $G$ of the system and the motion of the system relative to a moving frame attached to $G$.

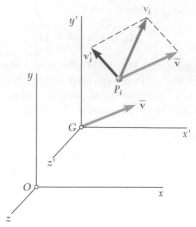

**Fig. 14.7**

Let $P_i$ be a particle of the system, $\mathbf{v}_i$ its velocity relative to the newtonian frame of reference $Oxyz$, and $\mathbf{v}_i'$ its velocity relative to the moving frame $Gx'y'z'$ which is in translation with respect to $Oxyz$ (Fig. 14.7). We recall from the preceding section that

$$\mathbf{v}_i = \overline{\mathbf{v}} + \mathbf{v}_i' \tag{14.22}$$

where $\overline{\mathbf{v}}$ denotes the velocity of the mass center $G$ relative to the newtonian frame $Oxyz$. Observing that $v_i^2$ is equal to the scalar product $\mathbf{v}_i \cdot \mathbf{v}_i$, we express the kinetic energy $T$ of the system relative to the newtonian frame $Oxyz$ as follows:

$$T = \frac{1}{2} \sum_{i=1}^{n} m_i v_i^2 = \frac{1}{2} \sum_{i=1}^{n} (m_i \mathbf{v}_i \cdot \mathbf{v}_i)$$

or, substituting for $\mathbf{v}_i$ from (14.22),

$$T = \frac{1}{2} \sum_{i=1}^{n} [m_i(\overline{\mathbf{v}} + \mathbf{v}_i') \cdot (\overline{\mathbf{v}} + \mathbf{v}_i')]$$

$$= \frac{1}{2} \left( \sum_{i=1}^{n} m_i \right) \overline{v}^2 + \overline{\mathbf{v}} \cdot \sum_{i=1}^{n} m_i \mathbf{v}_i' + \frac{1}{2} \sum_{i=1}^{n} m_i v_i'^2$$

The first sum represents the total mass $m$ of the system. Recalling Eq. (14.13), we note that the second sum is equal to $m\overline{\mathbf{v}}'$ and thus to zero, since $\overline{\mathbf{v}}'$, which represents the velocity of $G$ relative to the frame $Gx'y'z'$, is clearly zero. We therefore write

$$T = \tfrac{1}{2} m \overline{v}^2 + \frac{1}{2} \sum_{i=1}^{n} m_i v_i'^2 \tag{14.29}$$

This equation shows that the kinetic energy $T$ of a system of particles can be obtained *by adding the kinetic energy of the mass center $G$ (assuming the entire mass concentrated at $G$) and the kinetic energy of the system in its motion relative to the frame $Gx'y'z'$.*

## 14.8. WORK-ENERGY PRINCIPLE. CONSERVATION OF ENERGY FOR A SYSTEM OF PARTICLES

The principle of work and energy can be applied to each particle $P_i$ of a system of particles. We write

$$T_1 + U_{1 \to 2} = T_2 \qquad (14.30)$$

for each particle $P_i$, where $U_{1 \to 2}$ represents the work done by the internal forces $\mathbf{f}_{ij}$ and the resultant external force $\mathbf{F}_i$ acting on $P_i$. Adding the kinetic energies of the various particles of the system and considering the work of all the forces involved, we can apply Eq. (14.30) to the entire system. The quantities $T_1$ and $T_2$ now represent the kinetic energy of the entire system and can be computed from either Eq. (14.28) or Eq. (14.29). The quantity $U_{1 \to 2}$ represents the work of all the forces acting on the particles of the system. Note that while the internal forces $\mathbf{f}_{ij}$ and $\mathbf{f}_{ji}$ are equal and opposite, the work of these forces will not, in general, cancel out, since the particles $P_i$ and $P_j$ on which they act will, in general, undergo different displacements. Therefore, in computing $U_{1 \to 2}$, *we must consider the work of the internal forces $\mathbf{f}_{ij}$ as well as the work of the external forces $\mathbf{F}_i$.*

If all the forces acting on the particles of the system are conservative, Eq. (14.30) can be replaced by

$$T_1 + V_1 = T_2 + V_2 \qquad (14.31)$$

where $V$ represents the potential energy associated with the internal and external forces acting on the particles of the system. Equation (14.31) expresses the principle of *conservation of energy* for the system of particles.

## 14.9. PRINCIPLE OF IMPULSE AND MOMENTUM FOR A SYSTEM OF PARTICLES

Integrating Eqs. (14.10) and (14.11) in $t$ from $t_1$ to $t_2$, we write

$$\sum \int_{t_1}^{t_2} \mathbf{F}\, dt = \mathbf{L}_2 - \mathbf{L}_1 \qquad (14.32)$$

$$\sum \int_{t_1}^{t_2} \mathbf{M}_O\, dt = (\mathbf{H}_O)_2 - (\mathbf{H}_O)_1 \qquad (14.33)$$

Recalling the definition of the linear impulse of a force given in Sec. 13.10, we observe that the integrals in Eq. (14.32) represent the linear impulses of the external forces acting on the particles of the system. We shall refer in a similar way to the integrals in Eq. (14.33) as the *angular impulses* about $O$ of the external forces. Thus, Eq. (14.32) expresses that the sum of the linear impulses of the external forces acting on the system is equal to the change in linear momentum of the system. Similarly, Eq. (14.33) expresses that the sum of the angular impulses about $O$ of the external forces is equal to the change in angular momentum about $O$ of the system.

**Photo 14.2** Because the concrete block is a system of particles with no initial momentum, the linear momentum $\mathbf{L}_2$ equals the sum of the linear impulses of the external forces acting on the system.

In order to make clear the physical significance of Eqs. (14.32) and (14.33), we will rearrange the terms in these equations and write

$$\mathbf{L}_1 + \sum \int_{t_1}^{t_2} \mathbf{F}\, dt = \mathbf{L}_2 \qquad (14.34)$$

$$(\mathbf{H}_O)_1 + \sum \int_{t_1}^{t_2} \mathbf{M}_O\, dt = (\mathbf{H}_O)_2 \qquad (14.35)$$

In parts $a$ and $c$ of Fig. 14.8 we have sketched the momenta of the particles of the system at times $t_1$ and $t_2$, respectively. In part $b$ we have shown a vector equal to the sum of the linear impulses of the external forces and a couple of moment equal to the sum of the angular impulses about $O$ of the external forces. For simplicity, the particles have

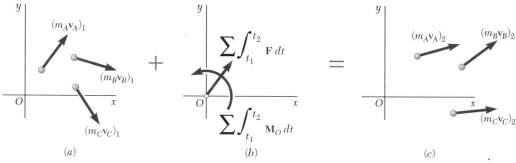

(a)          (b)          (c)

**Fig. 14.8**

been assumed to move in the plane of the figure, but the present discussion remains valid in the case of particles moving in space. Recalling from Eq. (14.6) that $\mathbf{L}$, by definition, is the resultant of the momenta $m_i\mathbf{v}_i$, we note that Eq. (14.34) expresses that the resultant of the vectors shown in parts $a$ and $b$ of Fig. 14.8 is equal to the resultant of the vectors shown in part $c$ of the same figure. Recalling from Eq. (14.7) that $\mathbf{H}_O$ is the moment resultant of the momenta $m_i\mathbf{v}_i$, we note that Eq. (14.35) similarly expresses that the moment resultant of the vectors in parts $a$ and $b$ of Fig. 14.8 is equal to the moment resultant of the vectors in part $c$. Together, Eqs. (14.34) and (14.35) thus express that *the momenta of the particles at time $t_1$ and the impulses of the external forces from $t_1$ to $t_2$ form a system of vectors equipollent to the system of the momenta of the particles at time $t_2$.* This has been indicated in Fig. 14.8 by the use of blue plus and equals signs.

If no external force acts on the particles of the system, the integrals in Eqs. (14.34) and (14.35) are zero, and these equations yield

$$\mathbf{L}_1 = \mathbf{L}_2 \qquad (14.36)$$
$$(\mathbf{H}_O)_1 = (\mathbf{H}_O)_2 \qquad (14.37)$$

We thus check the result obtained in Sec. 14.6: If no external force acts on the particles of a system, the linear momentum and the angular momentum about $O$ of the system of particles are conserved. The system of the initial momenta is equipollent to the system of the final momenta, and it follows that the angular momentum of the system of particles about *any* fixed point is conserved.

## SAMPLE PROBLEM 14.3

For the 200-kg space vehicle of Sample Prob. 14.1, it is known that at $t = 2.5$ s, the velocity of part $A$ is $\mathbf{v}_A = (270 \text{ m/s})\mathbf{i} - (120 \text{ m/s})\mathbf{j} + (160 \text{ m/s})\mathbf{k}$ and the velocity of part $B$ is parallel to the $xz$ plane. Determine the velocity of part $C$.

## SOLUTION

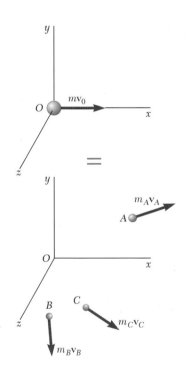

Since there is no external force, the initial momentum $m\mathbf{v}_0$ is equipollent to the system of the final momenta. Equating first the sums of the vectors in both parts of the adjoining sketch, and then the sums of their moments about $O$, we write

$\mathbf{L}_1 = \mathbf{L}_2$: $\qquad m\mathbf{v}_0 = m_A\mathbf{v}_A + m_B\mathbf{v}_B + m_C\mathbf{v}_C \qquad (1)$

$(\mathbf{H}_O)_1 = (\mathbf{H}_O)_2$: $\qquad 0 = \mathbf{r}_A \times m_A\mathbf{v}_A + \mathbf{r}_B \times m_B\mathbf{v}_B + \mathbf{r}_C \times m_C\mathbf{v}_C \qquad (2)$

Recalling from Sample Prob. 14.1 that $\mathbf{v}_0 = (150 \text{ m/s})\mathbf{i}$,

$$m_A = 100 \text{ kg} \qquad m_B = 60 \text{ kg} \qquad m_C = 40 \text{ kg}$$
$$\mathbf{r}_A = (555 \text{ m})\mathbf{i} - (180 \text{ m})\mathbf{j} + (240 \text{ m})\mathbf{k}$$
$$\mathbf{r}_B = (255 \text{ m})\mathbf{i} - (120 \text{ m})\mathbf{k}$$
$$\mathbf{r}_C = (105 \text{ m})\mathbf{i} + (450 \text{ m})\mathbf{j} - (420 \text{ m})\mathbf{k}$$

and using the information given in the statement of this problem, we rewrite Eqs. (1) and (2) as follows:

$$200(150\mathbf{i}) = 100(270\mathbf{i} - 120\mathbf{j} + 160\mathbf{k}) + 60[(v_B)_x\mathbf{i} + (v_B)_z\mathbf{k}]$$
$$+ 40[(v_C)_x\mathbf{i} + (v_C)_y\mathbf{j} + (v_C)_z\mathbf{k}] \qquad (1')$$

$$0 = 100\begin{vmatrix} \mathbf{i} & \mathbf{j} & \mathbf{k} \\ 555 & -180 & 240 \\ 270 & -120 & 160 \end{vmatrix} + 60\begin{vmatrix} \mathbf{i} & \mathbf{j} & \mathbf{k} \\ 255 & 0 & -120 \\ (v_B)_x & 0 & (v_B)_z \end{vmatrix}$$

$$+ 40\begin{vmatrix} \mathbf{i} & \mathbf{j} & \mathbf{k} \\ 105 & 450 & -420 \\ (v_C)_x & (v_C)_y & (v_C)_z \end{vmatrix} \qquad (2')$$

Equating to zero the coefficient of $\mathbf{j}$ in $(1')$ and the coefficients of $\mathbf{i}$ and $\mathbf{k}$ in $(2')$, we write, after reductions, the three scalar equations

$$(v_C)_y - 300 = 0$$
$$450(v_C)_z + 420(v_C)_y = 0$$
$$105(v_C)_y - 450(v_C)_x - 45\,000 = 0$$

which yield, respectively,

$$(v_C)_y = 300 \qquad (v_C)_z = -280 \qquad (v_C)_x = -30$$

The velocity of part $C$ is thus
$$\mathbf{v}_C = -(30 \text{ m/s})\mathbf{i} + (300 \text{ m/s})\mathbf{j} - (280 \text{ m/s})\mathbf{k} \qquad \blacktriangleleft$$

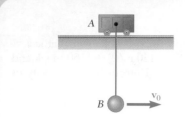

## SAMPLE PROBLEM 14.4

Ball $B$, of mass $m_B$, is suspended from a cord of length $l$ attached to cart $A$, of mass $m_A$, which can roll freely on a frictionless horizontal track. If the ball is given an initial horizontal velocity $\mathbf{v}_0$ while the cart is at rest, determine (*a*) the velocity of $B$ as it reaches its maximum elevation, (*b*) the maximum vertical distance $h$ through which $B$ will rise. (It is assumed that $v_0^2 < 2gl$.)

## SOLUTION

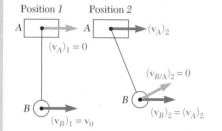

The impulse-momentum principle and the principle of conservation of energy will be applied to the cart-ball system between its initial position *1* and position *2*, when $B$ reaches its maximum elevation.

**Velocities   *Position 1:*** $\quad (\mathbf{v}_A)_1 = 0 \qquad (\mathbf{v}_B)_1 = \mathbf{v}_0$ $\qquad\qquad$ (1)

***Position 2:*** When ball $B$ reaches its maximum elevation, its velocity $(\mathbf{v}_{B/A})_2$ relative to its support $A$ is zero. Thus, at that instant, its absolute velocity is

$$(\mathbf{v}_B)_2 = (\mathbf{v}_A)_2 + (\mathbf{v}_{B/A})_2 = (\mathbf{v}_A)_2 \qquad (2)$$

**Impulse-Momentum Principle.** Noting that the external impulses consist of $\mathbf{W}_A t$, $\mathbf{W}_B t$, and $\mathbf{R} t$, where $\mathbf{R}$ is the reaction of the track on the cart, and recalling (1) and (2), we draw the impulse-momentum diagram and write

$$\Sigma m\mathbf{v}_1 + \Sigma\,\textbf{Ext Imp}_{1\rightarrow 2} = \Sigma m\mathbf{v}_2$$

$\xrightarrow{+}$ $x$ components: $\qquad m_B v_0 = (m_A + m_B)(v_A)_2$

which expresses that the linear momentum of the system is conserved in the horizontal direction. Solving for $(v_A)_2$:

$$(v_A)_2 = \frac{m_B}{m_A + m_B}\,v_0 \qquad (\mathbf{v}_B)_2 = (\mathbf{v}_A)_2 = \frac{m_B}{m_A + m_B}\,v_0 \rightarrow \quad \blacktriangleleft$$

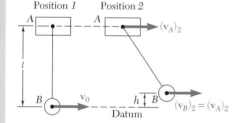

**Conservation of Energy**

***Position 1.*** ***Potential Energy:*** $\qquad V_1 = m_A gl$
$\qquad\qquad$ ***Kinetic Energy:*** $\qquad T_1 = \frac{1}{2} m_B v_0^2$

***Position 2.*** ***Potential Energy:*** $\qquad V_2 = m_A gl + m_B gh$
$\qquad\qquad$ ***Kinetic Energy:*** $\qquad T_2 = \frac{1}{2}(m_A + m_B)(v_A)_2^2$

$T_1 + V_1 = T_2 + V_2:\quad \frac{1}{2} m_B v_0^2 + m_A gl = \frac{1}{2}(m_A + m_B)(v_A)_2^2 + m_A gl + m_B gh$

Solving for $h$, we have

$$h = \frac{v_0^2}{2g} - \frac{m_A + m_B}{m_B}\frac{(v_A)_2^2}{2g}$$

or, substituting for $(v_A)_2$ the expression found above,

$$h = \frac{v_0^2}{2g} - \frac{m_B}{m_A + m_B}\frac{v_0^2}{2g} \qquad h = \frac{m_A}{m_A + m_B}\frac{v_0^2}{2g} \quad \blacktriangleleft$$

**Remarks.** (1) Recalling that $v_0^2 < 2gl$, it follows from the last equation that $h < l$; we thus check that $B$ stays below $A$ as assumed in our solution.

(2) For $m_A \gg m_B$, the answers obtained reduce to $(\mathbf{v}_B)_2 = (\mathbf{v}_A)_2 = 0$ and $h = v_0^2/2g$; $B$ oscillates as a simple pendulum with $A$ fixed. For $m_A \ll m_B$, they reduce to $(\mathbf{v}_B)_2 = (\mathbf{v}_A)_2 = \mathbf{v}_0$ and $h = 0$; $A$ and $B$ move with the same constant velocity $\mathbf{v}_0$.

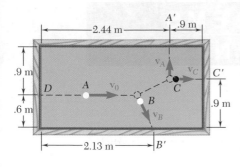

## SAMPLE PROBLEM 14.5

In a game of billiards, ball $A$ is given an initial velocity $\mathbf{v}_0$ of magnitude $v_0 = 3.05$ m/s along line $DA$ parallel to the axis of the table. It hits ball $B$ and then ball $C$, which are both at rest. Knowing that $A$ and $C$ hit the sides of the table squarely at points $A'$ and $C'$, respectively, that $B$ hits the side obliquely at $B'$, and assuming frictionless surfaces and perfectly elastic impacts, determine the velocities $\mathbf{v}_A$, $\mathbf{v}_B$, and $\mathbf{v}_C$ with which the balls hit the sides of the table. (*Remark.* In this sample problem and in several of the problems which follow, the billiard balls are assumed to be particles moving freely in a horizontal plane, rather than the rolling and sliding spheres they actually are.)

## SOLUTION

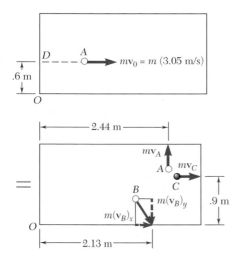

**Conservation of Momentum.** Since there is no external force, the initial momentum $m\mathbf{v}_0$ is equipollent to the system of momenta after the two collisions (and before any of the balls hits the side of the table). Referring to the adjoining sketch, we write

$\xrightarrow{+} x$ components: $\qquad m(3.05 \text{ m/s}) = m(v_B)_x + mv_C$ (1)

$+\uparrow y$ components: $\qquad\qquad\qquad 0 = mv_A - m(v_B)_y$ (2)

$+\text{\Large$\gamma$}$ moments about $O$: $-(.6 \text{ m})m(3.05 \text{ m/s}) = (2.44 \text{ m})mv_A$
$\qquad\qquad\qquad\qquad - (2.13 \text{ m})m(v_B)_y - (.9 \text{ m})mv_C$ (3)

Solving the three equations for $v_A$, $(v_B)_x$, and $(v_B)_y$ in terms of $v_C$,

$$v_A = (v_B)_y = 2.9v_C - 5.9 \qquad (v_B)_x = 3.05 - v_C \qquad (4)$$

**Conservation of Energy.** Since the surfaces are frictionless and the impacts are perfectly elastic, the initial kinetic energy $\frac{1}{2}mv_0^2$ is equal to the final kinetic energy of the system:

$$\tfrac{1}{2}mv_0^2 = \tfrac{1}{2}m_Av_A^2 + \tfrac{1}{2}m_Bv_B^2 + \tfrac{1}{2}m_Cv_C^2$$
$$v_A^2 + (v_B)_x^2 + (v_B)_y^2 + v_C^2 = (3.05 \text{ m/s})^2 \qquad (5)$$

Substituting for $v_A$, $(v_B)_x$, and $(v_B)_y$ from (4) into (5), we have

$$2(2.9v_C - 5.9)^2 + (3.05 - v_C)^2 + v_C^2 = 9.3$$
$$18.82v_C^2 - 74.54v_C + 69.62 = 0$$

Solving for $v_C$, we find $v_C = 2.45$ m/s and $v_C = 1.5$ m. Since only the first root yields a positive value for $v_A$ after substitution into Eqs. (4), we conclude that $v_C = 2.44$ m/s and

$$v_A = (v_B)_y - 2.9(2.45) - 5.9 = 1.205 \text{ m/s} \qquad (v_B)_x = 3.05 - 2.45 = 0.6 \text{ m/s}$$

$\mathbf{v}_A = 1.205 \text{ m/s}\uparrow \qquad \mathbf{v}_B = 1.346 \text{ m/s} \searrow 63.4° \qquad \mathbf{v}_C = 2.45 \text{ m/s} \rightarrow$ ◀

In the preceding lesson we defined the linear momentum and the angular momentum of a system of particles. In this lesson we defined the *kinetic energy T* of a system of particles:

$$T = \frac{1}{2} \sum_{i=1}^{n} m_i v_i^2 \qquad (14.28)$$

The solutions of the problems in the preceding lesson were based on the conservation of the linear momentum of a system of particles or on the observation of the motion of the mass center of a system of particles. In this lesson you will solve problems involving the following:

*1. Computation of the kinetic energy lost in collisions.* The kinetic energy $T_1$ of the system of particles before the collisions and its kinetic energy $T_2$ after the collisions are computed from Eq. (14.28) and are subtracted from each other. Keep in mind that, while linear momentum and angular momentum are vector quantities, kinetic energy is a *scalar* quantity.

*2. Conservation of linear momentum and conservation of energy.* As you saw in the preceding lesson, when the resultant of the external forces acting on a system of particles is zero, the linear momentum of the system is conserved. In problems involving two-dimensional motion, expressing that the initial linear momentum and the final linear momentum of the system are equipollent yields two algebraic equations. Equating the initial total energy of the system of particles (including potential energy as well as kinetic energy) to its final total energy yields an additional equation. Thus, you can write three equations which can be solved for three unknowns [Sample Prob. 14.5]. Note that if the resultant of the external forces is not zero but has a fixed direction, the component of the linear momentum in a direction perpendicular to the resultant is still conserved; the number of equations which can be used is then reduced to two [Sample Prob. 14.4].

*3. Conservation of linear and angular momentum.* When no external forces act on a system of particles, both the linear momentum of the system and its angular momentum about some arbitrary point are conserved. In the case of three-dimensional motion, this will enable you to write as many as six equations, although you may need to solve only some of them to obtain the desired answers [Sample Prob. 14.3]. In the case of two-dimensional motion, you will be able to write three equations which can be solved for three unknowns.

*4. Conservation of linear and angular momentum and conservation of energy.* In the case of the two-dimensional motion of a system of particles which are not subjected to any external forces, you will obtain two algebraic equations by expressing that the linear momentum of the system is conserved, one equation by writing that the angular momentum of the system about some arbitrary point is conserved, and a fourth equation by expressing that the total energy of the system is conserved. These equations can be solved for four unknowns.

# Problems

**14.31** In Prob. 14.6, determine the energy lost as the bullet (a) passes through block A, (b) becomes embedded in block B.

**14.32** In Prob. 14.1, determine the energy lost as a result of the first collision and verify that the total kinetic energy is unchanged as a result of the second collision.

**14.33** In Prob. 14.3, determine the total work done by the engines of cars A and B while the cars are accelerating to reach constant speeds.

**14.34** In Prob. 14.27, determine the increase in kinetic energy as a result of the explosion.

*14.35* Two automobiles A and B, of mass $m_A$ and $m_B$, respectively, are traveling in opposite directions when they collide head on. The impact is assumed perfectly plastic, and it is further assumed that the energy absorbed by each automobile is equal to its loss of kinetic energy with respect to a moving frame of reference attached to the mass center of the two-vehicle system. Denoting by $E_A$ and $E_B$, respectively, the energy absorbed by automobile A and by automobile B, (a) show that $E_A/E_B = m_B/m_A$, that is, the amount of energy absorbed by each vehicle is inversely proportional to its mass, (b) compute $E_A$ and $E_B$, knowing that $m_A = 2400$ kg and $m_B = 1350$ kg and that the speeds of A and B are, respectively, 135 km/h and 90 km/h.

**Fig. P14.35**

*14.36* It is assumed that each of the two automobiles involved in the collision described in Prob. 14.35 had been designed to safely withstand a test in which it crashed into a solid, immovable wall at the speed $v_0$. The severity of the collision of Prob. 14.35 may then be measured for each vehicle by the ratio of the energy absorbed in the collision to the energy it absorbed in the test. On that basis, show that the collision described in Prob. 14.35 is $(m_A/m_B)^2$ times more severe for automobile B than for automobile A.

**14.37** In a game of pool, ball A is moving with a velocity $\mathbf{v}_0 = v_0\mathbf{i}$ when it strikes balls B and C, which are at rest side by side. Assuming frictionless surfaces and perfectly elastic impact (that is, conservation of energy), determine the final velocity of each ball, assuming that the path of A is (a) perfectly centered and that A strikes B and C simultaneously, (b) not perfectly centered and that A strikes B slightly before it strikes C.

**Fig. P14.37**

**14.38** Solve Sample Prob. 14.4, assuming that cart $A$ is given an initial horizontal velocity $\mathbf{v}_0$ while ball $B$ is at rest.

**14.39 and 14.40** In a game of pool, ball $A$ is moving with a velocity $\mathbf{v}_0$ of magnitude $v_0 = 4.57$ m/s when it strikes balls $B$ and $C$, which are at rest and aligned as shown. Knowing that after the collision the three balls move in the directions indicated and assuming frictionless surfaces and perfectly elastic impact (that is, conservation of energy), determine the magnitudes of the velocities $\mathbf{v}_A$, $\mathbf{v}_B$, and $\mathbf{v}_C$.

**Fig. P14.39**

**Fig. P14.40**

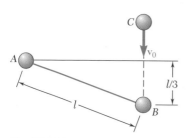

**Fig. P14.41**

**14.41** Three spheres, each of mass $m$, can slide freely on a frictionless, horizontal surface. Spheres $A$ and $B$ are attached to an inextensible, inelastic cord of length $l$ and are at rest in the position shown when sphere $B$ is struck squarely by sphere $C$ which is moving with a velocity $\mathbf{v}_0$. Knowing that the cord is taut when sphere $B$ is struck by sphere $C$ and assuming perfectly elastic impact between $B$ and $C$, and thus conservation of energy for the entire system, determine the velocity of each sphere immediately after impact.

**14.42** Three spheres, each of mass $m$, can slide freely on a frictionless, horizontal surface. Spheres $A$ and $B$ are attached to an inextensible, inelastic cord of length $l$ and are at rest in the position shown when sphere $B$ is struck squarely by sphere $C$ which is moving with a velocity $\mathbf{v}_0$. Knowing that the cord is slack when sphere $B$ is struck by sphere $C$ and assuming perfectly elastic impact between $B$ and $C$, determine (a) the velocity of each sphere immediately after the cord becomes taut, (b) the fraction of the initial kinetic energy of the system which is dissipated when the cord becomes taut.

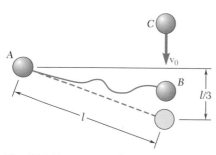

**Fig. P14.42**

**14.43** Ball $B$ is suspended from a cord of length $l$ attached to cart $A$, which can roll freely on a frictionless, horizontal track. The ball and the cart have the same mass $m$. If the ball is given an initial horizontal velocity $\mathbf{v}_0$ while the cart is at rest, describe the subsequent motion of the system, specifying the velocities of $A$ and $B$ for the following successive values of the angle $\theta$ (assume positive counterclockwise) that the cord will form with the vertical: (a) $\theta = \theta_{\max}$, (b) $\theta = 0$, (c) $\theta = \theta_{\min}$.

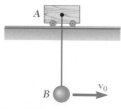

**Fig. P14.43**

**14.44** A 6-kg block $B$ starts from rest and slides on the 10-kg wedge $A$, which is supported by a horizontal surface. Neglecting friction, determine (a) the velocity of $B$ relative to $A$ after it has slid 1 m down the inclined surface of the wedge, (b) the corresponding velocity of $A$.

**Fig. *P14.44***

**14.45** Four small disks $A$, $B$, $C$, and $D$ can slide freely on a frictionless, horizontal surface. Disks $B$, $C$, and $D$ are connected by light rods and are at rest in the position shown when disk $B$ is struck squarely by disk $A$ which is moving to the right with a velocity $\mathbf{v}_0 = (12\ \text{m/s})\mathbf{i}$. The masses of the disks are $m_A = m_B = m_C = 7.5$ kg, and $m_D = 15$ kg. Knowing that the velocities of the disks immediately after the impact are $\mathbf{v}_A = \mathbf{v}_B = (2.5\ \text{m/s})\mathbf{i}$, $\mathbf{v}_C = v_C\mathbf{i}$, and $\mathbf{v}_D = v_D\mathbf{i}$, determine (a) the speeds $v_C$ and $v_D$, (b) the fraction of the initial kinetic energy of the system which is dissipated during the collision.

**14.46** Four small disks $A$, $B$, $C$, and $D$ can slide freely on a frictionless, horizontal surface. Disks $B$, $C$, and $D$ are connected by light rods and are at rest in the position shown when disk $B$ is struck squarely by disk $A$ which is moving to the right with a velocity $\mathbf{v}_0 = (12\ \text{m/s})\mathbf{i}$. The masses of the disks are $m_A = m_B = m_C = 7.5$ kg, and $m_D = 15$ kg. Knowing that the velocities of the disks immediately after the impact are $\mathbf{v}_A = 0$, $\mathbf{v}_B = (3.5\ \text{m/s})\mathbf{i}$, $\mathbf{v}_C = v_C\mathbf{i}$, and $\mathbf{v}_D = v_D\mathbf{i}$, determine (a) the speeds $v_C$ and $v_D$, (b) the fraction of the initial kinetic energy of the system which is dissipated during the collision.

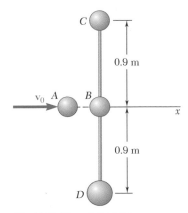

**Fig. P14.45 and P14.46**

**14.47** Two small spheres $A$ and $B$, respectively of mass $m$ and $2m$, are connected by a rigid rod of length $l$ and negligible mass. The two spheres are resting on a horizontal, frictionless surface when $A$ is suddenly given the velocity $\mathbf{v}_0 = v_0\mathbf{i}$. Determine (a) the linear momentum of the system and its angular momentum about its mass center $G$, (b) the velocities of $A$ and $B$ after the rod $AB$ has rotated through 180°.

**14.48** A 340.2 kg space vehicle traveling with a velocity $\mathbf{v}_0 = (457.2\ \text{m/s})\mathbf{k}$ passes through the origin $O$. Explosive charges then separate the vehicle into three parts $A$, $B$ and $C$, weighing, respectively, 556 N, 1112 N, and 1668 N. Knowing that shortly thereafter the positions of the three parts are, respectively, $A(73.15,\ 73.15,\ 658)$, $B(182.88,\ 402,\ 987.5)$, and $C(-146.3,\ -292.6,\ 585)$, where the coordinates are expressed in feet, that the velocity of $B$ is $\mathbf{v}_B = (152.4\ \text{m/s})\mathbf{i} + (335.28\ \text{m/s})\mathbf{j} + (670.56\ \text{m/s})\mathbf{k}$, and that the $x$ component of the velocity of $C$ is $-121.92$ m/s, determine the velocity of part $A$.

**Fig. P14.47**

**14.49** Three small spheres $A$, $B$, and $C$, each of mass $m$, are connected to a small ring $D$ of negligible mass by means of three inextensible, inelastic cords of length $l$. The spheres can slide freely on a frictionless horizontal surface and are rotating initially at a speed $v_0$ about ring $D$ which is at rest. Suddenly the cord $CD$ breaks. After the other two cords have again become taut, determine (a) the speed of ring $D$, (b) the relative speed at which spheres $A$ and $B$ rotate about $D$, (c) the fraction of the original energy of spheres $A$ and $B$ which is dissipated when cords $AD$ and $BD$ again became taut.

**14.50** Three small spheres $A$, $B$, and $C$, each of mass $m$, are connected to a small ring $D$ of mass $2m$ by means of three inextensible, inelastic cords of length $l$ which are equally spaced. The spheres can slide freely on a frictionless horizontal surface and are rotating initially at a speed $v_0$ about ring $D$ which is at rest. The cord $CD$ breaks at time $t = 0$ when the spheres and ring are in the position shown. Considering the system which consists of ring $D$ and spheres $A$ and $B$, determine for $t > 0$ (a) the position vector $\bar{\mathbf{r}}$ of the mass center $G$ of the system, (b) the angular momentum $\mathbf{H}_G$ of the system about $G$, (c) the kinetic energy $T$ of the system.

**Fig. P14.49 and P14.50**

**Fig. P14.51 and P14.52**

**14.51**   Three identical small spheres, each of weight .9 kg, can slide freely on a horizontal frictionless surface. Spheres $B$ and $C$ are connected by a light rod and are at rest in the position shown when sphere $B$ is struck squarely by sphere $A$ which is moving to the right with a velocity $\mathbf{v}_0 = (1.98 \text{ m/s})\mathbf{i}$. Knowing that $\theta = 30°$ and that the velocities of spheres $A$ and $B$ immediately after the impact are $\mathbf{v}_A = (.15 \text{ m/s})\mathbf{i}$ and $\mathbf{v}_B = (1.143 \text{ m/s})\mathbf{i} + (\mathbf{v}_B)_y\mathbf{j}$, determine $(\mathbf{v}_B)_y$ and the velocity of $C$ immediately after impact.

**14.52**   Three identical small spheres, each of weight .9 kg, can slide freely on a horizontal frictionless surface. Spheres $B$ and $C$ are connected by a light rod and are at rest in the position shown when sphere $B$ is struck squarely by sphere $A$ which is moving to the right with a velocity $\mathbf{v}_0 = (2.44 \text{ m/s})\mathbf{i}$. Knowing that $\theta = 45°$ and that the velocities of spheres $A$ and $B$ immediately after the impact are $\mathbf{v}_A = 0$ and $\mathbf{v}_B = (1.83 \text{ m/s})\mathbf{i} + (\mathbf{v}_B)_y\mathbf{j}$, determine $(v_B)_y$ and the velocity of $C$ immediately after impact.

**14.53**   Two small disks, $A$ and $B$, weighing 21.35 N and 10.68 N, respectively, can slide on a horizontal, frictionless surface. They are connected by a cord, .9 m long, and spin counterclockwise about their mass center $G$ at a rate of 8 rad/s. At $t = 0$, the coordinates of $G$ are $\bar{x}_0 = 0$, $\bar{y}_0 = 2.27$ m, and its velocity is $\bar{\mathbf{v}}_0 = (1.76 \text{ m/s})\mathbf{i} + (.44 \text{ m/s})\mathbf{j}$. Shortly thereafter the cord breaks; disk $A$ is then observed to move along a path parallel to the $y$ axis and disk $B$ along a path which intersects the $x$ axis at a distance $b = 7.32$ m from $O$. Determine (a) the velocities of $A$ and $B$ after the cord breaks, (b) the distance $a$ from the $y$ axis to the path of $A$.

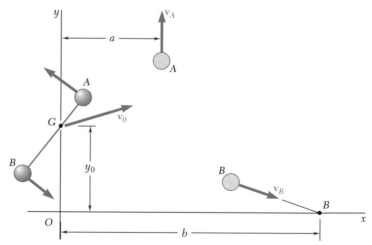

**Fig. P14.53 and P14.54**

**14.54**   Two small disks, $A$ and $B$, weighing 26.7 N and 13.3 N, respectively, can slide on a horizontal and frictionless surface. They are connected by a cord of negligible mass and spin about their mass center $G$. At $t = 0$, $G$ is moving with the velocity $\bar{\mathbf{v}}_0$ and its coordinates are $\bar{x}_0 = 0$, $\bar{y}_0 = 2.286$ m. Shortly thereafter, the cord breaks and disk $A$ is observed to move with a velocity $\bar{\mathbf{v}}_A = (2.34 \text{ m/s})\mathbf{j}$ in a straight line and at a distance $a = 1.7$ m from the $y$ axis, while $B$ moves with a velocity $\bar{\mathbf{v}}_B = (3.29 \text{ m/s})\mathbf{i} - (2.05 \text{ m/s})\mathbf{j}$ along a path intersecting the $x$ axis at a distance $b = 6.58$ m from the origin $O$. Determine (a) the initial velocity $\bar{\mathbf{v}}_0$ of the mass center $G$ of the two disks, (b) the length of the cord initially connecting the two disks, (c) the rate in rad/s at which the disks were spinning about $G$.

**14.55** In a game of billiards, ball $A$ is given an initial velocity $\mathbf{v}_0$ along the longitudinal axis of the table. It hits ball $B$ and then ball $C$, which are both at rest. Balls $A$ and $C$ are observed to hit the sides of the table squarely at $A'$ and $C'$, respectively, and ball $B$ is observed to hit the side obliquely at $B'$. Knowing that $v_0 = 4$ m/s, $v_A = 1.92$ m/s, and $a = 1.65$ m, determine (a) the velocities $\mathbf{v}_B$ and $\mathbf{v}_C$ of balls $B$ and $C$, (b) the point $C'$ where ball $C$ hits the side of the table. Assume frictionless surfaces and perfectly elastic impacts (that is, conservation of energy).

**14.56** For the game of billiards of Prob. 14.55, it is now assumed that $v_0 = 5$ m/s, $v_C = 3.2$ m/s, and $c = 1.22$ m. Determine (a) the velocities $\mathbf{v}_A$ and $\mathbf{v}_B$ of balls $A$ and $B$, (b) the point $A'$ where ball $A$ hits the side of the table.

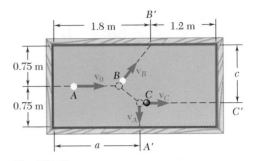

**Fig. P14.55**

**14.57** Three small identical spheres $A$, $B$, and $C$, which can slide on a horizontal, frictionless surface, are attached to three strings of length $l$ which are tied to a ring $G$. Initially the spheres rotate about the ring which moves along the $x$ axis with a velocity $\mathbf{v}_0$. Suddenly the ring breaks and the three spheres move freely in the $xy$ plane. Knowing that $\mathbf{v}_A = (2.6$ m/s$)\mathbf{j}$, $\mathbf{v}_C = (4.5$ m/s$)\mathbf{i}$, $a = 260$ mm, and $d = 150$ mm, determine (a) the initial velocity of the ring, (b) the length $l$ of the strings, (c) the rate in rad/s at which the spheres were rotating about $G$.

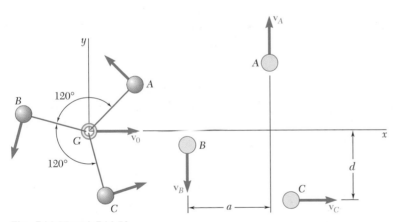

**Fig. *P14.57* and *P14.58***

**14.58** Three small identical spheres $A$, $B$, and $C$, which can slide on a horizontal, frictionless surface, are attached to three strings, 75 mm long, which are tied to a ring $G$. Initially the spheres rotate counterclockwise about the ring with a relative velocity of 0.75 m/s and the ring moves along the $x$ axis with a velocity $\mathbf{v}_0 = (0.4$ m/s$)\mathbf{i}$. Suddenly the ring breaks and the three spheres move freely in the $xy$ plane with $A$ and $B$ following paths parallel to the $y$ axis at a distance $a = 130$ mm from each other and $C$ following a path parallel to the $x$ axis. Determine (a) the velocity of each sphere, (b) the distance $d$.

## *14.10. VARIABLE SYSTEMS OF PARTICLES

All the systems of particles considered so far consisted of well-defined particles. These systems did not gain or lose any particles during their motion. In a large number of engineering applications, however, it is necessary to consider *variable systems of particles,* that is, systems which are continually gaining or losing particles, or doing both at the same time. Consider, for example, a hydraulic turbine. Its analysis involves the determination of the forces exerted by a stream of water on rotating blades, and we note that the particles of water in contact with the blades form an everchanging system which continually acquires and loses particles. Rockets furnish another example of variable systems, since their propulsion depends upon the continual ejection of fuel particles.

We recall that all the kinetics principles established so far were derived for constant systems of particles, which neither gain nor lose particles. We must therefore find a way to reduce the analysis of a variable system of particles to that of an auxiliary constant system. The procedure to follow is indicated in Secs. 14.11 and 14.12 for two broad categories of applications: a steady stream of particles and a system that is gaining or losing mass.

## *14.11. STEADY STREAM OF PARTICLES

**Fig. 14.9**

Consider a steady stream of particles, such as a stream of water diverted by a fixed vane or a flow of air through a duct or through a blower. In order to determine the resultant of the forces exerted on the particles in contact with the vane, duct, or blower, we isolate these particles and denote by $S$ the system thus defined (Fig. 14.9). We observe that $S$ is a variable system of particles, since it continually gains particles flowing in and loses an equal number of particles flowing out. Therefore, the kinetics principles that have been established so far cannot be directly applied to $S$.

However, we can easily define an auxiliary system of particles which does remain constant for a short interval of time $\Delta t$. Consider at time $t$ the system $S$ *plus* the particles which will enter $S$ during the interval at time $\Delta t$ (Fig. 14.10$a$). Next, consider at time $t + \Delta t$ the system $S$ *plus* the particles which have left $S$ during the interval $\Delta t$ (Fig. 14.10$c$). Clearly, *the same particles are involved in both cases,* and we can apply to those particles the principle of impulse and momentum. Since the total mass $m$ of the system $S$ remains constant, the particles entering the system and those leaving the system in the time $\Delta t$ must have the same mass $\Delta m$. Denoting by $\mathbf{v}_A$ and $\mathbf{v}_B$, respectively, the velocities of the particles entering $S$ at $A$ and leaving $S$ at $B$, we represent the momentum of the particles entering $S$ by $(\Delta m)\mathbf{v}_A$ (Fig. 14.10$a$) and the momentum of the particles leaving $S$ by $(\Delta m)\mathbf{v}_B$ (Fig. 14.10$c$). We also represent by appropriate vectors the momenta $m_i\mathbf{v}_i$ of the particles forming $S$ and the impulses of the forces exerted on $S$ and indicate by blue plus and equals signs that the system of the momenta and impulses in parts $a$ and $b$ of Fig. 14.10 is equipollent to the system of the momenta in part $c$ of the same figure.

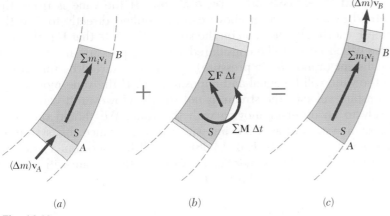

**Fig. 14.10**

The resultant $\Sigma m_i\mathbf{v}_i$ of the momenta of the particles of S is found on both sides of the equals sign and can thus be omitted. We conclude that *the system formed by the momentum $(\Delta m)\mathbf{v}_A$ of the particles entering S in the time $\Delta t$ and the impulses of the forces exerted on S during that time is equipollent to the momentum $(\Delta m)\mathbf{v}_B$ of the particles leaving S in the same time $\Delta t$.* We can therefore write

$$(\Delta m)\mathbf{v}_A + \Sigma\mathbf{F}\,\Delta t = (\Delta m)\mathbf{v}_B \tag{14.38}$$

A similar equation can be obtained by taking the moments of the vectors involved (see Sample Prob. 14.6). Dividing all terms of Eq. (14.38) by $\Delta t$ and letting $\Delta t$ approach zero, we obtain at the limit

$$\Sigma\mathbf{F} = \frac{dm}{dt}\,(\mathbf{v}_B - \mathbf{v}_A) \tag{14.39}$$

where $\mathbf{v}_B - \mathbf{v}_A$ represents the difference between the *vector* $\mathbf{v}_B$ and the *vector* $\mathbf{v}_A$.

If SI units are used, $dm/dt$ is expressed in kg/s and the velocities in m/s; we check that both members of Eq. (14.39) are expressed in the same units (newtons). If U.S. customary units are used, $dm/dt$ must be expressed in slugs/s and the velocities in ft/s; we check again that both members of the equation are expressed in the same units (pounds).†

The principle we have established can be used to analyze a large number of engineering applications. Some of the more common of these applications will be considered next.

---

†It is often convenient to express the mass rate of flow $dm/dt$ as the product $\rho Q$, where $\rho$ is the density of the stream (mass per unit volume) and $Q$ its volume rate of flow (volume per unit time). If SI units are used, $\rho$ is expressed in kg/m$^3$ (for instance, $\rho = 1000$ kg/m$^3$ for water) and $Q$ in m$^3$/s. However, if U.S. customary units are used, $\rho$ will generally have to be computed from the corresponding specific weight $\gamma$ (weight per unit volume), $\rho = \gamma/g$. Since $\gamma$ is expressed in lb/ft$^3$ (for instance, $\gamma = 62.4$ lb/ft$^3$ for water), $\rho$ is obtained in slugs/ft$^3$. The volume rate of flow $Q$ is expressed in ft$^3$/s.

**Fluid Stream Diverted by a Vane.** If the vane is fixed, the method of analysis given above can be applied directly to find the force **F** exerted by the vane on the stream. We note that **F** is the only force which needs to be considered since the pressure in the stream is constant (atmospheric pressure). The force exerted by the stream on the vane will be equal and opposite to **F**. If the vane moves with a constant velocity, the stream is not steady. However, it will appear steady to an observer moving with the vane. We should therefore choose a system of axes moving with the vane. Since this system of axes is not accelerated, Eq. (14.38) can still be used, but $\mathbf{v}_A$ and $\mathbf{v}_B$ must be replaced by the *relative velocities* of the stream with respect to the vane (see Sample Prob. 14.7).

**Fluid Flowing through a Pipe.** The force exerted by the fluid on a pipe transition such as a bend or a contraction can be determined by considering the system of particles $S$ in contact with the transition. Since, in general, the pressure in the flow will vary, the forces exerted on $S$ by the adjoining portions of the fluid should also be considered.

**Jet Engine.** In a jet engine, air enters with no velocity through the front of the engine and leaves through the rear with a high velocity. The energy required to accelerate the air particles is obtained by burning fuel. The mass of the burned fuel in the exhaust gases will usually be small enough compared with the mass of the air flowing through the engine that it can be neglected. Thus, the analysis of a jet engine reduces to that of an airstream. This stream can be considered as a steady stream if all velocities are measured with respect to the airplane. It will be assumed, therefore, that the airstream enters the engine with a velocity **v** of magnitude equal to the speed of the airplane and leaves with a velocity **u** equal to the relative velocity of

**Photo 14.3** The analysis of a jet engine reduces to that of a steady stream of air particles with all velocities measured with respect to the airplane.

v          u

**Fig. 14.11**

the exhaust gases (Fig. 14.11). Since the intake and exhaust pressures are nearly atmospheric, the only external force which needs to be considered is the force exerted by the engine on the airstream. This force is equal and opposite to the thrust.†

---

†Note that if the airplane is accelerated, it cannot be used as a newtonian frame of reference. The same result will be obtained for the thrust, however, by using a reference frame at rest with respect to the atmosphere, since the air particles will then be observed to enter the engine with no velocity and to leave it with a velocity of magnitude $u - v$.

**Fan.** We consider the system of particles $S$ shown in Fig. 14.12. The velocity $\mathbf{v}_A$ of the particles entering the system is assumed equal to zero, and the velocity $\mathbf{v}_B$ of the particles leaving the system is the velocity of the *slipstream*. The rate of flow can be obtained by multiplying $v_B$ by the cross-sectional area of the slipstream. Since the pressure all around $S$ is atmospheric, the only external force acting on $S$ is the thrust of the fan.

**Helicopter.** The determination of the thrust created by the rotating blades of a hovering helicopter is similar to the determination of the thrust of a fan. The velocity $\mathbf{v}_A$ of the air particles as they approach the blades is assumed to be zero, and the rate of flow is obtained by multiplying the magnitude of the velocity $\mathbf{v}_B$ of the slipstream by its cross-sectional area.

**Fig. 14.12**

## *14.12. SYSTEMS GAINING OR LOSING MASS

Let us now analyze a different type of variable system of particles, namely, a system which gains mass by continually absorbing particles or loses mass by continually expelling particles. Consider the system $S$ shown in Fig. 14.13. Its mass, equal to $m$ at the instant $t$, increases

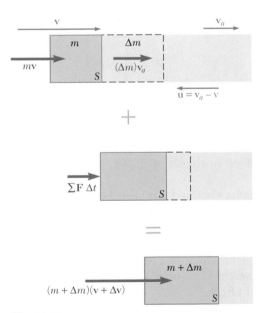

**Fig. 14.13**

by $\Delta m$ in the interval of time $\Delta t$. In order to apply the principle of impulse and momentum to the analysis of this system, we must consider at time $t$ the system $S$ *plus* the particles of mass $\Delta m$ which $S$ absorbs during the time interval $\Delta t$. The velocity of $S$ at time $t$ is denoted by $\mathbf{v}$, the velocity of $S$ at time $t + \Delta t$ is denoted by $\mathbf{v} + \Delta\mathbf{v}$, and the absolute velocity of the particles absorbed is denoted by $\mathbf{v}_a$. Applying the principle of impulse and momentum, we write

$$m\mathbf{v} + (\Delta m)\mathbf{v}_a + \Sigma\mathbf{F}\,\Delta t = (m + \Delta m)(\mathbf{v} + \Delta\mathbf{v}) \qquad (14.40)$$

Solving for the sum $\Sigma \mathbf{F} \, \Delta t$ of the impulses of the external forces acting on $S$ (excluding the forces exerted by the particles being absorbed), we have

$$\Sigma \mathbf{F} \, \Delta t = m \Delta \mathbf{v} + \Delta m (\mathbf{v} - \mathbf{v}_a) + (\Delta m)(\Delta \mathbf{v}) \qquad (14.41)$$

Introducing the *relative velocity* $\mathbf{u}$ with respect to $S$ of the particles which are absorbed, we write $\mathbf{u} = \mathbf{v}_a - \mathbf{v}$ and note, since $v_a < v$, that the relative velocity $\mathbf{u}$ is directed to the left, as shown in Fig. 14.13. Neglecting the last term in Eq. (14.41), which is of the second order, we write

$$\Sigma \mathbf{F} \, \Delta t = m \, \Delta \mathbf{v} - (\Delta m)\mathbf{u}$$

Dividing through by $\Delta t$ and letting $\Delta t$ approach zero, we have at the limit†

$$\Sigma \mathbf{F} = m \frac{d\mathbf{v}}{dt} - \frac{dm}{dt} \mathbf{u} \qquad (14.42)$$

Rearranging the terms and recalling that $d\mathbf{v}/dt = \mathbf{a}$, where $\mathbf{a}$ is the acceleration of the system $S$, we write

$$\Sigma \mathbf{F} + \frac{dm}{dt} \mathbf{u} = m\mathbf{a} \qquad (14.43)$$

which shows that the action on $S$ of the particles being absorbed is equivalent to a thrust

$$\mathbf{P} = \frac{dm}{dt} \mathbf{u} \qquad (14.44)$$

which tends to slow down the motion of $S$, since the relative velocity $\mathbf{u}$ of the particles is directed to the left. If SI units are used, $dm/dt$ is expressed in kg/s, the relative velocity $u$ in m/s, and the corresponding thrust in newtons. If U.S. customary units are used, $dm/dt$ must be expressed in slugs/s, $u$ in ft/s, and the corresponding thrust in pounds.‡

The equations obtained can also be used to determine the motion of a system $S$ losing mass. In this case, the rate of change of mass is negative, and the action on $S$ of the particles being expelled is equivalent to a thrust in the direction of $-\mathbf{u}$, that is, in the direction opposite to that in which the particles are being expelled. A *rocket* represents a typical case of a system continually losing mass (see Sample Prob. 14.8).

**Photo 14.4**   As the shuttle's booster rockets are fired, the gas particles they eject provide the thrust required for liftoff.

---

†When the absolute velocity $\mathbf{v}_a$ of the particles absorbed is zero, $\mathbf{u} = -\mathbf{v}$, and formula (14.42) becomes

$$\Sigma \mathbf{F} = \frac{d}{dt}(m\mathbf{v})$$

Comparing the formula obtained to Eq. (12.3) of Sec. 12.3, we observe that Newton's second law can be applied to a system gaining mass, *provided that the particles absorbed are initially at rest*. It may also be applied to a system losing mass, *provided that the velocity of the particles expelled is zero* with respect to the frame of reference selected.

‡See footnote on page 887.

## SAMPLE PROBLEM 14.6

Grain falls from a hopper onto a chute $CB$ at the rate of 108.86 kg/s. It hits the chute at $A$ with a velocity of 6.09 m/s and leaves at $B$ with a velocity of 4.57 m/s, forming an angle of 10° with the horizontal. Knowing that the combined weight of the chute and of the grain it supports is a force **W** of magnitude 2668.9 N applied at $G$, determine the reaction at the roller support $B$ and the components of the reaction at the hinge $C$.

## SOLUTION

We apply the principle of impulse and momentum for the time interval $\Delta t$ to the system consisting of the chute, the grain it supports, and the amount of grain which hits the chute in the interval $\Delta t$. Since the chute does not move, it has no momentum. We also note that the sum $\Sigma m_i\mathbf{v}_i$ of the momenta of the particles supported by the chute is the same at $t$ and $t + \Delta t$ and can thus be omitted.

Since the system formed by the momentum $(\Delta m)\mathbf{v}_A$ and the impulses is equipollent to the momentum $(\Delta m)\mathbf{v}_B$, we write

$\xrightarrow{+}$ $x$ components:
$$C_x \Delta t = (\Delta m)v_B \cos 10° \qquad (1)$$

$+\uparrow$ $y$ components:
$$-(\Delta m)v_A + C_y \Delta t - W \Delta t + B \Delta t$$
$$= -(\Delta m)v_B \sin 10° \quad (2)$$

$+\curvearrowleft$ moments about $C$:
$$-.9(\Delta m)v_A - 2.13(W \Delta t) + 3.66(B \Delta t)$$
$$= 1.8(\Delta m)v_B \cos 10° - 3.66(\Delta m)v_B \sin 10° \quad (3)$$

Using the given data, $W = 2668.9$ N, $v_A = 6.09$ m/s, $v_B = 4.57$ m/s, and $\Delta m/\Delta t = 108.86$ kg/s and solving Eq. (3) for $B$ and Eq. (1) for $C_x$,

$$3.66B = 2.13(2668.9) + .9(108.86)(6.09) +$$
$$1.8(108.86)(4.57)(\cos 10° - 2 \sin 10°)$$

$$3.66B = 6852.296 \qquad B = 1872.2 \text{ N} \qquad \mathbf{B} = 1872.2 \text{ N} \uparrow \blacktriangleleft$$

$$C_x = (108.86)(4.57) \cos 10° = 489.9 \text{ N} \qquad C_x = 489.9 \text{ N} \rightarrow \blacktriangleleft$$

Substituting for $B$ and solving Eq. (2) for $C_y$,

$$C_y = 2668.9 - 1872.2 + (108.86)(6.09 - 4.57 \sin 10°) = 1373 \text{ N}$$
$$C_y = 1373.3 \text{ N} \uparrow \blacktriangleleft$$

## SAMPLE PROBLEM 14.7

A nozzle discharges a stream of water of cross-sectional area $A$ with a velocity $\mathbf{v}_A$. The stream is deflected by a *single* blade which moves to the right with a constant velocity $\mathbf{V}$. Assuming that the water moves along the blade at constant speed, determine (*a*) the components of the force $\mathbf{F}$ exerted by the blade on the stream, (*b*) the velocity $\mathbf{V}$ for which maximum power is developed.

## SOLUTION

*a*. **Components of Force Exerted on Stream.** We choose a coordinate system which moves with the blade at a constant velocity $\mathbf{V}$. The particles of water strike the blade with a relative velocity $\mathbf{u}_A = \mathbf{v}_A - \mathbf{V}$ and leave the blade with a relative velocity $\mathbf{u}_B$. Since the particles move along the blade at a constant speed, the relative velocities $\mathbf{u}_A$ and $\mathbf{u}_B$ have the same magnitude $u$. Denoting the density of water by $\rho$, the mass of the particles striking the blade during the time interval $\Delta t$ is $\Delta m = A\rho(v_A - V)\,\Delta t$; an equal mass of particles leaves the blade during $\Delta t$. We apply the principle of impulse and momentum to the system formed by the particles in contact with the blade and the particles striking the blade in the time $\Delta t$.

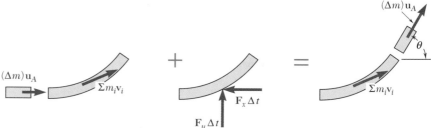

Recalling that $\mathbf{u}_A$ and $\mathbf{u}_B$ have the same magnitude $u$, and omitting the momentum $\Sigma m_i \mathbf{v}_i$ which appears on both sides, we write

$\xrightarrow{+}$ $x$ components: $\qquad (\Delta m)u - F_x\,\Delta t = (\Delta m)u\cos\theta$

$+\uparrow y$ components: $\qquad\qquad +F_y\,\Delta t = (\Delta m)u\sin\theta$

Substituting $\Delta m = A\rho(v_A - V)\,\Delta t$ and $u = v_A - V$, we obtain

$$\mathbf{F}_x = A\rho(v_A - V)^2(1 - \cos\theta) \leftarrow \qquad \mathbf{F}_y = A\rho(v_A - V)^2\sin\theta \uparrow \quad \blacktriangleleft$$

*b*. **Velocity of Blade for Maximum Power.** The power is obtained by multiplying the velocity $V$ of the blade by the component $F_x$ of the force exerted by the stream on the blade.

$$\text{Power} = F_x V = A\rho(v_A - V)^2(1 - \cos\theta)V$$

Differentiating the power with respect to $V$ and setting the derivative equal to zero, we obtain

$$\frac{d(\text{power})}{dV} = A\rho(v_A^2 - 4v_A V + 3V^2)(1 - \cos\theta) = 0$$

$$V = v_A \qquad V = \tfrac{1}{3}v_A \qquad \text{For maximum power } \mathbf{V} = \tfrac{1}{3}v_A \rightarrow \quad \blacktriangleleft$$

**Note.** These results are valid only when a *single* blade deflects the stream. Different results are obtained when a series of blades deflects the stream, as in a Pelton-wheel turbine. (See Prob. 14.81.)

## SAMPLE PROBLEM 14.8

A rocket of initial mass $m_0$ (including shell and fuel) is fired vertically at time $t = 0$. The fuel is consumed at a constant rate $q = dm/dt$ and is expelled at a constant speed $u$ relative to the rocket. Derive an expression for the magnitude of the velocity of the rocket at time $t$, neglecting the resistance of the air.

## SOLUTION

At time $t$, the mass of the rocket shell and remaining fuel is $m = m_0 - qt$, and the velocity is $\mathbf{v}$. During the time interval $\Delta t$, a mass of fuel $\Delta m = q \, \Delta t$ is expelled with a speed $u$ relative to the rocket. Denoting by $\mathbf{v}_e$ the absolute velocity of the expelled fuel, we apply the principle of impulse and momentum between time $t$ and time $t + \Delta t$.

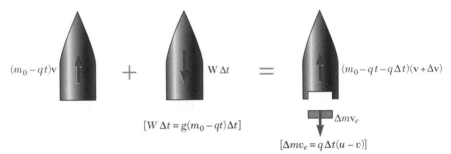

We write

$$(m_0 - qt)v - g(m_0 - qt) \, \Delta t = (m_0 - qt - q \, \Delta t)(v + \Delta v) - q \, \Delta t(u - v)$$

Dividing through by $\Delta t$ and letting $\Delta t$ approach zero, we obtain

$$-g(m_0 - qt) = (m_0 - qt)\frac{dv}{dt} - qu$$

Separating variables and integrating from $t = 0$, $v = 0$ to $t = t$, $v = v$,

$$dv = \left(\frac{qu}{m_0 - qt} - g\right) dt \qquad \int_0^v dv = \int_0^t \left(\frac{qu}{m_0 - qt} - g\right) dt$$

$$v = [-u \ln (m_0 - qt) - gt]_0^t \qquad v = u \ln \frac{m_0}{m_0 - qt} - gt \quad \blacktriangleleft$$

**Remark.** The mass remaining at time $t_f$, after all the fuel has been expended, is equal to the mass of the rocket shell $m_s = m_0 - qt_f$, and the maximum velocity attained by the rocket is $v_m = u \ln (m_0/m_s) - gt_f$. Assuming that the fuel is expelled in a relatively short period of time, the term $gt_f$ is small and we have $v_m \approx u \ln (m_0/m_s)$. In order to escape the gravitational field of the earth, a rocket must reach a velocity of 11.18 km/s. Assuming $u = 2200$ m/s and $v_m = 11.18$ km/s, we obtain $m_0/m_s = 161$. Thus, to project each kilogram of the rocket shell into space, it is necessary to consume more than 161 kg of fuel if a propellant yielding $u = 2200$ m/s is used.

This lesson is devoted to the study of the motion of *variable systems of particles,* that is, systems which are continually *gaining or losing particles* or doing both at the same time. The problems you will be asked to solve will involve (1) *steady streams of particles* and (2) *systems gaining or losing mass.*

*1.   To solve problems involving a steady stream of particles,* you will consider a portion S of the stream and express that the system formed by the momentum of the particles entering S at A in the time $\Delta t$ and the impulses of the forces exerted on S during that time is equipollent to the momentum of the particles leaving S at B in the same time $\Delta t$ (Fig. 14.10). Considering only the resultants of the vector systems involved, you can write the vector equation

$$(\Delta m)\mathbf{v}_A + \Sigma\mathbf{F}\,\Delta t = (\Delta m)\mathbf{v}_B \qquad (14.38)$$

You may want to consider as well the moments about a given point of the vector systems involved to obtain an additional equation [Sample Prob. 14.6], but many problems can be solved using Eq. (14.38) or the equation obtained by dividing all terms by $\Delta t$ and letting $\Delta t$ approach zero,

$$\Sigma\mathbf{F} = \frac{dm}{dt}(\mathbf{v}_B - \mathbf{v}_A) \qquad (14.39)$$

where $\mathbf{v}_B - \mathbf{v}_A$ represents a *vector subtraction* and where the mass rate of flow $dm/dt$ can be expressed as the product $\rho Q$ of the density $\rho$ of the stream (mass per unit volume) and the volume rate of flow $Q$ (volume per unit time). If U.S. customary units are used, $\rho$ is expressed as the ratio $\gamma/g$, where $\gamma$ is the specific weight of the stream and $g$ is the acceleration of gravity.

Typical problems involving a steady stream of particles have been described in Sec. 14.11. You may be asked to determine the following:

*a.   Thrust caused by a diverted flow.* Equation (14.39) is applicable, but you will get a better understanding of the problem if you use a solution based on Eq. (14.38).

*b.   Reactions at supports of vanes or conveyor belts.* First draw a diagram showing on one side of the equals sign the momentum $(\Delta m)\mathbf{v}_A$ of the particles impacting the vane or belt in the time $\Delta t$, as well as the impulses of the loads and reactions at the supports during that time, and showing on the other side the momentum $(\Delta m)\mathbf{v}_B$ of the particles leaving the vane or belt in the time $\Delta t$ [Sample Prob. 14.6]. Equating the $x$ components, $y$ components, and moments of the quantities on both sides of the equals sign will yield three scalar equations which can be solved for three unknowns.

*c. Thrust developed by a jet engine, a propeller, or a fan.* In most cases, a single unknown is involved, and that unknown can be obtained by solving the scalar equation derived from Eq. (14.38) or Eq. (14.39).

2. *To solve problems involving systems gaining mass,* you will consider the system $S$, which has a mass $m$ and is moving with a velocity $\mathbf{v}$ at time $t$, and the particles of mass $\Delta m$ with velocity $\mathbf{v}_a$ that $S$ will absorb in the time interval $\Delta t$ (Fig. 14.13). You will then express that the total momentum of $S$ and of the particles that will be absorbed, *plus* the impulse of the external forces exerted on $S$, are equipollent to the momentum of $S$ at time $t + \Delta t$. Noting that the mass of $S$ and its velocity at that time are, respectively, $m + \Delta m$ and $\mathbf{v} + \Delta \mathbf{v}$, you will write the vector equation

$$m\mathbf{v} + (\Delta m)\mathbf{v}_a + \Sigma \mathbf{F} \, \Delta t = (m + \Delta m)(\mathbf{v} + \Delta \mathbf{v}) \qquad (14.40)$$

As was shown in Sec. 14.12, if you introduce the relative velocity $\mathbf{u} = \mathbf{v}_a - \mathbf{v}$ of the particles being absorbed, you obtain the following expression for the resultant of the external forces applied to $S$:

$$\Sigma \mathbf{F} = m \frac{d\mathbf{v}}{dt} - \frac{dm}{dt}\mathbf{u} \qquad (14.42)$$

Furthermore, it was shown that the action on $S$ of the particles being absorbed is equivalent to a thrust

$$\mathbf{P} = \frac{dm}{dt}\mathbf{u} \qquad (14.44)$$

exerted in the direction of the relative velocity of the particles being absorbed.

Examples of systems gaining mass are conveyor belts and moving railroad cars being loaded with gravel or sand, and chains being pulled out of a pile.

3. *To solve problems involving systems losing mass,* such as rockets and rocket engines, you can use Eqs. (14.40) through (14.44), provided that you give negative values to the increment of mass $\Delta m$ and to the rate of change of mass $dm/dt$. It follows that the thrust defined by Eq. (14.44) will be exerted in a direction opposite to the direction of the relative velocity of the particles being ejected.

# Problems

**14.59**  A stream of water of cross-sectional area $A$ and velocity $\mathbf{v}_1$ strikes a plate which is held motionless by a force $\mathbf{P}$. Determine the magnitude of $\mathbf{P}$, knowing that $A = 500$ mm², $v_1 = 25$ m/s and $V = 0$.

**Fig. P14.59 and P14.60**

**14.60**  A stream of water of cross-sectional area $A$ and velocity $\mathbf{v}_1$ strikes a plate which moves to the right with a velocity $\mathbf{V}$. Determine the magnitude of $V$, knowing that $A = 600$ mm², $v_1 = 30$ m/s, and $P = 400$ N.

**14.61**  The nozzle shown discharges a stream of water at a flow rate $Q = 1.8$ m³/min with a velocity $\mathbf{v}$ of magnitude 18.29 m/s. The stream is split into two streams with equal flow rates by a wedge which is kept in a fixed position. Determine the components (drag and lift) of the force exerted by the stream on the wedge.

**14.62**  The nozzle shown discharges a stream of water at a flow rate $Q = 1.89$ m³/min with a velocity $\mathbf{v}$ of magnitude 14.6 m/s. The stream is split into two streams of equal flow rates by a wedge which is moving to the left at a constant speed of 3.66 m/s. Determine the components (drag and lift) of the force exerted by the stream on the wedge.

**Fig. P14.61 and P14.62**

**14.63**  Tree limbs and branches are being fed at $A$ at the rate of 4.54 kg/s into a shredder which spews the resulting wood chips at $C$ with a velocity of 18.29 m/s. Determine the horizontal component of the force exerted by the shredder on the truck hitch at $D$.

**Fig. P14.63**

**14.64**  A hose discharges water at a rate of 8 m³/min with a velocity of 50 m/s from the bow of a fireboat. Determine the thrust developed by the engine to keep the fireboat in a stationary position.

**14.65**  Sand falls from three hoppers onto a conveyor belt at a rate of 40 kg/s for each hopper. The sand hits the belt with a vertical velocity $v_1 = 3$ m/s and is discharged at $A$ with a horizontal velocity $v_2 = 4$ m/s. Knowing that the combined mass of the beam, belt system, and the sand it supports is 600 kg with a mass center at $G$, determine the reaction at $E$.

**Fig. _P14.64_**

**Fig. P14.65**

**14.66**  Knowing that the blade $AB$ of Sample Prob. 14.7 is in the shape of an arc of a circle, show that the resultant force **F** exerted by the blade on the stream is applied at the midpoint $C$ of the arc $AB$. (*Hint.* First show that the line of action of **F** must pass through the center $O$ of the circle.)

**14.67**  The nozzle shown discharges water at the rate of 800 L/min. Knowing that at both $B$ and $C$ the stream of water moves with a velocity of magnitude 30 m/s, and neglecting the weight of the vane, determine the force-couple system which must be applied at $A$ to hold the vane in place.

**14.68**  The nozzle shown discharges water at the rate of 1.13 m³/min. Knowing that at both $A$ and $B$ the stream of water moves with a velocity of magnitude 22.86 m/s and neglecting the weight of the vane, determine the components of the reactions at $C$ and $D$.

**Fig. P14.67**

**Fig. P14.68**

**Fig. *P14.69* and *P14.70***

**14.69** A stream of water flowing at a rate of 1.135 m³/min and moving with a velocity of magnitude 27.4 m/s at both *A* and *B* is deflected by a vane welded to a hinged plate. Knowing that the combined weight of the vane and plate is 178 N with the mass center at point *G*, determine (*a*) the angle *θ*, (*b*) the reaction at *C*.

**14.70** A stream of water flowing at a rate of 1.135 m³/min and moving with a velocity of magnitude *v* at both *A* and *B* is deflected by a vane welded to a hinged plate. The combined weight of the vane and plate is 178 N with the mass center at point *G*. Knowing that *θ* = 45°, determine (*a*) the speed *v* of the flow, (*b*) the reaction at *C*.

**14.71** The total drag due to air friction on a jet airplane traveling at 900 km/h is 35 kN. Knowing that the exhaust velocity is 600 m/s relative to the airplane, determine the mass of air which must pass through the engine per second to maintain the speed of 900 km/h in level flight.

**14.72** Prior to takeoff the pilot of a 2722 kg twin-engine airplane tests the reversible-pitch propellers by increasing the reverse thrust with the brakes at point *B* locked. Knowing that point *G* is the center of gravity of the airplane, determine the velocity of the air in the two 2 m diameter slipstreams when the nose wheel *A* begins to lift off the ground. Assume *γ* = 1.2 kg/m³ and neglect the approach velocity of the air.

**Fig. P14.72 and P14.73**

**14.73** Prior to takeoff the pilot of a 2722 kg twin-engine airplane tests the reversible-pitch propellers with the brakes at point *B* locked. Knowing that the velocity of the air in the two 2 m diameter slipstreams is 18.288 m/s and that point *G* is the center of gravity of the airplane, determine the reactions at points *A* and *B*. Assume *γ* = 1.2 kg/m³ and neglect the approach velocity of the air.

**14.74** The jet engine shown scoops in air at *A* at a rate of 90 kg/s and discharges it at *B* with a velocity of 600 m/s relative to the airplane. Determine the magnitude and line of action of the propulsive thrust developed by the engine when the speed of the airplane is (*a*) 480 km/h, (*b*) 960 km/h.

**Fig. P14.74**

**14.75** The helicopter shown can produce a maximum downward air speed of 24 m/s in a 9-m-diameter slipstream. Knowing that the weight of the helicopter and crew is 15 kN and assuming $\rho = 1.21$ kg/m$^3$ for air, determine the maximum load that the helicopter can lift while hovering in midair.

**Fig. P14.75**

**14.76** A jet airliner is cruising at a speed of 897.29 km/h with its engines scooping in air at the rate of 326.58 kg/s and discharging it with a velocity of 567 m/s relative to the plane when a control surface malfunction suddenly causes a 20 percent increase in drag. Knowing that the pilot maintains level flight with the same mass flow rate and discharge relative velocity, determine the new cruising speed. Assume that the drag due to air friction is proportional to the square of the speed of the plane.

**Fig. P14.76**

**14.77** In order to shorten the distance required for landing, a jet airplane is equipped with moveable vanes which partially reverse the direction of the air discharged by each of its engines. Each engine scoops in the air at a rate of 120 kg/s and discharges it with a velocity of 600 m/s relative to the engine. At an instant when the speed of the airplane is 270 km/h, determine the reverse thrust provided by each of the engines.

**Fig. *P14.77***

**14.78** A jetliner is cruising at a speed of 900 km/h with each of its three engines discharging air with a velocity of 800 m/s relative to the plane. Determine the speed of the airliner after it has lost the use of (*a*) one of its engines, (*b*) two of its engines. Assume that the drag due to air friction is proportional to the square of the speed and that the remaining engines keep operating at the same rate.

**Fig. *P14.78***

**14.79** In a Pelton-wheel turbine, a stream of water is deflected by a series of blades so that the rate at which water is deflected by the blades is equal to the rate at which water issues from the nozzle ($\Delta m/\Delta t = A\rho v_A$). Using the same notation as in Sample Prob. 14.7, (*a*) determine the velocity **V** of the blades for which maximum power is developed, (*b*) derive an expression for the maximum power, (*c*) derive an expression for the mechanical efficiency.

**Fig. P14.79**

**Fig. P14.81 and P14.82**

**14.80** While cruising in level flight at a speed of 913.3 km/h, a jet airplane scoops in air at a rate of 108.86 kg/s and discharges it with a velocity of 670.56 m/s relative to the airplane. Determine (a) the power actually used to propel the airplane, (b) the total power developed by the engine, (c) the mechanical efficiency of the airplane.

**14.81** The wind-turbine-generator shown operates at a wind speed of 30 km/h with an efficiency of 0.4. Knowing that the area swept out by the blades is a circle of diameter $d = 6.5$ m and that $\rho = 1.2$ kg/m³, determine (a) the kinetic energy of the air particles entering the 6.5-m-diameter circle per second, (b) the output power.

**14.82** The wind-turbine-generator shown has an output-power rating of 3.5 kW for a wind speed of 36 km/h and operates at an efficiency of 0.35. Knowing that $\rho = 1.2$ kg/m³, determine (a) the diameter $d$ of the circular area swept out by the blades, (b) the kinetic energy of the air particles entering the circular area per second.

**\*14.83** The depth of water flowing in a rectangular channel of width $b$ at a speed $v_1$ and a depth $d_1$ increases to a depth $d_2$ at a *hydraulic jump*. Express the rate of flow $Q$ in terms of $b$, $d_1$, and $d_2$.

**Fig. P14.83**

**\*14.84** Determine the rate of flow in the channel of Prob. 14.83, knowing that $b = 3$ m, $d_1 = 1.25$ m, and $d_2 = 1.5$ m.

**14.85** A circular reentrant orifice (also called Borda's mouthpiece) of diameter $D$ is placed at a depth $h$ below the surface of a tank. Knowing that the speed of the issuing stream is $v = \sqrt{2gh}$ and assuming that the speed of approach $v_1$ is zero, show that the diameter of the stream is $d = D/\sqrt{2}$. (*Hint:* Consider the section of water indicated, and note that $P$ is equal to the pressure at a depth $h$ multiplied by the area of the orifice.)

**Fig. P14.85**

**Fig. P14.86**

**14.86** A garden sprinkler has four rotating arms, each of which consists of two horizontal straight sections of pipe forming an angle of 120°. Each arm discharges water at a rate of .0189 m³/min with a velocity of 18.3 m/s relative to the arm. Knowing that the friction between the moving and stationary parts of the sprinkler is equivalent to a couple of magnitude $M = .3728$ N · m, determine the constant rate at which the sprinkler rotates.

**14.87**  A railroad car of length $L$ and mass $m_0$ when empty is moving freely on a horizontal track while being loaded with sand from a stationary chute at a rate $dm/dt = q$. Knowing that the car was approaching the chute at a speed $v_0$, determine ($a$) the mass of the car and its load after the car has cleared the chute, ($b$) the speed of the car at that time.

**Fig. P14.87**

**14.88**  The final component of a conveyor system receives sand at a rate of 100 kg/s at $A$ and discharges it at $B$. The sand is moving horizontally at $A$ and $B$ with a velocity of magnitude $v_A = v_B = 4.5$ m/s. Knowing that the combined weight of the component and of the sand it supports is $W = 4$ kN, determine the reactions at $C$ and $D$.

**Fig. P14.88**

**Fig. P14.89**

**14.89**  A chain of length $l$ and mass $m$ lies in a pile on the floor. If its end $A$ is raised vertically at a constant speed $v$, express in terms of the length $y$ of chain which is off the floor at any given instant ($a$) the magnitude of the force **P** applied at $A$, ($b$) the reaction of the floor.

**14.90**  Solve Prob. 14.89, assuming that the chain is being lowered to the floor at a constant speed $v$.

**14.91**  The ends of a chain lie in piles at $A$ and $C$. When given an initial speed $v$, the chain keeps moving freely at that speed over the pulley at $B$. Neglecting friction, determine the required value of $h$.

**Fig. P14.91**

**14.92** A chain of length $l$ and mass $m$ falls through a small hole in a plate. Initially, when $y$ is very small, the chain is at rest. In each case shown, determine (*a*) the acceleration of the first link $A$ as a function of $y$, (*b*) the velocity of the chain as the last link passes through the hole. In case *1* assume that the individual links are at rest until they fall through the hole; in case *2* assume that at any instant all links have the same speed. Ignore the effect of friction.

(1)                    (2)

**Fig. *P14.92***

**14.93** The main propulsion system of a space shuttle consists of three identical rocket engines, each of which burns the hydrogen-oxygen propellant at the rate of 340.2 kg/s and ejects it with a relative velocity of 3810 m/s. Determine the total thrust provided by the three engines.

**14.94** The main propulsion system of a space shuttle consists of three identical rocket engines which provide a total thrust of 5300 kN. Determine the rate at which the hydrogen-oxygen propellant is burned by each of the three engines, knowing that it is ejected with a relative velocity of 3810 m/s.

**Fig. P14.93 and P14.94**

**14.95** A rocket has a mass of 1500 kg, including 1200 kg of fuel, which is consumed at the rate of 15 kg/s. Knowing that the rocket is fired vertically from the ground and that its acceleration increases by 220 m/s² from the time it is fired to the time the last particle of fuel has been consumed, determine the relative velocity with which the fuel is being ejected.

**14.96** A rocket of mass 1500 kg (not including the fuel) is fired vertically at time $t = 0$. The fuel is consumed at a constant rate $q = dm/dt$ and is expelled at a constant speed of 450 m/s relative to the rocket. Knowing that the rocket is designed to achieve a maximum acceleration of 25 m/s² in 15.6 s, determine (*a*) the fuel consumption rate $q$, (*b*) the mass of fuel consumed.

**14.97** A weather satellite of mass 5000 kg, including fuel, has been ejected from a space shuttle describing a low circular orbit around the earth. After the satellite has slowly drifted to a safe distance from the shuttle, its engine is fired to increase its velocity by 2430 m/s as a first step to its transfer to a geosynchronous orbit. Knowing that the fuel is ejected with a relative velocity of 4200 m/s, determine the mass of fuel consumed in this maneuver.

**Fig. P14.97 and P14.98**

**14.98** Determine the increase in velocity of the weather satellite of Prob. 14.97 after 1500 kg of fuel has been consumed.

**14.99** The weight of a spacecraft, including fuel, is 51.6 kN when the rocket engines are fired to increase its velocity by 109.7 m/s. Knowing that 453.6 kg of fuel is consumed, determine the relative velocity of the fuel ejected.

**Fig. *P14.99* and *P14.100***

**14.100** The rocket engines of a spacecraft are fired to increase its velocity by 137 m/s. Knowing that 544.3 kg of fuel is ejected at a relative velocity of 1646 m/s, determine the weight of the spacecraft after the firing.

**14.101** A rocket of initial weight 32.47 kN, including 17.79 kN of fuel, is fired vertically at time $t = 0$. The fuel is consumed at a constant rate of 118 kg/s and is expelled at a constant speed of 457.2 m/s relative to the rocket. Determine the altitude of the rocket when all the fuel has been consumed.

**14.102** Determine the distance traveled by the spacecraft of Prob. 14.99 during the rocket engine firing, knowing that its initial speed was 2286 m/s and the duration of the firing was 60 s.

**14.103** A rocket has a mass of 960 kg, including 800 kg of fuel, which is consumed at the rate of 10 kg/s and ejected with a relative velocity of 3600 m/s. Knowing that the rocket is fired vertically from the ground, determine (*a*) the altitude at which all fuel has been consumed, (*b*) the velocity of the rocket at that time.

**14.104** In Prob. 14.97, determine the distance separating the weather satellite from the space shuttle 80 s after its engine has been fired, knowing that the fuel is consumed at a rate of 18.75 kg/s.

**14.105** In a jet airplane, the kinetic energy imparted to exhaust gases is wasted as far as propelling the airplane is concerned. The useful power is equal to the product of the force available to propel the airplane and the speed of the airplane. If $v$ is the speed of the airplane and $u$ is the relative speed of the expelled gases, show that the mechanical efficiency of the airplane is $\eta = 2v/(u + v)$. Explain why $\eta = 1$ when $u = v$.

**14.106** In a rocket, the kinetic energy imparted to the consumed and ejected fuel is wasted as far as propelling the rocket is concerned. The useful power is equal to the product of the force available to propel the rocket and the speed of the rocket. If $v$ is the speed of the rocket and $u$ is the relative speed of the expelled fuel, show that the mechanical efficiency of the rocket is $\eta = 2uv/(u^2 + v^2)$. Explain why $\eta = 1$ when $u = v$.

In this chapter we analyzed the motion of *systems of particles*, that is, the motion of a large number of particles considered together. In the first part of the chapter we considered systems consisting of well-defined particles, while in the second part we analyzed systems which are continually gaining or losing particles, or doing both at the same time.

**Effective forces**

We first defined the *effective force* of a particle $P_i$ of a given system as the product $m_i\mathbf{a}_i$ of its mass $m_i$ and its acceleration $\mathbf{a}_i$ with respect to a newtonian frame of reference centered at $O$ [Sec. 14.2]. We then showed that *the system of the external forces acting on the particles and the system of the effective forces of the particles are equipollent;* that is, both systems have the *same resultant* and the *same moment resultant* about $O$:

$$\sum_{i=1}^{n} \mathbf{F}_i = \sum_{i=1}^{n} m_i\mathbf{a}_i \tag{14.4}$$

$$\sum_{i=1}^{n} (\mathbf{r}_i \times \mathbf{F}_i) = \sum_{i=1}^{n} (\mathbf{r}_i \times m_i\mathbf{a}_i) \tag{14.5}$$

**Linear and angular momentum of a system of particles**

Defining the *linear momentum* $\mathbf{L}$ and the *angular momentum* $\mathbf{H}_O$ *about point* $O$ of the system of particles [Sec. 14.3] as

$$\mathbf{L} = \sum_{i=1}^{n} m_i\mathbf{v}_i \qquad \mathbf{H}_O = \sum_{i=1}^{n} (\mathbf{r}_i \times m_i\mathbf{v}_i) \tag{14.6, 14.7}$$

we showed that Eqs. (14.4) and (14.5) can be replaced by the equations

$$\Sigma\mathbf{F} = \dot{\mathbf{L}} \qquad \Sigma\mathbf{M}_O = \dot{\mathbf{H}}_O \tag{14.10, 14.11}$$

which express that *the resultant and the moment resultant about O of the external forces are, respectively, equal to the rates of change of the linear momentum and of the angular momentum about O of the system of particles.*

**Motion of the mass center of a system of particles**

In Sec. 14.4, we defined the mass center of a system of particles as the point $G$ whose position vector $\bar{\mathbf{r}}$ satisfies the equation

$$m\bar{\mathbf{r}} = \sum_{i=1}^{n} m_i\mathbf{r}_i \tag{14.12}$$

where $m$ represents the total mass $\sum_{i=1}^{n} m_i$ of the particles. Differentiating both members of Eq. (14.12) twice with respect to $t$, we obtained the relations

$$\mathbf{L} = m\overline{\mathbf{v}} \qquad \dot{\mathbf{L}} = m\overline{\mathbf{a}} \qquad (14.14, 14.15)$$

where $\overline{\mathbf{v}}$ and $\overline{\mathbf{a}}$ represent, respectively, the velocity and the acceleration of the mass center $G$. Substituting for $\dot{\mathbf{L}}$ from (14.15) into (14.10), we obtained the equation

$$\Sigma \mathbf{F} = m\overline{\mathbf{a}} \qquad (14.16)$$

from which we concluded that *the mass center of a system of particles moves as if the entire mass of the system and all the external forces were concentrated at that point* [Sample Prob. 14.1].

In Sec. 14.5 we considered the motion of the particles of a system with respect to a centroidal frame $Gx'y'z'$ attached to the mass center $G$ of the system and in translation with respect to the newtonian frame $Oxyz$ (Fig. 14.14). We defined the *angular momentum* of the system *about its mass center $G$* as the sum of the moments about $G$ of the momenta $m_i\mathbf{v}_i'$ of the particles in their motion relative to the frame $Gx'y'z'$. We also noted that the same result can be obtained by considering the moments about $G$ of the momenta $m_i\mathbf{v}_i$ of the particles in their absolute motion. We therefore wrote

$$\mathbf{H}_G = \sum_{i=1}^{n} (\mathbf{r}_i' \times m_i\mathbf{v}_i) = \sum_{i=1}^{n} (\mathbf{r}_i' \times m_i\mathbf{v}_i') \qquad (14.24)$$

and derived the relation

$$\Sigma \mathbf{M}_G = \dot{\mathbf{H}}_G \qquad (14.23)$$

which expresses that *the moment resultant about $G$ of the external forces is equal to the rate of change of the angular momentum about $G$ of the system of particles.* As will be seen later, this relation is fundamental to the study of the motion of rigid bodies.

When no external force acts on a system of particles [Sec. 14.6], it follows from Eqs. (14.10) and (14.11) that the linear momentum $\mathbf{L}$ and the angular momentum $\mathbf{H}_O$ of the system are conserved [Sample Probs. 14.2 and 14.3]. In problems involving central forces, the angular momentum of the system about the center of force $O$ will also be conserved.

The kinetic energy $T$ of a system of particles was defined as the sum of the kinetic energies of the particles [Sec. 14.7]:

$$T = \frac{1}{2} \sum_{i=1}^{n} m_i v_i^2 \qquad (14.28)$$

Angular momentum of a system of particles about its mass center

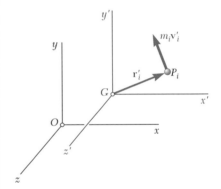

**Fig. 14.14**

Conservation of momentum

Kinetic energy of a system of particles

Using the centroidal frame of reference $Gx'y'z'$ of Fig. 14.14, we noted that the kinetic energy of the system can also be obtained by adding the kinetic energy $\frac{1}{2}m\bar{v}^2$ associated with the motion of the mass center $G$ and the kinetic energy of the system in its motion relative to the frame $Gx'y'z'$:

$$T = \tfrac{1}{2}m\bar{v}^2 + \frac{1}{2}\sum_{i=1}^{n} m_i v_i'^2 \qquad (14.29)$$

**Principle of work and energy**

The *principle of work and energy* can be applied to a system of particles as well as to individual particles [Sec. 14.8]. We wrote

$$T_1 + U_{1\to2} = T_2 \qquad (14.30)$$

and noted that $U_{1\to2}$ represents the work of *all* the forces acting on the particles of the system, internal as well as external.

**Conservation of energy**

If all the forces acting on the particles of the system are *conservative*, we can determine the potential energy $V$ of the system and write

$$T_1 + V_1 = T_2 + V_2 \qquad (14.31)$$

which expresses the *principle of conservation of energy* for a system of particles.

**Principle of impulse and momentum**

We saw in Sec. 14.9 that the *principle of impulse and momentum* for a system of particles can be expressed graphically as shown in Fig. 14.15. It states that the momenta of the particles at time $t_1$ and the impulses of the external forces from $t_1$ to $t_2$ form a system of vectors equipollent to the system of the momenta of the particles at time $t_2$.

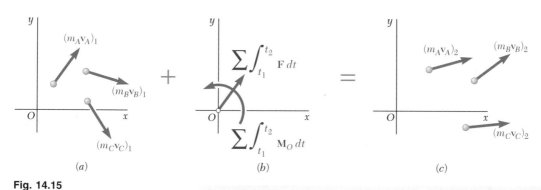

**Fig. 14.15**

If no external force acts on the particles of the system, the systems of momenta shown in parts *a* and *c* of Fig. 14.15 are equipollent and we have

$$\mathbf{L}_1 = \mathbf{L}_2 \qquad (\mathbf{H}_O)_1 = (\mathbf{H}_O)_2 \qquad (14.36,\ 14.37)$$

**Use of conservation principles in the solution of problems involving systems of particles**

Many problems involving the motion of systems of particles can be solved by applying simultaneously the principle of impulse and momentum and the principle of conservation of energy [Sample Prob. 14.4] or by expressing that the linear momentum, angular momentum, and energy of the system are conserved [Sample Prob. 14.5].

In the second part of the chapter, we considered *variable systems of particles*. First we considered a *steady stream of particles,* such as a stream of water diverted by a fixed vane or the flow of air through a jet engine [Sec. 14.11]. Applying the principle of impulse and momentum to a system $S$ of particles during a time interval $\Delta t$, and including the particles which enter the system at $A$ during that time interval and those (of the same mass $\Delta m$) which leave the system at $B$, we concluded that *the system formed by the momentum* $(\Delta m)\mathbf{v}_A$ *of the particles entering $S$ in the time $\Delta t$ and the impulses of the forces exerted on $S$ during that time is equipollent to the momentum* $(\Delta m)\mathbf{v}_B$ *of the particles leaving $S$ in the same time $\Delta t$* (Fig. 14.16). Equating the $x$ components, $y$ components,

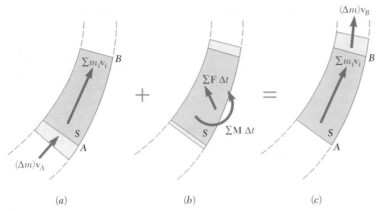

$(a)$        $(b)$        $(c)$

**Fig. 14.16**

and moments about a fixed point of the vectors involved, we could obtain as many as three equations, which could be solved for the desired unknowns [Sample Probs. 14.6 and 14.7]. From this result, we could also derive the following expression for the resultant $\Sigma\mathbf{F}$ of the forces exerted on $S$,

$$\Sigma\mathbf{F} = \frac{dm}{dt}(\mathbf{v}_B - \mathbf{v}_A) \qquad (14.39)$$

where $\mathbf{v}_B - \mathbf{v}_A$ represents the difference between the *vectors* $\mathbf{v}_B$ and $\mathbf{v}_A$ and where $dm/dt$ is the mass rate of flow of the stream (see footnote, page 887).

Considering next a system of particles gaining mass by continually absorbing particles or losing mass by continually expelling particles [Sec. 14.12], as in the case of a rocket, we applied the principle of impulse and momentum to the system during a time interval $\Delta t$, being careful to include the particles gained or lost during that time interval [Sample Prob. 14.8]. We also noted that the action on a system $S$ of the particles being *absorbed* by $S$ was equivalent to a thrust

$$\mathbf{P} = \frac{dm}{dt}\mathbf{u} \qquad (14.44)$$

where $dm/dt$ is the rate at which mass is being absorbed, and $\mathbf{u}$ is the velocity of the particles *relative to $S$*. In the case of particles being *expelled* by $S$, the rate $dm/dt$ is negative and the thrust $\mathbf{P}$ is exerted in a direction opposite to that in which the particles are being expelled.

**Fig. P14.107 and *P14.108***

**14.107** An airline employee tosses two suitcases, of weight 133.4 N and 178 N, respectively, onto a 22.68 kg baggage carrier in rapid succession. Knowing that the carrier is initially at rest and that the employee imparts a 2.74 m/s horizontal velocity to the 133.4 N suitcase and a 1.83 m/s horizontal velocity to the 178 N suitcase, determine the final velocity of the baggage carrier if the first suitcase tossed onto the carrier is (*a*) the 133.4 N suitcase, (*b*) the 178 N suitcase.

**14.108** An airline employee tosses two suitcases in rapid succession, with a horizontal velocity of 2.2 m/s, onto a 22.68 kg baggage carrier which is initially at rest. (*a*) Knowing that the final velocity of the baggage carrier is 1.09 m/s and that the first suitcase the employee tosses onto the carrier has a weight of 133.4 N, determine the weight of the other suitcase. (*b*) What would be the final velocity of the carrier if the employee reversed the order in which he tosses the suitcases?

**14.109** A system consists of three particles *A*, *B*, and *C*. We know that $m_A = 3$ kg, $m_B = 2$ kg, and $m_C = 4$ kg and that the velocities of the particles expressed in m/s are, respectively, $\mathbf{v}_A = 4\mathbf{i} + 2\mathbf{j} + 2\mathbf{k}$, $\mathbf{v}_B = 4\mathbf{i} + 3\mathbf{j}$, and $\mathbf{v}_C = -2\mathbf{i} + 4\mathbf{j} + 2\mathbf{k}$. Determine the angular momentum $\mathbf{H}_O$ of the system about *O*.

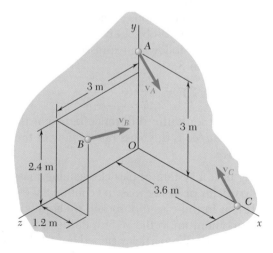

**Fig. P14.109**

**14.110** For the system of particles in Prob. 14.109, determine (*a*) the position vector $\bar{\mathbf{r}}$ of the mass center *G* of the system, (*b*) the linear momentum $m\bar{\mathbf{v}}$ of the system, (*c*) the angular momentum $\mathbf{H}_G$ of the system about *G*. Also verify that the answers to this problem and to Prob. 14.109 satisfy the equation given in Prob. 14.28.

**14.111**   A small 1361 kg airplane and a 2722 kg helicopter flying at an altitude of 1097 m are observed to collide directly above a tower located at $O$ in a wooded area. Four minutes earlier the helicopter had been sighted 8.8 km due west of the tower and the airplane 16 km west and 12 km north of the tower. As a result of the collision the helicopter was split into two pieces, $H_1$ and $H_2$, weighing 8896 N and 17.8 kN, respectively; the airplane remained in one piece as it fell to the ground. Knowing that the two fragments of the helicopter were located at points $H_1$ (457.2 m, $-91$ m) and $H_2$ (548.6 m, $-457.2$ m), respectively, and assuming that all pieces hit the ground at the same time, determine the coordinates of the point $A$ where the wreckage of the airplane will be found.

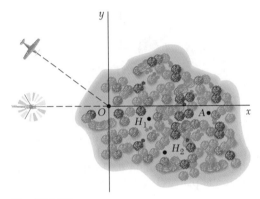

**Fig. P14.111**

*14.112*   In Prob. 14.111, knowing that the wreckage of the small airplane was found at $A$ (1097 m, 73.15 m) and the 907.18 kg fragment of the helicopter at point $H_1$ (365.76 m, $-182.88$ m), and assuming that all pieces hit the ground at the same time, determine the coordinates of the point $H_2$ where the other fragment of the helicopter will be found.

**14.113**   Two hemispheres are held together by a cord which maintains a spring under compression (the spring is not attached to the hemispheres). The potential energy of the compressed spring is 122 N · m and the assembly has an initial velocity $\mathbf{v}_0$ of magnitude $v_0 = 7.3$ m/s. Knowing that the cord is severed when $\theta = 30°$, causing the hemispheres to fly apart, determine the resulting velocity of each hemisphere.

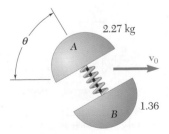

**Fig. P14.113**

**14.114** Two small spheres $A$ and $B$, with masses of 2.5 kg and 1 kg, respectively, are connected by a rigid rod of negligible mass. The two spheres are resting on a horizontal, frictionless surface when $A$ is suddenly given the velocity $\mathbf{v}_0 = (3.5 \text{ m/s})\mathbf{i}$. Determine ($a$) the linear momentum of the system and its angular momentum about its mass center $G$, ($b$) the velocities of $A$ and $B$ after the rod $AB$ has rotated through 180°.

**Fig. *P14.114***

**14.115** Water flows in a continuous sheet from between two plates $A$ and $B$ with a velocity $\mathbf{v}$ of magnitude 27.4 m/s. The stream is split into two parts by a smooth horizontal plate $C$. Knowing that the rates of flow in each of the two resulting streams are, respectively, $Q_1 = .098 \text{ m}^3/\text{min}$ and $Q_2 = .492 \text{ m}^3/\text{min}$, determine ($a$) the angle $\theta$, ($b$) the total force exerted by the stream on the horizontal plate. (1 ft$^3$ = 7.48 gal.)

**Fig. P14.115**

**14.116** A stream of water having a cross-sectional area of 9.68 cm$^2$ and moving with a velocity of magnitude 18.3 m/s at both $A$ and $B$ is deflected by two vanes which are welded as shown to a vertical plate. Knowing that the combined weight of the plate and vanes is 44.5 N, determine the reactions at $C$ and $D$.

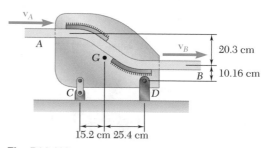

**Fig. P14.116**

**14.117** The propeller of a small airplane has a 2-m-diameter slipstream and produces a thrust of 3.6 kN when the airplane is at rest on the ground. Assuming $\rho = 1.21$ kg/m$^3$ for air, determine (a) the speed of the air in the slipstream, (b) the volume of air passing through the propeller per second, (c) the kinetic energy imparted per second to the air in the slipstream.

**14.118** A rocket has a mass of 960 kg, including 800 kg of fuel, which is consumed at the rate of 10 kg/s and ejected with a relative velocity of 3600 m/s. Knowing that the rocket is fired vertically from the ground, determine its acceleration (a) as it is fired, (b) as the last particle of fuel is being consumed.

# Computer Problems

**14.C1** A 9-kg block B starts from rest and slides down the inclined surface of a 15-kg wedge A which is supported by a horizontal surface. Neglecting friction and denoting by s the distance traveled by block B down the surface of the wedge, derive expressions for (a) the speed of block B relative to wedge A, (b) the speed of wedge A. (c) Plot the speed of B relative to A and the speed of A as functions of s for $0 \leq s \leq 0.6$ m.

**Fig. P14.C1**

**14.C2** A system of particles consists of n particles $A_i$ of weight $W_i$ and coordinates $x_i$, $y_i$, and $z_i$, having velocities of components $(v_x)_i$, $(v_y)_i$, and $(v_z)_i$. Derive expressions for the components of the angular momentum of the system about the origin O of the coordinates. Use computational software to solve Prob. 14.109.

**14.C3** A shell moving with a velocity of known components $v_x$, $v_y$, and $v_z$ explodes into three fragments of masses $m_1$, $m_2$, and $m_3$ at point $A_0$ at a distance $d$ from a vertical wall. Derive expressions for the speeds of the three fragments immediately after the explosion, knowing the coordinates $x_i$ and $y_i$ of the points $A_i$ ($i = 1, 2, 3$) where the fragments hit the wall. Use computational software to solve Prob. 14.26.

**Fig. P14.C3**

**14.C4** As a 5443 kg training plane lands on an aircraft carrier at a speed of 240.3 km/h, its tail hooks into the end of a 79.2 m long chain which lies in a pile below deck. The chain has a weight per unit length of 350.25 N/m and there is no other retarding force. Denoting by $x$ the distance traveled by the plane while the chain is being pulled out, use computational software to calculate and plot as functions of $x$ ($a$) the elapsed time, ($b$) the speed of the plane, ($c$) the deceleration of the plane.

**Fig. P14.C4**

**14.C5** A 14515 kg jet airplane maintains a constant speed of 1033.5 km/h while climbing at an angle $\alpha = 18°$. The airplane scoops in air at a rate of 272.2 kg/s and discharges it with a velocity of 3204.6 km/h relative to the airplane. Knowing that the pilot changes the angle of climb $\alpha$ while maintaining the same engine setting, calculate and plot as functions of $\alpha$ for $0 \le \alpha \le 20°$ (a) the initial acceleration of the plane, (b) the maximum speed that will be attained. Assume that the drag due to air friction is proportional to the square of the speed.

**Fig. P14.C5**

**14.C6** A rocket has a mass of 1200 kg, including 1000 kg of fuel, which is consumed at the rate of 12.5 kg/s and ejected with a relative velocity of 3.6 km/s. Knowing that the rocket is fired vertically from the ground and assuming a constant value for the acceleration of gravity, use computational software to calculate and plot from the time of ignition to the time when the last particle of fuel is being consumed (a) the acceleration $a$ of the rocket in m/s², (b) its velocity $v$ in m/s, (c) its elevation $h$ above the ground in kilometers. (*Hint:* Use for $v$ the expression derived in Sample Prob. 14.8, and integrate this expression analytically to obtain $h$.)

**14.C7** Two hemispheres are held together by a cord which maintains a spring under compression (the spring is not attached to the hemispheres). The potential energy of the compressed spring is 122 N · m and the assembly has an initial velocity $\mathbf{v}_0$ of magnitude $v_0 = 7.3$ m/s. Knowing that the cord is severed causing the hemispheres to fly apart, calculate and plot the magnitude of the resulting velocity for each hemisphere as a function of $\theta$ for $30° \le \theta \le 120°$.

**Fig. P14.C7**

## CHAPTER 15

# Kinematics of Rigid Bodies

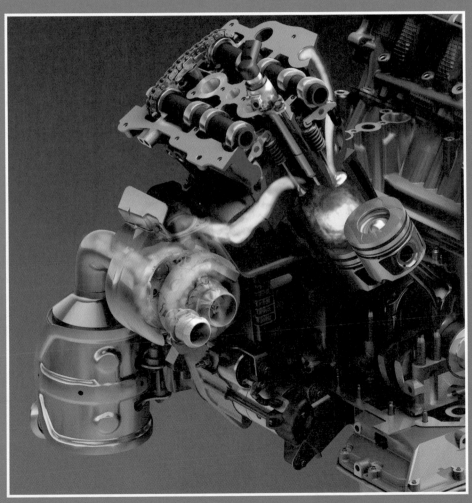

In this chapter you will learn to analyze the motion of mechanical systems such as the engine system shown. You may be asked, for example, to determine the angular velocity and angular acceleration of the connecting rod, knowing the velocity and acceleration of the piston.

## 15.1. INTRODUCTION

In this chapter, the kinematics of *rigid bodies* will be considered. You will investigate the relations existing between the time, the positions, the velocities, and the accelerations of the various particles forming a rigid body. As you will see, the various types of rigid-body motion can be conveniently grouped as follows:

**Fig. 15.1**

**Fig. 15.2**

**Fig. 15.3**

1. *Translation.* A motion is said to be a translation if any straight line inside the body keeps the same direction during the motion. It can also be observed that in a translation all the particles forming the body move along parallel paths. If these paths are straight lines, the motion is said to be a *rectilinear translation* (Fig. 15.1); if the paths are curved lines, the motion is a *curvilinear translation* (Fig. 15.2).

2. *Rotation about a Fixed Axis.* In this motion, the particles forming the rigid body move in parallel planes along circles centered on the same fixed axis (Fig. 15.3). If this axis, called the *axis of rotation,* intersects the rigid body, the particles located on the axis have zero velocity and zero acceleration.

   Rotation should not be confused with certain types of curvilinear translation. For example, the plate shown in Fig. 15.4*a* is in curvilinear translation, with all its particles moving along *parallel* circles, while the plate shown in Fig. 15.4*b* is in rotation, with all its particles moving along *concentric* circles.

(*a*) Curvilinear translation

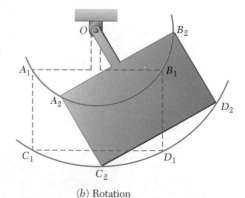

(*b*) Rotation

**Fig. 15.4**

In the first case, any given straight line drawn on the plate will maintain the same direction, whereas in the second case, point *O* remains fixed.

Because each particle moves in a given plane, the rotation of a body about a fixed axis is said to be a *plane motion*.

3. *General Plane Motion.* There are many other types of plane motion, that is, motions in which all the particles of the body move in parallel planes. Any plane motion which is neither a rotation nor a translation is referred to as a general plane motion. Two examples of general plane motion are given in Fig. 15.5.

**KINEMATICS OF RIGID BODIES**

| | |
|---|---|
| 15.1 | Introduction |
| 15.2 | Translation |
| 15.3 | Rotation about a Fixed Axis |
| 15.4 | Equations Defining the Rotation of a Rigid Body about a Fixed Axis |
| 15.5 | General Plane Motion |
| 15.6 | Absolute and Relative Velocity in Plane Motion |
| 15.7 | Instantaneous Center of Rotation in Plane Motion |
| 15.8 | Absolute and Relative Acceleration in Plane Motion |
| 15.9 | Analysis of Plane Motion in Terms of a Parameter |
| 15.10 | Rate of Change of a Vector with Respect to a Rotating Frame |
| 15.11 | Plane Motion of a Particle Relative to a Rotating Frame. Coriolis Acceleration |
| 15.12 | Motion about a Fixed Point |
| 15.13 | General Motion |
| 15.14 | Three-Dimensional Motion of a Particle Relative to a Rotating Frame. Coriolis Acceleration |
| 15.15 | Frame of Reference in General Motion |

(*a*) Rolling wheel      (*b*) Sliding rod

**Fig. 15.5**

4. *Motion about a Fixed Point.* The three-dimensional motion of a rigid body attached at a fixed point *O*, for example, the motion of a top on a rough floor (Fig. 15.6), is known as motion about a fixed point.

5. *General Motion.* Any motion of a rigid body which does not fall in any of the categories above is referred to as a general motion.

After a brief discussion in Sec. 15.2 of the motion of translation, the rotation of a rigid body about a fixed axis is considered in Sec. 15.3. The *angular velocity* and the *angular acceleration* of a rigid body about a fixed axis will be defined, and you will learn to express the velocity and the acceleration of a given point of the body in terms of its position vector and the angular velocity and angular acceleration of the body.

The following sections are devoted to the study of the general plane motion of a rigid body and to its application to the analysis of mechanisms such as gears, connecting rods, and pin-connected linkages. Resolving the plane motion of a slab into a translation and a rotation (Secs. 15.5 and 15.6), we will then express the velocity of a point *B* of the slab as the sum of the velocity of a reference point *A* and of the velocity of *B* relative to a frame of reference translating with *A* (that is, moving with *A* but not rotating). The same approach is used later in Sec. 15.8 to express the acceleration of *B* in terms of the acceleration of *A* and of the acceleration of *B* relative to a frame translating with *A*.

**Fig. 15.6**

An alternative method for the analysis of velocities in plane motion, based on the concept of *instantaneous center of rotation,* is given in Sec. 15.7; and still another method of analysis, based on the use of parametric expressions for the coordinates of a given point, is presented in Sec. 15.9.

The motion of a particle relative to a rotating frame of reference and the concept of *Coriolis acceleration* are discussed in Secs. 15.10 and 15.11, and the results obtained are applied to the analysis of the plane motion of mechanisms containing parts which slide on each other.

The remaining part of the chapter is devoted to the analysis of the three-dimensional motion of a rigid body, namely, the motion of a rigid body with a fixed point and the general motion of a rigid body. In Secs. 15.12 and 15.13, a fixed frame of reference or a frame of reference in translation will be used to carry out this analysis; in Secs. 15.14 and 15.15, the motion of the body relative to a rotating frame or to a frame in general motion will be considered, and the concept of Coriolis acceleration will again be used.

## 15.2. TRANSLATION

Consider a rigid body in translation (either rectilinear or curvilinear translation), and let $A$ and $B$ be any two of its particles (Fig. 15.7$a$). Denoting, respectively, by $\mathbf{r}_A$ and $\mathbf{r}_B$ the position vectors of $A$ and $B$ with respect to a fixed frame of reference and by $\mathbf{r}_{B/A}$ the vector joining $A$ and $B$, we write

$$\mathbf{r}_B = \mathbf{r}_A + \mathbf{r}_{B/A} \qquad (15.1)$$

Let us differentiate this relation with respect to $t$. We note that from the very definition of a translation, the vector $\mathbf{r}_{B/A}$ must maintain a constant direction; its magnitude must also be constant, since $A$ and

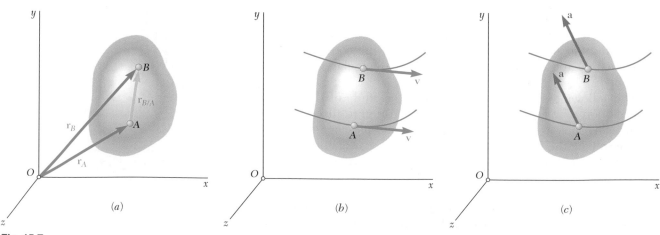

$(a)$ $(b)$ $(c)$

**Fig. 15.7**

$B$ belong to the same rigid body. Thus, the derivative of $\mathbf{r}_{B/A}$ is zero and we have

$$\mathbf{v}_B = \mathbf{v}_A \tag{15.2}$$

Differentiating once more, we write

$$\mathbf{a}_B = \mathbf{a}_A \tag{15.3}$$

Thus, *when a rigid body is in translation, all the points of the body have the same velocity and the same acceleration at any given instant* (Fig. 15.7*b* and *c*). In the case of curvilinear translation, the velocity and acceleration change in direction as well as in magnitude at every instant. In the case of rectilinear translation, all particles of the body move along parallel straight lines, and their velocity and acceleration keep the same direction during the entire motion.

## 15.3. ROTATION ABOUT A FIXED AXIS

Consider a rigid body which rotates about a fixed axis $AA'$. Let $P$ be a point of the body and $\mathbf{r}$ its position vector with respect to a fixed frame of reference. For convenience, let us assume that the frame is centered at point $O$ on $AA'$ and that the $z$ axis coincides with $AA'$ (Fig. 15.8). Let $B$ be the projection of $P$ on $AA'$; since $P$ must remain at a constant distance from $B$, it will describe a circle of center $B$ and of radius $r \sin \phi$, where $\phi$ denotes the angle formed by $\mathbf{r}$ and $AA'$.

The position of $P$ and of the entire body is completely defined by the angle $\theta$ the line $BP$ forms with the $zx$ plane. The angle $\theta$ is known as the *angular coordinate* of the body and is defined as positive when viewed as counterclockwise from $A'$. The angular coordinate will be expressed in radians (rad) or, occasionally, in degrees (°) or revolutions (rev). We recall that

$$1 \text{ rev} = 2\pi \text{ rad} = 360°$$

We recall from Sec. 11.9 that the velocity $\mathbf{v} = d\mathbf{r}/dt$ of a particle $P$ is a vector tangent to the path of $P$ and of magnitude $v = ds/dt$. Observing that the length $\Delta s$ of the arc described by $P$ when the body rotates through $\Delta\theta$ is

$$\Delta s = (BP)\,\Delta\theta = (r \sin \phi)\,\Delta\theta$$

and dividing both members by $\Delta t$, we obtain at the limit, as $\Delta t$ approaches zero,

$$v = \frac{ds}{dt} = r\dot{\theta} \sin \phi \tag{15.4}$$

where $\dot{\theta}$ denotes the time derivative of $\theta$. (Note that the angle $\theta$ depends on the position of $P$ within the body, but the rate of change $\dot{\theta}$ is itself independent of $P$.) We conclude that the velocity $\mathbf{v}$ of $P$ is

**Fig. 15.8**

**Fig. 15.9**

**Photo 15.1** For the central gear rotating about a fixed axis, the angular velocity and angular acceleration of that gear are vectors directed along the vertical axis of rotation.

a vector perpendicular to the plane containing $AA'$ and $\mathbf{r}$, and of magnitude $v$ defined by (15.4). But this is precisely the result we would obtain if we drew along $AA'$ a vector $\boldsymbol{\omega} = \dot{\theta}\mathbf{k}$ and formed the vector product $\boldsymbol{\omega} \times \mathbf{r}$ (Fig. 15.9). We thus write

$$\mathbf{v} = \frac{d\mathbf{r}}{dt} = \boldsymbol{\omega} \times \mathbf{r} \tag{15.5}$$

The vector

$$\boldsymbol{\omega} = \omega\mathbf{k} = \dot{\theta}\mathbf{k} \tag{15.6}$$

which is directed along the axis of rotation, is called the *angular velocity* of the body and is equal in magnitude to the rate of change $\dot{\theta}$ of the angular coordinate; its sense may be obtained by the right-hand rule (Sec. 3.6) from the sense of rotation of the body.†

The acceleration $\mathbf{a}$ of the particle $P$ will now be determined. Differentiating (15.5) and recalling the rule for the differentiation of a vector product (Sec. 11.10), we write

$$\begin{aligned}
\mathbf{a} &= \frac{d\mathbf{v}}{dt} = \frac{d}{dt}(\boldsymbol{\omega} \times \mathbf{r}) \\
&= \frac{d\boldsymbol{\omega}}{dt} \times \mathbf{r} + \boldsymbol{\omega} \times \frac{d\mathbf{r}}{dt} \\
&= \frac{d\boldsymbol{\omega}}{dt} \times \mathbf{r} + \boldsymbol{\omega} \times \mathbf{v} \tag{15.7}
\end{aligned}$$

The vector $d\boldsymbol{\omega}/dt$ is denoted by $\boldsymbol{\alpha}$ and is called the *angular acceleration* of the body. Substituting also for $\mathbf{v}$ from (15.5), we have

$$\mathbf{a} = \boldsymbol{\alpha} \times \mathbf{r} + \boldsymbol{\omega} \times (\boldsymbol{\omega} \times \mathbf{r}) \tag{15.8}$$

Differentiating (15.6) and recalling that $\mathbf{k}$ is constant in magnitude and direction, we have

$$\boldsymbol{\alpha} = \alpha\mathbf{k} = \dot{\omega}\mathbf{k} = \ddot{\theta}\mathbf{k} \tag{15.9}$$

Thus, the angular acceleration of a body rotating about a fixed axis is a vector directed along the axis of rotation, and is equal in magnitude to the rate of change $\dot{\omega}$ of the angular velocity. Returning to (15.8), we note that the acceleration of $P$ is the sum of two vectors. The first vector is equal to the vector product $\boldsymbol{\alpha} \times \mathbf{r}$; it is tangent to the circle described by $P$ and therefore represents the tangential component of the acceleration. The second vector is equal to the *vector triple product* $\boldsymbol{\omega} \times (\boldsymbol{\omega} \times \mathbf{r})$ obtained by forming the vector product of $\boldsymbol{\omega}$ and $\boldsymbol{\omega} \times \mathbf{r}$; since $\boldsymbol{\omega} \times \mathbf{r}$ is tangent to the circle described by $P$, the vector triple product is directed toward the center $B$ of the circle and therefore represents the normal component of the acceleration.

---

†It will be shown in Sec. 15.12 in the more general case of a rigid body rotating simultaneously about axes having different directions that angular velocities obey the parallelogram law of addition and thus are actually vector quantities.

**Rotation of a Representative Slab.** The rotation of a rigid body about a fixed axis can be defined by the motion of a representative slab in a reference plane perpendicular to the axis of rotation. Let us choose the $xy$ plane as the reference plane and assume that it coincides with the plane of the figure, with the $z$ axis pointing out of the paper (Fig. 15.10). Recalling from (15.6) that $\boldsymbol{\omega} = \omega\mathbf{k}$, we note

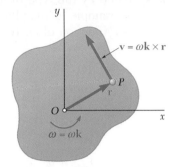

**Fig. 15.10**

that a positive value of the scalar $\omega$ corresponds to a counterclockwise rotation of the representative slab, and a negative value to a clockwise rotation. Substituting $\omega\mathbf{k}$ for $\boldsymbol{\omega}$ into Eq. (15.5), we express the velocity of any given point $P$ of the slab as

$$\mathbf{v} = \omega\mathbf{k} \times \mathbf{r} \qquad (15.10)$$

Since the vectors $\mathbf{k}$ and $\mathbf{r}$ are mutually perpendicular, the magnitude of the velocity $\mathbf{v}$ is

$$v = r\omega \qquad (15.10')$$

and its direction can be obtained by rotating $\mathbf{r}$ through 90° in the sense of rotation of the slab.

Substituting $\boldsymbol{\omega} = \omega\mathbf{k}$ and $\boldsymbol{\alpha} = \alpha\mathbf{k}$ into Eq. (15.8), and observing that cross-multiplying $\mathbf{r}$ twice by $\mathbf{k}$ results in a 180° rotation of the vector $\mathbf{r}$, we express the acceleration of point $P$ as

$$\mathbf{a} = \alpha\mathbf{k} \times \mathbf{r} - \omega^2\mathbf{r} \qquad (15.11)$$

Resolving $\mathbf{a}$ into tangential and normal components (Fig. 15.11), we write

$$\begin{aligned} \mathbf{a}_t &= \alpha\mathbf{k} \times \mathbf{r} & a_t &= r\alpha \\ \mathbf{a}_n &= -\omega^2\mathbf{r} & a_n &= r\omega^2 \end{aligned} \qquad (15.11')$$

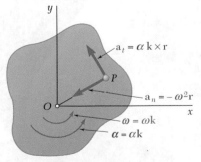

**Fig. 15.11**

The tangential component $\mathbf{a}_t$ points in the counterclockwise direction if the scalar $\alpha$ is positive, and in the clockwise direction if $\alpha$ is negative. The normal component $\mathbf{a}_n$ always points in the direction opposite to that of $\mathbf{r}$, that is, toward $O$.

## 15.4. EQUATIONS DEFINING THE ROTATION OF A RIGID BODY ABOUT A FIXED AXIS

The motion of a rigid body rotating about a fixed axis $AA'$ is said to be *known* when its angular coordinate $\theta$ can be expressed as a known function of $t$. In practice, however, the rotation of a rigid body is seldom defined by a relation between $\theta$ and $t$. More often, the conditions of motion will be specified by the type of angular acceleration that the body possesses. For example, $\alpha$ may be given as a function of $t$, as a function of $\theta$, or as a function of $\omega$. Recalling the relations (15.6) and (15.9), we write

$$\omega = \frac{d\theta}{dt} \tag{15.12}$$

$$\alpha = \frac{d\omega}{dt} = \frac{d^2\theta}{dt^2} \tag{15.13}$$

or, solving (15.12) for $dt$ and substituting into (15.13),

$$\alpha = \omega \frac{d\omega}{d\theta} \tag{15.14}$$

Since these equations are similar to those obtained in Chap. 11 for the rectilinear motion of a particle, their integration can be performed by following the procedure outlined in Sec. 11.3.

Two particular cases of rotation are frequently encountered:

1. *Uniform Rotation.* This case is characterized by the fact that the angular acceleration is zero. The angular velocity is thus constant, and the angular coordinate is given by the formula

$$\theta = \theta_0 + \omega t \tag{15.15}$$

2. *Uniformly Accelerated Rotation.* In this case, the angular acceleration is constant. The following formulas relating angular velocity, angular coordinate, and time can then be derived in a manner similar to that described in Sec. 11.5. The similarity between the formulas derived here and those obtained for the rectilinear uniformly accelerated motion of a particle is apparent.

$$
\begin{aligned}
\omega &= \omega_0 + \alpha t \\
\theta &= \theta_0 + \omega_0 t + \tfrac{1}{2}\alpha t^2 \\
\omega^2 &= \omega_0^2 + 2\alpha(\theta - \theta_0)
\end{aligned}
\tag{15.16}
$$

It should be emphasized that formula (15.15) can be used only when $\alpha = 0$, and formulas (15.16) can be used only when $\alpha = $ constant. In any other case, the general formulas (15.12) to (15.14) should be used.

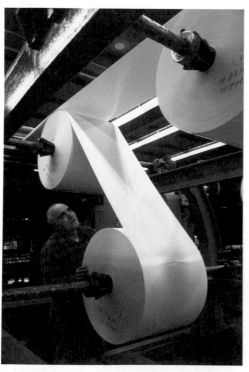

**Photo 15.2** If the lower roll has a constant angular velocity, the speed of the paper being wound onto it increases as the radius of the roll increases.

## SAMPLE PROBLEM 15.1

Load $B$ is connected to a double pulley by one of the two inextensible cables shown. The motion of the pulley is controlled by cable $C$, which has a constant acceleration of 22.86 cm/s² and an initial velocity of 30.48 cm/s, both directed to the right. Determine ($a$) the number of revolutions executed by the pulley in 2 s, ($b$) the velocity and change in position of the load $B$ after 2 s, and ($c$) the acceleration of point $D$ on the rim of the inner pulley at $t = 0$.

## SOLUTION

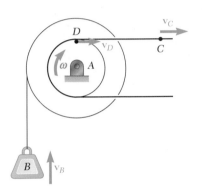

*a. Motion of Pulley.* Since the cable is inextensible, the velocity of point $D$ is equal to the velocity of point $C$ and the tangential component of the acceleration of $D$ is equal to the acceleration of $C$.

$$(\mathbf{v}_D)_0 = (\mathbf{v}_C)_0 = 30.48 \text{ cm/s} \rightarrow \qquad (\mathbf{a}_D)_t = \mathbf{a}_C = 22.8 \text{ cm/s}^2 \rightarrow$$

Noting that the distance from $D$ to the center of the pulley is 3 in., we write

$$(v_D)_0 = r\omega_0 \qquad 30.48 \text{ cm/s} = (7.62 \text{ cm})\omega_0 \qquad \boldsymbol{\omega}_0 = 4 \text{ rad/s} \downarrow$$
$$(a_D)_t = r\alpha \qquad 22.8 \text{ cm/s}^2 = (7.62 \text{ cm})\alpha \qquad \boldsymbol{\alpha} = 3 \text{ rad/s}^2 \downarrow$$

Using the equations of uniformly accelerated motion, we obtain, for $t = 2$ s,

$$\omega = \omega_0 + \alpha t = 4 \text{ rad/s} + (3 \text{ rad/s}^2)(2 \text{ s}) = 10 \text{ rad/s}$$
$$\boldsymbol{\omega} = 10 \text{ rad/s} \downarrow$$
$$\theta = \omega_0 t + \tfrac{1}{2}\alpha t^2 = (4 \text{ rad/s})(2 \text{ s}) + \tfrac{1}{2}(3 \text{ rad/s}^2)(2 \text{ s})^2 = 14 \text{ rad}$$
$$\theta = 14 \text{ rad} \downarrow$$
$$\text{Number of revolutions} = (14 \text{ rad})\left(\frac{1 \text{ rev}}{2\pi \text{ rad}}\right) = 2.23 \text{ rev} \quad \blacktriangleleft$$

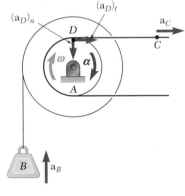

*b. Motion of Load B.* Using the following relations between linear and angular motion, with $r = 5$ in., we write

$$v_B = r\omega = (12.7 \text{ cm})(10 \text{ rad/s}) = 127 \text{ cm/s} \qquad \mathbf{v}_B = 127 \text{ cm/s} \uparrow \quad \blacktriangleleft$$
$$\Delta y_B = r\theta = (12.7 \text{ cm})(14 \text{ rad}) = 177.8 \text{ cm} \quad \Delta y_B = 177.8 \text{ cm upward} \quad \blacktriangleleft$$

*c. Acceleration of Point D at $t = 0$.* The tangential component of the acceleration is

$$(\mathbf{a}_D)_t = \mathbf{a}_C = 22.8 \text{ cm/s}^2 \rightarrow$$

Since, at $t = 0$, $\omega_0 = 4$ rad/s, the normal component of the acceleration is

$$(a_D)_n = r_D \omega_0^2 = (7.62 \text{ cm})(4 \text{ rad/s})^2 = 30.48 \text{ cm/s}^2 \quad (\mathbf{a}_D)_n = 30.48 \text{ cm/s}^2 \downarrow$$

The magnitude and direction of the total acceleration can be obtained by writing

$$\tan \phi = (30.48 \text{ cm/s}^2)/(22.8 \text{ cm/s}^2) \qquad \phi = 53.2°$$
$$a_D \sin 53.2° = 30.48 \text{ cm/s}^2 \qquad a_D = 38 \text{ cm/s}^2$$
$$\mathbf{a}_D = 38 \text{ cm/s}^2 \searrow 53.2° \quad \blacktriangleleft$$

# SOLVING PROBLEMS
# ON YOUR OWN

In this lesson we began the study of the motion of rigid bodies by considering two particular types of motion of rigid bodies: *translation* and *rotation* about a *fixed axis*.

**1. Rigid body in translation.** At any given instant, all the points of a rigid body in translation have the *same velocity* and the *same acceleration* (Fig. 15.7).

**2. Rigid body rotating about a fixed axis.** The position of a rigid body rotating about a fixed axis was defined at any given instant by the *angular coordinate* $\theta$, which is usually measured in *radians*. Selecting the unit vector $\mathbf{k}$ along the fixed axis and in such a way that the rotation of the body appears counterclockwise as seen from the tip of $\mathbf{k}$, we defined the *angular velocity* $\boldsymbol{\omega}$ and the *angular acceleration* $\boldsymbol{\alpha}$ of the body:

$$\boldsymbol{\omega} = \dot{\theta}\mathbf{k} \qquad \boldsymbol{\alpha} = \ddot{\theta}\mathbf{k} \qquad\qquad (15.6, 15.9)$$

In solving problems, keep in mind that the vectors $\boldsymbol{\omega}$ and $\boldsymbol{\alpha}$ are both directed along the fixed axis of rotation and that their sense can be obtained by the right-hand rule.

    *a. The velocity of a point P* of a body rotating about a fixed axis was found to be

$$\mathbf{v} = \boldsymbol{\omega} \times \mathbf{r} \qquad\qquad (15.5)$$

where $\boldsymbol{\omega}$ is the angular velocity of the body and $\mathbf{r}$ is the position vector drawn from any point on the axis of rotation to point $P$ (Fig. 15.9).

    *b. The acceleration of point P* was found to be

$$\mathbf{a} = \boldsymbol{\alpha} \times \mathbf{r} + \boldsymbol{\omega} \times (\boldsymbol{\omega} \times \mathbf{r}) \qquad\qquad (15.8)$$

Since vector products are not commutative, *be sure to write the vectors in the order shown* when using either of the above two equations.

**3. Rotation of a representative slab.** In many problems, you will be able to reduce the analysis of the rotation of a three-dimensional body about a fixed axis to the study of the rotation of a representative slab in a plane perpendicular to the fixed axis. The $z$ axis should be directed along the axis of rotation and point out of the paper. Thus, the representative slab will be rotating in the $xy$ plane about the origin $O$ of the coordinate system (Fig. 15.10).

To solve problems of this type you should do the following:

    *a. Draw a diagram of the representative slab,* showing its dimensions, its angular velocity and angular acceleration, as well as the vectors representing the velocities and accelerations of the points of the slab for which you have or seek information.

***b. Relate the rotation of the slab and the motion of points of the slab*** by writing the equations

$$v = r\omega \qquad\qquad (15.10')$$
$$a_t = r\alpha \qquad a_n = r\omega^2 \qquad\qquad (15.11')$$

Remember that the velocity **v** and the component $\mathbf{a}_t$ of the acceleration of a point $P$ of the slab are tangent to the circular path described by $P$. The directions of **v** and $\mathbf{a}_t$ are found by rotating the position vector **r** through 90° in the sense indicated by **ω** and **α**, respectively. The normal component $\mathbf{a}_n$ of the acceleration of $P$ is always directed toward the axis of rotation.

***4. Equations defining the rotation of a rigid body.*** You must have been pleased to note the similarity existing between the equations defining the rotation of a rigid body about a fixed axis [Eqs. (15.12) through (15.16)] and those in Chap. 11 defining the rectilinear motion of a particle [Eqs. (11.1) through (11.8)]. All you have to do to obtain the new set of equations is to substitute $\theta$, $\omega$, and $\alpha$ for $x$, $v$, and $a$ in the equations of Chap. 11.

# Problems

**15.1** The motion of a rotor is defined by the relation $\theta = 8t^3 - 6(t-2)^2$, where $\theta$ and $t$ are expressed in radians and seconds, respectively. Determine (a) when the angular acceleration is zero, (b) the angular coordinate and angular velocity at that time.

**15.2** The motion of an oscillating flywheel is defined by the relation $\theta = \theta_0 e^{-3\pi t} \cos 4\pi t$, where $\theta$ is expressed in radians and $t$ in seconds. Knowing that $\theta_0 = 0.5$ rad, determine the angular coordinate, the angular velocity, and the angular acceleration of the flywheel when (a) $t = 0$, (b) $t = 0.125$ s.

**15.3** The motion of an oscillating flywheel is defined by the relation $\theta = \theta_0 e^{-7\pi t/6} \sin 4\pi t$, where $\theta$ is expressed in radians and $t$ in seconds. Knowing that $\theta_0 = 0.4$ rad, determine the angular coordinate, the angular velocity, and the angular acceleration of the flywheel when (a) $t = 0.125$ s, (b) $t = \infty$.

**Fig. P15.2 and P15.3**

**15.4** A flywheel executes 1800 revolutions while it coasts to rest from a speed of 6000 rpm. Assuming uniformly accelerated motion, determine (a) the time required for the flywheel to coast to rest, (b) the time required for the flywheel to execute the first 900 revolutions.

**15.5** When the power to an electric motor is turned on the motor reaches its rated speed of 2400 rpm in 4 s, and when the power is turned off the motor coasts to rest in 40 s. Assuming uniformly accelerated motion, determine the number of revolutions that the motor executes (a) in reaching its rated speed, (b) in coasting to rest.

**Fig. P15.5**

**15.6** The angular acceleration of a flywheel is defined by the relation $\alpha = 30e^{-0.2t}$, where $\alpha$ and $t$ are expressed in rad/s$^2$ and seconds, respectively. Knowing that $\theta = 0$ and $\omega = 0$ at $t = 0$, determine the angular velocity and angular coordinate of the particle when $t = 0.5$ s.

**15.7** The angular acceleration of a shaft is defined by the relation $\alpha = -0.5\omega$, where $\alpha$ is expressed in rad/s$^2$ and $\omega$ in rad/s. Knowing that at $t = 0$ the angular velocity of the shaft is 30 rad/s, determine (a) the number of revolutions the shaft will execute before coming to rest, (b) the time required for the shaft to come to rest, (c) the time required for the angular velocity of the shaft to reduce to 2 percent of its initial value.

**15.8** The angular acceleration of an oscillating disk is defined by the relation $\alpha = -k\theta$. Determine (a) the value of $k$ for which $\omega = 12$ rad/s when $\theta = 0$ and $\theta = 6$ rad when $\omega = 0$, (b) the angular velocity of the disk when $\theta = 3$ rad.

**15.9** The rectangular block shown rotates about the diagonal $OA$ with a constant angular velocity of 6.76 rad/s. Knowing that the rotation is counterclockwise as viewed from $A$, determine the velocity and acceleration of point $B$ at the instant shown.

**15.10** In Prob. 15.9, determine the velocity and acceleration of point $B$ at the instant shown, assuming that the angular velocity is 3.38 rad/s and decreases at the rate of 5.07 rad/s$^2$.

**15.11** The assembly shown consists of two rods and a rectangular plate $BCDE$ which are welded together. The assembly rotates about the axis $AB$ with a constant angular velocity of 10 rad/s. Knowing that the rotation is counterclockwise as viewed from $B$, determine the velocity and acceleration of corner $E$.

**Fig. P15.9**

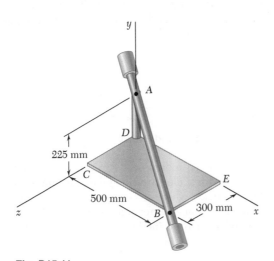

**Fig. P15.11**

**15.12** In Prob. 15.11, determine the velocity and acceleration of corner $C$, assuming that the angular velocity is 10 rad/s and decreases at the rate of 20 rad/s$^2$.

**15.13** The bent rod $ABCD$ rotates about a line joining points $A$ and $D$ with a constant angular velocity of 75 rad/s. Knowing that at the instant considered the velocity of corner $C$ is upward, determine the velocity and acceleration of corner $B$.

**15.14** In Prob. 15.13, determine the velocity and acceleration of corner $B$, assuming that the angular velocity is 75 rad/s and decreases at the rate of 600 rad/s$^2$.

**Fig. P15.13**

**15.15** The earth makes one complete revolution around the sun in 365.24 days. Assuming that the orbit of the earth is circular and has a radius of 149014581 km, determine the velocity and acceleration of the earth.

**Fig. P15.17 and P15.18**

*15.16* The earth makes one complete revolution on its axis in 23 h 56 min. Knowing that the mean radius of the earth is 6370 km, determine the linear velocity and acceleration of a point on the surface of the earth (*a*) at the equator, (*b*) at Philadelphia, latitude 40° north, (*c*) at the North Pole.

*15.17* A series of small machine components being moved by a conveyor belt passes over a 120-mm-radius idler pulley. At the instant shown, the velocity of point *A* is 300 mm/s to the left and its acceleration is 180 mm/s² to the right. Determine (*a*) the angular velocity and angular acceleration of the idler pulley, (*b*) the total acceleration of the machine component at *B*.

*15.18* A series of small machine components being moved by a conveyor belt passes over a 120-mm-radius idler pulley. At the instant shown, the angular velocity of the idler pulley is 4 rad/s clockwise. Determine the angular acceleration of the pulley for which the magnitude of the total acceleration of the machine component at *B* is 2400 mm/s².

*15.19* The belt shown moves over two pulleys without slipping. At the instant shown the pulleys are rotating clockwise and the speed of point *B* on the belt is 3.66 m/s, increasing at the rate of 29.3 m/s. Determine, at this instant, (*a*) the angular velocity and angular acceleration of each pulley, (*b*) the acceleration of point *P* on pulley *C*.

**Fig. P15.19 and P15.20**

*15.20* The belt shown moves over two pulleys without slipping. Pulley *A* starts from rest with a clockwise angular acceleration defined by the relation $\alpha = 120 - 0.002\omega^2$, where $\alpha$ is expressed in rad/s² and $\omega$ is expressed in rad/s. Determine, after one-half revolution of pulley *A*, (*a*) the magnitude of the acceleration of point *B* on the belt, (*b*) the acceleration of point *P* on pulley *C*.

*15.21* Three belts move over two pulleys without slipping in the speed reduction system shown. Point *A* on the input belt moves to the right with a constant speed of 0.6 m/s. Determine (*a*) the velocity of point *C* on the output belt, (*b*) the acceleration of point *B* on the output pulley.

**Fig. *P15.21* and *P15.22***

*15.22* Three belts move over two pulleys without slipping in the speed reduction system shown. At the instant shown the velocity of point *A* on the input belt is 0.6 m/s to the right, decreasing at the rate of 1.8 m/s². Determine, at this instant, (*a*) the velocity and acceleration of point *C* on the output belt, (*b*) the acceleration of point *B* on the output pulley.

**15.23** Ring $B$ has an inner radius $r_2$ and hangs from the horizontal shaft $A$ as shown. Knowing that shaft $A$ rotates with a constant angular velocity $\omega_A$ and that no slipping occurs, derive a relation in terms of $r_1$, $r_2$, $r_3$, and $\omega_A$ for ($a$) the angular velocity of ring $B$, ($b$) the acceleration of the points of shaft $A$ and ring $B$ which are in contact.

**15.24** Ring $B$ has an inner radius $r_2$ and hangs from the horizontal shaft $A$ as shown. Shaft $A$ rotates with a constant angular velocity of 25 rad/s and no slipping occurs. Knowing that $r_1 = 1.27$ cm, $r_2 = 6.35$ cm, and $r_3 = 8.9$ cm, determine ($a$) the angular velocity of ring $B$, ($b$) the acceleration of the points of shaft $A$ and ring $B$ which are in contact, ($c$) the magnitude of the acceleration of a point on the outside surface of ring $B$.

**15.25** A gear reduction system consists of three gears $A$, $B$, and $C$. Knowing that gear $A$ rotates clockwise with a constant angular velocity $\omega_A = 600$ rpm, determine ($a$) the angular velocities of gears $B$ and $C$, ($b$) the accelerations of the points on gears $B$ and $C$ which are in contact.

**Fig. P15.23 and P15.24**

**Fig. *P15.25* and *P15.26***

**15.26** A gear reduction system consists of three gears $A$, $B$, and $C$. Gear $A$ starts from rest at time $t = 0$ and rotates clockwise with constant angular acceleration. Knowing that the angular velocity of gear $A$ is 600 rpm at time $t = 2$ s, determine ($a$) the angular accelerations of gears $B$ and $C$, ($b$) the accelerations of the points on gears $B$ and $C$ which are in contact when $t = 0.5$ s.

**15.27** Two blocks and a pulley are connected by inextensible cords as shown. The relative velocity of block $A$ with respect to block $B$ is 0.8 m/s to the left at time $t = 0$ and 0.4 m/s to the left when $t = 0.25$ s. Knowing that the angular acceleration of the pulley is constant, find ($a$) the relative acceleration of block $A$ with respect to block $B$, ($b$) the distance block $A$ moves relative to block $B$ during the interval $0 \le t \le 0.25$ s.

**Fig. P15.27 and P15.28**

**15.28** Two blocks and a pulley are connected by inextensible cords as shown. The relative velocity of block $A$ with respect to block $B$ is 0.9 m/s to the left at time $t = 0$ and 0.45 m/s to the left after one-half revolution of the pulley. Knowing that the angular acceleration of the pulley is constant, find ($a$) the relative acceleration of block $A$ with respect to block $B$, ($b$) the distance block $A$ moves relative to block $B$ during the interval $0 \le t \le 0.3$ s.

**15.29** A pulley and two loads are connected by inextensible cords as shown. Load $A$ has a constant acceleration of 25.4 cm/s$^2$ and an initial velocity of 20.3 cm/s, both directed upward. Determine (*a*) the number of revolutions executed by the pulley in 3 s, (*b*) the velocity and position of load $B$ after 3 s, (*c*) the acceleration of point $D$ on the rim of the pulley at time $t = 0$.

**Fig. P15.29 and P15.30**

**15.30** A pulley and two loads are connected by inextensible cords as shown. The pulley starts from rest at $t = 0$ and is accelerated at the uniform rate of 2.4 rad/s$^2$ clockwise. At $t = 4$ s, determine the velocity and position (*a*) of load $A$, (*b*) of load $B$.

**15.31** Two friction disks $A$ and $B$ are to be brought into contact without slipping when the angular velocity of disk $A$ is 240 rpm counterclockwise. Disk $A$ starts from rest at time $t = 0$ and is given a constant angular acceleration of magnitude $\alpha$. Disk $B$ starts from rest at time $t = 2$ s and is given a constant clockwise angular acceleration, also of magnitude $\alpha$. Determine (*a*) the required angular acceleration magnitude $\alpha$, (*b*) the time at which the contact occurs.

**Fig. P15.31 and P15.32**

**15.32** Two friction disks $A$ and $B$ are brought into contact when the angular velocity of disk $A$ is 240 rpm counterclockwise and disk $B$ is at rest. A period of slipping follows and disk $B$ makes 2 revolutions before reaching its final angular velocity. Assuming that the angular acceleration of each disk is constant and inversely proportional to the cube of its radius, determine (a) the angular acceleration of each disk, (b) the time during which the disks slip.

**15.33** A simple friction drive consists of two disks $A$ and $B$. Initially, disk $B$ has a clockwise angular velocity of 500 rpm, and disk $A$ is at rest. It is known that disk $B$ will coast to rest in 60 s. However, rather than waiting until both disks are at rest to bring them together, disk $A$ is given a constant angular acceleration of 3 rad/s$^2$ counterclockwise. Determine (a) at what time the disks can be brought together if they are not to slip, (b) the angular velocity of each disk as contact is made.

6.35 cm    7.6 cm

**Fig. *P15.33* and *P15.34***

**15.34** Two friction wheels $A$ and $B$ are both rotating freely at 300 rpm counterclockwise when they are brought into contact. After 12 s of slippage, during which each wheel has a constant angular acceleration, wheel $B$ reaches a final angular velocity of 75 rpm counterclockwise. Determine (a) the angular acceleration of each wheel during the period of slippage, (b) the time at which the angular velocity of wheel $A$ is equal to zero.

*****15.35** Television recording tape is being rewound on a VCR reel which rotates with a constant angular velocity $\omega_0$. Denoting by $r$ the radius of the reel at any given time and by $b$ the thickness of the tape, derive an expression for the acceleration of the tape as it approaches the reel.

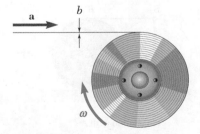

**Fig. P15.35**

**\*15.36** In a continuous printing process, paper is drawn into the presses at a constant speed $v$. Denoting by $r$ the radius of the paper roll at any given time and by $b$ the thickness of the paper, derive an expression for the angular acceleration of the paper roll.

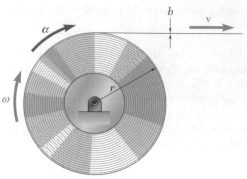

**Fig. P15.36**

## 15.5. GENERAL PLANE MOTION

As indicated in Sec. 15.1, we understand by general plane motion a plane motion which is neither a translation nor a rotation. As you will presently see, however, *a general plane motion can always be considered as the sum of a translation and a rotation.*

Consider, for example, a wheel rolling on a straight track (Fig. 15.12). Over a certain interval of time, two given points $A$ and $B$ will have moved, respectively, from $A_1$ to $A_2$ and from $B_1$ to $B_2$. The same result could be obtained through a translation which would bring $A$ and $B$ into $A_2$ and $B'_1$ (the line $AB$ remaining vertical), followed by a rotation about $A$ bringing $B$ into $B_2$. Although the original rolling motion differs from the combination of translation and rotation when these motions are taken in succession, the original motion can be exactly duplicated by a combination of simultaneous translation and rotation.

Another example of plane motion is given in Fig. 15.13, which represents a rod whose extremities slide along a horizontal and a vertical track, respectively. This motion can be replaced by a translation in a horizontal direction and a rotation about $A$ (Fig. 15.13a)

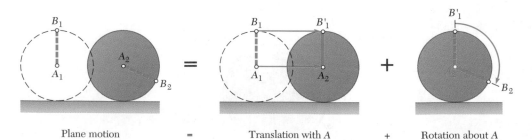

| Plane motion | = | Translation with $A$ | + | Rotation about $A$ |

**Fig. 15.12**

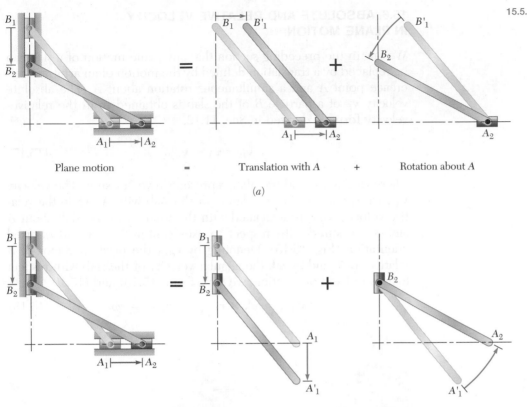

Plane motion = Translation with A + Rotation about A

(a)

Plane motion = Translation with B + Rotation about B

(b)

**Fig. 15.13**

or by a translation in a vertical direction and a rotation about $B$ (Fig. 15.13$b$).

In the general case of plane motion, we will consider a small displacement which brings two particles $A$ and $B$ of a representative slab, respectively, from $A_1$ and $B_1$ into $A_2$ and $B_2$ (Fig. 15.14). This displacement can be divided into two parts: in one, the particles move into $A_2$ and $B'_1$ while the line $AB$ maintains the same direction; in the other, $B$ moves into $B_2$ while $A$ remains fixed. The first part of the motion is clearly a translation and the second part a rotation about $A$.

Recalling from Sec. 11.12 the definition of the relative motion of a particle with respect to a moving frame of reference—as opposed to its absolute motion with respect to a fixed frame of reference—we can restate as follows the result obtained above: Given two particles $A$ and $B$ of a rigid slab in plane motion, the relative motion of $B$ with respect to a frame attached to $A$ and of fixed orientation is a rotation. To an observer moving with $A$ but not rotating, particle $B$ will appear to describe an arc of a circle centered at $A$.

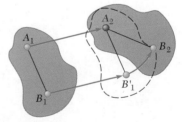

**Fig. 15.14**

## 15.6. ABSOLUTE AND RELATIVE VELOCITY IN PLANE MOTION

We saw in the preceding section that any plane motion of a slab can be replaced by a translation defined by the motion of an arbitrary reference point $A$ and a simultaneous rotation about $A$. The absolute velocity $\mathbf{v}_B$ of a particle $B$ of the slab is obtained from the relative-velocity formula derived in Sec. 11.12,

$$\mathbf{v}_B = \mathbf{v}_A + \mathbf{v}_{B/A} \qquad (15.17)$$

where the right-hand member represents a vector sum. The velocity $\mathbf{v}_A$ corresponds to the translation of the slab with $A$, while the relative velocity $\mathbf{v}_{B/A}$ is associated with the rotation of the slab about $A$ and is measured with respect to axes centered at $A$ and of fixed orientation (Fig. 15.15). Denoting by $\mathbf{r}_{B/A}$ the position vector of $B$ relative to $A$, and by $\omega\mathbf{k}$ the angular velocity of the slab with respect to axes of fixed orientation, we have from (15.10) and (15.10′)

$$\mathbf{v}_{B/A} = \omega\mathbf{k} \times \mathbf{r}_{B/A} \qquad v_{B/A} = r\omega \qquad (15.18)$$

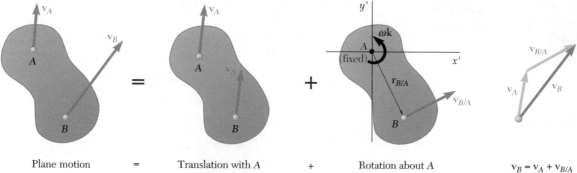

Plane motion $\quad = \quad$ Translation with $A$ $\quad + \quad$ Rotation about $A$ $\qquad$ $\mathbf{v}_B = \mathbf{v}_A + \mathbf{v}_{B/A}$

**Fig. 15.15**

where $r$ is the distance from $A$ to $B$. Substituting for $\mathbf{v}_{B/A}$ from (15.18) into (15.17), we can also write

$$\mathbf{v}_B = \mathbf{v}_A + \omega\mathbf{k} \times \mathbf{r}_{B/A} \qquad (15.17')$$

As an example, let us again consider the rod $AB$ of Fig. 15.13. Assuming that the velocity $\mathbf{v}_A$ of end $A$ is known, we propose to find the velocity $\mathbf{v}_B$ of end $B$ and the angular velocity $\boldsymbol{\omega}$ of the rod, in terms of the velocity $\mathbf{v}_A$, the length $l$, and the angle $\theta$. Choosing $A$ as a reference point, we express that the given motion is equivalent to a translation with $A$ and a simultaneous rotation about $A$ (Fig. 15.16). The absolute velocity of $B$ must therefore be equal to the vector sum

$$\mathbf{v}_B = \mathbf{v}_A + \mathbf{v}_{B/A} \qquad (15.17)$$

We note that while the direction of $\mathbf{v}_{B/A}$ is known, its magnitude $l\omega$ is unknown. However, this is compensated for by the fact that the direction of $\mathbf{v}_B$ is known. We can therefore complete the diagram of Fig. 15.16. Solving for the magnitudes $v_B$ and $\omega$, we write

$$v_B = v_A \tan \theta \qquad \omega = \frac{v_{B/A}}{l} = \frac{v_A}{l \cos \theta} \qquad (15.19)$$

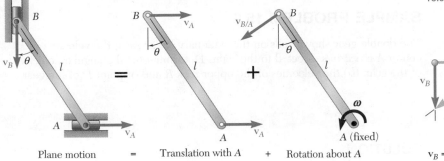

Plane motion   =   Translation with A   +   Rotation about A        $\mathbf{v}_B = \mathbf{v}_A + \mathbf{v}_{B/A}$

**Fig. 15.16**

The same result can be obtained by using $B$ as a point of reference. Resolving the given motion into a translation with $B$ and a simultaneous rotation about $B$ (Fig. 15.17), we write the equation

$$\mathbf{v}_A = \mathbf{v}_B + \mathbf{v}_{A/B} \tag{15.20}$$

which is represented graphically in Fig. 15.17. We note that $\mathbf{v}_{A/B}$ and $\mathbf{v}_{B/A}$ have the same magnitude $l\omega$ but opposite sense. The sense of the relative velocity depends, therefore, upon the point of reference which has been selected and should be carefully ascertained from the appropriate diagram (Fig. 15.16 or 15.17).

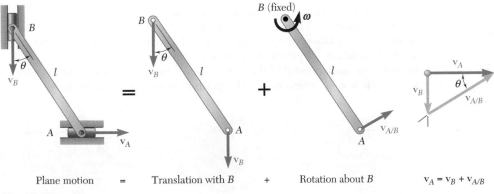

Plane motion   =   Translation with B   +   Rotation about B        $\mathbf{v}_A = \mathbf{v}_B + \mathbf{v}_{A/B}$

**Fig. 15.17**

Finally, we observe that the angular velocity $\boldsymbol{\omega}$ of the rod in its rotation about $B$ is the same as in its rotation about $A$. It is measured in both cases by the rate of change of the angle $\theta$. This result is quite general; we should therefore bear in mind that *the angular velocity $\boldsymbol{\omega}$ of a rigid body in plane motion is independent of the reference point.*

Most mechanisms consist not of one but of *several* moving parts. When the various parts of a mechanism are pin-connected, the analysis of the mechanism can be carried out by considering each part as a rigid body, keeping in mind that the points where two parts are connected must have the same absolute velocity (see Sample Prob. 15.3). A similar analysis can be used when gears are involved, since the teeth in contact must also have the same absolute velocity. However, when a mechanism contains parts which slide on each other, the relative velocity of the parts in contact must be taken into account (see Secs. 15.10 and 15.11).

The double gear shown rolls on the stationary lower rack; the velocity of its center A is 1.2 m/s directed to the right. Determine (a) the angular velocity of the gear, (b) the velocities of the upper rack R and of point D of the gear.

## SOLUTION

*a. Angular Velocity of the Gear.* Since the gear rolls on the lower rack, its center A moves through a distance equal to the outer circumference $2\pi r_1$ for each full revolution of the gear. Noting that 1 rev = $2\pi$ rad, and that when A moves to the right ($x_A > 0$) the gear rotates clockwise ($\theta < 0$), we write

$$\frac{x_A}{2\pi r_1} = -\frac{\theta}{2\pi} \qquad x_A = -r_1\theta$$

Differentiating with respect to the time $t$ and substituting the known values $v_A = 1.2$ m/s and $r_1 = 150$ mm = 0.150 m, we obtain

$$v_A = -r_1\omega \qquad 1.2 \text{ m/s} = -(0.150 \text{ m})\omega \qquad \omega = -8 \text{ rad/s}$$
$$\omega = \omega\mathbf{k} = -(8 \text{ rad/s})\mathbf{k} \quad \blacktriangleleft$$

where $\mathbf{k}$ is a unit vector pointing out of the paper.

*b. Velocities.* The rolling motion is resolved into two component motions: a translation with the center A and a rotation about the center A. In the translation, all points of the gear move with the same velocity $\mathbf{v}_A$. In the rotation, each point P of the gear moves about A with a relative velocity $\mathbf{v}_{P/A} = \omega\mathbf{k} \times \mathbf{r}_{P/A}$, where $\mathbf{r}_{P/A}$ is the position vector of P relative to A.

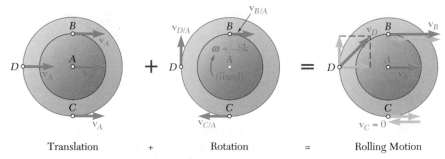

| Translation | + | Rotation | = | Rolling Motion |
|---|---|---|---|---|

*Velocity of Upper Rack.* The velocity of the upper rack is equal to the velocity of point B; we write

$$\mathbf{v}_R = \mathbf{v}_B = \mathbf{v}_A + \mathbf{v}_{B/A} = \mathbf{v}_A + \omega\mathbf{k} \times \mathbf{r}_{B/A}$$
$$= (1.2 \text{ m/s})\mathbf{i} - (8 \text{ rad/s})\mathbf{k} \times (0.100 \text{ m})\mathbf{j}$$
$$= (1.2 \text{ m/s})\mathbf{i} + (0.8 \text{ m/s})\mathbf{i} = (2 \text{ m/s})\mathbf{i}$$
$$\mathbf{v}_R = 2 \text{ m/s} \rightarrow \quad \blacktriangleleft$$

*Velocity of Point D.*

$$\mathbf{v}_D = \mathbf{v}_A + \mathbf{v}_{D/A} = \mathbf{v}_A + \omega\mathbf{k} \times \mathbf{r}_{D/A}$$
$$= (1.2 \text{ m/s})\mathbf{i} - (8 \text{ rad/s})\mathbf{k} \times (-0.150 \text{ m})\mathbf{i}$$
$$= (1.2 \text{ m/s})\mathbf{i} + (1.2 \text{ m/s})\mathbf{j}$$
$$\mathbf{v}_D = 1.697 \text{ m/s} \measuredangle 45° \quad \blacktriangleleft$$

# SAMPLE PROBLEM 15.3

In the engine system shown, the crank $AB$ has a constant clockwise angular velocity of 2000 rpm. For the crank position indicated, determine ($a$) the angular velocity of the connecting rod $BD$, ($b$) the velocity of the piston $P$.

## SOLUTION

**Motion of Crank $AB$.** The crank $AB$ rotates about point $A$. Expressing $\omega_{AB}$ in rad/s and writing $v_B = r\omega_{AB}$, we obtain

$$\omega_{AB} = \left(2000\frac{\text{rev}}{\text{min}}\right)\left(\frac{1\text{ min}}{60\text{ s}}\right)\left(\frac{2\pi\text{ rad}}{1\text{ rev}}\right) = 209.3\text{ rad/s}$$

$$v_B = (AB)\omega_{AB} = (7.6\text{ cm})(209.4\text{ rad/s}) = 1590.9\text{ cm/s}$$
$$\mathbf{v}_B = 1590.9\text{ cm/s} \, \text{⦢}\, 50°$$

**Motion of Connecting Rod $BD$.** We consider this motion as a general plane motion. Using the law of sines, we compute the angle $\beta$ between the connecting rod and the horizontal:

$$\frac{\sin 40°}{20.3\text{ cm}} = \frac{\sin\beta}{7.6\text{ cm}} \qquad \beta = 13.92°$$

The velocity $\mathbf{v}_D$ of the point $D$ where the rod is attached to the piston must be horizontal, while the velocity of point $B$ is equal to the velocity $\mathbf{v}_B$ obtained above. Resolving the motion of $BD$ into a translation with $B$ and a rotation about $B$, we obtain

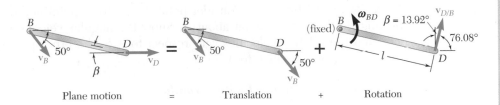

| Plane motion | = | Translation | + | Rotation |

Expressing the relation between the velocities $\mathbf{v}_D$, $\mathbf{v}_B$, and $\mathbf{v}_{D/B}$, we write

$$\mathbf{v}_D = \mathbf{v}_B + \mathbf{v}_{D/B}$$

We draw the vector diagram corresponding to this equation. Recalling that $\beta = 13.92°$, we determine the angles of the triangle and write

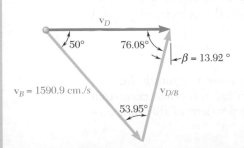

$$\frac{v_D}{\sin 53.95°} = \frac{v_{D/B}}{\sin 50°} = \frac{1590.9\text{ cm/s}}{\sin 76.08°}$$

$$v_{D/B} = 1256.16\text{ cm/s} \qquad \mathbf{v}_{D/B} = 1255.54\text{ cm/s} \,\measuredangle\, 76.08°$$
$$v_D = 1325.78\text{ cm/s} \qquad\qquad \mathbf{v}_D = 1325.14\text{ cm/s} \rightarrow$$
$$\mathbf{v}_P = \mathbf{v}_D = 1325.14\text{ cm/s} \rightarrow \quad ◄$$

Since $v_{D/B} = l\omega_{BD}$, we have

$$1255.54\text{ cm/s} = (20.3\text{ cm})\omega_{BD} \quad \boldsymbol{\omega}_{BD} = 61.83\text{ rad/s} \,\text{↑} \quad ◄$$

In this lesson you learned to analyze the velocity of bodies in *general plane motion.* You found that a general plane motion can always be considered as the sum of the two motions you studied in the last lesson, namely, *a translation and a rotation.*

To solve a problem involving the velocity of a body in plane motion you should take the following steps.

**1. *Whenever possible determine the velocity of the points of the body*** where the body is connected to another body whose motion is known. That other body may be an arm or crank rotating with a given angular velocity [Sample Prob. 15.3].

**2. *Next start drawing a "diagram equation"*** to use in your solution (Figs. 15.15 and 15.16). This "equation" will consist of the following diagrams.

    *a. Plane motion diagram:* Draw a diagram of the body including all dimensions and showing those points for which you know or seek the velocity.

    *b. Translation diagram:* Select a reference point $A$ for which you know the direction and/or the magnitude of the velocity $\mathbf{v}_A$, and draw a second diagram showing the body in translation with all of its points having the same velocity $\mathbf{v}_A$.

    *c. Rotation diagram:* Consider point $A$ as a fixed point and draw a diagram showing the body in rotation about $A$. Show the angular velocity $\boldsymbol{\omega} = \omega\mathbf{k}$ of the body and the relative velocities with respect to $A$ of the other points, such as the velocity $\mathbf{v}_{B/A}$ of $B$ relative to $A$.

**3. *Write the relative-velocity formula***

$$\mathbf{v}_B = \mathbf{v}_A + \mathbf{v}_{B/A}$$

While you can solve this vector equation analytically by writing the corresponding scalar equations, you will usually find it easier to solve it by using a vector triangle (Fig. 15.16).

**4. *A different reference point can be used to obtain an equivalent solution.*** For example, if point $B$ is selected as the reference point, the velocity of point $A$ is expressed as

$$\mathbf{v}_A = \mathbf{v}_B + \mathbf{v}_{A/B}$$

Note that the relative velocities $\mathbf{v}_{B/A}$ and $\mathbf{v}_{A/B}$ have the same magnitude but opposite sense. Relative velocities, therefore, depend upon the reference point that has been selected. The angular velocity, however, is independent of the choice of reference point.

# Problems

**15.37** The motion of rod $AB$ is guided by pins attached at $A$ and $B$ which slide in the slots shown. At the instant shown, $\theta = 40°$ and the pin at $B$ moves upward to the left with a constant velocity of 150 mm/s. Determine (a) the angular velocity of the rod, (b) the velocity of the pin at end $A$.

**15.38** The motion of rod $AB$ is guided by pins attached at $A$ and $B$ which slide in the slots shown. At the instant shown, $\theta = 30°$ and the pin at $A$ moves downward with a constant velocity of 225 mm/s. Determine (a) the angular velocity of the rod, (b) the velocity of the pin at end $B$.

**15.39** Rod $AB$ can slide freely along the floor and the inclined plane. At the instant shown, the velocity of end $A$ is 1.28 m/s to the left. Determine (a) the angular velocity of the rod, (b) the velocity of end $B$ of the rod.

**Fig. P15.37 and P15.38**

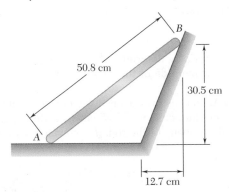

**Fig. P15.39 and P15.40**

**15.40** Rod $AB$ can slide freely along the floor and the inclined plane. At the instant shown the angular velocity of the rod is 4.2 rad/s counterclockwise. Determine (a) the velocity of end $A$ of the rod, (b) the velocity of end $B$ of the rod.

**15.41** The disk shown moves in the $xy$ plane. Knowing that $(v_A)_y = -7$ m/s, $(v_B)_x = -7.4$ m/s, and $(v_C)_x = -1.4$ m/s, determine (a) the angular velocity of the disk, (b) the velocity of point $B$.

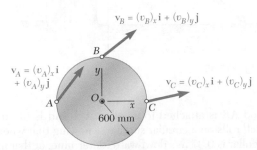

**Fig. P15.41**

**15.42**   In Prob. 15.41, determine (*a*) the velocity of point *O*, (*b*) the point of the disk with zero velocity.

**15.43**   The sheet metal form shown moves in the *xy* plane. Knowing that $(v_A)_x = 100$ mm/s, $(v_B)_y = -75$ mm/s, and $(v_C)_x = 400$ mm/s, determine (*a*) the angular velocity of the plate, (*b*) the velocity of point *A*.

**Fig. P15.43**

**15.44**   In Prob. 15.43, determine the locus of points of the sheet metal form for which the magnitude of the velocity is 200 mm/s.

**15.45**   Rod *AB* moves over a small wheel at *C* while end *A* moves to the right with a constant velocity of 63.5 cm/s. At the instant shown, determine (*a*) the angular velocity of the rod, (*b*) the velocity of end *B* of the rod.

**Fig. P15.45**

**Fig. P15.46**

**15.46**   Rod *AB* is attached to a collar at *A* and is fitted with a small wheel at *B* which rolls on a circular surface. Knowing that when $\theta = 60°$ the velocity of the collar is 0.37 m/s downward, determine, at that instant, (*a*) the angular velocity of rod *AB*, (*b*) the velocity of point *B*.

**15.47** The outer gear $C$ rotates with an angular velocity of 5 rad/s clockwise. Knowing that the inner gear $A$ is stationary, determine (a) the angular velocity of the intermediate gear $B$, (b) the angular velocity of the arm $AB$.

40 mm 80 mm

80 mm

A

B

C

**Fig. P15.47 and P15.48**

**15.48** The intermediate gear $B$ rotates with an angular velocity of 20 rad/s clockwise. Knowing that the outer gear $C$ is stationary, determine (a) the angular velocity of the inner gear $A$, (b) the angular velocity of the arm $AB$.

**15.49** In the simplified sketch of a ball bearing shown, the diameter of the inner race $A$ is 6.35 cm and the diameter of each ball is 1.3 cm. The outer race $B$ is stationary while the inner race has an angular velocity of 3600 rpm. Determine (a) the speed of the center of each ball, (b) the angular velocity of each ball, (c) the number of times per minute each ball describes a complete circle.

B   A

**Fig. P15.49**

12.7 cm

A   B   C 25.4 cm

38.1 cm — 38.1 cm

**Fig. P15.50**

**15.50** The outer gear $A$ rotates with an angular velocity of 3 rad/s counterclockwise. Knowing that the angular velocity of the intermediate gear $B$ is 6 rad/s clockwise, determine the angular velocity of (a) the arm $ABC$, (b) the outer gear $C$.

**15.51** In the planetary gear system shown, the radius of gears A, B, C, and D is 60 mm and the radius of the outer gear E is 180 mm. Knowing that gear E has an angular velocity of 120 rpm clockwise and that the central gear has an angular velocity of 150 rpm clockwise, determine (a) the angular velocity of each planetary gear, (b) the angular velocity of the spider connecting the planetary gears.

**Fig. P15.51 and P15.52**

**15.52** In the planetary gear system shown, the radius of the central gear A is a, the radius of the planetary gears is b, and the radius of the outer gear E is a + 2b. The angular velocity of gear A is $\omega_A$ clockwise, and the outer gear is stationary. If the angular velocity of the spider BCD is to be $\omega_A/5$ clockwise, determine (a) the required value of the ratio b/a, (b) the corresponding angular velocity of each planetary gear.

**15.53** Three gears A, B, and C are pinned at their centers to rod ABC. Knowing that end A of rod ABC is fixed and that gear A does not rotate, determine the angular velocity of gears B and C when rod ABC rotates clockwise with a constant angular velocity of 75 rpm.

**15.54** Three gears A, B, and C are pinned at their centers to rod ABC. Knowing that end C of rod ABC is fixed and that gear C does not rotate, determine the angular velocity of gears B and C when rod ABC rotates clockwise with a constant angular velocity of 80 rpm.

**Fig. *P15.53* and *P15.54***

**15.55** In the eccentric shown, a disk of 40-mm-radius revolves about shaft O that is located 10 mm from the center A of the disk. The distance between the center A of the disk and the pin at B is 160 mm. Knowing that the angular velocity of the disk is 900 rpm clockwise, determine the velocity of the block when $\theta = 30°$.

**Fig. P15.55**

**15.56** Determine the velocity of the block of Prob. 15.55 when $\theta = 120°$.

**15.57**   Knowing that the disk has a constant angular velocity of 15 rad/s clockwise, determine the angular velocity of bar *BD* and the velocity of collar *D* when (*a*) θ = 0, (*b*) θ = 90°, (*c*) θ = 180°.

7.1 cm

25.4 cm

**Fig. P15.57 and P15.58**

**15.58**   The disk has a constant angular velocity of 20 rad/s clockwise. (*a*) Determine the two values of the angle θ for which the velocity of collar *D* is zero. (*b*) For each of these values of θ, determine the corresponding value of the angular velocity of bar *BD*.

**15.59**   Knowing that at the instant shown the angular velocity of rod *BE* is 4 rad/s counterclockwise, determine (*a*) the angular velocity of rod *AD*, (*b*) the velocity of collar *D*, (*c*) the velocity of point *A*.

192 mm

A

240 mm

B

30°

360 mm

D

**Fig. P15.59 and P15.60**

**15.60**   Knowing that at the instant shown the velocity of collar *D* is 1.6 m/s upward, determine (*a*) the angular velocity of rod *AD*, (*b*) the velocity of point *B*, (*c*) the velocity of point *A*.

**15.61**   In the engine system shown, $l = 20.3$ cm and $b = 7.6$ cm. Knowing that the crank *AB* rotates with a constant angular velocity of 1000 rpm clockwise, determine the velocity of piston *P* and the angular velocity of the connecting rod when (*a*) θ = 0, (*b*) θ = 90°.

**15.62**   In the engine system shown, $l = 20.3$ cm and $b = 7.6$ cm. Knowing that the crank *AB* rotates with a constant angular velocity of 1000 rpm clockwise, determine the velocity of piston *P* and the angular velocity of the connecting rod when θ = 60°.

**Fig. P15.61 and P15.62**

**15.63** In the position shown, bar *AB* has an angular velocity of 4 rad/s clockwise. Determine the angular velocity of bars *BD* and *DE*.

**Fig. P15.63**

**15.64** Bar *AB* is rotating counterclockwise and, at the instant shown, the magnitude of the velocity of point *G* is 2.5 m/s. Determine the angular velocity of each of the three bars at that instant.

**15.65** At the instant shown, bar *AB* has a constant angular velocity of 25 rad/s counterclockwise. Determine at that instant (*a*) the angular velocity of the rectangular plate *FBDH*, (*b*) the velocity of point *F*.

**15.66** At the instant shown, bar *DE* has a constant angular velocity of 35 rad/s clockwise. Determine at that instant (*a*) the angular velocity of the rectangular plate *FBDH*, (*b*) the point on the plate *FBDH* with zero velocity.

**15.67** Knowing that at the instant shown the angular velocity of rod *AB* is 15 rad/s clockwise, determine (*a*) the angular velocity of rod *BD*, (*b*) the velocity of the midpoint of rod *BD*.

**Fig. P15.64**

**Fig. P15.65 and P15.66**

**Fig. P15.67**

**Fig. P15.68**

**15.68** In the position shown, bar *AB* has a constant angular velocity of 3 rad/s counterclockwise. Determine the angular velocity of bars *BD* and *DE*.

**15.69**   An automobile travels to the right at a constant speed of 80 km/h. If the diameter of the wheel is 560 mm, determine the velocities of points $B$, $C$, $D$, and $E$ on the rim of the wheel.

**Fig. P15.69**

**15.70**   The 10.2 cm-radius wheel shown rolls to the left with a velocity of 114.3 cm/s. Knowing that the distance $AD$ is 6.4 cm, determine the velocity of the collar and the angular velocity of rod $AB$ when (a) $\beta = 0$, (b) $\beta = 90°$.

**Fig. P15.70**

---

## 15.7. INSTANTANEOUS CENTER OF ROTATION IN PLANE MOTION

Consider the general plane motion of a slab. We propose to show that at any given instant the velocities of the various particles of the slab are the same as if the slab were rotating about a certain axis perpendicular to the plane of the slab, called the *instantaneous axis of rotation.* This axis intersects the plane of the slab at a point $C$, called the *instantaneous center of rotation* of the slab.

We first recall that the plane motion of a slab can always be replaced by a translation defined by the motion of an arbitrary reference point $A$ and by a rotation about $A$. As far as the velocities are concerned, the translation is characterized by the velocity $\mathbf{v}_A$ of the reference point $A$ and the rotation is characterized by the angular velocity $\boldsymbol{\omega}$ of the slab (which is independent of the choice of $A$). Thus, the velocity $\mathbf{v}_A$ of point $A$ and the angular velocity $\boldsymbol{\omega}$ of the slab define completely the

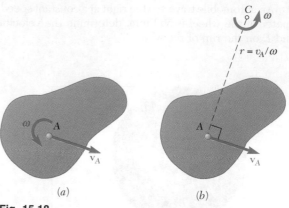

Fig. 15.18

velocities of all the other particles of the slab (Fig. 15.18$a$). Now let us assume that $\mathbf{v}_A$ and $\boldsymbol{\omega}$ are known and that they are both different from zero. (If $\mathbf{v}_A = 0$, point $A$ is itself the instantaneous center of rotation, and if $\boldsymbol{\omega} = 0$, all the particles have the same velocity $\mathbf{v}_A$.) These velocities could be obtained by letting the slab rotate with the angular velocity $\boldsymbol{\omega}$ about a point $C$ located on the perpendicular to $\mathbf{v}_A$ at a distance $r = v_A/\omega$ from $A$ as shown in Fig. 15.18$b$. We check that the velocity of $A$ would be perpendicular to $AC$ and that its magnitude would be $r\omega = (v_A/\omega)\omega = v_A$. Thus the velocities of all the other particles of the slab would be the same as originally defined. Therefore, *as far as the velocities are concerned, the slab seems to rotate about the instantaneous center $C$ at the instant considered.*

The position of the instantaneous center can be defined in two other ways. If the directions of the velocities of two particles $A$ and $B$ of the slab are known and if they are different, the instantaneous center $C$ is obtained by drawing the perpendicular to $\mathbf{v}_A$ through $A$ and the perpendicular to $\mathbf{v}_B$ through $B$ and determining the point in which these two lines intersect (Fig. 15.19$a$). If the velocities $\mathbf{v}_A$ and $\mathbf{v}_B$ of two particles $A$ and $B$ are perpendicular to the line $AB$ and if their magnitudes are known, the instantaneous center can be found by intersecting the line $AB$ with the line joining the extremities of the vectors $\mathbf{v}_A$ and $\mathbf{v}_B$ (Fig. 15.19$b$). Note that if $\mathbf{v}_A$ and $\mathbf{v}_B$ were parallel in Fig. 15.19$a$ or if $\mathbf{v}_A$ and $\mathbf{v}_B$ had the same magnitude in Fig. 15.19$b$,

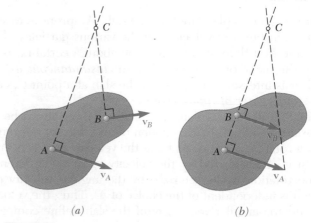

Fig. 15.19

the instantaneous center $C$ would be at an infinite distance and $\boldsymbol{\omega}$ would be zero; all points of the slab would have the same velocity.

To see how the concept of instantaneous center of rotation can be put to use, let us consider again the rod of Sec. 15.6. Drawing the perpendicular to $\mathbf{v}_A$ through $A$ and the perpendicular to $\mathbf{v}_B$ through $B$ (Fig. 15.20), we obtain the instantaneous center $C$. At the instant

**Fig. 15.20**

considered, the velocities of all the particles of the rod are thus the same as if the rod rotated about $C$. Now, if the magnitude $v_A$ of the velocity of $A$ is known, the magnitude $\omega$ of the angular velocity of the rod can be obtained by writing

$$\omega = \frac{v_A}{AC} = \frac{v_A}{l \cos \theta}$$

The magnitude of the velocity of $B$ can then be obtained by writing

$$v_B = (BC)\omega = l \sin \theta \frac{v_A}{l \cos \theta} = v_A \tan \theta$$

Note that only *absolute* velocities are involved in the computation.

The instantaneous center of a slab in plane motion can be located either on the slab or outside the slab. If it is located on the slab, the particle $C$ coinciding with the instantaneous center at a given instant $t$ must have zero velocity at that instant. However, it should be noted that the instantaneous center of rotation is valid only at a given instant. Thus, the particle $C$ of the slab which coincides with the instantaneous center at time $t$ will generally not coincide with the instantaneous center at time $t + \Delta t$; while its velocity is zero at time $t$, it will probably be different from zero at time $t + \Delta t$. This means that, in general, the particle $C$ *does not have zero acceleration* and, therefore, that the *accelerations* of the various particles of the slab *cannot* be determined as if the slab were rotating about $C$.

As the motion of the slab proceeds, the instantaneous center moves in space. But it was just pointed out that the position of the instantaneous center on the slab keeps changing. Thus, the instantaneous center describes one curve in space, called the *space centrode*, and another curve on the slab, called the *body centrode* (Fig. 15.21). It can be shown that at any instant, these two curves are tangent at $C$ and that as the slab moves, the body centrode appears to *roll* on the space centrode.

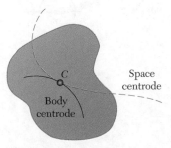

**Fig. 15.21**

## SAMPLE PROBLEM 15.4

Solve Sample Prob. 15.2, using the method of the instantaneous center of rotation.

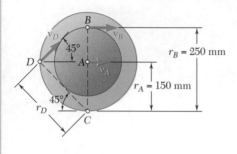

### SOLUTION

**a. Angular Velocity of the Gear.** Since the gear rolls on the stationary lower rack, the point of contact $C$ of the gear with the rack has no velocity; point $C$ is therefore the instantaneous center of rotation. We write

$$v_A = r_A \omega \qquad 1.2 \text{ m/s} = (0.150 \text{ m})\omega$$

$$\omega = 8 \text{ rad/s} \downarrow \quad \blacktriangleleft$$

**b. Velocities.** As far as velocities are concerned, all points of the gear seem to rotate about the instantaneous center.

*Velocity of Upper Rack.* Recalling that $v_R = v_B$, we write

$$v_R = v_B = r_B \omega \qquad v_R = (0.250 \text{ m})(8 \text{ rad/s}) = 2 \text{ m/s}$$

$$\mathbf{v}_R = 2 \text{ m/s} \rightarrow \quad \blacktriangleleft$$

*Velocity of Point D.* Since $r_D = (0.150 \text{ m})\sqrt{2} = 0.2121$ m, we write

$$v_D = r_D \omega \qquad v_D = (0.2121 \text{ m})(8 \text{ rad/s}) = 1.697 \text{ m/s}$$

$$\mathbf{v}_D = 1.697 \text{ m/s} \measuredangle 45° \quad \blacktriangleleft$$

## SAMPLE PROBLEM 15.5

Solve Sample Prob. 15.3, using the method of the instantaneous center of rotation.

### SOLUTION

**Motion of Crank AB.** Referring to Sample Prob. 15.3, we obtain the velocity of point $B$; $\mathbf{v}_B = 1591.44$ cm/s $\measuredangle 50°$.

**Motion of the Connecting Rod BD.** We first locate the instantaneous center $C$ by drawing lines perpendicular to the absolute velocities $\mathbf{v}_B$ and $\mathbf{v}_D$. Recalling from Sample Prob. 15.3 that $\beta = 13.95°$ and that $BD = 8$ in., we solve the triangle $BCD$.

$$\gamma_B = 40° + \beta = 53.92° \qquad \gamma_D = 90° - \beta = 76.06°$$

$$\frac{BC}{\sin 76.05°} = \frac{CD}{\sin 53.95°} = \frac{20.3 \text{ cm}}{\sin 50°}$$

$$BC = 25.7 \text{ cm} \qquad CD = 21.4 \text{ cm}.$$

Since the connecting rod $BD$ seems to rotate about point $C$, we write

$$v_B = (BC)\omega_{BD}$$
$$1591.44 \text{ cm/s} = (25.7 \text{ cm})\omega_{BD}$$
$$\omega_{BD} = 61.85 \text{ rad/s} \uparrow \quad \blacktriangleleft$$

$$v_D = (CD)\omega_{BD} = (21.4 \text{ cm})(61.9 \text{ rad/s})$$
$$= 13.25 \text{ m/s}$$

$$\mathbf{v}_P = \mathbf{v}_D = 13.24 \text{ m/s} \rightarrow \quad \blacktriangleleft$$

# SOLVING PROBLEMS
## ON YOUR OWN

In this lesson we introduced the *instantaneous center of rotation* in plane motion. This provides us with an alternative way for solving problems involving the *velocities* of the various points of a body in plane motion.

As its name suggests, the *instantaneous center of rotation* is the point about which you can assume a body is rotating at a given instant, as you determine the velocities of the points of the body at that instant.

**A. To determine the instantaneous center of rotation** of a body in plane motion, you should use one of the following procedures.

**1. If the velocity $\mathbf{v}_A$ of a point A and the angular velocity $\omega$ of the body are both known** (Fig. 15.18):

**a. Draw a sketch of the body,** showing point $A$, its velocity $\mathbf{v}_A$, and the angular velocity $\boldsymbol{\omega}$ of the body.

**b. From A draw a line perpendicular to $\mathbf{v}_A$** on the side of $\mathbf{v}_A$ from which this velocity is viewed as having *the same sense as $\boldsymbol{\omega}$.*

**c. Locate the instantaneous center C** on this line, at a distance $r = v_A/\omega$ from point $A$.

**2. If the directions of the velocities of two points A and B are known and are different** (Fig. 15.19*a*):

**a. Draw a sketch of the body,** showing points $A$ and $B$ and their velocities $\mathbf{v}_A$ and $\mathbf{v}_B$.

**b. From A and B draw lines perpendicular to $\mathbf{v}_A$ and $\mathbf{v}_B$, respectively.** The instantaneous center $C$ is located at the point where the two lines intersect.

**c. If the velocity of one of the two points is known,** you can determine the angular velocity of the body. For example, if you know $v_A$, you can write $\omega = v_A/AC$, where $AC$ is the distance from point $A$ to the instantaneous center $C$.

**3. If the velocities of two points A and B are known and are both perpendicular to the line AB** (Fig. 15.19*b*):

**a. Draw a sketch of the body,** showing points $A$ and $B$ with their velocities $\mathbf{v}_A$ and $\mathbf{v}_B$ *drawn to scale.*

**b. Draw a line through points A and B, and another line** through the tips of the vectors $\mathbf{v}_A$ and $\mathbf{v}_B$. The instantaneous center $C$ is located at the point where the two lines intersect.

(*continued*)

*c. The angular velocity of the body* is obtained by either dividing $\mathbf{v}_A$ by $AC$ or $\mathbf{v}_B$ by $BC$.

*d. If the velocities $\mathbf{v}_A$ and $\mathbf{v}_B$ have the same magnitude,* the two lines drawn in part $b$ do not intersect; the instantaneous center $C$ is at an infinite distance. The angular velocity $\boldsymbol{\omega}$ is zero and *the body is in translation.*

**B.   Once you have determined the instantaneous center and the angular velocity** of a body, you can determine the velocity $\mathbf{v}_P$ of any point $P$ of the body in the following way.

*1.   Draw a sketch of the body,* showing point $P$, the instantaneous center of rotation $C$, and the angular velocity $\boldsymbol{\omega}$.

*2.   Draw a line from $P$ to the instantaneous center $C$* and measure or calculate the distance from $P$ to $C$.

*3.   The velocity $\mathbf{v}_P$ is a vector perpendicular to the line $PC$,* of the same sense as $\boldsymbol{\omega}$, and of magnitude $v_P = (PC)\omega$.

*Finally, keep in mind* that the instantaneous center of rotation can be used *only* to determine velocities. *It cannot be used to determine accelerations.*

# Problems

**15.71**  A helicopter moves horizontally in the $x$ direction at a speed of 180 km/h. Knowing that the main blades rotate clockwise with an angular velocity of 180 rpm, determine the instantaneous axis of rotation of the main blades.

**Fig. P15.71**

**15.72**  A 5-m beam $AE$ is being lowered by means of two overhead cranes. At the instant shown it is known that the velocity of point $D$ is 1 m/s downward and the velocity of point $E$ is 1.5 m/s downward. Determine ($a$) the instantaneous center of rotation of the beam, ($b$) the velocity of point $A$.

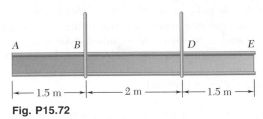

**Fig. P15.72**

**15.73**  The arm $ABC$ rotates with an angular velocity of 4 rad/s counterclockwise. Knowing that the angular velocity of the intermediate gear $B$ is 8 rad/s counterclockwise, determine ($a$) the instantaneous centers of rotation of gears $A$ and $C$, ($b$) the angular velocities of gears $A$ and $C$.

**Fig. P15.73**

**15.74**  A drum of radius 10.2 cm is mounted on a cylinder of radius 12.7 cm. A cord is wound around the drum, and its extremity $D$ is pulled to the left with a constant velocity of 15.24 cm/s, causing the cylinder to roll without sliding. Determine ($a$) the angular velocity of the cylinder, ($b$) the velocity of the center of the cylinder, ($c$) the length of cord which is wound or unwound per second.

**Fig. P15.74**

**15.75**  At the instant shown during deceleration, the velocity of an automobile is 12 m/s to the right. Knowing that the velocity of the contact point $A$ of the wheel with the ground is 1.5 m/s to the right, determine ($a$) the instantaneous center of rotation of the wheel, ($b$) the velocity of point $B$, ($c$) the velocity of point $D$.

**15.76**  At the instant shown during acceleration, the velocity of an automobile is 12 m/s to the right. Knowing that the velocity of the contact point $A$ of the wheel with the ground is 1.5 m/s to the left, determine ($a$) the instantaneous center of rotation of the wheel, ($b$) the velocity of point $B$, ($c$) the velocity of point $E$.

**Fig. P15.75 and P15.76**

25.4 cm/s

E

A

B

D

F

20.3 cm/s

**Fig. P15.78**

E

A

192 mm

B

240 mm

30°

360 mm

D

**Fig. P15.79**

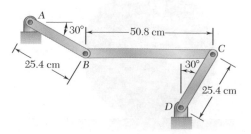

A

30°

50.8 cm

C

25.4 cm

B

30°

25.4 cm

D

**Fig. P15.81 and P15.82**

**15.77** A roll of paper rests on a horizontal surface as shown. End $D$ of the paper is then pulled to the right with a constant velocity of 20.3 cm/s and it is observed that the center $A$ of the roll moves to the right with a constant velocity of 7.6 cm/s. Determine ($a$) the instantaneous axis of rotation of the roll of paper, ($b$) the velocity of point $B$, ($c$) the number of inches of paper unwound per second.

7.6 cm

B

A

D

**Fig. P15.77**

**15.78** A double pulley is attached to a slider block by a pin at $A$. The 3.8 cm.-radius inner pulley is rigidly attached to the 7.6 cm-radius outer pulley. Knowing that each of the two cords is pulled at a constant speed as shown, determine ($a$) the instantaneous center of rotation of the double pulley, ($b$) the velocity of the slider block, ($c$) the number of inches of cord wrapped or unwrapped on each pulley per second.

**15.79** Knowing that at the instant shown the angular velocity of rod $BE$ is 4 rad/s counterclockwise, determine ($a$) the angular velocity of rod $AD$, ($b$) the velocity of collar $D$, ($c$) the velocity of point $A$.

**15.80** Knowing that at the instant shown the angular velocity of rod $AB$ is 15 rad/s clockwise, determine ($a$) the angular velocity of rod $BD$, ($b$) the velocity of the midpoint of rod $BD$.

A

200 mm

B

250 mm

D

E

200 mm

600 mm

**Fig. P15.80**

**15.81** Knowing that at the instant shown the angular velocity of bar $DC$ is 18 rad/s counterclockwise, determine ($a$) the angular velocity of bar $AB$, ($b$) the angular velocity of bar $BC$, ($c$) the velocity of the midpoint of bar $BC$.

**15.82** Knowing that at the instant shown bar $AB$ is rotating counterclockwise and that the magnitude of the velocity of the midpoint of bar $BC$ is 2.38 m/s, determine ($a$) the angular velocity of bar $AB$, ($b$) the angular velocity of bar $BC$, ($c$) the angular velocity of bar $DC$.

**15.83** Rod *ABD* is guided by wheels at *A* and *B* that roll in horizontal and vertical tracks. Knowing that at the instant shown $\beta = 60°$ and the velocity of wheel *B* is 800 mm/s downward, determine (*a*) the angular velocity of the rod, (*b*) the velocity of point *D*.

Fig. *P15.83*

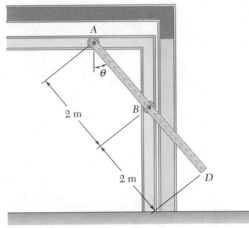

Fig. *P15.84*

**15.84** An overhead door is guided by wheels at *A* and *B* that roll in horizontal and vertical tracks. Knowing that when $\theta = 40°$ the velocity of wheel *B* is 0.6 m/s upward, determine (*a*) the angular velocity of the door, (*b*) the velocity of end *D* of the door.

**15.85** In the position shown, bar *DE* has a constant angular velocity of 15 rad/s clockwise. Determine (*a*) the distance *b* for which the velocity of point *F* is vertical, (*b*) the corresponding velocity of point *F*.

Fig. **P15.85**

**15.86** Collars *A* and *B* are pin-connected to bar *ABD* and can slide along fixed rods. Knowing that at the instant shown the velocity of *A* is 0.9 m/s to the right, determine (*a*) the angular velocity of *ABD*, (*b*) the velocity of point *D*.

Fig. **P15.86**

**Fig. P15.87**

**Fig. *P15.89***

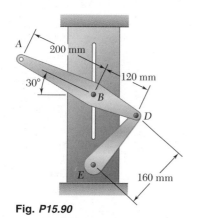

**Fig. *P15.90***

**15.87** Small wheels have been attached to the ends of bar *AB* and roll freely along the surfaces shown. Knowing that the velocity of wheel *B* is 2.286 m/s to the right at the instant shown, determine (*a*) the velocity of end *A* of the bar, (*b*) the angular velocity of the bar, (*c*) the velocity of the midpoint of the bar.

**15.88** At the instant shown, the angular velocity of bar *DE* is 8 rad/s counterclockwise. Determine (*a*) the angular velocity of bar *BD*, (*b*) the angular velocity of bar *AB*, (*c*) the velocity of the midpoint of bar *BD*.

**Fig. P15.88**

**15.89** Two identical rods *ABF* and *DBE* are connected by a pin at *B*. Knowing that at the instant shown the velocity of point *D* is 200 mm/s upward, determine the velocity of (*a*) point *E*, (*b*) point *F*.

**15.90** The pin at *B* is attached to member *ABD* and can slide freely along the slot cut in the fixed plate. Knowing that at the instant shown the angular velocity of arm *DE* is 3 rad/s clockwise, determine (*a*) the angular velocity of member *ABD*, (*b*) the velocity of point *A*.

**15.91** Two collars *C* and *D* move along the vertical rod shown. Knowing that the velocity of collar *C* is 83.8 cm/s downward, determine (*a*) the velocity of collar *D*, (*b*) the angular velocity of member *AB*.

**15.92** Two rods *AB* and *BD* are connected to three collars as shown. Knowing that collar *A* moves downward with a velocity of 15.2 cm/s, determine at the instant shown (*a*) the angular velocity of each rod, (*b*) the velocity of collar *D*.

**Fig. P15.92**

**Fig. P15.91**

**15.93**   At the instant shown, the velocity of collar $A$ is 0.4 m/s to the right and the velocity of collar $B$ is 1 m/s to the left. Determine (*a*) the angular velocity of bar $AD$, (*b*) the angular velocity of bar $BD$, (*c*) the velocity of point $D$.

**Fig. P15.93**

**Fig. P15.94**

**15.94**   At the instant shown, the angular velocity of bar $AB$ is 4.5 rad/s counterclockwise and the velocity of collar $E$ is 700 mm/s to the right. Determine (*a*) the angular velocity of bar $BD$, (*b*) the angular velocity of bar $DE$, (*c*) the velocity of point $D$.

**15.95**   Two rods $ABD$ and $DE$ are connected to three collars as shown. Knowing that the angular velocity of $ABD$ is 5 rad/s clockwise, determine at the instant shown (*a*) the angular velocity of $DE$, (*b*) the velocity of collar $E$.

**15.96**   Two 50.8 cm rods $AB$ and $DE$ are connected as shown. Point $D$ is the midpoint of rod $AB$, and at the instant shown rod $DE$ is horizontal. Knowing that the velocity of point $A$ is .3 m/s downward, determine (*a*) the angular velocity of rod $DE$, (*b*) the velocity of point $E$.

**Fig. P15.95**

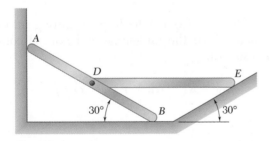

**Fig. P15.96**

**15.97**   Describe the space centrode and the body centrode of the gear of Sample Prob. 15.2 as the gear rolls on the stationary horizontal track.

**15.98**   Describe the space centrode and the body centrode of rod $ABD$ of Prob. 15.83. (*Hint:* The body centrode does not have to lie on a physical portion of the rod.)

**15.99**   Using the method of Sec. 15.7, solve Prob. 15.69.

**15.100**   Using the method of Sec. 15.7, solve Prob. 15.55.

**15.101**   Using the method of Sec. 15.7, solve Prob. 15.63.

**15.102**   Using the method of Sec. 15.7, solve Prob. 15.70.

## 15.8. ABSOLUTE AND RELATIVE ACCELERATION IN PLANE MOTION

We saw in Sec. 15.5 that any plane motion can be replaced by a translation defined by the motion of an arbitrary reference point $A$ and a simultaneous rotation about $A$. This property was used in Sec. 15.6 to determine the velocity of the various points of a moving slab. The same property will now be used to determine the acceleration of the points of the slab.

We first recall that the absolute acceleration $\mathbf{a}_B$ of a particle of the slab can be obtained from the relative-acceleration formula derived in Sec. 11.12,

$$\mathbf{a}_B = \mathbf{a}_A + \mathbf{a}_{B/A} \tag{15.21}$$

where the right-hand member represents a vector sum. The acceleration $\mathbf{a}_A$ corresponds to the translation of the slab with $A$, while the relative acceleration $\mathbf{a}_{B/A}$ is associated with the rotation of the slab about $A$ and is measured with respect to axes centered at $A$ and of fixed orientation. We recall from Sec. 15.3 that the relative acceleration $\mathbf{a}_{B/A}$ can be resolved into two components, a *tangential component* $(\mathbf{a}_{B/A})_t$ perpendicular to the line $AB$, and a *normal component* $(\mathbf{a}_{B/A})_n$ directed toward $A$ (Fig. 15.22). Denoting by $\mathbf{r}_{B/A}$ the position vector of $B$ relative to $A$ and, respectively, by $\omega\mathbf{k}$ and $\alpha\mathbf{k}$ the angular velocity and angular acceleration of the slab with respect to axes of fixed orientation, we have

$$(\mathbf{a}_{B/A})_t = \alpha\mathbf{k} \times \mathbf{r}_{B/A} \qquad (a_{B/A})_t = r\alpha$$
$$(\mathbf{a}_{B/A})_n = -\omega^2\mathbf{r}_{B/A} \qquad (a_{B/A})_n = r\omega^2 \tag{15.22}$$

where $r$ is the distance from $A$ to $B$. Substituting into (15.21) the expressions obtained for the tangential and normal components of $\mathbf{a}_{B/A}$, we can also write

$$\mathbf{a}_B = \mathbf{a}_A + \alpha\mathbf{k} \times \mathbf{r}_{B/A} - \omega^2\mathbf{r}_{B/A} \tag{15.21'}$$

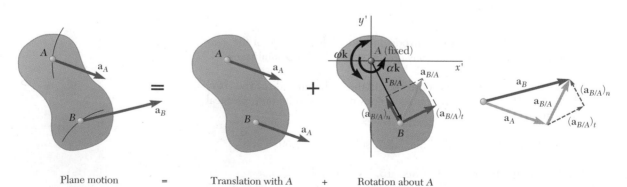

Plane motion   =   Translation with $A$   +   Rotation about $A$

**Fig. 15.22**

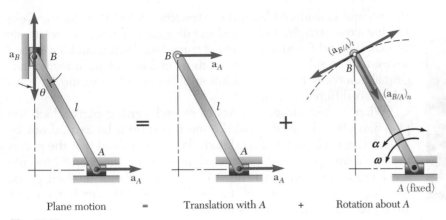

**Fig. 15.23**

As an example, let us again consider the rod $AB$ whose extremities slide, respectively, along a horizontal and a vertical track (Fig. 15.23). Assuming that the velocity $\mathbf{v}_A$ and the acceleration $\mathbf{a}_A$ of $A$ are known, we propose to determine the acceleration $\mathbf{a}_B$ of $B$ and the angular acceleration $\boldsymbol{\alpha}$ of the rod. Choosing $A$ as a reference point, we express that the given motion is equivalent to a translation with $A$ and a rotation about $A$. The absolute acceleration of $B$ must be equal to the sum

$$\mathbf{a}_B = \mathbf{a}_A + \mathbf{a}_{B/A}$$
$$= \mathbf{a}_A + (\mathbf{a}_{B/A})_n + (\mathbf{a}_{B/A})_t \qquad (15.23)$$

where $(\mathbf{a}_{B/A})_n$ has the magnitude $l\omega^2$ and is *directed toward A*, while $(\mathbf{a}_{B/A})_t$ has the magnitude $l\alpha$ and is perpendicular to $AB$. Students should note that there is no way to tell whether the tangential component $(\mathbf{a}_{B/A})_t$ is directed to the left or to the right, and therefore both possible directions for this component are indicated in Fig. 15.23. Similarly, both possible senses for $\mathbf{a}_B$ are indicated, since it is not known whether point $B$ is accelerated upward or downward.

Equation (15.23) has been expressed geometrically in Fig. 15.24. Four different vector polygons can be obtained, depending upon the sense of $\mathbf{a}_A$ and the relative magnitude of $a_A$ and $(a_{B/A})_n$. If we are to determine $a_B$ and $\alpha$ from one of these diagrams, we must know not only $a_A$ and $\theta$ but also $\omega$. The angular velocity of the rod should therefore be separately determined by one of the methods indicated in Secs. 15.6 and 15.7. The values of $a_B$ and $\alpha$ can then be obtained by considering successively the $x$ and $y$ components of the vectors shown in Fig. 15.24. In the case of polygon $a$, for example, we write

$\xrightarrow{+} x$ components:   $0 = a_A + l\omega^2 \sin\theta - l\alpha \cos\theta$

$+\uparrow y$ components:   $-a_B = -l\omega^2 \cos\theta - l\alpha \sin\theta$

and solve for $a_B$ and $\alpha$. The two unknowns can also be obtained by direct measurement on the vector polygon. In that case, care should be taken to draw first the known vectors $\mathbf{a}_A$ and $(\mathbf{a}_{B/A})_n$.

It is quite evident that the determination of accelerations is considerably more involved than the determination of velocities. Yet in

**Fig. 15.24**

**Photo 15.3** The central gear rotates about a fixed axis and is pin-connected to three bars which are in general plane motion.

the example considered here, the extremities $A$ and $B$ of the rod were moving along straight tracks, and the diagrams drawn were relatively simple. If $A$ and $B$ had moved along curved tracks, it would have been necessary to resolve the accelerations $\mathbf{a}_A$ and $\mathbf{a}_B$ into normal and tangential components and the solution of the problem would have involved six different vectors.

When a mechanism consists of several moving parts which are pin-connected, the analysis of the mechanism can be carried out by considering each part as a rigid body, keeping in mind that the points at which two parts are connected must have the same absolute acceleration (see Sample Prob. 15.7). In the case of meshed gears, the tangential components of the accelerations of the teeth in contact are equal, but their normal components are different.

## *15.9. ANALYSIS OF PLANE MOTION IN TERMS OF A PARAMETER

In the case of certain mechanisms, it is possible to express the coordinates $x$ and $y$ of all the significant points of the mechanism by means of simple analytic expressions containing a single parameter. It is sometimes advantageous in such a case to determine the absolute velocity and the absolute acceleration of the various points of the mechanism directly, since the components of the velocity and of the acceleration of a given point can be obtained by differentiating the coordinates $x$ and $y$ of that point.

Let us consider again the rod $AB$ whose extremities slide, respectively, in a horizontal and a vertical track (Fig. 15.25). The coordinates $x_A$ and $y_B$ of the extremities of the rod can be expressed in terms of the angle $\theta$ the rod forms with the vertical:

**Fig. 15.25**

$$x_A = l \sin \theta \qquad y_B = l \cos \theta \qquad (15.24)$$

Differentiating Eqs. (15.24) twice with respect to $t$, we write

$$v_A = \dot{x}_A = l\dot{\theta} \cos \theta$$
$$a_A = \ddot{x}_A = -l\dot{\theta}^2 \sin \theta + l\ddot{\theta} \cos \theta$$

$$v_B = \dot{y}_B = -l\dot{\theta} \sin \theta$$
$$a_B = \ddot{y}_B = -l\dot{\theta}^2 \cos \theta - l\ddot{\theta} \sin \theta$$

Recalling that $\dot{\theta} = \omega$ and $\ddot{\theta} = \alpha$, we obtain

$$v_A = l\omega \cos \theta \qquad\qquad v_B = -l\omega \sin \theta \qquad (15.25)$$

$$a_A = -l\omega^2 \sin \theta + l\alpha \cos \theta \qquad a_B = -l\omega^2 \cos \theta - l\alpha \sin \theta \qquad (15.26)$$

We note that a positive sign for $v_A$ or $a_A$ indicates that the velocity $\mathbf{v}_A$ or the acceleration $\mathbf{a}_A$ is directed to the right; a positive sign for $v_B$ or $a_B$ indicates that $\mathbf{v}_B$ or $\mathbf{a}_B$ is directed upward. Equations (15.25) can be used, for example to determine $v_B$ and $\omega$ when $v_A$ and $\theta$ are known. Substituting for $\omega$ in (15.26), we can then determine $a_B$ and $\alpha$ if $a_A$ is known.

## SAMPLE PROBLEM 15.6

The center of the double gear of Sample Prob. 15.2 has a velocity of 1.2 m/s to the right and an acceleration of 3 m/s² to the right. Recalling that the lower rack is stationary, determine (a) the angular acceleration of the gear, (b) the acceleration of points B, C, and D of the gear.

## SOLUTION

*a. Angular Acceleration of the Gear.* In Sample Prob. 15.2, we found that $x_A = -r_1\theta$ and $v_A = -r_1\omega$. Differentiating the latter with respect to time, we obtain $a_A = -r_1\alpha$.

$$v_A = -r_1\omega \qquad 1.2 \text{ m/s} = -(0.150 \text{ m})\omega \qquad \omega = -8 \text{ rad/s}$$
$$a_A = -r_1\alpha \qquad 3 \text{ m/s}^2 = -(0.150 \text{ m})\alpha \qquad \alpha = -20 \text{ rad/s}^2$$
$$\boldsymbol{\alpha} = \alpha\mathbf{k} = -(20 \text{ rad/s}^2)\mathbf{k} \quad \blacktriangleleft$$

*b. Accelerations.* The rolling motion of the gear is resolved into a translation with A and a rotation about A.

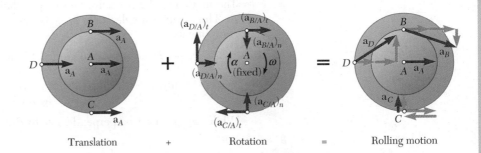

| Translation | + | Rotation | = | Rolling motion |

*Acceleration of Point B.* Adding vectorially the accelerations corresponding to the translation and to the rotation, we obtain

$$\mathbf{a}_B = \mathbf{a}_A + \mathbf{a}_{B/A} = \mathbf{a}_A + (\mathbf{a}_{B/A})_t + (\mathbf{a}_{B/A})_n$$
$$= \mathbf{a}_A + \alpha\mathbf{k} \times \mathbf{r}_{B/A} - \omega^2\mathbf{r}_{B/A}$$
$$= (3 \text{ m/s}^2)\mathbf{i} - (20 \text{ rad/s}^2)\mathbf{k} \times (0.100 \text{ m})\mathbf{j} - (8 \text{ rad/s})^2(0.100 \text{ m})\mathbf{j}$$
$$= (3 \text{ m/s}^2)\mathbf{i} + (2 \text{ m/s}^2)\mathbf{i} - (6.40 \text{ m/s}^2)\mathbf{j}$$
$$\mathbf{a}_B = 8.12 \text{ m/s}^2 \; \diagdown 52.0° \quad \blacktriangleleft$$

*Acceleration of Point C*

$$\mathbf{a}_C = \mathbf{a}_A + \mathbf{a}_{C/A} = \mathbf{a}_A + \alpha\mathbf{k} \times \mathbf{r}_{C/A} - \omega^2\mathbf{r}_{C/A}$$
$$= (3 \text{ m/s}^2)\mathbf{i} - (20 \text{ rad/s}^2)\mathbf{k} \times (-0.150 \text{ m})\mathbf{j} - (8 \text{ rad/s})^2(-0.150 \text{ m})\mathbf{j}$$
$$= (3 \text{ m/s}^2)\mathbf{i} - (3 \text{ m/s}^2)\mathbf{i} + (9.60 \text{ m/s}^2)\mathbf{j}$$
$$\mathbf{a}_C = 9.60 \text{ m/s}^2 \uparrow \quad \blacktriangleleft$$

*Acceleration of Point D*

$$\mathbf{a}_D = \mathbf{a}_A + \mathbf{a}_{D/A} = \mathbf{a}_A + \alpha\mathbf{k} \times \mathbf{r}_{D/A} - \omega^2\mathbf{r}_{D/A}$$
$$= (3 \text{ m/s}^2)\mathbf{i} - (20 \text{ rad/s}^2)\mathbf{k} \times (-0.150 \text{ m})\mathbf{i} - (8 \text{ rad/s})^2(-0.150 \text{ m})\mathbf{i}$$
$$= (3 \text{ m/s}^2)\mathbf{i} + (3 \text{ m/s}^2)\mathbf{j} + (9.60 \text{ m/s}^2)\mathbf{i}$$
$$\mathbf{a}_D = 12.95 \text{ m/s}^2 \; \measuredangle 13.4° \quad \blacktriangleleft$$

## SAMPLE PROBLEM 15.7

Crank *AB* of the engine system of Sample Prob. 15.3 has a constant clockwise angular velocity of 2000 rpm. For the crank position shown, determine the angular acceleration of the connecting rod *BD* and the acceleration of point *D*.

## SOLUTION

**Motion of Crank *AB*.** Since the crank rotates about *A* with constant $\omega_{AB} = 2000$ rpm $= \omega_{AB} = 209.3$ rads$^{-1}$, we have $\alpha_{AB} = 0$. The acceleration of *B* is therefore directed toward *A* and has a magnitude

$$a_B = r\omega_{AB}^2 = (7.6 \text{ cm})(209.3 \text{ rad/s})^2 = 3329.3 \text{ m/s}^2$$
$$\mathbf{a}_B = 3329.3 \text{ m/s}^2 \ \rlap{/}{\nearrow} \ 40°$$

**Motion of the Connecting Rod *BD*.** The angular velocity $\boldsymbol{\omega}_{BD}$ and the value of $\beta$ were obtained in Sample Prob. 15.3:

$$\boldsymbol{\omega}_{BD} = 61.83 \text{ rad/s} \uparrow \qquad \beta = 13.92°$$

The motion of *BD* is resolved into a translation with *B* and a rotation about *B*. The relative acceleration $\mathbf{a}_{D/B}$ is resolved into normal and tangential components:

$$(a_{D/B})_n = (BD)\omega_{BD}^2 = (20.3 \text{ cm})(61.83 \text{ rad/s})^2 = 776.06 \text{ m/s}^2$$
$$(\mathbf{a}_{D/B})_n = 776.06 \text{ m/s}^2 \ \searrow \ 13.92°$$
$$(a_{D/B})_t = (BD)\alpha_{BD} = (20.3)\alpha_{BD} = 20.3\alpha_{BD}$$
$$(\mathbf{a}_{D/B})_t = 20.3\alpha_{BD} \ \measuredangle \ 76.05°$$

While $(\mathbf{a}_{D/B})_t$ must be perpendicular to *BD*, its sense is not known.

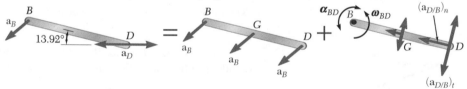

| Plane motion | = | Translation | + | Rotation |

Noting that the acceleration $\mathbf{a}_D$ must be horizontal, we write

$$\mathbf{a}_D = \mathbf{a}_B + \mathbf{a}_{D/B} = \mathbf{a}_B + (\mathbf{a}_{D/B})_n + (\mathbf{a}_{D/B})_t$$
$$[a_D \leftrightarrow] = [3329.3 \ \rlap{/}{\nearrow} \ 40°] + [776.06 \ \searrow \ 13.92°] + [20.3\alpha_{BD} \ \measuredangle \ 76.05°]$$

Equating *x* and *y* components, we obtain the following scalar equations:

$\overset{+}{\rightarrow}x$ components:
$$-a_D = -3329.3 \cos 40° - 776.06 \cos 13.92° + 20.3\alpha_{BD} \sin 13.92°$$
$+\uparrow y$ components:
$$0 = -3329.3 \sin 40° + 776.06 \sin 13.92° + 20.3\alpha_{BD} \cos 13.92°$$

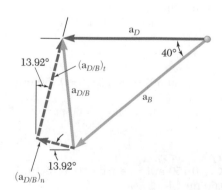

Solving the equations simultaneously, we obtain $\alpha_{BD} = +120$ rad/s$^2$ and $a_D = +3889.68$ rad/s$^2$. The positive signs indicate that the senses shown on the vector polygon are correct; we write

$$\boldsymbol{\alpha}_{BD} = 120 \text{ rad/s}^2 \uparrow \ \blacktriangleleft$$
$$\mathbf{a}_D = 3889.68 \text{ rad/s}^2 \leftarrow \ \blacktriangleleft$$

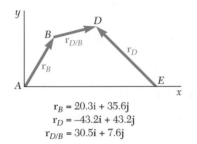

$r_B = 20.3i + 35.6j$
$r_D = -43.2i + 43.2j$
$r_{D/B} = 30.5i + 7.6j$

## SAMPLE PROBLEM 15.8

The linkage *ABDE* moves in the vertical plane. Knowing that in the position shown crank *AB* has a constant angular velocity $\omega_1$ of 20 rad/s counterclockwise, determine the angular velocities and angular accelerations of the connecting rod *BD* and of the crank *DE*.

## SOLUTION

This problem could be solved by the method used in Sample Prob. 15.7. In this case, however, the vector approach will be used. The position vectors $\mathbf{r}_B$, $\mathbf{r}_D$, and $\mathbf{r}_{D/B}$ are chosen as shown in the sketch.

**Velocities.** Since the motion of each element of the linkage is contained in the plane of the figure, we have

$$\boldsymbol{\omega}_{AB} = \omega_{AB}\mathbf{k} = (20 \text{ rad/s})\mathbf{k} \qquad \boldsymbol{\omega}_{BD} = \omega_{BD}\mathbf{k} \qquad \boldsymbol{\omega}_{DE} = \omega_{DE}\mathbf{k}$$

where **k** is a unit vector pointing out of the paper. We now write

$$\mathbf{v}_D = \mathbf{v}_B + \mathbf{v}_{D/B}$$
$$\omega_{DE}\mathbf{k} \times \mathbf{r}_D = \omega_{AB}\mathbf{k} \times \mathbf{r}_B + \omega_{BD}\mathbf{k} \times \mathbf{r}_{D/B}$$
$$\omega_{DE}\mathbf{k} \times (-43.2\mathbf{i} + 43.2\mathbf{j}) = 20\mathbf{k} \times (20.3\mathbf{i} + 35.8\mathbf{j}) + \omega_{BD}\mathbf{k} \times (30.5\mathbf{i} + 7.6\mathbf{j})$$
$$-43.2\omega_{DE}\mathbf{j} - 43.2\omega_{DE}\mathbf{i} = 406\mathbf{j} - 712\mathbf{i} + 30.5\omega_{BD}\mathbf{j} - 7.6\omega_{BD}\mathbf{i}$$

Equating the coefficients of the unit vectors **i** and **j**, we obtain the following two scalar equations:

$$-43.2\omega_{DE} = -712 - 7.6\omega_{BD}$$
$$-43.2\omega_{DE} = +406 + 30.5\omega_{BD}$$
$$\boldsymbol{\omega}_{BD} = -(29.34 \text{ rad/s})\mathbf{k} \qquad \boldsymbol{\omega}_{DE} = (21.64 \text{ rad/s})\mathbf{k} \quad \blacktriangleleft$$

**Accelerations.** Noting that at the instant considered crank *AB* has a constant angular velocity, we write

$$\boldsymbol{\alpha}_{AB} = 0 \qquad \boldsymbol{\alpha}_{BD} = \alpha_{BD}\mathbf{k} \qquad \boldsymbol{\alpha}_{DE} = \alpha_{DE}\mathbf{k} \tag{1}$$
$$\mathbf{a}_D = \mathbf{a}_B + \mathbf{a}_{D/B}$$

Each term of Eq. (1) is evaluated separately:

$$\mathbf{a}_D = \alpha_{DE}\mathbf{k} \times \mathbf{r}_D - \omega_{DE}^2\mathbf{r}_D$$
$$= \alpha_{DE}\mathbf{k} \times (-43.2\mathbf{i} + 43.2\mathbf{j}) - (21.64)^2(-43.2\mathbf{i} + 43.2\mathbf{j})$$
$$= -43.2\alpha_{DE}\mathbf{j} - 43.2\alpha_{DE}\mathbf{i} + 20230.1\mathbf{i} - 20230.1\mathbf{j}$$
$$\mathbf{a}_B = \alpha_{AB}\mathbf{k} \times \mathbf{r}_B - \omega_{AB}^2\mathbf{r}_B = 0 - (20)^2(20.3\mathbf{i} + 35.6\mathbf{j})$$
$$= -8120\mathbf{i} - 14240\mathbf{j}$$
$$\mathbf{a}_{D/B} = \alpha_{BD}\mathbf{k} \times \mathbf{r}_{D/B} - \omega_{BD}^2\mathbf{r}_{D/B}$$
$$= \alpha_{BD}\mathbf{k} \times (30.5\mathbf{i} + 7.6\mathbf{j}) - (29.34)^2(30.5\mathbf{i} + 7.6\mathbf{j})$$
$$= 30.5\alpha_{BD}\mathbf{j} - 7.6\alpha_{BD}\mathbf{i} - 26255.5\mathbf{i} - 6542.35\mathbf{j}$$

Substituting into Eq. (1) and equating the coefficients of **i** and **j**, we obtain

$$-43.2\alpha_{DE} + 7.6\alpha_{BD} = -14145.5$$
$$-43.2\alpha_{DE} - 30.5\alpha_{BD} = -41012.45$$
$$\boldsymbol{\alpha}_{BD} = +(705.2 \text{ rad/s}^2)\mathbf{k} \qquad \boldsymbol{\alpha}_{DE} = (451.5 \text{ rad/s}^2)\mathbf{k} \quad \blacktriangleleft$$

This lesson was devoted to the determination of the *accelerations* of the points of a *rigid body in plane motion.* As you did previously for velocities, you will again consider the plane motion of a rigid body as the sum of two motions, namely, *a translation and a rotation.*

To solve a problem involving accelerations in plane motion you should use the following steps:

1. **Determine the angular velocity of the body.** To find $\omega$ you can either

   *a.* Consider the motion of the body as the sum of a translation and a rotation as you did in See. 15.6, or

   *b.* Use the instantaneous center of rotation of the body as you did in Sec. 15.7. However, *keep in mind that you cannot use the instantaneous center to determine accelerations.*

2. **Start drawing a "diagram equation"** to use in your solution. This "equation" will involve the following diagrams (Fig. 15.44).

   *a. Plane motion diagram.* Draw a sketch of the body including all dimensions, as well as the angular velocity $\omega$. Show the angular acceleration $\alpha$ with its magnitude and sense if you know them. Also show those points for which you know or seek the accelerations, indicating all that you know about these accelerations.

   *b. Translation diagram.* Select a reference point $A$ for which you know the direction, the magnitude, or a component of the acceleration $\mathbf{a}_A$. Draw a second diagram showing the body in translation with each point having the same acceleration as point $A$.

   *c. Rotation diagram.* Considering point $A$ as a fixed reference point, draw a third diagram showing the body in rotation about $A$. Indicate the normal and tangential components of the relative accelerations of other points, such as the components $(\mathbf{a}_{B/A})_n$ and $(\mathbf{a}_{B/A})_t$ of the acceleration of point $B$ with respect to point $A$.

3. **Write the relative-acceleration formula**

$$\mathbf{a}_B = \mathbf{a}_A + \mathbf{a}_{B/A} \qquad \text{or} \qquad \mathbf{a}_B = \mathbf{a}_A + (\mathbf{a}_{B/A})_n + (\mathbf{a}_{B/A})_t$$

The sample problems illustrate three different ways to use this vector equation:

   *a. If $\alpha$ is given or can easily be determined,* you can use this equation to determine the accelerations of various points of the body [Sample Prob. 15.6].

***b. If α cannot easily be determined,*** select for point *B* a point for which you know the direction, the magnitude, or a component of the acceleration $\mathbf{a}_B$ and draw a vector diagram of the equation. Starting at the same point, draw all known acceleration components in tip-to-tail fashion for each member of the equation. Complete the diagram by drawing the two remaining vectors in appropriate directions and in such a way that the two sums of vectors end at a common point.

The magnitudes of the two remaining vectors can be found either graphically or analytically. Usually an analytic solution will require the solution of two simultaneous equations [Sample Prob. 15.7]. However, by first considering the components of the various vectors in a direction perpendicular to one of the unknown vectors, you may be able to obtain an equation in a single unknown.

One of the two vectors obtained by the method just described will be $(\mathbf{a}_{B/A})_t$, from which you can compute α. Once α has been found, the vector equation can be used to determine the acceleration of any other point of the body.

***c. A full vector approach*** can also be used to solve the vector equation. This is illustrated in Sample Prob. 15.8.

***4. The analysis of plane motion in terms of a parameter*** completed this lesson. This method should be used *only if it is possible* to express the coordinates $x$ and $y$ of all significant points of the body in terms of a single parameter (Sec. 15.9). By differentiating twice with respect to $t$ the coordinates $x$ and $y$ of a given point, you can determine the rectangular components of the absolute velocity and absolute acceleration of that point.

**15.103 and 15.104** Bar *BDE* is attached to two links *AB* and *CD*. Knowing that at the instant shown link *AB* rotates with a constant angular velocity of 3 rad/s clockwise, determine the acceleration (*a*) of point *D*, (*b*) of point *E*.

**Fig. P15.103**

**Fig. P15.104**

**15.105** Collars *B* and *D* are pin-connected to bar *ABD* and can slide along fixed rods. At the instant shown the angular velocity of bar *ABD* is zero and the acceleration of point *D* is 7.2 m/s² to the right. Determine (*a*) the angular acceleration of the bar, (*b*) the acceleration of point *B*, (*c*) the acceleration of point *A*.

**Fig. P15.105 and P15.106**

**15.106** Collars *B* and *D* are pin-connected to bar *ABD* and can slide along fixed rods. At the instant shown the angular velocity of bar *ABD* is zero and its angular acceleration is 12 rad/s² clockwise. Determine (*a*) the acceleration of point *D*, (*b*) the acceleration of point *B*, (*c*) the acceleration of point *A*.

**15.107** The acceleration of point *C* is .27 m/s² downward and the angular acceleration of the beam is 0.8 rad/s² clockwise. Knowing that the angular velocity of the beam is zero at the instant considered, determine the acceleration of each cable.

**Fig. P15.107**

**15.108** Knowing that at the instant shown the velocity of collar $A$ is zero and its acceleration is .24 m/s$^2$ to the left, determine ($a$) the angular acceleration of rod $ADB$, ($b$) the acceleration of point $B$.

**15.109** A heavy crate is being moved a short distance using three identical cylinders as rollers. Knowing that at the instant shown the crate has a velocity of 200 mm/s and an acceleration of 400 mm/s$^2$, both directed to the right, determine ($a$) the angular acceleration of the center cylinder, ($b$) the acceleration of point $A$ on the center cylinder.

**Fig. *P15.108***

**Fig. P15.109**

**15.110** At the instant shown the angular velocity of the wheel is 2 rad/s clockwise and its angular acceleration is 3 rad/s$^2$ counterclockwise. Knowing that the wheel rolls without slipping, determine the location of the point on the wheel with zero acceleration at this instant.

**15.111** A wheel rolls without slipping on a fixed cylinder. Knowing that at the instant shown the angular velocity of the wheel is 10 rad/s clockwise and its angular acceleration is 30 rad/s$^2$ counterclockwise, determine the acceleration of ($a$) point $A$, ($b$) point $B$, ($c$) point $C$.

**Fig. P15.110**

**Fig. P15.111**

**15.112** The 360-mm-radius flywheel is rigidly attached to a 30-mm-radius shaft that can roll along parallel rails. Knowing that at the instant shown the center of the shaft has a velocity of 24 mm/s and an acceleration of 10 mm/s$^2$, both directed down to the left, determine the acceleration ($a$) of point $A$, ($b$) of point $B$.

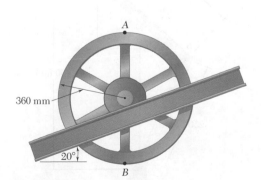

**Fig. P15.112**

**15.113 and 15.114**   A 60-mm-radius drum is rigidly attached to a 100-mm-radius drum as shown. One of the drums rolls without sliding on the surface shown, and a cord is wound around the other drum. Knowing that at the instant shown end $D$ of the cord has a velocity of 160 mm/s and an acceleration of 600 mm/s$^2$, both directed to the left, determine the accelerations of points $A$, $B$, and $C$ of the drums.

**Fig. P15.113**

**Fig. P15.114**

**15.115**   The motion of the 7.6-cm-radius cylinder is controlled by the cord shown. Knowing that end $E$ of the cord has a velocity of 30.5 cm/s and an acceleration of 48.8 cm/s$^2$, both directed upward, determine the accelerations of points $C$ and $D$ of the cylinder.

**15.116**   The 12.7-cm-radius drum rolls without slipping on a portion of a belt which moves downward to the left with a constant velocity of 15.2 cm/s. Knowing that at a given instant the velocity and acceleration of the center $A$ of the drum are as shown, determine the acceleration of point $D$.

**Fig. P15.115**

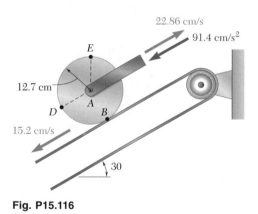

**Fig. P15.116**

**15.117**   Collar $D$ slides on a fixed horizontal rod with a constant velocity of 1.2 m/s to the right. Knowing that at the instant shown $x = 200$ mm, determine (a) the angular acceleration of bar $BD$, (b) the angular acceleration of bar $AB$.

**15.118**   Collar $D$ slides on a fixed horizontal rod with a constant velocity of 0.6 m/s to the right. Knowing that at the instant shown $x = 0$, determine (a) the angular acceleration of bar $BD$, (b) the angular acceleration of bar $AB$.

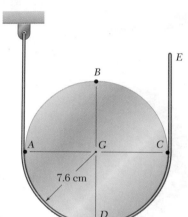

**Fig. P15.117 and P15.118**

**15.119**   Knowing that crank $AB$ rotates about point $A$ with a constant angular velocity of 900 rpm clockwise, determine the acceleration of the piston $P$ when $\theta = 60°$.

**15.120**   Knowing that crank $AB$ rotates about point $A$ with a constant angular velocity of 900 rpm clockwise, determine the acceleration of the piston $P$ when $\theta = 120°$.

**15.121**   Arm $AB$ has a constant angular velocity of 16 rad/s counterclockwise. At the instant when $\theta = 0$, determine the acceleration ($a$) of collar $D$, ($b$) of the midpoint $G$ of bar $BD$.

**Fig. P15.119 and P15.120**

**Fig. P15.121 and P15.122**

**15.122**   Arm $AB$ has a constant angular velocity of 16 rad/s counterclockwise. At the instant when $\theta = 90°$, determine the acceleration ($a$) of collar $D$, ($b$) of the midpoint $G$ of bar $BD$.

**15.123**   Collar $D$ slides on a fixed vertical rod. Knowing that the disk has a constant angular velocity of 15 rad/s clockwise, determine the angular acceleration of bar $BD$ and the acceleration of collar $D$ when ($a$) $\theta = 0$, ($b$) $\theta = 90°$, ($c$) $\theta = 180°$.

**15.124**   In the two-cylinder air compressor shown the connecting rods $BD$ and $BE$ are each 19.1 cm long and crank $AB$ rotates about the fixed point $A$ with a constant angular velocity of 1500 rpm clockwise. Determine the acceleration of each piston when $\theta = 0$.

**Fig. *P15.123***

**Fig. *P15.124***

**Fig. P15.125**

**Fig. P15.127 and P15.128**

**Fig. P15.130**

**15.125** Knowing that at the instant shown bar $AB$ has a constant angular velocity of 5 rad/s counterclockwise, determine (a) the angular acceleration of bar $BD$, (b) the angular acceleration of bar $DE$.

**15.126** Knowing that at the instant shown bar $AB$ has a constant angular velocity of 4 rad/s clockwise, determine the angular acceleration (a) of bar $BD$, (b) of bar $DE$.

**Fig. P15.126**

**15.127** Knowing that at the instant shown bar $AB$ has a constant angular velocity of 19 rad/s clockwise, determine (a) the angular acceleration of bar $BGD$, (b) the angular acceleration of bar $DE$.

**15.128** Knowing that at the instant shown bar $DE$ has a constant angular velocity of 18 rad/s clockwise, determine (a) the acceleration of point $B$, (b) the acceleration of point $G$.

**15.129** The cross $BHDF$ is supported by two links $AB$ and $DE$. Knowing that at the instant shown link $AB$ rotates with a constant angular velocity of 4 rad/s clockwise, determine (a) the angular velocity of the cross, (b) the angular acceleration of the cross, (c) the acceleration of point $H$.

**Fig. P15.129**

**15.130** At the instant shown the angular velocity of bar $DE$ is 4 rad/s clockwise and its angular acceleration is 10 rad/s² counterclockwise. Determine (a) the angular acceleration of bar $BGD$, (b) the acceleration of point $G$.

**15.131**   For the linkage of Prob. 15.68, determine the angular acceleration of (*a*) bar *BD*, (*b*) of bar *DE*.

**15.132**   Knowing that at the instant shown rod *AB* has a constant angular velocity of 15 rad/s counterclockwise, determine (*a*) the angular acceleration of rod *DE*, (*b*) the acceleration of point *D*.

10.2 cm   12.7 cm   12.7 cm   10.2 cm

*A*   *B*   *G*   *D*   *E*   7.6 cm

**Fig. P15.132**

**15.133 and** **15.134**   Solve the problem indicated making full use of the vector approach as done in Sample Prob. 15.8.

**15.133**   Prob. 15.126.
**15.134**   Prob. 15.132.

**15.135**   Denoting by $\mathbf{r}_A$ the position vector of point *A* of a rigid slab that is in plane motion, show that (*a*) the position vector $\mathbf{r}_C$ of the instantaneous center of rotation is

$$\mathbf{r}_C = \mathbf{r}_A + \frac{\boldsymbol{\omega} \times \mathbf{v}_A}{\omega^2}$$

where $\boldsymbol{\omega}$ is the angular velocity of the slab and $\mathbf{v}_A$ is the velocity of point *A*, (*b*) the acceleration of the instantaneous center of rotation is zero if, and only if,

$$\mathbf{a}_A = \frac{\alpha}{\omega}\mathbf{v}_A + \boldsymbol{\omega} \times \mathbf{v}_A$$

where $\boldsymbol{\alpha} = \alpha\mathbf{k}$ is the angular acceleration of the slab.

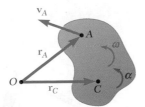

**Fig. P15.135**

**\*15.136**   Crank *AB* rotates counterclockwise with a constant angular velocity $\omega$. Using the method of Sec. 15.9, derive an expression for the angular velocity of rod *DE* in terms of $\theta$ and $\omega$.

**Fig. P15.136**

**\*15.137** Collar $D$ slides on a fixed vertical rod. The disk rotates with a constant clockwise angular velocity $\boldsymbol{\omega}$. Using the method of Sec. 15.9, derive an expression for the velocity of point $D$ in terms of $\theta$, $\omega$, $b$, and $l$.

**Fig. P15.137 and P15.138**

**\*15.138** Collar $D$ slides on a fixed vertical rod. The disk rotates with a constant clockwise angular velocity $\boldsymbol{\omega}$. Using the method of Sec. 15.9, derive an expression for the angular acceleration of rod $BD$ in terms of $\theta$, $\omega$, $b$, and $l$.

**\*15.139** A disk of radius $r$ rolls to the right with a constant velocity $\mathbf{v}$. Denoting by $P$ the point of the rim in contact with the ground at $t = 0$, derive expressions for the horizontal and vertical components of the velocity of $P$ at any time $t$.

**\*15.140** Rod $AB$ moves over a small wheel at $C$ while end $A$ moves to the right with a constant velocity $\mathbf{v}_A$. Using the method of Sec. 15.9, derive expressions for the angular velocity and angular acceleration of the rod.

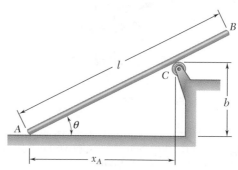

**Fig. P15.140 and P15.141**

**\*15.141** Rod $AB$ moves over a small wheel at $C$ while end $A$ moves to the right with a constant velocity $\mathbf{v}_A$. Using the method of Sec. 15.9, derive expressions for the horizontal and vertical components of the velocity of point $B$.

**\*15.142** Pin $C$ is attached to rod $CD$ and slides in a slot cut in arm $AB$. Knowing that rod $CD$ moves vertically upward with a constant velocity $\mathbf{v}_0$, derive an expression for the angular acceleration of arm $AB$.

**Fig. P15.142**

*15.143  The position of rod $AB$ is controlled by a disk of radius $r$ which is attached to yoke $CD$. Knowing that the yoke moves vertically upward with a constant velocity $\mathbf{v}_0$, derive an expression for the angular acceleration of rod $AB$.

*15.144  Crank $AB$ rotates with a constant clockwise angular velocity $\boldsymbol{\omega}$. Using the method of Sec. 15.9, derive expressions for the angular velocity of rod $BD$ and the velocity of the point on the rod coinciding with point $E$ in terms of $\theta$, $\omega$, $b$, and $l$.

Fig. *P15.143*

Fig. P15.144 and P15.145

*15.145  Crank $AB$ rotates with a constant clockwise angular velocity $\boldsymbol{\omega}$. Using the method of Sec. 15.9, derive an expression for the angular acceleration of rod $BD$ in terms of $\theta$, $\omega$, $b$, and $l$.

*15.146  A wheel of radius $r$ rolls without slipping along the inside of a fixed cylinder of radius $R$ with a constant angular velocity $\boldsymbol{\omega}$. Denoting by $P$ the point of the wheel in contact with the cylinder at $t = 0$, derive expressions for the horizontal and vertical components of the velocity of $P$ at any time $t$. (The curve described by point $P$ is a *hypocycloid*.)

*15.147  In Prob. 15.146, show that the path of $P$ is a vertical straight line when $r = R/2$. Derive expressions for the corresponding velocity and acceleration of $P$ at any time $t$.

Fig. P15.146

## 15.10. RATE OF CHANGE OF A VECTOR WITH RESPECT TO A ROTATING FRAME

We saw in Sec. 11.10 that the rate of change of a vector is the same with respect to a fixed frame and with respect to a frame in translation. In this section, the rates of change of a vector $\mathbf{Q}$ with respect to a fixed frame and with respect to a rotating frame of reference will be considered.† You will learn to determine the rate of change of $\mathbf{Q}$ with respect to one frame of reference when $\mathbf{Q}$ is defined by its components in another frame.

---

†It is recalled that the selection of a fixed frame of reference is arbitrary. Any frame may be designated as "fixed"; all others will then be considered as moving.

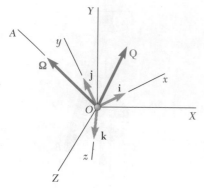

**Fig. 15.26**

Consider two frames of reference centered at $O$, a fixed frame $OXYZ$ and a frame $Oxyz$ which rotates about the fixed axis $OA$; let $\boldsymbol{\Omega}$ denote the angular velocity of the frame $Oxyz$ at a given instant (Fig. 15.26). Consider now a vector function $\mathbf{Q}(t)$ represented by the vector $\mathbf{Q}$ attached at $O$; as the time $t$ varies, both the direction and the magnitude of $\mathbf{Q}$ change. Since the variation of $\mathbf{Q}$ is viewed differently by an observer using $OXYZ$ as a frame of reference and by an observer using $Oxyz$, we should expect the rate of change of $\mathbf{Q}$ to depend upon the frame of reference which has been selected. Therefore, the rate of change of $\mathbf{Q}$ with respect to the fixed frame $OXYZ$ will be denoted by $(\dot{\mathbf{Q}})_{OXYZ}$, and the rate of change of $\mathbf{Q}$ with respect to the rotating frame $Oxyz$ will be denoted by $(\dot{\mathbf{Q}})_{Oxyz}$. We propose to determine the relation existing between these two rates of change.

Let us first resolve the vector $\mathbf{Q}$ into components along the $x$, $y$, and $z$ axes of the rotating frame. Denoting by $\mathbf{i}$, $\mathbf{j}$, and $\mathbf{k}$ the corresponding unit vectors, we write

$$\mathbf{Q} = Q_x\mathbf{i} + Q_y\mathbf{j} + Q_z\mathbf{k} \tag{15.27}$$

Differentiating (15.27) with respect to $t$ and considering the unit vectors $\mathbf{i}$, $\mathbf{j}$, $\mathbf{k}$ as fixed, we obtain the rate of change of $\mathbf{Q}$ *with respect to the rotating frame Oxyz:*

$$(\dot{\mathbf{Q}})_{Oxyz} = \dot{Q}_x\mathbf{i} + \dot{Q}_y\mathbf{j} + \dot{Q}_z\mathbf{k} \tag{15.28}$$

To obtain the rate of change of $\mathbf{Q}$ *with respect to the fixed frame OXYZ*, we must consider the unit vectors $\mathbf{i}$, $\mathbf{j}$, $\mathbf{k}$ as variable when differentiating (15.27). We therefore write

$$(\dot{\mathbf{Q}})_{OXYZ} = \dot{Q}_x\mathbf{i} + \dot{Q}_y\mathbf{j} + \dot{Q}_z\mathbf{k} + Q_x\frac{d\mathbf{i}}{dt} + Q_y\frac{d\mathbf{j}}{dt} + Q_z\frac{d\mathbf{k}}{dt} \tag{15.29}$$

Recalling (15.28), we observe that the sum of the first three terms in the right-hand member of (15.29) represents the rate of change $(\dot{\mathbf{Q}})_{Oxyz}$. We note, on the other hand, that the rate of change $(\dot{\mathbf{Q}})_{OXYZ}$ would reduce to the last three terms in (15.29) if the vector $\mathbf{Q}$ were fixed within the frame $Oxyz$, since $(\dot{\mathbf{Q}})_{Oxyz}$ would then be zero. But in that case, $(\dot{\mathbf{Q}})_{OXYZ}$ would represent the velocity of a particle located at the tip of $\mathbf{Q}$ and belonging to a body rigidly attached to the frame $Oxyz$. Thus, the last three terms in (15.29) represent the velocity of that particle; since the frame $Oxyz$ has an angular velocity $\boldsymbol{\Omega}$ with respect to $OXYZ$ at the instant considered, we write, by (15.5),

$$Q_x\frac{d\mathbf{i}}{dt} + Q_y\frac{d\mathbf{j}}{dt} + Q_z\frac{d\mathbf{k}}{dt} = \boldsymbol{\Omega} \times \mathbf{Q} \tag{15.30}$$

Substituting from (15.28) and (15.30) into (15.29), we obtain the fundamental relation

$$(\dot{\mathbf{Q}})_{OXYZ} = (\dot{\mathbf{Q}})_{Oxyz} + \boldsymbol{\Omega} \times \mathbf{Q} \tag{15.31}$$

We conclude that the rate of change of the vector $\mathbf{Q}$ with respect to the fixed frame $OXYZ$ is made of two parts: The first part represents the rate of change of $\mathbf{Q}$ with respect to the rotating frame $Oxyz$; the second part, $\boldsymbol{\Omega} \times \mathbf{Q}$, is induced by the rotation of the frame $Oxyz$.

The use of relation (15.31) simplifies the determination of the rate of change of a vector $\mathbf{Q}$ with respect to a fixed frame of reference $OXYZ$ when the vector $\mathbf{Q}$ is defined by its components along the axes of a rotating frame $Oxyz$, since this relation does not require the separate computation of the derivatives of the unit vectors defining the orientation of the rotating frame.

## 15.11. PLANE MOTION OF A PARTICLE RELATIVE TO A ROTATING FRAME. CORIOLIS ACCELERATION

Consider two frames of reference, both centered at $O$ and both in the plane of the figure, a fixed frame $OXY$ and a rotating frame $Oxy$ (Fig. 15.27). Let $P$ be a particle moving in the plane of the figure. The position vector $\mathbf{r}$ of $P$ is the same in both frames, but its rate of change depends upon the frame of reference which has been selected.

The absolute velocity $\mathbf{v}_P$ of the particle is defined as the velocity observed from the fixed frame $OXY$ and is equal to the rate of change $(\dot{\mathbf{r}})_{OXY}$ of $\mathbf{r}$ with respect to that frame. We can, however, express $\mathbf{v}_P$ in terms of the rate of change $(\dot{\mathbf{r}})_{Oxy}$ observed from the rotating frame if we make use of Eq. (15.31). Denoting by $\mathbf{\Omega}$ the angular velocity of the frame $Oxy$ with respect to $OXY$ at the instant considered, we write

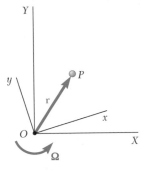

**Fig. 15.27**

$$\mathbf{v}_P = (\dot{\mathbf{r}})_{OXY} = \mathbf{\Omega} \times \mathbf{r} + (\dot{\mathbf{r}})_{Oxy} \qquad (15.32)$$

But $(\dot{\mathbf{r}})_{Oxy}$ defines the velocity of the particle $P$ *relative to the rotating* frame $Oxy$. Denoting the rotating frame by $\mathcal{F}$ for short, we represent the velocity $(\dot{\mathbf{r}})_{Oxy}$ of $P$ relative to the rotating frame by $\mathbf{v}_{P/\mathcal{F}}$. Let us imagine that a rigid slab has been attached to the rotating frame. Then $v_{P/\mathcal{F}}$ represents the velocity of $P$ along the path that it describes on that slab (Fig. 15.28), and the term $\mathbf{\Omega} \times \mathbf{r}$ in (15.32) represents the velocity $\mathbf{v}_{P'}$ of the point $P'$ of the slab—or rotating frame—which coincides with $P$ at the instant considered. Thus, we have

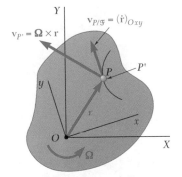

**Fig. 15.28**

$$\mathbf{v}_P = \mathbf{v}_{P'} + \mathbf{v}_{P/\mathcal{F}} \qquad (15.33)$$

where $\mathbf{v}_P$ = absolute velocity of particle $P$
  $\mathbf{v}_{P'}$ = velocity of point $P'$ of moving frame $\mathcal{F}$ coinciding with $P$
  $\mathbf{v}_{P/\mathcal{F}}$ = velocity of $P$ relative to moving frame $\mathcal{F}$

The absolute acceleration $\mathbf{a}_P$ of the particle is defined as the rate of change of $\mathbf{v}_P$ with respect to the fixed frame $OXY$. Computing the rates of change with respect to $OXY$ of the terms in (15.32), we write

$$\mathbf{a}_P = \dot{\mathbf{v}}_P = \dot{\mathbf{\Omega}} \times \mathbf{r} + \mathbf{\Omega} \times \dot{\mathbf{r}} + \frac{d}{dt}[(\dot{\mathbf{r}})_{Oxy}] \qquad (15.34)$$

where all derivatives are defined with respect to $OXY$, except where indicated otherwise. Referring to Eq. (15.31), we note that the last term in (15.34) can be expressed as

$$\frac{d}{dt}[(\dot{\mathbf{r}})_{Oxy}] = (\ddot{\mathbf{r}})_{Oxy} + \mathbf{\Omega} \times (\dot{\mathbf{r}})_{Oxy}$$

On the other hand, $\dot{\mathbf{r}}$ represents the velocity $\mathbf{v}_P$ and can be replaced by the right-hand member of Eq. (15.32). After completing these two substitutions into (15.34), we write

$$\mathbf{a}_P = \dot{\boldsymbol{\Omega}} \times \mathbf{r} + \boldsymbol{\Omega} \times (\boldsymbol{\Omega} \times \mathbf{r}) + 2\boldsymbol{\Omega} \times (\dot{\mathbf{r}})_{Oxy} + (\ddot{\mathbf{r}})_{Oxy} \qquad (15.35)$$

Referring to the expression (15.8) obtained in Sec. 15.3 for the acceleration of a particle in a rigid body rotating about a fixed axis, we note that the sum of the first two terms represents the acceleration $\mathbf{a}_{P'}$ of the point $P'$ of the rotating frame which coincides with $P$ at the instant considered. On the other hand, the last term defines the acceleration $\mathbf{a}_{P/\mathscr{F}}$ of $P$ relative to the rotating frame. If it were not for the third term, which has not been accounted for, a relation similar to (15.33) could be written for the accelerations, and $\mathbf{a}_P$ could be expressed as the sum of $\mathbf{a}_{P'}$ and $\mathbf{a}_{P/\mathscr{F}}$. However, it is clear that *such a relation would be incorrect* and that we must include the additional term. This term, which will be denoted by $\mathbf{a}_c$, is called the *complementary acceleration*, or *Coriolis acceleration*, after the French mathematician de Coriolis (1792–1843). We write

$$\mathbf{a}_P = \mathbf{a}_{P'} + \mathbf{a}_{P/\mathscr{F}} + \mathbf{a}_c \qquad (15.36)$$

where $\mathbf{a}_P$ = absolute acceleration of particle $P$
$\quad\;\; \mathbf{a}_{P'}$ = acceleration of point $P'$ of moving frame $\mathscr{F}$ coinciding with $P$
$\quad\;\; \mathbf{a}_{P/\mathscr{F}}$ = acceleration of $P$ relative to moving frame $\mathscr{F}$
$\quad\;\; \mathbf{a}_c = 2\boldsymbol{\Omega} \times (\dot{\mathbf{r}})_{Oxy} = 2\boldsymbol{\Omega} \times \mathbf{v}_{P/\mathscr{F}}$
$\qquad\quad$ = complementary, or Coriolis, acceleration†

We note that since point $P'$ moves in a circle about the origin $O$, its acceleration $\mathbf{a}_{P'}$ has, in general, two components: a component $(\mathbf{a}_{P'})_t$ tangent to the circle, and a component $(\mathbf{a}_{P'})_n$ directed toward $O$. Similarly, the acceleration $\mathbf{a}_{P/\mathscr{F}}$ generally has two components: a component $(\mathbf{a}_{P/\mathscr{F}})_t$ tangent to the path that $P$ describes on the rotating slab, and a component $(\mathbf{a}_{P/\mathscr{F}})_n$ directed toward the center of curvature of that path. We further note that since the vector $\boldsymbol{\Omega}$ is perpendicular to the plane of motion, and thus to $\mathbf{v}_{P/\mathscr{F}}$, the magnitude of the Coriolis acceleration $\mathbf{a}_c = 2\boldsymbol{\Omega} \times \mathbf{v}_{P/\mathscr{F}}$ is equal to $2\Omega v_{P/\mathscr{F}}$, and its direction can be obtained by rotating the vector $\mathbf{v}_{P/\mathscr{F}}$ through 90° in the sense of rotation of the moving frame (Fig. 15.29). The Coriolis acceleration reduces to zero when either $\boldsymbol{\Omega}$ or $\mathbf{v}_{P/\mathscr{F}}$ is zero.

The following example will help in understanding the physical meaning of the Coriolis acceleration. Consider a collar $P$ which is

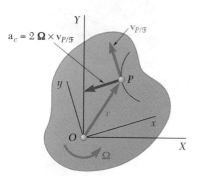

**Fig. 15.29**

---

†It is important to note the difference between Eq. (15.36) and Eq. (15.21) of Sec. 15.8. When we wrote

$$\mathbf{a}_B = \mathbf{a}_A + \mathbf{a}_{B/A} \qquad (15.21)$$

in Sec. 15.8, we were expressing the absolute acceleration of point $B$ as the sum of its acceleration $\mathbf{a}_{B/A}$ relative to a *frame in translation* and of the acceleration $\mathbf{a}_A$ of a point of that frame. We are now trying to relate the absolute acceleration of point $P$ to its acceleration $\mathbf{a}_{P/\mathscr{F}}$ relative to a *rotating frame $\mathscr{F}$* and to the acceleration $\mathbf{a}_{P'}$ of the point $P'$ of that frame which coincides with $P$; Eq. (15.36) shows that because the frame is rotating, it is necessary to include an additional term representing the Coriolis acceleration $\mathbf{a}_c$.

made to slide at a constant relative speed $u$ along a rod $OB$ rotating at a constant angular velocity $\boldsymbol{\omega}$ about $O$ (Fig. 15.30$a$). According to formula (15.36), the absolute acceleration of $P$ can be obtained by adding vectorially the acceleration $\mathbf{a}_A$ of the point $A$ of the rod coinciding with $P$, the relative acceleration $\mathbf{a}_{P/OB}$ of $P$ with respect to the rod, and the Coriolis acceleration $\mathbf{a}_c$. Since the angular velocity $\boldsymbol{\omega}$ of the rod is constant, $\mathbf{a}_A$ reduces to its normal component $(\mathbf{a}_A)_n$ of magnitude $r\omega^2$; and since $u$ is constant, the relative acceleration $\mathbf{a}_{P/OB}$ is zero. According to the definition given above, the Coriolis acceleration is a vector perpendicular to $OB$, of magnitude $2\omega u$, and directed as shown in the figure. The acceleration of the collar $P$ consists, therefore, of the two vectors shown in Fig. 15.30$a$. Note that the result obtained can be checked by applying the relation (11.44).

To understand better the significance of the Coriolis acceleration, let us consider the absolute velocity of $P$ at time $t$ and at time $t + \Delta t$ (Fig. 15.30$b$). The velocity at time $t$ can be resolved into its components $\mathbf{u}$ and $\mathbf{v}_A$; the velocity at time $t + \Delta t$ can be resolved into its components $\mathbf{u}'$ and $\mathbf{v}_{A'}$. Drawing these components from the same origin (Fig. 15.30$c$), we note that the change in velocity during the time $\Delta t$ can be represented by the sum of three vectors, $\overrightarrow{RR'}$, $\overrightarrow{TT''}$, and $\overrightarrow{T''T'}$. The vector $\overrightarrow{TT''}$ measures the change in direction of the velocity $\mathbf{v}_A$, and the quotient $\overrightarrow{TT''}/\Delta t$ represents the acceleration $\mathbf{a}_A$ when $\Delta t$ approaches zero. We check that the direction of $\overrightarrow{TT''}$ is that of $\mathbf{a}_A$ when $\Delta t$ approaches zero and that

$$\lim_{\Delta t \to 0} \frac{TT''}{\Delta t} = \lim_{\Delta t \to 0} v_A \frac{\Delta\theta}{\Delta t} = r\omega\omega = r\omega^2 = a_A$$

The vector $\overrightarrow{RR'}$ measures the change in direction of $\mathbf{u}$ due to the rotation of the rod; the vector $\overrightarrow{T''T'}$ measures the change in magnitude of $\mathbf{v}_A$ due to the motion of $P$ on the rod. The vectors $\overrightarrow{RR'}$ and $\overrightarrow{T''T'}$ result from the *combined effect* of the relative motion of $P$ and of the rotation of the rod; they would vanish if *either* of these two motions stopped. It is easily verified that the sum of these two vectors defines the Coriolis acceleration. Their direction is that of $\mathbf{a}_c$ when $\Delta t$ approaches zero, and since $RR' = u\,\Delta\theta$ and $T''T' = v_{A'} - v_A = (r + \Delta r)\omega - r\omega = \omega\,\Delta r$, we check that $a_c$ is equal to

$$\lim_{\Delta t \to 0} \left( \frac{RR'}{\Delta t} + \frac{T''T'}{\Delta t} \right) = \lim_{\Delta t \to 0} \left( u\frac{\Delta\theta}{\Delta t} + \omega\frac{\Delta r}{\Delta t} \right) = u\omega + \omega u = 2\omega u$$

Formulas (15.33) and (15.36) can be used to analyze the motion of mechanisms which contain parts sliding on each other. They make it possible, for example, to relate the absolute and relative motions of sliding pins and collars (see Sample Probs. 15.9 and 15.10). The concept of Coriolis acceleration is also very useful in the study of long-range projectiles and of other bodies whose motions are appreciably affected by the rotation of the earth. As was pointed out in Sec. 12.2, a system of axes attached to the earth does not truly constitute a newtonian frame of reference; such a system of axes should actually be considered as rotating. The formulas derived in this section will therefore facilitate the study of the motion of bodies with respect to axes attached to the earth.

(a)

(b)

(c)

**Fig. 15.30**

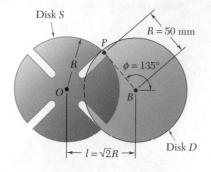

Disk S

P

R

$\phi = 135°$

O

B

R = 50 mm

l = √2R

Disk D

## SAMPLE PROBLEM 15.9

The Geneva mechanism shown is used in many counting instruments and in other applications where an intermittent rotary motion is required. Disk D rotates with a constant counterclockwise angular velocity $\omega_D$ of 10 rad/s. A pin P is attached to disk D and slides along one of several slots cut in disk S. It is desirable that the angular velocity of disk S be zero as the pin enters and leaves each slot; in the case of four slots, this will occur if the distance between the centers of the disks is $l = \sqrt{2}\,R$.

At the instant when $\phi = 150°$, determine (a) the angular velocity of disk S, (b) the velocity of pin P relative to disk S.

## SOLUTION

Disk S   P     Disk D

R

r

$\phi = 150°$

$\beta$

P'

O

B

l = √2R

We solve triangle OPB, which corresponds to the position $\phi = 150°$. Using the law of cosines, we write

$$r^2 = R^2 + l^2 - 2Rl \cos 30° = 0.551R^2 \qquad r = 0.742R = 37.1 \text{ mm}$$

From the law of sines,

$$\frac{\sin \beta}{R} = \frac{\sin 30°}{r} \qquad \sin \beta = \frac{\sin 30°}{0.742} \qquad \beta = 42.36°$$

Since pin P is attached to disk D, and since disk D rotates about point B, the magnitude of the absolute velocity of P is

$$v_P = R\omega_D = (50 \text{ mm})(10 \text{ rad/s}) = 500 \text{ mm/s}$$
$$\mathbf{v}_P = 500 \text{ mm/s} \; \angle 60° \blacktriangleleft$$

We consider now the motion of pin P along the slot in disk S. Denoting by P' the point of disk S which coincides with P at the instant considered and selecting a rotating frame $\mathscr{S}$ attached to disk S, we write

$$\mathbf{v}_P = \mathbf{v}_{P'} + \mathbf{v}_{P/\mathscr{S}}$$

$v_{P'}$

$v_P$

$\gamma$

$v_{P/\mathscr{S}}$

30°

$\beta = 42.4°$

Noting that $\mathbf{v}_{P'}$ is perpendicular to the radius OP and that $\mathbf{v}_{P/\mathscr{S}}$ is directed along the slot, we draw the velocity triangle corresponding to the equation above. From the triangle, we compute

$$\gamma = 90° - 42.4° - 30° = 17.64°$$
$$v_{P'} = v_P \sin \gamma = (500 \text{ mm/s}) \sin 17.6°$$
$$\mathbf{v}_{P'} = 151.2 \text{ mm/s} \; \nwarrow 42.4°$$
$$v_{P/\mathscr{S}} = v_P \cos \gamma = (500 \text{ mm/s}) \cos 17.6°$$
$$\mathbf{v}_{P/S} = \mathbf{v}_{P/\mathscr{S}} = 476.5 \text{ mm/s} \; \angle 42.4° \blacktriangleleft$$

Since $\mathbf{v}_{P'}$ is perpendicular to the radius OP, we write

$$v_{P'} = r\omega_{\mathscr{S}} \qquad 151.2 \text{ mm/s} = (37.1 \text{ mm})\omega_{\mathscr{S}}$$
$$\omega_S = \omega_{\mathscr{S}} = 4.08 \text{ rad/s} \; \circlearrowright \blacktriangleleft$$

## SAMPLE PROBLEM 15.10

In the Geneva mechanism of Sample Prob. 15.9, disk $D$ rotates with a constant counterclockwise angular velocity $\boldsymbol{\omega}_D$ of magnitude 10 rad/s. At the instant when $\phi = 150°$, determine the angular acceleration of disk $S$.

## SOLUTION

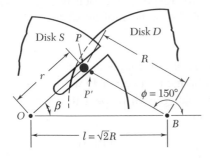

Referring to Sample Prob. 15.9, we obtain the angular velocity of the frame $\mathcal{S}$ attached to disk $S$ and the velocity of the pin relative to $\mathcal{S}$:

$$\omega_{\mathcal{S}} = 4.08 \text{ rad/s} \downarrow$$
$$\beta = 42.36° \qquad \mathbf{v}_{P/\mathcal{S}} = 476.5 \text{ mm/s} \nearrow 42.4°$$

Since pin $P$ moves with respect to the rotating frame $\mathcal{S}$, we write

$$\mathbf{a}_P = \mathbf{a}_{P'} + \mathbf{a}_{P/\mathcal{S}} + \mathbf{a}_c \qquad (1)$$

Each term of this vector equation is investigated separately.

**Absolute Acceleration $\mathbf{a}_P$.** Since disk $D$ rotates with a constant angular velocity, the absolute acceleration $\mathbf{a}_P$ is directed toward $B$. We have

$$a_P = R\omega_D^2 = (500 \text{ mm})(10 \text{ rad/s})^2 = 5000 \text{ mm/s}^2$$
$$\mathbf{a}_P = 5000 \text{ mm/s}^2 \; \searrow 30°$$

**Acceleration $\mathbf{a}_{P'}$ of the Coinciding Point $P'$.** The acceleration $\mathbf{a}_{P'}$ of the point $P'$ of the frame $\mathcal{S}$ which coincides with $P$ at the instant considered is resolved into normal and tangential components. (We recall from Sample Prob. 15.9 that $r = 37.1$ mm.)

$$(a_{P'})_n = r\omega_{\mathcal{S}}^2 = (37.1 \text{ mm})(4.08 \text{ rad/s})^2 = 617.6 \text{ mm/s}^2$$
$$(\mathbf{a}_{P'})_n = 618 \text{ mm/s}^2 \nearrow 42.4°$$
$$(\mathbf{a}_{P'})_t = r\alpha_{\mathcal{S}} \qquad (\mathbf{a}_{P'})_t = 37.1\alpha_{\mathcal{S}} \nwarrow 42.4°$$

**Relative Acceleration $\mathbf{a}_{P/\mathcal{S}}$.** Since the pin $P$ moves in a straight slot cut in disk $S$, the relative acceleration $\mathbf{a}_{P/\mathcal{S}}$ must be parallel to the slot; that is, its direction must be $\measuredangle 42.4°$.

**Coriolis Acceleration $\mathbf{a}_c$.** Rotating the relative velocity $\mathbf{v}_{P/\mathcal{S}}$ through 90° in the sense of $\boldsymbol{\omega}_{\mathcal{S}}$, we obtain the direction of the Coriolis component of the acceleration: $\nwarrow 42.4°$. We write

$$a_c = 2\omega_{\mathcal{S}}v_{P/\mathcal{S}} = 2(4.08 \text{ rad/s})(477 \text{ mm/s}) = 3890 \text{ mm/s}^2$$
$$\mathbf{a}_c = 3890 \text{ mm/s}^2 \nwarrow 42.4°$$

We rewrite Eq. (1) and substitute the accelerations found above:

$$\mathbf{a}_P = (\mathbf{a}_{P'})_n + (\mathbf{a}_{P'})_t + \mathbf{a}_{P/\mathcal{S}} + \mathbf{a}_c$$
$$[5000 \searrow 30°] = [618 \nearrow 42.4°] + [37.1\alpha_{\mathcal{S}} \nwarrow 42.4°]$$
$$+ [a_{P/\mathcal{S}} \measuredangle 42.4°] + [3890 \nwarrow 42.4°]$$

Equating components in a direction perpendicular to the slot,

$$5000 \cos 17.6° = 37.1\alpha_{\mathcal{S}} - 3890$$
$$\alpha_S = \alpha_{\mathcal{S}} = 233 \text{ rad/s}^2 \downarrow \quad \blacktriangleleft$$

In this lesson you studied the rate of change of a vector with respect to a rotating frame and then applied your knowledge to the analysis of the plane motion of a particle relative to a rotating frame.

**1.  Rate of change of a vector with respect to a fixed frame and with respect to a rotating frame.**   Denoting by $(\dot{\mathbf{Q}})_{OXYZ}$ the rate of change of a vector $\mathbf{Q}$ with respect to a fixed frame $OXYZ$ and by $(\dot{\mathbf{Q}})_{Oxyz}$ its rate of change with respect to a rotating frame $Oxyz$, we obtained the fundamental relation

$$(\dot{\mathbf{Q}})_{OXYZ} = (\dot{\mathbf{Q}})_{Oxyz} + \boldsymbol{\Omega} \times \mathbf{Q} \qquad (15.31)$$

where $\boldsymbol{\Omega}$ is the angular velocity of the rotating frame.

This fundamental relation will now be applied to the solution of two-dimensional problems.

**2.  Plane motion of a particle relative to a rotating frame.**   Using the above fundamental relation and designating by $\mathcal{F}$ the rotating frame, we obtained the following expressions for the velocity and the acceleration of a particle $P$:

$$\mathbf{v}_P = \mathbf{v}_{P'} + \mathbf{v}_{P/\mathcal{F}} \qquad (15.33)$$
$$\mathbf{a}_P = \mathbf{a}_{P'} + \mathbf{a}_{P/\mathcal{F}} + \mathbf{a}_c \qquad (15.36)$$

In these equations:

*a. The subscript P* refers to the absolute motion of the particle $P$, that is, to its motion with respect to a fixed frame of reference $OXY$.

*b. The subscript P′* refers to the motion of the point $P'$ of the rotating frame $\mathcal{F}$ which coincides with $P$ at the instant considered.

*c. The subscript P/$\mathcal{F}$* refers to the motion of the particle $P$ relative to the rotating frame $\mathcal{F}$.

*d. The term $\mathbf{a}_c$ represents the Coriolis acceleration of point P.* Its magnitude is $2\Omega v_{P/\mathcal{F}}$, and its direction is found by rotating $\mathbf{v}_{P/\mathcal{F}}$ through 90° in the sense of rotation of the frame $\mathcal{F}$.

You should keep in mind that the Coriolis acceleration should be taken into account whenever a part of the mechanism you are analyzing is moving with respect to another part that is rotating. The problems you will encounter in this lesson involve collars that slide on rotating rods, booms that extend from cranes rotating in a vertical plane, etc.

When solving a problem involving a rotating frame, you will find it convenient to draw vector diagrams representing Eqs. (15.33) and (15.36), respectively, and use these diagrams to obtain either an analytical or a graphical solution.

# Problems

**15.148**  Two rotating rods are connected by slider block *P*. The rod attached at *A* rotates with a constant clockwise angular velocity $\omega_A$. For $b = 300$ mm and $\omega_A = 10$ rad/s, determine for the position shown (*a*) the angular velocity of the rod attached at *B*, (*b*) the relative velocity of slider block *P* with respect to the rod on which it slides.

**15.149**  The motion of pin *P* is guided by slots cut in rods *AE* and *BD*. Knowing that the rods rotate with the constant angular velocities $\omega_A = 4$ rad/s clockwise and $\omega_B = 5$ rad/s clockwise, determine the velocity of pin *P* for the position shown.

**Fig. P15.148**

**Fig. P15.149**

**Fig. P15.150 and P15.151**

**15.150 and 15.151**  Pin *P* is attached to the wheel shown and slides in a slot cut in bar *BD*. The wheel rolls to the right without slipping with a constant angular velocity of 20 rad/s. Knowing that $x = 61$ cm when $\theta = 0$, determine the angular velocity of the bar and the relative velocity of pin *P* with respect to the rod for the given data.

     **15.150**  (*a*) $\theta = 0$, (*b*) $\theta = 90°$.
     **15.151**  $\theta = 30°$.

**15.152**  Knowing that at the instant shown the angular velocity of bar *AB* is 15 rad/s clockwise and the angular velocity of bar *EF* is 10 rad/s clockwise, determine (*a*) the angular velocity of rod *DE*, (*b*) the relative velocity of collar *B* with respect to rod *DE*.

**15.153**  Knowing that at the instant shown the angular velocity of rod *DE* is 10 rad/s clockwise and the angular velocity of bar *EF* is 15 rad/s counterclockwise, determine (*a*) the angular velocity of bar *AB*, (*b*) the relative velocity of collar *B* with respect to rod *DE*.

**Fig. P15.152 and P15.153**

**15.154** Bar $AB$ rotates clockwise with a constant angular velocity of 8 rad/s and rod $EF$ rotates clockwise with a constant angular velocity of 6 rad/s. Determine at the instant shown (*a*) the angular velocity of bar $BD$, (*b*) the relative velocity of collar $D$ with respect to rod $EF$.

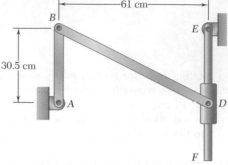

**Fig. *P15.154* and *P15.155***

**15.155** Bar $AB$ rotates clockwise with a constant angular velocity of 4 rad/s. Knowing that the magnitude of the velocity of collar $D$ is 6.1 m/s and that the angular velocity of bar $BD$ is counterclockwise at the instant shown, determine (*a*) the angular velocity of bar $EF$, (*b*) the relative velocity of collar $D$ with respect to rod $EF$.

**15.156** In the automated welding setup shown, the position of the two welding tips $G$ and $H$ is controlled by the hydraulic cylinder $D$ and rod $BC$. The cylinder is bolted to the vertical plate which at the instant shown rotates counterclockwise about $A$ with a constant angular velocity of 1.6 rad/s. Knowing that at the same instant the length $EF$ of the welding assembly is increasing at the constant rate of 300 mm/s, determine (*a*) the velocity of tip $H$, (*b*) the acceleration of tip $H$.

**Fig. *P15.156***

**15.157** In Prob. 15.156, determine (*a*) the velocity of tip $G$, (*b*) the acceleration of tip $G$.

**15.158** Four pins slide in four separate slots cut in a circular plate as shown. When the plate is at rest, each pin has a velocity directed as shown and of the same constant magnitude $u$. If each pin maintains the same velocity relative to the plate when the plate rotates about $O$ with a constant counterclockwise angular velocity $\boldsymbol{\omega}$, determine the acceleration of each pin.

**Fig. P15.158**

**15.159** Solve Prob. 15.158, assuming that the plate rotates about $O$ with a constant clockwise angular velocity $\boldsymbol{\omega}$.

**15.160** Pin $P$ is attached to the collar shown; the motion of the pin is guided by a slot cut in bar $BD$ and by the collar that slides on rod $AE$. Rod $AE$ rotates with a constant angular velocity of 6 rad/s clockwise and the distance from $A$ to $P$ increases at a constant rate of 2.4 m/s. Determine at the instant shown (*a*) the angular acceleration of bar $BD$, (*b*) the relative acceleration of pin $P$ with respect to bar $BD$.

**15.161** Pin $P$ is attached to the collar shown; the motion of the pin is guided by a slot cut in bar $BD$ and by the collar that slides on rod $AE$. Bar $BD$ rotates with a constant angular velocity of 6 rad/s counterclockwise and the distance from $B$ to $P$ decreases at a constant rate of 3.2 m/s. Determine at the instant shown (*a*) the angular acceleration of rod $AE$, (*b*) the relative acceleration of pin $P$ with respect to rod $AE$.

**15.162** A rocket sled is tested on a straight track that is built along a meridian. Knowing that the track is located at latitude 40° north, determine the Coriolis acceleration of the sled when it is moving north at a speed of 961.4 km/h.

*15.163* The cage of a mine elevator moves downward at a constant speed of 12.2 m/s. Determine the magnitude and direction of the Coriolis acceleration of the cage if the elevator is located (*a*) at the equator, (*b*) at latitude 40° north, (*c*) at latitude 40° south.

**15.164** Collar $P$ slides toward point $A$ at a constant relative speed of 1.8 m/s along rod $AB$ which rotates counterclockwise with a constant angular velocity of 5 rad/s. At the instant shown determine (*a*) the velocity and acceleration of point $P$, (*b*) the velocity and acceleration of point $D$.

**Fig. P15.160 and P15.161**

**Fig. P15.164 and P15.165**

**15.165** Collar $P$ slides toward point $A$ at a constant relative speed of 2.9 m/s along rod $AB$ which rotates clockwise with a constant angular velocity of 4 rad/s. At the instant shown determine (*a*) the angular velocities of bars $PD$ and $DE$, (*b*) the angular accelerations of bars $PD$ and $DE$.

**15.166** The motion of nozzle $D$ is controlled by arm $AB$. At the instant shown the arm is rotating counterclockwise at the constant rate $\omega = 2.4$ rad/s and portion $BC$ is being extended at the constant rate $u = 250$ mm/s with respect to the arm. For each of the arrangements shown, determine the acceleration of nozzle $D$.

**15.167** Solve Prob. 15.166, assuming that the direction of the relative velocity **u** is reversed so that portion $BD$ is being retracted.

**Fig. P15.166**

**15.168**  Collar E slides toward point A with a constant relative speed of 3 m/s along rod AB which rotates counterclockwise with a constant angular velocity of 15 rad/s. At the instant shown determine the angular velocity and angular acceleration of rod DE.

**Fig. *P15.168* and *P15.169***

**15.169**  Collar E slides away from point A with a constant relative speed of 2 m/s along rod AB which rotates clockwise with a constant angular velocity of 10 rad/s. At the instant shown determine the velocity and acceleration of collar D.

**Fig. P15.170**

**15.170**  Pin P slides in a circular slot cut in the plate shown at a constant relative speed u = 180 mm/s. Knowing that at the instant shown the plate rotates clockwise about A at the constant rate $\omega$ = 6 rad/s, determine the acceleration of the pin if it is located at (a) point A, (b) point B, (c) point C.

**15.171**  The collar P slides outward at a constant relative speed u along rod AB, which rotates counterclockwise with a constant angular velocity of 20 rpm. Knowing that r = 25.4 m. when $\theta$ = 0 and that the collar reaches B when $\theta$ = 90°, determine the magnitude of the acceleration of the collar P just as it reaches B.

**Fig. *P15.171***

**15.172**  Rod AD is bent in the shape of an arc of a circle of radius b = 150 mm. The position of the rod is controlled by pin B which slides in a horizontal slot and also slides along the rod. Knowing that at the instant shown pin B moves to the right at a constant speed of 75 mm/s, determine (a) the angular velocity of the rod, (b) the angular acceleration of the rod.

**Fig. P15.172**

**15.173**  Solve Prob. 15.172 when $\theta$ = 90°.

**15.174 and 15.175**  Pin *P* is attached to the wheel shown and slides in a slot cut in bar *BD*. The wheel rolls to the right without slipping with a constant angular velocity of 20 rad/s. Knowing that *x* = 61 cm when $\theta$ = 0, determine (*a*) the angular acceleration of the bar, (*b*) the relative acceleration of pin *P* with respect to the bar for the given data.

    **15.174**  $\theta$ = 0.
    **15.175**  $\theta$ = 90°.

**Fig. P15.174 and P15.175**

**15.176**  The disk shown rotates with a constant clockwise angular velocity of 12 rad/s. At the instant shown, determine (*a*) the angular velocity and angular acceleration of rod *BD*, (*b*) the velocity and acceleration of the point of the rod coinciding with *E*.

**Fig. P15.176**

**Fig. P15.177**

**15.177**  Collar *B* slides along rod *AC* and is attached to a block that moves in a vertical slot. Knowing that *R* = 45.7 cm, $\theta$ = 30°, $\omega$ = 6 rad/s, and $\alpha$ = 4 rad/s², determine the velocity and acceleration of collar *B*.

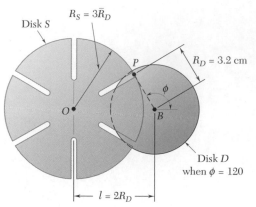

**15.178** The Geneva mechanism shown is used to provide an intermittent rotary motion of disk S. Disk D rotates with a constant counterclockwise angular velocity $\boldsymbol{\omega}_D$ of 8 rad/s. A pin P is attached to disk D and can slide in one of six equally spaced slots cut in disk S. It is desirable that the angular velocity of disk S be zero as the pin enters and leaves each of the six slots; this will occur if the distance between the centers of the disks and the radii of the disks are related as shown. Determine the angular velocity and angular acceleration of disk S at the instant when $\phi = 150°$.

**15.179** In Prob. 15.178, determine the angular velocity and angular acceleration of disk S at the instant when $\phi = 135°$.

**15.180** In Prob. 15.149, determine the acceleration of pin P.

## *15.12. MOTION ABOUT A FIXED POINT

In Sec. 15.3 the motion of a rigid body constrained to rotate about a fixed axis was considered. The more general case of the motion of a rigid body which has a fixed point O will now be examined.

First, it will be proved that *the most general displacement of a rigid body with a fixed point O is equivalent to a rotation of the body about an axis through O.*† Instead of considering the rigid body itself, we can detach a sphere of center O from the body and analyze the motion of that sphere. Clearly, the motion of the sphere completely characterizes the motion of the given body. Since three points define the position of a solid in space, the center O and two points A and B on the surface of the sphere will define the position of the sphere and thus the position of the body. Let $A_1$ and $B_1$ characterize the position of the sphere at one instant, and let $A_2$ and $B_2$ characterize its position at a later instant (Fig. 15.31a). Since the sphere is rigid, the lengths of the arcs of great circle $A_1B_1$ and $A_2B_2$ must be equal, but except for this requirement, the positions of $A_1$, $A_2$, $B_1$, and $B_2$ are arbitrary. We propose to prove that the points A and B can be brought, respectively, from $A_1$ and $B_1$ into $A_2$ and $B_2$ by a single rotation of the sphere about an axis.

For convenience, and without loss of generality, we select point B so that its initial position coincides with the final position of A; thus, $B_1 = A_2$ (Fig. 15.31b). We draw the arcs of great circle $A_1A_2$, $A_2B_2$ and the arcs bisecting, respectively, $A_1A_2$ and $A_2B_2$. Let C be the point of intersection of these last two arcs; we complete the construction by drawing $A_1C$, $A_2C$, and $B_2C$. As pointed out above, because of the rigidity of the sphere, $A_1B_1 = A_2B_2$. Since C is by construction equidistant from $A_1$, $A_2$, and $B_2$, we also have $A_1C = A_2C = B_2C$. As a result, the spherical triangles $A_1CA_2$ and $B_1CB_2$ are congruent and the

(a)

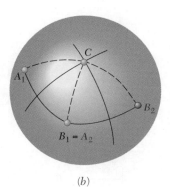

(b)

**Fig. 15.31**

† This is known as *Euler's theorem.*

angles $A_1CA_2$ and $B_1CB_2$ are equal. Denoting by $\theta$ the common value of these angles, we conclude that the sphere can be brought from its initial position into its final position by a single rotation through $\theta$ about the axis $OC$.

It follows that the motion during a time interval $\Delta t$ of a rigid body with a fixed point $O$ can be considered as a rotation through $\Delta\theta$ about a certain axis. Drawing along that axis a vector of magnitude $\Delta\theta/\Delta t$ and letting $\Delta t$ approach zero, we obtain at the limit the *instantaneous axis of rotation* and the angular velocity $\boldsymbol{\omega}$ of the body at the instant considered (Fig. 15.32). The velocity of a particle $P$ of the body can then be obtained, as in Sec. 15.3, by forming the vector product of $\boldsymbol{\omega}$ and of the position vector $\mathbf{r}$ of the particle:

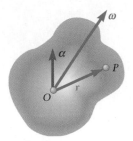

**Fig. 15.32**

$$\mathbf{v} = \frac{d\mathbf{r}}{dt} = \boldsymbol{\omega} \times \mathbf{r} \qquad (15.37)$$

The acceleration of the particle is obtained by differentiating (15.37) with respect to $t$. As in Sec. 15.3 we have

$$\mathbf{a} = \boldsymbol{\alpha} \times \mathbf{r} + \boldsymbol{\omega} \times (\boldsymbol{\omega} \times \mathbf{r}) \qquad (15.38)$$

where the angular acceleration $\boldsymbol{\alpha}$ is defined as the derivative

$$\boldsymbol{\alpha} = \frac{d\boldsymbol{\omega}}{dt} \qquad (15.39)$$

of the angular velocity $\boldsymbol{\omega}$.

In the case of the motion of a rigid body with a fixed point, the direction of $\boldsymbol{\omega}$ and of the instantaneous axis of rotation changes from one instant to the next. The angular acceleration $\boldsymbol{\alpha}$ therefore reflects the change in direction of $\boldsymbol{\omega}$ as well as its change in magnitude and, in general, *is not directed along the instantaneous axis of rotation.* While the particles of the body located on the instantaneous axis of rotation have zero velocity at the instant considered, they do not have zero acceleration. Also, the accelerations of the various particles of the body *cannot* be determined as if the body were rotating permanently about the instantaneous axis.

Recalling the definition of the velocity of a particle with position vector $\mathbf{r}$, we note that the angular acceleration $\boldsymbol{\alpha}$, as expressed in (15.39), represents the velocity of the tip of the vector $\boldsymbol{\omega}$. This property may be useful in the determination of the angular acceleration of a rigid body. For example, it follows that the vector $\boldsymbol{\alpha}$ is tangent to the curve described in space by the tip of the vector $\boldsymbol{\omega}$.

We should note that the vector $\boldsymbol{\omega}$ moves within the body, as well as in space. It thus generates two cones called, respectively, the *body cone* and the *space cone* (Fig. 15.33).† It can be shown that at any given instant, the two cones are tangent along the instantaneous axis of rotation and that as the body moves, the body cone appears to *roll* on the space cone.

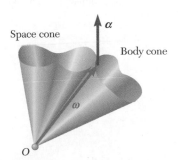

**Fig. 15.33**

†It is recalled that a *cone* is, by definition, a surface generated by a straight line passing through a fixed point. In general, the cones considered here *will not be circular cones.*

Before concluding our analysis of the motion of a rigid body with a fixed point, we should prove that angular velocities are actually vectors. As indicated in Sec. 2.3, some quantities, such as the *finite rotations* of a rigid body, have magnitude and direction but do not obey the parallelogram law of addition; these quantities cannot be considered as vectors. In contrast, angular velocities (and also *infinitesimal rotations*), as will be demonstrated presently, *do obey* the parallelogram law and thus are truly vector quantities.

(a)                                    (b)

**Fig. 15.34**

**Photo 15.4** When the ladder rotates about its fixed base, its angular velocity can be obtained by adding the angular velocities which correspond to simultaneous rotations about two different axes.

Consider a rigid body with a fixed point $O$ which at a given instant rotates simultaneously about the axes $OA$ and $OB$ with angular velocities $\boldsymbol{\omega}_1$ and $\boldsymbol{\omega}_2$ (Fig. 15.34$a$). We know that this motion must be equivalent at the instant considered to a single rotation of angular velocity $\boldsymbol{\omega}$. We propose to show that

$$\boldsymbol{\omega} = \boldsymbol{\omega}_1 + \boldsymbol{\omega}_2 \qquad (15.40)$$

that is, that the resulting angular velocity can be obtained by adding $\boldsymbol{\omega}_1$ and $\boldsymbol{\omega}_2$ by the parallelogram law (Fig. 15.34$b$).

Consider a particle $P$ of the body, defined by the position vector $\mathbf{r}$. Denoting, respectively, by $\mathbf{v}_1$, $\mathbf{v}_2$, and $\mathbf{v}$ the velocity of $P$ when the body rotates about $OA$ only, about $OB$ only, and about both axes simultaneously, we write

$$\mathbf{v} = \boldsymbol{\omega} \times \mathbf{r} \qquad \mathbf{v}_1 = \boldsymbol{\omega}_1 \times \mathbf{r} \qquad \mathbf{v}_2 = \boldsymbol{\omega}_2 \times \mathbf{r} \qquad (15.41)$$

But the vectorial character of *linear* velocities is well established (since they represent the derivatives of position vectors). We therefore have

$$\mathbf{v} = \mathbf{v}_1 + \mathbf{v}_2$$

where the plus sign indicates vector addition. Substituting from (15.41), we write

$$\boldsymbol{\omega} \times \mathbf{r} = \boldsymbol{\omega}_1 \times \mathbf{r} + \boldsymbol{\omega}_2 \times \mathbf{r}$$
$$\boldsymbol{\omega} \times \mathbf{r} = (\boldsymbol{\omega}_1 + \boldsymbol{\omega}_2) \times \mathbf{r}$$

where the plus sign still indicates vector addition. Since the relation obtained holds for an arbitrary $\mathbf{r}$, we conclude that (15.40) must be true.

## *15.13. GENERAL MOTION

The most general motion of a rigid body in space will now be considered. Let $A$ and $B$ be two particles of the body. We recall from Sec. 11.12 that the velocity of $B$ with respect to the fixed frame of reference $OXYZ$ can be expressed as

$$\mathbf{v}_B = \mathbf{v}_A + \mathbf{v}_{B/A} \tag{15.42}$$

where $\mathbf{v}_{B/A}$ is the velocity of $B$ relative to a frame $AX'Y'Z'$ attached to $A$ and of fixed orientation (Fig. 15.35). Since $A$ is fixed in this frame, the motion of the body relative to $AX'Y'Z'$ is the motion of a body with a fixed point. The relative velocity $\mathbf{v}_{B/A}$ can therefore be obtained from (15.37) after $\mathbf{r}$ has been replaced by the position vector $\mathbf{r}_{B/A}$ of $B$ relative to $A$. Substituting for $\mathbf{v}_{B/A}$ into (15.42), we write

$$\mathbf{v}_B = \mathbf{v}_A + \boldsymbol{\omega} \times \mathbf{r}_{B/A} \tag{15.43}$$

where $\boldsymbol{\omega}$ is the angular velocity of the body at the instant considered.

The acceleration of $B$ is obtained by a similar reasoning. We first write

$$\mathbf{a}_B = \mathbf{a}_A + \mathbf{a}_{B/A}$$

and, recalling Eq. (15.38),

$$\mathbf{a}_B = \mathbf{a}_A + \boldsymbol{\alpha} \times \mathbf{r}_{B/A} + \boldsymbol{\omega} \times (\boldsymbol{\omega} \times \mathbf{r}_{B/A}) \tag{15.44}$$

where $\boldsymbol{\alpha}$ is the angular acceleration of the body at the instant considered.

Equations (15.43) and (15.44) show that *the most general motion of a rigid body is equivalent, at any given instant, to the sum of a translation*, in which all the particles of the body have the same velocity and acceleration as a reference particle $A$, *and of a motion in which particle $A$ is assumed to be fixed*.†

It is easily shown, by solving (15.43) and (15.44) for $\mathbf{v}_A$ and $\mathbf{a}_A$, that the motion of the body with respect to a frame attached to $B$ would be characterized by the same vectors $\boldsymbol{\omega}$ and $\boldsymbol{\alpha}$ as its motion relative to $AX'Y'Z'$. The angular velocity and angular acceleration of a rigid body at a given instant are thus independent of the choice of reference point. On the other hand, one should keep in mind that whether the moving frame is attached to $A$ or to $B$, it should maintain a fixed orientation; that is, it should remain parallel to the fixed reference frame $OXYZ$ throughout the motion of the rigid body. In many problems it will be more convenient to use a moving frame which is allowed to rotate as well as to translate. The use of such moving frames will be discussed in Secs. 15.14 and 15.15.

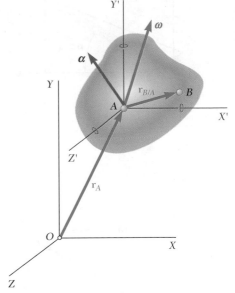

**Fig. 15.35**

†It is recalled from Sec. 15.12 that, in general, the vectors $\boldsymbol{\omega}$ and $\boldsymbol{\alpha}$ are not collinear, and that the accelerations of the particles of the body in their motion relative to the frame $AX'Y'Z'$ cannot be determined as if the body were rotating permanently about the instantaneous axis through $A$.

# SAMPLE PROBLEM 15.11

The crane shown rotates with a constant angular velocity $\boldsymbol{\omega}_1$ of 0.30 rad/s. Simultaneously, the boom is being raised with a constant angular velocity $\boldsymbol{\omega}_2$ of 0.50 rad/s relative to the cab. Knowing that the length of the boom $OP$ is $l = 12$ m, determine (a) the angular velocity $\boldsymbol{\omega}$ of the boom, (b) the angular acceleration $\boldsymbol{\alpha}$ of the boom, (c) the velocity $\mathbf{v}$ of the tip of the boom, (d) the acceleration $\mathbf{a}$ of the tip of the boom.

## SOLUTION

*a. Angular Velocity of Boom.* Adding the angular velocity $\boldsymbol{\omega}_1$ of the cab and the angular velocity $\boldsymbol{\omega}_2$ of the boom relative to the cab, we obtain the angular velocity $\boldsymbol{\omega}$ of the boom at the instant considered:

$$\boldsymbol{\omega} = \boldsymbol{\omega}_1 + \boldsymbol{\omega}_2 \qquad \boldsymbol{\omega} = (0.30 \text{ rad/s})\mathbf{j} + (0.50 \text{ rad/s})\mathbf{k} \quad \blacktriangleleft$$

*b. Angular Acceleration of Boom.* The angular acceleration $\boldsymbol{\alpha}$ of the boom is obtained by differentiating $\boldsymbol{\omega}$. Since the vector $\boldsymbol{\omega}_1$ is constant in magnitude and direction, we have

$$\boldsymbol{\alpha} = \dot{\boldsymbol{\omega}} = \dot{\boldsymbol{\omega}}_1 + \dot{\boldsymbol{\omega}}_2 = 0 + \dot{\boldsymbol{\omega}}_2$$

where the rate of change $\dot{\boldsymbol{\omega}}_2$ is to be computed with respect to the fixed frame $OXYZ$. However, it is more convenient to use a frame $Oxyz$ attached to the cab and rotating with it, since the vector $\boldsymbol{\omega}_2$ also rotates with the cab and therefore has zero rate of change with respect to that frame. Using Eq. (15.31) with $\mathbf{Q} = \boldsymbol{\omega}_2$ and $\boldsymbol{\Omega} = \boldsymbol{\omega}_1$, we write

$$(\dot{\mathbf{Q}})_{OXYZ} = (\dot{\mathbf{Q}})_{Oxyz} + \boldsymbol{\Omega} \times \mathbf{Q}$$
$$(\dot{\boldsymbol{\omega}}_2)_{OXYZ} = (\dot{\boldsymbol{\omega}}_2)_{Oxyz} + \boldsymbol{\omega}_1 \times \boldsymbol{\omega}_2$$
$$\boldsymbol{\alpha} = (\dot{\boldsymbol{\omega}}_2)_{OXYZ} = 0 + (0.30 \text{ rad/s})\mathbf{j} \times (0.50 \text{ rad/s})\mathbf{k}$$

$$\boldsymbol{\alpha} = (0.15 \text{ rad/s}^2)\mathbf{i} \quad \blacktriangleleft$$

*c. Velocity of Tip of Boom.* Noting that the position vector of point $P$ is $\mathbf{r} = (10.39 \text{ m})\mathbf{i} + (6 \text{ m})\mathbf{j}$ and using the expression found for $\boldsymbol{\omega}$ in part $a$, we write

$$\mathbf{v} = \boldsymbol{\omega} \times \mathbf{r} = \begin{vmatrix} \mathbf{i} & \mathbf{j} & \mathbf{k} \\ 0 & 0.30 \text{ rad/s} & 0.50 \text{ rad/s} \\ 10.39 \text{ m} & 6 \text{ m} & 0 \end{vmatrix}$$

$$\mathbf{v} = -(3 \text{ m/s})\mathbf{i} + (5.20 \text{ m/s})\mathbf{j} - (3.12 \text{ m/s})\mathbf{k} \quad \blacktriangleleft$$

*d. Acceleration of Tip of Boom.* Recalling that $\mathbf{v} = \boldsymbol{\omega} \times \mathbf{r}$, we write

$$\mathbf{a} = \boldsymbol{\alpha} \times \mathbf{r} + \boldsymbol{\omega} \times (\boldsymbol{\omega} \times \mathbf{r}) = \boldsymbol{\alpha} \times \mathbf{r} + \boldsymbol{\omega} \times \mathbf{v}$$

$$\mathbf{a} = \begin{vmatrix} \mathbf{i} & \mathbf{j} & \mathbf{k} \\ 0.15 & 0 & 0 \\ 10.39 & 6 & 0 \end{vmatrix} + \begin{vmatrix} \mathbf{i} & \mathbf{j} & \mathbf{k} \\ 0 & 0.30 & 0.50 \\ -3 & 5.20 & -3.12 \end{vmatrix}$$

$$= 0.90\mathbf{k} - 0.94\mathbf{i} - 2.60\mathbf{i} - 1.50\mathbf{j} + 0.90\mathbf{k}$$

$$\mathbf{a} = -(3.54 \text{ m/s}^2)\mathbf{i} - (1.50 \text{ m/s}^2)\mathbf{j} + (1.80 \text{ m/s}^2)\mathbf{k} \quad \blacktriangleleft$$

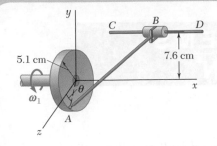

# SAMPLE PROBLEM 15.12

The rod $AB$, of length 17.8 cm, is attached to the disk by a ball-and-socket connection and to the collar $B$ by a clevis. The disk rotates in the $yz$ plane at a constant rate $\omega_1 = 12$ rad/s, while the collar is free to slide along the horizontal rod $CD$. For the position $\theta = 0$, determine (a) the velocity of the collar, (b) the angular velocity of the rod.

## SOLUTION

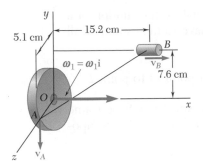

$\omega_1 = 12\,\mathbf{i}$
$\mathbf{r}_A = 5.1\,\mathbf{k}$
$\mathbf{r}_B = 15.2\,\mathbf{i} + 7.6\,\mathbf{j}$
$\mathbf{r}_{B/A} = 15.2\,\mathbf{i} + 7.6\,\mathbf{j} - 5.1\,\mathbf{k}$

**a. Velocity of Collar.** Since point $A$ is attached to the disk and since collar $B$ moves in a direction parallel to the $x$ axis, we have

$$\mathbf{v}_A = \boldsymbol{\omega}_1 \times \mathbf{r}_A = 12\mathbf{i} \times 5.1\mathbf{k} = -61.2\mathbf{j} \qquad \mathbf{v}_B = v_B\mathbf{i}$$

Denoting by $\boldsymbol{\omega}$ the angular velocity of the rod, we write

$$\mathbf{v}_B = \mathbf{v}_A + \mathbf{v}_{B/A} = \mathbf{v}_A + \boldsymbol{\omega} \times \mathbf{r}_{B/A}$$

$$v_B\mathbf{i} = -61.2\mathbf{j} + \begin{vmatrix} \mathbf{i} & \mathbf{j} & \mathbf{k} \\ \omega_x & \omega_y & \omega_z \\ 15.2 & 7.6 & -5.1 \end{vmatrix}$$

$$v_B\mathbf{i} = -24\mathbf{j} + (-5.1\omega_y - 7.6\omega_z)\mathbf{i} + (15.2\omega_z + 5.1\omega_x)\mathbf{j} + (7.6\omega_x - 15.2\omega_y)\mathbf{k}$$

Equating the coefficients of the unit vectors, we obtain

$$v_B = -5.1\omega_y - 7.6\omega_z \tag{1}$$
$$24 = 5.1\omega_x + 15.2\omega_z \tag{2}$$
$$0 = 7.6\omega_x - 15.2\omega_y \tag{3}$$

Clearly, the three equations obtained cannot be solved for the four unknowns $v_B$, $\omega_x$, $\omega_y$, and $\omega_z$. An additional equation reflecting the type of connection at $B$ will be obtained in part $b$. However, since the velocity of $B$ does not depend upon that connection, we should be able to obtain $v_B$ by eliminating the other unknowns from Eqs. (1), (2), and (3). We note that $\omega_x$ can be eliminated by multiplying Eq. (2) by 3, Eq. (3) by $-2$, and adding, and that $\omega_y$ and $\omega_z$ can be eliminated in a similar way. We verify that multiplying Eqs. (1), (2), and (3), respectively, by 6, 3, and $-2$ and adding eliminates all components of $\boldsymbol{\omega}$ and yields an equation that can be solved for $\mathbf{v}_B$:

$$6v_B + 72 = 0 \qquad v_B = -12 \qquad \mathbf{v}_B = -(12 \text{ m/s})\mathbf{i} \quad \blacktriangleleft$$

**b. Angular Velocity of Rod $AB$.** We noted that the angular velocity cannot be determined from Eqs. (1), (2), and (3) alone. An additional equation is obtained by considering the constraint imposed by the clevis at $B$.

The collar-clevis connection at $B$ permits rotation of $AB$ about the rod $CD$ and also about an axis perpendicular to the plane containing $AB$ and $CD$. It prevents rotation of $AB$ about the axis $EB$, which is perpendicular to $CD$ and lies in the plane containing $AB$ and $CD$. Thus the projection of $\boldsymbol{\omega}$ on $\mathbf{r}_{E/B}$ must be zero and we write†

$$\boldsymbol{\omega} \cdot \mathbf{r}_{E/B} = 0 \qquad (\omega_x\mathbf{i} + \omega_y\mathbf{j} + \omega_z\mathbf{k}) \cdot (-7.6\mathbf{j} + 5.1\mathbf{k}) = 0$$
$$-7.6\omega_y + 5.1\omega_z = 0 \tag{4}$$

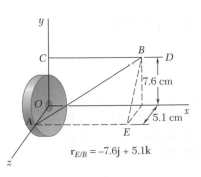

$\mathbf{r}_{E/B} = -7.6\mathbf{j} + 5.1\mathbf{k}$

Solving Eqs. (1) through (4) simultaneously, we obtain

$$v_B = -12 \qquad \omega_x = 1.46 \qquad \omega_y = .73 \qquad \omega_z = 1.09$$
$$\boldsymbol{\omega} = (1.46 \text{ rad/s})\mathbf{i} + (.73 \text{ rad/s})\mathbf{j} + (1.09 \text{ rad/s})\mathbf{k} \quad \blacktriangleleft$$

†We could also note that the direction of $EB$ is that of the vector triple product $\mathbf{r}_{B/C} \times (\mathbf{r}_{B/C} \times \mathbf{r}_{B/A})$ and write $\boldsymbol{\omega} \cdot [\mathbf{r}_{B/C} \times (\mathbf{r}_{B/C} \times \mathbf{r}_{B/A})] = 0$. This formulation would be particularly useful if the rod $CD$ were skew.

# SOLVING PROBLEMS
## ON YOUR OWN

In this lesson you started the study of the *kinematics of rigid bodies in three dimensions*. You first studied the *motion of a rigid body about a fixed point* and then the *general motion of a rigid body*.

**A.  Motion of a rigid body about a fixed point.**   To analyze the motion of a point $B$ of a body rotating about a fixed point $O$ you may have to take some or all of the following steps.

*1.  Determine the position vector* $\mathbf{r}$ connecting the fixed point $O$ to point $B$.

*2.  Determine the angular velocity* $\boldsymbol{\omega}$ *of the body* with respect to a fixed frame of reference. The angular velocity $\boldsymbol{\omega}$ will often be obtained by adding two component angular velocities $\boldsymbol{\omega}_1$ and $\boldsymbol{\omega}_2$ [Sample Prob. 15.11].

*3.  Compute the velocity of B* by using the equation

$$\mathbf{v} = \boldsymbol{\omega} \times \mathbf{r} \qquad (15.37)$$

Your computation will usually be facilitated if you express the vector product as a determinant.

*4.  Determine the angular acceleration* $\boldsymbol{\alpha}$ *of the body.* The angular acceleration $\boldsymbol{\alpha}$ represents the rate of change $(\dot{\boldsymbol{\omega}})_{OXYZ}$ of the vector $\boldsymbol{\omega}$ *with respect to a fixed frame of reference OXYZ* and reflects both *a change in magnitude and a change in direction* of the angular velocity. However, when computing $\boldsymbol{\alpha}$ you may find it convenient to first compute the rate of change $(\dot{\boldsymbol{\omega}})_{Oxyz}$ of $\boldsymbol{\omega}$ with respect to a rotating frame of reference $Oxyz$ of your choice and use Eq. (15.31) of the preceding lesson to obtain $\boldsymbol{\alpha}$. You will write

$$\boldsymbol{\alpha} = (\dot{\boldsymbol{\omega}})_{OXYZ} = (\dot{\boldsymbol{\omega}})_{Oxyz} + \boldsymbol{\Omega} \times \boldsymbol{\omega}$$

where $\boldsymbol{\Omega}$ is the angular velocity of the rotating frame $Oxyz$ [Sample Prob. 15.11].

*5.  Compute the acceleration of B* by using the equation

$$\mathbf{a} = \boldsymbol{\alpha} \times \mathbf{r} + \boldsymbol{\omega} \times (\boldsymbol{\omega} \times \mathbf{r}) \qquad (15.38)$$

Note that the vector product $(\boldsymbol{\omega} \times \mathbf{r})$ represents the velocity of point $B$ and was computed in step 3. Also, the computation of the first vector product in (15.38) will be facilitated if you express this product in determinant form. Remember that, as was the case with the plane motion of a rigid body, the instantaneous axis of rotation *cannot* be used to determine accelerations.

**B.  General motion of a rigid body.**  The general motion of a rigid body may be considered as *the sum of a translation and a rotation*. Keep the following in mind:

**a. In the translation part of the motion,** all the points of the body have the *same velocity* $\mathbf{v}_A$ *and the same acceleration* $\mathbf{a}_A$ as the point $A$ of the body that has been selected as the reference point.

**b. In the rotation part of the motion,** the same reference point $A$ is assumed to be a *fixed point*.

**1.  To determine the velocity of a point B** of the rigid body when you know the velocity $\mathbf{v}_A$ of the reference point $A$ and the angular velocity $\boldsymbol{\omega}$ of the body, you simply add $\mathbf{v}_A$ to the velocity $\mathbf{v}_{B/A} = \boldsymbol{\omega} \times \mathbf{r}_{B/A}$ of $B$ in its rotation about $A$:

$$\mathbf{v}_B = \mathbf{v}_A + \boldsymbol{\omega} \times \mathbf{r}_{B/A} \tag{15.43}$$

As indicated earlier, the computation of the vector product will usually be facilitated if you express this product in determinant form.

Equation (15.43) can also be used to determine the magnitude of $\mathbf{v}_B$ when its direction is known, even if $\boldsymbol{\omega}$ is not known. While the corresponding three scalar equations are linearly dependent and the components of $\boldsymbol{\omega}$ are indeterminate, these components can be eliminated and $\mathbf{v}_A$ can be found by using an appropriate linear combination of the three equations [Sample Prob. 15.12, part $a$]. Alternatively, you can assign an arbitrary value to one of the components of $\boldsymbol{\omega}$ and solve the equations for $\mathbf{v}_A$. However, an additional equation must be sought in order to determine the true values of the components of $\boldsymbol{\omega}$ [Sample Prob. 15.12, part $b$].

**2.  To determine the acceleration of a point B** of the rigid body when you know the acceleration $\mathbf{a}_A$ of the reference point $A$ and the angular acceleration $\boldsymbol{\alpha}$ of the body, you simply add $\mathbf{a}_A$ to the acceleration of $B$ in its rotation about $A$, as expressed by Eq. (15.38):

$$\mathbf{a}_B = \mathbf{a}_A + \boldsymbol{\alpha} \times \mathbf{r}_{B/A} + \boldsymbol{\omega} \times (\boldsymbol{\omega} \times \mathbf{r}_{B/A}) \tag{15.44}$$

Note that the vector product $(\boldsymbol{\omega} \times \mathbf{r}_{B/A})$ represents the velocity $\mathbf{v}_{B/A}$ of $B$ relative to $A$ and may already have been computed as part of your calculation of $\mathbf{v}_B$. We also remind you that the computation of the other two vector products will be facilitated if you express these products in determinant form.

The three scalar equations associated with Eq. (15.44) can also be used to determine the magnitude of $\mathbf{a}_B$ when its direction is known, even if $\boldsymbol{\omega}$ and $\boldsymbol{\alpha}$ are not known. While the components of $\boldsymbol{\omega}$ and $\boldsymbol{\alpha}$ are indeterminate, you can assign arbitrary values to one of the components of $\boldsymbol{\omega}$ and to one of the components of $\boldsymbol{\alpha}$ and solve the equations for $\mathbf{a}_B$.

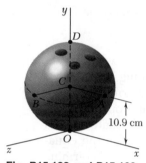

Dimensions in mm

**Fig. P15.181**

**Fig. P15.182 and P15.183**

# Problems

**15.181**  The plate and rods shown are welded together to form an assembly which rotates about the ball-and-socket joint $O$ with an angular velocity $\boldsymbol{\omega}$. Denoting the velocity of point $A$ by $\mathbf{v}_A = (v_A)_x\mathbf{i} + (v_A)_y\mathbf{j} + (v_A)_z\mathbf{k}$, and knowing that $(v_A)_x = 10$ mm/s and $(v_A)_z = 80$ mm/s, determine the velocity component $(v_A)_y$.

**15.182**  The bowling ball shown rolls without slipping on the horizontal $xz$ plane with an angular velocity $\boldsymbol{\omega} = \omega_x\mathbf{i} + \omega_y\mathbf{j} + \omega_z\mathbf{k}$. Knowing that $\mathbf{v}_A = (4.4\ \text{m/s})\mathbf{i} - (4.4\ \text{m/s})\mathbf{j} + (3.3\ \text{m/s})\mathbf{k}$ and $\mathbf{v}_D = (8.8\ \text{m/s})\mathbf{i} + (6.6\ \text{m/s})\mathbf{k}$, determine (a) the angular velocity of the bowling ball, (b) the velocity of its center $C$.

**15.183**  The bowling ball shown rolls without slipping on the horizontal $xz$ plane with an angular velocity $\boldsymbol{\omega} = \omega_x\mathbf{i} + \omega_y\mathbf{j} + \omega_z\mathbf{k}$. Knowing that $\mathbf{v}_B = (3.3\ \text{m/s})\mathbf{i} - (4.4\ \text{m/s})\mathbf{j} + (4.4\ \text{m/s})\mathbf{k}$ and $\mathbf{v}_D = (6.6\ \text{m/s})\mathbf{i} + (8.8\ \text{m/s})\mathbf{k}$, determine (a) the angular velocity of the bowling ball, (b) the velocity of its center $C$.

**15.184**  The L-shaped arm $BCD$ rotates about the $z$ axis with a constant angular velocity $\boldsymbol{\omega}_1$ of 5 rad/s. Knowing that the 19.1-cm-radius disk rotates about $BC$ with a constant angular velocity $\boldsymbol{\omega}_2$ of 4 rad/s, determine the angular acceleration of the disk.

**Fig. P15.184**

**Fig. P15.185**

**15.185**  The blade assembly of an oscillating fan rotates with a constant angular velocity $\boldsymbol{\omega}_1 = -(450\ \text{rpm})\mathbf{i}$ with respect to the motor housing. Determine the angular acceleration of the blade assembly, knowing that at the instant shown the angular velocity and the angular acceleration of the motor housing are, respectively, $\boldsymbol{\omega}_2 = -(3\ \text{rpm})\mathbf{j}$ and $\alpha_2 = 0$.

**15.186** Gear A is constrained to roll on the fixed gear B but is free to rotate about axle AD. Axle AD is connected by a clevis to the vertical shaft DE which rotates as shown with a constant angular velocity $\boldsymbol{\omega}_1$. Determine (a) the angular velocity of gear A, (b) the angular acceleration of gear A.

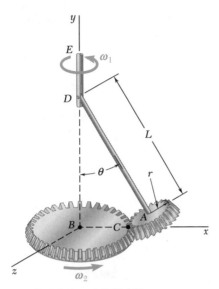

**Fig. P15.186 and P15.187**

**15.187** Gear A is constrained to roll on gear B but is free to rotate about axle AD. Axle AD is connected by a clevis to the vertical shaft DE which rotates as shown with a constant angular velocity $\boldsymbol{\omega}_1$. Knowing that gear B rotates with a constant angular velocity $\boldsymbol{\omega}_2$, determine (a) the angular velocity of gear A, (b) the angular acceleration of gear A.

**15.188** A 60-mm-radius disk spins at the constant rate $\omega_2 = 4$ rad/s about an axis held by a housing attached to a horizontal rod that rotates at the constant rate $\omega_1 = 5$ rad/s. For the position shown, determine (a) the angular acceleration of the disk, (b) the acceleration of point P on the rim of the disk if $\theta = 0$, (c) the acceleration of point P on the rim of the disk if $\theta = 90°$.

**Fig. P15.188 and P15.189**

**15.189** A 60-mm-radius disk spins at the constant rate $\omega_2 = 4$ rad/s about an axis held by a housing attached to a horizontal rod that rotates at the constant rate $\omega_1 = 5$ rad/s. Knowing that $\theta = 30°$, determine the acceleration of point P on the rim of the disk.

**15.190** A gun barrel of length $OP = 3.7$ m is mounted on a turret as shown. To keep the gun aimed at a moving target the azimuth angle $\beta$ is being increased at the rate $d\beta/dt = 30°/$s and the elevation angle $\gamma$ is being increased at the rate $d\gamma/dt = 10°/$s. For the position $\beta = 90°$ and $\gamma = 30°$, determine (a) the angular velocity of the barrel, (b) the angular acceleration of the barrel, (c) the velocity and acceleration of point P.

**Fig. P15.190**

**Fig. P15.191**

**15.191** In the planetary gear system shown, gears $A$ and $B$ are rigidly connected to each other and rotate as a unit about shaft $FG$. Gears $C$ and $D$ rotate with constant angular velocities of 15 rad/s and 30 rad/s, respectively, both counterclockwise when viewed from the right. Choosing the $z$ axis pointing out of the plane of the figure, determine ($a$) the common angular velocity of gears $A$ and $B$, ($b$) the common angular acceleration of gears $A$ and $B$, ($c$) the acceleration of the tooth of gear $B$ that is in contact with gear $D$ at point 2.

**15.192** In Prob. 15.186, determine the acceleration of the point on gear $A$ which is in contact with gear $B$.

**15.193** The cone shown rolls on the $zx$ plane with its apex at the origin of coordinates. Denoting by $\boldsymbol{\omega}_1$ the constant angular velocity of the axis $OB$ of the cone about the $y$ axis, determine ($a$) the rate of spin of the cone about the axis $OB$, ($b$) the total angular velocity of the cone, ($c$) the angular acceleration of the cone.

**Fig. P15.193**

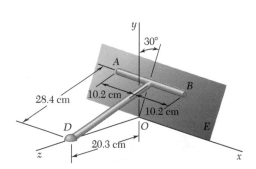

**Fig. P15.194**

**15.194** Two rods are welded together to form the assembly shown that is attached to a fixed ball-and-socket joint at $D$. Rod $AB$ moves on the inclined plane $E$ that is perpendicular to the $yz$ plane. Knowing that at the instant shown the speed of point $B$ is 10.2 cm./s and $\omega_y < 0$, determine ($a$) the angular velocity of the assembly, ($b$) the velocity of point $A$.

**15.195** In Prob. 15.194 the speed of point $B$ is known to be increasing at the rate of 20.3 cm/s². For the position shown, determine ($a$) the angular acceleration of the assembly, ($b$) the radius of curvature of the path of point $B$.

**Fig. P15.196 and P15.197**

**15.196** Rod $AB$ of length 325 mm is connected by ball-and-socket joints to collars $A$ and $B$, which slide along the two rods shown. Knowing that collar $B$ moves toward point $D$ at a constant speed of 1 m/s, determine the velocity of collar $A$ when $b = 100$ mm.

**15.197** Rod $AB$ of length 325 mm is connected by ball-and-socket joints to collars $A$ and $B$, which slide along the two rods shown. Knowing that collar $B$ moves away from point $D$ at a constant speed of 1.6 m/s, determine the velocity of collar $A$ when $b = 156$ mm.

**15.198 and 15.199**   Rod *BC*, of length 45.7 cm, is connected by ball-and-socket joints to the rotating arm *AB* and to collar *C* which can slide on the fixed rod *DE*. Arm *AB* is of length 5 cm and rotates at the constant rate $\omega_0 = 24$ rad/s. For the position shown determine the velocity of collar *C*.

**Fig. P15.198**                                                    **Fig. P15.199**

*15.200*   Rod *AB* of length 38.1 cm is connected by ball-and-socket joints to collars *A* and *B*, which slide along the two rods shown. Knowing that collar *B* moves toward point *D* at a constant speed of 6.35 cm/s, determine the velocity of collar *A* when $c = 10.2$ cm.

**Fig. *P15.200* and *P15.201***

*15.201*   Rod *AB* of length 38.1 cm is connected by ball-and-socket joints to collars *A* and *B*, which slide along the two rods shown. Knowing that collar *B* moves toward point *D* at a constant speed of 6.35 cm/s, determine the velocity of collar *A* when $c = 15.2$ cm.

**15.202** Rod $AB$ of length 500 mm is connected by ball-and-socket joints to collars $A$ and $B$, which slide along the two rods shown. Knowing that collar $B$ moves toward point $E$ at a constant speed of 200 mm/s, determine the velocity of collar $A$ as collar $B$ passes through point $D$.

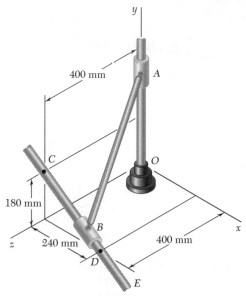

**Fig. P15.202 and P15.203**

**15.203** Rod $AB$ of length 500 mm is connected by ball-and-socket joints to collars $A$ and $B$, which slide along the two rods shown. Knowing that collar $B$ moves toward point $E$ at a constant speed of 200 mm/s, determine the velocity of collar $A$ as collar $B$ passes through point $C$.

**15.204** Two shafts $AC$ and $EG$, which lie in the vertical $yz$ plane, are connected by a universal joint at $D$. Shaft $AC$ rotates with a constant angular velocity $\boldsymbol{\omega}_1$ as shown. At a time when the arm of the crosspiece attached to shaft $AC$ is vertical, determine the angular velocity of shaft $EG$.

**Fig. P15.204**

**15.205** Solve Prob. 15.204 assuming that the arm of the crosspiece attached to shaft *AC* is horizontal.

**15.206** Rod *BC* of length 106.7 cm is connected by a ball-and-socket joint to collar *B* and by a clevis connection to collar *C*. Knowing that collar *B* moves toward *A* at a constant speed of 49.5 cm/s, determine at the instant shown (*a*) the angular velocity of the rod, (*b*) the velocity of collar *C*.

**Fig. P15.206**

**15.207** Rod *AB* of length 28 cm is connected by a ball-and-socket joint to collar *A* and by a clevis connection to collar *B*. Knowing that collar *B* moves down at a constant speed of 1.37 m/s, determine at the instant shown (*a*) the angular velocity of the rod, (*b*) the velocity of collar *A*.

**Fig. P15.207**

**15.208** In Prob. 15.196, determine the acceleration of collar *A*.

**15.209** In Prob. 15.197, determine the acceleration of collar *A*.

**15.210** In Prob. 15.198, determine the acceleration of collar *C*.

**15.211** In Prob. 15.199, determine the acceleration of collar *C*.

**15.212 and 15.213** For the mechanism of the problem indicated, determine the acceleration of collar *A*.

      **15.212** Mechanism of Prob. 15.200.
      **15.213** Mechanism of Prob. 15.201.

## *15.14. THREE-DIMENSIONAL MOTION OF A PARTICLE RELATIVE TO A ROTATING FRAME. CORIOLIS ACCELERATION

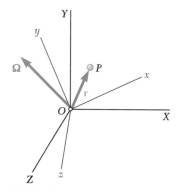

**Fig. 15.36**

**Fig. 15.37**

We saw in Sec. 15.10 that given a vector function $\mathbf{Q}(t)$ and two frames of reference centered at $O$—a fixed frame $OXYZ$ and a rotating frame $Oxyz$—the rates of change of $\mathbf{Q}$ with respect to the two frames satisfy the relation

$$(\dot{\mathbf{Q}})_{OXYZ} = (\dot{\mathbf{Q}})_{Oxyz} + \mathbf{\Omega} \times \mathbf{Q} \qquad (15.31)$$

We had assumed at the time that the frame $Oxyz$ was constrained to rotate about a fixed axis $OA$. However, the derivation given in Sec. 15.10 remains valid when the frame $Oxyz$ is constrained only to have a fixed point $O$. Under this more general assumption, the axis $OA$ represents the *instantaneous* axis of rotation of the frame $Oxyz$ (Sec. 15.12) and the vector $\mathbf{\Omega}$, its angular velocity at the instant considered (Fig. 15.36).

Let us now consider the three-dimensional motion of a particle $P$ relative to a rotating frame $Oxyz$ constrained to have a fixed origin $O$. Let $\mathbf{r}$ be the position vector of $P$ at a given instant and $\mathbf{\Omega}$ be the angular velocity of the frame $Oxyz$ with respect to the fixed frame $OXYZ$ at the same instant (Fig. 15.37). The derivations given in Sec. 15.11 for the two-dimensional motion of a particle can be readily extended to the three-dimensional case, and the absolute velocity $\mathbf{v}_P$ of $P$ (that is, its velocity with respect to the fixed frame $OXYZ$) can be expressed as

$$\mathbf{v}_P = \mathbf{\Omega} \times \mathbf{r} + (\dot{\mathbf{r}})_{Oxyz} \qquad (15.45)$$

Denoting by $\mathcal{F}$ the rotating frame $Oxyz$, we write this relation in the alternative form

$$\mathbf{v}_P = \mathbf{v}_{P'} + \mathbf{v}_{P/\mathcal{F}} \qquad (15.46)$$

where $\mathbf{v}_P$ = absolute velocity of particle $P$
$\quad\mathbf{v}_{P'}$ = velocity of point $P'$ of moving frame $\mathcal{F}$ coinciding with $P$
$\quad\mathbf{v}_{P/\mathcal{F}}$ = velocity of $P$ relative to moving frame $\mathcal{F}$

The absolute acceleration $\mathbf{a}_P$ of $P$ can be expressed as

$$\mathbf{a}_P = \dot{\mathbf{\Omega}} \times \mathbf{r} + \mathbf{\Omega} \times (\mathbf{\Omega} \times \mathbf{r}) + 2\mathbf{\Omega} \times (\dot{\mathbf{r}})_{Oxyz} + (\ddot{\mathbf{r}})_{Oxyz} \qquad (15.47)$$

An alternative form is

$$\mathbf{a}_P = \mathbf{a}_{P'} + \mathbf{a}_{P/\mathcal{F}} + \mathbf{a}_c \qquad (15.48)$$

where $\mathbf{a}_P$ = absolute acceleration of particle $P$
$\quad\mathbf{a}_{P'}$ = acceleration of point $P'$ of moving frame $\mathcal{F}$ coinciding with $P$

$\mathbf{a}_{P/\mathscr{F}}$ = acceleration of $P$ relative to moving frame $\mathscr{F}$

$\mathbf{a}_c = 2\mathbf{\Omega} \times (\dot{\mathbf{r}})_{Oxyz} = 2\mathbf{\Omega} \times \mathbf{v}_{P/\mathscr{F}}$
= complementary, or Coriolis, acceleration†

We note that the Coriolis acceleration is perpendicular to the vectors $\mathbf{\Omega}$ and $\mathbf{v}_{P/\mathscr{F}}$. However, since these vectors are usually not perpendicular to each other, the magnitude of $\mathbf{a}_c$ is in general *not* equal to $2\Omega v_{P/\mathscr{F}}$, as was the case for the plane motion of a particle. We further note that the Coriolis acceleration reduces to zero when the vectors $\mathbf{\Omega}$ and $\mathbf{v}_{P/\mathscr{F}}$ are parallel, or when either of them is zero.

Rotating frames of reference are particularly useful in the study of the three-dimensional motion of rigid bodies. If a rigid body has a fixed point $O$, as was the case for the crane of Sample Prob. 15.11, we can use a frame $Oxyz$ which is neither fixed nor rigidly attached to the rigid body. Denoting by $\mathbf{\Omega}$ the angular velocity of the frame $Oxyz$, we then resolve the angular velocity $\boldsymbol{\omega}$ of the body into the components $\mathbf{\Omega}$ and $\boldsymbol{\omega}_{B/\mathscr{F}}$, where the second component represents the angular velocity of the body relative to the frame $Oxyz$ (see Sample Prob. 15.14). An appropriate choice of the rotating frame often leads to a simpler analysis of the motion of the rigid body than would be possible with axes of fixed orientation. This is especially true in the case of the general three-dimensional motion of a rigid body, that is, when the rigid body under consideration has no fixed point (see Sample Prob. 15.15).

## *15.15. FRAME OF REFERENCE IN GENERAL MOTION

Consider a fixed frame of reference $OXYZ$ and a frame $Axyz$ which moves in a known, but arbitrary, fashion with respect to $OXYZ$ (Fig. 15.38). Let $P$ be a particle moving in space. The position of $P$ is defined at any instant by the vector $\mathbf{r}_P$ in the fixed frame, and by the vector $\mathbf{r}_{P/A}$ in the moving frame. Denoting by $\mathbf{r}_A$ the position vector of $A$ in the fixed frame, we have

$$\mathbf{r}_P = \mathbf{r}_A + \mathbf{r}_{P/A} \qquad (15.49)$$

The absolute velocity $\mathbf{v}_P$ of the particle is obtained by writing

$$\mathbf{v}_P = \dot{\mathbf{r}}_P = \dot{\mathbf{r}}_A + \dot{\mathbf{r}}_{P/A} \qquad (15.50)$$

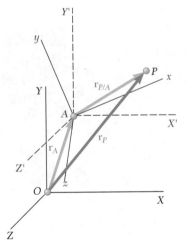

**Fig. 15.38**

where the derivatives are defined with respect to the fixed frame $OXYZ$. The first term in the right-hand member of (15.50) thus represents the velocity $\mathbf{v}_A$ of the origin $A$ of the moving axes. On the other hand, since the rate of change of a vector is the same with respect to a fixed frame and with respect to a frame in translation (Sec. 11.10), the second term can be regarded as the velocity $\mathbf{v}_{P/A}$ of $P$ relative to the frame $AX'Y'Z'$ of the same orientation as $OXYZ$ and the same origin as $Axyz$. We therefore have

$$\mathbf{v}_P = \mathbf{v}_A + \mathbf{v}_{P/A} \qquad (15.51)$$

---

†It is important to note the difference between Eq. (15.48) and Eq. (15.21) of Sec. 15.8. See the footnote on page 974.

But the velocity $\mathbf{v}_{P/A}$ of $P$ relative to $AX'Y'Z'$ can be obtained from (15.45) by substituting $\mathbf{r}_{P/A}$ for $\mathbf{r}$ in that equation. We write

$$\mathbf{v}_P = \mathbf{v}_A + \boldsymbol{\Omega} \times \mathbf{r}_{P/A} + (\dot{\mathbf{r}}_{P/A})_{Axyz} \tag{15.52}$$

where $\boldsymbol{\Omega}$ is the angular velocity of the frame $Axyz$ at the instant considered.

The absolute acceleration $\mathbf{a}_P$ of the particle is obtained by differentiating (15.51) and writing

$$\mathbf{a}_P = \dot{\mathbf{v}}_P = \dot{\mathbf{v}}_A + \dot{\mathbf{v}}_{P/A} \tag{15.53}$$

where the derivatives are defined with respect to either of the frames $OXYZ$ or $AX'Y'Z'$. Thus, the first term in the right-hand member of (15.53) represents the acceleration $\mathbf{a}_A$ of the origin $A$ of the moving axes and the second term represents the acceleration $\mathbf{a}_{P/A}$ of $P$ relative to the frame $AX'Y'Z'$. This acceleration can be obtained from (15.47) by substituting $\mathbf{r}_{P/A}$ for $\mathbf{r}$. We therefore write

$$\mathbf{a}_P = \mathbf{a}_A + \dot{\boldsymbol{\Omega}} \times \mathbf{r}_{P/A} + \boldsymbol{\Omega} \times (\boldsymbol{\Omega} \times \mathbf{r}_{P/A})$$
$$+ 2\boldsymbol{\Omega} \times (\dot{\mathbf{r}}_{P/A})_{Axyz} + (\ddot{\mathbf{r}}_{P/A})_{Axyz} \tag{15.54}$$

**Photo 15.5** The motion of air particles in a hurricane can be considered as motion relative to a frame of reference attached to the Earth and rotating with it.

Formulas (15.52) and (15.54) make it possible to determine the velocity and acceleration of a given particle with respect to a fixed frame of reference, when the motion of the particle is known with respect to a moving frame. These formulas become more significant, and considerably easier to remember, if we note that the sum of the first two terms in (15.52) represents the velocity of the point $P'$ of the moving frame which coincides with $P$ at the instant considered, and that the sum of the first three terms in (15.54) represents the acceleration of the same point. Thus, the relations (15.46) and (15.48) of the preceding section are still valid in the case of a reference frame in general motion, and we can write

$$\mathbf{v}_P = \mathbf{v}_{P'} + \mathbf{v}_{P/\mathscr{F}} \tag{15.46}$$
$$\mathbf{a}_P = \mathbf{a}_{P'} + \mathbf{a}_{P/\mathscr{F}} + \mathbf{a}_c \tag{15.48}$$

where the various vectors involved have been defined in Sec. 15.14.

It should be noted that if the moving reference frame $\mathscr{F}$ (or $Axyz$) is in translation, the velocity and acceleration of the point $P'$ of the frame which coincides with $P$ become, respectively, equal to the velocity and acceleration of the origin $A$ of the frame. On the other hand, since the frame maintains a fixed orientation, $\mathbf{a}_c$ is zero, and the relations (15.46) and (15.48) reduce, respectively, to the relations (11.33) and (11.34) derived in Sec. 11.12.

## SAMPLE PROBLEM 15.13

The bent rod $OAB$ rotates about the vertical $OB$. At the instant considered, its angular velocity and angular acceleration are, respectively, 20 rad/s and 200 rad/s², both clockwise when viewed from the positive $Y$ axis. The collar $D$ moves along the rod, and at the instant considered, $OD = 20.3$ cm. The velocity and acceleration of the collar relative to the rod are, respectively, 127 cm/s and 1524 cm/s², both upward. Determine (a) the velocity of the collar, (b) the acceleration of the collar.

## SOLUTION

**Frames of Reference.** The frame $OXYZ$ is fixed. We attach the rotating frame $Oxyz$ to the bent rod. Its angular velocity and angular acceleration relative to $OXYZ$ are therefore $\mathbf{\Omega} = (-20 \text{ rad/s})\mathbf{j}$ and $\dot{\mathbf{\Omega}} = (-200 \text{ rad/s}^2)\mathbf{j}$, respectively. The position vector of $D$ is

$$\mathbf{r} = (20.3 \text{ cm})(\sin 30°\mathbf{i} + \cos 30°\mathbf{j}) = (10.15 \text{ cm})\mathbf{i} + (17.58 \text{ cm})\mathbf{j}$$

*a. Velocity $\mathbf{v}_D$.* Denoting by $D'$ the point of the rod which coincides with $D$ and by $\mathscr{F}$ the rotating frame $Oxyz$, we write from Eq. (15.46)

$$\mathbf{v}_D = \mathbf{v}_{D'} + \mathbf{v}_{D/\mathscr{F}} \tag{1}$$

where

$$\mathbf{v}_{D'} = \mathbf{\Omega} \times \mathbf{r} = (-20 \text{ rad/s})\mathbf{j} \times [(10.15 \text{ cm})\mathbf{i} + (17.58 \text{ cm})\mathbf{j}] = (203 \text{ cm/s})\mathbf{k}$$
$$\mathbf{v}_{D/\mathscr{F}} = (127 \text{ cm/s})(\sin 30°\mathbf{i} + \cos 30°\mathbf{j}) = (63.5 \text{ cm/s})\mathbf{i} + (110 \text{ cm/s})\mathbf{j}$$

Substituting the values obtained for $\mathbf{v}_{D'}$ and $\mathbf{v}_{D/\mathscr{F}}$ into (1), we find

$$\mathbf{v}_D = (63.5 \text{ cm/s})\mathbf{i} + (110 \text{ cm/s})\mathbf{j} + (203 \text{ cm/s})\mathbf{k} \quad \blacktriangleleft$$

*b. Acceleration $\mathbf{a}_D$.* From Eq. (15.48) we write

$$\mathbf{a}_D = \mathbf{a}_{D'} + \mathbf{a}_{D/\mathscr{F}} + \mathbf{a}_c \tag{2}$$

where

$$\mathbf{a}_{D'} = \dot{\mathbf{\Omega}} \times \mathbf{r} + \mathbf{\Omega} \times (\mathbf{\Omega} \times \mathbf{r})$$
$$= (-200 \text{ rad/s}^2)\mathbf{j} \times [(10.15 \text{ cm})\mathbf{i} + (17.58 \text{ cm})\mathbf{j}] - (20 \text{ rad/s})\mathbf{j} \times (203 \text{ cm/s})\mathbf{k}$$
$$= +(2030 \text{ cm/s}^2)\mathbf{k} - (4060 \text{ cm/s}^2)\mathbf{i}$$
$$\mathbf{a}_{D/\mathscr{F}} = (1524 \text{ cm/s}^2)(\sin 30°\mathbf{i} + \cos 30°\mathbf{j}) = (762 \text{ cm/s}^2)\mathbf{i} + (1320 \text{ cm/s}^2)\mathbf{j}$$
$$\mathbf{a}_c = 2\mathbf{\Omega} \times \mathbf{v}_{D/\mathscr{F}}$$
$$= 2(-20 \text{ rad/s})\mathbf{j} \times [(63.5 \text{ cm/s})\mathbf{i} + (110 \text{ cm/s})\mathbf{j}] = (2540 \text{ cm/s}^2)\mathbf{k}$$

Substituting the values obtained for $\mathbf{a}_{D'}$, $\mathbf{a}_{D/\mathscr{F}}$, and $\mathbf{a}_c$ into (2),

$$\mathbf{a}_D = -(3298 \text{ cm/s}^2)\mathbf{i} + (1320 \text{ cm/s}^2)\mathbf{j} + (4570 \text{ cm/s}^2)\mathbf{k} \quad \blacktriangleleft$$

# SAMPLE PROBLEM 15.14

The crane shown rotates with a constant angular velocity $\boldsymbol{\omega}_1$ of 0.30 rad/s. Simultaneously, the boom is being raised with a constant angular velocity $\boldsymbol{\omega}_2$ of 0.50 rad/s relative to the cab. Knowing that the length of the boom $OP$ is $l = 12$ m, determine (a) the velocity of the tip of the boom, (b) the acceleration of the tip of the boom.

## SOLUTION

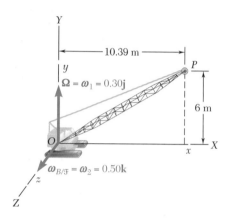

**Frames of Reference.** The frame $OXYZ$ is fixed. We attach the rotating frame $Oxyz$ to the cab. Its angular velocity with respect to the frame $OXYZ$ is therefore $\boldsymbol{\Omega} = \boldsymbol{\omega}_1 = (0.30 \text{ rad/s})\mathbf{j}$. The angular velocity of the boom relative to the cab and the rotating frame $Oxyz$ (or $\mathcal{F}$, for short) is $\boldsymbol{\omega}_{B/\mathcal{F}} = \boldsymbol{\omega}_2 = (0.50 \text{ rad/s})\mathbf{k}$.

*a. Velocity* $\mathbf{v}_P$. From Eq. (15.46) we write

$$\mathbf{v}_P = \mathbf{v}_{P'} + \mathbf{v}_{P/\mathcal{F}} \tag{1}$$

where $\mathbf{v}_{P'}$ is the velocity of the point $P'$ of the rotating frame which coincides with $P$:

$$\mathbf{v}_{P'} = \boldsymbol{\Omega} \times \mathbf{r} = (0.30 \text{ rad/s})\mathbf{j} \times [(10.39 \text{ m})\mathbf{i} + (6 \text{ m})\mathbf{j}] = -(3.12 \text{ m/s})\mathbf{k}$$

and where $\mathbf{v}_{P/\mathcal{F}}$ is the velocity of $P$ relative to the rotating frame $Oxyz$. But the angular velocity of the boom relative to $Oxyz$ was found to be $\boldsymbol{\omega}_{B/\mathcal{F}} = (0.50 \text{ rad/s})\mathbf{k}$. The velocity of its tip $P$ relative to $Oxyz$ is therefore

$$\begin{aligned} \mathbf{v}_{P/\mathcal{F}} = \boldsymbol{\omega}_{B/\mathcal{F}} \times \mathbf{r} &= (0.50 \text{ rad/s})\mathbf{k} \times [(10.39 \text{ m})\mathbf{i} + (6 \text{ m})\mathbf{j}] \\ &= -(3 \text{ m/s})\mathbf{i} + (5.20 \text{ m/s})\mathbf{j} \end{aligned}$$

Substituting the values obtained for $\mathbf{v}_{P'}$ and $\mathbf{v}_{P/\mathcal{F}}$ into (1), we find

$$\mathbf{v}_P = -(3 \text{ m/s})\mathbf{i} + (5.20 \text{ m/s})\mathbf{j} - (3.12 \text{ m/s})\mathbf{k} \quad \blacktriangleleft$$

*b. Acceleration* $\mathbf{a}_P$. From Eq. (15.48) we write

$$\mathbf{a}_P = \mathbf{a}_{P'} + \mathbf{a}_{P/\mathcal{F}} + \mathbf{a}_c \tag{2}$$

Since $\boldsymbol{\Omega}$ and $\boldsymbol{\omega}_{B/\mathcal{F}}$ are both constant, we have

$$\mathbf{a}_{P'} = \boldsymbol{\Omega} \times (\boldsymbol{\Omega} \times \mathbf{r}) = (0.30 \text{ rad/s})\mathbf{j} \times (-3.12 \text{ m/s})\mathbf{k} = -(0.94 \text{ m/s}^2)\mathbf{i}$$

$$\begin{aligned} \mathbf{a}_{P/\mathcal{F}} = \boldsymbol{\omega}_{B/\mathcal{F}} \times (\boldsymbol{\omega}_{B/\mathcal{F}} \times \mathbf{r}) \\ = (0.50 \text{ rad/s})\mathbf{k} \times [-(3 \text{ m/s})\mathbf{i} + (5.20 \text{ m/s})\mathbf{j}] \\ = -(1.50 \text{ m/s}^2)\mathbf{j} - (2.60 \text{ m/s}^2)\mathbf{i} \end{aligned}$$

$$\begin{aligned} \mathbf{a}_c = 2\boldsymbol{\Omega} \times \mathbf{v}_{P/\mathcal{F}} \\ = 2(0.30 \text{ rad/s})\mathbf{j} \times [-(3 \text{ m/s})\mathbf{i} + (5.20 \text{ m/s})\mathbf{j}] = (1.80 \text{ m/s}^2)\mathbf{k} \end{aligned}$$

Substituting for $\mathbf{a}_{P'}$, $\mathbf{a}_{P/\mathcal{F}}$, and $\mathbf{a}_c$ into (2), we find

$$\mathbf{a}_P = -(3.54 \text{ m/s}^2)\mathbf{i} - (1.50 \text{ m/s}^2)\mathbf{j} + (1.80 \text{ m/s}^2)\mathbf{k} \quad \blacktriangleleft$$

## SAMPLE PROBLEM 15.15

Disk $D$, of radius $R$, is pinned to end $A$ of the arm $OA$ of length $L$ located in the plane of the disk. The arm rotates about a vertical axis through $O$ at the constant rate $\omega_1$, and the disk rotates about $A$ at the constant rate $\omega_2$. Determine (a) the velocity of point $P$ located directly above $A$, (b) the acceleration of $P$, (c) the angular velocity and angular acceleration of the disk.

## SOLUTION

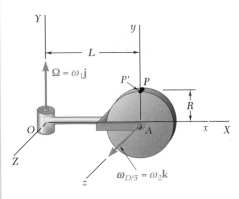

**Frames of Reference.** The frame $OXYZ$ is fixed. We attach the moving frame $Axyz$ to the arm $OA$. Its angular velocity with respect to the frame $OXYZ$ is therefore $\boldsymbol{\Omega} = \omega_1\mathbf{j}$. The angular velocity of disk $D$ relative to the moving frame $Axyz$ (or $\mathcal{F}$, for short) is $\boldsymbol{\omega}_{D/\mathcal{F}} = \omega_2\mathbf{k}$. The position vector of $P$ relative to $O$ is $\mathbf{r} = L\mathbf{i} + R\mathbf{j}$, and its position vector relative to $A$ is $\mathbf{r}_{P/A} = R\mathbf{j}$.

*a. Velocity $\mathbf{v}_P$.* Denoting by $P'$ the point of the moving frame which coincides with $P$, we write from Eq. (15.46)

$$\mathbf{v}_P = \mathbf{v}_{P'} + \mathbf{v}_{P/\mathcal{F}} \tag{1}$$

where $\mathbf{v}_{P'} = \boldsymbol{\Omega} \times \mathbf{r} = \omega_1\mathbf{j} \times (L\mathbf{i} + R\mathbf{j}) = -\omega_1 L\mathbf{k}$

$\mathbf{v}_{P/\mathcal{F}} = \boldsymbol{\omega}_{D/\mathcal{F}} \times \mathbf{r}_{P/A} = \omega_2\mathbf{k} \times R\mathbf{j} = -\omega_2 R\mathbf{i}$

Substituting the values obtained for $\mathbf{v}_{P'}$ and $\mathbf{v}_{P/\mathcal{F}}$ into (1), we find

$$\mathbf{v}_P = -\omega_2 R\mathbf{i} - \omega_1 L\mathbf{k} \quad \blacktriangleleft$$

*b. Acceleration $\mathbf{a}_P$.* From Eq. (15.48) we write

$$\mathbf{a}_P = \mathbf{a}_{P'} + \mathbf{a}_{P/\mathcal{F}} + \mathbf{a}_c \tag{2}$$

Since $\boldsymbol{\Omega}$ and $\boldsymbol{\omega}_{D/\mathcal{F}}$ are both constant, we have

$$\mathbf{a}_{P'} = \boldsymbol{\Omega} \times (\boldsymbol{\Omega} \times \mathbf{r}) = \omega_1\mathbf{j} \times (-\omega_1 L\mathbf{k}) = -\omega_1^2 L\mathbf{i}$$
$$\mathbf{a}_{P/\mathcal{F}} = \boldsymbol{\omega}_{D/\mathcal{F}} \times (\boldsymbol{\omega}_{D/\mathcal{F}} \times \mathbf{r}_{P/A}) = \omega_2\mathbf{k} \times (-\omega_2 R\mathbf{i}) = -\omega_2^2 R\mathbf{j}$$
$$\mathbf{a}_c = 2\boldsymbol{\Omega} \times \mathbf{v}_{P/\mathcal{F}} = 2\omega_1\mathbf{j} \times (-\omega_2 R\mathbf{i}) = 2\omega_1\omega_2 R\mathbf{k}$$

Substituting the values obtained into (2), we find

$$\mathbf{a}_P = -\omega_1^2 L\mathbf{i} - \omega_2^2 R\mathbf{j} + 2\omega_1\omega_2 R\mathbf{k} \quad \blacktriangleleft$$

*c. Angular Velocity and Angular Acceleration of Disk.*

$$\boldsymbol{\omega} = \boldsymbol{\Omega} + \boldsymbol{\omega}_{D/\mathcal{F}} \qquad \boldsymbol{\omega} = \omega_1\mathbf{j} + \omega_2\mathbf{k} \quad \blacktriangleleft$$

Using Eq. (15.31) with $\mathbf{Q} = \boldsymbol{\omega}$, we write

$$\boldsymbol{\alpha} = (\dot{\boldsymbol{\omega}})_{OXYZ} = (\dot{\boldsymbol{\omega}})_{Axyz} + \boldsymbol{\Omega} \times \boldsymbol{\omega}$$
$$= 0 + \omega_1\mathbf{j} \times (\omega_1\mathbf{j} + \omega_2\mathbf{k})$$

$$\boldsymbol{\alpha} = \omega_1\omega_2\mathbf{i} \quad \blacktriangleleft$$

In this lesson you concluded your study of the kinematics of rigid bodies by learning how to use an auxiliary frame of reference $\mathcal{F}$ to analyze the three-dimensional motion of a rigid body. This auxiliary frame may be a *rotating frame* with a fixed origin $O$, or it may be a *frame in general motion*.

**A. Using a rotating frame of reference.** As you approach a problem involving the use of a rotating frame $\mathcal{F}$ you should take the following steps.

**1. Select the rotating frame $\mathcal{F}$ that you wish to use** and draw the corresponding coordinate axes $x$, $y$, and $z$ from the fixed point $O$.

**2. Determine the angular velocity $\Omega$ of the frame $\mathcal{F}$** with respect to a fixed frame $OXYZ$. In most cases, you will have selected a frame which is attached to some rotating element of the system; $\Omega$ will then be the angular velocity of that element.

**3. Designate as $P'$ the point of the rotating frame $\mathcal{F}$** that coincides with the point $P$ of interest at the instant you are considering. Determine the velocity $\mathbf{v}_{P'}$ and the acceleration $\mathbf{a}_{P'}$ of point $P'$. Since $P'$ is part of $\mathcal{F}$ and has the same position vector $\mathbf{r}$ as $P$, you will find that

$$\mathbf{v}_{P'} = \boldsymbol{\Omega} \times \mathbf{r} \qquad \text{and} \qquad \mathbf{a}_{P'} = \boldsymbol{\alpha} \times \mathbf{r} + \boldsymbol{\Omega} \times (\boldsymbol{\Omega} \times \mathbf{r})$$

where $\boldsymbol{\alpha}$ is the angular acceleration of $\mathcal{F}$. However, in many of the problems that you will encounter, the angular velocity of $\mathcal{F}$ is constant in both magnitude and direction, and $\boldsymbol{\alpha} = 0$.

**4. Determine the velocity and acceleration of point $P$ with respect to the frame $\mathcal{F}$.** As you are trying to determine $\mathbf{v}_{P/\mathcal{F}}$ and $\mathbf{a}_{P/\mathcal{F}}$ you will find it useful to visualize the motion of $P$ on frame $\mathcal{F}$ when the frame is not rotating. If $P$ is a point of a rigid body $\mathcal{B}$ which has an angular velocity $\boldsymbol{\omega}_{\mathcal{B}}$ and an angular acceleration $\boldsymbol{\alpha}_{\mathcal{B}}$ relative to $\mathcal{F}$ [Sample Prob. 15.14], you will find that

$$\mathbf{v}_{P/\mathcal{F}} = \boldsymbol{\omega}_{\mathcal{B}} \times \mathbf{r} \qquad \text{and} \qquad \mathbf{a}_{P/\mathcal{F}} = \boldsymbol{\alpha}_{\mathcal{B}} \times \mathbf{r} + \boldsymbol{\omega}_{\mathcal{B}} \times (\boldsymbol{\omega}_{\mathcal{B}} \times \mathbf{r})$$

In many of the problems that you will encounter, the angular velocity of body $\mathcal{B}$ relative to frame $\mathcal{F}$ is constant in both magnitude and direction, and $\boldsymbol{\alpha}_{\mathcal{B}} = 0$.

**5. Determine the Coriolis acceleration.** Considering the angular velocity $\Omega$ of frame $\mathcal{F}$ and the velocity $\mathbf{v}_{P/\mathcal{F}}$ of point $P$ relative to that frame, which was computed in the previous step, you write

$$\mathbf{a}_c = 2\boldsymbol{\Omega} \times \mathbf{v}_{P/\mathcal{F}}$$

**6. The velocity and the acceleration of P with respect to the fixed frame OXYZ** can now be obtained by adding the expressions you have determined:

$$\mathbf{v}_P = \mathbf{v}_{P'} + \mathbf{v}_{P/\mathcal{F}} \tag{15.46}$$

$$\mathbf{a}_P = \mathbf{a}_{P'} + \mathbf{a}_{P/\mathcal{F}} + \mathbf{a}_c \tag{15.48}$$

**B. Using a frame of reference in general motion.** The steps that you will take differ only slightly from those listed under A. They consist of the following:

**1. Select the frame $\mathcal{F}$ that you wish to use and a reference point A in that frame,** from which you will draw the coordinate axes, $x$, $y$, and $z$ defining that frame. You will consider the motion of the frame as the sum of a *translation with A and a rotation about A*.

**2. Determine the velocity $\mathbf{v}_A$ of point A and the angular velocity $\Omega$ of the frame.** In most cases, you will have selected a frame which is attached to some element of the system; $\Omega$ will then be the angular velocity of that element.

**3. Designate as P′ the point of frame $\mathcal{F}$** that coincides with the point $P$ of interest at the instant you are considering, and determine the velocity $\mathbf{v}_{P'}$ and the acceleration $\mathbf{a}_{P'}$ of that point. In some cases, this can be done by visualizing the motion of $P$ if that point were prevented from moving with respect to $\mathcal{F}$ [Sample Prob. 15.15]. A more general approach is to recall that the motion of $P'$ is the sum of a translation with the reference point A and a rotation about A. The velocity $\mathbf{v}_{P'}$ and the acceleration $\mathbf{a}_{P'}$ of $P'$, therefore, can be obtained by adding $\mathbf{v}_A$ and $\mathbf{a}_A$, respectively, to the expressions found in paragraph A3 and replacing the position vector $\mathbf{r}$ by the vector $\mathbf{r}_{P/A}$ drawn from A to P:

$$\mathbf{v}_{P'} = \mathbf{v}_A + \Omega \times \mathbf{r}_{P/A} \qquad \mathbf{a}_{P'} = \mathbf{a}_A + \boldsymbol{\alpha} \times \mathbf{r}_{P/A} + \Omega \times (\Omega \times \mathbf{r}_{P/A})$$

*Steps 4, 5, and 6 are the same as in Part A,* except that the vector $\mathbf{r}$ should again be replaced by $\mathbf{r}_{P/A}$. Thus, Eqs. (15.46) and (15.48) can still be used to obtain the velocity and the acceleration of $P$ with respect to the fixed frame of reference OXYZ.

Fig. P15.214

Fig. P15.215

# Problems

**15.214** Rod $AB$ is welded to the 30.5 cm-radius plate which rotates at the constant rate $\omega_1 = 6$ rad/s. Knowing that collar $D$ moves toward end $B$ of the rod at a constant speed $u = 198$ cm/s, determine, for the position shown, (a) the velocity of $D$, (b) the acceleration of $D$.

**15.215** The circular plate shown rotates about its vertical diameter at the constant rate $\omega_1 = 10$ rad/s. Knowing that in the position shown the disk lies in the $XY$ plane and point $D$ of strap $CD$ moves upward at a constant relative speed $u = 1.37$ m/s, determine (a) the velocity of $D$, (b) the acceleration of $D$.

**15.216** The bent pipe shown rotates at the constant rate $\omega_1 = 10$ rad/s. Knowing that a ball bearing $D$ moves in portion $BC$ of the pipe toward end $C$ at a constant relative speed $u = 0.6$ m/s, determine at the instant shown (a) the velocity of $D$, (b) the acceleration of $D$.

**15.217** The bent rod shown rotates at the constant rate $\omega_1 = 5$ rad/s and collar $C$ moves toward point $B$ at a constant relative speed $u = 975$ mm/s. Knowing that collar $C$ is halfway between points $B$ and $D$ at the instant shown, determine its velocity and acceleration.

Fig. P15.216

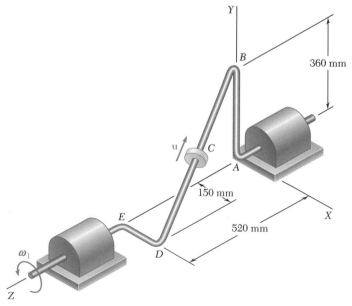

Fig. P15.217

**15.218** The rectangular plate shown rotates at the constant rate $\omega_2 = 12$ rad/s with respect to arm $AE$, which itself rotates at the constant rate $\omega_1 = 9$ rad/s about the Z axis. For the position shown, determine (a) the velocity of corner $B$, (b) the acceleration of corner $B$.

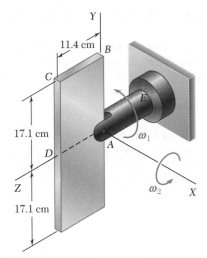

**Fig. *P15.218* and *P15.219***

**15.219** The rectangular plate shown rotates at the constant rate $\omega_2 = 12$ rad/s with respect to arm $AE$, which itself rotates at the constant rate $\omega_1 = 9$ rad/s about the Z axis. For the position shown, determine (a) the velocity of corner $C$, (b) the acceleration of corner $C$.

**15.220** A square plate of side 360 mm is hinged at $A$ and $B$ to a clevis. The plate rotates at the constant rate $\omega_2 = 4$ rad/s with respect to the clevis, which itself rotates at the constant rate $\omega_1 = 3$ rad/s about the $Y$ axis. For the position shown, determine (a) the velocity of point $C$, (b) the acceleration of point $C$.

**15.221** A square plate of side 360 mm is hinged at $A$ and $B$ to a clevis. The plate rotates at the constant rate $\omega_2 = 4$ rad/s with respect to the clevis, which itself rotates at the constant rate $\omega_1 = 3$ rad/s about the $Y$ axis. For the position shown, determine (a) the velocity of corner $D$, (b) the acceleration of corner $D$.

**Fig. P15.220 and P15.221**

**15.222** Solve Prob. 15.216 assuming that at the instant shown the angular velocity $\omega_1$ of the pipe is 10 rad/s and is decreasing at the rate of 15 rad/s², while the relative speed $u$ of the ball bearing is 0.6 m/s and is increasing at the rate of 3 m/s².

**15.223** Solve Prob. 15.217 assuming that at the instant shown the angular velocity $\omega_1$ of the rod is 5 rad/s and is increasing at the rate of 10 rad/s², while the relative speed $u$ of the collar $C$ is 975 mm/s and is decreasing at the rate of 6.5 m/s².

**15.224** Solve Prob. 15.219 assuming that at the instant shown the angular velocity $\omega_2$ of the plate with respect to arm $AE$ is 12 rad/s and is decreasing at the rate of 60 rad/s², while the angular velocity $\omega_1$ of the arm about the Z axis is 9 rad/s and is decreasing at the rate of 45 rad/s².

**Fig. P15.229**

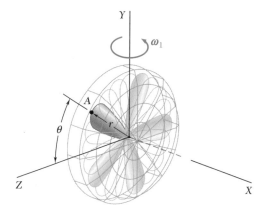

**Fig. P15.230**

**15.225**   Using the method of Sec. 15.14, solve Prob. 15.186.

**15.226**   Using the method of Sec. 15.14, solve Prob. 15.187.

*15.227*   Using the method of Sec. 15.14, solve Prob. 15.188.

*15.228*   Using the method of Sec. 15.14, solve Prob. 15.189.

**15.229**   The crane shown rotates at the constant rate $\omega_1 = 0.25$ rad/s; simultaneously, the telescoping boom is being lowered at the constant rate $\omega_2 = 0.40$ rad/s. Knowing that at the instant shown the length of the boom is 6 m and is increasing at the constant rate $u = 0.45$ m/s, determine the velocity and acceleration of point $B$.

**15.230**   The rotor of a fan rotates about the $Y$ axis at a constant rate $\omega_1 = 0.8$ rad/s. At the same time, the blades rotate at a constant rate $\omega_2 = d\theta/dt = 300$ rpm. Knowing that the distance between the rotor and point $A$ on the tip of the blade is $r = 160$ mm and that $\theta = 45°$ at the instant shown, determine the velocity and acceleration of point $A$.

**15.231**   The 40.6 cm. bar $AB$ is made to rotate at the constant rate $\omega_2 = d\theta/dt = 8$ rad/s with respect to the frame $CD$ which itself rotates at the constant rate $\omega_1 = 12$ rad/s about the $Y$ axis. Knowing that $\theta = 60°$ at the instant shown, determine the velocity and acceleration of point $A$.

**Fig. P15.231 and P15.232**

**15.232**   The 40.6 cm bar $AB$ is made to rotate at the rate $\omega_2 = d\theta/dt$ with respect to the frame $CD$ which itself rotates at the rate $\omega_1$ about the $Y$ axis. At the instant shown $\omega_1 = 12$ rad/s, $d\omega_1/dt = -16$ rad/s$^2$, $\omega_2 = 8$ rad/s, $d\omega_2/dt = 10$ rad/s$^2$, and $\theta = 60°$. Determine the velocity and acceleration of point $A$ at this instant.

**15.233** The arm $AB$ of length 5 m is used to provide an elevated platform for construction workers. In the position shown, arm $AB$ is being raised at the constant rate $d\theta/dt = 0.25$ rad/s; simultaneously, the unit is being rotated counterclockwise about the $Y$ axis at the constant rate $\omega_1 = 0.15$ rad/s. Knowing that $\theta = 20°$, determine the velocity and acceleration of point $B$.

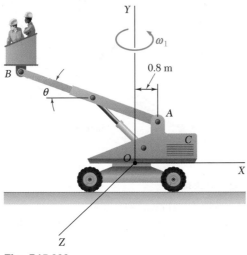

**Fig. P15.233**

**15.234** The rod $CD$ is attached to arm $ABC$ by means of a clevis at $C$ and is made to rotate at the rate $\omega_2 = d\theta/dt$ with respect to arm $ABC$. The entire assembly rotates about the fixed $Y$ axis at the constant rate $\omega_1 = 8$ rad/s. Knowing that $\theta = 45°$, $\omega_2 = 4$ rad/s, and $d\omega_2/dt = 5$ rad/s² at the instant shown, determine the velocity and acceleration of point $D$.

**15.235** A disk of 22.9 cm radius rotates at the constant rate $\omega_2 = 12$ rad/s with respect to arm $CD$, which itself rotates at the constant rate $\omega_1 = 8$ rad/s about the $Y$ axis. Determine at the instant shown the velocity and acceleration of point $A$ on the rim of the disk.

**Fig. P15.234**

**Fig. P15.235 and P15.236**

**15.236** A disk of 22.9 cm radius rotates at the constant rate $\omega_2 = 12$ rad/s with respect to arm $CD$, which itself rotates at the constant rate $\omega_1 = 8$ rad/s about the $Y$ axis. Determine at the instant shown the velocity and acceleration of point $B$ on the rim of the disk.

**Fig. P15.237**

**15.237 and 15.238**   In the position shown the thin rod moves at a constant speed $u$ = 60 mm/s out of the tube $BC$. At the same time tube $BC$ rotates at the constant rate $\omega_2$ = 1.5 rad/s with respect to arm $CD$. Knowing that the entire assembly rotates about the $X$ axis at the constant rate $\omega_1$ = 1.2 rad/s, determine the velocity and acceleration of end $A$ of the rod.

**Fig. P15.238**

**15.239**   The cylinder shown rotates at the constant rate $\omega_2$ = 8 rad/s with respect to rod $CD$, which itself rotates at the constant rate $\omega_1$ = 6 rad/s about the $X$ axis. For the position shown, determine the velocity and acceleration of point $A$ on the edge of the cylinder.

**Fig. P15.239 and P15.240**

**15.240**   The cylinder shown rotates at the constant rate $\omega_2$ = 8 rad/s with respect to rod $CD$, which itself rotates at the constant rate $\omega_1$ = 6 rad/s about the $X$ axis. For the position shown, determine the velocity and acceleration of point $B$ on the edge of the cylinder.

**15.241**   The vertical plate shown is welded to arm $EFG$, and the entire unit rotates at the constant rate $\omega_1$ = 1.6 rad/s about the $Y$ axis. At the same time, a continuous link belt moves around the perimeter of the plate at a constant speed $u$ = 90 mm/s. For the position shown, determine the acceleration of the link of the belt located ($a$) at point $A$, ($b$) at point $B$.

**15.242**   The vertical plate shown is welded to arm $EFG$, and the entire unit rotates at the constant rate $\omega_1$ = 1.6 rad/s about the $Y$ axis. At the same time, a continuous link belt moves around the perimeter of the plate at a constant speed $u$ = 90 mm/s. For the position shown, determine the acceleration of the link of the belt located ($a$) at point $C$, ($b$) at point $D$.

**Fig. P15.241 and P15.242**

This chapter was devoted to the study of the kinematics of rigid bodies.

We first considered the *translation* of a rigid body [Sec. 15.2] and observed that in such a motion, *all points of the body have the same velocity and the same acceleration at any given instant.*

We next considered the *rotation* of a rigid body about a fixed axis [Sec. 15.3]. The position of the body is defined by the angle $\theta$ that the line $BP$, drawn from the axis of rotation to a point $P$ of the body, forms with a fixed plane (Fig. 15.39). We found that the magnitude of the velocity of $P$ is

$$v = \frac{ds}{dt} = r\dot{\theta} \sin \phi \qquad (15.4)$$

where $\dot{\theta}$ is the time derivative of $\theta$. We then expressed the velocity of $P$ as

$$\mathbf{v} = \frac{d\mathbf{r}}{dt} = \boldsymbol{\omega} \times \mathbf{r} \qquad (15.5)$$

where the vector

$$\boldsymbol{\omega} = \omega\mathbf{k} = \dot{\theta}\mathbf{k} \qquad (15.6)$$

is directed along the fixed axis of rotation and represents the *angular velocity* of the body.

Denoting by $\boldsymbol{\alpha}$ the derivative $d\boldsymbol{\omega}/dt$ of the angular velocity, we expressed the acceleration of $P$ as

$$\mathbf{a} = \boldsymbol{\alpha} \times \mathbf{r} + \boldsymbol{\omega} \times (\boldsymbol{\omega} \times \mathbf{r}) \qquad (15.8)$$

Differentiating (15.6), and recalling that $\mathbf{k}$ is constant in magnitude and direction, we found that

$$\boldsymbol{\alpha} = \alpha\mathbf{k} = \dot{\omega}\mathbf{k} = \ddot{\theta}\mathbf{k} \qquad (15.9)$$

The vector $\boldsymbol{\alpha}$ represents the *angular acceleration* of the body and is directed along the fixed axis of rotation.

Rigid body in translation

Rigid body in rotation about a fixed axis

**Fig. 15.39**

**Fig. 15.40**

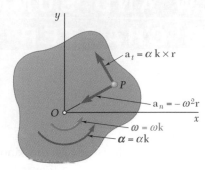

**Fig. 15.41**

Rotation of a representative slab

Next we considered the motion of a representative slab located in a plane perpendicular to the axis of rotation of the body (Fig. 15.40). Since the angular velocity is perpendicular to the slab, the velocity of a point $P$ of the slab was expressed as

$$\mathbf{v} = \omega\mathbf{k} \times \mathbf{r} \qquad (15.10)$$

where $\mathbf{v}$ is contained in the plane of the slab. Substituting $\boldsymbol{\omega} = \omega\mathbf{k}$ and $\boldsymbol{\alpha} = \alpha\mathbf{k}$ into (15.8), we found that the acceleration of $P$ could be resolved into tangential and normal components (Fig. 15.41) respectively equal to

Tangential and normal components

$$\begin{aligned} \mathbf{a}_t &= \alpha\mathbf{k} \times \mathbf{r} & a_t &= r\alpha \\ \mathbf{a}_n &= -\omega^2\mathbf{r} & a_n &= r\omega^2 \end{aligned} \qquad (15.11')$$

Angular velocity and angular acceleration of rotating slab

Recalling Eqs. (15.6) and (15.9), we obtained the following expressions for the *angular velocity* and the *angular acceleration* of the slab [Sec. 15.4]:

$$\omega = \frac{d\theta}{dt} \qquad (15.12)$$

$$\alpha = \frac{d\omega}{dt} = \frac{d^2\theta}{dt^2} \qquad (15.13)$$

or

$$\alpha = \omega\frac{d\omega}{d\theta} \qquad (15.14)$$

We noted that these expressions are similar to those obtained in Chap. 11 for the rectilinear motion of a particle.

Two particular cases of rotation are frequently encountered: *uniform rotation* and *uniformly accelerated rotation*. Problems involving either of these motions can be solved by using equations similar to those used in Secs. 11.4 and 11.5 for the uniform rectilinear motion and the uniformly accelerated rectilinear motion of a particle, but where $x$, $v$, and $a$ are replaced by $\theta$, $\omega$, and $\alpha$, respectively [Sample Prob. 15.1].

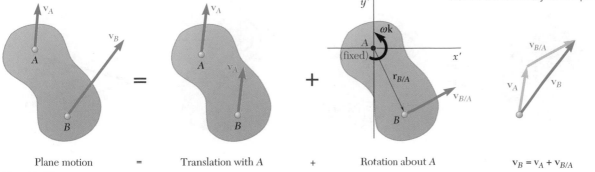

Plane motion    =    Translation with $A$    +    Rotation about $A$    $v_B = v_A + v_{B/A}$

**Fig. 15.42**

The *most general plane motion* of a rigid slab can be considered as the *sum of a translation and a rotation* [Sec. 15.5]. For example, the slab shown in Fig. 15.42 can be assumed to translate with point $A$, while simultaneously rotating about $A$. It follows [Sec. 15.6] that the velocity of any point $B$ of the slab can be expressed as

$$\mathbf{v}_B = \mathbf{v}_A + \mathbf{v}_{B/A} \qquad (15.17)$$

where $\mathbf{v}_A$ is the velocity of $A$ and $\mathbf{v}_{B/A}$ the relative velocity of $B$ with respect to $A$ or, more precisely, with respect to axes $x'y'$ translating with $A$. Denoting by $\mathbf{r}_{B/A}$ the position vector of $B$ relative to $A$, we found that

$$\mathbf{v}_{B/A} = \omega\mathbf{k} \times \mathbf{r}_{B/A} \qquad v_{B/A} = r\omega \qquad (15.18)$$

The fundamental equation (15.17) relating the absolute velocities of points $A$ and $B$ and the relative velocity of $B$ with respect to $A$ was expressed in the form of a vector diagram and used to solve problems involving the motion of various types of mechanisms [Sample Probs. 15.2 and 15.3].

Another approach to the solution of problems involving the velocities of the points of a rigid slab in plane motion was presented in Sec. 15.7 and used in Sample Probs. 15.4 and 15.5. It is based on the determination of the *instantaneous center of rotation C* of the slab (Fig. 15.43).

*Velocities in plane motion*

*Instantaneous center of rotation*

*(a)*          *(b)*

**Fig. 15.43**

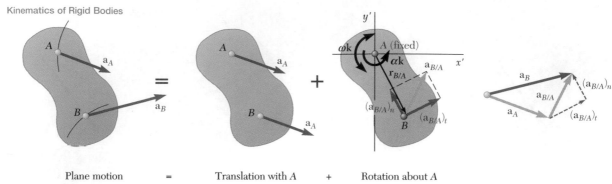

Plane motion     =     Translation with $A$     +     Rotation about $A$

**Fig. 15.44**

Accelerations in plane motion

The fact that any plane motion of a rigid slab can be considered as the sum of a translation of the slab with a reference point $A$ and a rotation about $A$ was used in Sec. 15.8 to relate the absolute accelerations of any two points $A$ and $B$ of the slab and the relative acceleration of $B$ with respect to $A$. We had

$$\mathbf{a}_B = \mathbf{a}_A + \mathbf{a}_{B/A} \tag{15.21}$$

where $\mathbf{a}_{B/A}$ consisted of a *normal component* $(\mathbf{a}_{B/A})_n$ of magnitude $r\omega^2$ directed toward $A$, and a *tangential component* $(\mathbf{a}_{B/A})_t$ of magnitude $r\alpha$ perpendicular to the line $AB$ (Fig. 15.44). The fundamental relation (15.21) was expressed in terms of vector diagrams or vector equations and used to determine the accelerations of given points of various mechanisms [Sample Probs. 15.6 through 15.8]. It should be noted that the instantaneous center of rotation $C$ considered in Sec. 15.7 cannot be used for the determination of accelerations, since point $C$, in general, does *not* have zero acceleration.

Coordinates expressed in terms of a parameter

In the case of certain mechanisms, it is possible to express the coordinates $x$ and $y$ of all significant points of the mechanism by means of simple analytic expressions containing a *single parameter*. The components of the absolute velocity and acceleration of a given point are then obtained by differentiating twice with respect to the time $t$ the coordinates $x$ and $y$ of that point [Sec. 15.9].

Rate of change of a vector with respect to a rotating frame

While the rate of change of a vector is the same with respect to a fixed frame of reference and with respect to a frame in translation, the rate of change of a vector with respect to a rotating frame is different. Therefore, in order to study the motion of a particle relative to a rotating frame we first had to compare the rates of change of a general vector $\mathbf{Q}$ with respect to a fixed frame $OXYZ$ and with respect to a frame $Oxyz$ rotating with an angular velocity $\mathbf{\Omega}$ [Sec. 15.10] (Fig. 15.45). We obtained the fundamental relation

$$(\dot{\mathbf{Q}})_{OXYZ} = (\dot{\mathbf{Q}})_{Oxyz} + \mathbf{\Omega} \times \mathbf{Q} \tag{15.31}$$

and we concluded that the rate of change of the vector $\mathbf{Q}$ with respect to the fixed frame $OXYZ$ is made of two parts: The first part represents the rate of change of $\mathbf{Q}$ with respect to the rotating frame $Oxyz$; the second part, $\mathbf{\Omega} \times \mathbf{Q}$, is induced by the rotation of the frame $Oxyz$.

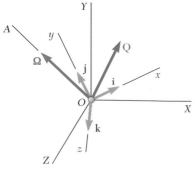

**Fig. 15.45**

The next part of the chapter [Sec. 15.11] was devoted to the two-dimensional kinematic analysis of a particle $P$ moving with respect to a frame $\mathscr{F}$ rotating with an angular velocity $\boldsymbol{\Omega}$ about a fixed axis (Fig. 15.46). We found that the absolute velocity of $P$ could be expressed as

$$\mathbf{v}_P = \mathbf{v}_{P'} + \mathbf{v}_{P/\mathscr{F}} \tag{15.33}$$

where $\mathbf{v}_P$ = absolute velocity of particle $P$

$\mathbf{v}_{P'}$ = velocity of point $P'$ of moving frame $\mathscr{F}$ coinciding with $P$

$\mathbf{v}_{P/\mathscr{F}}$ = velocity of $P$ relative to moving frame $\mathscr{F}$

We noted that the same expression for $\mathbf{v}_P$ is obtained if the frame is in translation rather than in rotation. However, when the frame is in rotation, the expression for the acceleration of $P$ is found to contain an additional term $\mathbf{a}_c$ called the *complementary acceleration* or *Coriolis acceleration*. We wrote

$$\mathbf{a}_P = \mathbf{a}_{P'} + \mathbf{a}_{P/\mathscr{F}} + \mathbf{a}_c \tag{15.36}$$

where $\mathbf{a}_P$ = absolute acceleration of particle $P$

$\mathbf{a}_{P'}$ = acceleration of point $P'$ of moving frame $\mathscr{F}$ coinciding with $P$

$\mathbf{a}_{P/\mathscr{F}}$ = acceleration of $P$ relative to moving frame $\mathscr{F}$

$\mathbf{a}_c = 2\boldsymbol{\Omega} \times (\dot{\mathbf{r}})_{Oxy} = 2\boldsymbol{\Omega} \times \mathbf{v}_{P/\mathscr{F}}$
= complementary, or Coriolis, acceleration

Since $\boldsymbol{\Omega}$ and $\mathbf{v}_{P/\mathscr{F}}$ are perpendicular to each other in the case of plane motion, the Coriolis acceleration was found to have a magnitude $a_c = 2\Omega v_{P/\mathscr{F}}$ and to point in the direction obtained by rotating the vector $\mathbf{v}_{P/\mathscr{F}}$ through 90° in the sense of rotation of the moving frame. Formulas (15.33) and (15.36) can be used to analyze the motion of mechanisms which contain parts sliding on each other [Sample Probs. 15.9 and 15.10].

The last part of the chapter was devoted to the study of the kinematics of rigid bodies in three dimensions. We first considered the motion of a rigid body with a fixed point [Sec. 15.12]. After proving that the most general displacement of a rigid body with a fixed point $O$ is equivalent to a rotation of the body about an axis through $O$, we were able to define the angular velocity $\boldsymbol{\omega}$ and the *instantaneous axis of rotation* of the body at a given instant. The velocity of a point $P$ of the body (Fig. 15.47) could again be expressed as

$$\mathbf{v} = \frac{d\mathbf{r}}{dt} = \boldsymbol{\omega} \times \mathbf{r} \tag{15.37}$$

Differentiating this expression, we also wrote

$$\mathbf{a} = \boldsymbol{\alpha} \times \mathbf{r} + \boldsymbol{\omega} \times (\boldsymbol{\omega} \times \mathbf{r}) \tag{15.38}$$

However, since the direction of $\boldsymbol{\omega}$ changes from one instant to the next, the angular acceleration $\boldsymbol{\alpha}$ is, in general, not directed along the instantaneous axis of rotation [Sample Prob. 15.11].

Plane motion of a particle relative to a rotating frame

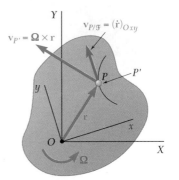

**Fig. 15.46**

Motion of a rigid body with a fixed point

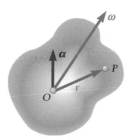

**Fig. 15.47**

General motion in space

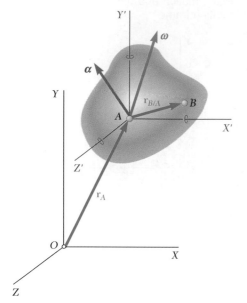

**Fig. 15.48**

Three-dimensional motion of a particle relative to a rotating frame

It was shown in Sec. 15.13 that *the most general motion of a rigid body in space is equivalent, at any given instant, to the sum of a translation and a rotation.* Considering two particles $A$ and $B$ of the body, we found that

$$\mathbf{v}_B = \mathbf{v}_A + \mathbf{v}_{B/A} \qquad (15.42)$$

where $\mathbf{v}_{B/A}$ is the velocity of $B$ relative to a frame $AX'Y'Z'$ attached to $A$ and of fixed orientation (Fig. 15.48). Denoting by $\mathbf{r}_{B/A}$ the position vector of $B$ relative to $A$, we wrote

$$\mathbf{v}_B = \mathbf{v}_A + \boldsymbol{\omega} \times \mathbf{r}_{B/A} \qquad (15.43)$$

where $\boldsymbol{\omega}$ is the angular velocity of the body at the instant considered [Sample Prob. 15.12]. The acceleration of $B$ was obtained by a similar reasoning. We first wrote

$$\mathbf{a}_B = \mathbf{a}_A + \mathbf{a}_{B/A}$$

and, recalling Eq. (15.38),

$$\mathbf{a}_B = \mathbf{a}_A + \boldsymbol{\alpha} \times \mathbf{r}_{B/A} + \boldsymbol{\omega} \times (\boldsymbol{\omega} \times \mathbf{r}_{B/A}) \qquad (15.44)$$

In the final two sections of the chapter we considered the three-dimensional motion of a particle $P$ relative to a frame $Oxyz$ rotating with an angular velocity $\boldsymbol{\Omega}$ with respect to a fixed frame $OXYZ$ (Fig. 15.49). In Sec. 15.14 we expressed the absolute velocity $\mathbf{v}_P$ of $P$ as

$$\mathbf{v}_P = \mathbf{v}_{P'} + \mathbf{v}_{P/\mathscr{F}} \qquad (15.46)$$

where $\mathbf{v}_P$ = absolute velocity of a particle $P$

$\qquad \mathbf{v}_{P'}$ = velocity of point $P'$ of moving frame $\mathscr{F}$ coinciding with $P$

$\qquad \mathbf{v}_{P/\mathscr{F}}$ = velocity of $P$ relative to moving frame $\mathscr{F}$

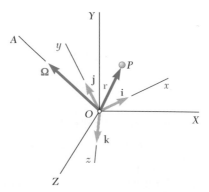

**Fig. 15.49**

The absolute acceleration $\mathbf{a}_P$ of $P$ was then expressed as

$$\mathbf{a}_P = \mathbf{a}_{P'} + \mathbf{a}_{P/\mathscr{F}} + \mathbf{a}_c \qquad (15.48)$$

where $\mathbf{a}_P$ = absolute acceleration of particle $P$

$\mathbf{a}_{P'}$ = acceleration of point $P'$ of moving frame $\mathscr{F}$ coinciding with $P$

$\mathbf{a}_{P/\mathscr{F}}$ = acceleration of $P$ relative to moving frame $\mathscr{F}$

$\mathbf{a}_c = 2\mathbf{\Omega} \times (\dot{\mathbf{r}})_{Oxyz} = 2\mathbf{\Omega} \times \mathbf{v}_{P/\mathscr{F}}$
= complementary, or Coriolis, acceleration

It was noted that the magnitude $a_c$ of the Coriolis acceleration is not equal to $2\Omega v_{P/\mathscr{F}}$ [Sample Prob. 15.13] except in the special case when $\mathbf{\Omega}$ and $\mathbf{v}_{P/\mathscr{F}}$ are perpendicular to each other.

We also observed [Sec. 15.15] that Eqs. (15.46) and (15.48) remain valid when the frame $Axyz$ moves in a known, but arbitrary, fashion with respect to the fixed frame $OXYZ$ (Fig. 15.50), provided that the motion of $A$ is included in the terms $\mathbf{v}_{P'}$ and $\mathbf{a}_{P'}$ representing the absolute velocity and acceleration of the coinciding point $P'$.

**Frame of reference in general motion**

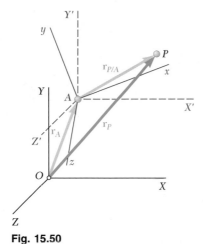

**Fig. 15.50**

Rotating frames of reference are particularly useful in the study of the three-dimensional motion of rigid bodies. Indeed, there are many cases where an appropriate choice of the rotating frame will lead to a simpler analysis of the motion of the rigid body than would be possible with axes of fixed orientation [Sample Probs. 15.14 and 15.15].

# Review Problems

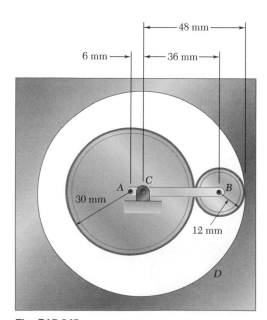

**Fig. P15.243**

**15.243** The bent rod *ABCDE* rotates about a line joining points *A* and *E* with a constant angular velocity of 12 rad/s. Knowing that the rotation is clockwise as viewed from *E*, determine the velocity and acceleration of corner *C*.

**15.244** The rated speed of drum *B* of the belt sander shown is 2400 rpm. When the power is turned off, it is observed that the sander coasts from its rated speed to rest in 10 s. Assuming uniformly decelerated motion, determine the velocity and acceleration of point *C* of the belt, (*a*) immediately before the power is turned off, (*b*) 9 s later.

**Fig. P15.244**

**15.245** Arm *ACB* rotates about point *C* with an angular velocity of 40 rad/s counterclockwise. Two friction disks *A* and *B* are pinned at their centers to arm *ACB* as shown. Knowing that the disks roll without slipping at surfaces of contact, determine the angular velocity of (*a*) disk *A*, (*b*) disk *B*.

**15.246** Bar *AB* is rotating clockwise and, at the instant shown, the magnitude of the velocity of point *G* is 3.6 m/s. Determine the angular velocity of each of the three bars at that instant.

**Fig. P15.246**

**Fig. *P15.245***

**15.247** A 60-mm-radius drum is rigidly attached to a 100-mm-radius drum as shown. One of the drums rolls without sliding on the surface shown, and a cord is wound around the other drum. Knowing that end $E$ of the cord is pulled to the left with a velocity of 120 mm/s, determine (*a*) the angular velocity of the drums, (*b*) the velocity of the center of the drums, (*c*) the length of cord wound or unwound per second.

**Fig. P15.247**

*15.248* Arm $ABD$ is connected by pins to a collar at $B$ and to crank $DE$. Knowing that the velocity of collar $B$ is 40.6 cm/s upward, determine (*a*) the angular velocity of arm $ABD$, (*b*) the velocity of point $A$.

**Fig. *P15.248***

**15.249** The 15.2 cm-radius drum rolls without slipping on a belt that moves to the left with a constant velocity of 30.5 cm/s. At an instant when the velocity and acceleration of the center $D$ of the drum are as shown, determine the accelerations of points $A$, $B$, and $C$ of the drum.

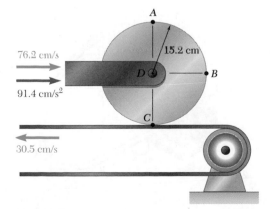

**Fig. P15.249**

**15.250** At the instant shown rod $AB$ has a constant angular velocity of 8 rad/s clockwise. Knowing that $l = 30.5$ cm determine the acceleration of the midpoint $C$ of member $BD$.

Fig. *P15.250*

**15.251** Pin $P$ is attached to the collar shown; the motion of the pin is guided by a slot cut in rod $BD$ and by the collar that slides on rod $AE$. Knowing that at the instant considered the rods rotate clockwise with $\omega_{AE} = 8$ rad/s and $\omega_{BD} = 3$ rad/s, determine the velocity of pin $P$.

**15.252** A chain is looped around two gears of radius 5.1 cm that can rotate freely with respect to the 40.6 cm arm $AB$. The chain moves about arm $AB$ in a clockwise direction at the constant rate of 10.2 cm/s relative to the arm. Knowing that in the position shown arm $AB$ rotates clockwise about $A$ at the constant rate $\omega = 0.75$ rad/s, determine the acceleration of each of the chain links $1$ and $2$.

**15.253** Rod $AB$ is connected by ball-and-socket joints to collar $A$ and to the 320-mm-diameter disk $C$. Knowing that disk $C$ rotates counterclockwise at the constant rate $\omega_0 = 3$ rad/s in the $zx$ plane, determine the velocity of collar $A$ for the position shown.

**15.254** The bent rod $EBD$ rotates at the constant rate $\omega_1 = 8$ rad/s. Knowing that collar $A$ moves upward along the rod at a constant relative speed $u = 600$ mm/s and that $\theta = 60°$, determine (a) the velocity of $A$, (b) the acceleration of $A$.

Fig. **P15.251**

Fig. **P15.252**

Fig. *P15.253*

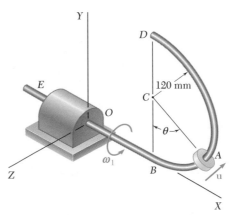

Fig. *P15.254*

# Computer Problems

**15.C1**  The disk shown has a constant angular velocity of 500 rpm counterclockwise. Knowing that rod *BD* is 25.4 cm long, use computational software to determine and plot, for values of $\theta$ from 0 to 360°, (*a*) the velocity of collar *D*, (*b*) the angular velocity of rod *BD*. (*c*) Determine the two values of $\theta$ for which the speed of collar *D* is zero.

**Fig. P15.C1**

**15.C2**  The 10.2 cm-radius wheel shown rolls to the left with a constant velocity of 114.3 cm/s. Knowing that the distance *AD* is 6.35 cm, determine and plot, for values of $\beta$ from 0 to 360°, (*a*) the velocity of collar *B*, (*b*) the angular velocity of rod *AB*.

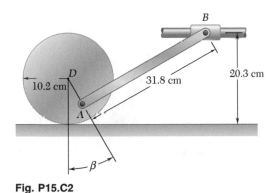

**Fig. P15.C2**

**15.C3**  In the engine system shown, $l = 20.3$ cm and $b = 7.6$ cm. Knowing that crank *AB* rotates with a constant angular velocity of 1000 rpm clockwise, use computational software to determine and plot, for values of $\theta$ from 0 to 180°, (*a*) the angular velocity and angular acceleration of rod *BD*, (*b*) the velocity and acceleration of the piston *P*.

**Fig. P15.C3**

**15.C4**   Knowing that crank $AB$ has a constant angular velocity of 160 rpm counterclockwise, determine and plot, for values of $\theta$ from 0 to 360°, (a) the angular velocity and angular acceleration of rod $BD$, (b) the velocity and acceleration of collar $D$.

**Fig. P15.C4**

**15.C5**   Two rotating rods are connected by a slider block $P$ as shown. Knowing that rod $BP$ rotates with a constant angular velocity of 6 rad/s counterclockwise, use computational software to determine and plot, for values of $\theta$ from 0 to 180°, (a) the angular velocity and angular acceleration of rod $AE$. (b) Determine the value of $\theta$ for which the angular acceleration $\alpha_{AE}$ of rod $AE$ is maximum and the corresponding value of $\alpha_{AE}$.

**Fig. P15.C5**

**Fig. P15.C6**

**15.C6**   Rod $AB$ moves over a small wheel at $C$ while end $A$ moves to the right with a constant velocity of 22.9 cm/s. Use computational software to determine and plot, for values of $\theta$ from 20° to 90°, (a) the velocity of point $B$, (b) the angular acceleration of the rod. (c) Determine the value of $\theta$ for which the angular acceleration $\alpha$ of the rod is maximum and the corresponding value of $\alpha$.

**15.C7**  Rod *AB* of length 500 mm is connected by ball-and-socket joints to collars *A* and *B*, which slide along the two rods shown. Collar *B* moves towards support *E* at a constant speed of 400 mm/s. Denoting by *d* the distance from point *C* to collar *B*, use computational software to determine and plot the velocity of collar *A* for values of *d* from 0 to 300 mm.

**Fig. P15.C7**

## CHAPTER 16

# Plane Motion of Rigid Bodies: Forces and Accelerations

Accelerometers attached at several points on a crash test dummy provide acceleration data which engineers can use to estimate the forces generated during the motion. In this chapter and in Chaps. 17 and 18 you will learn to analyze the motion of a rigid body by considering the motion of its mass center, the motion relative to its mass center, and the external forces acting on it.

## PLANE MOTION OF RIGID BODIES: FORCES AND ACCELERATIONS

16.1  Introduction
16.2  Equations of Motion for a Rigid Body
16.3  Angular Momentum of a Rigid Body in Plane Motion
16.4  Plane Motion of a Rigid Body. d'Alembert's Principle
16.5  A Remark on the Axioms of the Mechanics of Rigid Bodies
16.6  Solution of Problems Involving the Motion of a Rigid Body
16.7  Systems of Rigid Bodies
16.8  Constrained Plane Motion

## 16.1. INTRODUCTION

In this chapter and in Chaps. 17 and 18, you will study the *kinetics of rigid bodies,* that is, the relations existing between the forces acting on a rigid body, the shape and mass of the body, and the motion produced. In Chaps. 12 and 13, you studied similar relations, assuming then that the body could be considered as a particle, that is, that its mass could be concentrated in one point and that all forces acted at that point. The shape of the body, as well as the exact location of the points of application of the forces, will now be taken into account. You will also be concerned not only with the motion of the body as a whole but also with the motion of the body about its mass center.

Our approach will be to consider rigid bodies as made of large numbers of particles and to use the results obtained in Chap. 14 for the motion of systems of particles. Specifically, two equations from Chap. 14 will be used: Eq. (14.16), $\Sigma\mathbf{F} = m\overline{\mathbf{a}}$, which relates the resultant of the external forces and the acceleration of the mass center $G$ of the system of particles, and Eq. (14.23), $\Sigma\mathbf{M}_G = \dot{\mathbf{H}}_G$, which relates the moment resultant of the external forces and the angular momentum of the system of particles about $G$.

Except for Sec. 16.2, which applies to the most general case of the motion of a rigid body, the results derived in this chapter will be limited in two ways: (1) They will be restricted to the *plane motion* of rigid bodies, that is, to a motion in which each particle of the body remains at a constant distance from a fixed reference plane. (2) The rigid bodies considered will consist only of plane slabs and of bodies which are symmetrical with respect to the reference plane.† The study of the plane motion of nonsymmetrical three-dimensional bodies and, more generally, the motion of rigid bodies in three-dimensional space will be postponed until Chap. 18.

In Sec. 16.3, we define the angular momentum of a rigid body in plane motion and show that the rate of change of the angular momentum $\dot{\mathbf{H}}_G$ about the mass center is equal to the product $\overline{I}\boldsymbol{\alpha}$ of the centroidal mass moment of inertia $\overline{I}$ and the angular acceleration $\boldsymbol{\alpha}$ of the body. D'Alembert's principle, introduced in Sec. 16.4, is used to prove that the external forces acting on a rigid body are equivalent to a vector $m\overline{\mathbf{a}}$ attached at the mass center and a couple of moment $\overline{I}\boldsymbol{\alpha}$.

In Sec. 16.5, we derive the principle of transmissibility using only the parallelogram law and Newton's laws of motion, allowing us to remove this principle from the list of axioms (Sec. 1.2) required for the study of the statics and dynamics of rigid bodies.

Free-body-diagram equations are introduced in Sec. 16.6 and will be used in the solution of all problems involving the plane motion of rigid bodies.

After considering the plane motion of connected rigid bodies in Sec. 16.7, you will be prepared to solve a variety of problems involving the translation, centroidal rotation, and unconstrained motion of rigid bodies. In Sec. 16.8 and in the remaining part of the chapter, the solution of problems involving noncentroidal rotation, rolling motion, and other partially constrained plane motions of rigid bodies will be considered.

---

†Or, more generally, bodies which have a principal centroidal axis of inertia perpendicular to the reference plane.

## 16.2. EQUATIONS OF MOTION FOR A RIGID BODY

Consider a rigid body acted upon by several external forces $\mathbf{F}_1$, $\mathbf{F}_2$, $\mathbf{F}_3$, ... (Fig. 16.1). We can assume that the body is made of a large number $n$ of particles of mass $\Delta m_i$ ($i = 1, 2, \ldots, n$) and apply the results obtained in Chap. 14 for a system of particles (Fig. 16.2). Considering first the motion of the mass center $G$ of the body with respect to the newtonian frame of reference $Oxyz$, we recall Eq. (14.16) and write

$$\Sigma \mathbf{F} = m\overline{\mathbf{a}} \qquad (16.1)$$

where $m$ is the mass of the body and $\overline{\mathbf{a}}$ is the acceleration of the mass center $G$. Turning now to the motion of the body relative to the centroidal frame of reference $Gx'y'z'$, we recall Eq. (14.23) and write

$$\Sigma \mathbf{M}_G = \dot{\mathbf{H}}_G \qquad (16.2)$$

where $\dot{\mathbf{H}}_G$ represents the rate of change of $\mathbf{H}_G$, the angular momentum about $G$ of the system of particles forming the rigid body. In the following, $\mathbf{H}_G$ will simply be referred to as the *angular momentum of the rigid body about its mass center $G$*. Together Eqs. (16.1) and (16.2) express that *the system of the external forces is equipollent to the system consisting of the vector $m\overline{\mathbf{a}}$ attached at $G$ and the couple of moment $\dot{\mathbf{H}}_G$* (Fig. 16.3).†

**Fig. 16.1**

**Fig. 16.2**

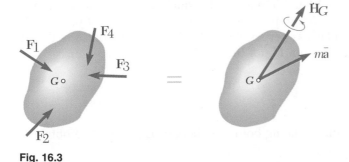

**Fig. 16.3**

Equations (16.1) and (16.2) apply in the most general case of the motion of a rigid body. In the rest of this chapter, however, our analysis will be limited to the *plane motion* of rigid bodies, that is, to a motion in which each particle remains at a constant distance from a fixed reference plane, and it will be assumed that the rigid bodies considered consist only of plane slabs and of bodies which are symmetrical with respect to the reference plane. Further study of the plane motion of nonsymmetrical three-dimensional bodies and of the motion of rigid bodies in three-dimensional space will be postponed until Chap. 18.

†Since the systems involved act on a rigid body, we could conclude at this point, by referring to Sec. 3.19, that the two systems are *equivalent* as well as equipollent and use red rather than blue equals signs in Fig. 16.3. However, by postponing this conclusion, we will be able to arrive at it independently (Secs. 16.4 and 18.5), thereby eliminating the necessity of including the principle of transmissibility among the axioms of mechanics (Sec. 16.5).

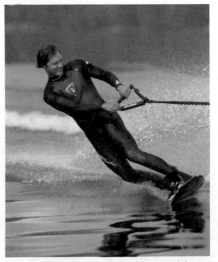

**Photo 16.1** The system of external forces acting on the man and wakeboard includes the weights, the tension in the tow rope, and the forces exerted by the water and the air.

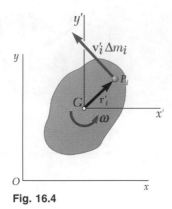

**Fig. 16.4**

## 16.3. ANGULAR MOMENTUM OF A RIGID BODY IN PLANE MOTION

Consider a rigid slab in plane motion. Assuming that the slab is made of a large number $n$ of particles $P_i$ of mass $\Delta m_i$ and recalling Eq. (14.24) of Sec. 14.5, we note that the angular momentum $\mathbf{H}_G$ of the slab about its mass center $G$ can be computed by taking the moments about $G$ of the momenta of the particles of the slab in their motion with respect to either of the frames $Oxy$ or $Gx'y'$ (Fig. 16.4). Choosing the latter course, we write

$$\mathbf{H}_G = \sum_{i=1}^{n} (\mathbf{r}'_i \times \mathbf{v}'_i \, \Delta m_i) \qquad (16.3)$$

where $\mathbf{r}'_i$ and $\mathbf{v}'_i \, \Delta m_i$ denote, respectively, the position vector and the linear momentum of the particle $P_i$ relative to the centroidal frame of reference $Gx'y'$. But since the particle belongs to the slab, we have $\mathbf{v}'_i = \boldsymbol{\omega} \times \mathbf{r}'_i$, where $\boldsymbol{\omega}$ is the angular velocity of the slab at the instant considered. We write

$$\mathbf{H}_G = \sum_{i=1}^{n} [\mathbf{r}'_i \times (\boldsymbol{\omega} \times \mathbf{r}'_i) \, \Delta m_i]$$

Referring to Fig. 16.4, we easily verify that the expression obtained represents a vector of the same direction as $\boldsymbol{\omega}$ (that is, perpendicular to the slab) and of magnitude equal to $\omega \Sigma r_i'^2 \, \Delta m_i$. Recalling that the sum $\Sigma r_i'^2 \, \Delta m_i$ represents the moment of inertia $\bar{I}$ of the slab about a centroidal axis perpendicular to the slab, we conclude that the angular momentum $\mathbf{H}_G$ of the slab about its mass center is

$$\mathbf{H}_G = \bar{I} \boldsymbol{\omega} \qquad (16.4)$$

Differentiating both members of Eq. (16.4) we obtain

$$\dot{\mathbf{H}}_G = \bar{I} \dot{\boldsymbol{\omega}} = \bar{I} \boldsymbol{\alpha} \qquad (16.5)$$

Thus the rate of change of the angular momentum of the slab is represented by a vector of the same direction as $\boldsymbol{\alpha}$ (that is, perpendicular to the slab) and of magnitude $\bar{I}\alpha$.

It should be kept in mind that the results obtained in this section have been derived for a rigid slab in plane motion. As you will see in Chap. 18, they remain valid in the case of the plane motion of rigid bodies which are symmetrical with respect to the reference plane.† However, they do not apply in the case of nonsymmetrical bodies or in the case of three-dimensional motion.

---

†Or, more generally, bodies which have a principal centroidal axis of inertia perpendicular to the reference plane.

# 16.4. PLANE MOTION OF A RIGID BODY. D'ALEMBERT'S PRINCIPLE

Consider a rigid slab of mass $m$ moving under the action of several external forces $\mathbf{F}_1$, $\mathbf{F}_2$, $\mathbf{F}_3$, . . . , contained in the plane of the slab (Fig. 16.5). Substituting for $\dot{\mathbf{H}}_G$ from Eq. (16.5) into Eq. (16.2) and writing the fundamental equations of motion (16.1) and (16.2) in scalar form, we have

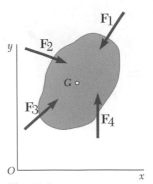

**Fig. 16.5**

$$\Sigma F_x = m\bar{a}_x \qquad \Sigma F_y = m\bar{a}_y \qquad \Sigma M_G = \bar{I}\alpha \qquad (16.6)$$

Equations (16.6) show that the acceleration of the mass center $G$ of the slab and its angular acceleration $\boldsymbol{\alpha}$ are easily obtained once the resultant of the external forces acting on the slab and their moment resultant about $G$ have been determined. Given appropriate initial conditions, the coordinates $\bar{x}$ and $\bar{y}$ of the mass center and the angular coordinate $\theta$ of the slab can then be obtained by integration at any instant $t$. Thus *the motion of the slab is completely defined by the resultant and moment resultant about G of the external forces acting on it.*

This property, which will be extended in Chap. 18 to the case of the three-dimensional motion of a rigid body, is characteristic of the motion of a rigid body. Indeed, as we saw in Chap. 14, the motion of a system of particles which are not rigidly connected will in general depend upon the specific external forces acting on the various particles, as well as upon the internal forces.

Since the motion of a rigid body depends only upon the resultant and moment resultant of the external forces acting on it, it follows that *two systems of forces which are equipollent,* that is, which have the same resultant and the same moment resultant, *are also equivalent;* that is, they have exactly the same effect on a given rigid body.†

Consider in particular the system of the external forces acting on a rigid body (Fig. 16.6a) and the system of the effective forces associated with the particles forming the rigid body (Fig. 16.6b). It was shown in Sec. 14.2 that the two systems thus defined are equipollent. But since the particles considered now form a rigid body, it follows from the discussion above that the two systems are also equivalent. We can thus state that *the external forces acting on a rigid body are equivalent to the effective forces of the various particles forming the body.* This statement is referred to as *d'Alembert's principle* after the French mathematician Jean le Rond d'Alembert (1717–1783), even though d'Alembert's original statement was written in a somewhat different form.

The fact that the system of external forces is *equivalent* to the system of the effective forces has been emphasized by the use of a red equals sign in Fig. 16.6 and also in Fig. 16.7, where using results obtained earlier in this section, we have replaced the effective forces by a vector $m\bar{\mathbf{a}}$ attached at the mass center $G$ of the slab and a couple of moment $\bar{I}\boldsymbol{\alpha}$.

(a)                    (b)

**Fig. 16.6**

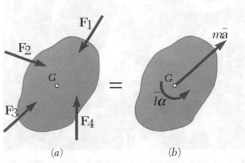

(a)                    (b)

**Fig. 16.7**

---

†This result has already been derived in Sec. 3.19 from the principle of transmissibility (Sec. 3.3). The present derivation is independent of that principle, however, and will make possible its elimination from the axioms of mechanics (Sec. 16.5).

**Fig. 16.7 (repeated)**

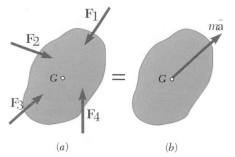

(a)                    (b)

**Fig. 16.8** Translation.

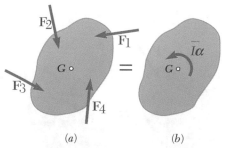

(a)                    (b)

**Fig. 16.9** Centroidal rotation.

**Photo 16.2** When the vehicle is at rest with the engine running, the motion of each gear is an example of centroidal rotation.

**Translation.** In the case of a body in translation, the angular acceleration of the body is identically equal to zero and its effective forces reduce to the vector $m\overline{\mathbf{a}}$ attached at $G$ (Fig. 16.8). Thus, the resultant of the external forces acting on a rigid body in translation passes through the mass center of the body and is equal to $m\overline{\mathbf{a}}$.

**Centroidal Rotation.** When a slab, or, more generally, a body symmetrical with respect to the reference plane, rotates about a fixed axis perpendicular to the reference plane and passing through its mass center $G$, we say that the body is in *centroidal rotation*. Since the acceleration $\overline{\mathbf{a}}$ is identically equal to zero, the effective forces of the body reduce to the couple $\overline{I}\boldsymbol{\alpha}$ (Fig. 16.9). Thus, the external forces acting on a body in centroidal rotation are equivalent to a couple of moment $\overline{I}\boldsymbol{\alpha}$.

**General Plane Motion.** Comparing Fig. 16.7 with Figs. 16.8 and 16.9, we observe that from the point of view of *kinetics*, the most general plane motion of a rigid body symmetrical with respect to the reference plane can be replaced by the sum of a translation and a centroidal rotation. We should note that this statement is more restrictive than the similar statement made earlier from the point of view of *kinematics* (Sec. 15.5), since we now require that the mass center of the body be selected as the reference point.

Referring to Eqs. (16.6), we observe that the first two equations are identical with the equations of motion of a particle of mass $m$ acted upon by the given forces $\mathbf{F}_1$, $\mathbf{F}_2$, $\mathbf{F}_3$, . . . We thus check that *the mass center $G$ of a rigid body in plane motion moves as if the entire mass of the body were concentrated at that point, and as if all the external forces acted on it*. We recall that this result has already been obtained in Sec. 14.4 in the general case of a system of particles, the particles being not necessarily rigidly connected. We also note, as we did in Sec. 14.4, that the system of the external forces does not, in general, reduce to a single vector $m\overline{\mathbf{a}}$ attached at $G$. Therefore, in the general case of the plane motion of a rigid body, *the resultant of the external forces acting on the body does not pass through the mass center of the body*.

Finally, it should be observed that the last of Eqs. (16.6) would still be valid if the rigid body, while subjected to the same applied forces, were constrained to rotate about a fixed axis through $G$. Thus, *a rigid body in plane motion rotates about its mass center as if this point were fixed*.

### *16.5. A REMARK ON THE AXIOMS OF THE MECHANICS OF RIGID BODIES

The fact that two equipollent systems of external forces acting on a rigid body are also equivalent, that is, have the same effect on that rigid body, has already been established in Sec. 3.19. But there it was derived from the *principle of transmissibility*, one of the axioms used in our study of the statics of rigid bodies. It should be noted that this axiom has not been used in the present chapter because Newton's second and third laws of motion make its use unnecessary in the study of the dynamics of rigid bodies.

In fact, the principle of transmissibility may now be *derived* from the other axioms used in the study of mechanics. This principle stated, without proof (Sec. 3.3), that the conditions of equilibrium or motion of a rigid body remain unchanged if a force **F** acting at a given point of the rigid body is replaced by a force **F'** of the same magnitude and same direction, but acting at a different point, provided that the two forces have the same line of action. But since **F** and **F'** have the same moment about any given point, it is clear that they form two equipollent systems of external forces. Thus, we may now *prove,* as a result of what we established in the preceding section, that **F** and **F'** have the same effect on the rigid body (Fig. 3.3).

The principle of transmissibility can therefore be removed from the list of axioms required for the study of the mechanics of rigid bodies. These axioms are reduced to the parallelogram law of addition of vectors and to Newton's laws of motion.

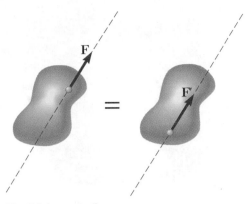

**Fig. 3.3 (*repeated*)**

## 16.6. SOLUTION OF PROBLEMS INVOLVING THE MOTION OF A RIGID BODY

We saw in Sec. 16.4 that when a rigid body is in plane motion, there exists a fundamental relation between the forces $\mathbf{F}_1$, $\mathbf{F}_2$, $\mathbf{F}_3$, . . . , acting on the body, the acceleration $\bar{\mathbf{a}}$ of its mass center, and the angular acceleration $\boldsymbol{\alpha}$ of the body. This relation, which is represented in Fig. 16.7 in the form of a *free-body-diagram equation,* can be used to determine the acceleration $\bar{\mathbf{a}}$ and the angular acceleration $\boldsymbol{\alpha}$ produced by a given system of forces acting on a rigid body or, conversely, to determine the forces which produce a given motion of the rigid body.

The three algebraic equations (16.6) can be used to solve problems of plane motion.† However, our experience in statics suggests that the solution of many problems involving rigid bodies could be simplified by an appropriate choice of the point about which the moments of the forces are computed. It is therefore preferable to remember the relation existing between the forces and the accelerations in the pictorial form shown in Fig. 16.7 and to derive from this fundamental relation the component or moment equations which fit best the solution of the problem under consideration.

The fundamental relation shown in Fig. 16.7 can be presented in an alternative form if we add to the external forces an inertia vector $-m\bar{\mathbf{a}}$ of sense opposite to that of $\bar{\mathbf{a}}$, attached at G, and an inertia couple $-\bar{I}\boldsymbol{\alpha}$ of moment equal in magnitude to $\bar{I}\alpha$ and of sense opposite to that of $\boldsymbol{\alpha}$ (Fig. 16.10). The system obtained is equivalent to zero, and the rigid body is said to be in *dynamic equilibrium.*

Whether the principle of equivalence of external and effective forces is directly applied, as in Fig. 16.7, or whether the concept of dynamic equilibrium is introduced, as in Fig. 16.10, the use of free-body-diagram equations showing vectorially the relationship existing between the forces applied on the rigid body and the resulting linear and angular accelerations presents considerable advantages over the blind application of formulas (16.6). These advantages can be summarized as follows:

**Fig. 16.10**

---

†We recall that the last of Eqs. (16.6) is valid only in the case of the plane motion of a rigid body symmetrical with respect to the reference plane. In all other cases, the methods of Chap. 18 should be used.

**Fig. 16.7 (repeated)**

**Fig. 16.10 (repeated)**

1.  The use of a pictorial representation provides a much clearer understanding of the effect of the forces on the motion of the body.

2.  This approach makes it possible to divide the solution of a dynamics problem into two parts: In the first part, the analysis of the kinematic and kinetic characteristics of the problem leads to the free-body diagrams of Fig. 16.7 or 16.10; in the second part, the diagram obtained is used to analyze the various forces and vectors involved by the methods of Chap. 3.

3.  A unified approach is provided for the analysis of the plane motion of a rigid body, regardless of the particular type of motion involved. While the kinematics of the various motions considered may vary from one case to the other, the approach to the kinetics of the motion is consistently the same. In every case a diagram will be drawn showing the external forces, the vector $m\overline{\mathbf{a}}$ associated with the motion of $G$, and the couple $\overline{I}\boldsymbol{\alpha}$ associated with the rotation of the body about $G$.

4.  The resolution of the plane motion of a rigid body into a translation and a centroidal rotation, which is used here, is a basic concept which can be applied effectively throughout the study of mechanics. It will be used again in Chap. 17 with the method of work and energy and the method of impulse and momentum.

5.  As you will see in Chap. 18, this approach can be extended to the study of the general three-dimensional motion of a rigid body. The motion of the body will again be resolved into a translation and a rotation about the mass center, and free-body-diagram equations will be used to indicate the relationship existing between the external forces and the rates of change of the linear and angular momentum of the body.

### 16.7. SYSTEMS OF RIGID BODIES

The method described in the preceding section can also be used in problems involving the plane motion of several connected rigid bodies. For each part of the system, a diagram similar to Fig. 16.7 or Fig. 16.10 can be drawn. The equations of motion obtained from these diagrams are solved simultaneously.

In some cases, as in Sample Prob. 16.3, a single diagram can be drawn for the entire system. This diagram should include all the external forces, as well as the vectors $m\overline{\mathbf{a}}$ and the couples $\overline{I}\boldsymbol{\alpha}$ associated with the various parts of the system. However, internal forces such as the forces exerted by connecting cables, can be omitted since they occur in pairs of equal and opposite forces and are thus equipollent to zero. The equations obtained by expressing that the system of the external forces is equipollent to the system of the effective forces can be solved for the remaining unknowns.†

It is not possible to use this second approach in problems involving more than three unknowns, since only three equations of motion are available when a single diagram is used. We need not elaborate upon this point, since the discussion involved would be completely similar to that given in Sec. 6.11 in the case of the equilibrium of a system of rigid bodies.

**Photo 16.3**   The forklift and moving load can be analyzed as a system of two connected rigid bodies in plane motion.

---

†Note that we cannot speak of *equivalent* systems since we are not dealing with a single rigid body.

## SAMPLE PROBLEM 16.1

When the forward speed of the truck shown was 9.1 m/s, the brakes were suddenly applied, causing all four wheels to stop rotating. It was observed that the truck skidded to rest in 6 m/s. Determine the magnitude of the normal reaction and of the friction force at each wheel as the truck skidded to rest.

## SOLUTION

**Kinematics of Motion.** Choosing the positive sense to the right and using the equations of uniformly accelerated motion, we write

$$\bar{v}_0 = +9.1 \text{ m/s} \qquad \bar{v}^2 = \bar{v}_0^2 + 2\bar{a}\bar{x} \qquad 0 = (9.1)^2 + 2\bar{a}(6)$$
$$\bar{a} = -6.9 \text{ m/s}^2 \qquad \mathbf{a} = 6.9 \text{ m/s}^2 \leftarrow$$

**Equations of Motion.** The external forces consist of the weight **W** of the truck and of the normal reactions and friction forces at the wheels. (The vectors $\mathbf{N}_A$ and $\mathbf{F}_A$ represent the sum of the reactions at the rear wheels, while $\mathbf{N}_B$ and $\mathbf{F}_B$ represent the sum of the reactions at the front wheels.) Since the truck is in translation, the effective forces reduce to the vector $m\bar{\mathbf{a}}$ attached at $G$. Three equations of motion are obtained by expressing that the system of the external forces is equivalent to the system of the effective forces.

$$+\uparrow\Sigma F_y = \Sigma(F_y)_{\text{eff}}: \qquad N_A + N_B - W = 0$$

Since $F_A = \mu_k N_A$ and $F_B = \mu_k N_B$, where $\mu_k$ is the coefficient of kinetic friction, we find that

$$F_A + F_B = \mu_k(N_A + N_B) = \mu_k W$$

$$\xrightarrow{+}\Sigma F_x = \Sigma(F_x)_{\text{eff}}: \qquad -(F_A + F_B) = -m\bar{a}$$

$$-\mu_k W = -\frac{W}{32.2 \text{ ft/s}^2}(6.9 \text{ m/s}^2)$$

$$\mu_k = 0.703$$

$$+\uparrow\Sigma M_A = \Sigma(M_A)_{\text{eff}}: \qquad -W(1.5 \text{ m}) + N_B(3.6 \text{ m}) = m\bar{a}(1.2 \text{ m})$$

$$-W(1.5 \text{ m}) + N_B(3.6 \text{ m}) = \frac{W}{32.2 \text{ ft/s}^2}(6.9 \text{ m/s}^2)(1.2 \text{ m})$$

$$N_B = 0.651W$$
$$F_B = \mu_k N_B = (0.703)(0.650W) \qquad F_B = 0.458W$$

$$+\uparrow\Sigma F_y = \Sigma(F_y)_{\text{eff}}: \qquad N_A + N_B - W = 0$$
$$N_A + 0.650W - W = 0$$
$$N_A = 0.350W$$
$$F_A = \mu_k N_A = (0.703)(0.350W) \qquad F_A = 0.246W$$

**Reactions at Each Wheel.** Recalling that the values computed above represent the sum of the reactions at the two front wheels or the two rear wheels, we obtain the magnitude of the reactions at each wheel by writing

$$N_{\text{front}} = \tfrac{1}{2}N_B = 0.325W \qquad N_{\text{rear}} = \tfrac{1}{2}N_A = 0.175W \quad \blacktriangleleft$$
$$F_{\text{front}} = \tfrac{1}{2}F_B = 0.229W \qquad F_{\text{rear}} = \tfrac{1}{2}F_A = 0.123W \quad \blacktriangleleft$$

## SAMPLE PROBLEM 16.2

The thin plate $ABCD$ of mass 8 kg is held in the position shown by the wire $BH$ and two links $AE$ and $DF$. Neglecting the mass of the links, determine immediately after wire $BH$ has been cut ($a$) the acceleration of the plate, ($b$) the force in each link.

## SOLUTION

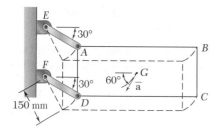

**Kinematics of Motion.** After wire $BH$ has been cut, we observe that corners $A$ and $D$ move along parallel circles of radius 150 mm centered, respectively, at $E$ and $F$. The motion of the plate is thus a curvilinear translation; the particles forming the plate move along parallel circles of radius 150 mm.

At the instant wire $BH$ is cut, the velocity of the plate is zero. Thus the acceleration $\bar{\mathbf{a}}$ of the mass center $G$ of the plate is tangent to the circular path which will be described by $G$.

**Equations of Motion.** The external forces consist of the weight $\mathbf{W}$ and the forces $\mathbf{F}_{AE}$ and $\mathbf{F}_{DF}$ exerted by the links. Since the plate is in translation, the effective forces reduce to the vector $m\bar{\mathbf{a}}$ attached at $G$ and directed along the $t$ axis. A free-body-diagram equation is drawn to show that the system of the external forces is equivalent to the system of the effective forces.

*a. Acceleration of the Plate.*

$$+\swarrow\Sigma F_t = \Sigma(F_t)_{\text{eff}}:$$
$$W\cos 30° = m\bar{a}$$
$$mg\cos 30° = m\bar{a}$$
$$\bar{a} = g\cos 30° = (9.81 \text{ m/s}^2)\cos 30° \tag{1}$$
$$\bar{\mathbf{a}} = 8.50 \text{ m/s}^2 \, \nearrow \, 60° \quad \blacktriangleleft$$

*b. Forces in Links AE and DF.*

$$+\nwarrow\Sigma F_n = \Sigma(F_n)_{\text{eff}}: \qquad F_{AE} + F_{DF} - W\sin 30° = 0 \tag{2}$$
$$+\downarrow\Sigma M_G = \Sigma(M_G)_{\text{eff}}:$$

$$(F_{AE}\sin 30°)(250 \text{ mm}) - (F_{AE}\cos 30°)(100 \text{ mm})$$
$$+ (F_{DF}\sin 30°)(250 \text{ mm}) + (F_{DF}\cos 30°)(100 \text{ mm}) = 0$$
$$38.4F_{AE} + 211.6F_{DF} = 0$$
$$F_{DF} = -0.1815F_{AE} \tag{3}$$

Substituting for $F_{DF}$ from (3) into (2), we write

$$F_{AE} - 0.1815F_{AE} - W\sin 30° = 0$$
$$F_{AE} = 0.6109W$$
$$F_{DF} = -0.1815(0.6109W) = -0.1109W$$

Noting that $W = mg = (8 \text{ kg})(9.81 \text{ m/s}^2) = 78.48$ N, we have

$$F_{AE} = 0.6109(78.48 \text{ N}) \qquad F_{AE} = 47.9 \text{ N } T \quad \blacktriangleleft$$
$$F_{DF} = -0.1109(78.48 \text{ N}) \qquad F_{DF} = 8.70 \text{ N } C \quad \blacktriangleleft$$

25.4 cm

G

15.2 cm

B

A

2.27 kg

4.5 kg

## SAMPLE PROBLEM 16.3

A pulley weighing 5.44 kg and having a radius of gyration of 20.3 kg is connected to two blocks as shown. Assuming no axle friction, determine the angular acceleration of the pulley and the acceleration of each block.

## SOLUTION

$\alpha$

$r_B$  $r_A$

G

B   A

$a_A$

$a_B$

**Sense of Motion.** Although an arbitrary sense of motion can be assumed (since no friction forces are involved) and later checked by the sign of the answer, we may prefer to determine the actual sense of rotation of the pulley first. The weight of block $B$ required to maintain the equilibrium of the pulley when it is acted upon by the 2.27 kg block $A$ is first determined. We write

$+\uparrow\Sigma M_G = 0:$    $W_B(15.2 \text{ cm}) - (2.27 \text{ kg})(25.4 \text{ cm}) = 0$    $W_B = 3.79 \text{ kg}$

Since block $B$ actually weighs 4.5 kg, the pulley will rotate counterclockwise.

**Kinematics of Motion.** Assuming $\boldsymbol{\alpha}$ counterclockwise and noting that $a_A = r_A\alpha$ and $a_B = r_B\alpha$, we obtain

$$\mathbf{a}_A = (25.4 \text{ cm})\alpha \uparrow \qquad \mathbf{a}_B = (15.2 \text{ cm})\alpha \downarrow$$

**Equations of Motion.** A single system consisting of the pulley and the two blocks is considered. Forces external to this system consist of the weights of the pulley and the two blocks and of the reaction at $G$. (The forces exerted by the cables on the pulley and on the blocks are internal to the system considered and cancel out.) Since the motion of the pulley is a centroidal rotation and the motion of each block is a translation, the effective forces reduce to the couple $\bar{I}\boldsymbol{\alpha}$ and the two vectors $m\mathbf{a}_A$ and $m\mathbf{a}_B$. The centroidal moment of inertia of the pulley is

$$\bar{I} = m\bar{k}^2 = 5.44 \text{ kg } (20.3 \text{ kg})^2 = 2241.77 \text{ kg} \cdot \text{cm}^2$$

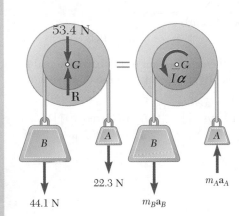

53.4 N

G

R

B   A

44.1 N

22.3 N

$m_B a_B$

G

$\bar{I}\alpha$

B   A

$m_A a_A$

Since the system of the external forces is equipollent to the system of the effective forces, we write

$+\uparrow\Sigma M_G = \Sigma(M_G)_{\text{eff}}:$

$(4.5 \text{ kg})(15.2 \text{ cm}) - (2.27 \text{ kg})(25.4 \text{ cm}) = +\bar{I}\alpha + m_B a_B(15.2 \text{ cm}) + m_A a_A(25.4 \text{ cm})$

$(4.5)(15.2) - (2.27)(25.4) = 2241.77\alpha + 4.5(15.2\alpha)(15.2) + 2.27(25.4\alpha)(25.4)$

$\alpha = +443.55 \text{ rad/s}^2$    $\boldsymbol{\alpha} = 443.55 \text{ rad/s}^2 \nwarrow$  ◄

$a_A = r_A\alpha = (25.4 \text{ cm})(443.55 \text{ rad/s}^2)$    $\mathbf{a}_A = 11266.17 \text{ cm/s}^2 \uparrow$  ◄

$a_B = r_B\alpha = (15.2 \text{ cm})(443.55 \text{ rad/s}^2)$    $\mathbf{a}_B = 6741.96 \text{ cm/s}^2 \downarrow$  ◄

# SAMPLE PROBLEM 16.4

A cord is wrapped around a homogeneous disk of radius $r = 0.5$ m and mass $m = 15$ kg. If the cord is pulled upward with a force $\mathbf{T}$ of magnitude 180 N, determine (a) the acceleration of the center of the disk, (b) the angular acceleration of the disk, (c) the acceleration of the cord.

## SOLUTION

**Equations of Motion.** We assume that the components $\bar{a}_x$ and $\bar{a}_y$ of the acceleration of the center are directed, respectively, to the right and upward and that the angular acceleration of the disk is counterclockwise. The external forces acting on the disk consist of the weight $\mathbf{W}$ and the force $\mathbf{T}$ exerted by the cord. This system is equivalent to the system of the effective forces, which consists of a vector of components $m\bar{a}_x$ and $m\bar{a}_y$ attached at $G$ and a couple $\bar{I}\alpha$. We write

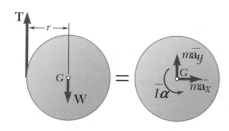

$$\xrightarrow{+}\Sigma F_x = \Sigma(F_x)_{\text{eff}}: \qquad 0 = m\bar{a}_x \qquad \bar{\mathbf{a}}_x = 0 \;\blacktriangleleft$$

$$+\uparrow\Sigma F_y = \Sigma(F_y)_{\text{eff}}: \qquad T - W = m\bar{a}_y$$

$$\bar{a}_y = \frac{T - W}{m}$$

Since $T = 180$ N, $m = 15$ kg, and $W = (15 \text{ kg})(9.81 \text{ m/s}^2) = 147.1$ N, we have

$$\bar{a}_y = \frac{180 \text{ N} - 147.1 \text{ N}}{15 \text{ kg}} = +2.19 \text{ m/s}^2 \qquad \bar{\mathbf{a}}_y = 2.19 \text{ m/s}^2 \uparrow \;\blacktriangleleft$$

$$+\!\!\uparrow\Sigma M_G = \Sigma(M_G)_{\text{eff}}: \qquad -Tr = \bar{I}\alpha$$

$$-Tr = (\tfrac{1}{2}mr^2)\alpha$$

$$\alpha = -\frac{2T}{mr} = -\frac{2(180 \text{ N})}{(15 \text{ kg})(0.5 \text{ m})} = -48.0 \text{ rad/s}^2$$

$$\boldsymbol{\alpha} = 48.0 \text{ rad/s}^2 \;\downarrow \;\blacktriangleleft$$

**Acceleration of Cord.** Since the acceleration of the cord is equal to the tangential component of the acceleration of point $A$ on the disk, we write

$$\mathbf{a}_{\text{cord}} = (\mathbf{a}_A)_t = \bar{\mathbf{a}} + (\mathbf{a}_{A/G})_t$$

$$= [2.19 \text{ m/s}^2 \uparrow] + [(0.5 \text{ m})(48 \text{ rad/s}^2)\uparrow]$$

$$\mathbf{a}_{\text{cord}} = 26.2 \text{ m/s}^2 \uparrow \;\blacktriangleleft$$

## SAMPLE PROBLEM 16.5

A uniform sphere of mass $m$ and radius $r$ is projected along a rough horizontal surface with a linear velocity $\bar{\mathbf{v}}_0$ and no angular velocity. Denoting by $\mu_k$ the coefficient of kinetic friction between the sphere and the floor, determine (a) the time $t_1$ at which the sphere will start rolling without sliding, (b) the linear velocity and angular velocity of the sphere at time $t_1$.

## SOLUTION

**Equations of Motion.** The positive sense is chosen to the right for $\bar{\mathbf{a}}$ and clockwise for $\boldsymbol{\alpha}$. The external forces acting on the sphere consist of the weight $\mathbf{W}$, the normal reaction $\mathbf{N}$, and the friction force $\mathbf{F}$. Since the point of the sphere in contact with the surface is sliding to the right, the friction force $\mathbf{F}$ is directed to the left. While the sphere is sliding, the magnitude of the friction force is $F = \mu_k N$. The effective forces consist of the vector $m\bar{\mathbf{a}}$ attached at $G$ and the couple $\bar{I}\boldsymbol{\alpha}$. Expressing that the system of the external forces is equivalent to the system of the effective forces, we write

$$+\uparrow \Sigma F_y = \Sigma(F_y)_{\text{eff}}: \qquad N - W = 0$$
$$N = W = mg \qquad F = \mu_k N = \mu_k mg$$
$$\xrightarrow{+} \Sigma F_x = \Sigma(F_x)_{\text{eff}}: \qquad -F = m\bar{a} \qquad -\mu_k mg = m\bar{a} \qquad \bar{a} = -\mu_k g$$
$$+\downarrow \Sigma M_G = \Sigma(M_G)_{\text{eff}}: \qquad Fr = \bar{I}\alpha$$

Noting that $\bar{I} = \frac{2}{5}mr^2$ and substituting the value obtained for $F$, we write

$$(\mu_k mg)r = \frac{2}{5}mr^2\alpha \qquad \alpha = \frac{5}{2}\frac{\mu_k g}{r}$$

**Kinematics of Motion.** As long as the sphere both rotates and slides, its linear and angular motions are uniformly accelerated.

$$t = 0, \bar{v} = \bar{v}_0 \qquad \bar{v} = \bar{v}_0 + \bar{a}t = \bar{v}_0 - \mu_k gt \tag{1}$$

$$t = 0, \omega_0 = 0 \qquad \omega = \omega_0 + \alpha t = 0 + \left(\frac{5}{2}\frac{\mu_k g}{r}\right)t \tag{2}$$

The sphere will start rolling without sliding when the velocity $\mathbf{v}_C$ of the point of contact $C$ is zero. At that time, $t = t_1$, point $C$ becomes the instantaneous center of rotation, and we have

$$\bar{v}_1 = r\omega_1 \tag{3}$$

Substituting in (3) the values obtained for $\bar{v}_1$ and $\omega_1$ by making $t = t_1$ in (1) and (2), respectively, we write

$$\bar{v}_0 - \mu_k gt_1 = r\left(\frac{5}{2}\frac{\mu_k g}{r}t_1\right) \qquad t_1 = \frac{2}{7}\frac{\bar{v}_0}{\mu_k g} \quad \blacktriangleleft$$

Substituting for $t_1$ into (2), we have

$$\omega_1 = \frac{5}{2}\frac{\mu_k g}{r}t_1 = \frac{5}{2}\frac{\mu_k g}{r}\left(\frac{2}{7}\frac{\bar{v}_0}{\mu_k g}\right) \qquad \omega_1 = \frac{5}{7}\frac{\bar{v}_0}{r} \qquad \omega_1 = \frac{5}{7}\frac{\bar{v}_0}{r}\downarrow \quad \blacktriangleleft$$

$$\bar{v}_1 = r\omega_1 = r\left(\frac{5}{7}\frac{\bar{v}_0}{r}\right) \qquad \bar{v}_1 = \frac{5}{7}\bar{v}_0 \qquad \mathbf{v}_1 = \frac{5}{7}\bar{v}_0 \rightarrow \quad \blacktriangleleft$$

# SOLVING PROBLEMS
## ON YOUR OWN

This chapter deals with the *plane motion* of rigid bodies, and in this first lesson we considered rigid bodies that are free to move under the action of applied forces.

**1. Effective forces.** We first recalled that a rigid body consists of a large number of particles. The effective forces of the particles forming the body were found to be equivalent to a vector $m\mathbf{a}$ attached at the mass center $G$ of the body and a couple of moment $\bar{I}\boldsymbol{\alpha}$ [Fig. 16.7]. Noting that the applied forces are equivalent to the effective forces, we wrote

$$\Sigma F_x = m\bar{a}_x \qquad \Sigma F_y = m\bar{a}_y \qquad \Sigma M_G = \bar{I}\alpha \qquad (16.5)$$

where $\bar{a}_x$ and $\bar{a}_y$ are the $x$ and $y$ components of the acceleration of the mass center $G$ of the body and $\alpha$ is the angular acceleration of the body. It is important to note that when these equations are used, *the moments of the applied forces must be computed with respect to the mass center of the body.* However, you learned a more efficient method of solution based on the use of a free-body-diagram equation.

**2. Free-body-diagram equation.** Your first step in the solution of a problem should be to draw a *free-body-diagram equation.*

 *a. A free-body-diagram equation consists* of two diagrams representing two equivalent systems of vectors. *In the first diagram* you should show *the forces exerted on the body,* including the applied forces, the reactions at the supports, and the weight of the body. *In the second diagram* you should show the vector $m\bar{\mathbf{a}}$ and the couple $\bar{I}\boldsymbol{\alpha}$ representing *the effective forces.*

 *b. Using a free-body-diagram equation* allows you to *sum components in any direction and to sum moments about any point.* When writing the three equations of motion needed to solve a given problem, you can therefore select one or more equations involving a single unknown. Solving these equations first and substituting the values obtained for the unknowns into the remaining equation(s) will yield a simpler solution.

**3. Plane motion of a rigid body.** The problems that you will be asked to solve will fall into one of the following categories.

 *a. Rigid body in translation.* For a body in translation, the angular acceleration is zero. The effective forces reduce to *the vector $m\bar{\mathbf{a}}$ applied at the mass center* [Sample Probs. 16.1 and 16.2].

***b. Rigid body in centroidal rotation.*** For a body in centroidal rotation, the acceleration of the mass center is zero. The effective forces reduce to *the couple* $\bar{I}\alpha$ [Sample Prob. 16.3].

***c. Rigid body in general plane motion.*** You can consider the general plane motion of a rigid body as the sum of a translation and a centroidal rotation. The effective forces are equivalent to the vector $m\bar{\mathbf{a}}$ and the couple $\bar{I}\alpha$ [Sample Probs. 16.4 and 16.5].

***4. Plane motion of a system of rigid bodies.*** You first should draw a free-body-diagram equation that includes all the rigid bodies of the system. A vector $m\bar{\mathbf{a}}$ and a couple $\bar{I}\alpha$ are attached to each body. However, the forces exerted on each other by the various bodies of the system can be omitted, since they occur in pairs of equal and opposite forces.

***a. If no more than three unknowns are involved,*** you can use this free-body-diagram equation and sum components in any direction and sum moments about any point to obtain equations that can be solved for the desired unknowns [Sample Prob. 16.3].

***b. If more than three unknowns are involved,*** you must draw a separate free-body-diagram equation for each of the rigid bodies of the system. Both internal forces and external forces should be included in each of the free-body-diagram equations, and care should be taken to represent with equal and opposite vectors the forces that two bodies exert on each other.

# Problems

**16.1** A 30-kg uniform thin panel is placed in a truck with end $A$ resting on a rough horizontal surface and end $B$ supported by a smooth vertical surface. Knowing that the deceleration of the truck is 4 m/s², determine (*a*) the reactions at ends $A$ and $B$, (*b*) the minimum required coefficient of static friction at end $A$.

**Fig. P16.1 and P16.2**

**16.2** A 30-kg uniform thin panel is placed in a truck with end $A$ resting on a rough horizontal surface and end $B$ supported by a smooth vertical surface. Knowing that the panel remains in the position shown, determine (*a*) the maximum allowable acceleration of the truck, (*b*) the corresponding minimum required coefficient of static friction at end $A$.

**16.3** Two identical 0.4 kg slender rods $AB$ and $BC$ are welded together to form an L-shaped assembly. The assembly is guided by two small wheels that roll freely in inclined parallel slots cut in a vertical plate. Knowing that $\theta = 30°$, determine (*a*) the acceleration of the assembly, (*b*) the reactions at $A$ and $C$.

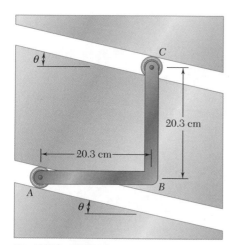

**Fig. P16.3 and P16.4**

**16.4** Two identical 0.4 kg slender rods $AB$ and $BC$ are welded together to form an L-shaped assembly. The assembly is guided by two small wheels that roll freely in inclined parallel slots cut in a vertical plate. Determine (*a*) the angle of inclination $\theta$ for which the reaction at $A$ is zero, (*b*) the corresponding acceleration of the assembly.

**16.5** For the truck of Sample Prob. 16.1, determine the distance through which the truck will skid if (*a*) the rear-wheel brakes fail to operate, (*b*) the front-wheel brakes fail to operate.

**16.6** Knowing that the coefficient of static friction between the tires and the road is 0.80 for the automobile shown, determine the maximum possible acceleration on a level road, assuming (*a*) four-wheel drive, (*b*) rear-wheel drive, (*c*) front-wheel drive.

**Fig. P16.6**

**16.7** The support bracket shown is used to transport a cylindrical can from one elevation to another. Knowing that $\mu_s = 0.30$ between the can and the bracket, determine (a) the magnitude of the upward acceleration **a** for which the can will slide on the bracket, (b) the smallest ratio $h/d$ for which the can will tip before it slides.

**16.8** Solve Prob. 16.7 assuming that the acceleration **a** of the bracket is downward.

**16.9** A 22.7 kg cabinet is mounted on casters that allow it to move freely ($\mu = 0$) on the floor. If a 11.3 kg force is applied as shown, determine (a) the acceleration of the cabinet, (b) the range of values of $h$ for which the cabinet will not tip.

**Fig. P16.7**

**Fig. *P16.9***

**16.10** Solve Prob. 16.9 assuming that the casters are locked and slide on the rough floor ($\mu_s = 0.25$).

**16.11** Bars $AB$ and $BE$, each of weight 3.6 kg, are welded together and are pin-connected to two links $AC$ and $BD$. Knowing that the assembly is released from rest in the position shown and neglecting the masses of the links, determine (a) the acceleration of the assembly, (b) the forces in the links.

**16.12** Members $ACE$ and $DCB$ are each 76.2 cm long and are connected by a pin at $C$. The mass center of the 9 kg member $AB$ is located at $G$. Determine (a) the acceleration of $AB$ immediately after the system has been released from rest in the position shown, (b) the corresponding force exerted by roller $A$ on member $AB$. Neglect the weight of members $ACE$ and $DCB$.

**Fig. P16.11**

**Fig. P16.12**

**Fig. P16.13 and P16.14**

**Fig. P16.15**

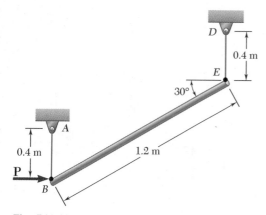

**Fig. P16.16**

**16.13** A 4-kg uniform slender rod *AB* is held in position by two ropes and the link *CA* which has a negligible weight. After rope *BD* is cut the assembly rotates in a vertical plane under the combined effect of gravity and a 6 N · m couple **M** applied to link *CA* as shown. Determine, immediately after rope *BD* has been cut, (*a*) the acceleration of rod *AB*, (*b*) the tension in rope *EB*.

**16.14** A 4-kg uniform slender rod *AB* is held in position by two ropes and the link *CA* which has a negligible weight. After rope *BD* is cut the assembly rotates in a vertical plane under the combined effect of gravity and a couple **M** applied to link *CA* as shown. Knowing that the magnitude of the acceleration of rod *AB* is 4 m/s² immediately after rope *BD* has been cut, determine (*a*) the magnitude of the couple **M**, (*b*) the tension in rope *EB*.

**16.15** The 7.5-kg rod *BC* connects a disk centered at *A* to crank *CD*. Knowing that the disk is made to rotate at the constant speed of 180 rpm, determine for the position shown the vertical components of the forces exerted on rod *BC* by the pins at *B* and *C*.

**16.16** At the instant shown the tensions in the vertical ropes *AB* and *DE* are 300 N and 200 N, respectively. Knowing that the mass of the uniform bar *BE* is 5 kg, determine, at this instant, (*a*) the force **P**, (*b*) the magnitude of the angular velocity of each rope, (*c*) the angular acceleration of each rope.

**\*16.17** Draw the shear and bending-moment diagrams for each of the bars *AB* and *BE* of Prob. 16.11.

**\*16.18** Draw the shear and bending-moment diagrams for the connecting rod *BC* of Prob. 16.15.

**16.19** For a rigid slab in translation, show that the system of the effective forces consists of vectors $(\Delta m_i)\overline{\mathbf{a}}$ attached to the various particles of the slab, where $\overline{\mathbf{a}}$ is the acceleration of the mass center *G* of the slab. Further show, by computing their sum and the sum of their moments about *G*, that the effective forces reduce to a single vector $m\overline{\mathbf{a}}$ attached at *G*.

**Fig. P16.19**          **Fig. P16.20**

**16.20** For a rigid slab in centroidal rotation, show that the system of the effective forces consists of vectors $-(\Delta m_i)\omega^2\mathbf{r}'_i$ and $(\Delta m_i)(\boldsymbol{\alpha} \times \mathbf{r}'_i)$ attached to the various particles $P_i$ of the slab, where $\boldsymbol{\omega}$ and $\boldsymbol{\alpha}$ are the angular velocity and angular acceleration of the slab, and where $\mathbf{r}'_i$ denotes the position of the particle $P_i$ relative to the mass center *G* of the slab. Further show, by computing their sum and the sum of their moments about *G*, that the effective forces reduce to a couple $\overline{I}\boldsymbol{\alpha}$.

**16.21**  A 200-kg flywheel is at rest when a constant 300 N · m couple **M** is applied at time $t = 0$. At $t = 28$ s the flywheel reaches its rated speed of 2400 rpm and the couple is removed. Knowing that the radius of gyration of the flywheel is 400 mm, determine (*a*) the average magnitude of the couple due to kinetic friction in the bearing, (*b*) the time at which the flywheel comes to rest.

**16.22**  A 226.8 kg flywheel is at rest when a constant 271.16 N.m couple **M** is applied at time $t = 0$. At $t = 30$ s the flywheel reaches its rated speed of 3000 rpm and the couple is removed. Assuming that the kinetic friction results in a constant couple of magnitude 16.9 N · m, determine (*a*) the radius of gyration of the flywheel, (*b*) the number of revolutions that the flywheel executes after the couple is removed and before it comes to rest.

**Fig. P16.21 and P16.22**

**16.23**  The flywheel shown consists of a 76.2 cm-diameter disk which weighs 108.9 kg. The coefficient of friction between the band and the flywheel is 0.35. If the initial angular velocity of the flywheel is 360 rpm clockwise, determine the magnitude of the force **P** required to stop the flywheel in 25 revolutions.

**16.24**  The 100-mm-radius brake drum is attached to a flywheel which is not shown. The drum and flywheel together have a mass of 300 kg and a radius of gyration of 600 mm. The coefficient of kinetic friction between the brake band and the drum is 0.30. Knowing that a force **P** of magnitude 50 N is applied at $A$ when the angular velocity is 180 rpm counterclockwise, determine the time required to stop the flywheel when $a = 200$ mm and $b = 160$ mm.

**Fig. P16.23**

**Fig. P16.24**

**16.25**  A 10-kg uniform disk is placed in contact with an inclined surface and a constant 11 N · m couple **M** is applied as shown. The weight of the link $AB$ is negligible. Knowing that the coefficient of kinetic friction at $D$ is 0.4, determine (*a*) the angular acceleration of the disk, (*b*) the force in the link $AB$.

**16.26**  A 10-kg uniform disk is placed in contact with an inclined surface and a constant couple **M** is applied as shown. The mass of the uniform link $AB$ is 5 kg and the coefficient of kinetic friction at $D$ is 0.25. Knowing that the angular acceleration of the disk is 16 rad/s$^2$ clockwise, determine (*a*) the magnitude of the couple **M**, (*b*) the force exerted on the disk at $D$.

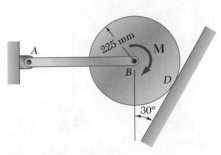

**Fig. P16.25 and P16.26**

**16.27**   Each of the double pulleys shown has a mass moment of inertia of 20 kg · m² and is initially at rest. The outside radius is 400 mm, and the inner radius is 200 mm. Determine (a) the angular acceleration of each pulley, (b) the angular velocity of each pulley after point A on the cord has moved 3 m.

785 N    80 kg    230 kg  150 kg    40 kg

(1)    (2)    (3)    (4)

**Fig. P16.27**

**16.28**   The flywheel shown has a radius of 600 mm, a mass of 144 kg, and a radius of gyration of 450 mm. An 18-kg block A is attached to a wire that is wrapped around the flywheel, and the system is released from rest. Neglecting the effect of friction, determine (a) the acceleration of block A, (b) the speed of block A after it has moved 1.8 m.

**16.29**   In order to determine the mass moment of inertia of a flywheel of radius .46 m, a 9 kg block is attached to a wire that is wrapped around the flywheel. The block is released and is observed to fall 3.7 m in 4.5 s. To eliminate bearing friction from the computations, a second block of weight 18.1 kg is used and is observed to fall 3.7 m in 2.8 s. Assuming that the moment of the couple due to friction remains constant, determine the mass moment of inertia of the flywheel.

$A$ $m$

**Fig. P16.28 and P16.29**

**16.30**   Two disks A and B, of mass $m_A = 2$ kg and $m_B = 4$ kg, are connected by a belt as shown. Assuming no slipping between the belt and the disks, determine the angular acceleration of each disk if a 2.70 N · m couple **M** is applied to disk A.

200 mm    300 mm

$A$    **M**    $B$

**Fig. P16.30**

**16.31**   The 11.3 kg double pulley shown is at rest and in equilibrium when a constant 4.75 N . m couple **M** is applied. Neglecting the effect of friction and knowing that the radius of gyration of the double pulley is 15.2 cm, determine (a) the angular acceleration of the double pulley, (b) the tension in each rope.

10.2 cm    20.3 cm

**M**

4.5 kg    2.27 kg

$A$    $B$

**Fig. P16.31**

**16.32**  The weight and radius of friction disk $A$ are $W_A = 5.4$ kg and $r_A = 15.2$ cm; the weight and radius of friction disk $B$ are $W_B = 2.7$ kg and $r_B = 10.2$ cm. The disks are at rest when a couple **M** of moment 84.7 N · cm is applied to disk $A$. Assuming that no slipping occurs between the disks, determine (*a*) the angular acceleration of each disk, (*b*) the friction force that disk $A$ exerts on disk $B$.

**16.33**  Solve Prob. 16.32 assuming that the couple **M** is applied to disk $B$.

**16.34**  Two identical 7.3 kg uniform cylinders are at rest when a constant couple **M** of magnitude 4.07 N · m is applied to cylinder $A$. Knowing that the coefficient of kinetic friction between cylinder $B$ and the horizontal surface is 0.2 and that no slipping occurs between the cylinders, determine (*a*) the angular acceleration of each cylinder, (*b*) the minimum allowable value of the coefficient of static friction between the cylinders.

**Fig. P16.32**

**Fig. P16.34 and P16.35**

**16.35**  Two identical 7.3 kg uniform cylinders are at rest when a constant couple **M** of magnitude 4.07 N · m is applied to cylinder $A$. Knowing that the coefficient of kinetic friction between the cylinders is 0.45 and the coefficient of kinetic friction between cylinder $B$ and the horizontal surface is 0.15, determine the angular acceleration of each cylinder.

**16.36**  A cylinder of radius $r$ and mass $m$ rests on two small casters $A$ and $B$ as shown. Initially, the cylinder is at rest and is set in motion by rotating caster $B$ clockwise at high speed so that slipping occurs between the cylinder and caster $B$. Denoting by $\mu_k$ the coefficient of kinetic friction and neglecting the moment of inertia of the free caster $A$, derive an expression for the angular acceleration of the cylinder.

**16.37**  In Prob. 16.36 assume that no slipping can occur between caster $B$ and the cylinder (such a case would exist if the cylinder and caster had gear teeth along their rims). Derive an expression for the maximum allowable counterclockwise acceleration $\alpha$ of the cylinder if it is not to lose contact with the caster at $A$.

**Fig. P16.36**

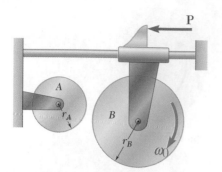

**Fig. P16.38, P16.39, and P16.40**

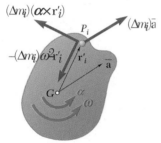

**Fig. P16.42**

**16.38** Disk $B$ has an angular velocity $\boldsymbol{\omega}_0$ when it is brought into contact with disk $A$, which is at rest. Show that ($a$) the final angular velocities of the disks are independent of the coefficient of friction $\mu_k$ between the disks as long as $\mu_k \neq 0$, ($b$) the final angular velocity of disk $B$ depends only upon $\boldsymbol{\omega}_0$ and the ratio of the masses $m_A$ and $m_B$ of the two disks.

**16.39** The 5.4 kg disk $A$ has a radius $r_A = 15.2$ cm and an initial angular velocity $\boldsymbol{\omega}_0 = 750$ rpm clockwise. The 13.6 kg disk $B$ has a radius $r_B = 25.4$ cm and is at rest. A force $\mathbf{P}$ of magnitude 22.2 N is then applied to bring the disks into contact. Knowing that $\mu_k = 0.25$ between the disks and neglecting bearing friction, determine ($a$) the angular acceleration of each disk, ($b$) the final angular velocity of each disk.

**16.40** Solve Prob. 16.39 assuming that disk $A$ is initially at rest and that disk $B$ has an angular velocity of 750 rpm clockwise.

**16.41** Show that the system of the effective forces for a rigid slab in plane motion reduces to a single vector, and express the distance from the mass center $G$ of the slab to the line of action of this vector in terms of the centroidal radius of gyration $\bar{k}$ of the slab, the magnitude $\bar{a}$ of the acceleration of $G$, and the angular acceleration $\alpha$.

**16.42** For a rigid slab in plane motion, show that the system of the effective forces consists of vectors $(\Delta m_i)\bar{\mathbf{a}}$, $-(\Delta m_i)\omega^2 \mathbf{r}'_i$, and $(\Delta m_i)(\boldsymbol{\alpha} \times \mathbf{r}'_i)$ attached to the various particles $P_i$ of the slab, where $\bar{\mathbf{a}}$ is the acceleration of the mass center $G$ of the slab, $\boldsymbol{\omega}$ is the angular velocity of the slab, $\boldsymbol{\alpha}$ is its angular acceleration, and $\mathbf{r}'_i$ denotes the position vector of the particle $P_i$ relative to $G$. Further show, by computing their sum and the sum of their moments about $G$, that the effective forces reduce to a vector $m\bar{\mathbf{a}}$ attached at $G$ and a couple $\bar{I}\alpha$.

**16.43** A force $\mathbf{P}$ of magnitude 3.34 N is applied to a tape wrapped around a thin 2.7 kg hoop. Knowing that the hoop rests on a frictionless horizontal surface, determine the acceleration of ($a$) point $A$, ($b$) point $B$.

**Fig. P16.43**

**Fig. P16.44**

**16.44** A force $\mathbf{P}$ is applied to a tape wrapped around a uniform disk that rests on a frictionless horizontal surface. Show that for each 360° rotation of the disk the center of the disk will move a distance $\pi r$.

**16.45** A uniform slender rod $AB$ rests on a frictionless horizontal surface, and a force **P** of magnitude 2 N is applied at $A$ in a direction perpendicular to the rod. Knowing that the rod has a mass of 3 kg, determine the acceleration of ($a$) point $A$, ($b$) point $B$.

**Fig. P16.45**

**16.46** A 120-kg satellite has a radius of gyration of 600 mm with respect to the $y$ axis and is symmetrical with respect to the $zx$ plane. Its orientation is changed by firing four small rockets $A$, $B$, $C$, and $D$, each of which produces a 16.20-N thrust **T** directed as shown. Determine the angular acceleration of the satellite and the acceleration of its mass center $G$ ($a$) when all four rockets are fired, ($b$) when all rockets except $D$ are fired.

**16.47** A uniform semicircular plate weighing 5.4 kg is suspended from three vertical wires at points $A$, $B$, and $C$, and a force **P** of magnitude 4.4 N is applied to point $B$. Immediately after **P** is applied, determine the acceleration of ($a$) the mass center of the plate, ($b$) point $C$.

**16.48** Immediately after the force **P** is applied to the plate of Prob. 16.47, determine the acceleration of ($a$) point $A$, ($b$) point $B$.

**16.49** A uniform slender L-shaped bar $ABC$ is at rest on a horizontal surface when a force **P** of magnitude 4 N is applied at point $A$. Neglecting friction between the bar and the surface and knowing that the mass of the bar is 2 kg, determine ($a$) the initial angular acceleration of the bar, ($b$) the initial acceleration of point $B$.

**Fig. P16.46**

**Fig. P16.47 and P16.48**

**Fig. P16.49**

**16.50**   The steel roll shown weighs 1270 kg, has a centroidal radius of
gyration of 15.2 cm, and is lifted by two cables looped around its shaft. Knowing that for each cable $T_A = 3069$ N and $T_B = 3247$ N, determine (*a*) the angular acceleration of the roll, (*b*) the acceleration of its mass center.

**Fig. P16.50 and P16.51**

**16.51**   The steel roll shown weighs 1270 kg, has a centroidal radius of
gyration of 15.2 cm, and is lifted by two cables looped around its shaft. Knowing that at the instant shown the acceleration of the roll is 15.2 cm/s² downward and that for each cable $T_A = 3024.8$ N, determine (*a*) the corresponding value of the tension $T_B$, (*b*) the angular acceleration of the roll.

**16.52**   A drum of 200-mm radius is attached to a disk of radius $r_A =$
150 mm. The disk and drum have a combined mass of 5 kg and a combined
radius of gyration of 120 mm and are suspended by two cords. Knowing that
$T_A = 35$ N and $T_B = 25$ N, determine the accelerations of points $A$ and $B$
on the cords.

**16.53**   A drum of 200-mm radius is attached to a disk of radius $r_A =$
140 mm. The disk and drum have a combined mass of 5 kg and are suspended by two cords. Knowing that the acceleration of point $B$ on the cord
is zero, $T_A = 40$ N, and $T_B = 20$ N, determine the combined radius of gyration of the disk and drum.

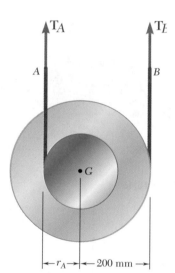

**Fig. P16.52 and P16.53**

**16.54**   The uniform disk shown, of mass $m$ and radius $r$, rotates counterclockwise. Its center $C$ is constrained to move in a slot cut in the vertical
member $AB$ and a horizontal force **P** is applied at $B$ to maintain contact at
$D$ between the disk and the vertical wall. The disk moves downward under
the influence of gravity and the friction at $D$. Denoting by $\mu_k$ the coefficient
of kinetic friction between the disk and the wall and neglecting friction in
the vertical slot, determine (*a*) the angular acceleration of the disk, (*b*) the
acceleration of the center $C$ of the disk.

**16.55**   The 5.4 kg uniform disk shown, of radius $r = 8.1$ cm, rotates
counterclockwise. Its center $C$ is constrained to move in a slot cut in the vertical member $AB$ and an 48.9 N horizontal force **P** is applied at $B$ to maintain contact at $D$ between the disk and the vertical wall. The disk moves
downward under the influence of gravity and the friction at $D$. Knowing that
the coefficient of kinetic friction between the disk and the wall is 0.12 and
neglecting friction in the vertical slot, determine (*a*) the angular acceleration
of the disk, (*b*) the acceleration of the center $C$ of the disk.

**Fig. *P16.54* and *P16.55***

**16.56 and 16.57**  A uniform slender L-shaped bar *ABC* is at rest in a vertical plane when two upward forces are applied as shown. Knowing that the mass of the bar is 20 kg, determine (*a*) the initial angular acceleration of the bar, (*b*) the initial acceleration of point *B*.

Fig. P16.56

Fig. P16.57

**16.58 and 16.59**  A beam *AB* of mass *m* and of uniform cross section is suspended from two springs as shown. If spring 2 breaks, determine at that instant (*a*) the angular acceleration of the beam, (*b*) the acceleration of point *A*, (*c*) the acceleration of point *B*.

Fig. P16.58

Fig. P16.59

**16.60**  A uniform slender bar *AB* of mass *m* is suspended from two springs as shown. If spring 2 breaks, determine at that instant (*a*) the angular acceleration of the bar, (*b*) the acceleration of point *A*, (*c*) the acceleration of point *B*.

Fig. P16.60

**16.61 through 16.63** A thin plate of the shape indicated and of mass $m$ is suspended from two springs as shown. If spring 2 breaks, determine the acceleration at that instant ($a$) of point $A$, ($b$) of point $B$.

**16.61** A square plate of side $b$.

**16.62** A thin hoop of radius $b$.

**16.63** A circular plate of radius $b$.

**Fig. P16.61**

**Fig. P16.62**

**Fig. P16.63**

**16.64** A sphere of radius $r$ and mass $m$ is projected along a rough horizontal surface with the initial velocities indicated. If the final velocity of the sphere is to be zero, express, in terms of $v_0$, $r$, and $\mu_k$, ($a$) the required magnitude of $\boldsymbol{\omega}_0$, ($b$) the time $t_1$ for the sphere to come to rest, ($c$) the distance the sphere will move before coming to rest.

**Fig. P16.64**

**16.65** A bowler projects a 200-mm-diameter ball of mass 5 kg along an alley with a forward velocity $\mathbf{v}_0$ of 5 m/s and a backspin $\boldsymbol{\omega}_0$ of 9 rad/s. Knowing that the coefficient of kinetic friction between the ball and the alley is 0.10, determine ($a$) the time $t_1$ at which the ball will start rolling without sliding, ($b$) the speed of the ball at time $t_1$, ($c$) the distance the ball will have traveled at time $t_1$.

**16.66** Solve Prob. 16.65 assuming that the bowler projects the ball with the same forward velocity but with a backspin of 18 rad/s.

**Fig. P16.65**

**16.67** A uniform sphere of radius $r$ and mass $m$ is placed with no initial velocity on a belt that moves to the right with a constant velocity $\mathbf{v}_1$. Denoting by $\mu_k$ the coefficient of kinetic friction between the sphere and the belt, determine ($a$) the time $t_1$ at which the sphere will start rolling without sliding, ($b$) the linear and angular velocities of the sphere at time $t_1$.

**16.68** Solve Prob. 16.67, assuming that the sphere is replaced by a wheel of radius $r$, mass $m$, and centroidal radius of gyration $\bar{k}$.

**Fig. P16.67**

**16.69** A sphere of radius $r$ and mass $m$ has a linear velocity $\mathbf{v}_0$ directed to the left and no angular velocity as it is placed on a belt moving to the right with a constant placed velocity $\mathbf{v}_1$. If after first sliding on the belt the sphere is to have no linear velocity relative to the ground as it starts rolling on the belt without sliding, determine in terms of $v_1$ and the coefficient of kinetic friction $\mu_k$ between the sphere and the belt ($a$) the required value of $v_0$, ($b$) time $t_1$ at which the sphere will start rolling on the belt, ($c$) the distance the sphere will have moved relative to the ground at time $t_1$.

**Fig. P16.69**

# 16.8. CONSTRAINED PLANE MOTION

Most engineering applications deal with rigid bodies which are moving under given constraints. For example, cranks must rotate about a fixed axis, wheels must roll without sliding, and connecting rods must describe certain prescribed motions. In all such cases, definite relations exist between the components of the acceleration $\bar{\mathbf{a}}$ of the mass center $G$ of the body considered and its angular acceleration $\boldsymbol{\alpha}$; the corresponding motion is said to be a *constrained motion*.

The solution of a problem involving a constrained plane motion calls first for a *kinematic analysis* of the problem. Consider, for example, a slender rod $AB$ of length $l$ and mass $m$ whose extremities are connected to blocks of negligible mass which slide along horizontal and vertical frictionless tracks. The rod is pulled by a force $\mathbf{P}$ applied at $A$ (Fig. 16.11). We know from Sec. 15.8 that the acceleration $\bar{\mathbf{a}}$ of the mass center $G$ of the rod can be determined at any given instant from the position of the rod, its angular velocity, and its angular acceleration at that instant. Suppose, for example, that the values of $\theta$, $\omega$, and $\alpha$ are known at a given instant and that we wish to determine the corresponding value of the force $\mathbf{P}$, as well as the reactions at $A$ and $B$. We should first *determine the components $\bar{a}_x$ and $\bar{a}_y$ of the acceleration of the mass center $G$* by the method of Sec. 15.8. We next apply d'Alembert's principle (Fig. 16.12), using the expressions obtained for $\bar{a}_x$ and $\bar{a}_y$. The unknown forces $\mathbf{P}$, $\mathbf{N}_A$, and $\mathbf{N}_B$ can then be determined by writing and solving the appropriate equations.

**Fig. 16.11**

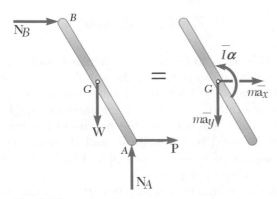

**Fig. 16.12**

Suppose now that the applied force $\mathbf{P}$, the angle $\theta$, and the angular velocity $\omega$ of the rod are known at a given instant and that we wish to find the angular acceleration $\alpha$ of the rod and the components $\bar{a}_x$ and $\bar{a}_y$ of the acceleration of its mass center at that instant, as well as the reactions at $A$ and $B$. The preliminary kinematic study of the problem will have for its object *to express the components $\bar{a}_x$ and $\bar{a}_y$ of the acceleration of $G$ in terms of the angular acceleration $\alpha$ of the rod*. This will be done by first expressing the acceleration of a suitable reference point such as $A$ in terms of the angular acceleration $\alpha$. The components $\bar{a}_x$ and $\bar{a}_y$ of the acceleration of $G$ can then be determined in terms of $\alpha$, and the expressions obtained carried into Fig. 16.12. Three equations can then be derived in terms of $\alpha$, $N_A$, and $N_B$ and solved for the three unknowns (see Sample Prob. 16.10). Note that the

**Fig. 16.13**

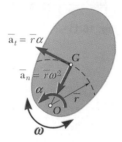

**Fig. 16.14**

method of dynamic equilibrium can also be used to carry out the solution of the two types of problems we have considered (Fig. 16.13).

When a mechanism consists of *several moving parts,* the approach just described can be used with each part of the mechanism. The procedure required to determine the various unknowns is then similar to the procedure followed in the case of the equilibrium of a system of connected rigid bodies (Sec. 6.11).

Earlier, we analyzed two particular cases of constrained plane motion: the translation of a rigid body, in which the angular acceleration of the body is constrained to be zero, and the centroidal rotation, in which the acceleration $\bar{\mathbf{a}}$ of the mass center of the body is constrained to be zero. Two other particular cases of constrained plane motion are of special interest: the *noncentroidal rotation* of a rigid body and the *rolling motion* of a disk or wheel. These two cases can be analyzed by one of the general methods described above. However, in view of the range of their applications, they deserve a few special comments.

**Noncentroidal Rotation.** The motion of a rigid body constrained to rotate about a fixed axis which does not pass through its mass center is called *noncentroidal rotation.* The mass center $G$ of the body moves along a circle of radius $\bar{r}$ centered at the point $O$, where the axis of rotation intersects the plane of reference (Fig. 16.14). Denoting, respectively, by $\boldsymbol{\omega}$ and $\boldsymbol{\alpha}$ the angular velocity and the angular acceleration of the line $OG$, we obtain the following expressions for the tangential and normal components of the acceleration of $G$:

$$\bar{a}_t = \bar{r}\alpha \qquad \bar{a}_n = \bar{r}\omega^2 \tag{16.7}$$

Since line $OG$ belongs to the body, its angular velocity $\boldsymbol{\omega}$ and its angular acceleration $\boldsymbol{\alpha}$ also represent the angular velocity and the angular acceleration of the body in its motion relative to $G$. Equations (16.7) define, therefore, the kinematic relation existing between the motion of the mass center $G$ and the motion of the body about $G$. They should be used to eliminate $\bar{a}_t$ and $\bar{a}_n$ from the equations obtained by applying d'Alembert's principle (Fig. 16.15) or the method of dynamic equilibrium (Fig. 16.16).

An interesting relation is obtained by equating the moments about the fixed point $O$ of the forces and vectors shown, respectively, in parts $a$ and $b$ of Fig. 16.15. We write

**Fig. 16.15**

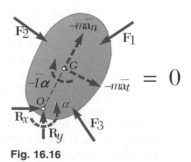

**Fig. 16.16**

$$+\curvearrowleft\Sigma M_O = \bar{I}\alpha + (m\bar{r}\alpha)\bar{r} = (\bar{I} + m\bar{r}^2)\alpha$$

But according to the parallel-axis theorem, we have $\bar{I} + m\bar{r}^2 = I_O$, where $I_O$ denotes the moment of inertia of the rigid body about the fixed axis. We therefore write

$$\Sigma M_O = I_O\alpha \qquad (16.8)$$

Although formula (16.8) expresses an important relation between the sum of the moments of the external forces about the fixed point $O$ and the product $I_O\alpha$, it should be clearly understood that this formula does not mean that the system of the external forces is equivalent to a couple of moment $I_O\alpha$. The system of the effective forces, and thus the system of the external forces, reduces to a couple only when $O$ coincides with $G$—that is, *only when the rotation is centroidal* (Sec. 16.4). In the more general case of noncentroidal rotation, the system of the external forces does not reduce to a couple.

A particular case of noncentroidal rotation is of special interest—the case of *uniform rotation,* in which the angular velocity $\boldsymbol{\omega}$ is constant. Since $\boldsymbol{\alpha}$ is zero, the inertia couple in Fig. 16.16 vanishes and the inertia vector reduces to its normal component. This component (also called *centrifugal force*) represents the tendency of the rigid body to break away from the axis of rotation.

**Rolling Motion.** Another important case of plane motion is the motion of a disk or wheel rolling on a plane surface. If the disk is constrained to roll without sliding, the acceleration $\bar{\mathbf{a}}$ of its mass center $G$ and its angular acceleration $\boldsymbol{\alpha}$ are not independent. Assuming that the disk is balanced, so that its mass center and its geometric center coincide, we first write that the distance $\bar{x}$ traveled by $G$ during a rotation $\theta$ of the disk is $\bar{x} = r\theta$, where $r$ is the radius of the disk. Differentiating this relation twice, we write

$$\bar{a} = r\alpha \qquad (16.9)$$

Recalling that the system of the effective forces in plane motion reduces to a vector $m\bar{\mathbf{a}}$ and a couple $\bar{I}\alpha$, we find that in the particular case of the rolling motion of a balanced disk, the effective forces reduce to a vector of magnitude $mr\alpha$ attached at $G$ and to a couple of magnitude $\bar{I}\alpha$. We may thus express that the external forces are equivalent to the vector and couple shown in Fig. 16.17.

**Fig. 16.17**

**Fig. 16.17 (repeated)**

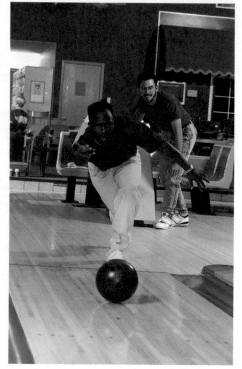

**Photo 16.4** As the ball hits the bowling alley, it first spins and slides, then rolls without sliding.

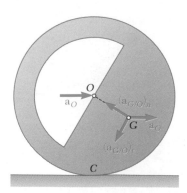

**Fig. 16.18**

When a disk *rolls without sliding,* there is no relative motion between the point of the disk in contact with the ground and the ground itself. Thus as far as the computation of the friction force **F** is concerned, a rolling disk can be compared with a block at rest on a surface. The magnitude $F$ of the friction force can have any value, as long as this value does not exceed the maximum value $F_m = \mu_s N$, where $\mu_s$ is the coefficient of static friction and $N$ is the magnitude of the normal force. In the case of a rolling disk, the magnitude $F$ of the friction force should therefore be determined independently of $N$ by solving the equation obtained from Fig. 16.17.

When *sliding is impending,* the friction force reaches its maximum value $F_m = \mu_s N$ and can be obtained from $N$.

When the disk *rotates and slides* at the same time, a relative motion exists between the point of the disk which is in contact with the ground and the ground itself, and the force of friction has the magnitude $F_k = \mu_k N$, where $\mu_k$ is the coefficient of kinetic friction. In this case, however, the motion of the mass center $G$ of the disk and the rotation of the disk about $G$ are independent, and $\bar{a}$ is not equal to $r\alpha$.

These three different cases can be summarized as follows:

| | | |
|---|---|---|
| Rolling, no sliding: | $F \leq \mu_s N$ | $\bar{a} = r\alpha$ |
| Rolling, sliding impending: | $F = \mu_s N$ | $\bar{a} = r\alpha$ |
| Rotating and sliding: | $F = \mu_k N$ | $\bar{a}$ and $\alpha$ independent |

When it is not known whether or not a disk slides, it should first be assumed that the disk rolls without sliding. If $F$ is found smaller than or equal to $\mu_s N$, the assumption is proved correct. If $F$ is found larger than $\mu_s N$, the assumption is incorrect and the problem should be started again, assuming rotating and sliding.

When a disk is *unbalanced,* that is, when its mass center $G$ does not coincide with its geometric center $O$, the relation (16.9) does not hold between $\bar{a}$ and $\alpha$. However, a similar relation holds between the magnitude $a_O$ of the acceleration of the geometric center and the angular acceleration $\alpha$ of an unbalanced disk which rolls without sliding. We have

$$a_O = r\alpha \tag{16.10}$$

To determine $\bar{a}$ in terms of the angular acceleration $\alpha$ and the angular velocity $\omega$ of the disk, we can use the relative-acceleration formula

$$\begin{aligned} \mathbf{\bar{a}} = \mathbf{\bar{a}}_G &= \mathbf{a}_O + \mathbf{a}_{G/O} \\ &= \mathbf{a}_O + (\mathbf{a}_{G/O})_t + (\mathbf{a}_{G/O})_n \end{aligned} \tag{16.11}$$

where the three component accelerations obtained have the directions indicated in Fig. 16.18 and the magnitudes $a_O = r\alpha$, $(a_{G/O})_t = (OG)\alpha$, and $(a_{G/O})_n = (OG)\omega^2$.

# SAMPLE PROBLEM 16.6

The portion $AOB$ of a mechanism consists of a 400-mm steel rod $OB$ welded to a gear $E$ of radius 120 mm which can rotate about a horizontal shaft $O$. It is actuated by a gear $D$ and, at the instant shown, has a clockwise angular velocity of 8 rad/s and a counterclockwise angular acceleration of 40 rad/s². Knowing that rod $OB$ has a mass of 3 kg and gear $E$ a mass of 4 kg and a radius of gyration of 85 mm, determine (a) the tangential force exerted by gear $D$ on gear $E$, (b) the components of the reaction at shaft $O$.

# SOLUTION

In determining the effective forces of the rigid body $AOB$, gear $E$ and rod $OB$ will be considered separately. Therefore, the components of the acceleration of the mass center $G_{OB}$ of the rod will be determined first:

$$(\bar{a}_{OB})_t = \bar{r}\alpha = (0.200 \text{ m})(40 \text{ rad/s}^2) = 8 \text{ m/s}^2$$
$$(\bar{a}_{OB})_n = \bar{r}\omega^2 = (0.200 \text{ m})(8 \text{ rad/s})^2 = 12.8 \text{ m/s}^2$$

**Equations of Motion.** Two sketches of the rigid body $AOB$ have been drawn. The first shows the external forces consisting of the weight $\mathbf{W}_E$ of gear $E$, the weight $\mathbf{W}_{OB}$ of the rod $OB$, the force $\mathbf{F}$ exerted by gear $D$, and the components $\mathbf{R}_x$ and $\mathbf{R}_y$ of the reaction at $O$. The magnitudes of the weights are, respectively,

$$W_E = m_E g = (4 \text{ kg})(9.81 \text{ m/s}^2) = 39.2 \text{ N}$$
$$W_{OB} = m_{OB} g = (3 \text{ kg})(9.81 \text{ m/s}^2) = 29.4 \text{ N}$$

The second sketch shows the effective forces, which consist of a couple $\bar{I}_E\boldsymbol{\alpha}$ (since gear $E$ is in centroidal rotation) and of a couple and two vector components at the mass center of $OB$. Since the accelerations are known, we compute the magnitudes of these components and couples:

$$\bar{I}_E\alpha = m_E \bar{k}_E^2 \alpha = (4 \text{ kg})(0.085 \text{ m})^2(40 \text{ rad/s}^2) = 1.156 \text{ N} \cdot \text{m}$$
$$m_{OB}(\bar{a}_{OB})_t = (3 \text{ kg})(8 \text{ m/s}^2) = 24.0 \text{ N}$$
$$m_{OB}(\bar{a}_{OB})_n = (3 \text{ kg})(12.8 \text{ m/s}^2) = 38.4 \text{ N}$$
$$\bar{I}_{OB}\alpha = (\tfrac{1}{12}m_{OB}L^2)\alpha = \tfrac{1}{12}(3 \text{ kg})(0.400 \text{ m})^2(40 \text{ rad/s}^2) = 1.600 \text{ N} \cdot \text{m}$$

Expressing that the system of the external forces is equivalent to the system of the effective forces, we write the following equations:

$+\curvearrowleft\Sigma M_O = \Sigma(M_O)_{\text{eff}}$:

$$F(0.120 \text{ m}) = \bar{I}_E\alpha + m_{OB}(\bar{a}_{OB})_t(0.200 \text{ m}) + \bar{I}_{OB}\alpha$$
$$F(0.120 \text{ m}) = 1.156 \text{ N} \cdot \text{m} + (24.0 \text{ N})(0.200 \text{ m}) + 1.600 \text{ N} \cdot \text{m}$$

$$F = 63.0 \text{ N} \qquad \mathbf{F} = 63.0 \text{ N} \downarrow \ \blacktriangleleft$$

$\xrightarrow{+}\Sigma F_x = \Sigma(F_x)_{\text{eff}}$: $\qquad R_x = m_{OB}(\bar{a}_{OB})_t$

$$R_x = 24.0 \text{ N} \qquad \mathbf{R}_x = 24.0 \text{ N} \rightarrow \ \blacktriangleleft$$

$+\uparrow\Sigma F_y = \Sigma(F_y)_{\text{eff}}$: $\quad R_y - F - W_E - W_{OB} = m_{OB}(\bar{a}_{OB})_n$
$$R_y - 63.0 \text{ N} - 39.2 \text{ N} - 29.4 \text{ N} = 38.4 \text{ N}$$

$$R_y = 170.0 \text{ N} \qquad \mathbf{R}_y \ 170.0 \text{ N} \uparrow \ \blacktriangleleft$$

## SAMPLE PROBLEM 16.7

A 15.2 × 20.3 cm rectangular plate weighing 27.2 kg is suspended from two pins $A$ and $B$. If pin $B$ is suddenly removed, determine (a) the angular acceleration of the plate, (b) the components of the reaction at pin $A$, immediately after pin $B$ has been removed.

## SOLUTION

*a.* **Angular Acceleration.** We observe that as the plate rotates about point $A$, its mass center $G$ describes a circle of radius $\bar{r}$ with center at $A$.

Since the plate is released from rest ($\omega = 0$), the normal component of the acceleration of $G$ is zero. The magnitude of the acceleration $\bar{a}$ of the mass center $G$ is thus $\bar{a} = \bar{r}\alpha$. We draw the diagram shown to express that the external forces are equivalent to the effective forces:

$$+\downarrow\Sigma M_A = \Sigma(M_A)_{\text{eff}}: \qquad W\bar{x} = (m\bar{a})\bar{r} + \bar{I}\alpha$$

Since $\bar{a} = \bar{r}\alpha$, we have

$$W\bar{x} = m(\bar{r}\alpha)\bar{r} + \bar{I}\alpha \qquad \alpha = \dfrac{W\bar{x}}{\dfrac{W}{g}\bar{r}^2 + \bar{I}} \qquad (1)$$

The centroidal moment of inertia of the plate is

$$\bar{I} = \dfrac{m}{12}(a^2 + b^2) = \dfrac{27.2 \text{ kg}}{12}\,[(20.3 \text{ cm})^2 + (15.2 \text{ cm})^2]$$

$$= 1457.76 \text{ kg cm}^2$$

Substituting this value of $\bar{I}$ together with $W = 266.8$ N, $\bar{r} = 12.7$ cm, and $\bar{x} = 10.2$ cm into Eq. (1), we obtain

$$\alpha = +.466 \text{ rad/s}^2 \qquad \boldsymbol{\alpha} = .466 \text{ rad/s}^2 \downarrow \quad \blacktriangleleft$$

*b.* **Reaction at A.** Using the computed value of $\alpha$, we determine the magnitude of the vector $m\bar{\mathbf{a}}$ attached at $G$.

$$m\bar{a} = m\bar{r}\alpha = 27.2 \text{ kg } (12.7 \text{ cm})(.466 \text{ rad/s}^2) = 1.6 \text{ N}$$

Showing this result on the diagram, we write the equations of motion

$$\xrightarrow{+}\Sigma F_x = \Sigma(F_x)_{\text{eff}}: \qquad A_x = -\tfrac{3}{5}(1.6 \text{ N}) = .96 \text{ N}$$
$$= -266.8 \text{ N} \qquad \mathbf{A}_x = .96 \text{ N} \leftarrow \quad \blacktriangleleft$$

$$+\uparrow\Sigma F_y = \Sigma(F_y)_{\text{eff}}: \qquad A_y - 266.8 \text{ N} = -\tfrac{4}{5}(1.6 \text{ N})$$
$$A_y = 265.5 \text{ N} \qquad \mathbf{A}_y = 265.5 \text{ N} \uparrow \quad \blacktriangleleft$$

The couple $\bar{I}\boldsymbol{\alpha}$ is not involved in the last two equations; nevertheless, it should be indicated on the diagram.

## SAMPLE PROBLEM 16.8

A sphere of radius $r$ and weight $W$ is released with no initial velocity on the incline and rolls without slipping. Determine (*a*) the minimum value of the coefficient of static friction compatible with the rolling motion, (*b*) the velocity of the center $G$ of the sphere after the sphere has rolled 10 ft, (*c*) the velocity of $G$ if the sphere were to move 10 ft down a frictionless 30° incline.

$\theta = 30°$

### SOLUTION

*a.* **Minimum $\mu_s$ for Rolling Motion.** The external forces **W**, **N**, and **F** form a system equivalent to the system of effective forces represented by the vector $m\bar{\mathbf{a}}$ and the couple $\bar{I}\boldsymbol{\alpha}$. Since the sphere rolls without sliding, we have $\bar{a} = r\alpha$.

$+\downarrow \Sigma M_C = \Sigma(M_C)_{\text{eff}}:$ 
$$(W \sin \theta)r = (m\bar{a})r + \bar{I}\alpha$$
$$(W \sin \theta)r = (mr\alpha)r + \bar{I}\alpha$$

Noting that $m = W/g$ and $\bar{I} = \frac{2}{5}mr^2$, we write

$$(W \sin \theta)r = \left(\frac{W}{g}r\alpha\right)r + \frac{2}{5}\frac{W}{g}r^2\alpha \qquad \alpha = +\frac{5g \sin \theta}{7r}$$

$$\bar{a} = r\alpha = \frac{5g \sin \theta}{7} = \frac{5(9.81 \text{ m/s}^2) \sin 30°}{7} = 3.5 \text{ m/s}^2$$

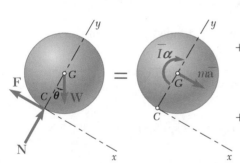

$+\searrow \Sigma F_x = \Sigma(F_x)_{\text{eff}}:$ 
$$W \sin \theta - F = m\bar{a}$$
$$W \sin \theta - F = \frac{W}{g}\frac{5g \sin \theta}{7}$$
$$F = +\frac{2}{7}W \sin \theta = \frac{2}{7}W \sin 30° \qquad \mathbf{F} = 0.143W \searrow 30°$$

$+\nearrow \Sigma F_y = \Sigma(F_y)_{\text{eff}}:$ 
$$N - W \cos \theta = 0$$
$$N = W \cos \theta = 0.866W \qquad \mathbf{N} = 0.866W \measuredangle 60°$$

$$\mu_s = \frac{F}{N} = \frac{0.143W}{0.866W} \qquad \qquad \mu_s = 0.165 \blacktriangleleft$$

*b.* **Velocity of Rolling Sphere.** We have uniformly accelerated motion:

$$\bar{v}_0 = 0 \qquad \bar{a} = 3.5 \text{ m/s}^2 \qquad \bar{x} = 3 \text{ m} \qquad \bar{x}_0 = 0$$
$$\bar{v}^2 = \bar{v}_0^2 + 2\bar{a}(\bar{x} - \bar{x}_0) \qquad \bar{v}^2 = 0 + 2(3.5 \text{ m/s}^2)(3 \text{ m})$$
$$\bar{v} = 4.58 \text{ m/s} \qquad \bar{\mathbf{v}} = 4.58 \text{ m/s} \searrow 30° \blacktriangleleft$$

*c.* **Velocity of Sliding Sphere.** Assuming now no friction, we have $F = 0$ and obtain

$+\downarrow \Sigma M_G = \Sigma(M_G)_{\text{eff}}:$ 
$$0 = \bar{I}\alpha \qquad \alpha = 0$$

$+\searrow \Sigma F_x = \Sigma(F_x)_{\text{eff}}:$ 
$$W \sin 30° = m\bar{a} \qquad 0.50W = \frac{W}{g}\bar{a}$$

$$\bar{a} = +4.9 \text{ m/s}^2 \qquad \bar{\mathbf{a}} = 4.9 \text{ m/s}^2 \searrow 30°$$

Substituting $\bar{a} = 4.9 \text{ m/s}^2$ into the equations for uniformly accelerated motion, we obtain

$$\bar{v}^2 = \bar{v}_0^2 + 2\bar{a}(\bar{x} - \bar{x}_0) \qquad \bar{v}^2 = 0 + 2(4.9 \text{ m/s}^2)(3 \text{ m})$$
$$\bar{v} = 5.4 \text{ m/s} \qquad \bar{\mathbf{v}} = 5.4 \text{ m/s} \searrow 30° \blacktriangleleft$$

# SAMPLE PROBLEM 16.9

A cord is wrapped around the inner drum of a wheel and pulled horizontally with a force of 200 N. The wheel has a mass of 50 kg and a radius of gyration of 70 mm. Knowing that $\mu_s = 0.20$ and $\mu_k = 0.15$, determine the acceleration of $G$ and the angular acceleration of the wheel.

## SOLUTION

*a.* **Assume Rolling without Sliding.** In this case, we have

$$\bar{a} = r\alpha = (0.100 \text{ m})\alpha$$

We can determine whether this assumption is justified by comparing the friction force obtained with the maximum available friction force. The moment of inertia of the wheel is

$$\bar{I} = m\bar{k}^2 = (50 \text{ kg})(0.070 \text{ m})^2 = 0.245 \text{ kg} \cdot \text{m}^2$$

*Equations of Motion*

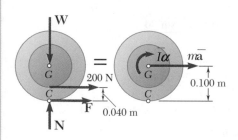

$$+\downarrow\Sigma M_C = \Sigma(M_C)_{\text{eff}}: \quad (200 \text{ N})(0.040 \text{ m}) = m\bar{a}(0.100 \text{ m}) + \bar{I}\alpha$$
$$8.00 \text{ N} \cdot \text{m} = (50 \text{ kg})(0.100 \text{ m})\alpha(0.100 \text{ m}) + (0.245 \text{ kg} \cdot \text{m}^2)\alpha$$
$$\alpha = +10.74 \text{ rad/s}^2$$
$$\bar{a} = r\alpha = (0.100 \text{ m})(10.74 \text{ rad/s}^2) = 1.074 \text{ m/s}^2$$

$$\xrightarrow{+}\Sigma F_x = \Sigma(F_x)_{\text{eff}}: \quad F + 200 \text{ N} = m\bar{a}$$
$$F + 200 \text{ N} = (50 \text{ kg})(1.074 \text{ m/s}^2)$$
$$F = -146.3 \text{ N} \qquad \mathbf{F} = 146.3 \text{ N} \leftarrow$$

$$+\uparrow\Sigma F_y = \Sigma(F_y)_{\text{eff}}:$$
$$N - W = 0 \qquad N = W = mg = (50 \text{ kg})(9.81 \text{ m/s}^2) = 490.5 \text{ N}$$
$$\mathbf{N} = 490.5 \text{ N} \uparrow$$

*Maximum Available Friction Force*

$$F_{\text{max}} = \mu_s N = 0.20(490.5 \text{ N}) = 98.1 \text{ N}$$

Since $F > F_{\text{max}}$, the assumed motion is impossible.

*b.* **Rotating and Sliding.** Since the wheel must rotate and slide at the same time, we draw a new diagram, where $\bar{a}$ and $\boldsymbol{\alpha}$ are independent and where

$$F = F_k = \mu_k N = 0.15(490.5 \text{ N}) = 73.6 \text{ N}$$

From the computation of part *a*, it appears that $\mathbf{F}$ should be directed to the left. We write the following equations of motion:

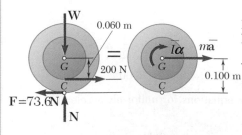

$$\xrightarrow{+}\Sigma F_x = \Sigma(F_x)_{\text{eff}}: \quad 200 \text{ N} - 73.6 \text{ N} = (50 \text{ kg})\bar{a}$$
$$\bar{a} = +2.53 \text{ m/s}^2 \qquad \bar{\mathbf{a}} = 2.53 \text{ m/s}^2 \rightarrow \blacktriangleleft$$

$$+\downarrow\Sigma M_G = \Sigma(M_G)_{\text{eff}}:$$
$$(73.6 \text{ N})(0.100 \text{ m}) - (200 \text{ N})(0.060 \text{ m}) = (0.245 \text{ kg} \cdot \text{m}^2)\alpha$$
$$\alpha = -18.94 \text{ rad/s}^2 \qquad \boldsymbol{\alpha} = 18.94 \text{ rad/s}^2 \curvearrowright \blacktriangleleft$$

## SAMPLE PROBLEM 16.10

The extremities of a 1.2 m rod weighing 22.7 kg can move freely and with no friction along two straight tracks as shown. If the rod is released with no velocity from the position shown, determine (a) the angular acceleration of the rod, (b) the reactions at A and B.

## SOLUTION

**Kinematics of Motion.** Since the motion is constrained, the acceleration of G must be related to the angular acceleration $\boldsymbol{\alpha}$. To obtain this relation, we first determine the magnitude of the acceleration $\mathbf{a}_A$ of point A in terms of $\alpha$. Assuming that $\boldsymbol{\alpha}$ is directed counterclockwise and noting that $a_{B/A} = 4\alpha$, we write

$$\mathbf{a}_B = \mathbf{a}_A + \mathbf{a}_{B/A}$$
$$[a_B \searrow 45°] = [a_A \rightarrow] + [4\alpha \nearrow 60°]$$

Noting that $\phi = 75°$ and using the law of sines, we obtain

$$a_A = 5.46\alpha \qquad a_B = 4.90\alpha$$

The acceleration of G is now obtained by writing

$$\overline{\mathbf{a}} = \mathbf{a}_G = \mathbf{a}_A + \mathbf{a}_{G/A}$$
$$\overline{\mathbf{a}} = [5.46\alpha \rightarrow] + [2\alpha \nearrow 60°]$$

Resolving $\overline{\mathbf{a}}$ into $x$ and $y$ components, we obtain

$$\overline{a}_x = 5.46\alpha - 2\alpha \cos 60° = 4.46\alpha \qquad \overline{\mathbf{a}}_x = 4.46\alpha \rightarrow$$
$$\overline{a}_y = -2\alpha \sin 60° = -1.732\alpha \qquad \overline{\mathbf{a}}_y = 1.732\alpha \downarrow$$

**Kinetics of Motion.** We draw a free-body-diagram equation expressing that the system of the external forces is equivalent to the system of the effective forces represented by the vector of components $m\overline{a}_x$ and $m\overline{a}_y$ attached at G and the couple $\overline{I}\alpha$. We compute the following magnitudes:

$$\overline{I} = \tfrac{1}{12}ml^2 = \frac{1}{12}22.7 \text{ kg } (1.2 \text{ m})^2 = 2.7 \text{ kg} \cdot \text{m}^2 \qquad \overline{I}\alpha = 2.7\alpha$$

$$m\overline{a}_x = 22.7 \text{ kg } (4.46\alpha) = 101.24\alpha \qquad m\overline{a}_y = -22.7 \, (1.732\alpha) = -39.3\alpha$$

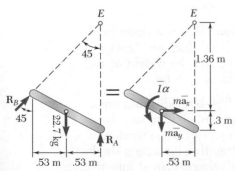

*Equations of Motion*

$+\curvearrowleft\Sigma M_E = \Sigma(M_E)_{\text{eff}}$:
$$(22.7)(.53) = (101.24\alpha)(1.36) + (39.3\alpha)(.53) + 2.7\alpha$$
$$\alpha = 0.075 \text{ rad/s}^2 \qquad \boldsymbol{\alpha} = 0.075 \text{ rad/s}^2 \curvearrowleft \blacktriangleleft$$

$\xrightarrow{+}\Sigma F_x = \Sigma(F_x)_{\text{eff}}$: $\qquad R_B \sin 45° = (101.24)(.73) = 7.6 \text{ kgf}$
$$R_B = 7.6 \text{ kgf} \qquad \mathbf{R}_B = 7.6 \text{ kgf} \nearrow 45° \blacktriangleleft$$

$+\uparrow\Sigma F_y = \Sigma(F_y)_{\text{eff}}$: $\qquad R_A + R_B \cos 45° - 222.70 = -(39.3)(.73)$
$$R_A = -28.7 - 52.26 + 222.7 = 167.49 \text{ kgf} \qquad \mathbf{R}_A = 167.49 \text{ kgf} \uparrow \blacktriangleleft$$

# SOLVING PROBLEMS
# ON YOUR OWN

In this lesson we considered the *plane motion of rigid bodies under constraints*. We found that the types of constraints involved in engineering problems vary widely. For example, a rigid body may be constrained to rotate about a fixed axis or to roll on a given surface, or it may be pin-connected to collars or to other bodies.

**1. Your solution of a problem involving the constrained motion of a rigid body,** will, in general, consist of two steps. First, you will consider the *kinematics of the motion*, and then you will solve the *kinetics portion of the problem*.

**2. The kinematic analysis of the motion** is done by using the methods you learned in Chap. 15. Due to the constraints, linear and angular accelerations will be related. (They will *not* be independent, as they were in the last lesson.) You should establish *relationships among the accelerations* (angular as well as linear), and your goal should be to express all accelerations in terms of a *single unknown acceleration*. This is the first step taken in the solution of each of the sample problems in this lesson.

 **a. For a body in noncentroidal rotation,** the components of the acceleration of the mass center are $\bar{a}_t = \bar{r}\alpha$ and $\bar{a}_n = \bar{r}\omega^2$, where $\omega$ will generally be known [Sample Probs. 16.6 and 16.7].

 **b. For a rolling disk or wheel,** the acceleration of the mass center is $\bar{a} = r\alpha$ [Sample Prob. 16.8].

 **c. For a body in general plane motion,** your best course of action, if neither $\bar{a}$ nor $\alpha$ is known or readily obtainable, is to express $\bar{a}$ in terms of $\alpha$ [Sample Prob. 16.10].

**3. The kinetic analysis of the motion** is carried out as follows.

 **a. Start by drawing a free-body-diagram equation.** This was done in all the sample problems of this lesson. In each case the left-hand diagram shows the external forces, including the applied forces, the reactions, and the weight of the body. The right-hand diagram shows the vector $m\bar{\mathbf{a}}$ and the couple $\bar{I}\boldsymbol{\alpha}$.

 **b. Next, reduce the number of unknowns** in the free-body-diagram equation by using the relationships among the accelerations that you found in your kinematic analysis. You will then be ready to consider equations that can be written by summing components or moments. Choose first an equation that involves a single unknown. After solving for that unknown, substitute the value obtained into the other equations, which you will then solve for the remaining unknowns.

**4. When solving problems involving rolling disks or wheels,** keep in mind the following.

**a. If sliding is impending,** the friction force exerted on the rolling body has reached its maximum value, $F_m = \mu_s N$, where $N$ is the normal force exerted on the body and $\mu_s$ is the coefficient of *static friction* between the surfaces of contact.

**b. If sliding is not impending,** the friction force $F$ can have *any value* smaller than $F_m$ and should, therefore, be considered as an independent unknown. After you have determined $F$, be sure to check that it is smaller than $F_m$; if it is not, *the body does not roll*, but rotates and slides as described in the next paragraph.

**c. If the body rotates and slides at the same time,** then the body is *not rolling* and the acceleration $\bar{a}$ of the mass center is *independent* of the angular acceleration $\alpha$ of the body: $\bar{a} \neq r\alpha$. On the other hand, the friction force has a well-defined value, $F = \mu_k N$, where $\mu_k$ is the coefficient of *kinetic friction* between the surfaces of contact.

**d. For an unbalanced rolling disk or wheel,** the relation $\bar{a} = r\alpha$ between the acceleration $\bar{a}$ of the mass center $G$ and the angular acceleration $\alpha$ of the disk or wheel *does not hold anymore*. However, a similar relation holds between the acceleration $a_O$ of the *geometric center $O$* and the angular acceleration $\alpha$ of the disk or wheel: $a_O = r\alpha$. This relation can be used to express $\bar{a}$ in terms of $\alpha$ and $\omega$ (Fig. 16.18).

**5. For a system of connected rigid bodies,** the goal of your *kinematic analysis* should be to determine all the accelerations from the given data, or to express them all in terms of a single unknown. (For systems with several degrees of freedom, you will need to use as many unknowns as there are degrees of freedom.)

Your *kinetic analysis* will generally be carried out by drawing a free-body-diagram equation for the entire system, as well as for one or several of the rigid bodies involved. In the latter case, both internal and external forces should be included, and care should be taken to represent with equal and opposite vectors the forces that two bodies exert on each other.

Fig. P16.70

# Problems

**16.70** Show that the couple $\bar{I}\alpha$ of Fig. 16.15 can be eliminated by attaching the vectors $m\bar{\mathbf{a}}_t$ and $m\bar{\mathbf{a}}_n$ at a point $P$ called the *center of percussion*, located on line $OG$ at a distance $GP = \bar{k}^2/\bar{r}$ from the mass center of the body.

**16.71** A uniform slender rod of length $L = 91.4$ cm and weight $W = 44.5$ N hangs freely from a hinge at $C$. A horizontal force $\mathbf{P}$ of magnitude 66.7 N is applied at end $B$. Knowing that $\bar{r} = 22.9$ cm, determine (*a*) the angular acceleration of the rod, (*b*) the components of the reaction at $C$.

**16.72** In Prob. 16.71, determine (*a*) the distance $\bar{r}$ for which the horizontal component of the reaction at $C$ is zero, (*b*) the corresponding angular acceleration of the rod.

**16.73** A uniform slender rod of length $L = 900$ mm and mass $m = 1$ kg hangs freely from a hinge at $A$. If a force $\mathbf{P}$ of magnitude 3.5 N is applied at $B$ horizontally to the left ($h = L$), determine (*a*) the angular acceleration of the rod, (*b*) the components of the reaction at $A$.

**16.74** In Prob. 16.73, determine (*a*) the distance $h$ for which the horizontal component of the reaction at $A$ is zero, (*b*) the corresponding angular acceleration of the rod.

**16.75** A portion of a circular cylindrical shell forms a small vane which is welded to the vertical shaft $AB$. The vane and shaft rotate about the $y$ axis with a constant angular velocity of 180 rpm counterclockwise. Knowing that the vane weighs 22.2 N, determine the horizontal components of the reaction at $A$.

Fig. P16.71                    Fig. P16.73

Fig. *P16.75*

**16.76** A uniform slender rod $AB$ of length $L$ and mass $m$ is pivoted at end $A$ and released from a horizontal position. The angular velocity of the rod as it passes through the vertical position is known to be $\omega = \sqrt{3g/L}$. (*a*) Express the tension in the rod at a distance $z$ from end $B$ in terms of $z$, $m$, $g$, and $L$ for the vertical position. (*b*) Knowing that the weight of the rod is 2 lb, determine the maximum tension in the rod for the vertical position.

Fig. *P16.76*

**16.77** A 120-mm-diameter hole is cut as shown in a thin disk of 600-mm diameter. The disk rotates in a horizontal plane about its geometric center $A$ at the constant rate of 480 rpm. Knowing that the disk has a mass of 30 kg after the hole has been cut, determine the horizontal component of the force exerted by the shaft on the disk at $A$.

**16.78** A uniform slender rod of mass 350 g/m is used to form the assembly shown. The assembly rotates clockwise at a constant rate of 120 rpm under the combined effect of gravity and the couple **M** which varies in magnitude and sense. Determine the magnitude and sense of the couple **M** and the reaction at point $A$ for (a) $\theta = 90°$, (b) $\theta = 180°$.

**Fig. P16.77**

**Fig. P16.78**

**16.79** A uniform rod of length $L$ and mass $m$ is supported as shown. If the cable attached at $B$ suddenly breaks, determine (a) the acceleration of end $B$, (b) the reaction at the pin support.

**Fig. P16.79**

**16.80** A uniform rod of length $L$ and mass $m$ is supported as shown. If the cable attached at $B$ suddenly breaks, determine (a) the distance $b$ for which the acceleration of end $A$ is maximum, (b) the corresponding acceleration of end $A$ and the reaction at $C$.

**16.81** A half cylinder of mass $m$ is released from rest in the position shown and swings freely about the horizontal diameter $AB$. Determine, at this instant, (a) the acceleration of point $D$, (b) the acceleration of the mass center of the half cylinder.

**Fig. P16.80**

**Fig. P16.81**

**16.82** A uniform rod of length $L$ and mass $m$ is supported as shown. Knowing that cable $AD$ suddenly breaks, determine at this instant (a) the angular acceleration of the rod, (b) the tensions in cables $AE$ and $BF$.

**Fig. P16.82**

**Fig. P16.83**

**16.83** A 4-kg slender rod is welded to the edge of a 3-kg uniform disk as shown. The assembly rotates about $A$ in a vertical plane under the combined effect of gravity and of the vertical force **P**. Knowing that at the instant shown the assembly has an angular velocity of 12 rad/s and an angular acceleration of 36 rad/s² both counterclockwise, determine (a) the force **P**, (b) the components of the reaction at $A$.

**16.84** Two identical 17.8-N slender rods $AB$ and $BC$ are connected by a pin at $B$ and by the cord $AC$. The assembly rotates in a vertical plane under the combined effect of gravity and a 8 N-m couple **M** applied to rod $AB$. Knowing that in the position shown the angular velocity of the assembly is zero, determine (a) the angular acceleration of the assembly, (b) the tension in cord $AC$.

**Fig. P16.84 and P16.85**

**16.85** Two identical 17.8-N slender rods $AB$ and $BC$ are connected by a pin at $B$ and by the cord $AC$. The assembly rotates in a vertical plane under the combined effect of gravity and a couple **M** applied to rod $AB$. Knowing that in the position shown the angular velocity of the assembly is zero and the tension in cord $AC$ is 2 N, determine (a) the angular acceleration of the assembly, (b) the magnitude of the couple **M**.

**16.86** Two slender rods, each of length $l$ and mass $m$, are released from rest in the position shown. Knowing that a small frictionless knob at end $B$ of rod $AB$ bears on rod $CD$, determine immediately after release (a) the acceleration of end $C$ of rod $CD$, (b) the force exerted on knob $B$.

**Fig. P16.86**

**16.87** Derive the equation $\Sigma M_C = I_C \alpha$ for the rolling disk of Fig. 16.17, where $\Sigma M_C$ represents the sum of the moments of the external forces about the instantaneous center $C$, and $I_C$ is the moment of inertia of the disk about $C$.

**16.88** Show that in the case of an unbalanced disk, the equation derived in Prob. 16.87 is valid only when the mass center $G$, the geometric center $O$, and the instantaneous center $C$ happen to lie in a straight line.

**16.89** A flywheel is rigidly attached to a shaft of 30-mm radius that can roll along parallel rails as shown. When released from rest, the system rolls 5 m in 40 s. Determine the centroidal radius of gyration of the system.

**Fig. P16.89 and P16.90**

**16.90** A flywheel of centroidal radius of gyration $\bar{k}$ is rigidly attached to a shaft that can roll along parallel rails. Denoting by $\mu_k$ the coefficient of static friction between the shaft and the rails, derive an expression for the largest angle of inclination $\beta$ for which no slipping will occur.

**16.91** A homogeneous sphere $S$, a uniform cylinder $C$, and a thin pipe $P$ are in contact when they are released from rest on the incline shown. Knowing that all three objects roll without slipping, determine, after 6 s of motion, the clear distance between ($a$) the pipe and the cylinder, ($b$) the cylinder and the sphere. Give the answers in both U.S. customary and SI units.

**16.92** An 355.8-N uniform cylinder is acted upon by a 222.4-N force as shown. Knowing that the cylinder rolls without slipping, determine ($a$) the acceleration of its center $G$, ($b$) the minimum value of the coefficient of static friction compatible with this motion.

**16.93 through 16.96** A drum of 10.2-cm. radius is attached to a disk of 20.3-cm. radius. The disk and drum have a total weight of 44.5 N and combined radius of gyration of 15.2 cm. A cord is attached as shown and pulled with a force **P** of magnitude 22.2 N. Knowing that the disk rolls without sliding, determine ($a$) the angular acceleration of the disk and the acceleration of $G$, ($b$) the minimum value of the coefficient of static friction compatible with this motion.

**16.97 through 16.100** A drum of 80-mm radius is attached to a disk of 160-mm radius. The disk and drum have a combined mass of 5 kg and combined radius of gyration of 120 mm. A cord is attached as shown and pulled with a force **P** of magnitude 20 N. Knowing that the coefficients of static and kinetic friction are $\mu_s = 0.25$ and $\mu_k = 0.20$, respectively, determine ($a$) whether or not the disk slides, ($b$) the angular acceleration of the disk and the acceleration of $G$.

**Fig. P16.91**

**Fig. P16.92**

**Fig. P16.93 and P16.97**

**Fig. P16.94 and P16.98**

**Fig. P16.95 and P16.99**

**Fig. P16.96 and P16.100**

**Fig. P16.101**

**16.101** A 44.5 N uniform square plate is supported by two identical 13.3 N uniform slender rods *AD* and *BE*. It is held in the position shown by rope *CF*. Determine, immediately after rope *CF* has been cut, (*a*) the acceleration of the plate, (*b*) the force exerted on the plate at point *B*.

**16.102** Solve Prob. 16.101 assuming that rod *AD* is replaced by a cable of negligible mass.

**16.103 and 16.104** Gear *C* has a mass of 5 kg and a centroidal radius of gyration of 75 mm. The uniform bar *AB* has a mass of 3 kg and gear *D* is stationary. If the system is released from rest in the position shown, determine (*a*) the angular acceleration of gear *C*, (*b*) the acceleration of point *B*.

**Fig. P16.103**　　　　　　　　　**Fig. P16.104**

**16.105** A hemisphere of mass *m* and radius *r* is released from rest in the position shown. Knowing that the hemisphere rolls without slipping, determine its initial angular acceleration. [*Hint:* Note that $OG = 3r/8$ and that, by the parallel-axis theorem, $\overline{I} = \frac{2}{5}mr^2 - m(OG)^2$.]

**Fig. P16.105 and P16.106**

**16.106** Solve Prob. 16.105, considering a half cylinder instead of a hemisphere. [*Hint:* Note that $OG = 4r/3\pi$ and that, by the parallel-axis theorem, $\overline{I} = \frac{1}{2}mr^2 - m(OG)^2$.]

**16.107** A half section of a uniform thin pipe of mass *m* is at rest when a force **P** is applied as shown. Assuming that the section rolls without sliding, determine (*a*) its initial angular acceleration, (*b*) the minimum value of the coefficient of static friction consistent with the motion.

**16.108** Solve Prob. 16.107 assuming that the force **P** applied at point *A* is directed horizontally to the right.

**Fig. P16.107**

**16.109**  A hemisphere of weight $W$ and radius $r$ is released from rest in the position shown. Determine (a) the minimum value of $\mu_s$ for which the hemisphere starts to roll without sliding, (b) the corresponding acceleration of point $B$. [*Hint*: Note that $OG = \frac{3}{8}r$ and that, by the parallel-axis theorem, $\bar{I} = \frac{2}{5}mr^2 - m(OG)^2$.]

**16.110**  The center of gravity $G$ of a 1.6 kg unbalanced tracking wheel is located at a distance $r = 2.3$ cm from its geometric center $B$. The radius of the wheel is $R = 7.6$ cm and its centroidal radius of gyration is 5.6 cm. At the instant shown the center $B$ of the wheel has a velocity of .32 m/s and an acceleration of 1.1 m/s², both directed to the left. Knowing that the wheel rolls without sliding and neglecting the mass of the driving yoke $AB$, determine the horizontal force **P** applied to the yoke.

**Fig. P16.109**

**Fig. P16.110**

**16.111**  End $A$ of a uniform 6.8 kg bar is attached to a horizontal rope and end $B$ contacts a floor with negligible friction. Knowing that the bar is released from rest in the position shown, determine immediately after release (a) the angular acceleration of the bar, (b) the tension in the rope, (c) the reaction at $B$.

**16.112**  The 7.3 kg uniform slender rod $AB$ is supported by a small wheel at $C$ and end $A$ can slide on the horizontal surface. Knowing that the rod is released from rest in the position shown and neglecting friction at $A$ and $C$, determine immediately after release (a) the angular acceleration of the rod, (b) the reaction at point $A$.

**Fig. P16.111**

**Fig. P16.112**

**Fig. P16.113 and P16.114**

**Fig. *P16.115* and *P16.116***

**16.113**   The 2-kg uniform rod *AB* is in the shape of a quarter circle and its ends are attached to collars of negligible weight that slide without friction along fixed rods. Knowing that rod *AB* is released from rest in the position shown, determine immediately after release (*a*) the angular acceleration of the rod, (*b*) the reaction at end *B*.

**16.114**   The 2-kg uniform rod *AB* is in the shape of a quarter circle and its ends are attached to collars of negligible weight that slide along fixed rods. Rod *AB* is released from rest in the position shown. Assuming no friction at *A* and a friction force of magnitude 0.9 N at *B*, determine the angular acceleration of the rod immediately after release.

**16.115**   End *A* of the 6-kg uniform rod *AB* rests on the inclined surface, while end *B* is attached to a collar of negligible mass which can slide along the vertical rod shown. Knowing that the rod is released from rest when $\theta = 35°$ and neglecting the effect of friction, determine immediately after release (*a*) the angular acceleration of the rod, (*b*) the reaction at *B*.

**16.116**   End *A* of the 6-kg uniform rod *AB* rests on the inclined surface, while end *B* is attached to a collar of negligible mass which can slide along the vertical rod shown. When the rod is at rest a vertical force **P** is applied at *B*, causing end *B* of the rod to start moving upward with an acceleration of 4 m/s². Knowing that $\theta = 35°$, determine the force **P**.

**16.117**   The 1.4 kg uniform rod *BD* is connected to crank *AB* and to a collar of negligible weight. A couple (not shown) is applied to crank *AB* causing it to rotate with an angular velocity of 12 rad/s counterclockwise and an angular acceleration of 80 rad/s² clockwise at the instant shown. Neglecting the effect of friction, determine the reaction at *D*.

**Fig. P16.117**

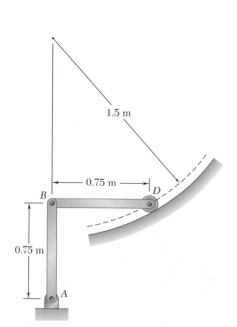

**Fig. P16.119**

**16.118**   In Prob. 16.117, determine the reaction at *D*, knowing that in the position shown crank *AB* has an angular velocity of 12 rad/s and an angular acceleration of 80 rad/s², both counterclockwise.

**16.119**   The 4-kg uniform slender bar *BD* is attached to bar *AB* and a wheel of negligible mass which rolls on a circular surface. Knowing that at the instant shown bar *AB* has an angular velocity of 6 rad/s and no angular acceleration, determine the reaction at point *D*.

**16.120** The collar $B$ of negligible mass can slide freely on the 4-kg uniform rod $CD$. Knowing that in the position shown crank $AB$ rotates with an angular velocity of 5 rad/s and an angular acceleration of 60 rad/s², both clockwise, determine the force exerted on rod $CD$ by collar $B$.

**Fig. P16.120 and P16.121**

**16.121** The collar $B$ of negligible mass can slide freely on the 4-kg uniform rod $CD$. Knowing that in the position shown crank $AB$ rotates with an angular velocity of 5 rad/s and an angular acceleration of 60 rad/s², both counterclockwise, determine the force exerted on rod $CD$ by collar $B$.

*16.122* The uniform 3.65 kg bar $AGB$ is attached to two collars of negligible weight that slide with negligible friction along fixed rods. Its motion is controlled by the force $P$ applied to collar $A$. Knowing that at the instant shown the speed and acceleration of collar $A$ are .6 m/s and zero, respectively, determine (*a*) the applied force $P$, (*b*) the force exerted by the fixed rod on the collar at $A$.

**Fig. *P16.122***

*16.123* The 4.5 kg uniform rod $ABD$ is attached to the crank $BC$ and is fitted with a small wheel that can roll without friction along a vertical slot. Knowing that at the instant shown crank $BC$ rotates with an angular velocity of 6 rad/s clockwise and an angular acceleration of 15 rad/s² counterclockwise, determine the reaction at $A$.

**16.124** A driver starts his car with the door on the passenger's side wide open ($\theta = 0$). The 36-kg door has a centroidal radius of gyration $\bar{k} = 250$ mm, and its mass center is located at a distance $r = 440$ mm from its vertical axis of rotation. Knowing that the driver maintains a constant acceleration of 2 m/s², determine the angular velocity of the door as it slams shut ($\theta = 90°$).

**Fig. *P16.123***

**Fig. P16.124**

**16.125** For the car of Prob. 16.124, determine the smallest constant acceleration that the driver can maintain if the door is to close and latch, knowing that as the door hits the frame its angular velocity must be at least 2 rad/s for the latching mechanism to operate.

**Fig. P16.126**

**Fig. P16.128 and P16.129**

**16.126**   The linkage *ABD* is formed by connecting two 4-kg bars and
a collar of negligible mass. The motion of the linkage is controlled by the
force **P** applied to the collar. Knowing that at the instant shown the angular
velocity and angular acceleration of bar *AB* are zero and 10 rad/s² counterclockwise, respectively, determine the force **P**.

**16.127**   Solve Prob. 16.126 assuming that at the instant shown the angular velocity and angular acceleration of bar *AB* are 5 rad/s clockwise and
zero, respectively.

**16.128**   The 1.8 kg uniform slender rod *AB*, the 3.6 kg uniform slender rod *BF*, and the 1.8 kg uniform thin sleeve *CE* are connected as shown
and move without friction in a vertical plane. The motion of the linkage is
controlled by the couple **M** applied to rod *AB*. Knowing that at the instant
shown the angular velocity of rod *AB* is 15 rad/s and the magnitude of the
couple **M** is 6.78 N determine (*a*) the angular acceleration of rod *AB*, (*b*) the
reaction at point *D*.

**16.129**   The 1.8 kg uniform slender rod *AB*, the 3.6 kg uniform slender rod *BF*, and the 1.8 kg uniform thin sleeve *CE* are connected as shown
and move without friction in a vertical plane. The motion of the linkage is
controlled by the couple **M** applied to rod *AB*. Knowing that at the instant
shown the angular velocity of rod *AB* is 30 rad/s and the angular acceleration of rod *AB* is 96 rad/s² clockwise, determine (*a*) the magnitude of the
couple **M**, (*b*) the reaction at point *D*.

**16.130**   The 2-kg rod *AB* and the 3-kg rod *BC* are connected as shown
to a disk that is made to rotate in a vertical plane at a constant angular velocity of 6 rad/s clockwise. For the position shown, determine the forces exerted at *A* and *B* on rod *AB*.

**Fig. P16.130 and P16.131**

**16.131**   The 2-kg rod *AB* and the 3-kg rod *BC* are connected as shown
to a disk that is made to rotate in a vertical plane. Knowing that at the instant
shown the disk has an angular acceleration of 18 rad/s² clockwise and no angular velocity, determine the components of the forces exerted at *A* and *B* on
rod *AB*.

**Fig. *P16.132***

**16.132**   In the engine system shown *l* = 25.4 cm and *b* = 10.2 cm. The
connecting rod *BD* is assumed to be a 1.36 kg uniform slender rod and is
attached to the 2 kg piston *P*. During a test of the system, crank *AB* is made
to rotate with a constant angular velocity of 600 rpm clockwise with no force
applied to the face of the piston. Determine the forces exerted on the connecting rod at *B* and *D* when *θ* = 180°. (Neglect the effect of the weight of
the rod.)

*16.133* Solve Prob. 16.132 when $\theta = 0$.

**16.134** Three uniform bars which have different lengths but equal masses are connected to form a linkage which lies in a vertical plane. Knowing that the system is released from rest in the position shown, determine the initial angular acceleration of each bar.

**Fig. P16.134**

*16.135* A uniform rod $AB$, of mass 4 kg and length $L = 1.5$ m, is released from rest in the position shown. Knowing that $\beta = 20°$, determine the values immediately after release of (a) the angular acceleration of the rod, (b) the acceleration of end $A$, (c) the reaction at $A$. Neglect the mass and friction of the roller at $A$.

**Fig. P16.135**

*16.136* The uniform slender 2-kg bar $BD$ is attached to the uniform 6-kg uniform disk by a pin at $B$ and released from rest in the position shown. Assuming that the disk rolls without slipping, determine (a) the initial reaction at the contact point $A$, (b) the corresponding smallest allowable value of the coefficient of static friction.

**Fig. P16.136**

**Fig. P16.137**

**Fig. P16.139 and P16.140**

**\*16.137**   Two disks, each of mass $m$ and radius $r$ are connected as shown by a continuous chain belt of negligible mass. If a pin at point $C$ of the chain belt is suddenly removed, determine ($a$) the angular acceleration of each disk, ($b$) the tension in the left-hand portion of the belt, ($c$) the acceleration of the center of disk $B$.

**\*16.138**   A uniform slender bar $AB$ of mass $m$ is suspended as shown from a uniform disk of the same mass $m$. Determine the accelerations of points $A$ and $B$ immediately after a horizontal force $\mathbf{P}$ has been applied at $B$.

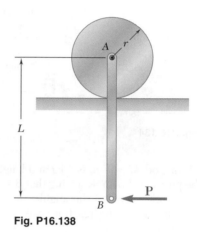

**Fig. P16.138**

**\*16.139**   Each of the 2.7 kg bars $AB$ and $BC$ is of length $L = 63.5$ cm. A horizontal force $\mathbf{P}$ of magnitude 22.2 N is applied to bar $BC$ as shown. Knowing that $b = L$ ($\mathbf{P}$ is applied at $C$), determine the angular acceleration of each bar.

**\*16.140**   Each of the 2.7 kg bars $AB$ and $BC$ is of length $L = 63.5$ cm. A horizontal force $\mathbf{P}$ of magnitude 22.2 N is applied to bar $BC$. For the position shown, determine ($a$) the distance $b$ for which the bars move as if they formed a single rigid body, ($b$) the corresponding angular acceleration of the bars.

**\*16.141**   Two identical uniform rods are connected by a pin at $B$ and are held in a horizontal position by three wires as shown. If the wires attached at $A$ and $B$ are cut simultaneously, determine at that instant the acceleration of ($a$) point $A$, ($b$) point $B$.

**Fig. *P16.141***

**\*16.142**   ($a$) Determine the magnitude and the location of the maximum bending moment in the rod of Prob. 16.73. ($b$) Show that the answer to part $a$ is independent of the mass of the rod.

**\*16.143**   Draw the shear and bending-moment diagrams for the rod of Prob. 16.79 immediately after the cable at $B$ breaks.

In this chapter, we studied the *kinetics of rigid bodies*, that is, the relations existing between the forces acting on a rigid body, the shape and mass of the body, and the motion produced. Except for the first two sections, which apply to the most general case of the motion of a rigid body, our analysis was restricted to the *plane motion of rigid slabs* and rigid bodies symmetrical with respect to the reference plane. The study of the plane motion of non-symmetrical rigid bodies and of the motion of rigid bodies in three-dimensional space will be considered in Chap. 18.

We first recalled [Sec. 16.2] the two fundamental equations derived in Chap. 14 for the motion of a system of particles and observed that they apply in the most general case of the motion of a rigid body. The first equation defines the motion of the mass center $G$ of the body; we have

$$\Sigma \mathbf{F} = m\overline{\mathbf{a}} \qquad (16.1)$$

where $m$ is the mass of the body and $\overline{\mathbf{a}}$ the acceleration of $G$. The second is related to the motion of the body relative to a centroidal frame of reference; we wrote

$$\Sigma \mathbf{M}_G = \dot{\mathbf{H}}_G \qquad (16.2)$$

where $\dot{\mathbf{H}}_G$ is the rate of change of the angular momentum $\mathbf{H}_G$ of the body about its mass center $G$. Together, Eqs. (16.1) and (16.2) express that *the system of the external forces is equipollent to the system consisting of the vector $m\overline{\mathbf{a}}$ attached at $G$ and the couple of moment $\dot{\mathbf{H}}_G$* (Fig. 16.19).

Restricting our analysis at this point and for the rest of the chapter to the plane motion of rigid slabs and rigid bodies symmetrical with respect to the reference plane, we showed [Sec. 16.3] that the angular momentum of the body could be expressed as

$$\mathbf{H}_G = \overline{I}\boldsymbol{\omega} \qquad (16.4)$$

where $\overline{I}$ is the moment of inertia of the body about a centroidal axis perpendicular to the reference plane and $\boldsymbol{\omega}$ is the angular velocity of the body. Differentiating both members of Eq. (16.4), we obtained

$$\dot{\mathbf{H}}_G = \overline{I}\dot{\boldsymbol{\omega}} = \overline{I}\boldsymbol{\alpha} \qquad (16.5)$$

which shows that in the restricted case considered here, the rate of change of the angular momentum of the rigid body can be

**Fundamental equations of motion for a rigid body**

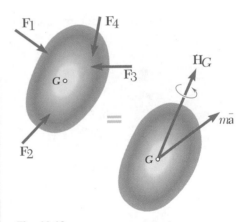

**Fig. 16.19**

**Angular momentum in plane motion**

Equations for the plane motion of a rigid
body

represented by a vector of the same direction as $\boldsymbol{\alpha}$ (that is, perpendicular to the plane of reference) and of magnitude $\bar{I}\alpha$.

It follows from the above [Sec. 16.4] that the plane motion of a rigid slab or of a rigid body symmetrical with respect to the reference plane is defined by the three scalar equations

$$\Sigma F_x = m\bar{a}_x \qquad \Sigma F_y = m\bar{a}_y \qquad \Sigma M_G = \bar{I}\alpha \qquad (16.6)$$

It further follows that *the external forces acting on the rigid body are actually **equivalent** to the effective forces of the various particles forming the body.* This statement, known as *d'Alembert's principle*, can be expressed in the form of the vector diagram shown in Fig. 16.20, where the effective forces have been represented by a vector $m\bar{a}$ attached at $G$ and a couple $\bar{I}\alpha$. In the particular case of a slab in *translation,* the effective forces shown in part $b$ of this figure reduce to the single vector $m\bar{a}$, while in the particular case of a slab in *centroidal rotation,* they reduce to the single couple $\bar{I}\alpha$; in any other case of plane motion, both the vector $m\bar{a}$ and the couple $\bar{I}\alpha$ should be included.

d'Alembert's principle

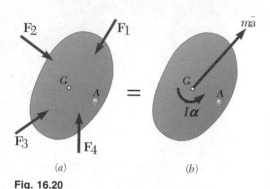

(a)             (b)

**Fig. 16.20**

Free-body-diagram equation

Any problem involving the plane motion of a rigid slab may be solved by drawing a *free-body-diagram equation* similar to that of Fig. 16.20 [Sec. 16.6]. Three equations of motion can then be obtained by equating the $x$ components, $y$ components, and moments about an arbitrary point $A$, of the forces and vectors involved [Sample Probs. 16.1, 16.2, 16.4, and 16.5]. An alternative solution can be obtained by adding to the external forces an *inertia vector* $-m\bar{a}$ of sense opposite to that of $\bar{a}$, attached at $G$, and an *inertia couple* $-\bar{I}\alpha$ of sense opposite to that of $\boldsymbol{\alpha}$. The system obtained in this way is equivalent to zero, and the slab is said to be in *dynamic equilibrium.*

Connected rigid bodies

The method described above can also be used to solve problems involving the plane motion of several connected rigid bodies [Sec. 16.7]. A free-body-diagram equation is drawn for each part of the system and the equations of motion obtained are solved simultaneously. In some cases, however, a single diagram can be drawn for the entire system, including all the external forces as well as the vectors $m\bar{a}$ and the couples $\bar{I}\alpha$ associated with the various parts of the system [Sample Prob. 16.3].

Constrained plane motion

In the second part of the chapter, we were concerned with rigid bodies *moving under given constraints* [Sec. 16.8]. While the kinetic analysis of the constrained plane motion of a rigid slab is the same as above, it must be supplemented by a *kinematic analysis* which has for its object to express the components $\bar{a}_x$ and $\bar{a}_y$ of the acceleration of the mass center $G$ of the slab in terms of its angular acceleration $\alpha$. Problems solved in this way included the *noncentroidal rotation* of rods and plates [Sample Probs. 16.6 and 16.7], the *rolling motion* of spheres and wheels [Sample Probs. 16.8 and 16.9], and the plane motion of *various types of linkages* [Sample Prob. 16.10].

**16.144** At the instant shown the angular velocity of links *BE* and *CF* is 6 rad/s counterclockwise and is decreasing at the rate of 12 rad/s². Knowing that the length of each link is 38.1 cm and neglecting the weight of the links, determine (*a*) the force **P**, (*b*) the corresponding force in each link. The weight of rod *AD* is 6.8 kg.

**Fig. P16.144**

**16.145** The 160-mm-radius brake drum is attached to a larger flywheel that is not shown. The total mass moment of inertia of the drum and the flywheel is 18 kg · m² and the coefficient of kinetic friction between the drum and the brake shoe is 0.35. Knowing that the angular velocity of the flywheel is 360 rpm counterclockwise when a force **P** of magnitude 300 N is applied to the pedal *C*, determine the number of revolutions executed by the flywheel before it comes to rest.

**Fig. P16.145**

**16.146** Disk *A* has a mass of 8 kg and an initial angular velocity of 480 rpm clockwise; disk *B* has a mass of 4 kg and is initially at rest. The disks are brought together by applying a horizontal force of magnitude 30 N to the axle of disk *A*. Knowing that $\mu_k = 0.15$ between the disks and neglecting bearing friction, determine (*a*) the angular acceleration of each disk, (*b*) the final angular velocity of each disk.

**Fig. P16.146**

**16.147** A force **P** of magnitude 3.3 N is applied to a tape wrapped around a uniform 2.7 kg disk. Knowing that the disk rests on a frictionless horizontal surface, determine the acceleration of (*a*) point *A*, (*b*) point *B*.

**Fig. P16.147**

**16.148**  The 181.4 kg crate shown is lowered by means of two overhead cranes. Knowing that at the instant shown the deceleration of cable $A$ is 0.9 m/s² and that of cable $B$ is 0.3 m/s², determine the tension in each cable.

**Fig. P16.148**

**16.149**  A beam $AB$ of mass $m$ and of uniform cross section is suspended from two springs as shown. If spring 2 breaks, determine at that instant (*a*) the angular acceleration of the beam, (*b*) the acceleration of point $A$, (*c*) the acceleration of point $B$.

**16.150**  A uniform rod of length $L$ and mass $m$ is supported as shown with $b = 0.2L$ when the cable attached to end $B$ suddenly breaks. Determine at this instant (*a*) the acceleration of end $B$, (*b*) the reaction at the pin support.

**Fig. P16.149**

**Fig. P16.150**

**16.151**  A wheel of radius $r$ and centroidal radius of gyration $\bar{k}$ is released from rest on the incline and rolls without sliding. Derive an expression for the acceleration of the center of the wheel in terms of $r$, $\bar{k}$, $\beta$, and $g$.

**Fig. P16.151**

**16.152**  A small clamp of mass $m_B$ is attached at $B$ to a hoop of mass $m_h$. The system is released from rest when $\theta = 90°$ and rolls without sliding. Knowing that $m_h = 5m_B$, determine (a) the angular acceleration of the hoop, (b) the horizontal and vertical components of the acceleration of $B$.

**Fig. P16.152**

**16.153**  The motion of the uniform rod $AB$ of mass 8 kg and length $L = 900$ mm is guided by small wheels of negligible mass that roll on the surface shown. If the rod is released from rest when $\theta = 20°$, determine (a) the angular acceleration of the rod, (b) the reaction at $A$.

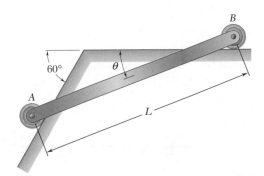

**Fig. P16.153**

**16.154**  The 300-mm uniform rod $BD$ of mass 3 kg is connected as shown to crank $AB$ and to collar $D$ of negligible mass, which can slide freely along a horizontal rod. Knowing that crank $AB$ rotates counterclockwise at the constant rate of 300 rpm, determine the reaction at $D$ when $\theta = 0$.

**Fig. P16.154**

**Fig. P16.155**

**16.155**  Two 3-kg uniform bars are connected to form the linkage shown. Neglecting the effect of friction, determine the reaction at $D$ immediately after the linkage is released from rest in the position shown.

# Computer Problems

**Fig. P16.C1**

**16.C1**  A 15-kg cylinder of diameter $b = 160$ mm and height $h = 140$ mm is placed on a 5-kg platform $CD$ that is held in the position shown by three cables. It is desired to determine the minimum value of $\mu_s$ between the cylinder and the platform for which the cylinder does not slip on the platform, immediately after cable $AB$ is cut. Using computational software calculate and plot the minimum allowable value of $\mu_s$ for values of $\theta$ from 0 to 40°. Knowing that the actual value of $\mu_s$ is 0.75, determine the value of $\theta$ at which slipping impends.

**16.C2**  The cranks $BE$ and $CF$ are made to rotate at a constant speed of 90 rpm counterclockwise. Determine and plot the vertical components of the forces exerted on the 5.4 kg uniform rod $ABCD$ by pins $B$ and $C$ as functions of the angle $\theta$ from 0 to 360°.

**Fig. P16.C2**

**16.C3**  End $A$ of the 4.5 kg rod $AB$ is moved to the left at a constant speed $v_A = 152.4$ cm/s. Using computational software calculate and plot the normal reactions at ends $A$ and $B$ of the rod for values of $\theta$ from 0 to 40°. Determine the value of $\theta$ at which end $B$ of the rod loses contact with the wall.

**16.C4**  A uniform slender bar $AB$ of weight $W$ is suspended from springs $AC$ and $BD$ as shown. Using computational software calculate and plot the accelerations of ends $A$ and $B$, immediately after spring $AC$ has broken, for values of $\theta$ from 0 to 90°.

**Fig. P16.C3**

**Fig. P16.C4**

**16.C5**  A 27.2 kg plank rests on two horizontal pipes *AB* and *CD* of a scaffolding. The pipes are 3 m apart, and the plank overhangs .6 m at each end. When pipe *CD* suddenly breaks, a 63.5 kg worker is standing on the plank at a distance *x* from pipe *AB*. Calculate and plot the initial acceleration of the worker for values of *x* from 0 to 3 m.

**Fig. P16.C5**

**16.C6**  In the engine system of Prob. 15.C3 of Chap. 15, the connecting rod *BD* is assumed to be a 2.26 kg slender rod and is attached to the 2.7 kg piston *P*. Knowing that during a test of the system crank *AB* is made to rotate with a constant angular velocity of 1000 rpm clockwise with no force applied to the face of the piston, use computational software to calculate and plot the horizontal and vertical components of the dynamic reactions exerted on the connecting rod at *B* and *D* for values of $\theta$ from 0 to 180°.

**16.C7**  The uniform slender bar *AB*, of mass 10 kg and length 720 mm, is suspended as shown from the 15 kg cart *C*. Knowing that the system is released from rest and neglecting the effect of friction, calculate and plot the initial acceleration of the cart and the initial angular acceleration of the bar as functions of $\theta$ from 0 to 90°.

**Fig. P16.C7**

## CHAPTER 17

# Plane Motion of Rigid Bodies: Energy and Momentum Methods

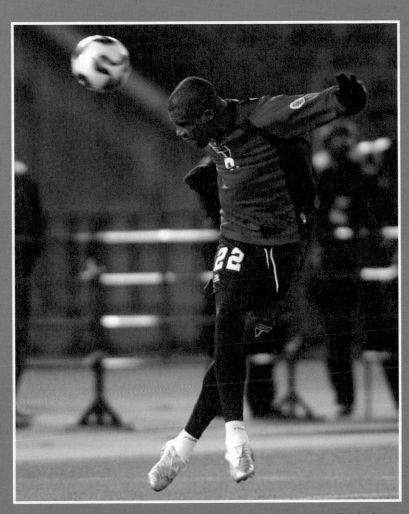

In this chapter the energy and momentum methods will be added to the tools available for your study of the motion of rigid bodies. For example, when an impulsive force is applied to a soccer ball, there is a period of deformation and restitution after which the ball moves with changed linear momentum and angular momentum.

## PLANE MOTION OF RIGID BODIES: ENERGY AND MOMENTUM METHODS

17.1    Introduction
17.2    Principle of Work and Energy for a Rigid Body
17.3    Work of Forces Acting on a Rigid Body
17.4    Kinetic Energy of a Rigid Body in Plane Motion
17.5    Systems of Rigid Bodies
17.6    Conservation of Energy
17.7    Power
17.8    Principle of Impulse and Momentum for the Plane Motion of a Rigid Body
17.9    Systems of Rigid Bodies
17.10   Conservation of Angular Momentum
17.11   Impulsive Motion
17.12   Eccentric Impact

## 17.1. INTRODUCTION

In this chapter the method of work and energy and the method of impulse and momentum will be used to analyze the plane motion of rigid bodies and of systems of rigid bodies.

The method of work and energy will be considered first. In Secs. 17.2 through 17.5, the work of a force and of a couple will be defined, and an expression for the kinetic energy of a rigid body in plane motion will be obtained. The principle of work and energy will then be used to solve problems involving displacements and velocities. In Sec. 17.6, the principle of conservation of energy will be applied to the solution of a variety of engineering problems.

In the second part of the chapter, the principle of impulse and momentum will be applied to the solution of problems involving velocities and time (Secs. 17.8 and 17.9) and the concept of conservation of angular momentum will be introduced and discussed (Sec. 17.10).

In the last part of the chapter (Secs. 17.11 and 17.12), problems involving the eccentric impact of rigid bodies will be considered. As was done in Chap. 13, where we analyzed the impact of particles, the coefficient of restitution between the colliding bodies will be used together with the principle of impulse and momentum in the solution of impact problems. It will also be shown that the method used is applicable not only when the colliding bodies move freely after the impact but also when the bodies are partially constrained in their motion.

## 17.2. PRINCIPLE OF WORK AND ENERGY FOR A RIGID BODY

The principle of work and energy will now be used to analyze the plane motion of rigid bodies. As was pointed out in Chap. 13, the method of work and energy is particularly well adapted to the solution of problems involving velocities and displacements. Its main advantage resides in the fact that the work of forces and the kinetic energy of particles are scalar quantities.

In order to apply the principle of work and energy to the analysis of the motion of a rigid body, it will again be assumed that the rigid body is made of a large number $n$ of particles of mass $\Delta m_i$. Recalling Eq. (14.30) of Sec. 14.8, we write

$$T_1 + U_{1 \to 2} = T_2 \tag{17.1}$$

where $T_1$, $T_2$ = initial and final values of total kinetic energy of particles forming the rigid body

$U_{1 \to 2}$ = work of all forces acting on various particles of body

The total kinetic energy

$$T = \frac{1}{2} \sum_{i=1}^{n} \Delta m_i v_i^2 \tag{17.2}$$

is obtained by adding positive scalar quantities and is itself a positive scalar quantity. You will see later how $T$ can be determined for various types of motion of a rigid body.

The expression $U_{1 \to 2}$ in (17.1) represents the work of all the forces acting on the various particles of the body, whether these forces are internal or external. However, as you will see presently, the total work of the internal forces holding together the particles of a rigid body is zero. Consider two particles $A$ and $B$ of a rigid body and the two equal and opposite forces $\mathbf{F}$ and $-\mathbf{F}$ they exert on each other (Fig. 17.1). While, in general, small displacements $d\mathbf{r}$ and $d\mathbf{r}'$ of the two particles are different, the components of these displacements along $AB$ must be equal; otherwise, the particles would not remain at the same distance from each other and the body would not be rigid. Therefore, the work of $\mathbf{F}$ is equal in magnitude and opposite in sign to the work of $-\mathbf{F}$, and their sum is zero. Thus, the total work of the internal forces acting on the particles of a rigid body is zero, and *the expression $U_{1 \to 2}$ in Eq. (17.1) reduces to the work of the external forces act-*ing on the body during the displacement considered.

**Fig. 17.1**

## 17.3. WORK OF FORCES ACTING ON A RIGID BODY

We saw in Sec. 13.2 that the work of a force $\mathbf{F}$ during a displacement of its point of application from $A_1$ to $A_2$ is

$$U_{1 \to 2} = \int_{A_1}^{A_2} \mathbf{F} \cdot d\mathbf{r} \qquad (17.3)$$

or

$$U_{1 \to 2} = \int_{s_1}^{s_2} (F \cos \alpha)\, ds \qquad (17.3')$$

where $F$ is the magnitude of the force, $\alpha$ is the angle it forms with the direction of motion of its point of application $A$, and $s$ is the variable of integration which measures the distance traveled by $A$ along its path.

In computing the work of the external forces acting on a rigid body, it is often convenient to determine the work of a couple without considering separately the work of each of the two forces forming the couple. Consider the two forces $\mathbf{F}$ and $-\mathbf{F}$ forming a couple of moment $\mathbf{M}$ and acting on a rigid body (Fig. 17.2). Any small displacement of the rigid body bringing $A$ and $B$, respectively, into $A'$ and $B''$ can be divided into two parts: in one part points $A$ and $B$ undergo equal displacements $d\mathbf{r}_1$; in the other part $A'$ remains fixed while $B'$ moves into $B''$ through a displacement $d\mathbf{r}_2$ of magnitude $ds_2 = r\, d\theta$. In the first part of the motion, the work of $\mathbf{F}$ is equal in magnitude and opposite in sign to the work of $-\mathbf{F}$ and their sum is zero. In the second part of the motion, only force $\mathbf{F}$ works, and its work is $dU = F\, ds_2 = Fr\, d\theta$. But the product $Fr$ is equal to the magnitude $M$ of the moment of the couple. Thus, the work of a couple of moment $\mathbf{M}$ acting on a rigid body is

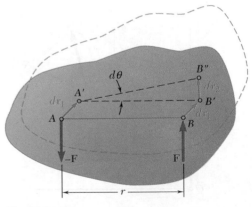

**Fig. 17.2**

$$dU = M\, d\theta \qquad (17.4)$$

where $d\theta$ is the small angle expressed in radians through which the body rotates. We again note that work should be expressed in units obtained by multiplying units of force by units of length. The work of the couple during a finite rotation of the rigid body is obtained by integrating both members of (17.4) from the initial value $\theta_1$ of the angle $\theta$ to its final value $\theta_2$. We write

$$U_{1 \to 2} = \int_{\theta_1}^{\theta_2} M \, d\theta \tag{17.5}$$

*When the moment* **M** *of the couple is constant,* formula (17.5) reduces to

$$U_{1 \to 2} = M(\theta_2 - \theta_1) \tag{17.6}$$

It was pointed out in Sec. 13.2 that a number of forces encountered in problems of kinetics *do no work.* They are forces applied to fixed points or acting in a direction perpendicular to the displacement of their point of application. Among the forces which do no work the following have been listed: the reaction at a frictionless pin when the body supported rotates about the pin, the reaction at a frictionless surface when the body in contact moves along the surface, and the weight of a body when its center of gravity moves horizontally. We can add now that *when a rigid body rolls without sliding on a fixed surface, the friction force* **F** *at the point of contact C does no work.* The velocity $\mathbf{v}_C$ of the point of contact $C$ is zero, and the work of the friction force **F** during a small displacement of the rigid body is

$$dU = F \, ds_C = F(v_C \, dt) = 0$$

## 17.4. KINETIC ENERGY OF A RIGID BODY IN PLANE MOTION

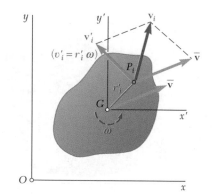

**Fig. 17.3**

Consider a rigid body of mass $m$ in plane motion. We recall from Sec. 14.7 that, if the absolute velocity $\mathbf{v}_i$ of each particle $P_i$ of the body is expressed as the sum of the velocity $\overline{\mathbf{v}}$ of the mass center $G$ of the body and of the velocity $\mathbf{v}'_i$ of the particle relative to a frame $Gx'y'$ attached to $G$ and of fixed orientation (Fig. 17.3), the kinetic energy of the system of particles forming the rigid body can be written in the form

$$T = \tfrac{1}{2}m\overline{v}^2 + \frac{1}{2}\sum_{i=1}^{n} \Delta m_i \, v_i'^2 \tag{17.7}$$

But the magnitude $v'_i$ of the relative velocity of $P_i$ is equal to the product $r'_i \omega$ of the distance $r'_i$ of $P_i$ from the axis through $G$ perpendicular to the plane of motion and of the magnitude $\omega$ of the angular velocity of the body at the instant considered. Substituting into (17.7), we have

$$T = \tfrac{1}{2}m\overline{v}^2 + \frac{1}{2}\left(\sum_{i=1}^{n} r_i'^2 \, \Delta m_i\right) \omega^2 \tag{17.8}$$

or, since the sum represents the moment of inertia $\overline{I}$ of the body about the axis through $G$,

$$T = \tfrac{1}{2}m\overline{v}^2 + \tfrac{1}{2}\overline{I}\omega^2 \tag{17.9}$$

We note that in the particular case of a body in translation ($\omega = 0$), the expression obtained reduces to $\frac{1}{2}m\bar{v}^2$, while in the case of a centroidal rotation ($\bar{v} = 0$), it reduces to $\frac{1}{2}\bar{I}\omega^2$. We conclude that the kinetic energy of a rigid body in plane motion can be separated into two parts: (1) the kinetic energy $\frac{1}{2}m\bar{v}^2$ associated with the motion of the mass center $G$ of the body, and (2) the kinetic energy $\frac{1}{2}\bar{I}\omega^2$ associated with the rotation of the body about $G$.

**Noncentroidal Rotation.** The relation (17.9) is valid for any type of plane motion and can therefore be used to express the kinetic energy of a rigid body rotating with an angular velocity $\boldsymbol{\omega}$ about a fixed axis through $O$ (Fig. 17.4). In that case, however, the kinetic energy of the body can be expressed more directly by noting that the speed $v_i$ of the particle $P_i$ is equal to the product $r_i\omega$ of the distance $r_i$ of $P_i$ from the fixed axis and the magnitude $\omega$ of the angular velocity of the body at the instant considered. Substituting into (17.2), we write

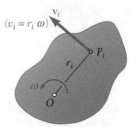

**Fig. 17.4**

$$T = \frac{1}{2}\sum_{i=1}^{n}\Delta m_i (r_i\omega)^2 = \frac{1}{2}\left(\sum_{i=1}^{n}r_i^2\,\Delta m_i\right)\omega^2$$

or, since the last sum represents the moment of inertia $I_O$ of the body about the fixed axis through $O$,

$$T = \tfrac{1}{2}I_O\omega^2 \tag{17.10}$$

We note that the results obtained are not limited to the motion of plane slabs or to the motion of bodies which are symmetrical with respect to the reference plane, and can be applied to the study of the plane motion of any rigid body, regardless of its shape. However, since Eq. (17.9) is applicable to any plane motion while Eq. (17.10) is applicable only in cases involving noncentroidal rotation, Eq. (17.9) will be used in the solution of all the sample problems.

## 17.5. SYSTEMS OF RIGID BODIES

When a problem involves several rigid bodies, each rigid body can be considered separately and the principle of work and energy can be applied to each body. Adding the kinetic energies of all the particles and considering the work of all the forces involved, we can also write the equation of work and energy for the entire system. We have

$$T_1 + U_{1\to 2} = T_2 \tag{17.11}$$

where $T$ represents the arithmetic sum of the kinetic energies of the rigid bodies forming the system (all terms are positive) and $U_{1\to 2}$ represents the work of all the forces acting on the various bodies, whether these forces are *internal* or *external* from the point of view of the system as a whole.

The method of work and energy is particularly useful in solving problems involving pin-connected members, blocks and pulleys connected by inextensible cords, and meshed gears. In all these cases, the internal forces occur by pairs of equal and opposite forces, and the points of application of the forces in each pair *move through equal distances* during a small displacement of the system. As a result, the work of the internal forces is zero and $U_{1\rightarrow2}$ reduces to the work of the *forces external to the system.*

## 17.6. CONSERVATION OF ENERGY

We saw in Sec. 13.6 that the work of conservative forces, such as the weight of a body or the force exerted by a spring, can be expressed as a change in potential energy. When a rigid body, or a system of rigid bodies, moves under the action of conservative forces, the principle of work and energy stated in Sec. 17.2 can be expressed in a modified form. Substituting for $U_{1\rightarrow2}$ from (13.19′) into (17.1), we write

$$T_1 + V_1 = T_2 + V_2 \tag{17.12}$$

Formula (17.12) indicates that when a rigid body, or a system of rigid bodies, moves under the action of conservative forces, *the sum of the kinetic energy and of the potential energy of the system remains constant.* It should be noted that in the case of the plane motion of a rigid body, the kinetic energy of the body should include both the *translational* term $\frac{1}{2}m\bar{v}^2$ and the *rotational* term $\frac{1}{2}\bar{I}\omega^2$.

As an example of application of the principle of conservation of energy, let us consider a slender rod $AB$, of length $l$ and mass $m$, whose extremities are connected to blocks of negligible mass sliding along horizontal and vertical tracks. We assume that the rod is released with no initial velocity from a horizontal position (Fig. 17.5$a$), and we wish to determine its angular velocity after it has rotated through an angle $\theta$ (Fig. 17.5$b$).

Since the initial velocity is zero, we have $T_1 = 0$. Measuring the potential energy from the level of the horizontal track, we write $V_1 = 0$. After the rod has rotated through $\theta$, the center of gravity $G$ of the

(a)    (b)

**Fig. 17.5**

rod is at a distance $\frac{1}{2}l \sin \theta$ below the reference level and we have

$$V_2 = -\frac{1}{2}Wl \sin \theta = -\frac{1}{2}mgl \sin \theta$$

Observing that in this position the instantaneous center of the rod is located at $C$ and that $CG = \frac{1}{2}l$, we write $\bar{v}_2 = \frac{1}{2}l\omega$ and obtain

$$T_2 = \frac{1}{2}m\bar{v}_2^2 + \frac{1}{2}\bar{I}\omega_2^2 = \frac{1}{2}m(\frac{1}{2}l\omega)^2 + \frac{1}{2}(\frac{1}{12}ml^2)\omega^2$$

$$= \frac{1}{2}\frac{ml^2}{3}\omega^2$$

Applying the principle of conservation of energy, we write

$$T_1 + V_1 = T_2 + V_2$$

$$0 = \frac{1}{2}\frac{ml^2}{3}\omega^2 - \frac{1}{2}mgl \sin \theta$$

$$\omega = \left(\frac{3g}{l} \sin \theta\right)^{1/2}$$

The advantages of the method of work and energy, as well as its shortcomings, were indicated in Sec. 13.4. Here we should add that the method of work and energy must be supplemented by the application of d'Alembert's principle when reactions at fixed axles, rollers, or sliding blocks are to be determined. For example, in order to compute the reactions at the extremities $A$ and $B$ of the rod of Fig. 17.5$b$, a diagram should be drawn to express that the system of the external forces applied to the rod is equivalent to the vector $m\bar{\mathbf{a}}$ and the couple $\bar{I}\boldsymbol{\alpha}$. The angular velocity $\boldsymbol{\omega}$ of the rod, however, is determined by the method of work and energy before the equations of motion are solved for the reactions. The complete analysis of the motion of the rod and of the forces exerted on the rod requires, therefore, the combined use of the method of work and energy and of the principle of equivalence of the external and effective forces.

## 17.7. POWER

*Power* was defined in Sec. 13.5 as the time rate at which work is done. In the case of a body acted upon by a force $\mathbf{F}$, and moving with a velocity $\mathbf{v}$, the power was expressed as follows:

$$\text{Power} = \frac{dU}{dt} = \mathbf{F} \cdot \mathbf{v} \tag{13.13}$$

In the case of a rigid body rotating with an angular velocity $\boldsymbol{\omega}$ and acted upon by a couple of moment $\mathbf{M}$ parallel to the axis of rotation, we have, by (17.4),

$$\text{Power} = \frac{dU}{dt} = \frac{M \, d\theta}{dt} = M\omega \tag{17.13}$$

The various units used to measure power, such as the watt and the horsepower, were defined in Sec. 13.5.

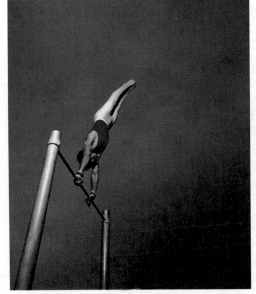

**Photo 17.1** By using the principle of conservation of energy and a free-body-diagram equation, you will find that a force larger than four times his own weight will be exerted on the hands of the gymnast if, keeping his body straight, he swings through 180° about the horizontal bar.

.38 m

A

108.9 kg

## SAMPLE PROBLEM 17.1

A 108.9 kg block is suspended from an inextensible cable which is wrapped around a drum of 0.38 m radius rigidly attached to a flywheel. The drum and flywheel have a combined centroidal moment of inertia $\bar{I} = 1.45$ kgm². At the instant shown, the velocity of the block is 1.8 m/s directed downward. Knowing that the bearing at $A$ is poorly lubricated and that the bearing friction is equivalent to a couple $\mathbf{M}$ of magnitude 81.3 N · m, determine the velocity of the block after it has moved 1.2 m downward.

## SOLUTION

$\omega_1$

$\mathbf{M} = 81.3$ N·m

$A_y$

$A_x$

$\bar{v}_1 = 1.8$ m/s

$s_1 = 0$

$W = 108.9$ kg

$\omega_2$

$\mathbf{M} = 81.3$ N·m

$A_y$

$A_x$

$s_1 = 0$

1.2 m

$s_2 = 1.2$ m

$\bar{v}_2$

$W = 108.9$ kg

We consider the system formed by the flywheel and the block. Since the cable is inextensible, the work done by the internal forces exerted by the cable cancels. The initial and final positions of the system and the external forces acting on the system are as shown.

**Kinetic Energy.**   *Position 1.*

Block:   $\bar{v}_1 = 1.8$ m/s

Flywheel:   $\omega_1 = \dfrac{\bar{v}_1}{r} = \dfrac{1.8 \text{ m/s}}{.38 \text{m}} = 4.7$ rad/s

$$T_1 = \tfrac{1}{2}m\bar{v}_1^2 + \tfrac{1}{2}\bar{I}\omega_1^2$$

$$= \frac{1}{2}108.9 \text{ kg}(1.8 \text{ m/s})^2 + \frac{1}{2}(1.45 \text{ kgm}^2)(4.7 \text{ rad/s})^2$$

$$= 192.4 \text{ kg} \cdot \text{m} = 1888.3 \text{ N} \cdot \text{m}$$

*Position 2.*   Noting that $\omega_2 = \bar{v}_2/.38$, we write

$$T_2 = \tfrac{1}{2}m\bar{v}_2^2 + \tfrac{1}{2}\bar{I}\omega_2^2$$

$$= \frac{1}{2}108.9 \, (\bar{v}_2)^2 + \left(\frac{1}{2}\right)(1.45)\left(\frac{v_2}{.38}\right)^2 = 59.47\bar{v}_2^2$$

**Work.**   During the motion, only the weight $\mathbf{W}$ of the block and the friction couple $\mathbf{M}$ do work. Noting that $\mathbf{W}$ does positive work and that the friction couple $\mathbf{M}$ does negative work, we write

$$s_1 = 0 \qquad s_2 = 1.2 \text{ m}$$

$$\theta_1 = 0 \qquad \theta_2 = \frac{s_2}{r} = \frac{1.2 \text{ m}}{.38 \text{ m}} = 3.16 \text{ rad}$$

$$U_{1\to2} = W(s_2 - s_1) - M(\theta_2 - \theta_1)$$
$$= (1068.3 \text{ N})(1.2 \text{ m}) - (81.3 \text{ N} \cdot \text{m})(3.16 \text{ rad})$$
$$= 1025.05 \text{ N} \cdot \text{m}$$

**Principle of Work and Energy**

$$T_1 + U_{1\to2} = T_2$$
$$1888.3 \text{ N} \cdot \text{m} + 1025.05 \text{ N} \cdot \text{m} = 59.47\bar{v}_2^2$$
$$\bar{v}_2 = 7 \text{ m/s} \qquad\qquad \bar{v}_2 = 7 \text{ m/s} \downarrow \; \blacktriangleleft$$

$r_A = 250$ mm

$r_B = 100$ mm

## SAMPLE PROBLEM 17.2

Gear A has a mass of 10 kg and a radius of gyration of 200 mm; gear B has a mass of 3 kg and a radius of gyration of 80 mm. The system is at rest when a couple **M** of magnitude 6 N · m is applied to gear B. Neglecting friction, determine (a) the number of revolutions executed by gear B before its angular velocity reaches 600 rpm, (b) the tangential force which gear B exerts on gear A.

## SOLUTION

**Motion of Entire System.** Noting that the peripheral speeds of the gears are equal, we write

$$r_A\omega_A = r_B\omega_B \qquad \omega_A = \omega_B\frac{r_B}{r_A} = \omega_B\frac{100 \text{ mm}}{250 \text{ mm}} = 0.40\omega_B$$

For $\omega_B = 600$ rpm, we have

$$\omega_B = 62.8 \text{ rad/s} \qquad \omega_A = 0.40\omega_B = 25.1 \text{ rad/s}$$
$$\bar{I}_A = m_A\bar{k}_A^2 = (10 \text{ kg})(0.200 \text{ m})^2 = 0.400 \text{ kg} \cdot \text{m}^2$$
$$\bar{I}_B = m_B\bar{k}_B^2 = (3 \text{ kg})(0.080 \text{ m})^2 = 0.0192 \text{ kg} \cdot \text{m}^2$$

*Kinetic Energy.* Since the system is initially at rest, $T_1 = 0$. Adding the kinetic energies of the two gears when $\omega_B = 600$ rpm, we obtain

$$T_2 = \tfrac{1}{2}\bar{I}_A\omega_A^2 + \tfrac{1}{2}\bar{I}_B\omega_B^2$$
$$= \tfrac{1}{2}(0.400 \text{ kg} \cdot \text{m}^2)(25.1 \text{ rad/s})^2 + \tfrac{1}{2}(0.0192 \text{ kg} \cdot \text{m}^2)(62.8 \text{ rad/s})^2$$
$$= 163.9 \text{ J}$$

*Work.* Denoting by $\theta_B$ the angular displacement of gear B, we have

$$U_{1\to2} = M\theta_B = (6 \text{ N} \cdot \text{m})(\theta_B \text{ rad}) = (6\theta_B) \text{ J}$$

**Principle of Work and Energy**

$$T_1 + U_{1\to2} = T_2$$
$$0 + (6\theta_B) \text{ J} = 163.9 \text{ J}$$
$$\theta_B = 27.32 \text{ rad} \qquad \theta_B = 4.35 \text{ rev} \quad \blacktriangleleft$$

**Motion of Gear A.** *Kinetic Energy.* Initially, gear A is at rest, so $T_1 = 0$. When $\omega_B = 600$ rpm, the kinetic energy of gear A is

$$T_2 = \tfrac{1}{2}\bar{I}_A\omega_A^2 = \tfrac{1}{2}(0.400 \text{ kg} \cdot \text{m}^2)(25.1 \text{ rad/s})^2 = 126.0 \text{ J}$$

*Work.* The forces acting on gear A are as shown. The tangential force **F** does work equal to the product of its magnitude and of the length $\theta_A r_A$ of the arc described by the point of contact. Since $\theta_A r_A = \theta_B r_B$, we have

$$U_{1\to2} = F(\theta_B r_B) = F(27.3 \text{ rad})(0.100 \text{ m}) = F(2.73 \text{ m})$$

**Principle of Work and Energy**

$$T_1 + U_{1\to2} = T_2$$
$$0 + F(2.73 \text{ m}) = 126.0 \text{ J}$$
$$F = +46.2 \text{ N} \qquad \mathbf{F} = 46.2 \text{ N} \swarrow \quad \blacktriangleleft$$

## SAMPLE PROBLEM 17.3

A sphere, a cylinder, and a hoop, each having the same mass and the same radius, are released from rest on an incline. Determine the velocity of each body after it has rolled through a distance corresponding to a change in elevation $h$.

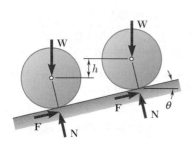

### SOLUTION

The problem will first be solved in general terms, and then results for each body will be found. We denote the mass by $m$, the centroidal moment of inertia by $\bar{I}$, the weight by $W$, and the radius by $r$.

**Kinematics.** Since each body rolls, the instantaneous center of rotation is located at $C$ and we write

$$\omega = \frac{\bar{v}}{r}$$

**Kinetic Energy**

$$T_1 = 0$$
$$T_2 = \tfrac{1}{2}m\bar{v}^2 + \tfrac{1}{2}\bar{I}\omega^2$$
$$= \tfrac{1}{2}m\bar{v}^2 + \tfrac{1}{2}\bar{I}\left(\frac{\bar{v}}{r}\right)^2 = \tfrac{1}{2}\left(m + \frac{\bar{I}}{r^2}\right)\bar{v}^2$$

**Work.** Since the friction force $\mathbf{F}$ in rolling motion does no work,

$$U_{1\to2} = Wh$$

**Principle of Work and Energy**

$$T_1 + U_{1\to2} = T_2$$
$$0 + Wh = \tfrac{1}{2}\left(m + \frac{\bar{I}}{r^2}\right)\bar{v}^2 \qquad \bar{v}^2 = \frac{2Wh}{m + \bar{I}/r^2}$$

Noting that $W = mg$, we rearrange the result and obtain

$$\bar{v}^2 = \frac{2gh}{1 + \bar{I}/mr^2}$$

**Velocities of Sphere, Cylinder, and Hoop.** Introducing successively the particular expression for $\bar{I}$, we obtain

| | | |
|---|---|---|
| *Sphere:* | $\bar{I} = \tfrac{2}{5}mr^2$ | $\bar{v} = 0.845\sqrt{2gh}$ ◀ |
| *Cylinder:* | $\bar{I} = \tfrac{1}{2}mr^2$ | $\bar{v} = 0.816\sqrt{2gh}$ ◀ |
| *Hoop:* | $\bar{I} = mr^2$ | $\bar{v} = 0.707\sqrt{2gh}$ ◀ |

*Remark.* Let us compare the results with the velocity attained by a frictionless block sliding through the same distance. The solution is identical to the above solution except that $\omega = 0$; we find $\bar{v} = \sqrt{2gh}$.

Comparing the results, we note that the velocity of the body is independent of both its mass and radius. However, the velocity does depend upon the quotient $\bar{I}/mr^2 = \bar{k}^2/r^2$, which measures the ratio of the rotational kinetic energy to the translational kinetic energy. Thus the hoop, which has the largest $\bar{k}$ for a given radius $r$, attains the smallest velocity, while the sliding block, which does not rotate, attains the largest velocity.

## SAMPLE PROBLEM 17.4

A 13.6-kg slender rod $AB$ is 1.5 m long and is pivoted about a point $O$ which is .3 m from end $B$. The other end is pressed against a spring of constant $k = 315$ kN/m. until the spring is compressed 2.54 cm. The rod is then in a horizontal position. If the rod is released from this position, determine its angular velocity and the reaction at the pivot $O$ as the rod passes through a vertical position.

## SOLUTION

**Position 1.** **Potential Energy.** Since the spring is compressed 1 in., we have $x_1 = 1$ in.

$$V_e = \tfrac{1}{2}kx_1^2 = \tfrac{1}{2}(315 \text{ kN/m})(2.54 \text{ cm})^2 = 101.6 \text{ N} \cdot \text{m}$$

Choosing the datum as shown, we have $V_g = 0$; therefore,

$$V_1 = V_e + V_g = 101.6 \text{ N} \cdot \text{m}$$

**Kinetic Energy.** Since the velocity in position $1$ is zero, we have $T_1 = 0$.

**Position 2.** **Potential Energy.** The elongation of the spring is zero, and we have $V_e = 0$. Since the center of gravity of the rod is now .46 m above the datum,

$$V_g = (133.4 \text{ N})(+.46 \text{ m}) = 61.3 \text{ N} \cdot \text{m}$$
$$V_2 = V_e + V_g = 61.3 \text{ N} \cdot \text{m}$$

**Kinetic Energy.** Denoting by $\boldsymbol{\omega}_2$ the angular velocity of the rod in position 2, we note that the rod rotates about $O$ and write $\bar{v}_2 = \bar{r}\omega_2 = .46\omega_2$.

$$\bar{I} = \tfrac{1}{12}ml^2 = \frac{1}{12} 13.6 \text{ kg}(1.5 \text{ m})^2 = 2.55 \text{ kgm}^2$$

$$T_2 = \tfrac{1}{2}m\bar{v}_2^2 + \tfrac{1}{2}\bar{I}\omega_2^2 = \frac{1}{2} 13.6(.46\omega_2)^2 + \tfrac{1}{2}(2.55)\omega_2^2 = 2.714\omega_2^2$$

**Conservation of Energy**

$$T_1 + V_1 = T_2 + V_2$$
$$0 + 101.6 \text{ N} \cdot \text{m} = 2.714\omega_2^2 + 61.3$$
$$\omega_2 = 3.85 \text{ rad/s} \; \downarrow \quad \blacktriangleleft$$

**Reaction in Position 2.** Since $\omega_2 = 3.85$ rad/s, the components of the acceleration of $G$ as the rod passes through position 2 are

$$\bar{a}_n = \bar{r}\omega_2^2 = (.46 \text{ m})(3.85 \text{ rad/s})^2 = 6.82 \text{ m/s}^2 \qquad \bar{\mathbf{a}}_n = 6.82 \text{ m/s}^2 \downarrow$$
$$a_t = \bar{r}\alpha \qquad\qquad\qquad\qquad\qquad\qquad \bar{\mathbf{a}}_t = \bar{r}\alpha \rightarrow$$

We express that the system of external forces is equivalent to the system of effective forces represented by the vector of components $m\bar{\mathbf{a}}_t$ and $m\bar{\mathbf{a}}_n$ attached at $G$ and the couple $\bar{I}\alpha$.

$+\downarrow\Sigma M_O = \Sigma(M_O)_{\text{eff}}$: $\qquad\qquad 0 = \bar{I}\alpha + m(\bar{r}\alpha)r \qquad \alpha = 0$

$\xrightarrow{+}\Sigma F_x = \Sigma(F_x)_{\text{eff}}$: $\qquad\qquad R_x = m(\bar{r}\alpha) \qquad\qquad R_x = 0$

$+\uparrow\Sigma F_y = \Sigma(F_y)_{\text{eff}}$: $\quad R_y - 133.4 \text{ N} = -m\bar{a}_n$

$$R_y - 133.4 \text{ N} = -13.6(6.82 \text{ m/s}^2)$$

$$R_y = +40.65 \text{ N} \qquad \mathbf{R} = 40.65 \text{ N} \uparrow \quad \blacktriangleleft$$

## SAMPLE PROBLEM 17.5

Each of the two slender rods shown is 0.75 m long and has a mass of 6 kg. If the system is released from rest with $\beta = 60°$, determine (a) the angular velocity of rod $AB$ when $\beta = 20°$, (b) the velocity of point $D$ at the same instant.

## SOLUTION

**Kinematics of Motion When $\beta = 20°$.** Since $\mathbf{v}_B$ is perpendicular to the rod $AB$ and $\mathbf{v}_D$ is horizontal, the instantaneous center of rotation of rod $BD$ is located at $C$. Considering the geometry of the figure, we obtain

$$BC = 0.75 \text{ m} \qquad CD = 2(0.75 \text{ m}) \sin 20° = 0.513 \text{ m}$$

Applying the law of cosines to triangle $CDE$, where $E$ is located at the mass center of rod $BD$, we find $EC = 0.522$ m. Denoting by $\omega$ the angular velocity of rod $AB$, we have

$$\bar{v}_{AB} = (0.375 \text{ m})\omega \qquad \bar{\mathbf{v}}_{AB} = 0.375\omega \searrow$$
$$v_B = (0.75 \text{ m})\omega \qquad \bar{\mathbf{v}}_B = 0.75\omega \searrow$$

Since rod $BD$ seems to rotate about point $C$, we write

$$v_B = (BC)\omega_{BD} \qquad (0.75 \text{ m})\omega = (0.75 \text{ m})\omega_{BD} \qquad \boldsymbol{\omega}_{BD} = \omega \uparrow$$
$$\bar{v}_{BD} = (EC)\omega_{BD} = (0.522 \text{ m})\omega \qquad \bar{\mathbf{v}}_{BD} = 0.522\omega \searrow$$

**Position 1. Potential Energy.** Choosing the datum as shown, and observing that $W = (6 \text{ kg})(9.81 \text{ m/s}^2) = 58.86$ N, we have

$$V_1 = 2W\bar{y}_1 = 2(58.86 \text{ N})(0.325 \text{ m}) = 38.26 \text{ J}$$

**Kinetic Energy.** Since the system is at rest, $T_1 = 0$.

**Position 2. Potential Energy**

$$V_2 = 2W\bar{y}_2 = 2(58.86 \text{ N})(0.1283 \text{ m}) = 15.10 \text{ J}$$

**Kinetic Energy**

$$I_{AB} = \bar{I}_{BD} = \tfrac{1}{12}ml^2 = \tfrac{1}{12}(6 \text{ kg})(0.75 \text{ m})^2 = 0.281 \text{ kg} \cdot \text{m}^2$$
$$T_2 = \tfrac{1}{2}m\bar{v}_{AB}^2 + \tfrac{1}{2}\bar{I}_{AB}\omega_{AB}^2 + \tfrac{1}{2}m\bar{v}_{BD}^2 + \tfrac{1}{2}\bar{I}_{BD}\omega_{BD}^2$$
$$= \tfrac{1}{2}(6)(0.375\omega)^2 + \tfrac{1}{2}(0.281)\omega^2 + \tfrac{1}{2}(6)(0.522\omega)^2 + \tfrac{1}{2}(0.281)\omega^2$$
$$= 1.520\omega^2$$

**Conservation of Energy**

$$T_1 + V_1 = T_2 + V_2$$
$$0 + 38.26 \text{ J} = 1.520\omega^2 + 15.10 \text{ J}$$
$$\omega = 3.90 \text{ rad/s} \qquad \boldsymbol{\omega}_{AB} = 3.90 \text{ rad/s} \downarrow \quad \blacktriangleleft$$

**Velocity of Point $D$**

$$v_D = (CD)\omega = (0.513 \text{ m})(3.90 \text{ rad/s}) = 2.00 \text{ m/s}$$
$$\mathbf{v}_D = 2.00 \text{ m/s} \rightarrow \quad \blacktriangleleft$$

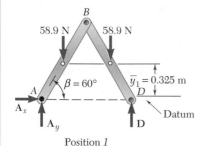

# SOLVING PROBLEMS
## ON YOUR OWN

In this lesson we introduced energy methods to determine the velocity of rigid bodies for various positions during their motion. As you found out previously in Chap. 13, energy methods should be considered for problems involving displacements and velocities.

**1. The method of work and energy,** when applied to all of the particles forming a rigid body, yields the equation

$$T_1 + U_{1 \rightarrow 2} = T_2 \qquad (17.1)$$

where $T_1$ and $T_2$ are, respectively, the initial and final values of the total kinetic energy of the particles forming the body and $U_{1 \rightarrow 2}$ is the *work done by the external forces* exerted on the rigid body.

 **a. Work of forces and couples.** To the expression for the work of a force (Chap. 13), we added the expression for the work of a couple and wrote

$$U_{1 \rightarrow 2} = \int_{A_1}^{A_2} \mathbf{F} \cdot d\mathbf{r} \qquad U_{1 \rightarrow 2} = \int_{\theta_1}^{\theta_2} M \, d\theta \qquad (17.3, 17.5)$$

When the moment of a couple is constant, the work of the couple is

$$U_{1 \rightarrow 2} = M(\theta_2 - \theta_1) \qquad (17.6)$$

where $\theta_1$ and $\theta_2$ are expressed in radians [Sample Probs. 17.1 and 17.2].

 **b. The kinetic energy of a rigid body in plane motion** was found by considering the motion of the body as the sum of a translation with its mass center and a rotation about the mass center.

$$T = \tfrac{1}{2}m\bar{v}^2 + \tfrac{1}{2}\bar{I}\omega^2 \qquad (17.9)$$

where $\bar{v}$ is the velocity of the mass center and $\omega$ is the angular velocity of the body [Sample Probs. 17.3 and 17.4].

**2. For a system of rigid bodies** we again used the equation

$$T_1 + U_{1 \rightarrow 2} = T_2 \qquad (17.1)$$

where $T$ is the sum of the kinetic energies of the bodies forming the system and $U$ is the work done by *all the forces acting on the bodies,* internal as well as external. Your computations will be simplified if you keep the following in mind.

 **a. The forces exerted on each other by pin-connected members or by meshed gears** are equal and opposite, and, since they have the same point of application, they undergo equal small displacements. Therefore, *their total work is zero* and can be omitted from your calculations [Sample Prob. 17.2].

*(continued)*

*b. The forces exerted by an inextensible cord* on the two bodies it connects have the same magnitude and their points of application move through equal distances, but the work of one force is positive and the work of the other is negative. Therefore, *their total work is zero* and can again be omitted from your calculations [Sample Prob. 17.1].

*c. The forces exerted by a spring* on the two bodies it connects also have the same magnitude, but their points of application will generally move through different distances. Therefore, *their total work is usually not zero* and should be taken into account in your calculations.

3. *The principle of conservation of energy* can be expressed as

$$T_1 + V_1 = T_2 + V_2 \qquad (17.12)$$

where $V$ represents the potential energy of the system. This principle can be used when a body or a system of bodies is acted upon by conservative forces, such as the force exerted by a spring or the force of gravity [Sample Probs. 17.4 and 17.5].

4. *The last section of this lesson was devoted to power,* which is the time rate at which work is done. For a body acted upon by a couple of moment $\mathbf{M}$, the power can be expressed as

$$\text{Power} = M\omega \qquad (17.13)$$

where $\omega$ is the angular velocity of the body expressed in rad/s. As you did in Chap. 13, you should express power either in watts or in horsepower (1 hp = 550 ft · lb/s).

# Problems

**17.1**  The rotor of an electric motor has an angular velocity of 3600 rpm when the load and power are cut off. The 50 kg rotor, which has a centroidal radius of gyration of 23 cm, then coasts to rest. Knowing that the kinetic friction of the rotor produces a couple of magnitude 3.4 N · m, determine the number of revolutions that the rotor executes before coming to rest.

**17.2**  A 200 kg flywheel is at rest when a constant 300 N · m couple is applied. After executing 560 revolutions, the flywheel reaches its rated speed of 2400 rpm. Knowing that the radius of gyration of the flywheel is 400 mm, determine the average magnitude of the couple due to kinetic friction in the bearing.

**17.3**  The flywheel of a small punching machine rotates at 360 rpm. Each punching operation requires 2034 J of work and it is desired that the speed of the flywheel after each punching be not less than 95 percent of the original speed. (a) Determine the required moment of inertia of the flywheel. (b) If a constant 24.4 N · m couple is applied to the shaft of the flywheel, determine the number of revolutions that must occur between two successive punchings, knowing that the initial velocity is to be 360 rpm at the start of each punching.

**17.4**  A 227 kg flywheel is at rest when a constant 271 N · m couple is applied. After executing 750 revolutions, the flywheel reaches its rated speed of 3000 rpm and the couple is removed. Assuming that the kinetic friction results in a constant couple of magnitude 17 N · m, determine (a) the radius of gyration of the flywheel, (b) the number of revolutions that the flywheel executes after the couple is removed and before it comes to rest.

**17.5**  The uniform cylinder $A$, of mass $m$ and radius $r$, has an angular velocity $\omega_0$ when it is brought into contact with an identical cylinder $B$ which is at rest. The coefficient of kinetic friction at the contact point $D$ is $\mu_k$. After a period of slipping, the cylinders attain constant angular velocities of equal magnitude and opposite direction at the same time. Knowing that the final kinetic energy of the system is one-half the initial kinetic energy of cylinder $A$, derive expressions for the number of revolutions executed by each cylinder before it attains a constant angular velocity.

**17.6**  The uniform 4 kg cylinder $A$, of radius $r = 150$ mm has an angular velocity $\omega_0 = 50$ rad/s when it is brought into contact with an identical cylinder $B$ which is at rest. The coefficient of kinetic friction at the contact point $D$ is $\mu_k$. After a period of slipping, the cylinders attain constant angular velocities of equal magnitude and opposite direction at the same time. Knowing that cylinder $A$ executes three revolutions before it attains a constant angular velocity and cylinder $B$ executes one revolution before it attains a constant angular velocity, determine (a) the final angular velocity of each cylinder, (b) the coefficient of kinetic friction $\mu_k$.

**Fig. P17.5 and P17.6**

**17.7**    Two uniform disks of the same material are attached to a shaft as shown. Disk $A$ has a weight of 4.5 kg and a radius $r = 15.2$ cm. Disk $B$ is twice as thick as disk $A$. Knowing that a couple $\mathbf{M}$ of magnitude 30 N · m is applied to disk $A$ when the system is at rest, determine the radius $nr$ of disk $B$ if the angular velocity of the system is to be 480 rpm after 5 revolutions.

**Fig. P17.7**

**17.8**    Two uniform slender rods $AB$ and $DE$ of weight $w$ per unit length are attached to a shaft $CF$ as shown. The length of rod $AB$ is $L$ and the length of rod $DE$ is $nL$. A couple $\mathbf{M}$ of constant moment is applied to shaft $CF$ when the system is at rest and removed after the system has executed one complete revolution. Neglecting the mass of the shaft, determine the length of rod $DE$ which results in the largest final speed of point $D$.

**17.9**    The double pulley shown has a mass of 14 kg and a centroidal radius of gyration of 165 mm. Cylinder $A$ and block $B$ are attached to cords that are wrapped on the pulleys as shown. The coefficient of kinetic friction between block $B$ and the surface is 0.25. Knowing that the system is released from rest in the position shown, determine ($a$) the velocity of cylinder $A$ as it strikes the ground, ($b$) the total distance that block $B$ moves before coming to rest.

**Fig. P17.8**

**Fig. P17.9**

**Fig. P17.10**

**17.10**    The 200-mm-radius brake drum is attached to a larger flywheel that is not shown. The total mass moment of inertia of the flywheel and drum is 19 kg · m² and the coefficient of kinetic friction between the drum and the brake shoe is 0.35. Knowing that the initial angular velocity of the flywheel is 360 rpm counterclockwise, determine the vertical force $\mathbf{P}$ that must be applied to the pedal $C$ if the system is to stop in 100 revolutions.

**17.11** Solve Prob. 17.10 assuming that the initial angular velocity of the flywheel is 360 rpm clockwise.

**17.12** Each of the gears $A$ and $B$ has a weight of 2.3 kg and a radius of gyration of 10.2 cm, while gear $C$ has a weight of 11.3 kg and a radius of gyration of 19 cm. A couple $\mathbf{M}$ of magnitude 9.15 N · m is applied to gear $C$. Determine (a) the number of revolutions of gear $C$ required for its angular velocity to increase from 100 to 450 rpm, (b) the corresponding tangential force acting on gear $A$.

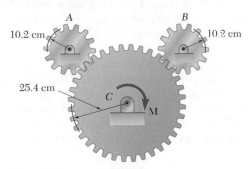

**Fig. P17.12**

**17.13** Solve Prob. 17.12 assuming that the 9.15 N · m couple is applied to gear $B$.

**17.14** A slender rod of length $l$ is pivoted about a point $C$ located at a distance $b$ from its center $G$. It is released from rest in a horizontal position and swings freely. Determine (a) the distance $b$ for which the angular velocity of the rod as it passes through a vertical position is maximum, (b) the corresponding values of its angular velocity and of the reaction at $C$.

**Fig. P17.14**

**17.15** A 3-kg slender rod rotates in a *vertical* plane about a pivot at $B$. A spring of constant $k = 300$ N/m and of unstretched length 120 mm is attached to the rod as shown. Knowing that in the position shown the rod has an angular velocity of 4 rad/s clockwise, determine the angular velocity of the rod after it has rotated through (a) 90°, (b) 180°.

**Fig. P17.15**

**17.16** A uniform sphere of radius $r$ is placed at corner $A$ and is given a slight clockwise motion. Assuming that the corner is sharp and becomes slightly embedded in the sphere, so that the coefficient of static friction at $A$ is very large, determine (a) the angle $\beta$ through which the sphere will have rotated when it loses contact with the corner, (b) the corresponding velocity of the center of the sphere.

**Fig. P17.16**

**Fig. P17.17**

**Fig. P17.18**

**17.17** Two identical slender rods $AB$ and $BC$ are welded together to form an L-shaped assembly. The assembly is pressed against a spring at $D$ and released from the position shown. Knowing that the maximum angle of rotation of the assembly in its subsequent motion is 90° counterclockwise, determine the magnitude of the angular velocity of the assembly as it passes through the position where rod $AB$ forms an angle of 30° with the horizontal.

**17.18** The 30-kg turbine disk has a centroidal radius of gyration of 175 mm and is rotating clockwise at a constant rate of 60 rpm when a small blade of weight 0.5 N at point $A$ becomes loose and is thrown off. Neglecting friction, determine the change in the angular velocity of the turbine disk after it has rotated through (a) 90°, (b) 270°.

**17.19** A 73-kg gymnast is executing a series of full-circle swings on the horizontal bar. In the position shown he has a small and negligible clockwise angular velocity and will maintain his body straight and rigid as he swings downward. Assuming that during the swing the centroidal radius of gyration of his body is 457 mm, determine his angular velocity and the force exerted on his hands after he has rotated through (a) 90°, (b) 180°.

**Fig. P17.19**

**Fig. P17.20**

**17.20** A rope is wrapped around a cylinder of radius $r$ and mass $m$ as shown. Knowing that the cylinder is released from rest, determine the velocity of the center of the cylinder after it has moved downward a distance $s$.

**17.21** A 24-kg uniform cylindrical roller, initially at rest, is acted upon by a 100 N force as shown. Knowing that the body rolls without slipping, determine (a) the velocity of its center $G$ after it has moved 1.8 m, (b) the friction force required to prevent slipping.

**Fig. P17.21**

**17.22**  A half-cylinder of mass $m$ and radius $r$ is released from rest in the position shown. Knowing that the half-cylinder rolls without sliding, determine (*a*) its angular velocity after it has rolled through 90°, (*b*) the reaction at the horizontal surface at the same instant. [*Hint:* Note that $GO = 4r/3\pi$ and that, by the parallel-axis theorem, $\bar{I} = \frac{1}{2}mr^2 - m(GO)^2$.]

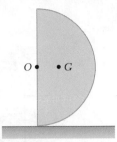

**Fig. P17.22**

**17.23**  A collar $B$, of mass $m$ and negligible dimensions, is attached to the rim of a hoop of the same mass $m$ and of radius $r$ that rolls without sliding on a horizontal surface. Determine the angular velocity $\boldsymbol{\omega}_1$ of the hoop in terms of $g$ and $r$ when $B$ is directly above the center $A$, knowing that the angular velocity of the hoop is $4\boldsymbol{\omega}_1$ when $B$ is directly below $A$.

**17.24**  The 9-kg rod $AB$ is attached by pins to two 6-kg uniform disks as shown. The assembly rolls without sliding on a horizontal surface. If the assembly is released from rest when $\theta = 60°$, determine (*a*) the angular velocity of the disks when $\theta = 180°$, (*b*) the force exerted by the surface on each disk at that instant.

**Fig. P17.23**

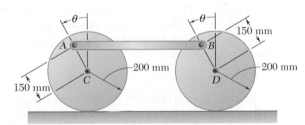

**Fig. P17.24**

**17.25**  A bar of mass $m = 5$ kg is held as shown between four disks each of mass $m' = 2$ kg and radius $r = 75$ mm. Knowing that the forces exerted on the disks are sufficient to prevent slipping and that the bar is released from rest, for each of the cases shown determine the velocity of the bar after it has moved through the distance $h$.

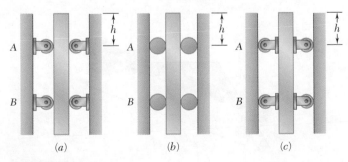

**Fig. P17.25**

**17.26**   The 1.36 kg uniform slender bar $AB$ is connected to the 2.7 kg gear $B$ which meshes with the stationary outer gear $C$. The centroidal radius of gyration of gear $B$ is 3 cm. Knowing that the system is released from rest in the position shown, determine (*a*) the angular velocity of the bar as it passes through the vertical position, (*b*) the corresponding angular velocity of gear $B$.

**Fig. P17.26 and P17.27**

**17.27**   The 1.36 kg uniform slender bar $AB$ is connected to the 2.7 kg gear $B$ which meshes with the stationary outer gear $C$. The centroidal radius of gyration of gear $B$ is 3 cm. The system is released in the position shown when the angular velocity of the bar is 4 rad/s counterclockwise. Knowing that the magnitude of the angular velocity of the bar as it passes through the vertical position is 12 rad/s, determine the amount of energy dissipated in friction.

**17.28**   Two identical uniform cylinders are connected by a cable of negligible mass and roll without sliding on the surfaces shown. Knowing that the system is released from rest when $\beta = 5°$, determine the angular velocity of each cylinder when $\beta = 15°$.

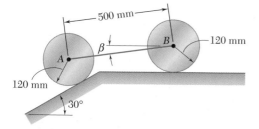

**Fig. P17.28**

**17.29**   The motion of the uniform rod $AB$ is guided by small wheels of negligible mass that roll on the surface shown. If the rod is released from rest when $\theta = 0$, determine the velocities of $A$ and $B$ when $\theta = 30°$.

**Fig. P17.29**

**17.30** The mechanism shown is one of two identical mechanisms attached to the two sides of a 90-kg uniform rectangular door. Edge $ABC$ of the door is guided by wheels of negligible mass that roll in horizontal and vertical tracks. A spring of constant $k = 600$ N/m is attached to wheel $B$. Knowing that the door is released from rest in the position $\theta = 30°$ with the spring unstretched, determine the velocity of wheel $A$ just as the door reaches the vertical position.

**Fig. P17.30 and P17.31**

**17.31** The mechanism shown is one of two identical mechanisms attached to the two sides of a 90-kg uniform rectangular door. Edge $ABC$ of the door is guided by wheels of negligible mass that roll in horizontal and vertical tracks. A spring of constant $k$ is attached to wheel $B$ in such a way that its tension is zero when $\theta = 30°$. Knowing that the door is released from rest in the position $\theta = 45°$ and reaches the vertical position with an angular velocity of 0.6 rad/s, determine the spring constant $k$.

**17.32** The motion of the uniform slender 2.3-kg rod $AB$ is guided at $A$ and $C$ by collars of negligible mass. The system is released from rest in the position $\theta = 45°$. Knowing that the applied force **P** is zero, determine the angular velocity of rod $AB$ when $\theta = 30°$.

**Fig. P17.32 and P17.33**

**17.33** The motion of the uniform slender 2.3 kg rod $AB$ is guided at $A$ and $C$ by collars of negligible mass. The system is released from rest in the position $\theta = 30°$. Knowing that the magnitude of the force **P** applied to collar $A$ is 9 N, determine the angular velocity of rod $AB$ when $\theta = 45°$.

**17.34** The 3.6 kg rod $AB$ is attached to a collar of negligible weight at $A$ and to a flywheel at $B$. The flywheel has a weight of 14.5 kg and a radius of gyration of 23 cm. Knowing that in the position shown the angular velocity of the flywheel is 60 rpm clockwise, determine the velocity of the flywheel when point $B$ is directly below $C$.

**Fig. *P17.34* and *P17.35***

**17.35** If in Prob. 17.34 the angular velocity of the flywheel is to be the same in the position shown and when point $B$ is directly above $C$, determine the required value of its angular velocity in the position shown.

**17.36** A 5-kg uniform square plate is supported by two identical 1.5-kg uniform slender rods $AD$ and $BE$ and is released from rest in the position $\theta = 45°$. Knowing that the angular velocity of $AD$ as it passes through the vertical position is 5.2 rad/s, determine the amount of energy dissipated in friction.

**Fig. P17.36**

**17.37** The uniform rods $AB$ and $BC$ are of mass 3 kg and 8 kg, respectively, and collar $C$ has a mass of 4 kg. Knowing that at the instant shown the velocity of collar $C$ is 0.9 m/s downward, determine the velocity of point $B$ after rod $AB$ has rotated through 90°.

**17.38** Wheel $A$ weighs 6.8 kg, has a centroidal radius of gyration of 15.2 cm, and rolls without sliding on the horizontal surface. Each of the uniform rods $AB$ and $BC$ is 50.8 cm long and weighs 3.6 kg. If point $A$ is moved slightly to the left and released, determine the velocity of point $A$ as rod $BC$ passes through a horizontal position.

**Fig. P17.37**

**Fig. P17.38**

**17.39** Knowing that the maximum allowable couple that can be applied to a shaft is 1.75 kN · m, determine the maximum power that can be transmitted by the shaft at (*a*) 180 rpm, (*b*) 480 rpm.

**17.40** The flywheel shown has a radius of 500 mm, a mass of 110 kg and a radius of gyration of 375 mm. A 20-kg block *A* is attached to a wire that is wrapped around the flywheel. Determine the power delivered by the electric motor attached to the shaft of the flywheel at the instant when the velocity of block *A* is 7.5 m/s up and its acceleration is (*a*) zero, (*b*) 0.9 m/s² up.

**Fig. P17.40**

**17.41** The shaft-disk-belt arrangement shown is used to transmit 2386.2 N · m/s from point *A* to point *D*. Knowing that the maximum allowable couples that can be applied to shafts *AB* and *CD* are 24.4 N · m and 78.6 N · m, respectively, determine the required minimum speed of shaft *AB*.

**Fig. P17.41**

**17.42** The experimental setup shown is used to measure the power output of a small turbine. When the turbine is operating at 200 rpm, the readings of the two spring scales are 4.5 kg and 10 kg, respectively. Determine the power being developed by the turbine.

**Fig. P17.42**

## 17.8. PRINCIPLE OF IMPULSE AND MOMENTUM FOR THE PLANE MOTION OF A RIGID BODY

The principle of impulse and momentum will now be applied to the analysis of the plane motion of rigid bodies and of systems of rigid bodies. As was pointed out in Chap. 13, the method of impulse and momentum is particularly well adapted to the solution of problems involving time and velocities. Moreover, the principle of impulse and momentum provides the only practicable method for the solution of problems involving impulsive motion or impact (Secs. 17.11 and 17.12).

Considering again a rigid body as made of a large number of particles $P_i$, we recall from Sec. 14.9 that the system formed by the momenta of the particles at time $t_1$ and the system of the impulses of the external forces applied from $t_1$ to $t_2$ are together equipollent to the system formed by the momenta of the particles at time $t_2$. Since the vectors associated with a rigid body can be considered as sliding vectors, it follows (Sec. 3.19) that the systems of vectors shown in Fig. 17.6 are not only equipollent but truly *equivalent* in the sense

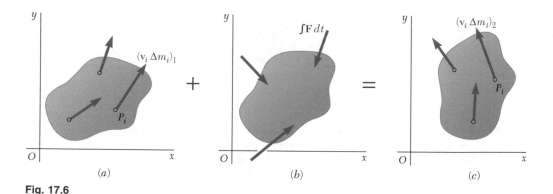

**Fig. 17.6**

that the vectors on the left-hand side of the equals sign can be transformed into the vectors on the right-hand side through the use of the fundamental operations listed in Sec. 3.13. We therefore write

$$\textbf{Syst Momenta}_1 + \textbf{Syst Ext Imp}_{1\to2} = \textbf{Syst Momenta}_2 \qquad (17.14)$$

But the momenta $\mathbf{v}_i \, \Delta m_i$ of the particles can be reduced to a vector attached at $G$, equal to their sum

$$\mathbf{L} = \sum_{i=1}^{n} \mathbf{v}_i \, \Delta m_i$$

and a couple of moment equal to the sum of their moments about $G$

$$\mathbf{H}_G = \sum_{i=1}^{n} \mathbf{r}_i' \times \mathbf{v}_i \, \Delta m_i$$

We recall from Sec. 14.3 that $\mathbf{L}$ and $\mathbf{H}_G$ define, respectively, the linear momentum and the angular momentum about $G$ of the system of particles forming the rigid body. We also note from Eq. (14.14) that $\mathbf{L} = m\overline{\mathbf{v}}$. On the other hand, restricting the present analysis to the plane motion of a rigid slab or of a rigid body symmetrical with respect to the reference plane, we recall from Eq. (16.4) that $\mathbf{H}_G = \overline{I}\boldsymbol{\omega}$. We thus conclude that the system of the momenta $\mathbf{v}_i \, \Delta m_i$ is equivalent to the *linear momentum vector* $m\overline{\mathbf{v}}$ attached at $G$ and to the *angular momentum couple* $\overline{I}\boldsymbol{\omega}$ (Fig. 17.7). Observing that the system of

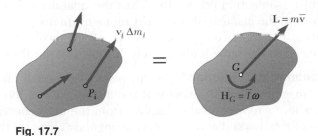

**Fig. 17.7**

momenta reduces to the vector $m\overline{\mathbf{v}}$ in the particular case of a translation ($\boldsymbol{\omega} = 0$) and to the couple $\overline{I}\boldsymbol{\omega}$ in the particular case of a centroidal rotation ($\overline{\mathbf{v}} = 0$), we verify once more that the plane motion of a rigid body symmetrical with respect to the reference plane can be resolved into a translation with the mass center $G$ and a rotation about $G$.

Replacing the system of momenta in parts $a$ and $c$ of Fig. 17.6 by the equivalent linear momentum vector and angular momentum couple, we obtain the three diagrams shown in Fig. 17.8. This figure

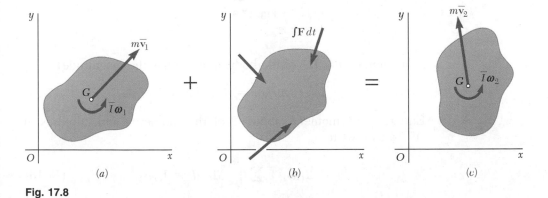

**Fig. 17.8**

expresses as a free-body-diagram equation the fundamental relation (17.14) in the case of the plane motion of a rigid slab or of a rigid body symmetrical with respect to the reference plane.

Three equations of motion can be derived from Fig. 17.8. Two equations are obtained by summing and equating the $x$ and $y$ *components* of the momenta and impulses, and the third equation is obtained by summing and equating the *moments* of these vectors *about any given point*. The coordinate axes can be chosen fixed in space, or allowed to move with the mass center of the body while maintaining a fixed direction. In either case, the point about which moments are taken should keep the same position relative to the coordinate axes during the interval of time considered.

In deriving the three equations of motion for a rigid body, care should be taken not to add linear and angular momenta indiscriminately. Confusion can be avoided by remembering that $m\bar{v}_x$ and $m\bar{v}_y$ represent the *components of a vector*, namely, the linear momentum vector $m\bar{\mathbf{v}}$, while $\bar{I}\omega$ represents the *magnitude of a couple*, namely, the angular momentum couple $\bar{I}\boldsymbol{\omega}$. Thus the quantity $\bar{I}\omega$ should be added only to the *moment* of the linear momentum $m\bar{\mathbf{v}}$, never to this vector itself nor to its components. All quantities involved will then be expressed in the same units, namely N · m · s or lb · ft · s.

**Noncentroidal Rotation.**  In this particular case of plane motion, the magnitude of the velocity of the mass center of the body is $\bar{v} = \bar{r}\omega$, where $\bar{r}$ represents the distance from the mass center to the fixed axis of rotation and $\boldsymbol{\omega}$ represents the angular velocity of the body at the instant considered; the magnitude of the momentum vector attached at $G$ is thus $m\bar{v} = m\bar{r}\omega$. Summing the moments about $O$ of the momentum vector and momentum couple (Fig. 17.9) and using the parallel-axis theorem for moments of inertia, we find that the an-

**Fig. 17.9**

gular momentum $\mathbf{H}_O$ of the body about $O$ has the magnitude†

$$\bar{I}\omega + (m\bar{r}\omega)\bar{r} = (\bar{I} + m\bar{r}^2)\omega = I_O\omega \tag{17.15}$$

Equating the moments about $O$ of the momenta and impulses in (17.14), we write

$$I_O\omega_1 + \sum \int_{t_1}^{t_2} M_O \, dt = I_O\omega_2 \tag{17.16}$$

In the general case of plane motion of a rigid body symmetrical with respect to the reference plane, Eq. (17.16) can be used with respect to the instantaneous axis of rotation under certain conditions. It is recommended, however, that all problems of plane motion be solved by the general method described earlier in this section.

---

†Note that the sum $\mathbf{H}_A$ of the moments about an arbitrary point $A$ of the momenta of the particles of a rigid slab is, in general, *not* equal to $I_A\boldsymbol{\omega}$. (See Prob. 17.67.)

## 17.9. SYSTEMS OF RIGID BODIES

The motion of several rigid bodies can be analyzed by applying the principle of impulse and momentum to each body separately (Sample Prob. 17.6). However, in solving problems involving no more than three unknowns (including the impulses of unknown reactions), it is often convenient to apply the principle of impulse and momentum to the system as a whole. The momentum and impulse diagrams are drawn for the entire system of bodies. For each moving part of the system, the diagrams of momenta should include a momentum vector, a momentum couple, or both. Impulses of forces internal to the system can be omitted from the impulse diagram, since they occur in pairs of equal and opposite vectors. Summing and equating successively the $x$ components, $y$ components, and moments of all vectors involved, one obtains three relations which express that the momenta at time $t_1$ and the impulses of the external forces form a system equipollent to the system of the momenta at time $t_2$.† Again, care should be taken not to add linear and angular momenta indiscriminately; each equation should be checked to make sure that consistent units have been used. This approach has been used in Sample Prob. 17.8 and, further on, in Sample Probs. 17.9 and 17.10.

## 17.10. CONSERVATION OF ANGULAR MOMENTUM

When no external force acts on a rigid body or a system of rigid bodies, the impulses of the external forces are zero and the system of the momenta at time $t_1$ is equipollent to the system of the momenta at time $t_2$. Summing and equating successively the $x$ components, $y$ components, and moments of the momenta at times $t_1$ and $t_2$, we conclude that the total linear momentum of the system is conserved in any direction and that its total angular momentum is conserved about any point.

There are many engineering applications, however, in which *the linear momentum is not conserved* yet *the angular momentum* $\mathbf{H}_O$ *of the system about a given point $O$ is conserved* that is, in which

$$(\mathbf{H}_O)_1 = (\mathbf{H}_O)_2 \tag{17.17}$$

Such cases occur when the lines of action of all external forces pass through $O$ or, more generally, when the sum of the angular impulses of the external forces about $O$ is zero.

Problems involving *conservation of angular momentum* about a point $O$ can be solved by the general method of impulse and momentum, that is, by drawing momentum and impulse diagrams as described in Secs. 17.8 and 17.9. Equation (17.17) is then obtained by summing and equating moments about $O$ (Sample Prob. 17.8). As you will see later in Sample Prob. 17.9, two additional equations can be written by summing and equating $x$ and $y$ components and these equations can be used to determine two unknown linear impulses, such as the impulses of the reaction components at a fixed point.

**Photo 17.2** A figure skater at the beginning and at the end of a spin. By using the principle of conservation of angular momentum you will find that her angular velocity is much higher at the end of the spin.

---

†Note that as in Sec. 16.7, we cannot speak of *equivalent* systems since we are not dealing with a single rigid body.

$r_A = 250$ mm

$r_B = 100$ mm

A

B

M

## SAMPLE PROBLEM 17.6

Gear $A$ has a mass of 10 kg and a radius of gyration of 200 mm, and gear $B$ has a mass of 3 kg and a radius of gyration of 80 mm. The system is at rest when a couple $\mathbf{M}$ of magnitude 6 N · m is applied to gear $B$. (These gears were considered in Sample Prob. 17.2.) Neglecting friction, determine (a) the time required for the angular velocity of gear $B$ to reach 600 rpm, (b) the tangential force which gear $B$ exerts on gear $A$.

## SOLUTION

We apply the principle of impulse and momentum to each gear separately. Since all forces and the couple are constant, their impulses are obtained by multiplying them by the unknown time $t$. We recall from Sample Prob. 17.2 that the centroidal moments of inertia and the final angular velocities are

$$\bar{I}_A = 0.400 \text{ kg} \cdot \text{m}^2 \qquad \bar{I}_B = 0.0192 \text{ kg} \cdot \text{m}^2$$
$$(\omega_A)_2 = 25.1 \text{ rad/s} \qquad (\omega_B)_2 = 62.8 \text{ rad/s}$$

**Principle of Impulse and Momentum for Gear $A$.** The systems of initial momenta, impulses, and final momenta are shown in three separate sketches.

$\bar{I}_A(\boldsymbol{\omega}_A)_1 = 0$

$A$

$A_x t$

$A_y t$

$r_A$

$Ft$

$\bar{I}_A(\boldsymbol{\omega}_A)_2$

**Syst Momenta$_1$ + Syst Ext Imp$_{1\to2}$ = Syst Momenta$_2$**

$+\curvearrowleft$ moments about $A$: $\qquad 0 - Ft r_A = -\bar{I}_A(\omega_A)_2$

$$Ft(0.250 \text{ m}) = (0.400 \text{ kg} \cdot \text{m}^2)(25.1 \text{ rad/s})$$
$$Ft = 40.2 \text{ N} \cdot \text{s}$$

**Principle of Impulse and Momentum for Gear $B$.**

$\bar{I}_B(\boldsymbol{\omega}_B)_1 = 0$

$B$

$B_x t$

$B_y t$

$r_B$

$Ft$

$Mt$

$\bar{I}_B(\boldsymbol{\omega}_B)_2$

$B$

**Syst Momenta$_1$ + Syst Ext Imp$_{1\to2}$ = Syst Momenta$_2$**

$+\curvearrowleft$ moments about $B$: $\qquad 0 + Mt - Ft r_B = \bar{I}_B(\omega_B)_2$

$$+(6 \text{ N} \cdot \text{m})t - (40.2 \text{ N} \cdot \text{s})(0.100 \text{ m}) = (0.0192 \text{ kg} \cdot \text{m}^2)(62.8 \text{ rad/s})$$
$$t = 0.871 \text{ s} \quad \blacktriangleleft$$

Recalling that $Ft = 40.2$ N · s, we write

$$F(0.871 \text{ s}) = 40.2 \text{ N} \cdot \text{s} \qquad F = +46.2 \text{ N}$$

Thus, the force exerted by gear $B$ on gear $A$ is $\qquad \mathbf{F} = 46.2 \text{ N} \swarrow \quad \blacktriangleleft$

## SAMPLE PROBLEM 17.7

A uniform sphere of mass $m$ and radius $r$ is projected along a rough horizontal surface with a linear velocity $\bar{\mathbf{v}}_1$ and no angular velocity. Denoting by $\mu_k$ the coefficient of kinetic friction between the sphere and the surface, determine (a) the time $t_2$ at which the sphere will start rolling without sliding, (b) the linear and angular velocities of the sphere at time $t_2$.

## SOLUTION

While the sphere is sliding relative to the surface, it is acted upon by the normal force $\mathbf{N}$, the friction force $\mathbf{F}$, and its weight $\mathbf{W}$ of magnitude $W = mg$.

**Principle of Impulse and Momentum.** We apply the principle of impulse and momentum to the sphere from the time $t_1 = 0$ when it is placed on the surface until the time $t_2 = t$ when it starts rolling without sliding.

$$\text{Syst Momenta}_1 + \text{Syst Ext Imp}_{1 \to 2} = \text{Syst Momenta}_2$$

$+\uparrow y$ components: $\qquad\qquad Nt - Wt = 0 \qquad\qquad (1)$

$\xrightarrow{+} x$ components: $\qquad\qquad m\bar{v}_1 - Ft = m\bar{v}_2 \qquad\qquad (2)$

$+\downarrow$ moments about $G$: $\qquad\qquad Ftr = \bar{I}\omega_2 \qquad\qquad (3)$

From (1) we obtain $N = W = mg$. During the entire time interval considered, sliding occurs at point $C$ and we have $F = \mu_k N = \mu_k mg$. Substituting $\mu_k mg$ for $F$ into (2), we write

$$m\bar{v}_1 - \mu_k mgt = m\bar{v}_2 \qquad\qquad \bar{v}_2 = \bar{v}_1 - \mu_k gt \qquad\qquad (4)$$

Substituting $F = \mu_k mg$ and $\bar{I} = \frac{2}{5}mr^2$ into (3),

$$\mu_k mgtr = \frac{2}{5}mr^2\omega_2 \qquad\qquad \omega_2 = \frac{5}{2}\frac{\mu_k g}{r}t \qquad\qquad (5)$$

The sphere will start rolling without sliding when the velocity $\mathbf{v}_C$ of the point of contact is zero. At that time, point $C$ becomes the instantaneous center of rotation, and we have $\bar{v}_2 = r\omega_2$. Substituting from (4) and (5), we write

$$\bar{v}_2 = r\omega_2 \qquad \bar{v}_1 - \mu_k gt = r\left(\frac{5}{2}\frac{\mu_k g}{r}t\right) \qquad t = \frac{2}{7}\frac{\bar{v}_1}{\mu_k g} \quad \blacktriangleleft$$

Substituting this expression for $t$ into (5),

$$\omega_2 = \frac{5}{2}\frac{\mu_k g}{r}\left(\frac{2}{7}\frac{\bar{v}_1}{\mu_k g}\right) \qquad \omega_2 = \frac{5}{7}\frac{\bar{v}_1}{r} \qquad \omega_2 = \frac{5}{7}\frac{\bar{v}_1}{r}\downarrow \quad \blacktriangleleft$$

$$\bar{v}_2 = r\omega_2 \qquad\qquad \bar{v}_2 = r\left(\frac{5}{7}\frac{v_1}{r}\right) \qquad \bar{\mathbf{v}}_2 = \frac{5}{7}\bar{v}_1 \rightarrow \quad \blacktriangleleft$$

# SAMPLE PROBLEM 17.8

Two solid spheres of radius 7.6 cm, weighing 0.9 kg each, are mounted at $A$ and $B$ on the horizontal rod $A'B'$, which rotates freely about the vertical with a counterclockwise angular velocity of 6 rad/s. The spheres are held in position by a cord which is suddenly cut. Knowing that the centroidal moment of inertia of the rod and pivot is $\bar{I}_R = .35$ kgm$^2$, determine (a) the angular velocity of the rod after the spheres have moved to positions $A'$ and $B'$, (b) the energy lost due to the plastic impact of the spheres and the stops at $A'$ and $B'$.

## SOLUTION

**a. Principle of Impulse and Momentum.** In order to determine the final angular velocity of the rod, we will express that the initial momenta of the various parts of the system and the impulses of the external forces are together equipollent to the final momenta of the system.

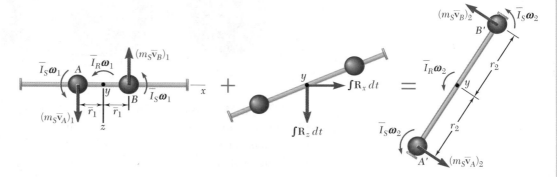

$$\textbf{Syst Momenta}_1 + \textbf{Syst Ext Imp}_{1\to2} = \textbf{Syst Momenta}_2$$

Observing that the external forces consist of the weights and the reaction at the pivot, which have no moment about the $y$ axis, and noting that $\bar{v}_A = \bar{v}_B = \bar{r}\omega$, we equate moments about the $y$ axis:

$$2(m_S\bar{r}_1\omega_1)\bar{r}_1 + 2\bar{I}_S\omega_1 + \bar{I}_R\omega_1 = 2(m_S\bar{r}_2\omega_2)\bar{r}_2 + 2\bar{I}_S\omega_2 + \bar{I}_R\omega_2$$
$$(2m_S\bar{r}_1^2 + 2\bar{I}_S + \bar{I}_R)\omega_1 = (2m_S\bar{r}_2^2 + 2\bar{I}_S + \bar{I}_R)\omega_2 \qquad (1)$$

which expresses that *the angular momentum of the system about the $y$ axis is conserved.* We now compute

$$\bar{I}_S = \tfrac{2}{5}m_Sa^2 = \tfrac{2}{5}(0.9 \text{ kg})(7.6 \text{ cm})^2 = 0.0021 \text{ kgm}^2$$
$$m_S\bar{r}_1^2 = (.9)(.127)^2 = 0.0145 \qquad m_S\bar{r}_2^2 = (.9)(.635)^2 = 0.363$$

Substituting these values, and $\bar{I}_R = .035$ and $\omega_1 = 6$ rad/s into (1):

$$0.0682(6 \text{ rad/s}) = .765\omega_2 \qquad \omega_2 = 0.535 \text{ rad/s} \uparrow \quad \blacktriangleleft$$

**b. Energy Lost.** The kinetic energy of the system at any instant is

$$T = 2(\tfrac{1}{2}m_S\bar{v}^2 + \tfrac{1}{2}\bar{I}_S\omega^2) + \tfrac{1}{2}\bar{I}_R\omega^2 = \tfrac{1}{2}(2m_S\bar{r}^2 + 2\bar{I}_S + \bar{I}_R)\omega^2$$

Recalling the numerical values found above, we have

$$T_1 = \tfrac{1}{2}(.0682)(6)^2 = 1.2276 \text{ N} \cdot \text{m} \qquad T_2 = \tfrac{1}{2}(.765)(.535)^2 = .1095 \text{ N} \cdot \text{m}$$
$$\Delta T = T_2 - T_1 = .1095 - 1.2276 \qquad \Delta T = -1.1181 \text{ N} \cdot \text{m} \quad \blacktriangleleft$$

# SOLVING PROBLEMS
# ON YOUR OWN

In this lesson you learned to use the method of impulse and momentum to solve problems involving the plane motion of rigid bodies. As you found out previously in Chap. 13, this method is most effective when used in the solution of problems involving velocities and time.

1. **The principle of impulse and momentum for the plane motion of a rigid body** is expressed by the following vector equation:

$$\textbf{Syst Momenta}_1 + \textbf{Syst Ext Imp}_{1\to2} = \textbf{Syst Momenta}_2 \quad (17.14)$$

where **Syst Momenta** represents the system of the momenta of the particles forming the rigid body, and **Syst Ext Imp** represents the system of all the external impulses exerted during the motion.

   *a. The system of the momenta of a rigid body* is equivalent to a linear momentum vector $m\bar{\textbf{v}}$ attached at the mass center of the body and an angular momentum couple $\bar{I}\boldsymbol{\omega}$ (Fig. 17.7).

   *b. You should draw a free-body-diagram equation for the rigid body* to express graphically the above vector equation. Your diagram equation will consist of three sketches of the body, representing respectively the initial momenta, the impulses of the external forces, and the final momenta. It will show that the system of the initial momenta and the system of the impulses of the external forces are together equivalent to the system of the final momenta (Fig. 17.8).

   *c. By using the free-body-diagram equation,* you can sum components in any direction and sum moments about any point. When summing moments about a point, remember to include the *angular momentum* $\bar{I}\omega$ of the body, as well as the *moments* of the components of its *linear momentum*. In most cases you will be able to select and solve an equation that involves only one unknown. This was done in all the sample problems of this lesson.

2. **In problems involving a system of rigid bodies,** you can apply the principle of impulse and momentum to the system as a whole. Since internal forces occur in equal and opposite pairs, they will not be part of your solution [Sample Prob. 17.8].

3. *Conservation of angular momentum about a given axis* occurs when, for a system of rigid bodies, *the sum of the moments of the external impulses about that axis is zero.* You can indeed easily observe from the free-body-diagram equation that the initial and final angular momenta of the system about that axis are equal and, thus, that *the angular momentum of the system about the given axis is conserved.* You can then sum the angular momenta of the various bodies of the system and the moments of their linear momenta about that axis to obtain an equation which can be solved for one unknown [Sample Prob. 17.8].

**Fig. P17.45**

**Fig. P17.46 and P17.47**

**Fig. P17.48 and P17.49**

**Fig. P17.50**

**17.43** A 181 kg flywheel is at rest when a constant 24.4 N · m couple is applied. It is observed that 4.3 min is required for the flywheel to reach its rated speed of 2400 rpm. Knowing that the radius of gyration of the flywheel is 35.6 cm, determine the average magnitude of the couple due to kinetic friction in the bearing.

**17.44** A 250-kg flywheel is at rest when a constant 300 N · m couple is applied at time $t = 0$. At $t = 28$ s the flywheel reaches its rated speed of 3000 rpm and the couple is removed. Assuming that the kinetic friction results in a constant couple of magnitude 12.5 N · m, determine (a) the radius of gyration of the flywheel, (b) the time at which the flywheel comes to rest.

**17.45** A bolt located 5 cm from the center of an automobile wheel is tightened by applying the couple shown for 0.10 s. Assuming that the wheel is free to rotate and is initially at rest, determine the resulting angular velocity of the wheel. The wheel weighs 19 kg and has a radius of gyration of 27.4 cm.

**17.46** A uniform 65.3 kg cube is attached to a uniform 61.7 kg circular shaft as shown and a couple **M** of constant magnitude is applied to the shaft when the system is at rest. Knowing that $r = 10.2$ cm, $L = 30.5$ cm, and the angular velocity of the system is 960 rpm after 4 s, determine the magnitude of the couple **M**.

**17.47** A uniform 75-kg cube is attached to a uniform 70-kg circular shaft as shown and a couple **M** of constant magnitude 20 N · m is applied to the shaft. Knowing that $r = 100$ mm and $L = 300$ mm, determine the time required for the angular velocity of the system to increase from 1000 rpm to 2000 rpm.

**17.48** A disk of constant thickness, initially at rest, is placed in contact with a belt that moves with a constant velocity **v**. Denoting by $\mu_k$ the coefficient of kinetic friction between the disk and the belt, derive an expression for the time required for the disk to reach a constant angular velocity.

**17.49** Disk A, of mass 2 kg and radius $r = 60$ mm, is at rest when it is placed in contact with a belt which moves at a constant speed $v = 15$ m/s. Knowing that $\mu_k = 0.20$ between the disk and the belt, determine the time required for the disk to reach a constant angular velocity.

**17.50** A sphere of radius $r$ and weight $W$ with an initial clockwise angular velocity $\boldsymbol{\omega}_0$ is placed in the corner formed by the floor and a vertical wall. Denoting by $\mu_k$ the coefficient of kinetic friction at $A$ and $B$, derive an expression for the time required for the sphere to come to rest.

**17.51** Two identical uniform cylinders of mass $m$ and radius $r$ are at rest at time $t = 0$ when a couple **M** of constant magnitude M $<$ $mgr$ is applied to cylinder $A$. Knowing that the coefficient of kinetic friction between cylinder $B$ and the horizontal surface is $\mu_k < \dfrac{M}{2mgr}$ and that no slipping occurs between the two cylinders, derive an expression for the angular velocity of cylinder $B$ at time $t$.

**Fig. P17.51 and P17.52**

**17.52** Two identical 7.3 kg uniform cylinders of radius $r = 10.2$ cm are at rest when a couple **M** of constant magnitude 5.4 N · m is applied to cylinder $A$. Slipping occurs between the two cylinders and between cylinder $B$ and the horizontal surface. Knowing that the coefficient of kinetic friction is 0.5 between the two cylinders and 0.2 between cylinder $B$ and the horizontal surface, determine the angular velocity of each cylinder after 5 s.

**17.53** Each of the double pulleys shown has a centroidal mass moment of inertia of 0.25 kg · m², an inner radius of 100 mm, and an outer radius of 150 mm. Neglecting bearing friction, determine (*a*) the velocity of the cylinder 3 s after the system is released from rest, (*b*) the tension in the cord connecting the pulleys.

**17.54** Each of the gears $A$ and $B$ has a mass of 675 g and has a radius of gyration of 40 mm, while gear $C$ has a mass of 3.6 kg and has a radius of gyration of 100 mm. Assume that kinetic friction in the bearings of gears $A$, $B$, and $C$ produces couples of constant magnitude 0.15 N · m, 0.15 N · m, and 0.3 N · m, respectively. Knowing that the initial angular velocity of gear $C$ is 2000 rpm, determine the time required for the system to come to rest.

**17.55** A computer tape moves over the two drums shown. Drum $A$ weighs .64 kg and has a radius of gyration of 1.9 cm, while drum $B$ weighs 1.6 kg and has a radius of gyration of 3.2 cm. In the lower portion of the tape the tension is constant and equal to $T_A = .34$ kg. Knowing that the tape is initially at rest, determine (*a*) the required constant tension $T_B$ if the velocity of the tape is to be $v = 3$ m/s after 0.24 s, (*b*) the corresponding tension in the portion of tape between the drums.

**17.56** Show that the system of momenta for a rigid slab in plane motion reduces to a single vector, and express the distance from the mass center $G$ to the line of action of this vector in terms of the centroidal radius of gyration $\bar{k}$ of the slab, the magnitude $\bar{v}$ of the velocity of $G$, and the angular velocity $\omega$.

**Fig. P17.53**

**Fig. P17.54**

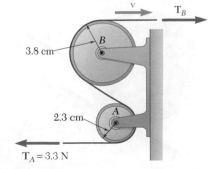

**Fig. P17.55**

**17.57** Show that, when a rigid slab rotates about a fixed axis through $O$ perpendicular to the slab, the system of the momenta of its particles is equivalent to a single vector of magnitude $m\bar{r}\omega$, perpendicular to the line $OG$, and applied to a point $P$ on this line, called the *center of percussion*, at a distance $GP = \bar{k}^2/\bar{r}$ from the mass center of the slab.

**Fig. P17.57**

**17.58** Show that the sum $\mathbf{H}_A$ of the moments about a point $A$ of the momenta of the particles of a rigid slab in plane motion is equal to $I_A\boldsymbol{\omega}$, where $\boldsymbol{\omega}$ is the angular velocity of the slab at the instant considered and $I_A$ is the moment of inertia of the slab about $A$, if and only if one of the following conditions is satisfied: ($a$) $A$ is the mass center of the slab, ($b$) $A$ is the instantaneous center of rotation, ($c$) the velocity of $A$ is directed along a line joining point $A$ and the mass center $G$.

**17.59** Consider a rigid slab initially at rest and subjected to an impulsive force $\mathbf{F}$ contained in the plane of the slab. We define the *center of percussion* $P$ as the point of intersection of the line of action of $\mathbf{F}$ with the perpendicular drawn from $G$. ($a$) Show that the instantaneous center of rotation $C$ of the slab is located on line $GP$ at a distance $GC = \bar{k}^2/GP$ on the opposite side of $G$. ($b$) Show that if the center of percussion were located at $C$ the instantaneous center of rotation would be located at $P$.

**Fig. P17.59**

**17.60** A flywheel is rigidly attached to a 38-mm-radius shaft that rolls without sliding along parallel rails. Knowing that after being released from rest the system attains a speed of 152 mm/s in 30 s, determine the centroidal radius of gyration of the system.

**Fig. P17.60**

**17.61** A drum of 100-mm radius is attached to a disk of 200-mm radius. The disk and drum have a combined mass of 5 kg and a combined radius of gyration of 150 mm. A cord is attached to the drum at $A$ and pulled with a constant force $\mathbf{P}$ of magnitude 25 N. Knowing that the disk rolls without sliding and that its initial angular velocity is 10 rad/s clockwise, determine ($a$) the angular velocity of the disk after 5 s, ($b$) the corresponding total impulse of the friction force exerted on the disk at $B$.

**Fig. P17.61**

**17.62** Cords are wrapped around a thin-walled pipe and a solid cylinder as shown. Knowing that the pipe and the cylinder are each released from rest at time $t = 0$, determine at time $t$ the velocity of the center of (a) the pipe, (b) the cylinder.

**17.63** A 30.5 cm-radius cylinder of weight 7.3 kg rests on a 2.7 kg carriage. The system is at rest when a force **P** of magnitude 11.2 N is applied as shown for 1.2 s. Knowing that the cylinder rolls without sliding on the carriage and neglecting the mass of the wheels of the carriage, determine the resulting velocity of (a) the carriage, (b) the center of the cylinder.

**17.64** The bar $AB$ of negligible mass is attached by pins to two disks and the system is released from rest in the position shown. Disk $A$ has a weight of 5.4 kg and a radius of gyration of 9 cm. Disk $B$ has a weight of 2.7 kg and a radius of gyration of 8.1 cm. Assuming that disk $A$ does not reach the corner $D$, determine (a) the velocity of the bar $AB$ after 0.6 s, (b) the corresponding total impulse of the force exerted by disk $A$ on the bar.

(a)       (b)

**Fig. P17.62**

**Fig. P17.63**

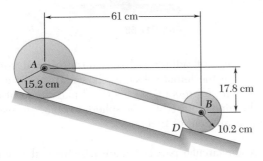

**Fig. *P17.64* and *P17.65***

**17.65** The 4.5 kg uniform bar $AB$ is attached by pins to two uniform disks and the system is released from rest in the position shown. The weight of disk $A$ is 5.4 kg and that of disk $B$ is 2.7 kg. Assuming that disk $A$ does not reach the corner $D$, determine the velocity of bar $AB$ after 0.6 s.

**17.66** A 15-kg double pulley has a radius of gyration of 125 mm and is attached to a 10-kg slider block by a pin at point $G$. The system is at rest when constant forces $P_A$ and $P_B$ are applied to the cords as shown. Knowing that after 2 s the velocity of point $A$ is 3.3 m/s to the left and the velocity of point $B$ is 5.7 m/s to the right, determine the magnitudes of the forces $P_A$ and $P_B$.

**Fig. P17.66**

**Fig. P17.67**

**17.67** A 160-mm-diameter pipe of mass 6 kg rests on a 1.5-kg plate. The pipe and plate are initially at rest when a force **P** of magnitude 25 N is applied for 0.75 s. Knowing that $\mu_s = 0.25$ and $\mu_k = 0.20$ between the plate and both the pipe and the floor, determine (a) whether the pipe slides with respect to the plate, (b) the resulting velocities of the pipe and of the plate.

**17.68** A sphere of radius $r$ and mass $m$ is placed on a horizontal floor with no linear velocity but with a clockwise angular velocity $\boldsymbol{\omega}_0$. Denoting by $\mu_k$ the coefficient of kinetic friction between the sphere and the floor, determine (a) the time $t_1$ at which the sphere will start rolling without sliding, (b) the linear and angular velocities of the sphere at time $t_1$.

**Fig. P17.68**

**Fig. P17.69**

**17.69** A sphere of radius $r$ and mass $m$ is projected along a rough horizontal surface with the initial velocities shown. If the final velocity of the sphere is to be zero, express (a) the required magnitude of $\boldsymbol{\omega}_0$ in terms of $v_0$ and $r$, (b) the time required for the sphere to come to rest in terms of $v_0$ and the kinetic coefficient of friction $\mu_k$.

**17.70** A semicircular panel of radius $r$ is attached with hinges to a circular plate of radius $r$ and initially held in the vertical position as shown. The plate and the panel are made of the same material and have the same thickness. Knowing that the entire assembly is rotating freely with an initial angular velocity $\boldsymbol{\omega}_0$, determine the angular velocity of the assembly after the panel has been released and comes to rest against the plate.

**Fig. P17.70**

**17.71** A 1.6-kg tube $AB$ can slide freely on rod $DE$ which in turn can rotate freely in a horizontal plane. Initially the assembly is rotating with an angular velocity $\omega = 5$ rad/s and the tube is held in position by a cord. The moment of inertia of the rod and bracket about the vertical axis of rotation is 0.30 kg · m² and the centroidal moment of inertia of the tube about a vertical axis is 0.0025 kg · m². If the cord suddenly breaks, determine (a) the angular velocity of the assembly after the tube has moved to end $E$, (b) the energy lost during the plastic impact at $E$.

**Fig. P17.71**

**17.72** Two 4.5 kg disks and a small motor are mounted on a 6.8 kg rectangular platform which is free to rotate about a central vertical spindle. The normal operating speed of the motor is 180 rpm. If the motor is started when the system is at rest, determine the angular velocity of all elements of the system after the motor has attained its normal operating speed. Neglect the mass of the motor and of the belt.

**17.73** A 4.5 kg disk is attached to the shaft of a motor mounted on arm *AB* which is free to rotate about the vertical axle *CD*. The arm-and-motor unit has a moment of inertia of 0.0044 kgm² with respect to axle *CD*, and the normal operating speed of the motor is 360 rpm. Knowing that the system is initially at rest, determine the angular velocities of the arm and of the disk when the motor reaches a speed of 360 rpm.

**Fig. P17.72**

**Fig. P17.73**

**17.74** Two 0.36-kg balls are put successively into the center *C* of the slender 1.8-kg tube *AB*. Knowing that when the first ball is put into the tube the initial angular velocity of the tube is 8 rad/s and neglecting the effect of friction, determine the angular velocity of the tube just after (*a*) the first ball has left the tube, (*b*) the second ball has left the tube.

**17.75** The 30-kg uniform disk *A* and the bar *BC* are at rest and the 5-kg uniform disk *D* has an initial angular velocity $\omega_1$ of magnitude 440 rpm when the compressed spring is released and disk *D* contacts disk *A*. The system rotates freely about the vertical spindle *BE*. After a period of slippage, disk *D* rolls without slipping. Knowing that the magnitude of the final angular velocity of disk *D* is 176 rpm, determine the final angular velocities of bar *BC* and disk *A*. Neglect the mass of bar *BC*.

**Fig. P17.74**

**Fig. P17.75**

11.4 cm

11.4 cm

A

23 cm

B

Motor

C

**Fig. P17.76**

**17.76** The 3.6 kg disk $B$ is attached to the shaft of a motor mounted on plate $A$, which can rotate freely about the vertical shaft $C$. The motor-plate-shaft unit has a moment of inertia of 0.02 kgm² with respect to the axis of the shaft. If the motor is started when the system is at rest, determine the angular velocities of the disk and of the plate after the motor has attained its normal operating speed of 360 rpm.

**17.77** A 0.54 kg slender bar $B$ fits inside a slot cut in a horizontal triangular plate which is free to rotate about a vertical axis through point $O$. Initially the angular velocity of the plate is 10 rad/s clockwise and the bar is at rest relative to the plate in the position shown. After the bar is given a slight nudge it moves to the end of the slot. Knowing that the mass moment of inertia of the plate about $O$ is 0.03 kgm², determine the angular velocity of the system after the plastic impact of the bar with the end of the slot.

A

10.2 cm

10.2 cm

B

O  40.6 cm  45°  D

**Fig. P17.77**

**17.78** The 4 kg rod $AB$ can slide freely inside the 6-kg tube $CD$. The rod was entirely within the tube ($x = 0$) and released with no initial velocity relative to the tube when the angular velocity of the assembly was 5 rad/s. Neglecting the effect of friction, determine the speed of the rod relative to the tube when $x = 400$ mm.

**17.79** A small 3-kg collar $C$ can slide freely on a thin ring of mass 4.5 kg and radius 325 mm. The ring is welded to a short vertical shaft, which can rotate freely in a fixed bearing. Initially the ring has an angular velocity of 35 rad/s and the collar is at the top of the ring ($\theta = 0$) when it is given a slight nudge. Neglecting the effect of friction, determine (*a*) the angular velocity of the ring as the collar passes through the position $\theta = 90°$, (*b*) the corresponding velocity of the collar relative to the ring.

800 mm

D

$\omega$

B

x

A

800 mm

C

**Fig. P17.78**

C

$\theta$

R

**Fig. P17.79**

**17.80**  A 1.6 kg collar $A$ and a 0.64 kg collar $B$ can slide without friction on a frame, consisting of the horizontal rod $OE$ and the vertical rod $CD$, which is free to rotate about its vertical axis of symmetry. The two collars are connected by a cord running over a pulley that is attached to the frame at $O$. At the instant shown, the velocity $\mathbf{v}_A$ of collar $A$ has a magnitude of 1.9 m/s and a stop prevents collar $B$ from moving. The stop is suddenly removed and collar $A$ moves toward $E$. As it reaches a distance of 12.2 cm from $O$, the magnitude of its velocity is observed to be 2.3 m/s. Determine at that instant the magnitude of the angular velocity of the frame and the moment of inertia of the frame and pulley system about $CD$.

**17.81**  Collar $C$ has a weight of 8.2 kg and can slide freely on rod $AB$, which in turn can rotate freely in a horizontal plane. The assembly is rotating with an angular velocity $\boldsymbol{\omega}$ of 1.5 rad/s when a spring located between $A$ and $C$ is released, projecting the collar along the rod with an initial relative speed of $v_r = 1.4$ m/s. Knowing that the combined mass moment of inertia about $B$ of the rod and spring is 0.116 kgm², determine (a) the minimum distance between the collar and point $B$ in the ensuing motion, (b) the corresponding angular velocity of the assembly.

**Fig. P17.80**

**Fig. P17.81**

**17.82**  In Prob. 17.71 determine the velocity of the tube relative to the rod as the tube strikes end $E$ of the assembly.

**17.83**  In Prob. 17.74, determine the velocity of each ball relative to the tube as it leaves the tube.

**17.84**  A 2.7 kg uniform cylinder $A$ can roll without sliding on a 4.5 kg cart $C$ and is attached to a spring $AB$ of constant $k = 103.6$ N/m as shown. The system is released from rest when the spring is stretched 2 cm. Neglecting wheel friction, determine the velocity of the cart and the angular velocity of the cylinder when the spring first reaches its undeformed state.

**17.85**  Rod $AB$ has a weight of 2.7 kg and is attached to a 4.5 kg cart $C$. Knowing that the system is released from rest in the position shown and neglecting friction, determine (a) the velocity of point $B$ as rod $AB$ passes through a vertical position, (b) the corresponding velocity of the cart $C$.

**Fig. P17.84**

**Fig. P17.85**

## 17.11. IMPULSIVE MOTION

You saw in Chap. 13 that the method of impulse and momentum is
the only practicable method for the solution of problems involving
the impulsive motion of a particle. Now you will find that problems
involving the impulsive motion of a rigid body are particularly well
suited to a solution by the method of impulse and momentum. Since
the time interval considered in the computation of linear impulses
and angular impulses is very short, the bodies involved can be assumed to occupy the same position during that time interval, making
the computation quite simple.

## 17.12. ECCENTRIC IMPACT

In Secs. 13.13 and 13.14, you learned to solve problems of *central impact,* that is, problems in which the mass centers of the two colliding
bodies are located on the line of impact. You will now analyze the *eccentric impact* of two rigid bodies. Consider two bodies which collide,
and denote by $\mathbf{v}_A$ and $\mathbf{v}_B$ the velocities before impact *of the two points
of contact A and B* (Fig. 17.10*a*). Under the impact, the two bodies

(*a*)

**Fig. 17.10**

(*b*)

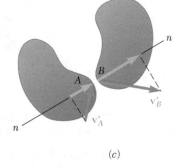

(*c*)

will *deform,* and at the end of the period of deformation, the velocities $\mathbf{u}_A$ and $\mathbf{u}_B$ of A and B will have equal components along the line
of impact *nn* (Fig. 17.10*b*). A period of *restitution* will then take place,
at the end of which A and B will have velocities $\mathbf{v}'_A$ and $\mathbf{v}'_B$ (Fig. 17.10*c*).
Assuming that the bodies are frictionless, we find that the forces they
exert on each other are directed along the line of impact. Denoting
the magnitude of the impulse of one of these forces during the period of deformation by $\int P \, dt$ and the magnitude of its impulse during the period of restitution by $\int R \, dt$, we recall that the coefficient
of restitution $e$ is defined as the ratio

$$e = \frac{\int R \, dt}{\int P \, dt} \tag{17.18}$$

We propose to show that the relation established in Sec. 13.13 between the relative velocities of two particles before and after impact
also holds between the components along the line of impact of the

relative velocities of the two points of contact $A$ and $B$. We propose to show, therefore, that

$$(v'_B)_n - (v'_A)_n = e[(v_A)_n - (v_B)_n] \tag{17.19}$$

It will first be assumed that the motion of each of the two colliding bodies of Fig. 17.10 is unconstrained. Thus the only impulsive forces exerted on the bodies during the impact are applied at $A$ and $B$, respectively. Consider the body to which point $A$ belongs and draw the three momentum and impulse diagrams corresponding to the period of deformation (Fig. 17.11). We denote by $\bar{\mathbf{v}}$ and $\bar{\mathbf{u}}$, respectively,

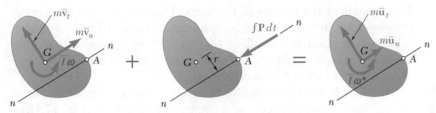

**Fig. 17.11**

the velocity of the mass center at the beginning and at the end of the period of deformation, and we denote by $\boldsymbol{\omega}$ and $\boldsymbol{\omega}^\circ$ the angular velocity of the body at the same instants. Summing and equating the components of the momenta and impulses along the line of impact $nn$, we write

$$m\bar{v}_n - \int P\,dt = m\bar{u}_n \tag{17.20}$$

Summing and equating the moments about $G$ of the momenta and impulses, we also write

$$\bar{I}\omega - r\int P\,dt = \bar{I}\omega^\circ \tag{17.21}$$

where $r$ represents the perpendicular distance from $G$ to the line of impact. Considering now the period of restitution, we obtain in a similar way

$$m\bar{u}_n - \int R\,dt = m\bar{v}'_n \tag{17.22}$$
$$\bar{I}\omega^\circ - r\int R\,dt = \bar{I}\omega' \tag{17.23}$$

where $\bar{\mathbf{v}}'$ and $\boldsymbol{\omega}'$ represent, respectively, the velocity of the mass center and the angular velocity of the body after impact. Solving (17.20) and (17.22) for the two impulses and substituting into (17.18), and then solving (17.21) and (17.23) for the same two impulses and substituting again into (17.18), we obtain the following two alternative expressions for the coefficient of restitution:

$$e = \frac{\bar{u}_n - \bar{v}'_n}{\bar{v}_n - \bar{u}_n} \qquad e = \frac{\omega^\circ - \omega'}{\omega - \omega^\circ} \tag{17.24}$$

Multiplying by $r$ the numerator and denominator of the second expression obtained for $e$, and adding respectively to the numerator and denominator of the first expression, we have

$$e = \frac{\bar{u}_n + r\omega^* - (\bar{v}_n' + r\omega')}{\bar{v}_n + r\omega - (\bar{u}_n + r\omega^*)} \tag{17.25}$$

Observing that $\bar{v}_n + r\omega$ represents the component $(v_A)_n$ along $nn$ of the velocity of the point of contact $A$ and that, similarly, $\bar{u}_n + r\omega^*$ and $\bar{v}_n' + r\omega'$ represent, respectively, the components $(u_A)_n$ and $(v_A')_n$, we write

$$e = \frac{(u_A)_n - (v_A')_n}{(v_A)_n - (u_A)_n} \tag{17.26}$$

The analysis of the motion of the second body leads to a similar expression for $e$ in terms of the components along $nn$ of the successive velocities of point $B$. Recalling that $(u_A)_n = (u_B)_n$, and eliminating these two velocity components by a manipulation similar to the one used in Sec. 13.13, we obtain relation (17.19).

If one or both of the colliding bodies is constrained to rotate about a fixed point $O$, as in the case of a compound pendulum (Fig. 17.12$a$), an impulsive reaction will be exerted at $O$ (Fig. 17.12$b$). Let us ver-

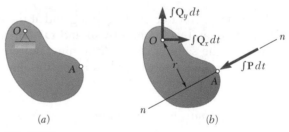

(a)  (b)

**Fig. 17.12**

**Photo 17.3** The method of impulse and momentum is used in the analysis of data from a pendulum impact test. The system shown was developed to test for compliance with safety standards for seat backs, instrument panels, and other vehicle interior components.

ify that while their derivation must be modified, Eqs. (17.26) and (17.19) remain valid. Applying formula (17.16) to the period of deformation and to the period of restitution, we write

$$I_O\omega - r\!\int\! P\,dt = I_O\omega^* \tag{17.27}$$
$$I_O\omega^* - r\!\int\! R\,dt = I_O\omega' \tag{17.28}$$

where $r$ represents the perpendicular distance from the fixed point $O$ to the line of impact. Solving (17.27) and (17.28) for the two impulses and substituting into (17.18), and then observing that $r\omega$, $r\omega^*$, and $r\omega'$ represent the components along $nn$ of the successive velocities of point $A$, we write

$$e = \frac{\omega^* - \omega'}{\omega - \omega^*} = \frac{r\omega^* - r\omega'}{r\omega - r\omega^*} = \frac{(u_A)_n - (v_A')_n}{(v_A)_n - (u_A)_n}$$

and check that Eq. (17.26) still holds. Thus Eq. (17.19) remains valid when one or both of the colliding bodies is constrained to rotate about a fixed point $O$.

In order to determine the velocities of the two colliding bodies after impact, relation (17.19) should be used in conjunction with one or several other equations obtained by applying the principle of impulse and momentum (Sample Prob. 17.10).

# SAMPLE PROBLEM 17.9

A 0.02 kg bullet $B$ is fired with a horizontal velocity of 457.2 m/s into the side of a 9 kg square panel suspended from a hinge at $A$. Knowing that the panel is initially at rest, determine (a) the angular velocity of the panel immediately after the bullet becomes embedded, (b) the impulsive reaction at $A$, assuming that the bullet becomes embedded in 0.0006 s.

## SOLUTION

**Principle of Impulse and Momentum.** We consider the bullet and the panel as a single system and express that the initial momenta of the bullet and panel and the impulses of the external forces are together equipollent to the final momenta of the system. Since the time interval $\Delta t = 0.0006$ s is very short, we neglect all nonimpulsive forces and consider only the external impulses $\mathbf{A}_x \, \Delta t$ and $\mathbf{A}_y \, \Delta t$.

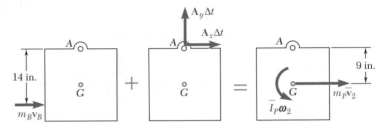

$$\textbf{Syst Momenta}_1 + \textbf{Syst Ext Imp}_{1\to2} = \textbf{Syst Momenta}_2$$

$+\uparrow$ moments about $A$: $\qquad m_B v_B(0.36 \text{ m}) + 0 = m_P \bar{v}_2(0.23 \text{ m}) + \bar{I}_P \omega_2 \quad (1)$

$\xrightarrow{+} x$ components: $\qquad\qquad m_B v_B + A_x \, \Delta t = m_P \bar{v}_2 \qquad\qquad\qquad (2)$

$+\uparrow y$ components: $\qquad\qquad\qquad 0 + A_y \, \Delta t = 0 \qquad\qquad\qquad\qquad (3)$

The centroidal mass moment of inertia of the square panel is

$$\bar{I}_P = \tfrac{1}{6} m_P b^2 = \frac{1}{6}(9 \text{ kg})(45.7 \text{ m})^2 = 3132.7 \text{ kgm}^2$$

Substituting this value as well as the given data into (1) and noting that

$$\bar{v}_2 = (.23 \text{ m})\omega_2$$

we write

$$(0.02)(457.2)(.36) = 3132.7\omega_2 + (9)(.23\omega_2)(.23)$$

$$\omega_2 = 10^{-3} \text{ rad/s} \qquad\qquad \omega_2 = 10^{-3} \text{ rad/s} \uparrow \quad \blacktriangleleft$$
$$\bar{v}_2 = (.23)\omega_2 = (.23)(10^{-3} \text{ rad/s}) = 2.4 \times 10^{-4} \text{m/s}$$

Substituting $\bar{v}_2 = 2.3 \times 10^{-4}$ m/s, $\Delta t = 0.0006$ s, and the given data into Eq. (2), we have

$$(.02)(457.2) + A_x(0.0006) = (9)(2.3 \times 10^{-4})$$

$$A_x = -15236.4 \text{ N} \qquad A_x = 15236.4 \text{ N} \leftarrow \quad \blacktriangleleft$$

From Eq. (3), we find $\qquad\qquad A_y = 0 \qquad\qquad\qquad A_y = 0 \quad \blacktriangleleft$

# SAMPLE PROBLEM 17.10

A 2 kg sphere moving horizontally to the right with an initial velocity of 5 m/s strikes the lower end of an 8-kg rigid rod $AB$. The rod is suspended from a hinge at $A$ and is initially at rest. Knowing that the coefficient of restitution between the rod and the sphere is 0.80, determine the angular velocity of the rod and the velocity of the sphere immediately after the impact.

## SOLUTION

**Principle of Impulse and Momentum.** We consider the rod and sphere as a single system and express that the initial momenta of the rod and sphere and the impulses of the external forces are together equipollent to the final momenta of the system. We note that the only impulsive force external to the system is the impulsive reaction at $A$.

$$\textbf{Syst Momenta}_1 + \textbf{Syst Ext Imp}_{1 \to 2} = \textbf{Syst Momenta}_2$$

$+\curvearrowleft$ moments about $A$:

$$m_s v_s (1.2 \text{ m}) = m_s v_s'(1.2 \text{ m}) + m_R \bar{v}_R' \,(0.6 \text{ m}) + \bar{I}\omega' \qquad (1)$$

Since the rod rotates about $A$, we have $\bar{v}_R' = \bar{r}\omega' = (0.6 \text{ m})\omega'$. Also,

$$\bar{I} = \tfrac{1}{12}mL^2 = \tfrac{1}{12}(8 \text{ kg})(1.2 \text{ m})^2 = 0.96 \text{ kg} \cdot \text{m}^2$$

Substituting these values and the given data into Eq. (1), we have

$$(2 \text{ kg})(5 \text{ m/s})(1.2 \text{ m}) = (2 \text{ kg})v_s'(1.2 \text{ m}) + (8 \text{ kg})(0.6 \text{ m})\omega'(0.6 \text{ m})$$
$$+ \; (0.96 \text{ kg} \cdot \text{m}^2)\omega'$$
$$12 = 2.4v_s' + 3.84\omega' \qquad (2)$$

**Relative Velocities.** Choosing positive to the right, we write

$$v_B' - v_s' = e(v_s - v_B)$$

Substituting $v_s = 5$ m/s, $v_B = 0$, and $e = 0.80$, we obtain

$$v_B' - v_s' = 0.80(5 \text{ m/s}) \qquad (3)$$

Again noting that the rod rotates about $A$, we write

$$v_B' = (1.2 \text{ m})\omega' \qquad (4)$$

Solving Eqs. (2) to (4) simultaneously, we obtain

$$\omega' = 3.21 \text{ rad/s} \qquad\qquad \omega' = 3.21 \text{ rad/s} \;\curvearrowleft \quad \blacktriangleleft$$
$$v_s' = -0.143 \text{ m/s} \qquad\qquad \mathbf{v}_s' = 0.143 \text{ m/s} \leftarrow \quad \blacktriangleleft$$

## SAMPLE PROBLEM 17.11

A square package of side $a$ and mass $m$ moves down a conveyor belt $A$ with a constant velocity $\bar{\mathbf{v}}_1$. At the end of the conveyor belt, the corner of the package strikes a rigid support at $B$. Assuming that the impact at $B$ is perfectly plastic, derive an expression for the smallest magnitude of the velocity $\bar{\mathbf{v}}_1$ for which the package will rotate about $B$ and reach conveyor belt $C$.

## SOLUTION

**Principle of Impulse and Momentum.** Since the impact between the package and the support is perfectly plastic, the package rotates about $B$ during the impact. We apply the principle of impulse and momentum to the package and note that the only impulsive force external to the package is the impulsive reaction at $B$.

$$\textbf{Syst Momenta}_1 + \textbf{Syst Ext Imp}_{1\rightarrow2} = \textbf{Syst Momenta}_2$$

$+\!\uparrow$ moments about $B$: $\qquad (m\bar{v}_1)(\tfrac{1}{2}a) + 0 = (m\bar{v}_2)(\tfrac{1}{2}\sqrt{2}a) + \bar{I}\omega_2 \qquad (1)$

Since the package rotates about $B$, we have $\bar{v}_2 = (GB)\omega_2 = \tfrac{1}{2}\sqrt{2}a\omega_2$. We substitute this expression, together with $\bar{I} = \tfrac{1}{6}ma^2$, into Eq. (1):

$$(m\bar{v}_1)(\tfrac{1}{2}a) = m(\tfrac{1}{2}\sqrt{2}a\omega_2)(\tfrac{1}{2}\sqrt{2}a) + \tfrac{1}{6}ma^2\omega_2 \qquad \bar{v}_1 = \tfrac{4}{3}a\omega_2 \qquad (2)$$

**Principle of Conservation of Energy.** We apply the principle of conservation of energy between position 2 and position 3.

Position 2

$GB = \tfrac{1}{2}\sqrt{2}a = 0.707a$

$h_2 = GB \sin (45° + 15°)$

$\qquad = 0.612a$

**Position 2.** $V_2 = Wh_2$. Recalling that $\bar{v}_2 = \tfrac{1}{2}\sqrt{2}a\omega_2$, we write

$$T_2 = \tfrac{1}{2}m\bar{v}_2^2 + \tfrac{1}{2}\bar{I}\omega_2^2 = \tfrac{1}{2}m(\tfrac{1}{2}\sqrt{2}a\omega_2)^2 + \tfrac{1}{2}(\tfrac{1}{6}ma^2)\omega_2^2 = \tfrac{1}{3}ma^2\omega_2^2$$

**Position 3.** Since the package must reach conveyor belt $C$, it must pass through position 3 where $G$ is directly above $B$. Also, since we wish to determine the smallest velocity for which the package will reach this position, we choose $\bar{v}_3 = \omega_3 = 0$. Therefore $T_3 = 0$ and $V_3 = Wh_3$.

Position 3

$h_3 = GB = 0.707a$

**Conservation of Energy**

$$T_2 + V_2 = T_3 + V_3$$
$$\tfrac{1}{3}ma^2\omega_2^2 + Wh_2 = 0 + Wh_3$$
$$\omega_2^2 = \frac{3W}{ma^2}(h_3 - h_2) = \frac{3g}{a^2}(h_3 - h_2) \qquad (3)$$

Substituting the computed values of $h_2$ and $h_3$ into Eq. (3), we obtain

$$\omega_2^2 = \frac{3g}{a^2}(0.707a - 0.612a) = \frac{3g}{a^2}(0.095a) \qquad \omega_2 = \sqrt{0.285\, g/a}$$

$$\bar{v}_1 = \tfrac{4}{3}a\omega_2 = \tfrac{4}{3}a\sqrt{0.285g/a} \qquad \bar{v}_1 = 0.712\sqrt{ga} \quad \blacktriangleleft$$

# SOLVING PROBLEMS
# ON YOUR OWN

This lesson was devoted to the *impulsive motion* and to the *eccentric impact of rigid bodies.*

*1. Impulsive motion* occurs when a rigid body is subjected to a very large force **F** for a very short interval of time $\Delta t$; the resulting impulse **F** $\Delta t$ is both finite and different from zero. Such forces are referred to as *impulsive forces* and are encountered whenever there is an impact between two rigid bodies. Forces for which the impulse is zero are referred to as *nonimpulsive forces.* As you saw in Chap. 13, the following forces can be assumed to be nonimpulsive: the *weight* of a body, the force exerted by a *spring,* and any other force which is *known* to be small by comparison with the impulsive forces. Unknown reactions, however, *cannot be assumed* to be nonimpulsive.

*2. Eccentric impact of rigid bodies.* You saw that when two bodies collide, the velocity components along the line of impact of the *points of contact A and B* before and after impact satisfy the following equation:

$$(v'_B)_n - (v'_A)_n = e[(v_A)_n - (v_B)_n] \qquad (17.19)$$

where the left-hand member is the *relative velocity after the impact,* and the right-hand member is the product of the coefficient of restitution and the *relative velocity before the impact.*

   This equation expresses the same relation between the velocity components of the points of contact before and after an impact that you used for particles in Chap. 13.

*3. To solve a problem involving an impact* you should use the *method of impulse and momentum* and take the following steps.

   *a. Draw a free-body-diagram equation of the body* that will express that the system consisting of the momenta immediately before impact and of the impulses of the external forces is equivalent to the system of the momenta immediately after impact.

   *b. The free-body-diagram equation* will relate the velocities before and after impact and the impulsive forces and reactions. In some cases, you will be able to determine the unknown velocities and impulsive reactions by solving equations obtained by summing components and moments [Sample Prob. 17.9].

   *c. In the case of an impact in which $e > 0$,* the number of unknowns will be greater than the number of equations that you can write by summing components and moments, and you should supplement the equations obtained from the free-body-diagram equation with Eq. (17.19), which relates the relative velocities of the points of contact before and after impact [Sample Prob. 17.10].

   *d. During an impact you must use the method of impulse and momentum.* However, *before and after the impact* you can, if necessary, use some of the other methods of solution that you have learned, such as the method of work and energy [Sample Prob. 17.11].

# Problems

**17.86** Two identical 4.6 kg slender rods $AB$ and $BC$ are welded together to form an L-shaped assembly which is suspended from a hinge at $B$ and is at rest in a vertical plane. A .014 kg bullet strikes the assembly with a velocity of 548.64 m/s perpendicular to $BC$. Knowing that $d = 0$, determine the angular velocity of the assembly immediately after the bullet becomes embedded.

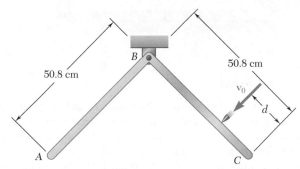

**Fig. P17.86 and P17.87**

**17.87** Two identical 4.6 kg slender rods $AB$ and $BC$ are welded together to form an L-shaped assembly which is suspended from a hinge at $B$ and is at rest in a vertical plane. A 0.014 kg bullet strikes the assembly with a velocity of 548.64 m/s perpendicular to $BC$. Knowing that the angular velocity of the assembly is 2 rad/s immediately after the bullet becomes embedded, determine the distance $d$.

**17.88** A 15-g magnet $D$ is released from rest in the position shown, falls a distance of 320 mm, and becomes attached at $A$ to the 200-g steel bar $AB$. Assuming that the impact is perfectly plastic, determine the angular velocity of the bar and the velocity of the magnet immediately after the impact.

**Fig. P17.88**

**17.89** A 30-kg uniform circular plate of radius $r$ is supported by a ball-and-socket joint at point $A$ and is at rest in the vertical $xy$ plane when a bullet with a mass of 15 g is fired with the velocity $\mathbf{v}_0 = -(210 \text{ m/s})\mathbf{k}$ and hits the plate at point $C$. Knowing that $r = 400$ mm and $h = 700$ mm, determine (a) the angular velocity of the plate immediately after the bullet becomes embedded, (b) the impulsive reaction at point $A$, assuming that the bullet becomes embedded in 1.1 ms.

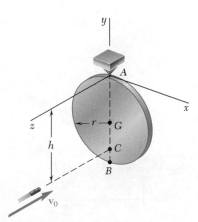

**Fig. P17.89**

**17.90**   In Prob. 17.89, determine (*a*) the required distance *h* if the impulsive reaction at point *A* is to be zero, (*b*) the corresponding velocity of the mass center *G* of the plate immediately after the bullet becomes embedded at *C*.

**17.91**   A uniform slender rod *AB* of mass *m* is at rest on a frictionless horizontal surface when hook *C* engages a small pin at *A*. Knowing that the hook is pulled upward with a constant velocity $\mathbf{v}_0$, determine the impulse exerted on the rod (*a*) at *A*, (*b*) at *B*. Assume that the velocity of the hook is unchanged and that the impact is perfectly plastic.

**Fig. P17.91**

**17.92**   A uniform slender rod *AB* of mass *m* and length *L* has a vertical velocity of magnitude $v_1$ and no angular velocity when it strikes a rigid frictionless support at point *C*. Knowing that $h = L/4$ and assuming perfectly plastic impact, determine (*a*) the angular velocity of the rod and the velocity of its mass center immediately after the impact, (*b*) the impulse exerted on the rod at point *C*.

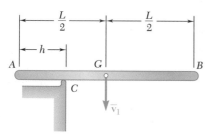

**Fig. P17.92**

**17.93**   The uniform slender rod *AB* of mass 3 kg and length 750 mm forms an angle $\beta = 30°$ with the vertical as it strikes the smooth corner shown with a vertical velocity $\bar{\mathbf{v}}_1$ of magnitude 2.4 m/s and no angular velocity. Assuming that the impact is perfectly plastic, determine the angular velocity of the rod immediately after the impact.

**Fig. P17.93 and P17.94**

**17.94**   The uniform slender rod *AB* of mass 3 kg and length 750 mm forms an angle $\beta$ with the vertical as it strikes the smooth corner shown with a vertical velocity $\bar{\mathbf{v}}_1$ of magnitude 2.4 m/s and no angular velocity. Knowing that the magnitude of the velocity of the mass center *G* of the rod immediately after the impact is 0.6 m/s and assuming that the impact is perfectly plastic, determine (*a*) the angle $\beta$, (*b*) the impulse exerted on the rod at point *A*.

**17.95**   A slender rod of mass *m* and length *L* is released from rest in the position shown and hits edge *D*. Assuming perfectly elastic impact ($e = 1$) at *D*, determine the distance *b* for which the rod will rebound with no angular velocity.

**Fig. P17.95**

**17.96** A uniform sphere of radius $r$ rolls down the incline shown without slipping. It hits a horizontal surface and, after slipping for a while, it starts rolling again. Assuming that the sphere does not bounce as it hits the horizontal surface, determine its angular velocity and the velocity of its mass center after it has resumed rolling.

**Fig. P17.96**

**17.97** A uniform slender rod $AB$ is at rest on a frictionless horizontal table when end $A$ of the rod is struck by a hammer which delivers an impulse that is perpendicular to the rod. In the subsequent motion, determine the distance $b$ through which the rod will move each time it completes a full revolution.

**17.98** A uniform slender rod of length $L$ is dropped onto rigid supports at $A$ and $B$. Immediately before striking $A$ the velocity of the rod is $\bar{\mathbf{v}}_1$. Since support $B$ is slightly lower than support $A$, the rod strikes $A$ before it strikes $B$. Assuming perfectly elastic impact at both $A$ and $B$, determine the angular velocity of the rod and the velocity of its mass center immediately after the rod $(a)$ strikes support $A$, $(b)$ strikes support $B$, $(c)$ again strikes support $A$.

**Fig. P17.97**

**Fig. P17.98**

**17.99** The slender rod $AB$ of length $L$ forms an angle $\beta$ with the vertical axis as it strikes the frictionless surface shown with a vertical velocity $\bar{\mathbf{v}}_1$ and no angular velocity. Assuming that the impact is perfectly elastic, derive an expression for the angular velocity of the rod immediately after the impact.

**17.100** A uniform slender rod $AB$ of mass $m$ and length $L$ is falling freely with a velocity $\mathbf{v}_0$ when end $B$ strikes a smooth inclined surface as shown. Assuming that the impact is perfectly elastic, determine the angular velocity of the rod and the velocity of its mass center immediately after the impact.

**Fig. P17.99**

**Fig. P17.100**

**17.101** A uniformly loaded square crate is falling freely with a velocity $\mathbf{v}_0$ when cable $AB$ suddenly becomes taut. Assuming that the impact is perfectly plastic, determine the angular velocity of the crate and the velocity of its mass center immediately after the cable becomes taut.

**Fig. P17.101**

**17.102** A uniform slender rod $AB$ of mass $m$ and length $L$ is released from rest in the position shown. Knowing that the impact between knob $B$ and the horizontal surface is perfectly elastic, determine ($a$) the angular velocity of the rod immediately after the impact, ($b$) the impulses exerted on the rod at points $A$ and $B$. (Neglect the masses of the knobs at points $A$ and $B$ and the friction between the knobs and the horizontal surface.)

**Fig. P17.102**

**17.103** A slender 5-kg rod is released from rest in the position shown. It is observed that after the rod strikes the vertical surface it rebounds to a horizontal position. ($a$) Determine the coefficient of restitution between knob $K$ and the surface. ($b$) Show that the same rebound can be expected for any position of knob $K$. (Neglect the mass of knob $K$.)

**17.104** A 0.03 kg bullet is fired with a horizontal velocity into the lower end of a 20.4 kg slender bar which is initially at rest in a vertical plane. Knowing that the bullet becomes embedded in the bar and that the maximum angle of rotation of the bar in its subsequent motion is 45° clockwise, determine the speed of the bullet before the impact.

**Fig. P17.103**

**Fig. P17.104 and P17.105**

**17.105** A .03 kg bullet is fired with a horizontal velocity of magnitude 457.2 m/s into the lower end of an 101.6 kg slender bar which is initially at rest in a vertical plane. Knowing that the bullet becomes embedded in the bar, determine the maximum clockwise angle of rotation of the bar in its subsequent motion.

**17.106** The uniform slender rod $AB$ is in equilibrium in the position shown when end $A$ is given a slight nudge, causing the rod to rotate counterclockwise and hit the horizontal surface. Knowing that the coefficient of restitution between the knob at $A$ and the horizontal surface is 0.50, determine the maximum angle of rebound $\theta$ of the rod.

**Fig. P17.106**

**17.107** A uniform slender rod $AB$ of length $L = 600$ mm is placed with its center equidistant from two supports that are located at a distance $b = 100$ mm from each other. End $B$ of the rod is raised a distance $h_0 = 80$ mm and released; the rod then rocks on the supports as shown. Assuming that the impact at each support is perfectly plastic and that no slipping occurs between the rod and the supports, determine ($a$) the height $h_1$ reached by end $A$ after the first impact, ($b$) the height $h_2$ reached by end $B$ after the second impact.

**Fig. P17.107**

**17.108** A uniformly loaded square crate is released from rest with its corner $D$ directly above $A$; it rotates about $A$ until its corner $B$ strikes the floor, and then rotates about $B$. The floor is sufficiently rough to prevent slipping and the impact at $B$ is perfectly plastic. Denoting by $\boldsymbol{\omega}_0$ the angular velocity of the crate immediately after $B$ strikes the floor, determine ($a$) the angular velocity of the crate immediately before $B$ strikes the floor, ($b$) the fraction of the kinetic energy of the crate lost during the impact, ($c$) the angle $\theta$ through which the crate will rotate after $B$ strikes the floor.

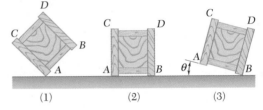

**Fig. P17.108**

**17.109** A slender rod $AB$ is released from rest in the position shown. It swings down to a vertical position and strikes a second and identical rod $CD$ which is resting on a frictionless surface. Assuming that the coefficient of restitution between the rods is 0.4, determine the velocity of rod $CD$ immediately after the impact.

**17.110** Solve Prob. 17.109 assuming that the impact between the rods is perfectly elastic.

**Fig. P17.109**

**17.111** The 8.16 kg rigid body $BD$ consists of two identical 6-cm-radius spheres and the rod which connects them and has a centroidal radius of gyration of 25.4 cm. The body is at rest on a horizontal frictionless surface when it is struck by the 2.7 kg sphere $A$ which has a radius of 6 cm and is moving as shown with a velocity $\mathbf{v}_1$ of magnitude 3.7 m/s. Assuming a perfectly plastic impact, determine immediately after the impact, ($a$) the angular velocity of body $BD$, ($b$) the velocity of point $G$.

**Fig. P17.111**

**Fig. P17.113**

**17.112** Solve Prob. 17.111, assuming that the impact is perfectly elastic.

**17.113** Block *A* of mass *m* is attached to a cord which is wrapped around a uniform disk of mass *M*. The block is released from rest and falls through a distance *h* before the cord becomes taut. Derive expressions for the velocity of the block and the angular velocity of the disk immediately after the impact. Assume that the impact is (*a*) perfectly plastic, (*b*) perfectly elastic.

**17.114** The plank *CDE* has a weight of 13.6 kg and rests on a small pivot at *D*. The 50 kg gymnast *A* is standing on the plank at *C* when the 63.5 kg gymnast *B* jumps from a height of 2.3 m and strikes the plank at *E*. Assuming perfectly plastic impact and that gymnast *A* is standing absolutely straight, determine the height to which gymnast *A* will rise.

**Fig. P17.114**

**17.115** Solve Prob. 17.114 assuming that the gymnasts change places so that gymnast *A* jumps onto the plank while gymnast *B* stands at *C*.

**17.116** The 2.5 kg slender rod *AB* is released from rest in the position shown and swings to a vertical position where it strikes the 1.5 kg slender rod *CD*. Knowing that the coefficient of restitution between the knob *K* attached to rod *AB* and rod *CD* is 0.8, determine the maximum angle $\theta_m$ through which rod *CD* will rotate after the impact.

**17.117** The uniform slender rod *AB* of mass $m_{AB}$ is attached by a pin to collar *C* of mass $m_C$ and the system is falling freely with a velocity $\mathbf{v}_0$ when collar *C* strikes a horizontal surface as shown. Denoting by *e* the coefficient of restitution between the collar and the surface, determine (*a*) the angular velocity of the rod immediately after the impact, (*b*) the impulse exerted on collar *C* by the surface.

**Fig. P17.116**

**Fig. P17.117**

**17.118** Sphere $A$ of mass $m$ and radius $r$ rolls without slipping with a velocity $\overline{\mathbf{v}}_1$ on a horizontal surface when it hits squarely an identical sphere $B$ that is at rest. Denoting by $\mu_k$ the coefficient of kinetic friction between the spheres and the surface, neglecting friction between the spheres, and assuming perfectly elastic impact, determine (a) the linear and angular velocities of each sphere immediately after the impact, (b) the velocity of each sphere after it has started rolling uniformly.

**Fig. P17.118**

**17.119** A small rubber ball of radius $r$ is thrown against a rough floor with a velocity $\overline{\mathbf{v}}_A$ of magnitude $v_0$ and a backspin $\boldsymbol{\omega}_A$ of magnitude $\omega_0$. It is observed that the ball bounced from $A$ to $B$, then from $B$ to $A$, then from $A$ to $B$, etc. Assuming perfectly elastic impact, determine the required magnitude $\omega_0$ of the backspin in terms of $\overline{v}_0$ and $r$.

**Fig. P17.119**

**17.120** In a game of pool, ball $A$ is rolling without slipping with a velocity $\overline{\mathbf{v}}_0$ as it hits obliquely ball $B$, which is at rest. Denoting by $r$ the radius of each ball and by $\mu_k$ the coefficient of kinetic friction between the balls and the table surface, and assuming perfectly elastic impact, determine (a) the linear and angular velocity of each ball immediately after the impact, (b) the velocity of ball $B$ after it has started rolling uniformly.

**Fig. P17.120**

In this chapter we again considered the method of work and energy and the method of impulse and momentum. In the first part of the chapter we studied the method of work and energy and its application to the analysis of the motion of rigid bodies and systems of rigid bodies.

**Principle of work and energy for a rigid body**

In Sec. 17.2, we first expressed the principle of work and energy for a rigid body in the form

$$T_1 + U_{1\rightarrow2} = T_2 \qquad (17.1)$$

where $T_1$ and $T_2$ represent the initial and final values of the kinetic energy of the rigid body and $U_{1\rightarrow2}$ represents the work of the *external forces* acting on the rigid body.

**Work of a force or a couple**

In Sec. 17.3, we recalled the expression found in Chap. 13 for the work of a force **F** applied at a point $A$, namely

$$U_{1\rightarrow2} = \int_{s_1}^{s_2} (F \cos \alpha)\, ds \qquad (17.3')$$

where $F$ was the magnitude of the force, $\alpha$ the angle it formed with the direction of motion of $A$, and $s$ the variable of integration measuring the distance traveled by $A$ along its path. We also derived the expression for the *work of a couple of moment* **M** applied to a rigid body during a rotation in $\theta$ of the rigid body:

$$U_{1\rightarrow2} = \int_{\theta_1}^{\theta_2} M\, d\theta \qquad (17.5)$$

**Kinetic energy in plane motion**

We then derived an expression for the kinetic energy of a rigid body in plane motion [Sec. 17.4]. We wrote

$$T = \tfrac{1}{2}m\bar{v}^2 + \tfrac{1}{2}\bar{I}\omega^2 \qquad (17.9)$$

where $\bar{v}$ is the velocity of the mass center $G$ of the body, $\omega$ is the angular velocity of the body, and $\bar{I}$ is its moment of inertia about an axis through $G$ perpendicular to the plane of reference (Fig. 17.13) [Sample Prob. 17.3]. We noted that the kinetic energy of a rigid body in plane motion can be separated into two parts: (1) the kinetic energy $\tfrac{1}{2}m\bar{v}^2$ associated with the motion of the mass center $G$ of the body, and (2) the kinetic energy $\tfrac{1}{2}\bar{I}\omega^2$ associated with the rotation of the body about $G$.

**Fig. 17.13**

For a rigid body rotating about a fixed axis through $O$ with an angular velocity $\boldsymbol{\omega}$, we had

$$T = \tfrac{1}{2}I_O\omega^2 \qquad (17.10)$$

where $I_O$ was the moment of inertia of the body about the fixed axis. We noted that the result obtained is not limited to the rotation of plane slabs or of bodies symmetrical with respect to the reference plane, but is valid regardless of the shape of the body or of the location of the axis of rotation.

Equation (17.1) can be applied to the motion of systems of rigid bodies [Sec. 17.5] as long as all the forces acting on the various bodies involved—internal as well as external to the system— are included in the computation of $U_{1\rightarrow2}$. However, in the case of systems consisting of pin-connected members, or blocks and pulleys connected by inextensible cords, or meshed gears, the points of application of the internal forces move through equal distances and the work of these forces cancels out [Sample Probs. 17.1 and 17.2].

When a rigid body, or a system of rigid bodies, moves under the action of conservative forces, the principle of work and energy can be expressed in the form

$$T_1 + V_1 = T_2 + V_2 \qquad (17.12)$$

which is referred to as the *principle of conservation of energy* [Sec. 17.6]. This principle can be used to solve problems involving conservative forces such as the force of gravity or the force exerted by a spring [Sample Probs. 17.4 and 17.5]. However, when a reaction is to be determined, the principle of conservation of energy must be supplemented by the application of d'Alembert's principle [Sample Prob. 17.4].

In Sec. 17.7, we extended the concept of power to a rotating body subjected to a couple, writing

$$\text{Power} = \frac{dU}{dt} = \frac{M\,d\theta}{dt} = M\omega \qquad (17.13)$$

where $M$ is the magnitude of the couple and $\omega$ the angular velocity of the body.

The middle part of the chapter was devoted to the method of impulse and momentum and its application to the solution of various types of problems involving the plane motion of rigid slabs and rigid bodies symmetrical with respect to the reference plane.

We first recalled the *principle of impulse and momentum* as it was derived in Sec. 14.9 for a system of particles and applied it to the *motion of a rigid body* [Sec. 17.8]. We wrote

**Syst Momenta**$_1$ + **Syst Ext Imp**$_{1\rightarrow2}$ = **Syst Momenta**$_2$ (17.14)

Next we showed that for a rigid slab or a rigid body symmetrical
with respect to the reference plane, the system of the momenta of
the particles forming the body is equivalent to a vector $m\bar{\mathbf{v}}$ attached
at the mass center $G$ of the body and a couple $\bar{I}\boldsymbol{\omega}$ (Fig. 17.14). The

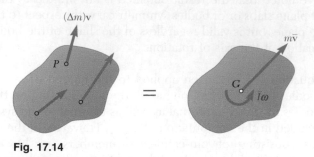

**Fig. 17.14**

vector $m\bar{\mathbf{v}}$ is associated with the translation of the body with $G$ and
represents the *linear momentum* of the body, while the couple $\bar{I}\boldsymbol{\omega}$
corresponds to the rotation of the body about $G$ and represents the
*angular momentum* of the body about an axis through $G$.

Equation (17.14) can be expressed graphically as shown in
Fig. 17.15 by drawing three diagrams representing respectively the
system of the initial momenta of the body, the impulses of the
external forces acting on the body, and the system of the final momenta of the body.

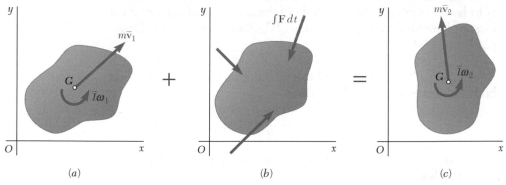

**Fig. 17.15**

Summing and equating respectively the *x components*, the *y
components*, and the *moments about any given point* of the vectors
shown in that figure, we obtain three equations of motion which can
be solved for the desired unknowns [Sample Probs. 17.6 and 17.7].

In problems dealing with several connected rigid bodies [Sec.
17.9], each body can be considered separately [Sample Prob. 17.6],
or, if no more than three unknowns are involved, the principle of

impulse and momentum can be applied to the entire system, considering the impulses of the external forces only [Sample Prob. 17.8].

Conservation of angular momentum

When the lines of action of all the external forces acting on a system of rigid bodies pass through a given point $O$, the angular momentum of the system about $O$ is conserved [Sec. 17.10]. It was suggested that problems involving conservation of angular momentum be solved by the general method described above [Sample Prob. 17.8].

Impulsive motion

The last part of the chapter was devoted to the *impulsive motion* and the *eccentric impact* of rigid bodies. In Sec. 17.11, we recalled that the method of impulse and momentum is the only practicable method for the solution of problems involving impulsive motion and that the computation of impulses in such problems is particularly simple [Sample Prob. 17.9].

Eccentric impact

In Sec. 17.12, we recalled that the eccentric impact of two rigid bodies is defined as an impact in which the mass centers of the colliding bodies are *not* located on the line of impact. It was shown that in such a situation a relation similar to that derived in Chap. 13 for the central impact of two particles and involving the coefficient of restitution $e$ still holds, but that *the velocities of points A and B where contact occurs during the impact should be used*. We have

$$(v'_B)_n - (v'_A)_n = e[(v_A)_n - (v_B)_n] \qquad (17.19)$$

where $(v_A)_n$ and $(v_B)_n$ are the components along the line of impact of the velocities of $A$ and $B$ before the impact, and $(v'_A)_n$ and $(v'_B)_n$ are their components after the impact (Fig. 17.16). Equation

 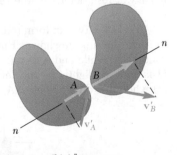

(a) Before impact  (b) After impact

**Fig. 17.16**

(17.19) is applicable not only when the colliding bodies move freely after the impact but also when the bodies are partially constrained in their motion. It should be used in conjunction with one or several other equations obtained by applying the principle of impulse and momentum [Sample Prob. 17.10]. We also considered problems where the method of impulse and momentum and the method of work and energy can be combined [Sample Prob. 17.11].

# Review Problems

**17.121 and 17.122** A slender 6-kg rod can rotate in a vertical plane about a pivot at $B$. A spring of constant $k = 600$ N/m and an unstretched length of 225 mm is attached to the rod as shown. Knowing that the rod is released from rest in the position shown, determine its angular velocity after it has rotated through 90°.

**Fig. P17.121**

**Fig. *P17.123***

**Fig. P17.122**

*17.123* The 8.2 kg cradle is supported as shown by two uniform disks that roll without sliding at all surfaces of contact. The weight of each disk is $W = 5.4$ kg and the radius of each disk is $r = 10.2$ cm. Knowing that the system is initially at rest, determine the velocity of the cradle after it has moved 38 cm.

**17.124** Two uniform rods, each of mass $m$ and length $L$, are connected to form the linkage shown. End $D$ of rod $BD$ can slide freely in the horizontal slot, while end $A$ of rod $AB$ is supported by a pin and bracket. If end $D$ is moved slightly to the left and then released, determine its velocity (*a*) when it is directly below $A$, (*b*) when rod $AB$ is vertical.

**Fig. P17.124**

**Fig. P17.125**

**17.125** The 317.5 kg flywheel of a small hoisting engine has a radius of gyration of 61 cm. If the power is cut off when the angular velocity of the flywheel is 100 rpm clockwise, determine the time required for the system to come to rest.

**17.126** A wheel of radius $r$ and centroidal radius of gyration $\bar{k}$ is released from rest on the incline shown at time $t = 0$. Assuming that the wheel rolls without sliding, determine (a) the velocity of its center at time $t$, (b) the coefficient of static friction required to prevent slipping.

Fig. *P17.126*

**17.127** A 1.134 kg disk of radius 100 mm is attached to the yoke $BCD$ by means of short shafts fitted in bearings at $B$ and $D$. The 0.68 kg yoke has a radius of gyration of 75 mm about the $x$ axis. Initially the assembly is rotating at 120 rpm with the disk in the plane of the yoke ($\theta = 0$). If the disk is slightly disturbed and rotates with respect to the yoke until $\theta = 90°$, where it is stopped by a small bar at $D$, determine the final angular velocity of the assembly.

Fig. **P17.127**

**17.128** In the helicopter shown, a vertical tail rotor is used to prevent rotation of the cab as the speed of the main blades is changed. Assuming that the tail rotor is not operating, determine the final angular velocity of the cab after the speed of the main blades has been changed from 180 to 240 rpm. (The speed of the main blades is measured relative to the cab, and the cab has a centroidal moment of inertia of 881 kg · m². Each of the four main blades is assumed to be a slender 4.27-m rod of mass 24.9 kg.)

Fig. **P17.128**

**17.129** A 40-g bullet is fired with a horizontal velocity of 600 m/s into the lower end of a slender 7 kg bar of length $L = 600$ mm. Knowing that $h = 240$ mm and that the bar is initially at rest, determine (a) the angular velocity of the bar immediately after the bullet becomes embedded, (b) the impulsive reaction at $C$, assuming that the bullet becomes embedded in 0.001 s.

**Fig. P17.129**

**17.130** A uniform slender rod of mass $m$ and length $L$ strikes a rigid frictionless support at point $C$ with an angular velocity of magnitude $\omega_1$ when the velocity of its mass center $G$ is zero. Knowing that the angular velocity of the rod immediately after the impact is $\omega_1/2$, counterclockwise, and assuming perfectly elastic impact, determine (a) the ratio $h/L$, (b) the velocity of the mass center of the rod immediately after the impact, (c) the impulse exerted on the rod at point $C$.

**Fig. P17.130**

**17.131** A .04 kg bullet is fired with a horizontal velocity of 290 m/s into the 8.2 kg wooden beam $AB$. The beam is suspended from a collar of negligible weight that can slide along a horizontal rod. Neglecting friction between the collar and the rod, determine the maximum angle of rotation of the beam during its subsequent motion.

**Fig. P17.131**

**17.132** Member $ABC$ has a weight of 2.3 kg and is attached to a pin support at $B$. A 0.68 kg sphere $D$ strikes end $C$ of member $ABC$ with a vertical velocity $\mathbf{v}_1$ of 2.7 kg. Knowing that $L = 76.2$ cm and that the coefficient of restitution between the sphere and member $ABC$ is 0.5, determine immediately after the impact (a) the angular velocity of member $ABC$, (b) the velocity of the sphere.

**Fig. P17.132**

# Computer Problems

**17.C1** The 4.5 kg slender rod $AB$ is welded to the 2.7-kg uniform disk, which rotates about a pivot at $A$. A spring of constant $k = 65.7$ N/m is attached to the disk and is unstretched when rod $AB$ is horizontal. The assembly is released from rest in the position shown. Calculate and plot the angular velocity of the assembly after it has rotated through an angle $\theta$, for values of $\theta$ from 0 to $\theta_m$, the angle of maximum rotation. Determine the maximum angular velocity of the assembly and the corresponding value of $\theta$.

**17.C2** Two 3-kg slender rods are welded to the edge of a 4-kg uniform disk as shown. The assembly is released from rest in the position shown and swings freely about the pivot $C$. Calculate and plot the angular velocity of the assembly after it has rotated through an angle $\theta$, for values of $\theta$ from 0 to $\theta_m$, the angle of maximum rotation. Determine the maximum angular velocity of the assembly and the corresponding value of $\theta$.

**Fig. P17.C1**

**Fig. P17.C2**

**17.C3** The uniform slender rod $AB$ of length $L = 102$ cm and weight 4.5 kg rests on a small wheel at $D$ and is attached to a collar of negligible mass that can slide freely on the vertical rod $EF$. Knowing that $a = 25.4$ cm and that the rod is released from rest when $\theta = 0$, use computational software to calculate and plot the angular velocity of the rod and the velocity of end $A$ for values of $\theta$ from 0 to 50°. Determine the maximum angular velocity of the rod and the corresponding value of $\theta$.

**Fig. P17.C3**

**Fig. P17.C4**

**17.C4** Each of the two identical slender bars shown has a length $L = 600$ mm. Knowing that the system is released from rest when the bars are horizontal, use computational software to calculate and plot the angular velocity of rod $AB$ and the velocity of point $D$ for values of $\theta$ from 0 to 90°.

**17.C5** Collar $C$ has a weight of 2.3 kg and can slide without friction on rod $AB$. A spring of constant 656.7 N/m and an unstretched length $r_0 = 63.5$ cm is attached as shown to the collar and to the hub $B$. The total mass moment of inertia of the rod, hub, and spring is known to be 0.03 kgm² about $B$. Initially the collar is held at a distance of 63.5 cm from the axis of rotation by a small pin protruding from the rod. The pin is suddenly removed as the assembly is rotating in a horizontal plane with angular velocity $\omega_0$ of 10 rad/s. Denoting by $r$ the distance of the collar from the axis of rotation, use computational software to calculate and plot the angular velocity of the assembly and the velocity of the collar relative to the rod for values of $r$ from 63.5 cm to 89 cm. Determine the maximum value of $r$ in the ensuing motion.

**Fig. P17.C5**

**17.C6** Rod $AB$ has a weight of 2.7 kg and is attached at $A$ to a 4.5 kg cart $C$. Knowing that the system is released from rest when $\theta = 45°$ and neglecting friction, use computational software to calculate the velocity of the cart and the velocity of end $B$ of the rod for values of $\theta$ from $+45°$ to $-90°$ using 15° decrements. Using appropriate smaller decrements, determine the value of $\theta$ for which the velocity of the cart to the left is maximum and the corresponding value of the velocity.

**Fig. P17.C6**

**17.C7** A uniform 200-mm-radius sphere rolls over a series of parallel horizontal bars equally spaced at a distance $d$. As it rotates without slipping about a given bar, the sphere strikes the next bar and starts rotating about that bar without slipping, until it strikes the next bar, and so on. Assuming perfectly plastic impact and knowing that the sphere has an angular velocity $\omega_0$ of 1.5 rad/s as its mass center $G$ is directly above bar $A$, use computational software to calculate, for values of the spacing $d$ from 20 to 120 mm and using 10-mm increments, (a) the angular velocity $\omega_1$ of the sphere as $G$ passes directly above bar $B$, (b) the number of bars over which the sphere will roll after leaving bar $A$.

Fig. P17.C7

## CHAPTER 18

# Kinetics of Rigid Bodies in Three Dimensions

While the general principles that you learned in earlier chapters can be used to solve problems involving the three-dimensional motion of rigid bodies, the solution of these problems requires a new approach and is considerably more involved than the solution of two-dimensional problems. One example is the study of the motion of a spinning top with a fixed point and an axis, which can assume any orientation in space.

**KINETICS OF RIGID BODIES IN THREE DIMENSIONS**

18.1 Introduction
18.2 Angular Momentum of a Rigid Body in Three Dimensions
18.3 Application of the Principle of Impulse and Momentum to the Three-Dimensional Motion of a Rigid Body
18.4 Kinetic Energy of a Rigid Body in Three Dimensions
18.5 Motion of a Rigid Body in Three Dimensions
18.6 Euler's Equations of Motion. Extension of d'Alembert's Principle to the Motion of a Rigid Body in Three Dimensions
18.7 Motion of a Rigid Body about a Fixed Point
18.8 Rotation of a Rigid Body about a Fixed Axis
18.9 Motion of a Gyroscope. Eulerian Angles
18.10 Steady Precession of a Gyroscope
18.11 Motion of an Axisymmetrical Body under No Force

## *18.1. INTRODUCTION

In Chaps. 16 and 17 we were concerned with the plane motion of rigid bodies and of systems of rigid bodies. In Chap. 16 and in the second half of Chap. 17 (momentum method), our study was further restricted to that of plane slabs and of bodies symmetrical with respect to the reference plane. However, many of the fundamental results obtained in these two chapters remain valid in the case of the motion of a rigid body in three dimensions.

For example, the two fundamental equations

$$\Sigma \mathbf{F} = m\bar{\mathbf{a}} \qquad (18.1)$$

$$\Sigma \mathbf{M}_G = \dot{\mathbf{H}}_G \qquad (18.2)$$

on which the analysis of the plane motion of a rigid body was based, remain valid in the most general case of motion of a rigid body. As indicated in Sec. 16.2, these equations express that the system of the external forces is equipollent to the system consisting of the vector $m\bar{\mathbf{a}}$ attached at $G$ and the couple of moment $\dot{\mathbf{H}}_G$ (Fig. 18.1). How-

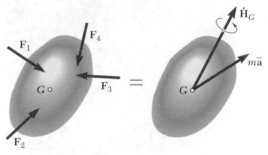

**Fig. 18.1**

ever, the relation $\mathbf{H}_G = \bar{I}\boldsymbol{\omega}$, which enabled us to determine the angular momentum of a rigid slab and which played an important part in the solution of problems involving the plane motion of slabs and bodies symmetrical with respect to the reference plane, ceases to be valid in the case of nonsymmetrical bodies or three-dimensional motion. Thus in the first part of the chapter, in Sec. 18.2, a more general method for computing the angular momentum $\mathbf{H}_G$ of a rigid body in three dimensions will be developed.

Similarly, although the main feature of the impulse-momentum method discussed in Sec. 17.7, namely, the reduction of the momenta of the particles of a rigid body to a linear momentum vector $m\bar{\mathbf{v}}$ attached at the mass center $G$ of the body and an angular momentum couple $\mathbf{H}_G$, remains valid, the relation $\mathbf{H}_G = \bar{I}\boldsymbol{\omega}$ must be discarded and replaced by the more general relation developed in Sec. 18.2 before this method can be applied to the three-dimensional motion of a rigid body (Sec. 18.3).

We also note that the work-energy principle (Sec. 17.2) and the principle of conservation of energy (Sec. 17.6) still apply in the case of the motion of a rigid body in three dimensions. However, the expression obtained in Sec. 17.4 for the kinetic energy of a rigid body in plane motion will be replaced by a new expression developed in Sec. 18.4 for a rigid body in three-dimensional motion.

In the second part of the chapter, you will first learn to determine the rate of change $\dot{\mathbf{H}}_G$ of the angular momentum $\mathbf{H}_G$ of a three-dimensional rigid body, using a rotating frame of reference with respect to which the moments and products of inertia of the body remain constant (Sec. 18.5). Equations (18.1) and (18.2) will then be expressed in the form of free-body-diagram equations, which can be used to solve various problems involving the three-dimensional motion of rigid bodies (Secs. 18.6 through 18.8).

The last part of the chapter (Secs. 18.9 through 18.11) is devoted to the study of the motion of the gyroscope or, more generally, of an axisymmetrical body with a fixed point located on its axis of symmetry. In Sec. 18.10, the particular case of the steady precession of a gyroscope will be considered, and, in Sec. 18.11, the motion of an axisymmetrical body subjected to no force, except its own weight, will be analyzed.

## *18.2. ANGULAR MOMENTUM OF A RIGID BODY IN THREE DIMENSIONS

In this section you will see how the angular momentum $\mathbf{H}_G$ of a body about its mass center $G$ can be determined from the angular velocity $\boldsymbol{\omega}$ of the body in the case of three-dimensional motion.

According to Eq. (14.24), the angular momentum of the body about $G$ can be expressed as

$$\mathbf{H}_G = \sum_{i=1}^{n} (\mathbf{r}'_i \times \mathbf{v}'_i \, \Delta m_i) \qquad (18.3)$$

where $\mathbf{r}'_i$ and $\mathbf{v}'_i$ denote, respectively, the position vector and the velocity of the particle $P_i$, of mass $\Delta m_i$, relative to the centroidal frame $Gxyz$ (Fig. 18.2). But $\mathbf{v}'_i = \boldsymbol{\omega} \times \mathbf{r}'_i$, where $\boldsymbol{\omega}$ is the angular

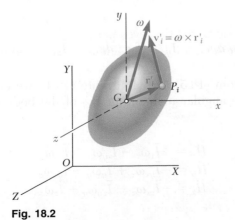

**Fig. 18.2**

velocity of the body at the instant considered. Substituting into (18.3), we have

$$\mathbf{H}_G = \sum_{i=1}^{n} [\mathbf{r}'_i \times (\boldsymbol{\omega} \times \mathbf{r}'_i) \, \Delta m_i]$$

Recalling the rule for determining the rectangular components of a vector product (Sec. 3.5), we obtain the following expression for the $x$ component of the angular momentum:

$$\begin{aligned}
H_x &= \sum_{i=1}^{n} [y_i(\boldsymbol{\omega} \times \mathbf{r}'_i)_z - z_i(\boldsymbol{\omega} \times \mathbf{r}'_i)_y] \, \Delta m_i \\
&= \sum_{i=1}^{n} [y_i(\omega_x y_i - \omega_y x_i) - z_i(\omega_z x_i - \omega_x z_i)] \, \Delta m_i \\
&= \omega_x \sum_i (y_i^2 + z_i^2) \, \Delta m_i - \omega_y \sum_i x_i y_i \, \Delta m_i - \omega_z \sum_i z_i x_i \, \Delta m_i
\end{aligned}$$

Replacing the sums by integrals in this expression and in the two similar expressions which are obtained for $H_y$ and $H_z$, we have

$$\begin{aligned}
H_x &= \omega_x \!\int\! (y^2 + z^2) \, dm - \omega_y \!\int\! xy \, dm - \omega_z \!\int\! zx \, dm \\
H_y &= -\omega_x \!\int\! xy \, dm + \omega_y \!\int\! (z^2 + x^2) \, dm - \omega_z \!\int\! yz \, dm \\
H_z &= -\omega_x \!\int\! zx \, dm - \omega_y \!\int\! yz \, dm + \omega_z \!\int\! (x^2 + y^2) \, dm
\end{aligned} \qquad (18.4)$$

We note that the integrals containing squares represent the *centroidal mass moments of inertia* of the body about the $x$, $y$, and $z$ axes, respectively (Sec. 9.11); we have

$$\bar{I}_x = \int (y^2 + z^2) \, dm \qquad \bar{I}_y = \int (z^2 + x^2) \, dm \qquad (18.5)$$

$$\bar{I}_z = \int (x^2 + y^2) \, dm$$

Similarly, the integrals containing products of coordinates represent the *centroidal mass products of inertia* of the body (Sec. 9.16); we have

$$\bar{I}_{xy} = \int xy \, dm \qquad \bar{I}_{yz} = \int yz \, dm \qquad \bar{I}_{zx} = \int zx \, dm \qquad (18.6)$$

Substituting from (18.5) and (18.6) into (18.4), we obtain the components of the angular momentum $\mathbf{H}_G$ of the body about its mass center:

$$\begin{aligned}
H_x &= +\bar{I}_x \omega_x - \bar{I}_{xy} \omega_y - \bar{I}_{xz} \omega_z \\
H_y &= -\bar{I}_{yx} \omega_x + \bar{I}_y \omega_y - \bar{I}_{yz} \omega_z \\
H_z &= -\bar{I}_{zx} \omega_x - \bar{I}_{zy} \omega_y + \bar{I}_z \omega_z
\end{aligned} \qquad (18.7)$$

The relations (18.7) show that the operation which transforms the vector $\boldsymbol{\omega}$ into the vector $\mathbf{H}_G$ (Fig. 18.3) is characterized by the array of moments and products of inertia

$$\begin{pmatrix} \bar{I}_x & -\bar{I}_{xy} & -\bar{I}_{xz} \\ -\bar{I}_{yx} & \bar{I}_y & -\bar{I}_{yz} \\ -\bar{I}_{zx} & -\bar{I}_{zy} & \bar{I}_z \end{pmatrix} \tag{18.8}$$

The array (18.8) defines the *inertia tensor* of the body at its mass center $G$.† A new array of moments and products of inertia would be obtained if a different system of axes were used. The transformation characterized by this new array, however, would still be the same. Clearly, the angular momentum $\mathbf{H}_G$ corresponding to a given angular velocity $\boldsymbol{\omega}$ is independent of the choice of the coordinate axes. As was shown in Secs. 9.17 and 9.18, it is always possible to select a system of axes $Gx'y'z'$, called *principal axes of inertia*, with respect to which all the products of inertia of a given body are zero. The array (18.8) takes then the diagonalized form

$$\begin{pmatrix} \bar{I}_{x'} & 0 & 0 \\ 0 & \bar{I}_{y'} & 0 \\ 0 & 0 & \bar{I}_{z'} \end{pmatrix} \tag{18.9}$$

where $\bar{I}_{x'}, \bar{I}_{y'}, \bar{I}_{z'}$ represent the *principal centroidal moments of inertia* of the body, and the relations (18.7) reduce to

$$H_{x'} = \bar{I}_{x'}\omega_{x'} \qquad H_{y'} = \bar{I}_{y'}\omega_{y'} \qquad H_{z'} = \bar{I}_{z'}\omega_{z'} \tag{18.10}$$

We note that if the three principal centroidal moments of inertia $\bar{I}_{x'}, \bar{I}_{y'}, \bar{I}_{z'}$ are equal, the components $H_{x'}, H_{y'}, H_{z'}$ of the angular momentum about $G$ are proportional to the components $\omega_{x'}, \omega_{y'}, \omega_{z'}$ of the angular velocity, and the vectors $\mathbf{H}_G$ and $\boldsymbol{\omega}$ are collinear. In general, however, the principal moments of inertia will be different, and the vectors $\mathbf{H}_G$ and $\boldsymbol{\omega}$ *will have different directions*, except when two of the three components of $\boldsymbol{\omega}$ happen to be zero, that is, when $\boldsymbol{\omega}$ is directed along one of the coordinate axes. Thus, *the angular momentum $\mathbf{H}_G$ of a rigid body and its angular velocity $\boldsymbol{\omega}$ have the same direction if, and only if, $\boldsymbol{\omega}$ is directed along a principal axis of inertia.*‡

---

†Setting $\bar{I}_x = I_{11}, \bar{I}_y = I_{22}, \bar{I}_z = I_{33}$, and $-\bar{I}_{xy} = I_{12}, -\bar{I}_{xz} = I_{13}$, etc., we may write the inertia tensor (18.8) in the standard form

$$\begin{pmatrix} I_{11} & I_{12} & I_{13} \\ I_{21} & I_{22} & I_{23} \\ I_{31} & I_{32} & I_{33} \end{pmatrix}$$

Denoting by $H_1, H_2, H_3$ the components of the angular momentum $\mathbf{H}_G$ and by $\omega_1, \omega_2, \omega_3$ the components of the angular velocity $\boldsymbol{\omega}$, we can write the relations (18.7) in the form

$$H_i = \sum_j I_{ij}\omega_j$$

where $i$ and $j$ take the values 1, 2, 3. The quantities $I_{ij}$ are said to be the *components* of the inertia tensor. Since $I_{ij} = I_{ji}$, the inertia tensor is a *symmetric tensor of the second order*.

‡In the particular case when $\bar{I}_{x'} = \bar{I}_{y'} = \bar{I}_{z'}$, any line through $G$ can be considered as a principal axis of inertia, and the vectors $\mathbf{H}_G$ and $\boldsymbol{\omega}$ are always collinear.

**Fig. 18.3**

**Photo 18.1** The design of a robotic welder for
an automobile assembly line requires a three-
dimensional study of both kinematics and
kinetics.

Since this condition is satisfied in the case of the plane motion of a
rigid body symmetrical with respect to the reference plane, we were
able in Secs. 16.3 and 17.8 to represent the angular momentum $\mathbf{H}_G$
of such a body by the vector $\bar{I}\boldsymbol{\omega}$. We must realize, however, that this
result cannot be extended to the case of the plane motion of a nonsymmetrical body, or to the case of the three-dimensional motion of
a rigid body. Except when $\boldsymbol{\omega}$ happens to be directed along a principal axis of inertia, the angular momentum and angular velocity of a
rigid body have different directions, and the relation (18.7) or (18.10)
must be used to determine $\mathbf{H}_G$ from $\boldsymbol{\omega}$.

**Reduction of the Momenta of the Particles of a Rigid Body
to a Momentum Vector and a Couple at _G._** We saw in Sec. 17.8
that the system formed by the momenta of the various particles of a
rigid body can be reduced to a vector $\mathbf{L}$ attached at the mass center
$G$ of the body, representing the linear momentum of the body, and
to a couple $\mathbf{H}_G$, representing the angular momentum of the body
about $G$ (Fig. 18.4). We are now in a position to determine the vector

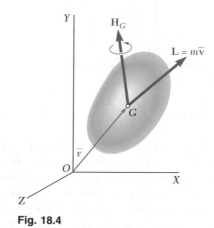

**Fig. 18.4**

$\mathbf{L}$ and the couple $\mathbf{H}_G$ in the most general case of three-dimensional
motion of a rigid body. As in the case of the two-dimensional motion
considered in Sec. 17.8, the linear momentum $\mathbf{L}$ of the body is equal
to the product $m\bar{\mathbf{v}}$ of its mass $m$ and the velocity $\bar{\mathbf{v}}$ of its mass center
$G$. The angular momentum $\mathbf{H}_G$, however, can no longer be obtained
by simply multiplying the angular velocity $\boldsymbol{\omega}$ of the body by the scalar
$\bar{I}$; it must now be obtained from the components of $\boldsymbol{\omega}$ and from the
centroidal moments and products of inertia of the body through the
use of Eq. (18.7) or (18.10).

We should also note that once the linear momentum $m\bar{\mathbf{v}}$ and the
angular momentum $\mathbf{H}_G$ of a rigid body have been determined, its angular momentum $\mathbf{H}_O$ about any given point $O$ can be obtained by
adding the moments about $O$ of the vector $m\bar{\mathbf{v}}$ and of the couple $\mathbf{H}_G$.
We write

$$\mathbf{H}_O = \bar{\mathbf{r}} \times m\bar{\mathbf{v}} + \mathbf{H}_G \tag{18.11}$$

**Angular Momentum of a Rigid Body Constrained to Rotate about a Fixed Point.** In the particular case of a rigid body constrained to rotate in three-dimensional space about a fixed point $O$ (Fig. 18.5a), it is sometimes convenient to determine the angular

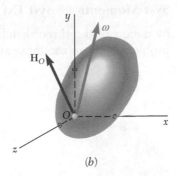

(a)   (b)

**Fig. 18.5**

momentum $\mathbf{H}_O$ of the body about the fixed point $O$. While $\mathbf{H}_O$ could be obtained by first computing $\mathbf{H}_G$ as indicated above and then using Eq. (18.11), it is often advantageous to determine $\mathbf{H}_O$ directly from the angular velocity $\boldsymbol{\omega}$ of the body and its moments and products of inertia with respect to a frame $Oxyz$ centered at the fixed point $O$. Recalling Eq. (14.7), we write

$$\mathbf{H}_O = \sum_{i=1}^{n} (\mathbf{r}_i \times \mathbf{v}_i \, \Delta m_i) \tag{18.12}$$

where $\mathbf{r}_i$ and $\mathbf{v}_i$ denote, respectively, the position vector and the velocity of the particle $P_i$ with respect to the fixed frame $Oxyz$. Substituting $\mathbf{v}_i = \boldsymbol{\omega} \times \mathbf{r}_i$, and after manipulations similar to those used in the earlier part of this section, we find that the components of the angular momentum $\mathbf{H}_O$ (Fig. 18.5b) are given by the relations

$$\begin{aligned}
H_x &= +I_x\omega_x - I_{xy}\omega_y - I_{xz}\omega_z \\
H_y &= -I_{yx}\omega_x + I_y\omega_y - I_{yz}\omega_z \\
H_z &= -I_{zx}\omega_x - I_{zy}\omega_y + I_z\omega_z
\end{aligned} \tag{18.13}$$

where the moments of inertia $I_x$, $I_y$, $I_z$ and the products of inertia $I_{xy}$, $I_{yz}$, $I_{zx}$ are computed with respect to the frame $Oxyz$ centered at the fixed point $O$.

## *18.3. APPLICATION OF THE PRINCIPLE OF IMPULSE AND MOMENTUM TO THE THREE-DIMENSIONAL MOTION OF A RIGID BODY

Before we can apply the fundamental Eq. (18.2) to the solution of problems involving the three-dimensional motion of a rigid body, we must learn to compute the derivative of the vector $\mathbf{H}_G$. This will be done in Sec. 18.5. The results obtained in the preceding section can, however, be used right away to solve problems by the impulse-momentum method.

Photo 18.2 As a result of the impulsive force applied by the bowling ball, a pin acquires both linear momentum and angular momentum.

Recalling that the system formed by the momenta of the particles of a rigid body reduces to a linear momentum vector $m\bar{\mathbf{v}}$ attached at the mass center $G$ of the body and an angular momentum couple $\mathbf{H}_G$, we represent graphically the fundamental relation

**Syst Momenta$_1$ + Syst Ext Imp$_{1\to2}$ = Syst Momenta$_2$**     (17.4)

by means of the three sketches shown in Fig. 18.6. To solve a given problem, we can use these sketches to write appropriate component

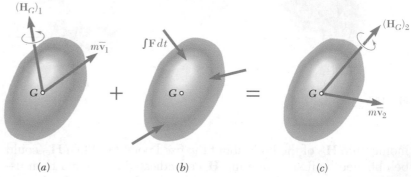

**Fig. 18.6**

and moment equations, keeping in mind that the components of the angular momentum $\mathbf{H}_G$ are related to the components of the angular velocity $\boldsymbol{\omega}$ by Eqs. (18.7) of the preceding section.

In solving problems dealing with the motion of a body rotating about a fixed point $O$, it will be convenient to eliminate the impulse of the reaction at $O$ by writing an equation involving the moments of the momenta and impulses about $O$. We recall that the angular momentum $\mathbf{H}_O$ of the body about the fixed point $O$ can be obtained either directly from Eqs. (18.13) or by first computing its linear momentum $m\bar{\mathbf{v}}$ and its angular momentum $\mathbf{H}_G$ and then using Eq. (18.11).

### *18.4. KINETIC ENERGY OF A RIGID BODY IN THREE DIMENSIONS

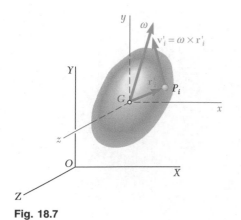

**Fig. 18.7**

Consider a rigid body of mass $m$ in three-dimensional motion. We recall from Sec. 14.6 that if the absolute velocity $\mathbf{v}_i$ of each particle $P_i$ of the body is expressed as the sum of the velocity $\bar{\mathbf{v}}$ of the mass center $G$ of the body and of the velocity $\mathbf{v}_i'$ of the particle relative to a frame $Gxyz$ attached to $G$ and of fixed orientation (Fig. 18.7), the kinetic energy of the system of particles forming the rigid body can be written in the form

$$T = \tfrac{1}{2}m\bar{v}^2 + \frac{1}{2}\sum_{i=1}^{n}\Delta m_i v_i'^2 \qquad (18.14)$$

where the last term represents the kinetic energy $T'$ of the body relative to the centroidal frame $Gxyz$. Since $v_i' = |\mathbf{v}_i'| = |\boldsymbol{\omega} \times \mathbf{r}_i'|$, we write

$$T' = \frac{1}{2}\sum_{i=1}^{n}\Delta m_i v_i'^2 = \frac{1}{2}\sum_{i=1}^{n}|\boldsymbol{\omega} \times \mathbf{r}_i'|^2\,\Delta m_i$$

Expressing the square in terms of the rectangular components of the vector product, and replacing the sums by integrals, we have

$$T' = \tfrac{1}{2}\int[(\omega_x y - \omega_y x)^2 + (\omega_y z - \omega_z y)^2 + (\omega_z x - \omega_x z)^2]\, dm$$
$$= \tfrac{1}{2}[\omega_x^2\int(y^2 + z^2)\, dm + \omega_y^2\int(z^2 + x^2)\, dm + \omega_z^2\int(x^2 + y^2)\, dm$$
$$- 2\omega_x\omega_y\int xy\, dm - 2\omega_y\omega_z\int yz\, dm - 2\omega_z\omega_x\int zx\, dm]$$

or, recalling the relations (18.5) and (18.6),

$$T' = \tfrac{1}{2}(\bar{I}_x\omega_x^2 + \bar{I}_y\omega_y^2 + \bar{I}_z\omega_z^2 - 2\bar{I}_{xy}\omega_x\omega_y - 2\bar{I}_{yz}\omega_y\omega_z - 2\bar{I}_{zx}\omega_z\omega_x)$$

$$(18.15)$$

Substituting into (18.14) the expression (18.15) we have just obtained for the kinetic energy of the body relative to centroidal axes, we write

$$T = \tfrac{1}{2}m\bar{v}^2 + \tfrac{1}{2}(\bar{I}_x\omega_x^2 + \bar{I}_y\omega_y^2 + \bar{I}_z\omega_z^2 - 2\bar{I}_{xy}\omega_x\omega_y$$
$$- 2\bar{I}_{yz}\omega_y\omega_z - 2\bar{I}_{zx}\omega_z\omega_x) \qquad (18.16)$$

If the axes of coordinates are chosen so that they coincide at the instant considered with the principal axes $x'$, $y'$, $z'$ of the body, the relation obtained reduces to

$$T = \tfrac{1}{2}m\bar{v}^2 + \tfrac{1}{2}(\bar{I}_{x'}\omega_{x'}^2 + \bar{I}_{y'}\omega_{y'}^2 + \bar{I}_{z'}\omega_{z'}^2) \qquad (18.17)$$

where $\bar{\mathbf{v}}$ = velocity of mass center
$\boldsymbol{\omega}$ = angular velocity
$m$ = mass of rigid body
$\bar{I}_{x'}, \bar{I}_{y'}, \bar{I}_{z'}$ = principal centroidal moments of inertia

The results we have obtained enable us to apply to the three-dimensional motion of a rigid body the principles of work and energy (Sec. 17.2) and conservation of energy (Sec. 17.6).

**Kinetic Energy of a Rigid Body with a Fixed Point.** In the particular case of a rigid body rotating in three-dimensional space about a fixed point $O$, the kinetic energy of the body can be expressed in terms of its moments and products of inertia with respect to axes attached at $O$ (Fig. 18.8). Recalling the definition of the kinetic energy of a system of particles, and substituting $v_i = |\mathbf{v}_i| = |\boldsymbol{\omega} \times \mathbf{r}_i|$, we write

$$T = \frac{1}{2}\sum_{i=1}^{n}\Delta m_i v_i^2 = \frac{1}{2}\sum_{i=1}^{n}|\boldsymbol{\omega} \times \mathbf{r}_i|^2\, \Delta m_i \qquad (18.18)$$

Manipulations similar to those used to derive Eq. (18.15) yield

$$T = \tfrac{1}{2}(I_x\omega_x^2 + I_y\omega_y^2 + I_z\omega_z^2 - 2I_{xy}\omega_x\omega_y - 2I_{yz}\omega_y\omega_z - 2I_{zx}\omega_z\omega_x)$$

$$(18.19)$$

or, if the principal axes $x'$, $y'$, $z'$ of the body at the origin $O$ are chosen as coordinate axes,

$$T = \tfrac{1}{2}(I_{x'}\omega_{x'}^2 + I_{y'}\omega_{y'}^2 + I_{z'}\omega_{z'}^2) \qquad (18.20)$$

**Fig. 18.8**

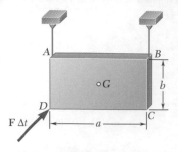

## SAMPLE PROBLEM 18.1

A rectangular plate of mass $m$ suspended from two wires at $A$ and $B$ is hit at $D$ in a direction perpendicular to the plate. Denoting by $\mathbf{F}\,\Delta t$ the impulse applied at $D$, determine immediately after the impact ($a$) the velocity of the mass center $G$, ($b$) the angular velocity of the plate.

## SOLUTION

Assuming that the wires remain taut and thus that the components $\bar{v}_y$ of $\bar{\mathbf{v}}$ and $\omega_z$ of $\boldsymbol{\omega}$ are zero after the impact, we have

$$\bar{\mathbf{v}} = \bar{v}_x\mathbf{i} + \bar{v}_z\mathbf{k} \qquad \boldsymbol{\omega} = \omega_x\mathbf{i} + \omega_y\mathbf{j}$$

and since the $x$, $y$, $z$ axes are principal axes of inertia,

$$\mathbf{H}_G = \bar{I}_x\omega_x\mathbf{i} + \bar{I}_y\omega_y\mathbf{j} \qquad \mathbf{H}_G = \tfrac{1}{12}mb^2\omega_x\mathbf{i} + \tfrac{1}{12}ma^2\omega_y\mathbf{j} \qquad (1)$$

**Principle of Impulse and Momentum.** Since the initial momenta are zero, the system of the impulses must be equivalent to the system of the final momenta:

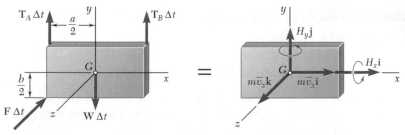

**$a$. Velocity of Mass Center.** Equating the components of the impulses and momenta in the $x$ and $z$ directions:

$x$ components: $\qquad 0 = m\bar{v}_x \qquad \bar{v}_x = 0$

$z$ components: $\qquad -F\,\Delta t = m\bar{v}_z \qquad \bar{v}_z = -F\,\Delta t/m$

$$\bar{\mathbf{v}} = \bar{v}_x\mathbf{i} + \bar{v}_z\mathbf{k} \qquad \bar{\mathbf{v}} = -(F\,\Delta t/m)\mathbf{k} \quad \blacktriangleleft$$

**$b$. Angular Velocity.** Equating the moments of the impulses and momenta about the $x$ and $y$ axes:

About $x$ axis: $\qquad \tfrac{1}{2}bF\,\Delta t = H_x$

About $y$ axis: $\qquad -\tfrac{1}{2}aF\,\Delta t = H_y$

$$\mathbf{H}_G = H_x\mathbf{i} + H_y\mathbf{j} \qquad \mathbf{H}_G = \tfrac{1}{2}bF\,\Delta t\mathbf{i} - \tfrac{1}{2}aF\,\Delta t\mathbf{j} \qquad (2)$$

Comparing Eqs. (1) and (2), we conclude that

$$\omega_x = 6F\,\Delta t/mb \qquad \omega_y = -6F\,\Delta t/ma$$
$$\boldsymbol{\omega} = \omega_x\mathbf{i} + \omega_y\mathbf{j} \qquad \boldsymbol{\omega} = (6F\,\Delta t/mab)(a\mathbf{i} - b\mathbf{j}) \quad \blacktriangleleft$$

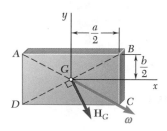

We note that $\boldsymbol{\omega}$ is directed along the diagonal $AC$.

**Remark:** Equating the $y$ components of the impulses and momenta, and their moments about the $z$ axis, we obtain two additional equations which yield $T_A = T_B = \tfrac{1}{2}W$. We thus verify that the wires remain taut and that our assumption was correct.

# SAMPLE PROBLEM 18.2

A homogeneous disk of radius $r$ and mass $m$ is mounted on an axle $OG$ of length $L$ and negligible mass. The axle is pivoted at the fixed point $O$, and the disk is constrained to roll on a horizontal floor. Knowing that the disk rotates counterclockwise at the rate $\omega_1$ about the axle $OG$, determine (a) the angular velocity of the disk, (b) its angular momentum about $O$, (c) its kinetic energy, (d) the vector and couple at $G$ equivalent to the momenta of the particles of the disk.

# SOLUTION

*a. Angular Velocity.* As the disk rotates about the axle $OG$ it also rotates with the axle about the $y$ axis at a rate $\omega_2$ clockwise. The total angular velocity of the disk is therefore

$$\boldsymbol{\omega} = \omega_1\mathbf{i} - \omega_2\mathbf{j} \qquad (1)$$

To determine $\omega_2$ we write that the velocity of $C$ is zero:

$$\mathbf{v}_C = \boldsymbol{\omega} \times \mathbf{r}_C = 0$$
$$(\omega_1\mathbf{i} - \omega_2\mathbf{j}) \times (L\mathbf{i} - r\mathbf{j}) = 0$$
$$(L\omega_2 - r\omega_1)\mathbf{k} = 0 \qquad \omega_2 = r\omega_1/L$$

Substituting into (1) for $\omega_2$: $\qquad \boldsymbol{\omega} = \omega_1\mathbf{i} = (r\omega_1/L)\mathbf{j}$ ◀

*b. Angular Momentum about $O$.* Assuming the axle to be part of the disk, we can consider the disk to have a fixed point at $O$. Since the $x$, $y$, and $z$ axes are principal axes of inertia for the disk,

$$H_x = I_x\omega_x = (\tfrac{1}{2}mr^2)\omega_1$$
$$H_y = I_y\omega_y = (mL^2 + \tfrac{1}{4}mr^2)(-r\omega_1/L)$$
$$H_z = I_z\omega_z = (mL^2 + \tfrac{1}{4}mr^2)0 = 0$$
$$\mathbf{H}_O = \tfrac{1}{2}mr^2\omega_1\mathbf{i} - m(L^2 + \tfrac{1}{4}r^2)(r\omega_1/L)\mathbf{j} \quad ◀$$

*c. Kinetic Energy.* Using the values obtained for the moments of inertia and the components of $\boldsymbol{\omega}$, we have

$$T = \tfrac{1}{2}(I_x\omega_x^2 + I_y\omega_y^2 + I_z\omega_z^2) = \tfrac{1}{2}[\tfrac{1}{2}mr^2\omega_1^2 + m(L^2 + \tfrac{1}{4}r^2)(-r\omega_1/L)^2]$$

$$T = \tfrac{1}{8}mr^2\left(6 + \frac{r^2}{L^2}\right)\omega_1^2 \quad ◀$$

*d. Momentum Vector and Couple at $G$.* The linear momentum vector $m\bar{\mathbf{v}}$ and the angular momentum couple $\mathbf{H}_G$ are

$$m\bar{\mathbf{v}} = mr\omega_1\mathbf{k} \quad ◀$$

and

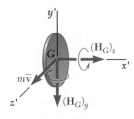

$$\mathbf{H}_G = \bar{I}_{x'}\omega_x\mathbf{i} + \bar{I}_{y'}\omega_y\mathbf{j} + \bar{I}_{z'}\omega_z\mathbf{k} = \tfrac{1}{2}mr^2\omega_1\mathbf{i} + \tfrac{1}{4}mr^2(-r\omega_1/L)\mathbf{j}$$

$$\mathbf{H}_G = \tfrac{1}{2}mr^2\omega_1\left(\mathbf{i} - \frac{r}{2L}\mathbf{j}\right) \quad ◀$$

# SOLVING PROBLEMS
## ON YOUR OWN

In this lesson you learned to compute the *angular momentum of a rigid body in three dimensions* and to apply the principle of impulse and momentum to the three-dimensional motion of a rigid body. You also learned to compute the *kinetic energy of a rigid body in three dimensions.* It is important for you to keep in mind that, except for very special situations, the angular momentum of a rigid body in three dimensions *cannot* be expressed as the product $\bar{I}\boldsymbol{\omega}$ and, therefore, *will not have the same direction as the angular velocity* $\boldsymbol{\omega}$ (Fig. 18.3).

1. *To compute the angular momentum $\mathbf{H}_G$ of a rigid body about its mass center G,* you must first determine the angular velocity $\boldsymbol{\omega}$ of the body with respect to a system of axes *centered at G and of fixed orientation.* Since you will be asked in this lesson to determine the angular momentum of the body *at a given instant only,* select the system of axes which will be most convenient for your computations.

   *a. If the principal axes of inertia of the body at G are known,* use these axes as coordinate axes $x'$, $y'$, and $z'$, since the corresponding products of inertia of the body will be equal to zero. Resolve $\boldsymbol{\omega}$ into components $\omega_{x'}$, $\omega_{y'}$, and $\omega_{z'}$ along these axes and compute the principal moments of inertia $\bar{I}_{x'}$, $\bar{I}_{y'}$, and $\bar{I}_{z'}$. The corresponding components of the angular momentum $\mathbf{H}_G$ are

   $$H_{x'} = \bar{I}_{x'}\omega_{x'} \qquad H_{y'} = \bar{I}_{y'}\omega_{y'} \qquad H_{z'} = \bar{I}_{z'}\omega_{z'} \qquad (18.10)$$

   *b. If the principal axes of inertia of the body at G are not known,* you must use Eqs. (18.7) to determine the components of the angular momentum $\mathbf{H}_G$. These equations require prior computation of the *product of inertia* of the body as well as prior computation of its moments of inertia with respect to the selected axes.

   *c. The magnitude and direction cosines of $\mathbf{H}_G$* are obtained from formulas similar to those used in Statics [Sec. 2.12]. We have

   $$H_G = \sqrt{H_x^2 + H_y^2 + H_z^2}$$
   $$\cos\theta_x = \frac{H_x}{H_G} \qquad \cos\theta_y = \frac{H_y}{H_G} \qquad \cos\theta_z = \frac{H_z}{H_G}$$

   *d. Once $\mathbf{H}_G$ has been determined,* you can obtain the angular momentum of the body *about any given point O* by observing from Fig. 18.4 that

   $$\mathbf{H}_O = \bar{\mathbf{r}} \times m\bar{\mathbf{v}} + \mathbf{H}_G \qquad (18.11)$$

   where $\bar{\mathbf{r}}$ is the position vector of G relative to O, and $m\bar{\mathbf{v}}$ is the linear momentum of the body.

2. *To compute the angular momentum $\mathbf{H}_O$ of a rigid body with a fixed point O,* follow the procedure described in paragraph 1, except that you should now use axes centered at the fixed point O.

   *a. If the principal axes of inertia of the body at O are known,* resolve $\boldsymbol{\omega}$ into components along these axes [Sample Prob. 18.2]. The corresponding components of the angular momentum $\mathbf{H}_G$ are obtained from equations similar to Eqs. (18.10).

*b. If the principal axes of inertia of the body at O are not known,* you must compute the products as well as the moments of inertia of the body with respect to the axes that you have selected and use Eqs. (18.13) to determine the components of the angular momentum $\mathbf{H}_O$.

3. *To apply the principle of impulse and momentum* to the solution of a problem involving the three-dimensional motion of a rigid body, you will use the same vector equation that you used for plane motion in Chap. 17,

$$\textbf{Syst Momenta}_1 + \textbf{Syst Ext Imp}_{1\rightarrow 2} = \textbf{Syst Momenta}_2 \qquad (17.4)$$

where the initial and final systems of momenta are each represented by a *linear-momentum vector* $m\overline{\mathbf{v}}$ and an *angular-momentum couple* $\mathbf{H}_G$. Now, however, these vector-and-couple systems should be represented in three dimensions as shown in Fig. 18.6, and $\mathbf{H}_G$ should be determined as explained in paragraph 1.

*a. In problems involving the application of a known impulse to a rigid body,* draw the free-body-diagram equation corresponding to Eq. (17.4). Equating the components of the vectors involved, you will determine the final linear momentum $m\overline{\mathbf{v}}$ of the body and, thus, the corresponding velocity $\overline{\mathbf{v}}$ of its mass center. Equating moments about $G$, you will determine the final angular momentum $\mathbf{H}_G$ of the body. You will then substitute the values obtained for the components of $\mathbf{H}_G$ into Eqs. (18.10) or (18.7) and solve these equations for the corresponding values of the components of the angular velocity $\boldsymbol{\omega}$ of the body [Sample Prob. 18.1].

*b. In problems involving unknown impulses,* draw the free-body-diagram equation corresponding to Eq. (17.4) and write equations which do not involve the unknown impulses. Such equations can be obtained by equating moments about the point or line of impact.

4. *To compute the kinetic energy of a rigid body with a fixed point O,* resolve the angular velocity $\boldsymbol{\omega}$ into components along axes of your choice and compute the moments and products of inertia of the body with respect to these axes. As was the case for the computation of the angular momentum, use the principal axes of inertia $x'$, $y'$, and $z'$ if you can easily determine them. The products of inertia will then be zero [Sample Prob. 18.2], and the expression for the kinetic energy will reduce to

$$T = \tfrac{1}{2}(I_{x'}\omega_{x'}^2 + I_{y'}\omega_{y'}^2 + I_{z'}\omega_{z'}^2) \qquad (18.20)$$

If you must use axes other than the principal axes of inertia, the kinetic energy of the body should be expressed as shown in Eq. (18.19).

5. *To compute the kinetic energy of a rigid body in general motion,* consider the motion as the sum of a *translation with the mass center G and a rotation about G.* The kinetic energy associated with the translation is $\tfrac{1}{2}m\overline{v}^2$. If principal axes of inertia can be used, the kinetic energy associated with the rotation about $G$ can be expressed in the form used in Eq. (18.20). The total kinetic energy of the rigid body is then

$$T = \tfrac{1}{2}m\overline{v}^2 + \tfrac{1}{2}(\overline{I}_{x'}\omega_{x'}^2 + \overline{I}_{y'}\omega_{y'}^2 + \overline{I}_{z'}\omega_{z'}^2) \qquad (18.17)$$

If you must use axes other than the principal axes of inertia to determine the kinetic energy associated with the rotation about $G$, the total kinetic energy of the body should be expressed as shown in Eq. (18.16).

**18.1** A 5.4-kg slender rod is bent to form a rectangular frame which is attached to a shaft and rotates about its diagonal as shown. Knowing that the assembly has an angular velocity of constant magnitude $\omega = 10$ rad/s, determine the angular momentum $\mathbf{H}_G$ of the frame about its mass center $G$.

Fig. P18.2

**Fig. P18.1**

**18.2** A thin homogeneous square plate of mass $m$ and side $a$ is welded to a vertical shaft $AB$ with which it forms an angle of 45°. Knowing that the shaft rotates with a constant angular velocity $\boldsymbol{\omega}$, determine the angular momentum $\mathbf{H}_A$ of the plate about point $A$.

**18.3** A uniform 1.6 kg rod $AB$ is welded at its midpoint $G$ to a vertical shaft $GD$. Knowing that the shaft rotates with an angular velocity of constant magnitude $\omega = 1200$ rpm, determine the angular momentum $\mathbf{H}_G$ of the rod about $G$.

**18.4** A thin homogeneous disk of mass $m$ and radius $r$ is mounted on the horizontal axle $AB$. The plane of the disk forms an angle $\beta = 20°$ with the vertical. Knowing that the axle rotates with an angular velocity $\boldsymbol{\omega}$, determine the angle $\theta$ formed by the axle and the angular momentum of the disk about $G$.

**Fig. P18.3**

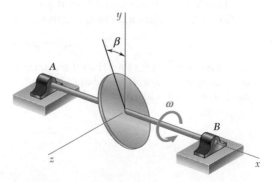

Fig. P18.4

**1158**

**18.5** A solid rectangular parallelepiped of mass $m$ has a square base of side $a$ and a length $2a$. Knowing that it rotates at the constant rate $\omega$ about its diagonal $AC'$ and that its rotation is observed from $A$ as counterclockwise, determine (a) the magnitude of the angular momentum $\mathbf{H}_G$ of the parallelepiped about its mass center $G$, (b) the angle that $\mathbf{H}_G$ forms with the diagonal $AC'$.

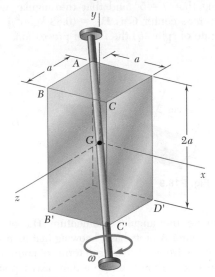

**Fig. P18.5**

**18.6** Solve Prob. 18.5, assuming that the solid rectangular parallelepiped has been replaced by a hollow one consisting of six thin metal plates welded together.

**18.7** A homogeneous disk of mass $m = 8$ kg rotates at the constant rate $\omega_1 = 12$ rad/s with respect to arm $OA$, which itself rotates at the constant rate $\omega_2 = 4$ rad/s about the $y$ axis. Determine the angular momentum $\mathbf{H}_A$ of the disk about its center $A$.

**Fig. P18.7**

**18.8** A homogeneous disk of mass $m = 6$ kg rotates at the constant rate $\omega_1 = 16$ rad/s with respect to arm $ABC$, which is welded to a shaft $DCE$ rotating at the constant rate $\omega_2 = 8$ rad/s. Determine the angular momentum $\mathbf{H}_A$ of the disk about its center $A$.

**Fig. P18.8**

**18.9** The 27.2 kg projectile shown has a radius of gyration of 6 cm about its axis of symmetry $Gx$ and a radius of gyration of 25.4 cm about the transverse axis $Gy$. Its angular velocity $\boldsymbol{\omega}$ can be resolved into two components: one component, directed along $Gx$, measures the *rate of spin* of the projectile, while the other component, directed along $GD$, measures its *rate of precession*. Knowing that $\theta = 5°$ and that the angular momentum of the projectile about its mass center $G$ is $\mathbf{H}_G = (0.88 \text{ kg.m}^2)\mathbf{i} - (0.0025 \text{ kgm}^2)\mathbf{j}$, determine ($a$) the rate of spin, ($b$) the rate of precession.

**Fig. P18.9**

**18.10** Determine the angular momentum $\mathbf{H}_A$ of the projectile of Prob. 18.9 about the center $A$ of its base, knowing that its mass center $G$ has a velocity $\overline{\mathbf{v}}$ of 594.4 m/s. Give your answer in terms of components respectively parallel to the $x$ and $y$ axes shown and to a third axis $z$ pointing toward you.

**18.11** Determine the angular momentum $\mathbf{H}_O$ of the disk of Sample Prob. 18.2 from the expressions obtained for its linear momentum $m\overline{\mathbf{v}}$ and its angular momentum $\mathbf{H}_G$, using Eqs. (18.11). Verify that the result obtained is the same as that obtained by direct computation.

**18.12** ($a$) Show that the angular momentum $\mathbf{H}_B$ of a rigid body about point $B$ can be obtained by adding to the angular momentum $\mathbf{H}_A$ of that body about point $A$ the vector product of the vector $\mathbf{r}_{A/B}$ drawn from $B$ to $A$ and the linear momentum $m\overline{\mathbf{v}}$ of the body:

$$\mathbf{H}_B = \mathbf{H}_A + \mathbf{r}_{A/B} \times m\overline{\mathbf{v}}$$

($b$) Further show that when a rigid body rotates about a fixed axis, its angular momentum is the same about any two points $A$ and $B$ located on the fixed axis ($\mathbf{H}_A = \mathbf{H}_B$) if, and only if, the mass center $G$ of the body is located on the fixed axis.

**18.13** Determine the angular momentum $\mathbf{H}_O$ of the disk of Prob. 18.7 about the fixed point $O$.

**18.14** Determine the angular momentum $\mathbf{H}_D$ of the disk of Prob. 18.8 about point $D$.

**18.15** Two L-shaped arms, each of mass 5 kg, are welded at the one-third points of the 600 mm shaft $AB$ to form the assembly shown. Knowing that the assembly rotates at the constant rate of 360 rpm, determine ($a$) the angular momentum $\mathbf{H}_A$ of the assembly about point $A$, ($b$) the angle formed by $\mathbf{H}_A$ and $AB$.

**18.16** For the assembly of Prob. 18.15, determine ($a$) the angular momentum $\mathbf{H}_B$ of the assembly about point $B$, ($b$) the angle formed by $\mathbf{H}_B$ and $BA$.

**Fig. P18.15**

**18.17** A 4.5 kg rod of uniform cross section is used to form the shaft shown. Knowing that the shaft rotates with a constant angular velocity $\boldsymbol{\omega}$ of magnitude 12 rad/s, determine (a) the angular momentum $\mathbf{H}_G$ of the shaft about its mass center G, (b) the angle formed by $\mathbf{H}_G$ and the axis AB.

**Fig. P18.17**

**18.18** Determine the angular momentum of the shaft of Prob. 18.17 about (a) point A, (b) point B.

**18.19** Two triangular plates, each of mass 8 kg, are welded to a vertical shaft AB. Knowing that the system rotates at the constant rate $\omega = 6$ rad/s, determine its angular momentum about G.

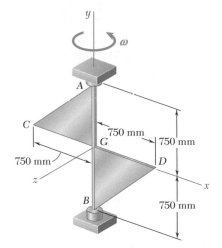

**Fig. P18.19**

**18.20** The assembly shown consists of two pieces of sheet aluminum of uniform thickness and total mass 1.6 kg welded to a light axle supported by bearings A and B. Knowing that the assembly rotates with an angular velocity of constant magnitude $\omega = 20$ rad/s, determine the angular momentum $\mathbf{H}_G$ of the assembly about point G.

**Fig. P18.20**

**18.21** A wire of uniform cross section having a mass of 1.2 kg/m is used to form the wire figure shown, which is suspended from cord AD. Knowing that an impulse $\mathbf{F}\Delta t = (5.4 \text{ N} \cdot \text{s})\mathbf{j}$ is applied to the wire figure at point E, determine (a) the velocity of the mass center G of the wire figure, (b) the angular velocity of the wire figure.

**Fig. P18.21 and P18.22**

**18.22** A wire of uniform cross section with a mass of 1.2 kg/m is used to form the wire figure shown, which is suspended from cord AD. Knowing that an impulse $\mathbf{F}\Delta t = -(5.4 \text{ N} \cdot \text{s})\mathbf{i}$ is applied to the wire figure at point E, determine (a) the velocity of the mass center G of the wire figure, (b) the angular velocity of the wire figure.

**18.23**   Three slender homogeneous rods, each of mass $m$ and length $d$, are welded together to form the assembly shown, which hangs from a wire attached at $G$. The assembly is hit at $A$ in a vertical downward direction. Denoting the corresponding impulse by $\mathbf{F}\,\Delta t$, determine immediately after the impact (a) the velocity of the mass center $G$, (b) the angular velocity of the assembly.

**Fig. P18.23**

**18.24**   Solve Prob. 18.23, assuming that the assembly is hit at $B$ in a direction opposite to that of the $z$ axis.

**18.25**   A uniform rod of mass $m$ and length $5a$ is bent into the shape shown and is suspended from a wire attached at point $B$. Knowing that the rod is hit at point $A$ in the negative $y$ direction and denoting the corresponding impulse by $-(F\Delta t)\mathbf{j}$, determine immediately after the impact (a) the velocity of the mass center $G$, (b) the angular velocity of the rod.

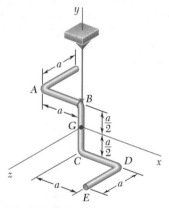

**Fig. P18.25**

**18.26**   Solve Prob. 18.25, assuming that the rod is hit at point $C$ in the negative $z$ direction.

**18.27**   A uniform rod of mass $m$ is bent into the shape shown and is suspended from a wire attached to its mass center $G$. The bent rod is hit at $A$ in a direction perpendicular to the plane containing the rod (in the positive $x$ direction). Denoting the corresponding impulse by $\mathbf{F}\,\Delta t$, determine immediately after the impact (a) the velocity of the mass center $G$, (b) the angular velocity of the rod.

**18.28**   Solve Prob. 18.27 assuming that the bent rod is hit at $B$.

**Fig. P18.27**

**18.29**  The coordinate axes shown represent the principal centroidal axes of inertia of a 1500-kg space probe whose radii of gyration are $k_x = 0.55$ m, $k_y = 0.57$ m, and $k_z = 0.50$ m. The probe has no angular velocity when a 0.14-kg meteorite strikes one of its solar panels at $A$ with a velocity $\mathbf{v}_0 = (720$ m/s$)\mathbf{i} - (900$ m/s$)\mathbf{j} + (960$ m/s$)\mathbf{k}$ relative to the probe. Knowing that the meteorite emerges on the other side of the panel with no change in the direction of its velocity, but with a speed reduced by 20 percent, determine the final angular velocity of the probe.

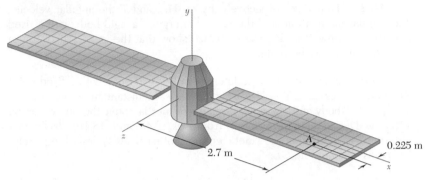

**Fig. P18.29 and P18.30**

**18.30**  The coordinate axes shown represent the principal centroidal axes of inertia of a 1500-kg space probe whose radii of gyration are $k_x = 0.55$ m, $k_y = 0.57$ m, and $k_z = 0.50$ m. The probe has no angular velocity when a 0.14-kg meteorite strikes one of its solar panels at $A$ and emerges on the other side of the panel with no change in the direction of its velocity, but with a speed reduced by 25 percent. Knowing that the final angular velocity of the probe is $\boldsymbol{\omega} = (0.05$ rad/s$)\mathbf{i} - (0.12$ rad/s$)\mathbf{j} + \omega_z\mathbf{k}$ and that the $x$ component of the resulting change in the velocity of the mass center of the probe is $-17$ mm/s, determine (a) the component $\omega_z$ of the final angular velocity of the probe, (b) the relative velocity $\mathbf{v}_0$ with which the meteorite strikes the panel.

**18.31**  A circular plate of radius $a$ and mass $m$ supported by a ball-and-socket joint at point $A$ is rotating about the $y$ axis with a constant angular velocity $\boldsymbol{\omega} = \omega_0\mathbf{j}$ when an obstruction is suddenly introduced at point $B$ in the $xy$ plane. Assuming the impact at point $B$ to be perfectly plastic ($e = 0$), determine immediately after the impact (a) the angular velocity of the plate, (b) the velocity of its mass center $G$.

**Fig. P18.31**

**18.32**  Determine the impulse exerted on the plate of Prob. 18.31 during the impact by (a) the obstruction at point $B$, (b) the support at point $A$.

**18.33**  A 1200-kg satellite designed to study the sun has an angular velocity $\boldsymbol{\omega}_0 = (0.050$ rad/s$)\mathbf{i} + (0.075$ rad/s$)\mathbf{k}$ when two small jets are activated at $A$ and $B$ in a direction parallel to the $y$ axis. Knowing that the coordinate axes are principal centroidal axes, that the radii of gyration of the satellite are $\bar{k}_x = 1.120$ m, $\bar{k}_y = 1.200$ m, $\bar{k}_z = 0.900$ m, and that each jet produces a 50-N thrust, determine (a) the required operating time of each jet if the angular velocity of the satellite is to be reduced to zero, (b) the resulting change in the velocity of the mass center $G$.

**18.34**  If jet $A$ in Prob. 18.33 is inoperative, determine (a) the required operating time of jet $B$ to reduce to zero the $x$ component of the angular velocity of the satellite, (b) the resulting final angular velocity, (c) the resulting change in the velocity of the mass center $G$.

**Fig. P18.33**

**Fig. P18.35**

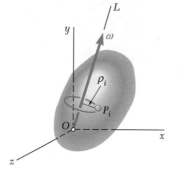

**Fig. P18.38**

**18.35** A rectangular plate of mass $m$ is falling with a velocity $\bar{\mathbf{v}}_0$ and no angular velocity when its corner $C$ strikes an obstruction. Assuming the impact to be perfectly plastic ($e = 0$), determine the angular velocity of the plate immediately after the impact.

**18.36** For the plate of Prob. 18.35, determine (a) the velocity of its mass center $G$ immediately after the impact, (b) the impulse exerted on the plate by the obstruction during the impact.

**18.37** Denoting, respectively, by $\boldsymbol{\omega}$, $\mathbf{H}_O$, and $T$ the angular velocity, the angular momentum, and the kinetic energy of a rigid body with a fixed point $O$, (a) prove that $\mathbf{H}_O \cdot \boldsymbol{\omega} = 2T$; (b) show that the angle $\theta$ between $\boldsymbol{\omega}$ and $\mathbf{H}_O$ will always be acute.

**18.38** Show that the kinetic energy of a rigid body with a fixed point $O$ can be expressed as $T = \frac{1}{2}I_{OL}\omega^2$, where $\omega$ is the instantaneous angular velocity of the body and $I_{OL}$ is its moment of inertia about the line of action $OL$ of $\boldsymbol{\omega}$. Derive this expression (a) from Eqs. (9.46) and (18.19), (b) by considering $T$ as the sum of the kinetic energies of particles $P_i$ describing circles of radius $\rho_i$ about line $OL$.

**18.39** Determine the kinetic energy of the rectangular frame of Prob. 18.1.

**18.40** Determine the kinetic energy of the square plate of Prob. 18.2.

**18.41** Determine the kinetic energy of rod $AB$ of Prob. 18.3.

**18.42** Determine the kinetic energy of the disk of Prob. 18.4.

**18.43** Determine the kinetic energy of the solid parallelepiped of Prob. 18.5.

**18.44** Determine the kinetic energy of the hollow parallelepiped of Prob. 18.6.

**18.45** Determine the kinetic energy of the disk of Prob. 18.7.

**18.46** Determine the kinetic energy of the disk of Prob. 18.8.

**18.47** Determine the kinetic energy of the assembly of Prob. 18.15.

**18.48** Determine the kinetic energy of the shaft of Prob. 18.17.

**18.49** Determine the kinetic energy of the assembly of Prob. 18.19.

**18.50** Determine the kinetic energy imparted to the plate of Sample Prob. 18.1.

**18.51** Determine the kinetic energy of the space probe of Prob. 18.29 in its motion about its mass center after its collision with the meteorite.

**18.52** Determine the kinetic energy of the space probe of Prob. 18.30 in its motion about its mass center after its collision with the meteorite.

**18.53** Determine the kinetic energy lost when the plate of Prob. 18.31 hits the obstruction at point $B$.

**18.54** Determine the kinetic energy lost when corner $C$ of the plate of Prob. 18.35 hits the obstruction.

As indicated in Sec. 18.2, the fundamental equations

$$\Sigma \mathbf{F} = m\overline{\mathbf{a}} \qquad (18.1)$$

$$\Sigma \mathbf{M}_G = \dot{\mathbf{H}}_G \qquad (18.2)$$

remain valid in the most general case of the motion of a rigid body. Before Eq. (18.2) could be applied to the three-dimensional motion of a rigid body, however, it was necessary to derive Eqs. (18.7), which relate the components of the angular momentum $\mathbf{H}_G$ and those of the angular velocity $\boldsymbol{\omega}$. It still remains for us to find an effective and convenient way for computing the components of the derivative $\dot{\mathbf{H}}_G$ of the angular momentum.

Since $\mathbf{H}_G$ represents the angular momentum of the body in its motion relative to centroidal axes $GX'Y'Z'$ of fixed orientation (Fig. 18.9), and since $\dot{\mathbf{H}}_G$ represents the rate of change of $\mathbf{H}_G$ with respect to the same axes, it would seem natural to use components of $\boldsymbol{\omega}$ and $\mathbf{H}_G$ along the axes $X'$, $Y'$, $Z'$ in writing the relations (18.7). But since the body rotates, its moments and products of inertia would change continually, and it would be necessary to determine their values as functions of the time. It is therefore more convenient to use axes $x$, $y$, $z$ attached to the body, ensuring that its moments and products of inertia will maintain the same values during the motion. This is permissible since, as indicated earlier, the transformation of $\boldsymbol{\omega}$ into $\mathbf{H}_G$ is independent of the system of coordinate axes selected. The angular velocity $\boldsymbol{\omega}$, however, should still be *defined* with respect to the frame $GX'Y'Z'$ of fixed orientation. The vector $\boldsymbol{\omega}$ may then be *resolved* into components along the rotating $x$, $y$, and $z$ axes. Applying the relations (18.7), we obtain the *components* of the vector $\mathbf{H}_G$ along the rotating axes. The vector $\mathbf{H}_G$, however, represents the angular momentum about $G$ of the body *in its motion relative to the frame $GX'Y'Z'$*.

Differentiating with respect to $t$ the components of the angular momentum in (18.7), we define the rate of change of the vector $\mathbf{H}_G$ with respect to the rotating frame $Gxyz$:

$$(\dot{\mathbf{H}}_G)_{Gxyz} = \dot{H}_x\mathbf{i} + \dot{H}_y\mathbf{j} + \dot{H}_z\mathbf{k} \qquad (18.21)$$

where $\mathbf{i}$, $\mathbf{j}$, $\mathbf{k}$ are the unit vectors along the rotating axes. Recalling from Sec. 15.10 that the rate of change $\dot{\mathbf{H}}_G$ of the vector $\mathbf{H}_G$ with respect to the frame $GX'Y'Z'$ is found by adding to $(\dot{\mathbf{H}}_G)_{Gxyz}$ the vector product $\boldsymbol{\Omega} \times \mathbf{H}_G$, where $\boldsymbol{\Omega}$ denotes the angular velocity of the rotating frame, we write

$$\dot{\mathbf{H}}_G = (\dot{\mathbf{H}}_G)_{Gxyz} + \boldsymbol{\Omega} \times \mathbf{H}_G \qquad (18.22)$$

where $\mathbf{H}_G$ = angular momentum of body with respect to frame $GX'Y'Z'$ of fixed orientation

$(\dot{\mathbf{H}}_G)_{Gxyz}$ = rate of change of $\mathbf{H}_G$ with respect to rotating frame $Gxyz$, to be computed from the relations (18.7) and (18.21)

$\boldsymbol{\Omega}$ = angular velocity of rotating frame $Gxyz$

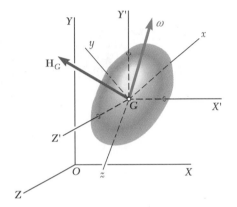

**Fig. 18.9**

Substituting for $\dot{\mathbf{H}}_G$ from (18.22) into (18.2), we have

$$\Sigma \mathbf{M}_G = (\dot{\mathbf{H}}_G)_{Gxyz} + \boldsymbol{\Omega} \times \mathbf{H}_G \qquad (18.23)$$

If the rotating frame is attached to the body, as has been assumed in this discussion, its angular velocity $\boldsymbol{\Omega}$ is identically equal to the angular velocity $\boldsymbol{\omega}$ of the body. There are many applications, however, where it is advantageous to use a frame of reference which is not actually attached to the body but rotates in an independent manner. For example, if the body considered is axisymmetrical, as in Sample Prob. 18.5 or Sec. 18.9, it is possible to select a frame of reference with respect to which the moments and products of inertia of the body remain constant, but which rotates less than the body itself.[†] As a result, it is possible to obtain simpler expressions for the angular velocity $\boldsymbol{\omega}$ and the angular momentum $\mathbf{H}_G$ of the body than could have been obtained if the frame of reference had actually been attached to the body. It is clear that in such cases the angular velocity $\boldsymbol{\Omega}$ of the rotating frame and the angular velocity $\boldsymbol{\omega}$ of the body are different.

## *18.6. EULER'S EQUATIONS OF MOTION. EXTENSION OF D'ALEMBERT'S PRINCIPLE TO THE MOTION OF A RIGID BODY IN THREE DIMENSIONS

If the $x$, $y$, and $z$ axes are chosen to coincide with the principal axes of inertia of the body, the simplified relations (18.10) can be used to determine the components of the angular momentum $\mathbf{H}_G$. Omitting the primes from the subscripts, we write

$$\mathbf{H}_G = \bar{I}_x \omega_x \mathbf{i} + \bar{I}_y \omega_y \mathbf{j} + \bar{I}_z \omega_z \mathbf{k} \qquad (18.24)$$

where $\bar{I}_x$, $\bar{I}_y$, and $\bar{I}_z$ denote the principal centroidal moments of inertia of the body. Substituting for $\mathbf{H}_G$ from (18.24) into (18.23) and setting $\boldsymbol{\Omega} = \boldsymbol{\omega}$, we obtain the three scalar equations

$$\Sigma M_x = \bar{I}_x \dot{\omega}_x - (\bar{I}_y - \bar{I}_z)\omega_y \omega_z$$
$$\Sigma M_y = \bar{I}_y \dot{\omega}_y - (\bar{I}_z - \bar{I}_x)\omega_z \omega_x \qquad (18.25)$$
$$\Sigma M_z = \bar{I}_z \dot{\omega}_z - (\bar{I}_x - \bar{I}_y)\omega_x \omega_y$$

These equations, called *Euler's equations of motion* after the Swiss mathematician Leonhard Euler (1707–1783), can be used to analyze the motion of a rigid body about its mass center. In the following sections, however, Eq. (18.23) will be used in preference to Eqs. (18.25), since the former is more general and the compact vectorial form in which it is expressed is easier to remember.

Writing Eq. (18.1) in scalar form, we obtain the three additional equations

$$\Sigma F_x = m\bar{a}_x \qquad \Sigma F_y = m\bar{a}_y \qquad \Sigma F_z = m\bar{a}_z \qquad (18.26)$$

which, together with Euler's equations, form a system of six differential equations. Given appropriate initial conditions, these differential

---

[†]More specifically, the frame of reference will have no spin (see Sec. 18.9).

equations have a unique solution. Thus, the motion of a rigid body in three dimensions is completely defined by the resultant and the moment resultant of the external forces acting on it. This result will be recognized as a generalization of a similar result obtained in Sec. 16.4 in the case of the plane motion of a rigid slab. It follows that in three as well as two dimensions, two systems of forces which are equipollent are also equivalent; that is, they have the same effect on a given rigid body.

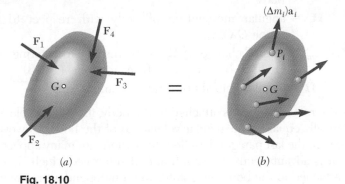

**Fig. 18.10**

Considering in particular the system of the external forces acting on a rigid body (Fig. 18.10a) and the system of the effective forces associated with the particles forming the rigid body (Fig. 18.10b), we can state that the two systems—which were shown in Sec. 14.2 to be equipollent—are also equivalent. This is the extension of d'Alembert's principle to the three-dimensional motion of a rigid body. Replacing the effective forces in Fig. 18.10b by an equivalent force-couple system, we verify that the system of the external forces acting on a rigid body in three-dimensional motion is equivalent to the system consisting of the vector $m\bar{\mathbf{a}}$ attached at the mass center $G$ of the body and the couple of moment $\dot{\mathbf{H}}_G$ (Fig. 18.11), where $\dot{\mathbf{H}}_G$ is obtained from the relations (18.7) and (18.22). Note that the equivalence of the systems of vectors shown in Fig. 18.10 and in Fig. 18.11 has been indicated by *red* equals signs. Problems involving the three-dimensional motion of a rigid body can be solved by considering the free-body-diagram equation represented in Fig. 18.11 and writing appropriate scalar equations relating the components or moments of the external and effective forces (see Sample Prob. 18.3).

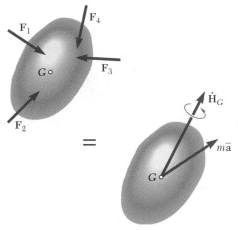

**Fig. 18.11**

### *18.7. MOTION OF A RIGID BODY ABOUT A FIXED POINT

When a rigid body is constrained to rotate about a fixed point $O$, it is desirable to write an equation involving the moments about $O$ of the external and effective forces, since this equation will not contain the unknown reaction at $O$. While such an equation can be obtained from Fig. 18.11, it may be more convenient to write it by considering the rate of change of the angular momentum $\mathbf{H}_O$ of the body about the fixed point $O$ (Fig. 18.12). Recalling Eq. (14.11), we write

$$\Sigma \mathbf{M}_O = \dot{\mathbf{H}}_O \qquad (18.27)$$

where $\dot{\mathbf{H}}_O$ denotes the rate of change of the vector $\mathbf{H}_O$ with respect to the fixed frame $OXYZ$. A derivation similar to that used in Sec. 18.5

**Fig. 18.12**

**Photo 18.3** The revolving radio telescope is an
example of a structure constrained to rotate
about a fixed point.

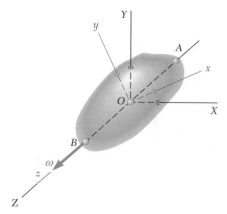

**Fig. 18.13**

enables us to relate $\dot{\mathbf{H}}_O$ to the rate of change $(\dot{\mathbf{H}}_O)_{Oxyz}$ of $\mathbf{H}_O$ with
respect to the rotating frame $Oxyz$. Substitution into (18.27) leads to
the equation

$$\Sigma\mathbf{M}_O = (\dot{\mathbf{H}}_O)_{Oxyz} + \boldsymbol{\Omega} \times \mathbf{H}_O \qquad (18.28)$$

where  $\Sigma\mathbf{M}_O$ = sum of moments about $O$ of forces applied to rigid
body

$\mathbf{H}_O$ = angular momentum of body with respect to fixed
frame $OXYZ$

$(\dot{\mathbf{H}}_O)_{Oxyz}$ = rate of change of $\mathbf{H}_O$ with respect to rotating frame
$Oxyz$, to be computed from relations (18.13)

$\boldsymbol{\Omega}$ = angular velocity of rotating frame $Oxyz$

If the rotating frame is attached to the body, its angular velocity $\boldsymbol{\Omega}$
is identically equal to the angular velocity $\boldsymbol{\omega}$ of the body. However, as
indicated in the last paragraph of Sec. 18.5, there are many applications
where it is advantageous to use a frame of reference which is not actually attached to the body but rotates in an independent manner.

## *18.8. ROTATION OF A RIGID BODY ABOUT A FIXED AXIS

Equation (18.28), which was derived in the preceding section, will be
used to analyze the motion of a rigid body constrained to rotate about
a fixed axis $AB$ (Fig. 18.13). First, we note that the angular velocity
of the body with respect to the fixed frame $OXYZ$ is represented by
the vector $\boldsymbol{\omega}$ directed along the axis of rotation. Attaching the moving frame of reference $Oxyz$ to the body, with the $z$ axis along $AB$, we
have $\boldsymbol{\omega} = \omega\mathbf{k}$. Substituting $\omega_x = 0$, $\omega_y = 0$, $\omega_z = \omega$ into the relations
(18.13), we obtain the components along the rotating axes of the
angular momentum $\mathbf{H}_O$ of the body about $O$:

$$H_x = -I_{xz}\omega \qquad H_y = -I_{yz}\omega \qquad H_z = I_z\omega$$

Since the frame $Oxyz$ is attached to the body, we have $\boldsymbol{\Omega} = \boldsymbol{\omega}$ and
Eq. (18.28) yields

$$\begin{aligned}
\Sigma\mathbf{M}_O &= (\dot{\mathbf{H}}_O)_{Oxyz} + \boldsymbol{\omega} \times \mathbf{H}_O \\
&= (-I_{xz}\mathbf{i} - I_{yz}\mathbf{j} + I_z\mathbf{k})\dot{\omega} + \omega\mathbf{k} \times (-I_{xz}\mathbf{i} - I_{yz}\mathbf{j} + I_z\mathbf{k})\omega \\
&= (-I_{xz}\mathbf{i} - I_{yz}\mathbf{j} + I_z\mathbf{k})\alpha + (-I_{xz}\mathbf{j} + I_{yz}\mathbf{i})\omega^2
\end{aligned}$$

The result obtained can be expressed by the three scalar equations

$$\begin{aligned}
\Sigma M_x &= -I_{xz}\alpha + I_{yz}\omega^2 \\
\Sigma M_y &= -I_{yz}\alpha - I_{xz}\omega^2 \qquad (18.29) \\
\Sigma M_z &= I_z\alpha
\end{aligned}$$

When the forces applied to the body are known, the angular
acceleration $\alpha$ can be obtained from the last of Eqs. (18.29). The angular velocity $\omega$ is then determined by integration and the values
obtained for $\alpha$ and $\omega$ substituted into the first two equations (18.29).
These equations plus the three equations (18.26) which define the
motion of the mass center of the body can then be used to determine
the reactions at the bearings $A$ and $B$.

It is possible to select axes other than the ones shown in Fig. 18.13 to analyze the rotation of a rigid body about a fixed axis. In many cases, the principal axes of inertia of the body will be found more advantageous. It is therefore wise to revert to Eq. (18.28) and to select the system of axes which best fits the problem under consideration.

If the rotating body is symmetrical with respect to the $xy$ plane, the products of inertia $I_{xz}$ and $I_{yz}$ are equal to zero and Eqs. (18.29) reduce to

$$\Sigma M_x = 0 \qquad \Sigma M_y = 0 \qquad \Sigma M_z = I_z \alpha \qquad (18.30)$$

which is in accord with the results obtained in Chap. 16. If, on the other hand, the products of inertia $I_{xz}$ and $I_{yz}$ are different from zero, the sum of the moments of the external forces about the $x$ and $y$ axes will also be different from zero, even when the body rotates at a constant rate $\omega$. Indeed, in the latter case, Eqs. (18.29) yield

$$\Sigma M_x = I_{yz}\omega^2 \qquad \Sigma M_y = -I_{xz}\omega^2 \qquad \Sigma M_z = 0 \qquad (18.31)$$

This last observation leads us to discuss the *balancing of rotating shafts*. Consider, for instance, the crankshaft shown in Fig. 18.14*a*, which is symmetrical about its mass center $G$. We first observe that when the crankshaft is at rest, it exerts no lateral thrust on its supports, since its center of gravity $G$ is located directly above $A$. The shaft is said to be *statically balanced*. The reaction at $A$, often referred to as a *static reaction*, is vertical and its magnitude is equal to the weight $W$ of the shaft. Let us now assume that the shaft rotates with a constant angular velocity $\omega$. Attaching our frame of reference to the shaft, with its origin at $G$, the $z$ axis along $AB$, and the $y$ axis in the plane of symmetry of the shaft (Fig. 18.14*b*), we note that $I_{xz}$ is zero and that $I_{yz}$ is positive. According to Eqs. (18.31), the external forces include a couple of moment $I_{yz}\omega^2\mathbf{i}$. Since this couple is formed by the reaction at $B$ and the horizontal component of the reaction at $A$, we have

$$\mathbf{A}_y = \frac{I_{yz}\omega^2}{l}\mathbf{j} \qquad \mathbf{B} = -\frac{I_{yz}\omega^2}{l}\mathbf{j} \qquad (18.32)$$

Since the bearing reactions are proportional to $\omega^2$, the shaft will have a tendency to tear away from its bearings when rotating at high speeds. Moreover, since the bearing reactions $\mathbf{A}_y$ and $\mathbf{B}$, called *dynamic reactions*, are contained in the $yz$ plane, they rotate with the shaft and cause the structure supporting it to vibrate. These undesirable effects will be avoided if, by rearranging the distribution of mass around the shaft or by adding corrective masses, we let $I_{yz}$ become equal to zero. The dynamic reactions $\mathbf{A}_y$ and $\mathbf{B}$ will vanish and the reactions at the bearings will reduce to the static reaction $\mathbf{A}_z$, the direction of which is fixed. The shaft will then be *dynamically as well as statically balanced*.

**Photo 18.4**   The forces exerted by a rotating automobile crankshaft on its bearings are the static and dynamic reactions. The crankshaft can be designed to be dynamically as well as statically balanced.

(a)

(b)

**Fig. 18.14**

## SAMPLE PROBLEM 18.3

A slender rod $AB$ of length $L = 2.4$ m and weight $W = 18$ kg is pinned at $A$ to a vertical axle $DE$ which rotates with a constant angular velocity $\boldsymbol{\omega}$ of 15 rad/s. The rod is maintained in position by means of a horizontal wire $BC$ attached to the axle and to the end $B$ of the rod. Determine the tension in the wire and the reaction at $A$.

## SOLUTION

The effective forces reduce to the vector $m\bar{\mathbf{a}}$ attached at $G$ and the couple $\dot{\mathbf{H}}_G$. Since $G$ describes a horizontal circle of radius $\bar{r} = \frac{1}{2}L \cos \beta$ at the constant rate $\omega$, we have

$$\bar{\mathbf{a}} = \mathbf{a}_n = -\bar{r}\omega^2 \mathbf{I} = -(\tfrac{1}{2}L \cos \beta)\omega^2 \mathbf{I} = -(135 \text{ m/s}^2)\mathbf{I}$$

$$m\bar{\mathbf{a}} = 18 \text{ kg}(-135) = -(2430 \text{ N})\mathbf{I}$$

**Determination of $\dot{\mathbf{H}}_G$.** We first compute the angular momentum $\mathbf{H}_G$. Using the principal centroidal axes of inertia $x$, $y$, $z$, we write

$$\bar{I}_x = \tfrac{1}{12}mL^2 \qquad \bar{I}_y = 0 \qquad \bar{I}_z = \tfrac{1}{12}mL^2$$
$$\omega_x = -\omega \cos \beta \quad \omega_y = \omega \sin \beta \quad \omega_z = 0$$
$$\mathbf{H}_G = \bar{I}_x\omega_x\mathbf{i} + \bar{I}_y\omega_y\mathbf{j} + \bar{I}_z\omega_z\mathbf{k}$$
$$\mathbf{H}_G = -\tfrac{1}{12}mL^2\omega \cos \beta \, \mathbf{i}$$

The rate of change $\dot{\mathbf{H}}_G$ of $\mathbf{H}_G$ with respect to axes of fixed orientation is obtained from Eq. (18.22). Observing that the rate of change $(\dot{\mathbf{H}}_G)_{Gxyz}$ of $\mathbf{H}_G$ with respect to the rotating frame $Gxyz$ is zero, and that the angular velocity $\boldsymbol{\Omega}$ of that frame is equal to the angular velocity $\boldsymbol{\omega}$ of the rod, we have

$$\dot{\mathbf{H}}_G = (\dot{\mathbf{H}}_G)_{Gxyz} + \boldsymbol{\omega} \times \mathbf{H}_G$$
$$\dot{\mathbf{H}}_G = 0 + (-\omega \cos \beta \, \mathbf{i} + \omega \sin \beta \, \mathbf{j}) \times (-\tfrac{1}{12}mL^2\omega \cos \beta \, \mathbf{i})$$
$$\dot{\mathbf{H}}_G = \tfrac{1}{12}mL^2\omega^2 \sin \beta \cos \beta \, \mathbf{k} = (842 \text{ N} \cdot \text{m})\mathbf{k}$$

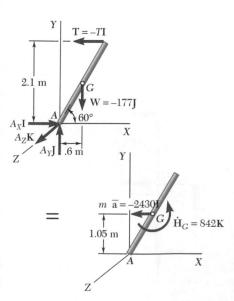

**Equations of Motion.** Expressing that the system of the external forces is equivalent to the system of the effective forces, we write

$\Sigma \mathbf{M}_A = \Sigma(\mathbf{M}_A)_{\text{eff}}$:
$$2.1\mathbf{J} \times (-T\mathbf{I}) + .6\mathbf{I} \times (-177\mathbf{J}) = 1.05\mathbf{J} \times (-2430\mathbf{I}) + 842\mathbf{K}$$
$$(2.1T - 106.2)\mathbf{K} = (2551.5 + 842)\mathbf{K} \qquad T = 1666.5 \text{ N} \quad \blacktriangleleft$$

$\Sigma \mathbf{F} = \Sigma \mathbf{F}_{\text{eff}}$: $\quad A_X\mathbf{I} + A_Y\mathbf{J} + A_Z\mathbf{K} - 1666.5\mathbf{I} - 177\mathbf{J} = -2430\mathbf{I}$
$$\mathbf{A} = -(763.5 \text{ N})\mathbf{I} + (177 \text{ N})\mathbf{J} \quad \blacktriangleleft$$

**Remark.** The value of $T$ could have been obtained from $\mathbf{H}_A$ and Eq. (18.28). However, the method used here also yields the reaction at $A$. Moreover, it draws attention to the effect of the asymmetry of the rod on the solution of the problem by clearly showing that both the vector $m\bar{\mathbf{a}}$ and the couple $\dot{\mathbf{H}}_G$ must be used to represent the effective forces.

# SAMPLE PROBLEM 18.4

Two 100-mm rods $A$ and $B$, each of mass 300 g, are welded to shaft $CD$ which is supported by bearings at $C$ and $D$. If a couple $\mathbf{M}$ of magnitude equal to 6 N · m is applied to the shaft, determine the components of the dynamic reactions at $C$ and $D$ at the instant when the shaft has reached an angular velocity at 1200 rpm. Neglect the moment of inertia of the shaft itself.

## SOLUTION

**Angular Momentum about $O$.** We attach to the body the frame of reference $Oxyz$ and note that the axes chosen are not principal axes of inertia for the body. Since the body rotates about the $x$ axis, we have $\omega_x = \omega$ and $\omega_y = \omega_z = 0$. Substituting into Eqs. (18.13),

$$H_x = I_x\omega \qquad H_y = -I_{xy}\omega \qquad H_z = -I_{xz}\omega$$
$$\mathbf{H}_O = (I_x\mathbf{i} - I_{xy}\mathbf{j} - I_{xz}\mathbf{k})\omega$$

**Moments of the External Forces about $O$.** Since the frame of reference rotates with the angular velocity $\boldsymbol{\omega}$, Eq. (18.28) yields

$$\Sigma\mathbf{M}_O = (\dot{\mathbf{H}}_O)_{Oxyz} + \boldsymbol{\omega} \times \mathbf{H}_O$$
$$= (I_x\mathbf{i} - I_{xy}\mathbf{j} - I_{xz}\mathbf{k})\alpha + \omega\mathbf{i} \times (I_x\mathbf{i} - I_{xy}\mathbf{j} - I_{xz}\mathbf{k})\omega$$
$$= I_x\alpha\mathbf{i} - (I_{xy}\alpha - I_{xz}\omega^2)\mathbf{j} - (I_{xz}\alpha + I_{xy}\omega^2)\mathbf{k} \qquad (1)$$

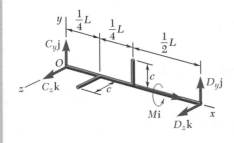

**Dynamic Reaction at $D$.** The external forces consist of the weights of the shaft and rods, the couple $\mathbf{M}$, the static reactions at $C$ and $D$, and the dynamic reactions at $C$ and $D$. Since the weights and static reactions are balanced, the external forces reduce to the couple $\mathbf{M}$ and the dynamic reactions $\mathbf{C}$ and $\mathbf{D}$ as shown in the figure. Taking moments about $O$, we have

$$\Sigma\mathbf{M}_O = L\mathbf{i} \times (D_y\mathbf{j} + D_z\mathbf{k}) + M\mathbf{i} = M\mathbf{i} - D_zL\mathbf{j} + D_yL\mathbf{k} \qquad (2)$$

Equating the coefficients of the unit vector $\mathbf{i}$ in (1) and (2):

$$M = I_x\alpha \qquad M = 2(\tfrac{1}{3}mc^2)\alpha \qquad \alpha = 3M/2mc^2$$

Equating the coefficients of $\mathbf{k}$ and $\mathbf{j}$ in (1) and (2):

$$D_y = -(I_{xz}\alpha + I_{xy}\omega^2)/L \qquad D_z = (I_{xy}\alpha - I_{xz}\omega^2)/L \qquad (3)$$

Using the parallel-axis theorem, and noting that the product of inertia of each rod is zero with respect to centroidal axes, we have

$$I_{xy} = \Sigma m\bar{x}\bar{y} = m(\tfrac{1}{2}L)(\tfrac{1}{2}c) = \tfrac{1}{4}mLc$$
$$I_{xz} = \Sigma m\bar{x}\bar{z} = m(\tfrac{1}{4}L)(\tfrac{1}{2}c) = \tfrac{1}{8}mLc$$

Substituting into (3) the values found for $I_{xy}$, $I_{xz}$, and $\alpha$:

$$D_y = -\tfrac{3}{16}(M/c) - \tfrac{1}{4}mc\omega^2 \qquad D_z = \tfrac{3}{8}(M/c) - \tfrac{1}{8}mc\omega^2$$

Substituting $\omega = 1200$ rpm $= 125.7$ rad/s, $c = 0.100$ m, $M = 6$ N · m, and $m = 0.300$ kg, we have

$$D_y = -129.8 \text{ N} \qquad D_z = -36.8 \text{ N} \quad \blacktriangleleft$$

**Dynamic Reaction at $C$.** Using a frame of reference attached at $D$, we obtain equations similar to Eqs. (3), which yield

$$C_y = -152.2 \text{ N} \qquad C_z = -155.2 \text{ N} \quad \blacktriangleleft$$

## SAMPLE PROBLEM 18.5

A homogeneous disk of radius $r$ and mass $m$ is mounted on an axle $OG$ of length $L$ and negligible mass. The axle is pivoted at the fixed point $O$ and the disk is constrained to roll on a horizontal floor. Knowing that the disk rotates counterclockwise at the constant rate $\omega_1$ about the axle, determine (a) the force (assumed vertical) exerted by the floor on the disk, (b) the reaction at the pivot $O$.

## SOLUTION

The effective forces reduce to the vector $m\bar{\mathbf{a}}$ attached at $G$ and the couple $\dot{\mathbf{H}}_G$. Recalling from Sample Prob. 18.2 that the axle rotates about the $y$ axis at the rate $\omega_2 = r\omega_1/L$, we write

$$m\bar{\mathbf{a}} = -mL\omega_2^2\mathbf{i} = -mL(r\omega_1/L)^2\mathbf{i} = -(mr^2\omega_1^2/L)\mathbf{i} \qquad (1)$$

**Determination of $\dot{\mathbf{H}}_G$.** We recall from Sample Prob. 18.2 that the angular momentum of the disk about $G$ is

$$\mathbf{H}_G = \tfrac{1}{2}mr^2\omega_1\left(\mathbf{i} - \frac{r}{2L}\mathbf{j}\right)$$

where $\mathbf{H}_G$ is resolved into components along the rotating axes $x'$, $y'$, $z'$, with $x'$ along $OG$ and $y'$ vertical. The rate of change $\dot{\mathbf{H}}_G$ of $\mathbf{H}_G$ with respect to axes of fixed orientation is obtained from Eq. (18.22). Noting that the rate of change $(\dot{\mathbf{H}}_G)_{Gx'y'z'}$ of $\mathbf{H}_G$ with respect to the rotating frame is zero, and that the angular velocity $\boldsymbol{\Omega}$ of that frame is

$$\boldsymbol{\Omega} = -\omega_2\mathbf{j} = -\frac{r\omega_1}{L}\mathbf{j}$$

we have

$$\dot{\mathbf{H}}_G = (\dot{\mathbf{H}}_G)_{Gx'y'z'} + \boldsymbol{\Omega} \times \mathbf{H}_G$$
$$= 0 - \frac{r\omega_1}{L}\mathbf{j} \times \tfrac{1}{2}mr^2\omega_1\left(\mathbf{i} - \frac{r}{2L}\mathbf{j}\right)$$
$$= \tfrac{1}{2}mr^2(r/L)\omega_1^2\mathbf{k} \qquad (2)$$

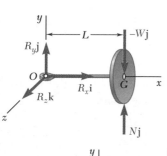

**Equations of Motion.** Expressing that the system of the external forces is equivalent to the system of the effective forces, we write

$$\Sigma\mathbf{M}_O = \Sigma(\mathbf{M}_O)_{\text{eff}}: \qquad L\mathbf{i} \times (N\mathbf{j} - W\mathbf{j}) = \dot{\mathbf{H}}_G$$
$$(N - W)L\mathbf{k} = \tfrac{1}{2}mr^2(r/L)\omega_1^2\mathbf{k}$$

$$N = W + \tfrac{1}{2}mr(r/L)^2\omega_1^2 \qquad \mathbf{N} = [W + \tfrac{1}{2}mr(r/L)^2\omega_1^2]\mathbf{j} \quad (3) \quad \blacktriangleleft$$

$$\Sigma\mathbf{F} = \Sigma\mathbf{F}_{\text{eff}}: \qquad \mathbf{R} + N\mathbf{j} - W\mathbf{j} = m\bar{\mathbf{a}}$$

Substituting for $N$ from (3), for $m\bar{\mathbf{a}}$ from (1), and solving for $\mathbf{R}$, we have

$$\mathbf{R} = -(mr^2\omega_1^2/L)\mathbf{i} - \tfrac{1}{2}mr(r/L)^2\omega_1^2\mathbf{j}$$

$$\mathbf{R} = -\frac{mr^2\omega_1^2}{L}\left(\mathbf{i} + \frac{r}{2L}\mathbf{j}\right) \quad \blacktriangleleft$$

In this lesson you will be asked to solve problems involving the *three-dimensional motion of rigid bodies*. The method you will use is basically the same that you used in Chap. 16 in your study of the plane motion of rigid bodies. You will draw a free-body-diagram equation showing that the system of the external forces is equivalent to the system of the effective forces, and you will equate sums of components and sums of moments on both sides of this equation. Now, however, the system of the effective forces will be represented by the vector $m\overline{\mathbf{a}}$ and a couple vector $\dot{\mathbf{H}}_G$, the determination of which will be explained in paragraphs 1 and 2 below.

To solve a problem involving the three-dimensional motion of a rigid body, you should take the following steps:

1. **Determine the angular momentum $\mathbf{H}_G$ of the body about its mass center $G$** from its angular velocity $\boldsymbol{\omega}$ with respect to a frame of reference $GX'Y'Z'$ of fixed orientation. This is an operation you learned to perform in the preceding lesson. However, since the configuration of the body will be changing with time, it will now be necessary for you to use an auxiliary system of axes $Gx'y'z'$ (Fig. 18.9) to compute the components of $\boldsymbol{\omega}$ and the moments and products of inertia of the body. These axes may be rigidly attached to the body, in which case their angular velocity is equal to $\boldsymbol{\omega}$ [Sample Probs. 18.3 and 18.4], or they may have an angular velocity $\boldsymbol{\Omega}$ of their own [Sample Prob. 18.5].

Recall the following from the preceding lesson:

   *a. If the principal axes of inertia of the body at $G$ are known,* use these axes as coordinate axes $x'$, $y'$, and $z'$, since the corresponding products of inertia of the body will be equal to zero. (Note that if the body is axisymmetric, these axes do not need to be rigidly attached to the body.) Resolve $\boldsymbol{\omega}$ into components $\omega_{x'}$, $\omega_{y'}$, and $\omega_{z'}$ along these axes and compute the principal moments of inertia $\overline{I}_{x'}$, $\overline{I}_{y'}$, and $\overline{I}_{z'}$. The corresponding components of the angular momentum $\mathbf{H}_G$ are

$$H_{x'} = \overline{I}_{x'}\omega_{x'} \qquad H_{y'} = \overline{I}_{y'}\omega_{y'} \qquad H_{z'} = \overline{I}_{z'}\omega_{z'} \qquad (18.10)$$

   *b. If the principal axes of inertia of the body at $G$ are not known,* you must use Eqs. (18.7) to determine the components of the angular momentum $\mathbf{H}_G$. These equations require your prior computation of the *products of inertia* of the body, as well as of its moments of inertia, with respect to the selected axes.

2. **Compute the rate of change $\dot{\mathbf{H}}_G$ of the angular momentum $\mathbf{H}_G$ with respect to the frame $GX'Y'Z'$.** Note that this frame has a *fixed orientation*, while the frame $Gx'y'z'$ you used when you calculated the components of the vector $\boldsymbol{\omega}$ was a *rotating frame*. We refer you to our discussion in Sec. 15.10 of the rate of change of a vector with respect to a rotating frame. Recalling Eq. (15.31), you will express the rate of change $\dot{\mathbf{H}}_G$ as follows:

$$\dot{\mathbf{H}}_G = (\dot{\mathbf{H}}_G)_{Gx'y'z'} + \boldsymbol{\Omega} \times \mathbf{H}_G \qquad (18.22)$$

The first term in the right-hand member of Eq. (18.22) represents the rate of change of $\mathbf{H}_G$ with respect to the rotating frame $Gx'y'z'$. This term will drop out if $\boldsymbol{\omega}$—and thus, $\mathbf{H}_G$—

*(continued)*

remain constant in both magnitude and direction when viewed from that frame. On the other hand, if any of the time derivatives $\dot{\omega}_{x'}$, $\dot{\omega}_{y'}$, and $\dot{\omega}_{z'}$ is different from zero, $(\dot{\mathbf{H}}_G)_{Gx'y'z'}$ will also be different from zero, and its components should be determined by differentiating Eqs. (18.10) with respect to $t$. Finally, we remind you that if the rotating frame is rigidly attached to the body, its angular velocity will be the same as that of the body, and $\boldsymbol{\Omega}$ can be replaced by $\boldsymbol{\omega}$.

3. *Draw the free-body-diagram equation for the rigid body,* showing that the system of the external forces exerted on the body is equivalent to the vector $m\overline{\mathbf{a}}$ applied at $G$ and the couple vector $\dot{\mathbf{H}}_G$ (Fig. 18.11). By equating components in any direction and moments about any point, you can write as many as six independent scalar equations of motion [Sample Probs. 18.3 and 18.5].

4. *When solving problems involving the motion of a rigid body about a fixed point O,* you may find it convenient to use the following equation, derived in Sec. 18.7, which eliminates the components of the reaction at the support $O$,

$$\Sigma\mathbf{M}_O = (\dot{\mathbf{H}}_O)_{Oxyz} + \boldsymbol{\Omega} \times \mathbf{H}_O \qquad (18.28)$$

where the first term in the right-hand member represents the rate of change of $\mathbf{H}_O$ with respect to the rotating frame $Oxyz$, and where $\boldsymbol{\Omega}$ is the angular velocity of that frame.

5. *When determining the reactions at the bearings of a rotating shaft,* use Eq. (18.28) and take the following steps:

   *a. Place the fixed point O at one of the two bearings supporting the shaft* and attach the rotating frame $Oxyz$ to the shaft, with one of the axes directed along it. Assuming, for instance, that the $x$ axis has been aligned with the shaft, you will have $\boldsymbol{\Omega} = \boldsymbol{\omega} = \omega\mathbf{i}$ [Sample Prob. 18.4].

   *b. Since the selected axes, usually, will not be the principal axes of inertia at O,* you must compute the *products of inertia* of the shaft, as well as its moments of inertia, with respect to these axes, and use Eqs. (18.13) to determine $\mathbf{H}_O$. Assuming again that the $x$ axis has been aligned with the shaft, Eqs. (18.13) reduce to

$$H_x = I_x\omega \qquad H_y = -I_{yz}\omega \qquad H_z = -I_{zx}\omega \qquad (18.13')$$

which shows that $\mathbf{H}_O$ *will not be directed along the shaft.*

   *c. To obtain $\dot{\mathbf{H}}_O$, substitute the expressions obtained into Eq. (18.28), and let $\boldsymbol{\Omega} = \boldsymbol{\omega} = \omega\mathbf{i}$. If the angular velocity of the shaft is constant, the first term in the right-hand member of the equation will drop out. However, if the shaft has an angular acceleration $\boldsymbol{\alpha} = \alpha\mathbf{i}$, the first term will not be zero and must be determined by differentiating with respect to $t$ the expressions in (18.13'). The result will be equations similar to Eqs. (18.13'), with $\omega$ replaced by $\alpha$.

   *d. Since point O coincides with one of the bearings,* the three scalar equations corresponding to Eq. (18.28) can be solved for the components of the dynamic reaction at the other bearing. If the mass center $G$ of the shaft is located on the line joining the two bearings, the effective force $m\overline{\mathbf{a}}$ will be zero. Drawing the free-body-diagram equation of the shaft, you will then observe that the components of the dynamic reaction at the first bearing must be equal and opposite to those you have just determined. If $G$ is not located on the line joining the two bearings, you can determine the reaction at the first bearing by placing the fixed point $O$ at the second bearing and repeating the earlier procedure [Sample Prob. 18.4]; or you can obtain additional equations of motion from the free-body-diagram equation of the shaft, making sure to first determine and include the effective force $m\overline{\mathbf{a}}$ applied at $G$.

   *e. Most problems call for the determination of the "dynamic reactions"* at the bearings, that is, for the *additional forces* exerted by the bearings on the shaft when the shaft is rotating. When determining dynamic reactions, ignore the effect of static loads, such as the weight of the shaft.

# Problems

**18.55**  Determine the rate of change $\dot{\mathbf{H}}_G$ of the angular momentum $\mathbf{H}_G$ of the rectangular frame of Prob. 18.1.

**18.56**  Determine the rate of change $\dot{\mathbf{H}}_A$ of the angular momentum $\mathbf{H}_A$ of the square plate of Prob. 18.2.

**18.57**  Determine the rate of change $\dot{\mathbf{H}}_G$ of the angular momentum $\mathbf{H}_G$ of rod $AB$ of Prob. 18.3.

**18.58**  Determine the rate of change $\dot{\mathbf{H}}_G$ of the angular momentum $\mathbf{H}_G$ of the disk of Prob. 18.4 for an arbitrary value of $\beta$, knowing that its angular velocity $\boldsymbol{\omega}$ remains constant.

**18.59**  Determine the rate of change $\dot{\mathbf{H}}_A$ of the angular momentum $\mathbf{H}_A$ of the disk of Prob. 18.7.

**18.60**  Determine the rate of change $\dot{\mathbf{H}}_A$ of the angular momentum $\mathbf{H}_A$ of the disk of Prob. 18.8.

***18.61***  Determine the rate of change $\dot{\mathbf{H}}_A$ of the angular momentum $\mathbf{H}_A$ of the plate of Prob. 18.2, knowing that it has an angular velocity $\boldsymbol{\omega} = \omega\mathbf{j}$ and an angular acceleration $\boldsymbol{\alpha} = \alpha\mathbf{j}$.

***18.62***  Determine the rate of change $\dot{\mathbf{H}}_G$ of the angular momentum $\mathbf{H}_G$ of the disk of Prob. 18.4 for an arbitrary value of $\beta$, knowing that the disk has an angular velocity $\boldsymbol{\omega} = \omega\mathbf{i}$ and an angular acceleration $\boldsymbol{\alpha} = \alpha\mathbf{i}$.

**18.63**  The assembly shown consists of pieces of sheet aluminum of uniform thickness and of total mass 1.5 kg welded to a light axle supported by bearings $A$ and $B$. Knowing that the assembly rotates at the constant rate $\omega = 240$ rpm, determine the dynamic reactions at $A$ and $B$.

120 mm
120 mm
120 mm
120 mm
120 mm
120 mm
120 mm
120 mm
120 mm
$A$
$C$
$D$
$E$
$B$
$\omega$
$y$
$z$
$x$

**Fig. P18.63**

**18.64**  Solve Prob. 18.63 assuming $\omega = 360$ rpm.

**18.65** A homogeneous 3.6 kg disk is mounted on the horizontal shaft *AB*. The plane of the disk forms a 20° angle with the *yz* plane as shown. Knowing that the shaft rotates with a constant angular velocity **ω** of magnitude 10 rad/s, determine the dynamic reactions at points *A* and *B*.

**Fig. P18.65**

**18.66** A thin homogeneous 0.73 kg rod is used to form the shaft shown. Knowing that the shaft rotates with a constant angular velocity **ω** of magnitude 20 rad/s, determine the dynamic reactions at points *A* and *B*.

**Fig. P18.66**

**18.67** Two 2.4-kg slender rods are bent to form two square frames which are welded at the one-third points of the 600-mm shaft *AB* to form the assembly shown. Knowing that the assembly rotates with an angular velocity of constant magnitude *ω* = 20 rad/s, determine the dynamic reactions at the bearings *A* and *B*.

**Fig. *P18.67***

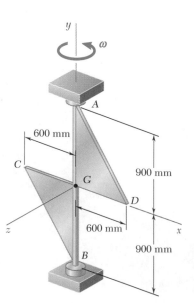

**Fig. *P18.68***

**18.68** Each of the two triangular plates shown has a mass of 5 kg and is welded to a vertical shaft *AB*. Knowing that the assembly rotates at the constant rate *ω* = 8 rpm, determine the dynamic reactions at *A* and *B*.

**18.69** Knowing that the assembly of Prob. 18.68 is initially at rest (*ω* = 0) when a couple of moment **M**₀ = (36 N · m)**j** is applied to the shaft, determine (*a*) the resulting angular acceleration of the assembly, (*b*) the dynamic reactions at *A* and *B* immediately after the couple is applied.

**18.70** The sheet-metal component shown is of uniform thickness and has a mass of 600 g. It is attached to a light axle supported by bearings at $A$ and $B$ located 150 mm apart. The component is at rest when it is subjected to a couple $\mathbf{M}_0$ as shown. If the resulting angular acceleration is $\boldsymbol{\alpha} = (12 \text{ rad/s}^2)\mathbf{k}$, determine (a) the couple $\mathbf{M}_0$, (b) the dynamic reactions at $A$ and $B$ immediately after the couple has been applied.

**Fig. P18.70**

**18.71** Knowing that the assembly of Prob. 18.65 is initially at rest ($\omega = 0$) when a couple of moment $\mathbf{M}_0 = (24.4 \text{ N} \cdot \text{m})\mathbf{i}$ is applied to the shaft, determine (a) the resulting angular acceleration of the assembly, (b) the dynamic reactions at points $A$ and $B$ immediately after the couple has been applied.

**18.72** Knowing that the shaft of Prob. 18.66 is initially at rest ($\omega = 0$), determine (a) the magnitude of the couple $\mathbf{M}_0 = M_0\mathbf{i}$ required to impart to the shaft an angular acceleration $\boldsymbol{\alpha} = (150 \text{ rad/s}^2)\mathbf{i}$, (b) the dynamic reactions at points $A$ and $B$ immediately after the couple has been applied.

**18.73** The assembly shown has a mass of 6 kg and consists of four thin 400-mm-diameter semicircular aluminum plates welded to a light 1-m-long shaft $AB$. The assembly is at rest ($\omega = 0$) at time $t = 0$ when a couple $\mathbf{M}_0$ is applied to it as shown, causing the assembly to complete one full revolution in 2 s. Determine (a) the couple $\mathbf{M}_0$, (b) the dynamic reactions at $A$ and $B$ at $t = 0$.

**18.74** For the assembly of Prob. 18.73, determine the dynamic reactions at $A$ and $B$ at $t = 2$ s.

**18.75** The assembly of Prob. 18.63 is initially at rest ($\omega = 0$) when a couple $\mathbf{M}_0$ is applied to axle $AB$. Knowing that the resulting angular acceleration of the assembly is $\boldsymbol{\alpha} = (150 \text{ rad/s}^2)\mathbf{i}$, determine (a) the couple $\mathbf{M}_0$, (b) the dynamic reactions at $A$ and $B$ immediately after the couple is applied.

**18.76** For the sheet-metal component of Prob. 18.70, determine (a) the angular velocity of the component 0.6 s after the couple $\mathbf{M}_0$ has been applied to it, (b) the magnitude of the dynamic reactions at $A$ and $B$ at that time.

**Fig. P18.73**

**Fig. P18.78**

**18.77**  Each wheel of an automobile has a weight of 21.8 kg, a diameter of 58.4 cm, and a radius of gyration of 23 cm. The automobile travels around an unbanked curve of radius 137.2 m at a speed of 104 km/h. Knowing that the transverse distance between the wheels is 1.4 m, determine the additional normal force exerted by the ground on each outside wheel due to the motion of the car.

**18.78**  The essential structure of a certain type of aircraft turn indicator is shown. Each spring has a constant of 584 N/m, and the 0.2 kg uniform disk of 5 cm. radius spins at the rate of 10,000 rpm. The springs are stretched and exert equal vertical forces on yoke $AB$ when the airplane is traveling in a straight path. Determine the angle through which the yoke will rotate when the pilot executes a horizontal turn of 686 m radius to the right at a speed of 801.2 km/h. Indicate whether point $A$ will move up or down.

**18.79**  The blade of an oscillating fan and the rotor of its motor have a total mass of 300 g and a combined radius of gyration of 75 mm. They are supported by bearings at $A$ and $B$, 125 mm apart, and rotate at the rate $\omega_1 = 1800$ rpm. Determine the dynamic reactions at $A$ and $B$ when the motor casing has an angular velocity $\omega_2 = (0.6 \text{ rad/s})\mathbf{j}$.

**Fig. P18.79**

**18.80**  An airplane propeller has a mass of 160 kg and has a radius of gyration of 800 mm. Knowing that the propeller rotates at 1600 rpm as the airplane is traveling on a vertical circular path of 600-m radius at 540 km/h, determine the magnitude of the couple exerted by the propeller on its shaft due to the rotation of the airplane.

**Fig. P18.80**

**18.81**  The flywheel of an automobile engine, which is rigidly attached to the crankshaft, is equivalent to a 51 cm-diameter, 1.9 cm-thick steel plate. Determine the magnitude of the couple exerted by the flywheel on the horizontal crankshaft as the automobile travels around an unbanked curve of 183 m radius at a speed of 96 km/h, with the flywheel rotating at 2700 rpm. Assume the automobile to have ($a$) a rear-wheel drive with the engine mounted longitudinally, ($b$) a front-wheel drive with the engine mounted transversely. (Specific weight of steel = 7861 kg/m³.)

**18.82**  The uniform thin 2.3 kg disk spins at a constant rate $\omega_2 = 6$ rad/s about an axis held by a housing attached to a horizontal rod that rotates at the constant rate $\omega_1 = 3$ rad/s. Determine the couple which represents the dynamic reaction at the support $A$.

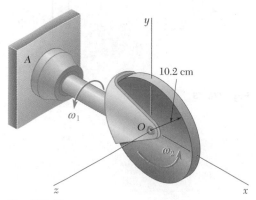

**Fig. P18.82**

**18.83**  A uniform square plate of side $a = 225$ mm is hinged at points $A$ and $B$ to a clevis which rotates with a constant angular velocity $\boldsymbol{\omega}$ about a vertical axis. Determine (a) the constant angle $\beta$ that the plate forms with the horizontal $x$ axis when $\omega = 12$ rad/s, (b) the largest value of $\omega$ for which the plate remains vertical ($\beta = 90°$).

**Fig. P18.83 and P18.84**

**18.84**  A uniform square plate of side $a = 300$ mm is hinged at points $A$ and $B$ to a clevis which rotates with a constant angular velocity $\boldsymbol{\omega}$ about a vertical axis. Determine (a) the value of $\omega$ for which the plate forms a constant angle $\beta = 60°$ with the horizontal $x$ axis, (b) the largest value of $\omega$ for which the plate remains vertical ($\beta = 90°$).

**18.85**  A thin ring of radius $r = 20.3$ cm is attached by a collar at point $A$ to a vertical shaft which rotates with a constant angular velocity $\boldsymbol{\omega}$. Determine (a) the constant angle $\beta$ that the plane of the ring forms with the vertical when $\omega = 10$ rad/s, (b) the largest value of $\omega$ for which the ring will remain vertical ($\beta = 0$).

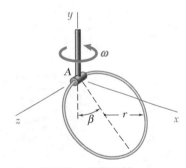

**Fig. P18.85 and P18.86**

**18.86**  A thin ring of radius $r = 30.5$ cm is attached by a collar at point $A$ to a vertical shaft which rotates with a constant angular velocity $\boldsymbol{\omega}$. Determine (a) the value of $\omega$ for which the ring forms a constant angle $\beta = 45°$ with the vertical $y$ axis, (b) the largest value of $\omega$ for which the ring will remain vertical ($\beta = 0$).

**18.87**  The 0.9-kg gear $A$ is constrained to roll on the fixed gear $B$, but is free to rotate about axle $AD$. Axle $AD$, of length 50.8 cm and negligible weight, is connected by a clevis to the vertical shaft $DE$ which rotates as shown with a constant angular velocity $\boldsymbol{\omega}_1$. Assuming that gear $A$ can be approximated by a thin disk of radius 10.1 cm, determine the largest allowable value of $\omega_1$ if gear $A$ is not to lose contact with gear $B$.

**18.88**  Determine the force $\mathbf{F}$ exerted by gear $B$ on gear $A$ of Prob. 18.87 when shaft $DE$ rotates with the constant angular velocity $\boldsymbol{\omega}_1 = 4$ rad/s. (*Hint:* The force $\mathbf{F}$ must be perpendicular to the line drawn from $D$ to $C$.)

**Fig. *P18.87***

**18.89 and 18.90**   The slender rod $AB$ is attached by a clevis to arm
$BCD$ which rotates with a constant angular velocity $\boldsymbol{\omega}$ about the centerline
of its vertical portion $CD$. Determine the magnitude of the angular velocity $\boldsymbol{\omega}$.

**Fig. P18.89**                                  **Fig. P18.90**

**18.91**   A thin homogeneous disk of mass $m$ and radius $a$ is held by a
fork-ended horizontal rod $ABC$. The disk and the rod rotate with the angu-
lar velocities shown. Assuming that both $\boldsymbol{\omega}_1$ and $\boldsymbol{\omega}_2$ are constant, determine
the dynamic reactions at $A$ and $B$.

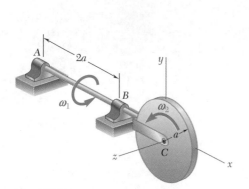

**Fig. P18.91**

**18.92**   A thin disk of weight $W = 4.5$ kg rotates with an angular velocity $\boldsymbol{\omega}_2$ with respect to arm $ABC$, which itself rotates with an angular velocity $\boldsymbol{\omega}_1$ about the $y$ axis. Knowing that $\omega_1 = 5$ rad/s and $\omega_2 = 15$ rad/s and
that both are constant, determine the force-couple system representing the
dynamic reaction at the support at $A$.

**18.93**   The 300-g disk shown spins at the rate $\omega_1 = 750$ rpm, while axle
$AB$ rotates as shown with an angular velocity $\boldsymbol{\omega}_2$ of 6 rad/s. Determine the
dynamic reactions at $A$ and $B$.

**Fig. P18.92**

**Fig. P18.93 and P18.94**

**18.94**   The 300-g disk shown spins at the rate $\omega_1 = 750$ rpm, while axle
$AB$ rotates as shown with an angular velocity $\boldsymbol{\omega}_2$. Determine the maximum
allowable magnitude of $\boldsymbol{\omega}_2$ if the dynamic reactions at $A$ and $B$ are not to exceed 1 N each.

**18.95** Two disks, each of mass 5 kg and radius 300 mm, spin as shown at the rate $\omega_1 = 1200$ rpm about a rod $AB$ of negligible mass which rotates about the horizontal $z$ axis at the rate $\omega_2 = 60$ rpm. ($a$) Determine the dynamic reactions at points $C$ and $D$. ($b$) Solve part $a$ assuming that the direction of spin of disk $A$ is reversed.

**Fig. P18.95 and P18.96**

**18.96** Two disks, each of mass 5 kg and radius 300 mm, spin as shown at the rate $\omega_1 = 1200$ rpm about a rod $AB$ of negligible mass which rotates about the horizontal $z$ axis at the rate $\omega_2$. Determine the maximum allowable value of $\omega_2$ if the magnitudes of the dynamic reactions at points $C$ and $D$ are not to exceed 350 N each.

**\*18.97** A 45.4 kg advertising panel of length $2a = 2.2$ m and width $2b = 1.5$ m is kept rotating at a constant rate $\omega_1$ about its horizontal axis by a small electric motor attached at $A$ to frame $ACB$. This frame itself is kept rotating at a constant rate $\omega_2$ about a vertical axis by a second motor attached at $C$ to the column $CD$. Knowing that the panel and the frame complete a full revolution in 6 s and 12 s, respectively, express, as a function of the angle $\theta$, the dynamic reaction exerted on column $CD$ by its support at $D$.

**Fig. P18.97**

**\*18.98** For the system of Prob. 18.97, show that ($a$) the dynamic reaction at $D$ is independent of the length $2a$ of the panel, ($b$) the ratio $M_1/M_2$ of the magnitudes of the couples exerted by the motors at $A$ and $C$, respectively, is independent of the dimensions and mass of the panel and is equal to $\omega_2/2\omega_1$ at any given instant.

**18.99** For the disk of Prob. 18.91, determine ($a$) the couple $M_1\mathbf{i}$ which should be applied to rod $AB$ to give it an angular acceleration $\alpha_1\mathbf{i}$ when the rod has an angular velocity $\omega_1\mathbf{i}$, knowing that the disk rotates at the constant rate $\omega_2$, ($b$) the dynamic reactions at $A$ and $B$ at that instant.

**18.100** At the instant shown the 300-g disk of Prob. 18.93 has an angular velocity of magnitude $\omega_1 = 12$ rad/s, which is decreasing at the rate of 4 rad/s$^2$ due to friction in the bearing at $C$. Knowing that $AB$ rotates with a constant angular velocity $\omega_2$ of 6 rad/s, determine the dynamic reactions at $A$ and $B$.

**18.101**  An 36-kg disk of radius 25.4 cm spins as shown at the constant rate $\omega_1 = 80$ rad/s with respect to the bent axle $ABC$. The system is at rest when a couple $\mathbf{M}_0 = (2 \; \text{N} \cdot \text{m})\mathbf{j}$ is applied for 5 s and then removed. Determine the force-couple system representing the dynamic reaction at the support at point $A$ after the couple was removed. Assume that the bent axle $ABC$ has a negligible mass.

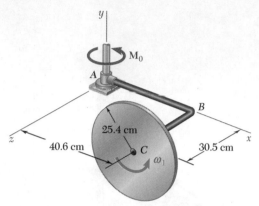

**Fig. P18.101 and P18.102**

**18.102**  An 36 kg disk of radius 25.4 cm spins as shown at the constant rate $\omega_1 = 80$ rad/s with respect to the bent axle $ABC$. The system is at rest when a couple $\mathbf{M}_0$ is applied as shown for 3 s and then removed. Knowing that the maximum angular velocity reached by the bent axle $ABC$ is 2.4 rad/s, determine (*a*) the couple $\mathbf{M}_0$, (*b*) the force-couple system representing the dynamic reaction at the support at point $A$ just before the couple was removed. Assume that the bent axle $ABC$ has a negligible mass.

*18.103*  A 2.5 kg homogeneous disk of radius 100 mm rotates with an angular velocity $\boldsymbol{\omega}_1$ with respect to arm $ABC$, which is welded to a shaft $DCE$ rotating as shown at the constant rate $\omega_2 = 12$ rad/s. Friction in the bearing at $A$ causes $\omega_1$ to decrease at the rate of 15 rad/s². Determine the dynamic reactions at $D$ and $E$ at a time when $\omega_1$ has decreased to 50 rad/s.

**Fig. *P18.103* and *P18.104***

*18.104*  A 2.5 kg homogeneous disk of radius 100 mm rotates at the constant rate $\omega_1 = 50$ rad/s with respect to arm $ABC$, which is welded to a shaft $DCE$. Knowing that at the instant shown shaft $DCE$ has an angular velocity of $\boldsymbol{\omega}_2 = (12 \; \text{rad/s})\mathbf{k}$ and an angular acceleration $\boldsymbol{\alpha}_2 = (8 \; \text{rad/s}^2)\mathbf{k}$, determine (*a*) the couple which must be applied to shaft $DCE$ to produce that angular acceleration, (*b*) the corresponding dynamic reactions at $D$ and $E$.

## *18.9. MOTION OF A GYROSCOPE. EULERIAN ANGLES

A *gyroscope* consists essentially of a rotor which can spin freely about its geometric axis. When mounted in a Cardan's suspension (Fig. 18.15), a gyroscope can assume any orientation, but its mass center must remain fixed in space. In order to define the position of a gyroscope at a given instant, let us select a fixed frame of reference $OXYZ$, with the origin $O$ located at the mass center of the gyroscope and the $Z$ axis directed along the line defined by the bearings $A$ and $A'$ of the outer gimbal. We will consider a reference position of the gyroscope in which the two gimbals and a given diameter $DD'$ of the rotor are located in the fixed $YZ$ plane (Fig. 18.15a). The gyroscope can be brought from this reference position into any arbitrary position (Fig. 18.15b) by means of the following steps: (1) a rotation of the outer gimbal through an angle $\phi$ about the axis $AA'$, (2) a rotation of the inner gimbal through $\theta$ about $BB'$, and (3) a rotation of the rotor through $\psi$ about $CC'$. The angles $\phi$, $\theta$, and $\psi$ are called the *Eulerian angles*; they completely characterize the position of the gyroscope at any given instant. Their derivatives $\dot{\phi}$, $\dot{\theta}$, and $\dot{\psi}$ define, respectively, the rate of *precession*, the rate of *nutation,* and the rate of *spin* of the gyroscope at the instant considered.

In order to compute the components of the angular velocity and of the angular momentum of the gyroscope, we will use a rotating system of axes *Oxyz attached to the inner gimbal*, with the $y$ axis along $BB'$ and the $z$ axis along $CC'$ (Fig. 18.16). These axes are principal axes of inertia for the gyroscope. While they follow it in its precession and nutation, however, they do not spin; for that reason, they are more convenient to use than axes actually attached to the gyroscope. The angular velocity $\boldsymbol{\omega}$ of the gyroscope with respect to the fixed frame of reference $OXYZ$ will now be expressed as the sum of three partial angular velocities corresponding, respectively, to the precession, the nutation, and the spin of the gyroscope. Denoting by $\mathbf{i}$, $\mathbf{j}$, and $\mathbf{k}$ the unit vectors along the rotating axes, and by $\mathbf{K}$ the unit vector along the fixed $Z$ axis, we have

$$\boldsymbol{\omega} = \dot{\phi}\mathbf{K} + \dot{\theta}\mathbf{j} + \dot{\psi}\mathbf{k} \qquad (18.33)$$

Since the vector components obtained for $\boldsymbol{\omega}$ in (18.33) are not orthogonal (Fig. 18.16), the unit vector $\mathbf{K}$ will be resolved into components along the $x$ and $z$ axes; we write

$$\mathbf{K} = -\sin\theta\,\mathbf{i} + \cos\theta\,\mathbf{k} \qquad (18.34)$$

and, substituting for $\mathbf{K}$ into (18.33),

$$\boldsymbol{\omega} = -\dot{\phi}\sin\theta\,\mathbf{i} + \dot{\theta}\mathbf{j} + (\dot{\psi} + \dot{\phi}\,\cos\theta)\mathbf{k} \qquad (18.35)$$

Since the coordinate axes are principal axes of inertia, the components

(a)

(b)

**Fig. 18.15**

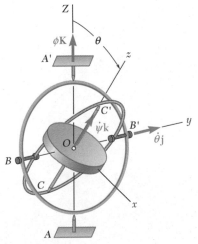

**Fig. 18.16**

of the angular momentum $\mathbf{H}_O$ can be obtained by multiplying the components of $\boldsymbol{\omega}$ by the moments of inertia of the rotor about the $x$, $y$, and $z$ axes, respectively. Denoting by $I$ the moment of inertia of the rotor about its spin axis, by $I'$ its moment of inertia about a transverse axis through $O$, and neglecting the mass of the gimbals, we write

$$\mathbf{H}_O = -I'\dot{\phi} \sin \theta \, \mathbf{i} + I'\dot{\theta}\mathbf{j} + I(\dot{\psi} + \dot{\phi} \cos \theta)\mathbf{k} \qquad (18.36)$$

Recalling that the rotating axes are attached to the inner gimbal and thus do not spin, we express their angular velocity as the sum

$$\boldsymbol{\Omega} = \dot{\phi}\mathbf{K} + \dot{\theta}\mathbf{j} \qquad (18.37)$$

or, substituting for $\mathbf{K}$ from (18.34),

$$\boldsymbol{\Omega} = -\dot{\phi} \sin \theta \, \mathbf{i} + \dot{\theta}\mathbf{j} + \dot{\phi} \cos \theta \, \mathbf{k} \qquad (18.38)$$

Substituting for $\mathbf{H}_O$ and $\boldsymbol{\Omega}$ from (18.36) and (18.38) into the equation

$$\Sigma\mathbf{M}_O = (\dot{\mathbf{H}}_O)_{Oxyz} + \boldsymbol{\Omega} \times \mathbf{H}_O \qquad (18.28)$$

we obtain the three differential equations

$$\Sigma M_x = -I'(\ddot{\phi} \sin \theta + 2\dot{\theta}\dot{\phi} \cos \theta) + I\dot{\theta} \, (\dot{\psi} + \dot{\phi} \cos \theta)$$
$$\Sigma M_y = I'(\ddot{\theta} - \dot{\phi}^2 \sin \theta \cos \theta) + I\dot{\phi} \sin \theta \, (\dot{\psi} + \dot{\phi} \cos \theta) \qquad (18.39)$$
$$\Sigma M_z = I \frac{d}{dt} \, (\dot{\psi} + \dot{\phi} \cos \theta)$$

The Eq. (18.39) define the motion of a gyroscope subjected to a given system of forces when the mass of its gimbals is neglected. They can also be used to define the motion of an *axisymmetrical body* (or body of revolution) attached at a point on its axis of symmetry, and to define the motion of an axisymmetrical body about its mass center. While the gimbals of the gyroscope helped us visualize the Eulerian angles, it is clear that these angles can be used to define the position of any rigid body with respect to axes centered at a point of the body, regardless of the way in which the body is actually supported.

Since the Eq. (18.39) are nonlinear, it will not be possible, in general, to express the Eulerian angles $\phi$, $\theta$, and $\psi$ as analytical functions of the time $t$, and numerical methods of solution may have to be used. However, as you will see in the following sections, there are several particular cases of interest which can be analyzed easily.

## *18.10. STEADY PRECESSION OF A GYROSCOPE

Let us now investigate the particular case of gyroscopic motion in which the angle $\theta$, the rate of precession $\dot{\phi}$, and the rate of spin $\dot{\psi}$ remain constant. We propose to determine the forces which must be applied to the gyroscope to maintain this motion, known as the *steady precession* of a gyroscope.

Instead of applying the general Eq. (18.39), we will determine the sum of the moments of the required forces by computing the rate of change of the angular momentum of the gyroscope in the particular case considered. We first note that the angular velocity $\boldsymbol{\omega}$ of the gyroscope, its angular momentum $\mathbf{H}_O$, and the angular velocity $\boldsymbol{\Omega}$ of the rotating frame of reference (Fig. 18.17) reduce, respectively, to

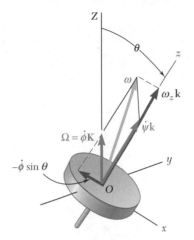

$$\boldsymbol{\omega} = -\dot{\phi}\sin\theta\,\mathbf{i} + \omega_z\mathbf{k} \tag{18.40}$$
$$\mathbf{H}_O = -I'\dot{\phi}\sin\theta\,\mathbf{i} + I\omega_z\mathbf{k} \tag{18.41}$$
$$\boldsymbol{\Omega} = -\dot{\phi}\sin\theta\,\mathbf{i} + \dot{\phi}\cos\theta\,\mathbf{k} \tag{18.42}$$

**Fig. 18.17**

where $\omega_z = \dot{\psi} + \dot{\phi}\cos\theta =$ rectangular component along spin axis of total angular velocity of gyroscope

Since $\theta$, $\dot{\phi}$, and $\dot{\psi}$ are constant, the vector $\mathbf{H}_O$ is constant in magnitude and direction with respect to the rotating frame of reference and its rate of change $(\dot{\mathbf{H}}_O)_{Oxyz}$ with respect to that frame is zero. Thus Eq. (18.28) reduces to

$$\Sigma\mathbf{M}_O = \boldsymbol{\Omega} \times \mathbf{H}_O \tag{18.43}$$

which yields, after substitutions from (18.41) and (18.42),

$$\Sigma\mathbf{M}_O = (I\omega_z - I'\dot{\phi}\cos\theta)\dot{\phi}\sin\theta\,\mathbf{j} \tag{18.44}$$

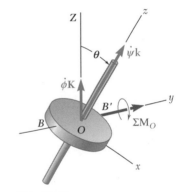

**Fig. 18.18**

Since the mass center of the gyroscope is fixed in space, we have, by (18.1), $\Sigma\mathbf{F} = 0$; thus, the forces which must be applied to the gyroscope to maintain its steady precession reduce to a couple of moment equal to the right-hand member of Eq. (18.44). We note that *this couple should be applied about an axis perpendicular to the precession axis and to the spin axis of the gyroscope* (Fig. 18.18).

In the particular case when the precession axis and the spin axis are at a right angle to each other, we have $\theta = 90°$ and Eq. (18.44) reduces to

$$\Sigma\mathbf{M}_O = I\dot{\psi}\dot{\phi}\mathbf{j} \tag{18.45}$$

Thus, if we apply to the gyroscope a couple $\mathbf{M}_O$ about an axis perpendicular to its axis of spin, the gyroscope will precess about an axis perpendicular to both the spin axis and the couple axis, in a sense such that the vectors representing the spin, the couple, and the precession, respectively, form a right-handed triad (Fig. 18.19).

Because of the relatively large couples required to change the orientation of their axles, gyroscopes are used as stabilizers in torpedoes

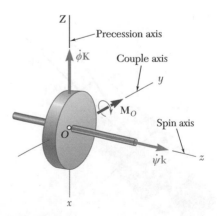

**Fig. 18.19**

and ships. Spinning bullets and shells remain tangent to their trajectory because of gyroscopic action. And a bicycle is easier to keep balanced at high speeds because of the stabilizing effect of its spinning wheels. However, gyroscopic action is not always welcome and must be taken into account in the design of bearings supporting rotating shafts subjected to forced precession. The reactions exerted by its propellers on an airplane which changes its direction of flight must also be taken into consideration and compensated for whenever possible.

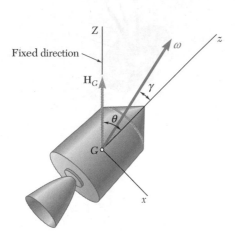

**Fig. 18.20**

## *18.11. MOTION OF AN AXISYMMETRICAL BODY UNDER NO FORCE

In this section you will analyze the motion about its mass center of an axisymmetrical body under no force except its own weight. Examples of such a motion are furnished by projectiles, if air resistance is neglected, and by artificial satellites and space vehicles after burnout of their launching rockets.

Since the sum of the moments of the external forces about the mass center $G$ of the body is zero, Eq. (18.2) yields $\dot{\mathbf{H}}_G = 0$. It follows that the angular momentum $\mathbf{H}_G$ of the body about $G$ is constant. Thus, the direction of $\mathbf{H}_G$ is fixed in space and can be used to define the $Z$ axis, or axis of precession (Fig. 18.20). Selecting a rotating system of axes $Gxyz$ with the $z$ axis along the axis of symmetry of the body, the $x$ axis in the plane defined by the $Z$ and $z$ axes, and the $y$ axis pointing away from you, we have

$$H_x = -H_G \sin \theta \qquad H_y = 0 \qquad H_z = H_G \cos \theta \quad (18.46)$$

where $\theta$ represents the angle formed by the $Z$ and $z$ axes, and $H_G$ denotes the constant magnitude of the angular momentum of the body about $G$. Since the $x$, $y$, and $z$ axes are principal axes of inertia for the body considered, we can write

$$H_x = I'\omega_x \qquad H_y = I'\omega_y \qquad H_z = I\omega_z \quad (18.47)$$

where $I$ denotes the moment of inertia of the body about its axis of symmetry and $I'$ denotes its moment of inertia about a transverse axis through $G$. It follows from Eqs. (18.46) and (18.47) that

$$\omega_x = -\frac{H_G \sin \theta}{I'} \qquad \omega_y = 0 \qquad \omega_z = \frac{H_G \cos \theta}{I} \quad (18.48)$$

The second of the relations obtained shows that the angular velocity $\boldsymbol{\omega}$ has no component along the $y$ axis, that is, along an axis perpendicular to the $Zz$ plane. Thus, the angle $\theta$ formed by the $Z$ and $z$ axes remains constant and *the body is in steady precession about the Z axis.*

Dividing the first and third of the relations (18.48) member by member, and observing from Fig. 18.21 that $-\omega_x/\omega_z = \tan \gamma$, we obtain the following relation between the angles $\gamma$ and $\theta$ that the vectors $\boldsymbol{\omega}$ and $\mathbf{H}_G$, respectively, form with the axis of symmetry of the body:

$$\tan \gamma = \frac{I}{I'} \tan \theta \quad (18.49)$$

**Fig. 18.21**

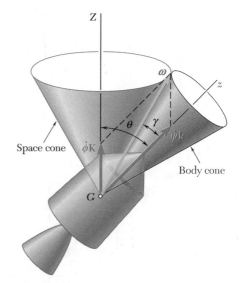

**Fig. 18.22**

There are two particular cases of motion of an axisymmetrical body
under no force which involve no precession: (1) If the body is set to
spin about its axis of symmetry, we have $\omega_x = 0$ and, by (18.47),
$H_x = 0$; the vectors $\boldsymbol{\omega}$ and $\mathbf{H}_G$ have the same orientation and the body
keeps spinning about its axis of symmetry (Fig. 18.22a). (2) If the body
is set to spin about a transverse axis, we have $\omega_z = 0$ and, by (18.47),
$H_z = 0$; again $\boldsymbol{\omega}$ and $\mathbf{H}_G$ have the same orientation and the body keeps
spinning about the given transverse axis (Fig. 18.22b).

Considering now the general case represented in Fig. 18.21, we
recall from Sec. 15.12 that the motion of a body about a fixed point—
or about its mass center—can be represented by the motion of a body
cone rolling on a space cone. In the case of steady precession, the two
cones are circular, since the angles $\gamma$ and $\theta - \gamma$ that the angular velocity $\boldsymbol{\omega}$ forms, respectively, with the axis of symmetry of the body
and with the precession axis are constant. Two cases should be
distinguished:

**Fig. 18.23**

1. $I < I'$. This is the case of an elongated body, such as the space
   vehicle of Fig. 18.23. By (18.49) we have $\gamma < \theta$; the vector $\boldsymbol{\omega}$
   lies inside the angle $ZGz$; the space cone and the body cone
   are tangent externally; the spin and the precession are both
   observed as counterclockwise from the positive $z$ axis. The
   precession is said to be *direct*.

2. $I > I'$. This is the case of a flattened body, such as the satellite of Fig. 18.24. By (18.49) we have $\gamma > \theta$; since the vector
   $\boldsymbol{\omega}$ must lie outside the angle $ZGz$, the vector $\dot{\psi}\mathbf{k}$ has a sense
   opposite to that of the $z$ axis; the space cone is inside the body
   cone; the precession and the spin have opposite senses. The
   precession is said to be *retrograde*.

**Fig. 18.24**

## SAMPLE PROBLEM 18.6

A space satellite of mass $m$ is known to be dynamically equivalent to two thin disks of equal mass. The disks are of radius $a = 800$ mm and are rigidly connected by a light rod of length $2a$. Initially the satellite is spinning freely about its axis of symmetry at the rate $\omega_0 = 60$ rpm. A meteorite, of mass $m_0 = m/1000$ and traveling with a velocity $\mathbf{v}_0$ of 2000 m/s relative to the satellite, strikes the satellite and becomes embedded at $C$. Determine ($a$) the angular velocity of the statellite immediately after impact, ($b$) the precession axis of the ensuing motion, ($c$) the rates of precession and spin of the ensuing motion.

## SOLUTION

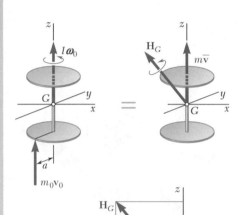

**Moments of Inertia.** We note that the axes shown are principal axes of inertia for the satellite and write

$$I = I_z = \tfrac{1}{2}ma^2 \qquad I' = I_x = I_y = 2[\tfrac{1}{4}(\tfrac{1}{2}m)a^2 + (\tfrac{1}{2}m)a^2] = \tfrac{5}{4}ma^2$$

**Principle of Impulse and Momentum.** We consider the satellite and the meteorite as a single system. Since no external force acts on this system, the momenta before and after impact are equipollent. Taking moments about $G$, we write

$$-a\mathbf{j} \times m_0v_0\mathbf{k} + I\omega_0\mathbf{k} = \mathbf{H}_G$$
$$\mathbf{H}_G = -m_0v_0a\mathbf{i} + I\omega_0\mathbf{k} \qquad (1)$$

**Angular Velocity after Impact.** Substituting the values obtained for the components of $\mathbf{H}_G$ and for the moments of inertia into

$$H_x = I_x\omega_x \qquad H_y = I_y\omega_y \qquad H_z = I_z\omega_z$$

we write

$$-m_0v_0a = I'\omega_x = \tfrac{5}{4}ma^2\omega_x \qquad 0 = I'\omega_y \qquad I\omega_0 = I\omega_z$$
$$\omega_x = -\frac{4}{5}\frac{m_0v_0}{ma} \qquad \omega_y = 0 \qquad \omega_z = \omega_0 \qquad (2)$$

For the satellite considered we have $\omega_0 = 60$ rpm $= 6.283$ rad/s, $m_0/m = \frac{1}{1000}$, $a = 0.800$ m, and $v_0 = 2000$ m/s; we find

$$\omega_x = -2 \text{ rad/s} \qquad \omega_y = 0 \qquad \omega_z = 6.283 \text{ rad/s}$$
$$\omega = \sqrt{\omega_x^2 + \omega_z^2} = 6.594 \text{ rad/s} \qquad \tan\gamma = \frac{-\omega_x}{\omega_z} = +0.3183$$
$$\omega = 63.0 \text{ rpm} \qquad \gamma = 17.7° \quad \blacktriangleleft$$

**Precession Axis.** Since in free motion the direction of the angular momentum $\mathbf{H}_G$ is fixed in space, the satellite will precess about this direction. The angle $\theta$ formed by the precession axis and the $z$ axis is

$$\tan\theta = \frac{-H_x}{H_z} = \frac{m_0v_0a}{I\omega_0} = \frac{2m_0v_0}{ma\omega_0} = 0.796 \qquad \theta = 38.5° \quad \blacktriangleleft$$

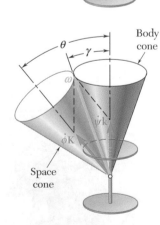

**Rates of Precession and Spin.** We sketch the space and body cones for the free motion of the satellite. Using the law of sines, we compute the rates of precession and spin.

$$\frac{\omega}{\sin\theta} = \frac{\dot{\phi}}{\sin\gamma} = \frac{\dot{\psi}}{\sin(\theta - \gamma)}$$
$$\dot{\phi} = 30.8 \text{ rpm} \qquad \dot{\psi} = 35.9 \text{ rpm} \quad \blacktriangleleft$$

In this lesson we analyzed the motion of *gyroscopes* and of other *axisymmetrical bodies* with a fixed point $O$. In order to define the position of these bodies at any given instant, we introduced the three *Eulerian angles* $\phi$, $\theta$, and $\psi$ (Fig. 18.15), and noted that their time derivatives define, respectively, the rate of *precession*, the rate of *nutation* and the rate of *spin* (Fig. 18.16). The problems you will encounter fall into one of the following categories.

**1.   *Steady precession.*** This is the motion of a gyroscope or other axisymmetrical body with a fixed point located on its axis of symmetry, in which the angle $\theta$, the rate of precession $\dot{\phi}$, and the rate of spin $\dot{\psi}$ all remain constant.

   ***a. Using the rotating frame of reference Oxyz*** shown in Fig. 18.17, which *precesses with the body, but does not spin* with it, we obtained the following expressions for the angular velocity $\boldsymbol{\omega}$ of the body, its angular momentum $\mathbf{H}_O$, and the angular velocity $\boldsymbol{\Omega}$ of the frame $Oxyz$:

$$\boldsymbol{\omega} = -\dot{\phi}\sin\theta\,\mathbf{i} + \omega_z\mathbf{k} \tag{18.40}$$

$$\mathbf{H}_O = -I'\dot{\phi}\sin\theta\,\mathbf{i} + I\,\omega_z\mathbf{k} \tag{18.41}$$

$$\boldsymbol{\Omega} = -\dot{\phi}\sin\theta\,\mathbf{i} + \dot{\phi}\cos\theta\,\mathbf{k} \tag{18.42}$$

where $I$ = moment of inertia of body about its axis of symmetry
   $I'$ = moment of inertia of body about a transverse axis through $O$
   $\omega_z$ = *rectangular* component of $\boldsymbol{\omega}$ along $z$ axis = $\dot{\psi} + \dot{\phi}\cos\theta$

   ***b. The sum of the moments about O of the forces applied to the body is equal to the rate of change of its angular momentum,*** as expressed by Eq. (18.28). But, since $\theta$ and the rates of change $\dot{\phi}$ and $\dot{\psi}$ are constant, it follows from Eq. (18.41) that $\mathbf{H}_O$ remains constant in magnitude and direction when viewed from the frame $Oxyz$. Thus, its rate of change is zero with respect to that frame and you can write

$$\Sigma\mathbf{M}_O = \boldsymbol{\Omega} \times \mathbf{H}_O \tag{18.43}$$

where $\boldsymbol{\Omega}$ and $\mathbf{H}_O$ are defined, respectively by Eq. (18.42) and Eq. (18.41). The equation obtained shows that the moment resultant at $O$ of the forces applied to the body is perpendicular to both the axis of precession and the axis of spin (Fig. 18.18).

   ***c. Keep in mind that the method described applies,*** not only to gyroscopes, where the fixed point $O$ coincides with the mass center $G$, but also *to any axisymmetrical body with a fixed point O located on its axis of symmetry*. This method, therefore, can be used to analyze the *steady precession of a top on a rough floor.*

   ***d. When an axisymmetrical body has no fixed point, but is in steady precession about its mass center G,*** you should draw a *free-body-diagram equation* showing that the system of the external forces exerted on the body (including the body's weight) is equivalent to the vector $m\overline{\mathbf{a}}$ applied at $G$ and the couple vector $\dot{\mathbf{H}}_G$. You can use Eqs. (18.40) through (18.42), replacing $\mathbf{H}_O$ with $\mathbf{H}_G$, and express the moment of the couple as

$$\dot{\mathbf{H}}_G = \boldsymbol{\Omega} \times \mathbf{H}_G$$

You can then use the free-body-diagram equation to write as many as six independent scalar equations.

*(continued)*

**2. Motion of an axisymmetrical body under no force, except its own weight.** We have $\Sigma\mathbf{M}_G = 0$ and, thus, $\dot{\mathbf{H}}_G = 0$; it follows that *the angular momentum* $\mathbf{H}_G$ *is constant in magnitude and direction* (Sec. 18.11). The body is in *steady precession* with the precession axis $GZ$ directed along $\mathbf{H}_G$ (Fig. 18.20). Using the rotating frame $Gxyz$ and denoting by $\gamma$ the angle that $\boldsymbol{\omega}$ forms with the spin axis $Gz$ (Fig. 18.21), we obtained the following relation between $\gamma$ and the angle $\theta$ formed by the precession and spin axes:

$$\tan\gamma = \frac{I}{I'}\tan\theta \tag{18.49}$$

The precession is said to be *direct* if $I < I'$ (Fig. 18.23) and *retrograde* if $I > I'$ (Fig. 18.24).

  *a. In many of the problems* dealing with the motion of an axisymmetrical body under no force, you will be asked to determine the *precession axis* and the *rates of precession and spin* of the body, knowing the magnitude of its *angular velocity* $\boldsymbol{\omega}$ and the angle $\gamma$ that it forms with the axis of symmetry $Gz$ (Fig. 18.21). From Eq. (18.49) you will determine the angle $\theta$ that the precession axis $GZ$ forms with $Gz$ and resolve $\boldsymbol{\omega}$ into its two *oblique components* $\dot{\phi}\mathbf{K}$ and $\dot{\psi}\mathbf{k}$. Using the law of sines, you will then determine the rate of precession $\dot{\phi}$ and the rate of spin $\dot{\psi}$.

  *b. In other problems,* the body will be subjected to *a given impulse* and you will first determine the resulting *angular momentum* $\mathbf{H}_G$. Using Eqs. (18.10), you will calculate the rectangular components of the angular velocity $\boldsymbol{\omega}$, its magnitude $\omega$, and the angle $\gamma$ that it forms with the axis of symmetry. You will then determine the *precession axis* and the *rates of precession and spin* as described above [Sample Prob. 18.6].

**3. General motion of an axisymmetric body with a fixed point $O$ located on its axis of symmetry, and subjected only to its own weight.** This is a motion in which the angle $\theta$ is allowed to vary. At any given instant you should take into account the rate of precession $\dot{\phi}$, the rate of spin $\dot{\psi}$, *and the rate of nutation* $\dot{\theta}$, none of which will remain constant. An example of such a motion is the motion of a top, which is discussed in Probs. 18.139 and 18.140. The rotating frame of reference $Oxyz$ that you will use is still the one shown in Fig. 18.18, but this frame will now rotate about the $y$ axis at the rate $\dot{\theta}$. Equations (18.40), (18.41), and (18.42), therefore, should be replaced by the following equations:

$$\boldsymbol{\omega} = -\dot{\phi}\sin\theta\,\mathbf{i} + \dot{\theta}\mathbf{j} + (\dot{\psi} + \dot{\phi}\cos\theta)\,\mathbf{k} \tag{18.40$'$}$$

$$\mathbf{H}_O = -I'\dot{\phi}\sin\theta\,\mathbf{i} + I'\dot{\theta}\mathbf{j} + I(\dot{\psi} + \dot{\phi}\cos\theta)\,\mathbf{k} \tag{18.41$'$}$$

$$\boldsymbol{\Omega} = -\dot{\phi}\sin\theta\,\mathbf{i} + \dot{\theta}\mathbf{j} + \dot{\phi}\cos\theta\,\mathbf{k} \tag{18.42$'$}$$

Since substituting these expressions into Eq. (18.44) would lead to nonlinear differential equations, it is preferable, whenever feasible, to apply the following conservation principles.

  *a. Conservation of energy.* Denoting by $c$ the distance between the fixed point $O$ and the mass center $G$ of the body, and by $E$ the total energy, you will write

$$T + V = E: \qquad \tfrac{1}{2}(I'\omega_x^2 + I'\omega_y^2 + I\omega_z^2) + mgc\cos\theta = E$$

and substitute for the components of $\boldsymbol{\omega}$ the expressions obtained in Eq. (18.40$'$). Note that $c$ will be positive or negative, depending upon the position of $G$ relative to $O$. Also, $c = 0$ if $G$ coincides with $O$; the *kinetic energy* is then conserved.

  *b. Conservation of the angular momentum about the axis of precession.* Since the support at $O$ is located on the $Z$ axis, and since the weight of the body and the $Z$ axis are both vertical and, thus, parallel to each other, it follows that $\Sigma M_Z = 0$ and, thus, that $H_Z$ remains constant. This can be expressed by writing that the scalar product $\mathbf{K}\cdot\mathbf{H}_O$ is constant, where $\mathbf{K}$ is the unit vector along the $Z$ axis.

  *c. Conservation of the angular momentum about the axis of spin.* Since the support at $O$ and the center of gravity $G$ are both located on the $z$ axis, it follows that $\Sigma M_z = 0$ and, thus, that $H_z$ remains constant. This is expressed by writing that the coefficient of the unit vector $\mathbf{k}$ in Eq. (18.41$'$) is constant. Note that this last conservation principle cannot be applied when the body is restrained from spinning about its axis of symmetry, but in that case the only variables are $\theta$ and $\phi$.

# Problems

**18.105** The 0.09 kg top shown is supported at the fixed point $O$. The radii of gyration of the top with respect to its axis of symmetry and with respect to a transverse axis through $O$ are 2.7 cm and 5.7 cm, respectively. Knowing that $c = 4.8$ cm and that the rate of spin of the top about its axis of symmetry is 1800 rpm, determine the two possible rates of steady precession corresponding to $\theta = 30°$.

**18.106** The top shown is supported at the fixed point $O$ and its moments of inertia about its axis of symmetry and about a transverse axis through $O$ are denoted, respectively, by $I$ and $I'$. (a) Show that the condition for steady precession of the top is

$$(I\omega_z - I'\dot{\phi} \cos \theta)\dot{\phi} = Wc$$

where $\dot{\phi}$ is the rate of precession and $\omega_z$ is the rectangular component of the angular velocity along the axis of symmetry of the top. (b) Show that if the rate of spin $\dot{\psi}$ of the top is very large compared with its rate of precession $\dot{\phi}$, the condition for steady precession is $I\dot{\psi}\dot{\phi} \approx Wc$. (c) Determine the percentage error introduced when this last relation is used to approximate the slower of the two rates of precession obtained for the top of Prob. 18.105.

**18.107** A uniform thin disk of 15.2 cm diameter is attached to the end of a rod $AB$ of negligible mass which is supported by a ball-and-socket joint at point $A$. Knowing that the disk is observed to precess about the vertical axis $AC$ at the constant rate of 36 rpm in the sense indicated and that its axis of symmetry $AB$ forms an angle $\beta = 60°$ with $AC$, determine the rate at which the disk spins about rod $AB$.

**18.108** A uniform thin disk of 15.2 cm diameter is attached to the end of a rod $AB$ of negligible mass which is supported by a ball-and-socket joint at point $A$. Knowing that the disk is spinning about its axis of symmetry $AB$ at the rate of 2100 rpm in the sense indicated and that $AB$ forms an angle $\beta = 45°$ with the vertical axis $AC$, determine the two possible rates of steady precession of the disk about the axis $AC$.

**18.109** A solid cone of height 240 mm with a circular base of radius 80 mm is supported by a ball-and-socket joint at $A$. The cone spins about its axis of symmetry $AB$ at the constant rate of 1600 rpm in the sense indicated. Knowing that the cone is observed to precess about the vertical axis $AC$ at the constant rate of 40 rpm in the sense indicated, determine the angle $\beta$.

**18.110** A solid cone of height 240 mm with a circular base of radius 80 mm is supported by a ball-and-socket joint at $A$. Knowing that the cone is observed to precess about the vertical axis $AC$ at the constant rate of 30 rpm in the sense indicated and that its axis of symmetry forms an angle $\beta = 30°$ with $AC$, determine the rate at which the cone spins about its axis $AB$.

**Fig. P18.105 and P18.106**

**Fig. P18.107 and P18.108**

**Fig. P18.109 and P18.110**

**Fig. P18.112 and P18.113**

**Fig. P18.111**

**18.111** A high-speed photographic record shows that a certain projectile was fired with a horizontal velocity $\bar{v}$ of 550 m/s and with its axis of symmetry forming an angle $\beta = 3°$ with the horizontal. The rate of spin $\dot{\psi}$ of the projectile was 6000 rpm, and the atmospheric drag was equivalent to a force **D** of 110 N acting at the center of pressure $C_P$ located at a distance $c = 175$ mm from $G$. (a) Knowing that the projectile has a mass of 18 kg and a radius of gyration of 60 mm with respect to its axis of symmetry, determine its approximate rate of steady precession. (b) If it is further known that the radius of gyration of the projectile with respect to a transverse axis through $G$ is 250 mm, determine the exact values of the two possible rates of precession.

**18.112** A homogeneous cone of height $h$ and with a base of diameter $d < h$ is attached as shown to a cord $AB$. The cone spins about its axis $BC$ at the constant rate $\dot{\psi}$ and precesses about the vertical through $A$ at the constant rate $\dot{\phi}$. Determine the angle $\beta$ for which the axis $BC$ of the cone is aligned with the cord $AB$ ($\theta = \beta$).

**18.113** A homogeneous cone of height $h = 30.5$ cm and with a base diameter $d = 15.2$ cm is attached as shown to a cord $AB$. Knowing that the angles that cord $AB$ and the axis $BC$ of the cone form with the vertical are, respectively, $\beta = 45°$ and $\theta = 30°$, and that the cone precesses at the constant rate $\dot{\phi} = 8$ rad/s in the sense indicated, determine (a) the rate of spin $\dot{\psi}$ of the cone about its axis $BC$, (b) the length of cord $AB$.

**18.114** If the earth were a sphere, the gravitational attraction of the sun, moon, and planets would at all times be equivalent to a single force **R** acting at the mass center of the earth. However, the earth is actually an oblate spheroid and the gravitational system acting on the earth is equivalent to a force **R** and a couple **M**. Knowing that the effect of the couple **M** is to cause the axis of the earth to precess about the axis $GA$ at the rate of one revolution in 25,800 years, determine the average magnitude of the couple **M** applied to the earth. Assume that the average specific weight of the earth is 5510 kg/m$^3$, that the average radius of the earth is 6345 km, and that $\bar{I} = \frac{2}{5} mR^2$. (*Note:* This forced precession is known as the precession of the equinoxes and is not to be confused with the free precession discussed in Prob. 18.121.)

**18.115** A homogeneous sphere of radius $c$ is attached as shown to a cord $AB$. The cord forms an angle $\beta$ with the vertical and precesses at the constant rate $\dot{\phi}$, while the sphere spins at the constant rate $\dot{\psi}$ about its diameter $BC$. Determine the angle $\theta$ that $BC$ forms with the vertical.

**18.116** A homogeneous sphere of radius $c = 40$ mm is attached as shown to a cord $AB$. The cord forms an angle $\beta = 30°$ with the vertical and is observed to precess at the constant rate $\dot{\phi} = 5$ rad/s about the vertical through $A$. Determine the angle $\theta$ that the diameter $BC$ forms with the vertical, knowing that the sphere (a) has no spin, (b) spins about its diameter $BC$ at the rate $\dot{\psi} = 30$ rad/s, (c) spins about $BC$ at the rate $\dot{\psi} = -30$ rad/s.

**Fig. P18.114**

**Fig. P18.115 and P18.116**

**18.117**   Show that for an axisymmetrical body under no force, the rates of precession and spin can be expressed, respectively, as

$$\dot{\phi} = \frac{H_G}{I'}$$

and

$$\dot{\psi} = \frac{H_G \cos \theta (I' - I)}{II'}$$

where $H_G$ is the constant value of the angular momentum of the body.

**18.118**   (a) Show that for an axisymmetrical body under no force, the rate of precession can be expressed as

$$\dot{\phi} = \frac{I\omega_2}{I' \cos \theta}$$

where $\omega_2$ is the rectangular component of $\boldsymbol{\omega}$ along the axis of symmetry of the body. (b) Use this result to check that the condition (18.44) for steady precession is satisfied by an axisymmetrical body under no force.

**18.119**   Show that the angular velocity vector $\boldsymbol{\omega}$ of an axisymmetrical body under no force is observed from the body itself to rotate about the axis of symmetry at the constant rate

$$n = \frac{I' - I}{I'} \omega_2$$

where $\omega_2$ is the rectangular component of $\boldsymbol{\omega}$ along the axis of symmetry of the body.

**18.120**   For an axisymmetrical body under no force, prove (a) that the rate of retrograde precession can never be less than twice the rate of spin of the body about its axis of symmetry, (b) that in Fig. 18.24 the axis of symmetry of the body can never lie within the space cone.

**18.121**   Using the relation given in Prob. 18.119, determine the period of precession of the north pole of the earth about the axis of symmetry of the earth. The earth may be approximated by an oblate spheroid of axial moment of inertia $I$ and of transverse moment of inertia $I' = 0.99671I$. (*Note:* Actual observations show a period of precession of the north pole of about 432.5 mean solar days; the difference between the observed and computed periods is due to the fact that the earth is not a perfectly rigid body. The free precession considered here should not be confused with the much slower precession of the equinoxes, which is a forced precession. See Prob. 18.114.)

**18.122**   The angular velocity vector of a football which has just been kicked is horizontal, and its axis of symmetry $OC$ is oriented as shown. Knowing that the magnitude of the angular velocity is 200 rpm and that the ratio of the axial and transverse moments of inertia is $I/I' = \frac{1}{3}$, determine (a) the orientation of the axis of precession $OA$, (b) the rates of precession and spin.

15°

$\omega$

$\beta$

**Fig. P18.122**

**Fig. P18.124**

**Fig. *P18.126***

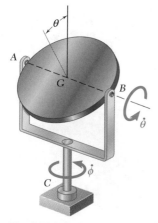

**Fig. P18.129**

**18.123**   A coin is tossed into the air. It is observed to spin at the rate of 600 rpm about an axis $GC$ perpendicular to the coin and to precess about the vertical direction $GD$. Knowing that $GC$ forms an angle of 15° with $GD$, determine (*a*) the angle that the angular velocity $\boldsymbol{\omega}$ of the coin forms with $GD$, (*b*) the rate of precession of the coin about $GD$.

**Fig. P18.123**

**18.124**   An 363 kg geostationary satellite is spinning with an angular velocity $\boldsymbol{\omega}_0 = (1.5 \text{ rad/s})\mathbf{j}$ when it is hit at $B$ by a 0.18 kg meteorite traveling with a velocity $\mathbf{v}_0 = -(488 \text{ m/s})\mathbf{i} + (396 \text{ m/s})\mathbf{j} + (1219 \text{ m/s})\mathbf{k}$ relative to the satellite. Knowing that $b = 50.8$ cm and that the radii of gyration of the satellite are $\overline{k}_x = \overline{k}_z = 73.2$ cm and $\overline{k}_y = 82.3$ cm, determine the precession axis and the rates of precession and spin of the satellite after the impact.

**18.125**   Solve Prob. 18.124, assuming that the meteorite hits the satellite at $A$ instead of $B$.

***18.126***   A space station consists of two sections $A$ and $B$ of equal masses, which are rigidly connected. Each section is dynamically equivalent to a homogeneous cylinder of length 15 m and radius 3 m. Knowing that the station is precessing about the fixed direction $GD$ at the constant rate of 2 rev/h, determine the rate of spin of the station about its axis of symmetry $CC'$.

***18.127***   Solve Sample Prob. 18.6 assuming that the meteorite strikes the satellite at $C$ with a velocity $\mathbf{v}_0 = (2000 \text{ m/s})\mathbf{i}$.

**18.128**   After the motion determined in Sample Prob. 18.6 has been established, the rod connecting disks $A$ and $B$ of the satellite breaks, and disk $A$ moves freely as a separate body. Knowing that the rod and the $z$ axis coincide when the rod breaks, determine the precession axis, the rate of precession, and the rate of spin for the ensuing motion of disk $A$.

**18.129**   A homogeneous disk of mass $m$ is connected at $A$ and $B$ to a fork-ended shaft of negligible mass which is supported by a bearing at $C$. The disk is free to rotate about its horizontal diameter $AB$ and the shaft is free to rotate about a vertical axis through $C$. Initially the disk lies in a vertical plane ($\theta_0 = 90°$) and the shaft has an angular velocity $\dot{\phi}_0 = 16$ rad/s. If the disk is slightly disturbed, determine for the ensuing motion (*a*) the minimum value of $\dot{\phi}$, (*b*) the maximum value of $\dot{\theta}$.

**18.130**  A homogeneous square plate of mass $m$ and side $c$ is held at points $A$ and $B$ by a frame of negligible mass which is supported by bearings at points $C$ and $D$. The plate is free to rotate about $AB$, and the frame is free to rotate about the vertical $CD$. Knowing that, initially, $\theta_0 = 45°$, $\dot{\theta}_0 = 0$, and $\dot{\phi}_0 = 8$ rad/s, determine for the ensuing motion (a) the range of values of $\theta$, (b) the minimum value of $\dot{\phi}$, (c) the maximum value of $\dot{\theta}$.

**18.131**  A homogeneous square plate of mass $m$ and side $c$ is held at points $A$ and $B$ by a frame of negligible mass which is supported by bearings at points $C$ and $D$. The plate is free to rotate about $AB$, and the frame is free to rotate about the vertical $CD$. Initially the plate lies in the plane of the frame ($\theta_0 = 90°$) and the frame has an angular velocity $\dot{\phi}_0 = 8$ rad/s. If the plate is slightly disturbed, determine for the ensuing motion (a) the minimum value of $\dot{\phi}$, (b) the maximum value of $\dot{\theta}$.

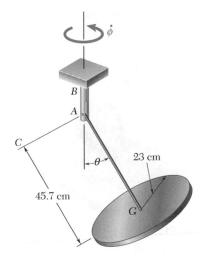

**Fig. P18.130 and P18.131**

**18.132**  A homogeneous disk of radius 23 cm is welded to a rod $AG$ of length 45.7 cm and of negligible weight which is connected by a clevis to a vertical shaft $AB$. The rod and disk can rotate freely about a horizontal axis $AC$, and shaft $AB$ can rotate freely about a vertical axis. Initially rod $AG$ is horizontal ($\theta_0 = 90°$) and has no angular velocity about $AC$. Knowing that the maximum value $\dot{\phi}_m$ of the angular velocity of shaft $AB$ in the ensuing motion is twice its initial value $\dot{\phi}_0$, determine (a) the minimum value of $\theta$, (b) the initial angular velocity $\dot{\phi}_0$ of shaft $AB$.

**18.133**  A homogeneous disk of radius 23 cm is welded to a rod $AG$ of length 45.7 cm and of negligible weight which is connected by a clevis to a vertical shaft $AB$. The rod and disk can rotate freely about a horizontal axis $AC$, and shaft $AB$ can rotate freely about a vertical axis. Initially rod $AG$ is horizontal ($\theta_0 = 90°$) and has no angular velocity about $AC$. Knowing that the smallest value of $\theta$ in the ensuing motion is $30°$, determine (a) the initial angular velocity of shaft $AB$, (b) its maximum angular velocity.

**18.134**  A homogeneous sphere of mass $m$ and radius $a$ is welded to a rod $AB$ of negligible mass, which is connected by a clevis to a vertical shaft $AC$. The rod and sphere can rotate freely about a horizontal axis at $A$, and shaft $AC$ can rotate freely about a vertical axis. The system is released in the position $\beta = 0$ with an angular velocity $\dot{\phi}_0$ about the vertical axis and no angular velocity about the horizontal axis. Knowing that the largest value of $\beta$ in the ensuing motion is $30°$, determine (a) the initial angular velocity $\dot{\phi}_0$, (b) the value of $\dot{\phi}$ when $\beta = 30°$.

**Fig. P18.132 and P18.133**

**Fig. P18.134**

**Fig. P18.135 and P18.136**

**Fig. P18.137 and P18.138**

*18.135    A homogeneous disk of radius 23 cm is welded to a rod $AG$ of length 45.7 cm and of negligible weight which is supported by a ball and socket at $A$. The disk is released with a rate of spin $\dot{\psi}_0 = 50$ rad/s, with zero rates of precession and nutation, and with rod $AG$ horizontal ($\theta_0 = 90°$). Determine (a) the smallest value of $\theta$ in the ensuing motion, (b) the rates of precession and spin as the disk passes through its lowest position.

*18.136    A homogeneous disk of radius 23 cm is welded to a rod $AG$ of length 45.7 cm and of negligible weight which is supported by a ball and socket at $A$. The disk is released with a rate of spin $\dot{\psi}_0$, counterclockwise as seen from $A$, with zero rates of precession and nutation, and with rod $AG$ horizontal ($\theta_0 = 90°$). Knowing that the smallest value of $\theta$ in the ensuing motion is 30°, determine (a) the rate of spin $\dot{\psi}_0$ of the disk in its initial position, (b) the rates of precession and spin as the disk passes through its lowest position.

*18.137    The top shown is supported at the fixed point $O$. Denoting by $\phi$, $\theta$, and $\psi$ the Eulerian angles defining the position of the top with respect to a fixed frame of reference, consider the general motion of the top in which all Eulerian angles vary.

(a) Observing that $\Sigma M_Z = 0$ and $\Sigma M_z = 0$, and denoting by $I$ and $I'$, respectively, the moments of inertia of the top about its axis of symmetry and about a transverse axis through $O$, derive the two first-order differential equations of motion

$$I'\dot{\phi}\sin^2\theta + I(\dot{\psi} + \dot{\phi}\cos\theta)\cos\theta = \alpha \qquad (1)$$
$$I(\dot{\psi} + \dot{\phi}\cos\theta) = \beta \qquad (2)$$

where $\alpha$ and $\beta$ are constants depending upon the initial conditions. These equations express that the angular momentum of the top is conserved about both the $Z$ and $z$ axes, that is, that the rectangular component of $\mathbf{H}_O$ along each of these axes is constant.

(b) Use Eqs. (1) and (2) to show that the rectangular component $\omega_z$ of the angular velocity of the top is constant and that the rate of precession $\dot{\phi}$ depends upon the value of the angle of nutation $\theta$.

*18.138    (a) Applying the principle of conservation of energy, derive a third differential equation for the general motion of the top of Prob. 18.137.

(b) Eliminating the derivatives $\dot{\phi}$ and $\dot{\psi}$ from the equation obtained and from the two equations of Prob. 18.137, show that the rate of nutation $\dot{\theta}$ is defined by the differential equation $\dot{\theta}^2 = f(\theta)$, where

$$f(\theta) = \frac{1}{I'}\left(2E - \frac{\beta^2}{I} - 2mgc\cos\theta\right) - \left(\frac{\alpha - \beta\cos\theta}{I'\sin\theta}\right)^2 \qquad (1)$$

(c) Further show, by introducing the auxiliary variable $x = \cos\theta$, that the maximum and minimum values of $\theta$ can be obtained by solving for $x$ the cubic equation

$$\left(2E - \frac{\beta^2}{I} - 2mgcx\right)(1 - x^2) - \frac{1}{I'}(\alpha - \beta x)^2 = 0 \qquad (2)$$

*18.139 A solid cone of height 180 mm with a circular base of radius 60 mm is supported by a ball and socket at $A$. The cone is released from the position $\theta_0 = 30°$ with a rate of spin $\dot{\psi}_0 = 300$ rad/s, a rate of precession $\dot{\phi}_0 = 20$ rad/s, and a zero rate of nutation. Determine (a) the maximum value of $\theta$ in the ensuing motion, (b) the corresponding values of the rates of spin and precession. [*Hint:* Use Eq. (2) of Prob. 18.138; you can either solve this equation numerically or reduce it to a quadratic equation, since one of its roots is known.]

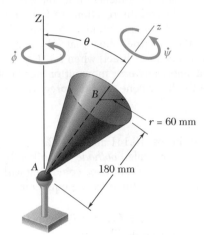

**Fig. P18.139 and P18.140**

*18.140 A solid cone of height 180 mm with a circular base of radius 60 mm is supported by a ball and socket at $A$. The cone is released from the position $\theta_0 = 30°$ with a rate of spin $\dot{\psi}_0 = 300$ rad/s, a rate of precession $\dot{\phi}_0 = -4$ rad/s, and a zero rate of nutation. Determine (a) the maximum value of $\theta$ in the ensuing motion, (b) the corresponding values of the rates of spin and precession, (c) the value of $\theta$ for which the sense of the precession is reversed. (See hint of Prob. 18.139.)

*18.141 Consider a rigid body of arbitrary shape which is attached at its mass center $O$ and subjected to no force other than its weight and the reaction of the support at $O$.

(a) Prove that the angular momentum $\mathbf{H}_O$ of the body about the fixed point $O$ is constant in magnitude and direction, that the kinetic energy $T$ of the body is constant, and that the projection along $\mathbf{H}_O$ of the angular velocity $\boldsymbol{\omega}$ of the body is constant.

(b) Show that the tip of the vector $\boldsymbol{\omega}$ describes a curve on a fixed plane in space (called the *invariable plane*), which is perpendicular to $\mathbf{H}_O$ and at a distance $2T/H_O$ from $O$.

(c) Show that with respect to a frame of reference attached to the body and coinciding with its principal axes of inertia, the tip of the vector $\boldsymbol{\omega}$ appears to describe a curve on an ellipsoid of equation

$$I_x\omega_x^2 + I_y\omega_y^2 + I_z\omega_z^2 = 2T = \text{constant}$$

The ellipsoid (called the *Poinsot ellipsoid*) is rigidly attached to the body and is of the same shape as the ellipsoid of inertia, but of a different size.

**Fig. P18.141**

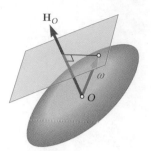

$\mathbf{H}_O$

$\omega$

$O$

**Fig. P18.142**

**\*18.142** Referring to Prob. 18.141, (a) prove that the Poinsot ellipsoid is tangent to the invariable plane, (b) show that the motion of the rigid body must be such that the Poinsot ellipsoid appears to roll on the invariable plane. [*Hint:* In part *a*, show that the normal to the Poinsot ellipsoid at the tip of **ω** is parallel to **H**$_O$. It is recalled that the direction of the normal to a surface of equation $F(x, y, z) = $ constant at a point *P* is the same as that of **grad** *F* at point *P*.]

**\*18.143** Using the results obtained in Probs. 18.141 and 18.142, show that for an axisymmetrical body attached at its mass center *O* and under no force other than its weight and the reaction at *O*, the Poinsot ellipsoid is an ellipsoid of revolution and the space and body cones are both circular and are tangent to each other. Further show that (a) the two cones are tangent externally, and the precession is direct, when $I < I'$, where *I* and *I'* denote, respectively, the axial and transverse moment of inertia of the body, (b) the space cone is inside the body cone, and the precession is retrograde, when $I > I'$.

**\*18.144** Refer to Probs. 18.141 and 18.142.
(a) Show that the curve (called *polhode*) described by the tip of the vector **ω** with respect to a frame of reference coinciding with the principal axes of inertia of the rigid body is defined by the equations

$$I_x\omega_x^2 + I_y\omega_y^2 + I_z\omega_z^2 = 2T = \text{constant} \tag{1}$$
$$I_x^2\omega_x^2 + I_y^2\omega_y^2 + I_z^2\omega_z^2 = H_O^2 = \text{constant} \tag{2}$$

and that the curve can, therefore, be obtained by intersecting the Poinsot ellipsoid with the ellipsoid defined by Eq. (2).
(b) Further show, assuming $I_x > I_y > I_z$, that the polhodes obtained for various values of $H_O$ have the shapes indicated in the figure.
(c) Using the result obtained in part *b*, show that a rigid body under no force can rotate about a fixed centroidal axis if, and only if, that axis coincides with one of the principal axes of inertia of the body, and that the motion will be stable if the axis of rotation coincides with the major or minor axis of the Poinsot ellipsoid (*z* or *x* axis in the figure) and unstable if it coincides with the intermediate axis (*y* axis).

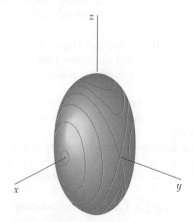

$z$

$x$

$y$

**Fig. P18.144**

This chapter was devoted to the kinetic analysis of the motion of rigid bodies in three dimensions.

We first noted [Sec. 18.1] that the two fundamental equations derived in Chap. 14 for the motion of a system of particles

$$\Sigma \mathbf{F} = m\overline{\mathbf{a}} \tag{18.1}$$

$$\Sigma \mathbf{M}_G = \dot{\mathbf{H}}_G \tag{18.2}$$

**Fundamental equations of motion for a rigid body**

provide the foundation of our analysis, just as they did in Chap. 16 in the case of the plane motion of rigid bodies. The computation of the angular momentum $\mathbf{H}_G$ of the body and of its derivative $\dot{\mathbf{H}}_G$, however, are now considerably more involved.

In Sec. 18.2, we saw that the rectangular components of the angular momentum $\mathbf{H}_G$ of a rigid body can be expressed as follows in terms of the components of its angular velocity $\boldsymbol{\omega}$ and of its centroidal moments and products of inertia:

**Angular momentum of a rigid body in three dimensions**

$$
\begin{aligned}
H_x &= +\overline{I}_x\omega_x - \overline{I}_{xy}\omega_y - \overline{I}_{xz}\omega_z \\
H_y &= -\overline{I}_{yx}\omega_x + \overline{I}_y\omega_y - \overline{I}_{yz}\omega_z \\
H_z &= -\overline{I}_{zx}\omega_x - \overline{I}_{zy}\omega_y + \overline{I}_z\omega_z
\end{aligned} \tag{18.7}
$$

If *principal axes of inertia Gx'y'z'* are used, these relations reduce to

$$H_{x'} = \overline{I}_{x'}\omega_{x'} \qquad H_{y'} = \overline{I}_{y'}\omega_{y'} \qquad H_{z'} = \overline{I}_{z'}\omega_{z'} \tag{18.10}$$

We observed that, in general, *the angular momentum $\mathbf{H}_G$ and the angular velocity $\boldsymbol{\omega}$ do not have the same direction* (Fig. 18.25). They will, however, have the same direction if $\boldsymbol{\omega}$ is directed along one of the principal axes of inertia of the body.

**Fig. 18.25**

Angular momentum about a given point

Recalling that the system of the momenta of the particles forming a rigid body can be reduced to the vector $m\bar{\mathbf{v}}$ attached at $G$ and the couple $\mathbf{H}_G$ (Fig. 18.26), we noted that, once the linear momentum $m\bar{\mathbf{v}}$ and the angular momentum $\mathbf{H}_G$ of a rigid body have been determined, the angular momentum $\mathbf{H}_O$ of the body about any given point $O$ can be obtained by writing

$$\mathbf{H}_O = \bar{\mathbf{r}} \times m\bar{\mathbf{v}} + \mathbf{H}_G \tag{18.11}$$

Rigid body with a fixed point

In the particular case of a rigid body *constrained to rotate about a fixed point O*, the components of the angular momentum $\mathbf{H}_O$ of the body about $O$ can be obtained directly from the components of its angular velocity and from its moments and products of inertia with respect to axes through $O$. We wrote

$$H_x = +I_x\omega_x - I_{xy}\omega_y - I_{xz}\omega_z$$
$$H_y = -I_{yx}\omega_x + I_y\omega_y - I_{yz}\omega_z \tag{18.13}$$
$$H_z = -I_{zx}\omega_x - I_{zy}\omega_y + I_z\omega_z$$

Principle of impulse and momentum

The *principle of impulse and momentum* for a rigid body in three-dimensional motion [Sec. 18.3] is expressed by the same fundamental formula that was used in Chap. 17 for a rigid body in plane motion,

**Syst Momenta$_1$ + Syst Ext Imp$_{1\rightarrow2}$ = Syst Momenta$_2$** (17.4)

but the systems of the initial and final momenta should now be represented as shown in Fig. 18.26, and $\mathbf{H}_G$ should be computed from the relations (18.7) or (18.10) [Sample Probs. 18.1 and 18.2].

Kinetic energy of a rigid body in three dimensions

The *kinetic energy* of a rigid body in three-dimensional motion can be divided into two parts [Sec. 18.4], one associated with the motion of its mass center $G$ and the other with its motion about $G$. Using principal centroidal axes $x'$, $y'$, $z'$, we wrote

$$T = \tfrac{1}{2}m\bar{v}^2 + \tfrac{1}{2}(\bar{I}_{x'}\omega_{x'}^2 + \bar{I}_{y'}\omega_{y'}^2 + \bar{I}_{z'}\omega_{z'}^2) \tag{18.17}$$

where
$\bar{\mathbf{v}}$ = velocity of mass center
$\boldsymbol{\omega}$ = angular velocity
$m$ = mass of rigid body
$\bar{I}_{x'}, \bar{I}_{y'}, \bar{I}_{z'}$ = principal centroidal moments of inertia

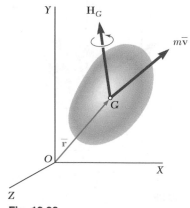

**Fig. 18.26**

We also noted that, in the case of a rigid body *constrained to rotate about a fixed point O*, the kinetic energy of the body can be expressed as

$$T = \tfrac{1}{2}(I_{x'}\omega_{x'}^2 + I_{y'}\omega_{y'}^2 + I_{z'}\omega_{z'}^2) \qquad (18.20)$$

where the $x'$, $y'$, and $z'$ axes are the principal axes of inertia of the body at $O$. The results obtained in Sec. 18.4 make it possible to extend to the three-dimensional motion of a rigid body the application of the *principle of work and energy* and of the *principle of conservation of energy*.

The second part of the chapter was devoted to the application of the fundamental equations

$$\Sigma \mathbf{F} = m\overline{\mathbf{a}} \qquad (18.1)$$
$$\Sigma \mathbf{M}_G = \dot{\mathbf{H}}_G \qquad (18.2)$$

Using a rotating frame to write the equations of motion of a rigid body in space

to the motion of a rigid body in three dimensions. We first recalled [Sec. 18.5] that $\mathbf{H}_G$ represents the angular momentum of the body relative to a centroidal frame $GX'Y'Z'$ of fixed orientation (Fig. 18.27)

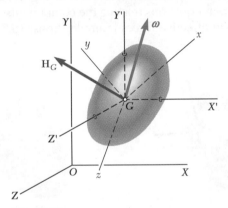

**Fig. 18.27**

and that $\dot{\mathbf{H}}_G$ in Eq. (18.2) represents the rate of change of $\mathbf{H}_G$ with respect to that frame. We noted that, as the body rotates, its moments and products of inertia with respect to the frame $GX'Y'Z'$ change continually. Therefore, it is more convenient to use a rotating frame $Gxyz$ when resolving $\boldsymbol{\omega}$ into components and computing the moments and products of inertia that will be used to determine $\mathbf{H}_G$ from Eqs. (18.7) or (18.10). However, since $\dot{\mathbf{H}}_G$ in Eq. (18.2) represents the rate of change of $\mathbf{H}_G$ with respect to the frame $GX'Y'Z'$ of fixed orientation, we must use the method of Sec. 15.10 to determine its value. Recalling Eq. (15.31), we wrote

$$\dot{\mathbf{H}}_G = (\dot{\mathbf{H}}_G)_{Gxyz} + \boldsymbol{\Omega} \times \mathbf{H}_G \qquad (18.22)$$

where $\mathbf{H}_G$ = angular momentum of body with respect to frame $GX'Y'Z'$ of fixed orientation

$(\dot{\mathbf{H}}_G)_{Gxyz}$ = rate of change of $\mathbf{H}_G$ with respect to rotating frame $Gxyz$, to be computed from relations (18.7)

$\boldsymbol{\Omega}$ = angular velocity of the rotating frame $Gxyz$

Substituting for $\dot{\mathbf{H}}_G$ from (18.22) into (18.2), we obtained

$$\Sigma\mathbf{M}_G = (\dot{\mathbf{H}}_G)_{Gxyz} + \boldsymbol{\Omega} \times \mathbf{H}_G \qquad (18.23)$$

If the rotating frame is actually attached to the body, its angular velocity $\boldsymbol{\Omega}$ is identically equal to the angular velocity $\boldsymbol{\omega}$ of the body. There are many applications, however, where it is advantageous to use a frame of reference which is not attached to the body but rotates in an independent manner [Sample Prob. 18.5].

**Euler's equations of motion. d'Alembert's principle**

Setting $\boldsymbol{\Omega} = \boldsymbol{\omega}$ in Eq. (18.23), using principal axes, and writing this equation in scalar form, we obtained *Euler's equations of motion* [Sec. 18.6]. A discussion of the solution of these equations and of the scalar equations corresponding to Eq. (18.1) led us to extend d'Alembert's principle to the three-dimensional motion of a rigid body and to conclude that the system of the external forces acting on the rigid body is not only equipollent, but actually *equivalent* to the effective forces of the body represented by the vector $m\bar{\mathbf{a}}$ and the couple $\dot{\mathbf{H}}_G$ (Fig. 18.28). Problems involving the three-dimensional motion of a rigid body can be solved by considering the free-body-diagram equation represented in Fig. 18.28 and writing appropriate scalar equations relating the components or moments of the external and effective forces [Sample Probs. 18.3 and 18.5].

**Free-body-diagram equation**

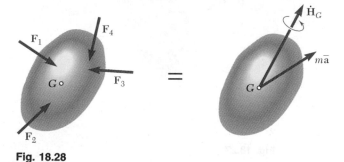

**Fig. 18.28**

**Rigid body with a fixed point**

In the case of a rigid body *constrained to rotate about a fixed point O*, an alternative method of solution, involving the moments of the forces and the rate of change of the angular momentum about point $O$, can be used. We wrote [Sec. 18.7]:

$$\Sigma\mathbf{M}_O = (\dot{\mathbf{H}}_O)_{Oxyz} + \boldsymbol{\Omega} \times \mathbf{H}_O \qquad (18.28)$$

where $\Sigma\mathbf{M}_O$ = sum of moments about $O$ of forces applied to rigid body

$\mathbf{H}_O$ = angular momentum of body with respect to fixed frame $OXYZ$

$(\dot{\mathbf{H}}_O)_{Oxyz}$ = rate of change of $\mathbf{H}_O$ with respect to rotating frame $Oxyz$, to be computed from relations (18.13)

$\boldsymbol{\Omega}$ = angular velocity of rotating frame $Oxyz$

This approach can be used to solve certain problems involving the rotation of a rigid body about a fixed axis [Sec. 18.8], for example, an unbalanced rotating shaft [Sample Prob. 18.4].

In the last part of the chapter, we considered the motion of *gyroscopes* and other *axisymmetrical bodies.* Introducing the *Eulerian angles* $\phi$, $\theta$, and $\psi$ to define the position of a gyroscope (Fig. 18.29), we observed that their derivatives $\dot{\phi}$, $\dot{\theta}$, and $\dot{\psi}$ represent, respectively, the rates of *precession, nutation,* and *spin* of the gyroscope [Sec. 18.9]. Expressing the angular velocity $\boldsymbol{\omega}$ in terms of these derivatives, we wrote

Motion of a gyroscope

$$\boldsymbol{\omega} = -\dot{\phi} \sin \theta \, \mathbf{i} + \dot{\theta}\mathbf{j} + (\dot{\psi} + \dot{\phi} \cos \theta)\mathbf{k} \quad (18.35)$$

Fig. 18.29

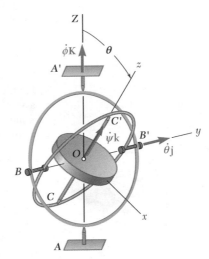

Fig. 18.30

where the unit vectors are associated with a frame $Oxyz$ attached to the inner gimbal of the gyroscope (Fig. 18.30) and rotate, therefore, with the angular velocity

$$\boldsymbol{\Omega} = -\dot{\phi} \sin \theta \, \mathbf{i} + \dot{\theta}\mathbf{j} + \dot{\phi} \cos \theta \, \mathbf{k} \quad (18.38)$$

Denoting by $I$ the moment of inertia of the gyroscope with respect to its spin axis $z$ and by $I'$ its moment of inertia with respect to a transverse axis through $O$, we wrote

$$\mathbf{H}_O = -I'\dot{\phi} \sin \theta \, \mathbf{i} + I'\dot{\theta}\mathbf{j} + I(\dot{\psi} + \dot{\phi} \cos \theta)\mathbf{k} \quad (18.36)$$

Substituting for $\mathbf{H}_O$ and $\boldsymbol{\Omega}$ into Eq. (18.28) led us to the differential equations defining the motion of the gyroscope.

In the particular case of the *steady precession* of a gyroscope [Sec. 18.10], the angle $\theta$, the rate of precession $\dot{\phi}$, and the rate of spin $\dot{\psi}$ remain constant. We saw that such a motion is possible only if the moments of the external forces about $O$ satisfy the relation

Steady precession

$$\Sigma \mathbf{M}_O = (I\omega_z - I'\dot{\phi} \cos \theta)\dot{\phi} \sin \theta \mathbf{j} \quad (18.44)$$

that is, if the external forces reduce to a couple of moment equal to the right-hand member of Eq. (18.44) and applied *about an axis perpendicular to the precession axis and to the spin axis* (Fig. 18.31). The chapter ended with a discussion of the motion of an axisymmetrical body spinning and precessing *under no force* [Sec. 18.11; Sample Prob. 18.6].

Fig. 18.31

# Review Problems

**Fig. P18.145**

**18.145** Two uniform rods $AB$ and $CE$, each of weight 1.36 kg and length 61 cm, are welded to each other at their midpoints. Knowing that this assembly has an angular velocity of constant magnitude $\omega = 12$ rad/s, determine the magnitude and direction of the angular momentum $\mathbf{H}_D$ of the assembly about $D$.

**18.146** Determine the kinetic energy of the assembly of Prob. 18.145.

**18.147** One of the sculptures displayed on a university campus consists of a hollow cube made of six aluminum sheets, each $1.5 \times 1.5$ m, welded together and reinforced with internal braces of negligible mass. The cube is mounted on a fixed base $A$ and can rotate freely about its vertical diagonal $AB$. As she passes by this display on the way to a class in mechanics, an engineering student grabs corner $C$ of the cube and pushes it for 1.2 s in a direction perpendicular to the plane $ABC$ with an average force of 50 N. Having observed that it takes 5 s for the cube to complete one full revolution, she flips out her calculator and proceeds to determine the mass of the cube. What is the result of her calculation? (*Hint:* The perpendicular distance from the diagonal joining two vertices of a cube to any of its other six vertices can be obtained by multiplying the side of the cube by $\sqrt{2/3}$.)

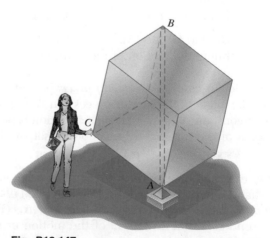

**Fig. *P18.147***

**18.148**   A 2268 kg probe in orbit about the moon is 2.4 m high and has octagonal bases of sides 1.2 m. The coordinate axes shown are the principal centroidal axes of inertia of the probe, and its radii of gyration are $k_x = 0.75$ m, $k_y = 0.8$ m, and $k_z = 0.77$ m. The probe is equipped with a main 556 N thruster $E$ and with four 22.2 N thrusters $A$, $B$, $C$, and $D$ which can expel fuel in the positive $y$ direction. The probe has an angular velocity $\boldsymbol{\omega} = (0.040 \text{ rad/s})\mathbf{i} + (0.060 \text{ rad/s})\mathbf{k}$ when two of the 22.2 N thrusters are used to reduce the angular velocity to zero. Determine ($a$) which of the thrusters should be used, ($b$) the operating time of each of these thrusters, ($c$) for how long the main thruster $E$ should be activated if the velocity of the mass center of the probe is to remain unchanged.

**Fig. P18.148**

**18.149**   A square plate of side $a$ and mass $m$ supported by a ball-and-socket joint at $A$ is rotating about the $y$ axis with a constant angular velocity $\boldsymbol{\omega} = \omega_0\mathbf{j}$ when an obstruction is suddenly introduced at $B$ in the $xy$ plane. Assuming the impact at $B$ is perfectly plastic ($e = 0$), determine immediately after impact ($a$) the angular velocity of the plate, ($b$) the velocity of its mass center $G$.

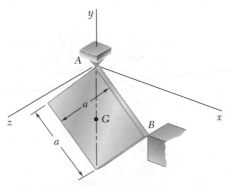

**Fig. P18.149**

*18.150*   Determine the impulse exerted on the plate of Prob. 18.149 during the impact by ($a$) the obstruction at $B$, ($b$) the support at $A$.

**18.151**   Determine the rate of change $\dot{\mathbf{H}}_D$ of the angular momentum $\mathbf{H}_D$ of the assembly of Prob. 18.145, assuming that at the instant considered the assembly has an angular velocity $\boldsymbol{\omega} = (12 \text{ rad/s})\mathbf{i}$ and an angular acceleration $\boldsymbol{\alpha} = -(96 \text{ rad/s}^2)\mathbf{i}$.

**18.152** When the 18 kg wheel shown is attached to a balancing machine and made to spin at a rate of 750 rpm, it is found that the forces exerted by the wheel on the machine are equivalent to a force-couple system consisting of a force $\mathbf{F} = (161 \text{ N})\mathbf{j}$ applied at $C$ and a couple $\mathbf{M}_C = (14.7 \text{ N} \cdot \text{m})\mathbf{k}$, where the unit vectors form a triad which rotates with the wheel. (*a*) Determine the distance from the axis of rotation to the mass center of the wheel and the products of inertia $I_{xy}$ and $I_{zx}$, (*b*) If only two corrective weights are to be used to balance the wheel statically and dynamically, what should these weights be and at which of the points $A$, $B$, $D$, or $E$ should they be placed?

18.4 cm

18.4 cm

7.6 cm   7.6 cm

**Fig. P18.152**

*18.153* The blade of a portable saw and the rotor of its motor have a total mass of 1.75 kg and a combined radius of gyration of 30 mm. Knowing that the blade rotates as shown at the rate $\omega_1 = 2500$ rpm, determine the magnitude and direction of the couple $\mathbf{M}$ that a worker must exert on the handle of the saw to rotate it with a constant angular velocity $\omega_2 = -(2.4 \text{ rad/s})\mathbf{j}$.

**Fig. *P18.153***

**18.154**  A homogeneous disk of mass $m = 3$ kg rotates at the constant rate $\omega_1 = 16$ rad/s with respect to arm $ABC$, which is welded to shaft $DCE$ rotating at the constant rate $\omega_2 = 8$ rad/s. Determine the dynamic reactions at $D$ and $E$.

**Fig. P18.154**

**18.155**  It is assumed that at the instant shown shaft $DCE$ of Prob. 18.154 has an angular velocity $\boldsymbol{\omega}_2 = (8 \text{ rad/s})\mathbf{i}$ and an angular acceleration $\boldsymbol{\alpha}_2 = (6 \text{ rad/s}^2)\mathbf{i}$. Recalling that the disk rotates with a constant angular velocity $\boldsymbol{\omega}_1 = (16 \text{ rad/s})\mathbf{j}$, determine (a) the couple that must be applied to shaft $DCE$ to produce the given angular acceleration, (b) the corresponding dynamic reactions at $D$ and $E$.

**18.156**  A 22.2 kN satellite is 2.2 m high and has octagonal bases of sides 1.1 m. The coordinate axes shown are the principal centroidal axes of inertia of the satellite, and its radii of gyration are $k_x = k_z = 0.8$ m and $k_y = 0.9$ m. The satellite is equipped with a main 556 N thruster $E$ and four 22.2 N thrusters $A$, $B$, $C$, and $D$, which can expel fuel in the positive $y$ direction. The satellite is spinning at the rate of 36 rev/h about its axis of symmetry $Gy$, which maintains a fixed direction in space, when thrusters $A$ and $B$ are activated for 2 s. Determine (a) the precession axis of the satellite, (b) its rate of precession, (c) its rate of spin.

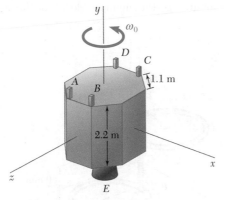

**Fig. P18.156**

# Computer Problems

**18.C1** A thin homogeneous disk of weight $W$ and radius $r$ is mounted on the horizontal axle $AB$. The plane of the disk forms an angle $\beta$ with the plane normal to the axle. The axle rotates with a constant angular velocity $\omega$. For values of $\beta$ from 0 to 90°, determine and plot (a) the angle $\theta$ formed by the axle and the angular momentum of the disk about $G$, (b) the kinetic energy of the disk, and (c) the magnitude of the rate of change of the angular momentum of the disk about $G$.

**Fig. P18.C1**

**18.C2** A wire of uniform cross section and mass per unit length $m' = 0.065$ kg/m is used to form the wire figure shown, which is suspended from cord $AD$. An impulse $\mathbf{F}\,\Delta t = (2\text{ N}\cdot\text{s})\mathbf{j}$ is applied to the wire figure at point $E$. Use computational software to calculate and plot, immediately after the impact, for values of $\theta$ from 0 to 180°, (a) the velocity of the mass center of the wire figure, (b) the angular velocity of the figure.

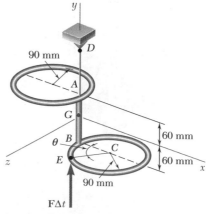

**Fig. P18.C2**

**18.C3** Two L-shaped arms, each of mass 2 kg, are welded to the one-third points of the 600-mm shaft $AB$ of negligible mass. The assembly is at rest ($\omega = 0$) at time $t = 0$ when a couple of moment $\mathbf{M}_0 = (1.6 \text{ N} \cdot \text{m})\mathbf{k}$ is applied to the shaft. Determine the dynamic reactions exerted on the shaft by the bearings at $A$ and $B$ at any time $t$. Resolve these reactions into components directed along $x$ and $y$ axes rotating with the assembly. Plot the components of the reactions and the magnitudes of the reactions from $t = 0$ to $t = 0.5$ s.

**Fig. P18.C3**

**18.C4** A couple $\mathbf{M}_0 = (0.036 \text{ N} \cdot \text{m})\mathbf{i}$ is applied to an assembly consisting of pieces of sheet aluminum of uniform thickness and of total mass 1.35 kg, which are welded to a light axle supported by bearings at $A$ and $B$. Determine the dynamic reactions exerted by the bearings on the axle at any time $t$ after the couple has been applied. Resolve these reactions into components directed along $y$ and $z$ axes rotating with the assembly. Use computational software (a) to calculate and plot the components of the reactions from $t = 0$ to $t = 2$ s, (b) to determine the time at which the $z$ components of the reactions at $A$ and $B$ are equal to zero.

**Fig. P18.C4**

**18.C5**  A 2.3 kg homogeneous disk of radius 4 in. can rotate with respect to arm $ABC$, which is welded to a shaft $DCE$ supported by bearings at $D$ and $E$. Both the arm and the shaft are of negligible mass. At time $t = 0$ a couple $\mathbf{M}_0 = (57.3 \text{ N} \cdot \text{cm})\mathbf{k}$ is applied to shaft $DCE$. Knowing that at $t = 0$ the angular velocity of the disk is $\omega_1 = (60 \text{ rad/s})\mathbf{j}$ and that friction in the bearing at $A$ causes the magnitude of $\omega_1$ to decrease at the rate of 15 rad/s$^2$, determine the dynamic reactions exerted on the shaft by the bearings at $D$ and $E$ at any time $t$. Resolve these reactions into components directed along $x$ and $y$ axes rotating with the shaft. Use computational software (a) to calculate and plot the components of the reactions from $t = 0$ to $t = 4$ s, (b) to determine the times $t_1$ and $t_2$ at which the $x$ and $y$ components of the reactions at $E$ are respectively equal to zero.

**Fig. P18.C5**

**18.C6**  A homogeneous disk of radius 23 cm is welded to a rod $AG$ of length 45.7 cm and of negligible mass which is connected by a clevis to a vertical shaft $AB$. The rod and disk can rotate freely about a horizontal axis $AC$, and shaft $AB$ can rotate freely about a vertical axis. Initially rod $AG$ forms a given angle $\theta_0$ with the downward vertical and its angular velocity $\dot{\theta}_0$ about $AC$ is zero. Shaft $AB$ is then given an angular velocity $\dot{\phi}_0$ about the vertical. Use computational software (a) to calculate the minimum value $\theta_m$ of the angle $\theta$ in the ensuing motion and the period of oscillation in $\theta$, that is, the time required for $\theta$ to regain its initial value $\theta_0$, (b) to compute and plot the angular velocity $\dot{\phi}$ of shaft $AB$ for values of $\theta$ from $\theta_0$ to $\theta_m$. Apply this software with the initial conditions (i) $\theta_0 = 90°$, $\dot{\phi}_0 = 5$ rad/s, (ii) $\theta_0 = 90°$, $\dot{\phi}_0 = 10$ rad/s, (iii) $\theta_0 = 60°$, $\dot{\phi}_0 = 5$ rad/s. [*Hint:* Use the principle of conservation of energy and the fact that the angular momentum of the body about the vertical through $A$ is conserved to obtain an equation of the form $\dot{\theta}^2 = f(\theta)$. This equation can be integrated by a numerical method.]

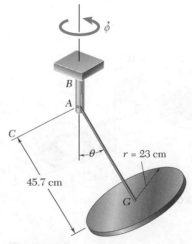

**Fig. P18.C6**

**18.C7** A homogeneous disk of radius 23 cm is welded to a rod $AG$ of length 45.7 cm and of negligible mass which is supported by a ball-and-socket joint at $A$. The disk is released in the position $\theta = \theta_0$ with a rate of spin $\dot{\psi}_0$, a rate of precession $\dot{\phi}_0$, and a zero rate of nutation. Use computational software (a) to calculate the minimum value $\theta_m$ of the angle $\theta$ in the ensuing motion and the period of oscillation in $\theta$, that is, the time required for $\theta$ to regain its initial value $\theta_0$, (b) to compute and plot the rate of spin $\dot{\psi}$ and the rate of precession $\dot{\phi}$ for values of $\theta$ from $\theta_0$ to $\theta_m$. Apply this software with the initial conditions (i) $\theta_0 = 90°$, $\dot{\psi}_0 = 50$ rad/s, $\dot{\phi}_0 = 0$, (ii) $\theta_0 = 90°$, $\dot{\psi}_0 = 0$, $\dot{\phi}_0 = 5$ rad/s, (iii) $\theta_0 = 90°$, $\dot{\psi}_0 = 50$ rad/s, $\dot{\phi}_0 = 5$ rad/s, (iv) $\theta_0 = 90°$, $\dot{\psi}_0 = 10$ rad/s, $\dot{\phi}_0 = 5$ rad/s, (v) $\theta_0 = 60°$, $\dot{\psi}_0 = 0$, $\dot{\phi}_0 = 5$ rad/s, (vi) $\theta_0 = 60°$, $\dot{\psi}_0 = 50$ rad/s, $\dot{\phi}_0 = 5$ rad/s. [*Hint:* Use the principle of conservation of energy and the fact that the angular momentum of the body is conserved about both the $Z$ and $z$ axes to obtain an equation of the form $\dot{\theta}^2 = f(\theta)$. This equation can be integrated by a numerical method.]

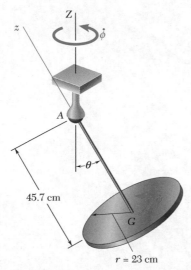

**Fig. P18.C7**

## CHAPTER 19

# Mechanical Vibrations

Mechanical systems may undergo *free vibrations* or they may be subjected to *forced vibrations*. The vibrations are *damped* when friction forces are present and *undamped* otherwise. The automobile suspension shown consists essentially of a spring and a shock absorber, which will cause the body of the car to undergo *damped forced vibrations* when the car is driven over an uneven road.

## MECHANICAL VIBRATIONS

19.1 Introduction
     **Vibrations without Damping**
19.2 Free Vibrations of Particles.
     Simple Harmonic Motion
19.3 Simple Pendulum (Approximate
     Solution)
19.4 Simple Pendulum (Exact
     Solution)
19.5 Free Vibrations of Rigid Bodies
19.6 Application of the Principle of
     Conservation of Energy
19.7 Forced Vibrations
     **Damped Vibrations**
19.8 Damped Free Vibrations
19.9 Damped Forced Vibrations
19.10 Electrical Analogues

## 19.1. INTRODUCTION

A *mechanical vibration* is the motion of a particle or a body which oscillates about a position of equilibrium. Most vibrations in machines and structures are undesirable because of the increased stresses and energy losses which accompany them. They should therefore be eliminated or reduced as much as possible by appropriate design. The analysis of vibrations has become increasingly important in recent years owing to the current trend toward higher-speed machines and lighter structures. There is every reason to expect that this trend will continue and that an even greater need for vibration analysis will develop in the future.

The analysis of vibrations is a very extensive subject to which entire texts have been devoted. Our present study will therefore be limited to the simpler types of vibrations, namely, the vibrations of a body or a system of bodies with one degree of freedom.

A mechanical vibration generally results when a system is displaced from a position of stable equilibrium. The system tends to return to this position under the action of restoring forces (either elastic forces, as in the case of a mass attached to a spring, or gravitational forces, as in the case of a pendulum). But the system generally reaches its original position with a certain acquired velocity which carries it beyond that position. Since the process can be repeated indefinitely, the system keeps moving back and forth across its position of equilibrium. The time interval required for the system to complete a full cycle of motion is called the *period* of the vibration. The number of cycles per unit time defines the *frequency*, and the maximum displacement of the system from its position of equilibrium is called the *amplitude* of the vibration.

When the motion is maintained by the restoring forces only, the vibration is said to be a *free vibration* (Secs. 19.2 to 19.6). When a periodic force is applied to the system, the resulting motion is described as a *forced vibration* (Sec. 19.7). When the effects of friction can be neglected, the vibrations are said to be *undamped*. However, all vibrations are actually *damped* to some degree. If a free vibration is only slightly damped, its amplitude slowly decreases until, after a certain time, the motion comes to a stop. But if damping is large enough to prevent any true vibration, the system then slowly regains its original position (Sec. 19.8). A damped forced vibration is maintained as long as the periodic force which produces the vibration is applied. The amplitude of the vibration, however, is affected by the magnitude of the damping forces (Sec. 19.9).

## VIBRATIONS WITHOUT DAMPING

### 19.2. FREE VIBRATIONS OF PARTICLES. SIMPLE HARMONIC MOTION

Consider a body of mass $m$ attached to a spring of constant $k$ (Fig. 19.1$a$). Since at the present time we are concerned only with the motion of its mass center, we will refer to this body as a particle. When the particle is in static equilibrium, the forces acting on it are its weight $\mathbf{W}$ and the force $\mathbf{T}$ exerted by the spring, of magnitude

$T = k\delta_{st}$, where $\delta_{st}$ denotes the elongation of the spring. We have, therefore,

$$W = k\delta_{st}$$

Suppose now that the particle is displaced through a distance $x_m$ from its equilibrium position and released with no initial velocity. If $x_m$ has been chosen smaller than $\delta_{st}$, the particle will move back and forth through its equilibrium position; a vibration of amplitude $x_m$ has been generated. Note that the vibration can also be produced by imparting a certain initial velocity to the particle when it is in its equilibrium position $x = 0$ or, more generally, by starting the particle from any given position $x = x_0$ with a given initial velocity $\mathbf{v}_0$.

To analyze the vibration, let us consider the particle in a position $P$ at some arbitrary time $t$ (Fig. 19.1b). Denoting by $x$ the displacement $OP$ measured from the equilibrium position $O$ (positive downward), we note that the forces acting on the particle are its weight $\mathbf{W}$ and the force $\mathbf{T}$ exerted by the spring which, in this position, has a magnitude $T = k(\delta_{st} + x)$. Recalling that $W = k\delta_{st}$, we find that the magnitude of the resultant $\mathbf{F}$ of the two forces (positive downward) is

$$F = W - k(\delta_{st} + x) = -kx \quad (19.1)$$

Thus the *resultant* of the forces exerted on the particle is proportional to the displacement $OP$ *measured from the equilibrium position*. Recalling the sign convention, we note that $\mathbf{F}$ is always directed *toward* the equilibrium position $O$. Substituting for $F$ into the fundamental equation $F = ma$ and recalling that $a$ is the second derivative $\ddot{x}$ of $x$ with respect to $t$, we write

$$m\ddot{x} + kx = 0 \quad (19.2)$$

Note that the same sign convention should be used for the acceleration $\ddot{x}$ and for the displacement $x$, namely, positive downward.

The motion defined by Eq. (19.2) is called a *simple harmonic motion*. It is characterized by the fact that *the acceleration is proportional to the displacement and of opposite direction*. We can verify that each of the functions $x_1 = \sin(\sqrt{k/m}\ t)$ and $x_2 = \cos(\sqrt{k/m}\ t)$ satisfies Eq. (19.2). These functions, therefore, constitute two *particular solutions* of the differential equation (19.2). The *general solution* of Eq. (19.2) is obtained by multiplying each of the particular solutions by an arbitrary constant and adding. Thus, the general solution is expressed as

$$x = C_1 x_1 + C_2 x_2 = C_1 \sin\left(\sqrt{\frac{k}{m}}t\right) + C_2 \cos\left(\sqrt{\frac{k}{m}}t\right) \quad (19.3)$$

We note that $x$ is a *periodic function* of the time $t$ and does, therefore, represent a vibration of the particle $P$. The coefficient of $t$ in the expression we have obtained is referred to as the *natural circular frequency* of the vibration and is denoted by $\omega_n$. We have

$$\text{Natural circular frequency} = \omega_n = \sqrt{\frac{k}{m}} \quad (19.4)$$

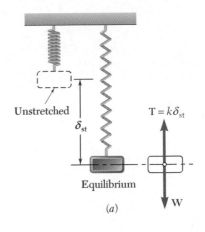

**Fig. 19.1**

Substituting for $\sqrt{k/m}$ into Eq. (19.3), we write

$$x = C_1 \sin \omega_n t + C_2 \cos \omega_n t \qquad (19.5)$$

This is the general solution of the differential equation

$$\ddot{x} + \omega_n^2 x = 0 \qquad (19.6)$$

which can be obtained from Eq. (19.2) by dividing both terms by $m$ and observing that $k/m = \omega_n^2$. Differentiating twice both members of Eq. (19.5) with respect to $t$, we obtain the following expressions for the velocity and the acceleration at time $t$:

$$v = \dot{x} = C_1 \omega_n \cos \omega_n t - C_2 \omega_n \sin \omega_n t \qquad (19.7)$$

$$a = \ddot{x} = -C_1 \omega_n^2 \sin \omega_n t - C_2 \omega_n^2 \cos \omega_n t \qquad (19.8)$$

The values of the constants $C_1$ and $C_2$ depend upon the *initial conditions* of the motion. For example, we have $C_1 = 0$ if the particle is displaced from its equilibrium position and released at $t = 0$ with no initial velocity, and we have $C_2 = 0$ if the particle is started from $O$ at $t = 0$ with a certain initial velocity. In general, substituting $t = 0$ and the initial values $x_0$ and $v_0$ of the displacement and the velocity into Eqs. (19.5) and (19.7), we find that $C_1 = v_0/\omega_n$ and $C_2 = x_0$.

The expressions obtained for the displacement, velocity, and acceleration of a particle can be written in a more compact form if we observe that Eq. (19.5) expresses that the displacement $x = OP$ is the sum of the $x$ components of two vectors $\mathbf{C}_1$ and $\mathbf{C}_2$, respectively, of magnitude $C_1$ and $C_2$, directed as shown in Fig. 19.2$a$. As $t$ varies, both vectors rotate clockwise; we also note that the magnitude of their resultant $\overrightarrow{OQ}$ is equal to the maximum displacement $x_m$. The simple harmonic motion of $P$ along the $x$ axis can thus be obtained by projecting on this axis the motion of a point $Q$ describing an *auxiliary circle* of radius $x_m$ *with a constant angular velocity* $\omega_n$ (which explains the name of natural *circular* frequency given to $\omega_n$). Denoting by $\phi$ the angle formed by the vectors $\overrightarrow{OQ}$ and $\mathbf{C}_1$, we write

$$OP = OQ \sin (\omega_n t + \phi) \qquad (19.9)$$

which leads to new expressions for the displacement, velocity, and acceleration of $P$:

$$x = x_m \sin (\omega_n t + \phi) \qquad (19.10)$$

$$v = \dot{x} = x_m \omega_n \cos (\omega_n t + \phi) \qquad (19.11)$$

$$a = \ddot{x} = -x_m \omega_n^2 \sin (\omega_n t + \phi) \qquad (19.12)$$

The displacement-time curve is represented by a sine curve (Fig. 19.2$b$); the maximum value $x_m$ of the displacement is called the *amplitude* of the vibration, and the angle $\phi$ which defines the initial position of $Q$ on the circle is called the *phase angle*. We note from Fig. 19.2 that a full *cycle* is described as the angle $\omega_n t$ increases by $2\pi$ rad. The corresponding value of $t$, denoted by $\tau_n$, is called the *period* of the free vibration and is measured in seconds. We have

(a)                                    (b)

**Fig. 19.2**

$$\text{Period} = \tau_n = \frac{2\pi}{\omega_n} \qquad (19.13)$$

The number of cycles described per unit of time is denoted by $f_n$ and is known as the *natural frequency* of the vibration. We write

$$\text{Natural frequency} = f_n = \frac{1}{\tau_n} = \frac{\omega_n}{2\pi} \qquad (19.14)$$

The unit of frequency is a frequency of 1 cycle per second, corresponding to a period of l s. In terms of base units the unit of frequency is thus $1/s$ or $s^{-1}$. It is called a *hertz* (Hz) in the SI system of units. It also follows from Eq. (19.14) that a frequency of 1 $s^{-1}$ or 1 Hz corresponds to a circular frequency of $2\pi$ rad/s. In problems involving angular velocities expressed in revolutions per minute (rpm), we have 1 rpm $= \frac{1}{60}$ $s^{-1} = \frac{1}{60}$ Hz, or 1 rpm $= (2\pi/60)$ rad/s.

Recalling that $\omega_n$ was defined in (19.4) in terms of the constant $k$ of the spring and the mass $m$ of the particle, we observe that the period and the frequency are independent of the initial conditions and of the amplitude of the vibration. Note that $\tau_n$ and $f_n$ depend on the *mass* rather than on the *weight* of the particle and thus are independent of the value of $g$.

The velocity-time and acceleration-time curves can be represented by sine curves of the same period as the displacement-time curve, but with different phase angles. From Eqs. (19.11) and (19.12), we note that the maximum values of the magnitudes of the velocity and acceleration are

$$v_m = x_m \omega_n \qquad a_m = x_m \omega_n^2 \qquad (19.15)$$

Since the point $Q$ describes the auxiliary circle, of radius $x_m$, at the constant angular velocity $\omega_n$, its velocity and acceleration are equal, respectively, to the expressions (19.15). Recalling Eqs. (19.11) and (19.12), we find, therefore, that the velocity and acceleration of $P$ can be obtained at any instant by projecting on the $x$ axis vectors of magnitudes $v_m = x_m \omega_n$ and $a_m = x_m \omega_n^2$ representing, respectively, the velocity and acceleration of $Q$ at the same instant (Fig. 19.3).

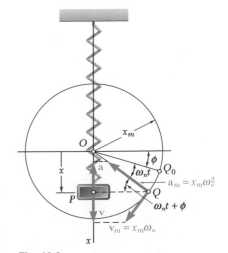

**Fig. 19.3**

The results obtained are not limited to the solution of the problem of a mass attached to a spring. They can be used to analyze the rectilinear motion of a particle *whenever the resultant* **F** *of the forces acting on the particle is proportional to the displacement x and directed toward O.* The fundamental equation of motion $F = ma$ can then be written in the form of Eq. (19.6), which is characteristic of a simple harmonic motion. Observing that the coefficient of $x$ must be equal to $\omega_n^2$, we can easily determine the natural circular frequency $\omega_n$ of the motion. Substituting the value obtained for $\omega_n$ into Eqs. (19.13) and (19.14), we then obtain the period $\tau_n$ and the natural frequency $f_n$ of the motion.

*(a)*

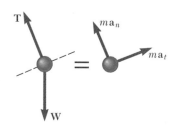

*(b)*

**Fig. 19.4**

### 19.3. SIMPLE PENDULUM (APPROXIMATE SOLUTION)

Most of the vibrations encountered in engineering applications can be represented by a simple harmonic motion. Many others, although of a different type, can be *approximated* by a simple harmonic motion, provided that their amplitude remains small. Consider, for example, a *simple pendulum,* consisting of a bob of mass $m$ attached to a cord of length $l$, which can oscillate in a vertical plane (Fig. 19.4a). At a given time $t$, the cord forms an angle $\theta$ with the vertical. The forces acting on the bob are its weight **W** and the force **T** exerted by the cord (Fig. 19.4b). Resolving the vector $m\mathbf{a}$ into tangential and normal components, with $m\mathbf{a}_t$ directed to the right, that is, in the direction corresponding to increasing values of $\theta$, and observing that $a_t = l\alpha = l\ddot{\theta}$, we write

$$\Sigma F_t = ma_t: \qquad\qquad -W \sin \theta = ml\ddot{\theta}$$

Noting that $W = mg$ and dividing through by $ml$, we obtain

$$\ddot{\theta} + \frac{g}{l} \sin \theta = 0 \tag{19.16}$$

For oscillations of small amplitude, we can replace $\sin \theta$ by $\theta$, expressed in radians, and write

$$\ddot{\theta} + \frac{g}{l} \theta = 0 \tag{19.17}$$

Comparison with Eq. (19.6) shows that the differential Eq. (19.17) is that of a simple harmonic motion with a natural circular frequency $\omega_n$ equal to $(g/l)^{1/2}$. The general solution of Eq. (19.17) can, therefore, be expressed as

$$\theta = \theta_m \sin (\omega_n t + \phi)$$

where $\theta_m$ is the amplitude of the oscillations and $\phi$ is a phase angle. Substituting into Eq. (19.13) the value obtained for $\omega_n$, we get the following expression for the period of the small oscillations of a pendulum of length $l$:

$$\tau_n = \frac{2\pi}{\omega_n} = 2\pi \sqrt{\frac{l}{g}} \tag{19.18}$$

# *19.4. SIMPLE PENDULUM (EXACT SOLUTION)

Formula (19.18) is only approximate. To obtain an exact expression for the period of the oscillations of a simple pendulum, we must return to Eq. (19.16). Multiplying both terms by $2\dot{\theta}$ and integrating from an initial position corresponding to the maximum deflection, that is, $\theta = \theta_m$ and $\dot{\theta} = 0$, we write

$$\left(\frac{d\theta}{dt}\right)^2 = \frac{2g}{l}(\cos\theta - \cos\theta_m)$$

Replacing $\cos\theta$ by $1 - 2\sin^2(\theta/2)$ and $\cos\theta_m$ by a similar expression, solving for $dt$, and integrating over a quarter period from $t = 0$, $\theta = 0$ to $t = \tau_n/4$, $\theta = \theta_m$, we have

$$\tau_n = 2\sqrt{\frac{l}{g}}\int_0^{\theta_m}\frac{d\theta}{\sqrt{\sin^2(\theta_m/2) - \sin^2(\theta/2)}}$$

The integral in the right-hand member is known as an *elliptic integral;* it cannot be expressed in terms of the usual algebraic or trigonometric functions. However, setting

$$\sin(\theta/2) = \sin(\theta_m/2)\sin\phi$$

we can write

$$\tau_n = 4\sqrt{\frac{l}{g}}\int_0^{\pi/2}\frac{d\phi}{\sqrt{1 - \sin^2(\theta_m/2)\sin^2\phi}} \qquad (19.19)$$

where the integral obtained, commonly denoted by $K$, can be calculated by using a numerical method of integration. It can also be found in *tables of elliptic integrals* for various values of $\theta_m/2$.† In order to compare the result just obtained with that of the preceding section, we write Eq. (19.19) in the form

$$\tau_n = \frac{2K}{\pi}\left(2\pi\sqrt{\frac{l}{g}}\right) \qquad (19.20)$$

Formula (19.20) shows that the actual value of the period of a simple pendulum can be obtained by multiplying the approximate value given in Eq. (19.18) by the correction factor $2K/\pi$. Values of the correction factor are given in Table 19.1 for various values of the amplitude $\theta_m$. We note that for ordinary engineering computations the correction factor can be omitted as long as the amplitude does not exceed $10°$.

**Table 19.1.** Correction Factor for the Period of a Simple Pendulum

| $\theta_m$ | 0° | 10° | 20° | 30° | 60° | 90° | 120° | 150° | 180° |
|---|---|---|---|---|---|---|---|---|---|
| $K$ | 1.571 | 1.574 | 1.583 | 1.598 | 1.686 | 1.854 | 2.157 | 2.768 | ∞ |
| $2K/\pi$ | 1.000 | 1.002 | 1.008 | 1.017 | 1.073 | 1.180 | 1.373 | 1.762 | ∞ |

†See, for example, *Standard Mathematical Tables,* Chemical Rubber Publishing Company, Cleveland, Ohio.

(a)

(b)

## SAMPLE PROBLEM 19.1

A 50-kg block moves between vertical guides as shown. The block is pulled 40 mm down from its equilibrium position and released. For each spring arrangement, determine the period of the vibration, the maximum velocity of the block, and the maximum acceleration of the block.

## SOLUTION

*a.* **Springs Attached in Parallel.** We first determine the constant $k$ of a single spring equivalent to the two springs *by finding the magnitude of the force* **P** *required to cause a given deflection* $\delta$. Since for a deflection $\delta$ the magnitudes of the forces exerted by the springs are, respectively, $k_1\delta$ and $k_2\delta$, we have

$$P = k_1\delta + k_2\delta = (k_1 + k_2)\delta$$

The constant $k$ of the single equivalent spring is

$$k = \frac{P}{\delta} = k_1 + k_2 = 4 \text{ kN/m} + 6 \text{ kN/m} = 10 \text{ kN/m} = 10^4 \text{ N/m}$$

*Period of Vibration:* Since $m = 50$ kg, Eq. (19.4) yields

$$\omega_n^2 = \frac{k}{m} = \frac{10^4 \text{ N/m}}{50 \text{ kg}} \qquad \omega_n = 14.14 \text{ rad/s}$$

$$\tau_n = 2\pi/\omega_n \qquad\qquad \tau_n = 0.444 \text{ s} \quad \blacktriangleleft$$

*Maximum Velocity:* $v_m = x_m\omega_n = (0.040 \text{ m})(14.14 \text{ rad/s})$

$$v_m = 0.566 \text{ m/s} \qquad \mathbf{v}_m = 0.566 \text{ m/s} \updownarrow \quad \blacktriangleleft$$

*Maximum Acceleration:* $a_m = x_m\omega_n^2 = (0.040 \text{ m})(14.14 \text{ rad/s})^2$

$$a_m = 8.00 \text{ m/s}^2 \qquad \mathbf{a}_m = 8.00 \text{ m/s}^2 \updownarrow \quad \blacktriangleleft$$

*b.* **Springs Attached in Series.** We first determine the constant $k$ of a single spring equivalent to the two springs *by finding the total elongation* $\delta$ of the springs under a given static load **P**. To facilitate the computation, a static load of magnitude $P = 12$ kN is used.

$$\delta = \delta_1 + \delta_2 = \frac{P}{k_1} + \frac{P}{k_2} = \frac{12 \text{ kN}}{4 \text{ kN/m}} + \frac{12 \text{ kN}}{6 \text{ kN/m}} = 5 \text{ m}$$

$$k = \frac{P}{\delta} = \frac{12 \text{ kN}}{5 \text{ m}} = 2.4 \text{ kN/m} = 2400 \text{ N/m}$$

*Period of Vibration:* $\omega_n^2 = \dfrac{k}{m} = \dfrac{2400 \text{ N/m}}{50 \text{ kg}} \qquad \omega_n = 6.93 \text{ rad/s}$

$$\tau_n = \frac{2\pi}{\omega_n} \qquad\qquad \tau_n = 0.907 \text{ s} \quad \blacktriangleleft$$

*Maximum Velocity:* $v_m = x_m\omega_n = (0.040 \text{ m})(6.93 \text{ rad/s})$

$$v_m = 0.277 \text{ m/s} \qquad \mathbf{v}_m = 0.277 \text{ m/s} \updownarrow \quad \blacktriangleleft$$

*Maximum Acceleration:* $a_m = x_m\omega_n^2 = (0.040 \text{ m})(6.93 \text{ rad/s})^2$

$$a_m = 1.920 \text{ m/s}^2 \qquad \mathbf{a}_m = 1.920 \text{ m/s}^2 \updownarrow \quad \blacktriangleleft$$

# SOLVING PROBLEMS ON YOUR OWN

This chapter deals with *mechanical vibrations,* that is, with the motion of a particle or body oscillating about a position of equilibrium.

In this first lesson, we saw that a *free vibration* of a particle occurs when the particle is subjected to a force proportional to its displacement and of opposite direction, such as the force exerted by a spring (Fig. 19.1). The resulting motion, called a *simple harmonic motion,* is characterized by the differential equation

$$m\ddot{x} + kx = 0 \tag{19.2}$$

where $x$ is the displacement of the particle, $\ddot{x}$ is its acceleration, $m$ is its mass, and $k$ is the constant of the spring. The solution of this differential equation was found to be

$$x = x_m \sin(\omega_n t + \phi) \tag{19.10}$$

where $x_m$ = amplitude of the vibration
$\omega_n = \sqrt{k/m}$ = natural circular frequency (rad/s)
$\phi$ = phase angle (rad)

We also defined the *period* of the vibration as the time $\tau_n = 2\pi/\omega_n$ needed for the particle to complete one cycle, and the *natural frequency* as the number of cycles per second, $f_n = 1/\tau_n = \omega_n/2\pi$, expressed in Hz or $s^{-1}$. Differentiating Eq. (19.10) twice yields the velocity and the acceleration of the particle at any time. The maximum values of the velocity and acceleration were found to be

$$v_m = x_m \omega_n \qquad a_m = x_m \omega_n^2 \tag{19.15}$$

To determine the parameters in Eq. (19.10) you can follow these steps.

*1. Draw a free-body diagram showing the forces exerted on the particle* when the particle is at a distance $x$ from its position of equilibrium. The resultant of these forces will be proportional to $x$ and its direction will be opposite to the positive direction of $x$ [Eq. (19.1)].

*2. Write the differential equation of motion* by equating to $m\ddot{x}$ the resultant of the forces found in step 1. Note that once a positive direction for $x$ has been chosen, the same sign convention should be used for the acceleration $\ddot{x}$. After transposition, you will obtain an equation of the form of Eq. (19.2).

*3. Determine the natural circular frequency $\omega_n$* by dividing the coefficient of $x$ by the coefficient of $\ddot{x}$ in this equation and taking the square root of the result obtained. Make sure that $\omega_n$ is expressed in rad/s.

*(continued)*

**4. *Determine the amplitude $x_m$ and the phase angle $\phi$*** by substituting the value obtained for $\omega_n$ and the initial values of $x$ and $\dot{x}$ into Eq. (19.10) and the equation obtained by differentiating Eq. (19.10) with respect to $t$.

Equation (19.10) and the two equations obtained by differentiating Eq. (19.10) twice with respect to $t$ can now be used to find the displacement, velocity, and acceleration of the particle at any time. Equations (19.15) yield the maximum velocity $v_m$ and the maximum acceleration $a_m$.

**5. *You also saw that for the small oscillations of a simple pendulum,*** the angle $\theta$ that the cord of the pendulum forms with the vertical satisfies the differential equation

$$\ddot{\theta} + \frac{g}{l}\,\theta = 0 \qquad\qquad (19.17)$$

where $l$ is the length of the cord and where $\theta$ is expressed in radians [Sec. 19.3]. This equation defines again a *simple harmonic motion,* and its solution is of the same form as Eq. (19.10),

$$\theta = \theta_m \sin\,(\omega_n t + \phi)$$

where the natural circular frequency $\omega_n = \sqrt{g/l}$ is expressed in rad/s. The determination of the various constants in this expression is carried out in a manner similar to that described above. Remember that the velocity of the bob is tangent to the path and that its magnitude is $v = l\dot{\theta}$, while the acceleration of the bob has a tangential component $\mathbf{a}_t$, of magnitude $a_t = l\ddot{\theta}$, and a normal component $\mathbf{a}_n$ directed toward the center of the path and of magnitude $a_n = l\dot{\theta}^2$.

# Problems

**19.1**  A particle moves in simple harmonic motion. Knowing that the amplitude is 45.7 cm and the maximum velocity is 1.8 m/s, determine the maximum acceleration of the particle and the period of its motion.

**19.2**  A particle moves in simple harmonic motion. Knowing that the maximum velocity is 200 mm/s and the maximum acceleration is 4 m/s$^2$, determine the amplitude and frequency of the motion.

**19.3**  Determine the amplitude and maximum acceleration of a particle which moves in simple harmonic motion with a maximum velocity of 1.2 m/s and a frequency of 6 Hz.

**19.4**  A 9 kg block is initially held so that the vertical spring attached as shown is undeformed. Knowing that the block is suddenly released from rest, determine (*a*) the amplitude and frequency of the resulting motion, (*b*) the maximum velocity and maximum acceleration of the block.

15761 N/m

9 kg

**Fig. P19.4**

**19.5**  A 32 kg block is attached to a spring and can move without friction in a slot as shown. The block is in its equilibrium position when it is struck by a hammer which imparts to the block an initial velocity of 3 m/s. Determine (*a*) the period and frequency of the resulting motion, (*b*) the amplitude of the motion and the maximum acceleration of the block.

32 kg

$k = 131$ kN/m

**Fig. P19.5**

**19.6**  An instrument package *A* is bolted to a shaker table as shown. The table moves vertically in simple harmonic motion at the same frequency as the variable-speed motor which drives it. The package is to be tested at a peak acceleration of 50 m/s$^2$. Knowing that the amplitude of the shaker table is 58 mm, determine (*a*) the required speed of the motor in rpm, (*b*) the maximum velocity of the table.

**19.7**  A simple pendulum consisting of a bob attached to a cord oscillates in a vertical plane with a period of 1.3 s. Assuming simple harmonic motion and knowing that the maximum velocity of the bob is 0.4 m/s, determine (*a*) the amplitude of the motion in degrees, (*b*) the maximum tangential acceleration of the bob.

A

**Fig. P19.6**

**19.8**  A 4.5 kg block *A* rests on a 18 kg plate *B* which is attached to an unstretched spring of constant $k = 875.6$ N/m. Plate *B* is slowly moved 6 cm. to the left and released from rest. Assuming that block *A* does not slip on the plate, determine (*a*) the amplitude and frequency of the resulting motion, (*b*) the corresponding smallest allowable value of the coefficient of static friction.

A

B        k

**Fig. *P19.8***

**1223**

**Fig. P19.9**

**Fig. P19.11**

**Fig. P19.12**

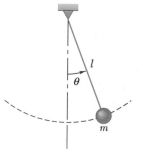

**Fig. P19.16**

**19.9**  A 1.8 kg collar is attached to a spring of constant 1051 N/m and can slide without friction on a horizontal rod. The collar is at rest when it is struck with a mallet and given an initial velocity of 1.4 m/s. Determine the amplitude and the maximum acceleration of the collar during the resulting motion.

**19.10**  The motion of a particle is described by the equation $x = 60 \cos 10\pi t + 45 \sin(10\pi t - \pi/3)$, where $x$ is expressed in millimeters and $t$ in seconds. Determine (a) the period of the resulting motion, (b) its amplitude, (c) its phase angle.

**19.11**  A variable-speed motor is rigidly attached to beam $BC$. The rotor is slightly unbalanced and causes the beam to vibrate with a frequency equal to the motor speed. When the speed of the motor is less than 600 rpm or more than 1200 rpm, a small object placed at $A$ is observed to remain in contact with the beam. For speeds between 600 rpm and 1200 rpm the object is observed to "dance" and actually to lose contact with the beam. Determine the amplitude of the motion of $A$ when the speed of the motor is (a) 600 rpm, (b) 1200 rpm. Give answers in both SI and U.S. customary units.

**19.12**  A 1.4-kg block is supported as shown by a spring of constant $k = 400$ N/m which can act in tension or compression. The block is in its equilibrium position when it is struck from below by a hammer which imparts to the block an upward velocity of 2.5 m/s. Determine (a) the time required for the block to move 60 mm upward, (b) the corresponding velocity and acceleration of the block.

**19.13**  In Prob. 19.12, determine the position, velocity, and acceleration of the block 0.90 s after it has been struck by the hammer.

**19.14**  A 32 kg block attached to a spring of constant $k = 131$ kN/m can move without friction in a slot as shown. The block is given an initial 38 cm displacement downward from its equilibrium position and released. Determine 1.5 s after the block has been released (a) the total distance traveled by the block, (b) the acceleration of the block.

**Fig. P19.14**          **Fig. P19.15**

**19.15**  A 4.5 kg collar $C$ is released from rest in the position shown and slides without friction on a vertical rod until it hits a spring of constant $k = 730$ N/m which it compresses. The velocity of the collar is reduced to zero, and the collar reverses the direction of its motion and returns to its initial position. The cycle is then repeated. Determine (a) the period of the motion of the collar, (b) the velocity of the collar 0.4 s after it was released. (*Note:* This is a periodic motion, but not simple harmonic motion.)

**19.16**  The bob of a simple pendulum of length $l = 1.2$ m is moving with a velocity of 180 mm/s to the right at time $t = 0$ when $\theta = 0$. Assuming simple harmonic motion, determine at $t = 1.5$ s (a) the angle $\theta$, (b) the magnitudes of the velocity and acceleration of the bob.

**19.17** A 4.5 kg collar rests on but is not attached to the spring shown. It is observed that when the collar is pushed down 23 cm or more and released, it loses contact with the spring. Determine (*a*) the spring constant, (*b*) the position, velocity, and acceleration of the collar 0.16 s after it has been pushed down 23 cm and released.

**Fig. *P19.17***     **Fig. P19.18**

**19.18** A 13.6 kg block is supported by the spring arrangement shown. If the block is moved from its equilibrium 44 mm vertically downward and released, determine (*a*) the period and frequency of the resulting motion, (*b*) the maximum velocity and acceleration of the block.

**19.19** A 50 kg block is supported by the spring arrangement shown. The block is moved vertically downward from its equilibrium position and released. Knowing that the amplitude of the resulting motion is 60 mm, determine (*a*) the period and frequency of the motion, (*b*) the maximum velocity and maximum acceleration of the block.

**19.20** A 4.5 kg block, attached to the lower end of a spring whose upper end is fixed, vibrates with a period of 6.8 s. Knowing that the constant $k$ of the spring is inversely proportional to its length, determine the period of a 2.7 kg block which is attached to the center of the same spring if the upper and lower ends of the spring are fixed.

**19.21** Two springs of constants $k_1$ and $k_2$ are connected in series to a block *A* that vibrates in simple harmonic motion with a period of 5 s. When the same two springs are connected in parallel to the same block, the block vibrates with a period of 2 s. Determine the ratio $k_1/k_2$ of the two spring constants.

**Fig. P19.19**

**Fig. P19.21**

**19.22** A 25 kg block is supported by the spring arrangement shown. If the block is moved vertically downward from its equilibrium position and released, determine (*a*) the period and frequency of the resulting motion, (*b*) the maximum velocity and acceleration of the block if the amplitude of the motion is 30 mm.

**Fig. *P19.22***

**Fig. P19.23**

**19.23** A 4.08 kg block is supported as shown by three springs, each of which has a constant $k$. In the equilibrium position the tensions in springs $A$, $B$, and $C$ are 22 N, 16 N, and 12 N, respectively. The block is moved vertically downward 12.5 mm from its equilibrium position and released from rest. Knowing that in the resulting motion the minimum tension in spring $B$ is zero, determine ($a$) the spring constant $k$, ($b$) the frequency of the motion, ($c$) the maximum compression in spring $A$.

**19.24** The period of vibration of the system of three springs and a block is observed to be 0.2 s. After the lower spring of constant $k_2 = 16$ kN/m is removed from the system the period is observed to be 0.25 s. Determine ($a$) the mass $m$, ($b$) the spring constant $k_1$.

**Fig. P19.24**          **Fig. P19.25**

**19.25** The period of vibration of the system shown is observed to be 0.2 s. After the spring of constant $k_2 = 3.5$ kN/m is removed and block $A$ is connected to the spring of constant $k_1$, the period is observed to be 0.12 s. Determine ($a$) the constant $k_1$ of the remaining spring, ($b$) the mass of block $A$.

**Fig. P19.26**

**19.26** The 45.3 kg block $D$ is supported as shown by three springs, each of which has a constant $k$. The period of vertical vibration is to remain unchanged when the block is replaced by a 54.4-kg block, spring $A$ is replaced by a spring of constant $k_A$, and the other two springs are unchanged. Knowing that the period of vibration is 0.4 s, determine the values of $k$ and $k_A$.

**19.27** The period of small oscillations of the system shown is observed to be 1.6 s. After a 6.4 kg collar is placed on top of collar $A$, the period of oscillation is observed to be 2.1 s. Determine ($a$) the weight of collar $A$, ($b$) the spring constant $k$.

**Fig. P19.27**

**19.28** Rod $AB$ is attached to a hinge at $A$ and to two springs, each of constant $k$. If $h = 700$ mm, $d = 300$ mm, and $m = 20$ kg, determine the value of $k$ for which the period of small oscillations is (*a*) 1 s, (*b*) infinite. Neglect the mass of the rod and assume that each spring can act in either tension or compression.

**19.29** Rod $AB$ is attached to a hinge at $A$ and to two springs, each of constant $k = 1.35$ kN/m. (*a*) Determine the mass $m$ of the block $C$ for which the period of small oscillations is 4 s. (*b*) If end $B$ is depressed 60 mm and released, determine the maximum velocity of block $C$. Neglect the mass of the rod and assume that each spring can act in either tension or compression.

**Fig. P19.28**

**Fig. P19.29**

**Fig. *P19.30***

**19.30** From mechanics of materials it is known that for a simply supported beam of uniform cross section a static load $\mathbf{P}$ applied at the center will cause a deflection $\delta_A = PL^3/48EI$, where $L$ is the length of the beam, $E$ is the modulus of elasticity, and $I$ is the moment of inertia of the cross-sectional area of the beam. Knowing that $L = 4.6$ m, $E = 207 \times 10^6$ kN/m$^2$, and $I = 1.73 \times 10^{-5}$ m$^4$, determine (*a*) the equivalent spring constant of the beam, (*b*) the frequency of vibration of a 680-kg block attached to the center of the beam. Neglect the mass of the beam and assume that the load remains in contact with the beam.

**19.31** From mechanics of materials it is known that when a static load $\mathbf{P}$ is applied at the end $B$ of a uniform metal rod fixed at end $A$, the length of the rod will increase by an amount $\delta = PL/AE$, where $L$ is the length of the undeformed rod, $A$ is its cross-sectional area, and $E$ is the modulus of elasticity of the material. Knowing that $L = 45.7$ cm and $E = 200 \times 10^6$ kPa and that the diameter of the rod is 0.8 cm, and neglecting the weight of the rod, determine (*a*) the equivalent spring constant of the rod, (*b*) the frequency of the vertical vibrations of a block of weight $W = 7.3$ kg attached to end $B$ of the same rod.

**19.32** Denoting by $\delta_{st}$ the static deflection of a beam under a given load, show that the frequency of vibration of the load is

$$f = \frac{1}{2\pi} \sqrt{\frac{g}{\delta_{st}}}$$

Neglect the mass of the beam, and assume that the load remains in contact with the beam.

**Fig. P19.31**

**\*19.33** The force-deflection equation for a nonlinear spring fixed at one end is $F = 4x^{1/2}$ where $F$ is the force, expressed in newtons, applied at the other end and $x$ is the deflection expressed in meters. (*a*) Determine the deflection $x_0$ if a 100-g block is suspended from the spring and is at rest. (*b*) Assuming that the slope of the force-deflection curve at the point corresponding to this loading can be used as an equivalent spring constant, determine the frequency of vibration of the block if it is given a very small downward displacement from its equilibrium position and released.

**\*19.34** Expanding the integrand in Eq. (19.19) of Sec. 19.4 into a series of even powers of sin $\phi$ and integrating, show that the period of a simple pendulum of length $l$ may be approximated by the formula

$$\tau = 2\pi\sqrt{\frac{l}{g}}\left(1 + \frac{1}{4}\,\sin^2\frac{\theta_m}{2}\right)$$

where $\theta_m$ is the amplitude of the oscillations.

**\*19.35** Using the formula given in Prob. 19.34, determine the amplitude $\theta_m$ for which the period of a simple pendulum is 1 percent longer than the period of the same pendulum for small oscillations.

**\*19.36** Using the data of Table 19.1, determine the period of a simple pendulum of length $l = 800$ mm (*a*) for small oscillations, (*b*) for oscillations of amplitude $\theta_m = 30°$, (*c*) for oscillations of amplitude $\theta_m = 90°$.

**\*19.37** Using the data of Table 19.1, determine the length in meters of a simple pendulum which oscillates with a period of 3 s and an amplitude of $\theta_m = 60°$.

---

## 19.5. FREE VIBRATIONS OF RIGID BODIES

The analysis of the vibrations of a rigid body or of a system of rigid bodies possessing a single degree of freedom is similar to the analysis of the vibrations of a particle. An appropriate variable, such as a distance $x$ or an angle $\theta$, is chosen to define the position of the body or system of bodies, and an equation relating this variable and its second derivative with respect to $t$ is written. If the equation obtained is of the same form as (19.6), that is, if we have

$$\ddot{x} + \omega_n^2 x = 0 \qquad \text{or} \qquad \ddot{\theta} + \omega_n^2 \theta = 0 \qquad (19.21)$$

the vibration considered is a simple harmonic motion. The period and natural frequency of the vibration can then be obtained by identifying $\omega_n$ and substituting its value into Eqs. (19.13) and (19.14).

In general, a simple way to obtain one of Eqs. (19.21) is to express that the system of the external forces is equivalent to the system of the effective forces by drawing a free-body-diagram equation for an arbitrary value of the variable and writing the appropriate equation of motion. We recall that our goal should be *the determination of the coefficient* of the variable $x$ or $\theta$, *not* the determination of the variable itself or of the derivative $\ddot{x}$ or $\ddot{\theta}$. Setting this coefficient equal to $\omega_n^2$, we obtain the natural circular frequency $\omega_n$, from which $\tau_n$ and $f_n$ can be determined.

The method we have outlined can be used to analyze vibrations which are truly represented by a simple harmonic motion, or vibrations of small amplitude which can be *approximated* by a simple harmonic motion. As an example, let us determine the period of the small oscillations of a square plate of side $2b$ which is suspended from the midpoint $O$ of one of its sides (Fig. 19.5$a$). We consider the plate in an arbitrary position defined by the angle $\theta$ that the line $OG$ forms with the vertical and draw a free-body-diagram equation to express that the weight $\mathbf{W}$ of the plate and the components $\mathbf{R}_x$ and $\mathbf{R}_y$ of the reaction at $O$ are equivalent to the vectors $m\mathbf{a}_t$ and $m\mathbf{a}_n$ and to the couple $\bar{I}\boldsymbol{\alpha}$ (Fig. 19.5$b$). Since the angular velocity and angular acceleration of the plate are equal, respectively, to $\dot{\theta}$ and $\ddot{\theta}$, the magnitudes of the two vectors are, respectively, $mb\ddot{\theta}$ and $mb\dot{\theta}^2$, while the moment of the couple is $\bar{I}\ddot{\theta}$. In previous applications of this method (Chap. 16), we tried whenever possible to assume the correct sense for the acceleration. Here, however, we must assume the same positive sense for $\theta$ and $\ddot{\theta}$ in order to obtain an equation of the form (19.21). Consequently, the angular acceleration $\ddot{\theta}$ will be assumed positive counterclockwise, even though this assumption is obviously unrealistic. Equating moments about $O$, we write

$+\uparrow\qquad\qquad -W(b\sin\theta) = (mb\ddot{\theta})b + \bar{I}\ddot{\theta}$

Noting that $\bar{I} = \frac{1}{12}m[(2b)^2 + (2b)^2] = \frac{2}{3}mb^2$ and $W = mg$, we obtain

$$\ddot{\theta} + \frac{3}{5}\frac{g}{b}\sin\theta = 0 \qquad (19.22)$$

For oscillations of small amplitude, we can replace $\sin\theta$ by $\theta$, expressed in radians, and write

$$\ddot{\theta} + \frac{3}{5}\frac{g}{b}\theta = 0 \qquad (19.23)$$

Comparison with (19.21) shows that the equation obtained is that of a simple harmonic motion and that the natural circular frequency $\omega_n$ of the oscillations is equal to $(3g/5b)^{1/2}$. Substituting into (19.13), we find that the period of the oscillations is

$$\tau_n = \frac{2\pi}{\omega_n} = 2\pi\sqrt{\frac{5b}{3g}} \qquad (19.24)$$

The result obtained is valid only for oscillations of small amplitude. A more accurate description of the motion of the plate is obtained by comparing Eqs. (19.16) and (19.22). We note that the two equations are identical if we choose $l$ equal to $5b/3$. This means that the plate will oscillate as a simple pendulum of length $l = 5b/3$, and the results of Sec. 19.4 can be used to correct the value of the period given in (19.24). The point $A$ of the plate located on line $OG$ at a distance $l = 5b/3$ from $O$ is defined as the *center of oscillation* corresponding to $O$ (Fig. 19.5$a$).

**Fig. 19.5**

## SAMPLE PROBLEM 19.2

A cylinder of weight $W$ and radius $r$ is suspended from a looped cord as shown. One end of the cord is attached directly to a rigid support, while the other end is attached to a spring of constant $k$. Determine the period and natural frequency of the vibrations of the cylinder.

## SOLUTION

**Kinematics of Motion.** We express the linear displacement and the acceleration of the cylinder in terms of the angular displacement $\theta$. Choosing the positive sense clockwise and measuring the displacements from the equilibrium position, we write

$$\bar{x} = r\theta \qquad \delta = 2\bar{x} = 2r\theta$$

$$\alpha = \ddot{\theta}\downarrow \qquad \bar{a} = r\alpha = r\ddot{\theta} \qquad \bar{\mathbf{a}} = r\ddot{\theta}\downarrow \qquad (1)$$

**Equations of Motion.** The system of external forces acting on the cylinder consists of the weight $\mathbf{W}$ and of the forces $\mathbf{T}_1$ and $\mathbf{T}_2$ exerted by the cord. We express that this system is equivalent to the system of effective forces represented by the vector $m\bar{\mathbf{a}}$ attached at $G$ and the couple $\bar{I}\alpha$.

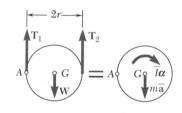

$$+\downarrow \Sigma M_A = \Sigma(M_A)_{\text{eff}}: \qquad Wr - T_2(2r) = m\bar{a}r + \bar{I}\alpha \qquad (2)$$

When the cylinder is in its position of equilibrium, the tension in the cord is $T_0 = \frac{1}{2}W$. We note that for an angular displacement $\theta$, the magnitude of $\mathbf{T}_2$ is

$$T_2 = T_0 + k\delta = \tfrac{1}{2}W + k\delta = \tfrac{1}{2}W + k(2r\theta) \qquad (3)$$

Substituting from (1) and (3) into (2), and recalling that $\bar{I} = \frac{1}{2}mr^2$, we write

$$Wr - (\tfrac{1}{2}W + 2kr\theta)(2r) = m(r\ddot{\theta})r + \tfrac{1}{2}mr^2\ddot{\theta}$$

$$\ddot{\theta} + \frac{8}{3}\frac{k}{m}\theta = 0$$

The motion is seen to be simple harmonic, and we have

$$\omega_n^2 = \frac{8}{3}\frac{k}{m} \qquad \omega_n = \sqrt{\frac{8}{3}\frac{k}{m}}$$

$$\tau_n = \frac{2\pi}{\omega_n} \qquad \tau_n = 2\pi\sqrt{\frac{3}{8}\frac{m}{k}} \qquad \blacktriangleleft$$

$$f_n = \frac{\omega_n}{2\pi} \qquad f_n = \frac{1}{2\pi}\sqrt{\frac{8}{3}\frac{k}{m}} \qquad \blacktriangleleft$$

## SAMPLE PROBLEM 19.3

A circular disk, weighing 9 kg and of radius 20.3 cm, is suspended from a wire as shown. The disk is rotated (thus twisting the wire) and then released; the period of the torsional vibration is observed to be 1.13 s. A gear is then suspended from the same wire, and the period of torsional vibration for the gear is observed to be 1.93 s. Assuming that the moment of the couple exerted by the wire is proportional to the angle of twist, determine (a) the torsional spring constant of the wire, (b) the centroidal moment of inertia of the gear, (c) the maximum angular velocity reached by the gear if it is rotated through 90° and released.

## SOLUTION

**a. Vibration of Disk.** Denoting by $\theta$ the angular displacement of the disk, we express that the magnitude of the couple exerted by the wire is $M = K\theta$, where $K$ is the torsional spring constant of the wire. Since this couple must be equivalent to the couple $\bar{I}\alpha$ representing the effective forces of the disk, we write

$$+\curvearrowleft\Sigma M_O = \Sigma(M_O)_{\text{eff}}: \qquad\qquad +K\theta = -\bar{I}\ddot{\theta}$$

$$\ddot{\theta} + \frac{K}{\bar{I}}\theta = 0$$

The motion is seen to be simple harmonic, and we have

$$\omega_n^2 = \frac{K}{\bar{I}} \qquad \tau_n = \frac{2\pi}{\omega_n} \qquad \tau_n = 2\pi\sqrt{\frac{\bar{I}}{K}} \qquad (1)$$

For the disk, we have

$$\tau_n = 1.13 \text{ s} \qquad \bar{I} = \tfrac{1}{2}mr^2 = \frac{1}{2}\left(9 \text{ kg}\right)\left(20.3\right)^2 = 0.185 \text{ kg m}^2$$

Substituting into (1), we obtain

$$1.13 = 2\pi\sqrt{\frac{0.185}{K}} \qquad K = 5.72 \text{ N} \cdot \text{m/rad} \quad \blacktriangleleft$$

**b. Vibration of Gear.** Since the period of vibration of the gear is 1.93 s and $K = 5.72$ N · m/rad, Eq. (1) yields

$$1.93 = 2\pi\sqrt{\frac{\bar{I}}{5.72}} \qquad \bar{I}_{\text{gear}} = 0.539 \text{ kg m}^2 \quad \blacktriangleleft$$

**c. Maximum Angular Velocity of Gear.** Since the motion is simple harmonic, we have

$$\theta = \theta_m \sin \omega_n t \qquad \omega = \theta_m\omega_n \cos \omega_n t \qquad \omega_m = \theta_m\omega_n$$

Recalling that $\theta_m = 90° = 1.571$ rad and $\tau = 1.93$ s, we write

$$\omega_m = \theta_m\omega_n = \theta_m\left(\frac{2\pi}{\tau}\right) = (1.571 \text{ rad})\left(\frac{2\pi}{1.93 \text{ s}}\right)$$

$$\omega_m = 5.11 \text{ rad/s} \quad \blacktriangleleft$$

In this lesson you saw that a rigid body, or a system of rigid bodies, whose position can be defined by a single coordinate $x$ or $\theta$, will execute a simple harmonic motion if the differential equation obtained by applying Newton's second law is of the form

$$\ddot{x} + \omega_n^2 x = 0 \qquad \text{or} \qquad \ddot{\theta} + \omega_n^2 \theta = 0 \qquad (19.21)$$

Your goal should be to determine $\omega_n$, from which you can obtain the period $\tau_n$ and the natural frequency $f_n$. Taking into account the initial conditions, you can then write an equation of the form

$$x = x_m \sin(\omega_n t + \phi) \qquad (19.10)$$

where $x$ should be replaced by $\theta$ if a rotation is involved. To solve the problems in this lesson, you will follow these steps:

1. *Choose a coordinate which will measure the displacement of the body* from its equilibrium position. You will find that many of the problems in this lesson involve the rotation of a body about a fixed axis and that the angle measuring the rotation of the body from its equilibrium position is the most convenient coordinate to use. In problems involving the general plane motion of a body, where a coordinate $x$ (and possibly a coordinate $y$) is used to define the position of the mass center $G$ of the body, and a coordinate $\theta$ is used to measure its rotation about $G$, find kinematic relations which will allow you to express $x$ (and $y$) in terms of $\theta$ [Sample Prob. 19.2].

2. *Draw a free-body-diagram equation* to express that the system of the external forces is equivalent to the system of the effective forces, which consists of the vector $m\bar{\mathbf{a}}$ and the couple $\bar{I}\boldsymbol{\alpha}$, where $\bar{a} = \ddot{x}$ and $\alpha = \ddot{\theta}$. Be sure that each applied force or couple is drawn in a direction consistent with the assumed displacement and that the senses of $\bar{\mathbf{a}}$ and $\boldsymbol{\alpha}$ are, respectively, those in which the coordinates $x$ and $\theta$ are increasing.

3. *Write the differential equations of motion* by equating the sums of the components of the external and effective forces in the $x$ and $y$ directions and the sums of their moments about a given point. If necessary, use the kinematic relations developed in step 1 to obtain equations involving only the coordinate $\theta$. If $\theta$ is a small angle, replace $\sin\theta$ by $\theta$ and $\cos\theta$ by 1, if these functions appear in your equations. Eliminating any unknown reactions, you will obtain an equation of the type of Eqs. (19.21). Note that in problems involving a body rotating about a fixed axis, you can immediately obtain such an equation by equating the moments of the external and effective forces about the fixed axis.

**4. Comparing the equation you have obtained with one of Eqs. (19.21),** you can identify $\omega_n^2$ and, thus, determine the natural circular frequency $\omega_n$. Remember that the object of your analysis is *not to solve* the differential equation you have obtained, *but to identify* $\omega_n^2$.

**5. Determine the amplitude and the phase angle $\phi$** by substituting the value obtained for $\omega_n$ and the initial values of the coordinate and its first derivative into Eq. (19.10) and the equation obtained by differentiating (19.10) with respect to $t$. From Eq. (19.10) and the two equations obtained by differentiating (19.10) twice with respect to $t$, and using the kinematic relations developed in step 1, you will be able to determine the position, velocity, and acceleration of any point of the body at any given time.

**6. In problems involving torsional vibrations,** the torsional spring constant $K$ is expressed in $N \cdot m/rad$ or $lb \cdot ft/rad$. The product of $K$ and the angle of twist $\theta$, expressed in radians, yields the moment of the restoring couple, which should be equated to the sum of the moments of the effective forces or couples about the axis of rotation [Sample Prob. 19.3].

**Fig. P19.38**

**19.38** The 9 kg uniform rod $AB$ is attached to springs at $A$ and $B$, each of constant 850 N/m, which can act in tension or compression. If the end $A$ of the rod is depressed slightly and released, determine (a) the frequency of vibration, (b) the amplitude of the angular motion of the rod, knowing that the maximum velocity of point $A$ is 1.1 mm/s.

**19.39** A 7.5 kg slender rod $AB$ is riveted to a 6 kg uniform disk as shown. A belt is attached to the rim of the disk and to a spring which holds the rod at rest in the position shown. If end $A$ of the rod is moved 20 mm down and released, determine (a) the period of vibration, (b) the maximum velocity of end $A$.

**Fig. P19.39**

**19.40** A 9 kg uniform cylinder can roll without sliding on an incline and is attached to a spring $AB$ as shown. If the center of the cylinder is moved 1.27 cm down the incline from the equilibrium position and released from rest, determine (a) the period of vibration, (b) the maximum velocity of the center of the cylinder.

**Fig. P19.40**

**19.41** Two identical 2.7 kg uniform rods are hinged at point $B$ and attached to a spring of constant $k = 4028$ N/m. The rods are guided by small wheels of negligible weight and the system is in equilibrium when the rods are horizontal as shown. Knowing that point $B$ is depressed 2 cm and released, determine (a) the frequency of the resulting vibration, (b) the magnitude of the maximum angular acceleration of rod $AB$.

**Fig. P19.41**

**19.42** A 1.2 kg uniform rod $AB$ is attached to a hinge at $A$ and to two springs, each of constant $k = 450$ N/m (a) Determine the mass $m$ of block $C$ for which the period of small oscillations is 0.6 s. (b) If end $B$ is depressed 60 mm and released, determine the maximum velocity of block $C$.

**19.43** A 6 kg uniform cylinder can roll without sliding on a horizontal surface and is attached by a pin at point $C$ to the 4 kg horizontal bar $AB$. The bar is attached to two springs, each of constant $k = 5$ kN/m as shown. Knowing that the bar is moved 12 mm to the right of the equilibrium position and released, determine (a) the period of vibration of the system, (b) the magnitude of the maximum velocity of bar $AB$.

**Fig. P19.42**

**Fig. *P19.43* and *P19.44***

**19.44** A 6 kg uniform cylinder is assumed to roll without sliding on a horizontal surface and is attached by a pin at point $C$ to the 4 kg horizontal bar $AB$. The bar is attached to two springs, each of constant $k = 3.5$ kN/m as shown. Knowing that the coefficient of static friction between the cylinder and the surface is 0.5, determine the maximum amplitude of the motion of point $C$ which is compatible with the assumption of rolling.

**19.45** A semicircular hole is cut in a uniform square plate which is attached to a frictionless pin at its geometric center $O$. Determine (a) the period of small oscillations of the plate, (b) the length of a simple pendulum which has the same period.

**Fig. P19.45**

**19.46** Determine the period of small oscillations of a uniform quarter-circular cylinder of radius $r = 0.3$ m which rolls without slipping. [*Hint:* Note that $GO = 4\sqrt{2}r/3\pi$ and that, by the parallel-axis theorem, $I = \frac{1}{2}mr^2 - m(GO)^2$.]

**19.47** For the uniform semicircular plate of radius $r$, determine the period of small oscillations if the plate (a) is suspended from point $A$ as shown, (b) is suspended from a pin located at point $B$.

**Fig. P19.46**

**Fig. P19.47**

**19.48** A thin homogeneous wire is bent into the shape of an isosceles triangle of sides $b$, $b$, and $1.6b$. Determine the period of small oscillations if the wire (a) is suspended from point $A$ as shown, (b) is suspended from point $B$.

**Fig. P19.48**

**Fig. P19.49**

$r = 25.4$ cm

**Fig. P19.51**

**19.49**   A uniform rectangular plate is suspended from a pin located at the midpoint of one edge as shown. Considering the dimension $b$ constant, determine (*a*) the ratio $c/b$ for which the period of oscillation of the plate is minimum, (*b*) the ratio $c/b$ for which the period of oscillation of the plate is the same as the period of a simple pendulum of length $c$.

**19.50**   The period of small oscillations about $A$ of a connecting rod is observed to be 1.03 s. Knowing that the distance $r_a = 160$ mm, determine the centroidal radius of gyration of the connecting rod.

**Fig. P19.50**

**19.51**   A uniform disk of radius $r = 25.4$ cm is attached at $A$ to a 66 cm rod $AB$ of negligible weight which can rotate freely in a vertical plane about $B$. Determine the period of small oscillations (*a*) if the disk is free to rotate in a bearing at $A$, (*b*) if the rod is riveted to the disk at $A$.

**19.52**   A *compound pendulum* is defined as a rigid slab which oscillates about a fixed point $O$, called the center of suspension. Show that the period of oscillation of a compound pendulum is equal to the period of a simple pendulum of length $OA$, where the distance from $A$ to the mass center $G$ is $GA = \bar{k}^2 \sqrt{r}$. Point $A$ is defined as the center of oscillation and coincides with the center of percussion defined in Prob. 17.57.

**Fig. P19.52 and P19.53**

**19.53**   A rigid slab oscillates about a fixed point $O$. Show that the smallest period of oscillation occurs when the distance $\bar{r}$ from point $O$ to the mass center $G$ is equal to $\bar{k}$.

**19.54**   Show that if the compound pendulum of Prob. 19.52 is suspended from $A$ instead of $O$, the period of oscillation is the same as before and the new center of oscillation is located at $O$.

**19.55** Two uniform rods, each of mass $m$ and length $l$ are welded together to form an L-shaped assembly. The assembly is constrained by two springs, each of constant $k$, and is in equilibrium in a vertical plane in the position shown. Determine the frequency of small oscillations of the system.

**19.56** A uniform disk of radius $r$ and mass $m$ can roll without slipping on a cylindrical surface and is attached to bar $ABC$ of length $L$ and negligible mass. The bar is attached at point $A$ to a spring of constant $k$ and can rotate freely about point $B$ in the vertical plane. Knowing that end $A$ is given a small displacement and released, determine the frequency of the resulting vibration in terms of $m$, $L$, $k$, and $g$.

**Fig. P19.55**

**Fig. P19.56**

**19.57** The 7.3 kg uniform bar $AB$ is hinged at $C$ and is attached at $A$ to a spring of constant $k = 730$ N/m. If end $A$ is given a small displacement and released, determine ($a$) the frequency of small oscillations, ($b$) the smallest value of the spring constant $k$ for which oscillations will occur.

**Fig. P19.57**

**Fig. P19.58**

**19.58**  A 9 kg uniform equilateral triangular plate is suspended from a pin located at one of its vertices and is attached to two springs, each of constant $k = 204$ N/m. Knowing that the plate is given a small angular displacement and released, determine the frequency of the resulting vibration.

**19.59**  A 3 kg uniform arm $ABC$ is supported by a pin at $B$ and is attached to identical springs at $A$ and $C$. The system is in equilibrium in a vertical plane in the position shown. Knowing that point $C$ is pushed down 15 mm and released, determine the angular acceleration of the arm 0.7 s later.

**Fig. P19.59**

**19.60**  A homogeneous wire bent to form the figure shown is attached to a pin support at $A$. Knowing that $r = 320$ mm and that point $B$ is pushed down 30 mm and released, determine the magnitude of the acceleration of $B$, 10 s later.

**Fig. P19.60**

**19.61**  A uniform disk of radius $r = 25.4$ cm is attached at $A$ to a 66 cm rod $AB$ of negligible weight which can rotate freely in a vertical plane about $B$. If the rod is displaced 2° from the position shown and released, determine the magnitude of the maximum velocity of point $A$, assuming that the disk ($a$) is free to rotate in a bearing at $A$, ($b$) is riveted to the rod at $A$.

**19.62**  A homogeneous rod of weight per unit length equal to 0.45 kg/m is used to form the assembly shown, which rotates freely about pivot $A$ in a vertical plane. Knowing that the assembly is displaced 2° clockwise from its equilibrium position and released, determine its angular velocity and angular acceleration 5 s later.

**Fig. P19.61**

**Fig. P19.62**

**19.63** A period of 4.1 s is observed for the angular oscillations of a 450-g gyroscope rotor suspended from a wire as shown. Knowing that a period of 6.2 s is obtained when a 50-mm-diameter steel sphere is suspended in the same fashion, determine the centroidal radius of gyration of the rotor. (Density of steel = 7850 kg/m³.)

**Fig. P19.63**

**19.64** A 3 kg slender rod is suspended from a steel wire which is known to have a torsional spring constant $K = 1.95$ N · m/rad. If the rod is rotated through 180° about the vertical and then released, determine (a) the period of oscillation, (b) the maximum velocity of end A of the rod.

**19.65** A 1.8 kg circular disk of radius $r = 102$ cm is suspended at its center C from wires AB and BC soldered together at B. The torsional spring constants of the wires are $K_1 = 4$ N · m/rad for AB and $K_2 = 2$ N · m/rad for BC. Determine the period of oscillation of the disk about the axis AC.

**Fig. P19.64**

**Fig. P19.65**

**19.66** A 54.4 kg uniform circular plate is welded to two elastic rods which have fixed ends at supports A and B as shown. The torsional spring constant of each rod is 203 N · m/rad, and the system is in equilibrium when the plate is vertical. Knowing that the plate is rotated 2° about axis AB and released, determine (a) the period of oscillation, (b) the magnitude of the maximum velocity of the mass center G of the plate.

**Fig. P19.66**

**Fig. P19.67**

**19.67** A horizontal platform $P$ is held by several rigid bars which are connected to a vertical wire. The period of oscillation of the platform is found to be 2.2 s when the platform is empty and 3.8 s when an object $A$ of uniform moment of inertia is placed on the platform with its mass center directly above the center of the plate. Knowing that the wire has a torsional constant $K = 27$ N · m/rad, determine the centroidal moment of inertia of object $A$.

**19.68** A uniform equilateral triangular plate of side $b$ is suspended from three vertical wires of the same length $l$. Determine the period of small oscillations of the plate when ($a$) it is rotated through a small angle about a vertical axis through its mass center $G$, ($b$) it is given a small horizontal displacement in a direction perpendicular to $AB$.

**Fig. P19.68**

## 19.6. APPLICATION OF THE PRINCIPLE OF CONSERVATION OF ENERGY

We saw in Sec. 19.2 that when a particle of mass $m$ is in simple harmonic motion, the resultant **F** of the forces exerted on the particle has a magnitude proportional to the displacement $x$ measured from the position of equilibrium $O$ and is directed toward $O$; we write $F = -kx$. Referring to Sec. 13.6, we note that **F** is a *conservative force* and that the corresponding potential energy is $V = \frac{1}{2}kx^2$, where $V$ is assumed equal to zero in the equilibrium position $x = 0$. Since the velocity of the particle is equal to $\dot{x}$, its kinetic energy is $T = \frac{1}{2}m\dot{x}^2$ and we can express that the total energy of the particle is conserved by writing

$$T + V = \text{constant} \qquad \tfrac{1}{2}m\dot{x}^2 + \tfrac{1}{2}kx^2 = \text{constant}$$

Dividing through by $m/2$ and recalling from Sec. 19.2 that $k/m = \omega_n^2$, where $\omega_n$ is the natural circular frequency of the vibration, we have

$$\dot{x}^2 + \omega_n^2 x^2 = \text{constant} \qquad (19.25)$$

Equation (19.25) is characteristic of a simple harmonic motion, since it can be obtained from Eq. (19.6) by multiplying both terms by $2\dot{x}$ and integrating.

The principle of conservation of energy provides a convenient way for determining the period of vibration of a rigid body or of a system of rigid bodies possessing a single degree of freedom, once it has been established that the motion of the system is a simple harmonic motion or that it can be approximated by a simple harmonic motion. Choosing an appropriate variable, such as a distance $x$ or an angle $\theta$, we consider two particular positions of the system:

1. *The displacement of the system is maximum;* we have $T_1 = 0$, and $V_1$ can be expressed in terms of the amplitude $x_m$ or $\theta_m$ (choosing $V = 0$ in the equilibrium position).
2. *The system passes through its equilibrium position;* we have $V_2 = 0$, and $T_2$ can be expressed in terms of the maximum velocity $\dot{x}_m$ or the maximum angular velocity $\dot{\theta}_m$.

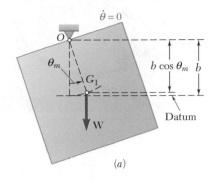

We then express that the total energy of the system is conserved and write $T_1 + V_1 = T_2 + V_2$. Recalling from (19.15) that for simple harmonic motion the maximum velocity is equal to the product of the amplitude and of the natural circular frequency $\omega_n$, we find that the equation obtained can be solved for $\omega_n$.

As an example, let us consider again the square plate of Sec. 19.5. In the position of maximum displacement (Fig. 19.6a), we have

$$T_1 = 0 \qquad V_1 = W(b - b \cos \theta_m) = Wb(1 - \cos \theta_m)$$

or, since $1 - \cos \theta_m = 2 \sin^2 (\theta_m/2) \approx 2(\theta_m/2)^2 = \theta_m^2/2$ for oscillations of small amplitude,

$$T_1 = 0 \qquad V_1 = \tfrac{1}{2}Wb\theta_m^2 \qquad (19.26)$$

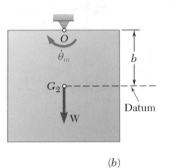

**Fig. 19.6**

As the plate passes through its position of equilibrium (Fig. 19.6b), its velocity is maximum and we have

$$T_2 = \tfrac{1}{2}m\bar{v}_m^2 + \tfrac{1}{2}\bar{I}\omega_m^2 = \tfrac{1}{2}mb^2\dot{\theta}_m^2 + \tfrac{1}{2}\bar{I}\dot{\theta}_m^2 \qquad V_2 = 0$$

or, recalling from Sec. 19.5 that $\bar{I} = \tfrac{2}{3}mb^2$,

$$T_2 = \tfrac{1}{2}(\tfrac{5}{3}mb^2)\dot{\theta}_m^2 \qquad V_2 = 0 \qquad (19.27)$$

Substituting from (19.26) and (19.27) into $T_1 + V_1 = T_2 + V_2$, and noting that the maximum velocity $\dot{\theta}_m$ is equal to the product $\theta_m\omega_n$, we write

$$\tfrac{1}{2}Wb\theta_m^2 = \tfrac{1}{2}(\tfrac{5}{3}mb^2)\theta_m^2\omega_n^2 \qquad (19.28)$$

which yields $\omega_n^2 = 3g/5b$ and

$$\tau_n = \frac{2\pi}{\omega_n} = 2\pi\sqrt{\frac{5b}{3g}} \qquad (19.29)$$

as previously obtained.

# SAMPLE PROBLEM 19.4

Determine the period of small oscillations of a cylinder of radius $r$ which rolls without slipping inside a curved surface of radius $R$.

## SOLUTION

Position 1

Position 2

We denote by $\theta$ the angle which line $OG$ forms with the vertical. Since the cylinder rolls without slipping, we may apply the principle of conservation of energy between position 1, where $\theta = \theta_m$, and position 2, where $\theta = 0$.

**Position 1**
*Kinetic Energy.*   Since the velocity of the cylinder is zero, $T_1 = 0$.
*Potential Energy.*   Choosing a datum as shown and denoting by $W$ the weight of the cylinder, we have

$$V_1 = Wh = W(R - r)(1 - \cos \theta)$$

Noting that for small oscillations $(1 - \cos \theta) = 2 \sin^2 (\theta/2) \approx \theta^2/2$, we have

$$V_1 = W(R - r)\frac{\theta_m^2}{2}$$

**Position 2.**   Denoting by $\dot\theta_m$ the angular velocity of line $OG$ as the cylinder passes through position 2, and observing that point $C$ is the instantaneous center of rotation of the cylinder, we write

$$\bar{v}_m = (R - r)\dot\theta_m \qquad \omega_m = \frac{\bar{v}_m}{r} = \frac{R - r}{r}\dot\theta_m$$

*Kinetic Energy*

$$T_2 = \tfrac{1}{2}m\bar{v}_m^2 + \tfrac{1}{2}\bar{I}\omega_m^2$$

$$= \tfrac{1}{2}m(R - r)^2\dot\theta_m^2 + \tfrac{1}{2}(\tfrac{1}{2}mr^2)\left(\frac{R - r}{r}\right)^2\dot\theta_m^2$$

$$= \tfrac{3}{4}m(R - r)^2\dot\theta_m^2$$

*Potential Energy*

$$V_2 = 0$$

**Conservation of Energy**

$$T_1 + V_1 = T_2 + V_2$$

$$0 + W(R - r)\frac{\theta_m^2}{2} = \tfrac{3}{4}m(R - r)^2\dot\theta_m^2 + 0$$

Since $\dot\theta_m = \omega_n\theta_m$ and $W = mg$, we write

$$mg(R - r)\frac{\theta_m^2}{2} = \tfrac{3}{4}m(R - r)^2(\omega_n\theta_m)^2 \qquad \omega_n^2 = \frac{2}{3}\frac{g}{R - r}$$

$$\tau_n = \frac{2\pi}{\omega_n} \qquad \tau_n = 2\pi\sqrt{\frac{3}{2}\frac{R - r}{g}} \quad \blacktriangleleft$$

In the problems which follow you will be asked to use the *principle of conservation of energy* to determine the period or natural frequency of the simple harmonic motion of a particle or rigid body. Assuming that you choose an angle $\theta$ to define the position of the system (with $\theta = 0$ in the equilibrium position), as you will in most of the problems in this lesson, you will express that the total energy of the system is conserved, $T_1 + V_1 = T_2 + V_2$, between the position 1 of maximum displacement ($\theta_1 = \theta_m$, $\dot{\theta}_1 = 0$) and the position 2 of maximum velocity ($\dot{\theta}_2 = \dot{\theta}_m$, $\theta_2 = 0$). It follows that $T_1$ and $V_2$ will both be zero, and the energy equation will reduce to $V_1 = T_2$, where $V_1$ and $T_2$ are homogeneous quadratic expressions in $\theta_m$ and $\dot{\theta}_m$, respectively. Recalling that, for a simple harmonic motion, $\dot{\theta}_m = \theta_m \omega_n$ and substituting this product into the energy equation, you will obtain, after reduction, an equation that you can solve for $\omega_n^2$. Once you have determined the natural circular frequency $\omega_n$, you can obtain the period $\tau_n$ and the natural frequency $f_n$ of the vibration.

The steps that you should take are as follows:

1. *Calculate the potential energy $V_1$ of the system in its position of maximum displacement.* Draw a sketch of the system in its position of maximum displacement and express the potential energy of all the forces involved (internal as well as external) in terms of the maximum displacement $x_m$ or $\theta_m$.

   *a. The potential energy associated with the weight $W$ of a body is $V_g = Wy$,* where $y$ is the elevation of the center of gravity $G$ of the body above its equilibrium position. If the problem you are solving involves the oscillation of a rigid body about a horizontal axis through a point $O$ located at a distance $b$ from $G$ (Fig. 19.6), express $y$ in terms of the angle $\theta$ that the line $OG$ forms with the vertical: $y = b(1 - \cos\theta)$. But, for small values of $\theta$, you can replace this expression with $y = \frac{1}{2}b\theta^2$ [Sample Prob. 19.4]. Therefore, when $\theta$ reaches its maximum value $\theta_m$, and for oscillations of small amplitude, you can express $V_g$ as

$$V_g = \tfrac{1}{2}Wb\theta_m^2$$

Note that *if $G$ is located above $O$* in its equilibrium position (instead of below $O$, as we have assumed), the vertical displacement $y$ will be negative and should be approximated as $y = -\frac{1}{2}b\theta^2$, which will result in a negative value for $V_g$. In the absence of other forces, the equilibrium position will be unstable, and the system will not oscillate. (See, for instance, Prob. 19.91.)

   *b. The potential energy associated with the elastic force exerted by a spring is $V_e = \frac{1}{2}kx^2$,* where $k$ is the constant of the spring and $x$ its deflection. In problems involving the rotation of a body about an axis, you will generally have $x = a\theta$, where $a$ is the distance from the axis of rotation to the point of the body

(*continued*)

where the spring is attached, and where $\theta$ is the angle of rotation. Therefore, when $x$ reaches its maximum value $x_m$ and $\theta$ reaches its maximum value $\theta_m$, you can express $V_e$ as

$$V_e = \tfrac{1}{2}kx_m^2 = \tfrac{1}{2}ka^2\theta_m^2$$

*c. The potential energy $V_1$ of the system in its position of maximum displacement* is obtained by adding the various potential energies that you have computed. It will be equal to the product of a constant and $\theta_m^2$.

**2. *Calculate the kinetic energy $T_2$ of the system in its position of maximum velocity.*** Note that this position is also the equilibrium position of the system.

*a. If the system consists of a single rigid body,* the kinetic energy $T_2$ of the system will be the sum of the kinetic energy associated with the motion of the mass center $G$ of the body and the kinetic energy associated with the rotation of the body about $G$. You will write, therefore,

$$T_2 = \tfrac{1}{2}m\bar{v}_m^2 + \tfrac{1}{2}\bar{I}\omega_m^2$$

Assuming that the position of the body has been defined by an angle $\theta$, express $\bar{v}_m$ and $\omega_m$ in terms of the rate of change $\dot{\theta}_m$ of $\theta$ as the body passes through its equilibrium position. The kinetic energy of the body will thus be expressed as the product of a constant and $\dot{\theta}_m^2$. Note that if $\theta$ measures the rotation of the body about its mass center, as was the case for the plate of Fig. 19.6, then $\omega_m = \dot{\theta}_m$. In other cases, however, the kinematics of the motion should be used to derive a relation between $\omega_m$ and $\dot{\theta}_m$ [Sample Prob. 19.4].

*b. If the system consists of several rigid bodies,* repeat the above computation for each of the bodies, using the same coordinate $\theta$, and add the results obtained.

**3. *Equate the potential energy $V_1$ of the system to its kinetic energy $T_2$,***

$$V_1 = T_2$$

and, recalling the first of Eqs. (19.15), replace $\dot{\theta}_m$ in the right-hand term by the product of the amplitude $\theta_m$ and the circular frequency $\omega_n$. Since both terms now contain the factor $\theta_m^2$, this factor can be canceled and the resulting equation can be solved for the circular frequency $\omega_n$.

# Problems

**19.69** Two blocks, each of mass 1.5 kg, are attached to links which are pin-connected to bar $BC$ as shown. The masses of the links and bar are negligible, and the blocks can slide without friction. Block $D$ is attached to a spring of constant $k = 720$ N/m. Knowing that block $A$ is moved 15 mm from its equilibrium position and released, determine the magnitude of the maximum velocity of block $D$ during the resulting motion.

**19.70** Two blocks, each of mass 1.5 kg, are attached to links which are pin-connected to bar $BC$ as shown. The masses of the links and bar are negligible, and the blocks can slide without friction. Block $D$ is attached to a spring of constant $k = 720$ N/m. Knowing that block $A$ is at rest when it is struck horizontally with a mallet and given an initial velocity of 250 mm/s, determine the magnitude of the maximum displacement of block $D$ during the resulting motion.

**Fig. P19.69 and P19.70**

**19.71** Two small spheres, $A$ and $C$, each of mass $m$, are attached to rod $AB$, which is supported by a pin and bracket at $B$ and by a spring $CD$ of constant $k$. Knowing that the mass of the rod is negligible and that the system is in equilibrium when the rod is horizontal, determine the frequency of the small oscillations of the system.

**19.72** A 9 kg block is attached to spring $A$ and connected to spring $B$ by a cord and pulley. The block is held in the position shown with both springs unstretched when the support is removed and the block is released with no initial velocity. Neglecting friction and the masses of the pulley and the springs, determine (a) the period of the resulting vibration, (b) the magnitude of the maximum velocity achieved by the block.

**Fig. P19.71**

**Fig. P19.72**

**Fig. P19.73**

**Fig. *P19.75***

$k_B = 525$ N/m   $k_C = 876$ N/m

A

B   C

$\longleftarrow$ 45.7 cm $\longrightarrow$ $\longleftarrow$ 45.7 cm $\longrightarrow$

**Fig. P19.77**

**19.73**  The inner rim of a 38 kg flywheel is placed on a knife edge, and the period of its small oscillations is found to be 1.26 s. Determine the centroidal moment of inertia of the flywheel.

**19.74**  A uniform rod $AB$ can rotate in a vertical plane about a horizontal axis at $C$ located at a distance $c$ above the mass center $G$ of the rod. For small oscillations determine the value of $c$ for which the frequency of the motion will be maximum.

**Fig. P19.74**

**19.75**  A connecting rod is supported by a knife edge at point $A$; the period of its small oscillations is observed to be 0.895 s. The rod is then inverted and supported by a knife edge at point $B$ and the period of its small oscillations is observed to be 0.805 s. Knowing that $r_a + r_b = 27$ cm, determine (*a*) the location of the mass center $G$, (*b*) the centroidal radius of gyration $\bar{k}$.

**19.76**  A thin uniform plate cut into the shape of a quarter circle can rotate in a vertical plane about a horizontal axis at point $O$. Determine the period of small oscillations of the plate.

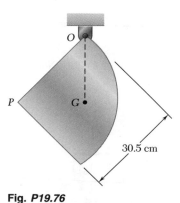

30.5 cm

**Fig. *P19.76***

**19.77**  A uniform rod $ABC$ weighs 2.7 kg and is attached to two springs as shown. If end $C$ is given a small displacement and released, determine the frequency of vibration of the rod.

**19.78** A uniform disk of radius $r$ and mass $m$ can roll without slipping on a cylindrical surface and is attached to bar $ABC$ of length $L$ and negligible mass. The bar is attached to a spring of constant $k$ and can rotate freely in the vertical plane about point $B$. Knowing that end $A$ is given a small displacement and released, determine the frequency of the resulting oscillations in terms of $m$, $L$, $k$, and $g$.

**Fig. P19.78**

**19.79** A 7 kg uniform cylinder can roll without sliding on an incline and is attached to a spring $AB$ as shown. If the center of the cylinder is moved 10 mm down the incline and released, determine (*a*) the period of vibration, (*b*) the maximum velocity of the center of the cylinder.

**Fig. P19.79**

**19.80** Two uniform rods, each of mass $m = 0.6$ kg and length $l = 160$ mm, are welded together to form the assembly shown. Knowing that the constant of each spring is $k = 120$ N/m and that end $A$ is given a small displacement and released, determine the frequency of the resulting motion.

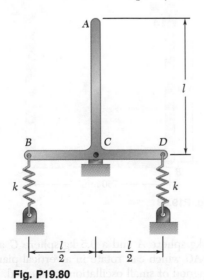

**Fig. P19.80**

**19.81**  A slender 10 kg bar $AB$ of length $l = 0.6$ m is connected to two collars of negligible weight. Collar $A$ is attached to a spring of constant $k = 1.5$ kN/m and can slide on a horizontal rod, while collar $B$ can slide freely on a vertical rod. Knowing that the system is in equilibrium when bar $AB$ is vertical and that collar $A$ is given a small displacement and released, determine the period of the resulting vibrations.

**Fig. *P19.81* and *P19.82***

**19.82**  A slender 5 kg bar $AB$ of length $l = 0.6$ m is connected to two collars, each of mass 2.5 kg. Collar $A$ is attached to a spring of constant $k = 1.5$ kN/m and can slide on a horizontal rod, while collar $B$ can slide freely on a vertical rod. Knowing that the system is in equilibrium when bar $AB$ is vertical and that collar $A$ is given a small displacement and released, determine the period of the resulting vibrations.

**19.83**  Three identical 3.6 kg uniform slender bars are connected by pins as shown and can move in a vertical plane. Knowing that bar $BC$ is given a small displacement and released, determine the period of vibration of the system.

**Fig. P19.83**

**Fig. P19.84**

**19.84**  A 0.7 kg sphere $A$ and a 0.5 kg sphere $C$ are attached to the ends of a 1 kg rod $AC$ which can rotate in a vertical plane about an axis at $B$. Determine the period of small oscillations of the rod.

**19.85**  Spheres $A$ and $C$, each of weight $W$, are attached to the ends of a homogeneous rod of the same weight $W$ and of length $2l$ which is bent as shown. The system is allowed to oscillate about a frictionless pin at $B$. Knowing that $\beta = 40°$ and $l = 63.5$ cm, determine the frequency of small oscillations.

**Fig. P19.85**

**19.86**  The 1.4 kg rod $AB$ is bolted to a 2.3 kg disk. Knowing that the disk rolls without sliding, determine the period of small oscillations of the system.

**19.87**  Two 6 kg uniform disks are attached to the 9 kg rod $AB$ as shown. Knowing that the constant of the spring is 5 kN/m and that the disks roll without sliding, determine the frequency of vibration of the system.

**Fig. P19.86**

**Fig. *P19.87***

**19.88**  A half section of a uniform cylinder of radius $r$ and mass $m$ rests on two casters $A$ and $B$, each of which is a uniform cylinder of radius $r/4$ and mass $m/8$. Knowing that the half cylinder is rotated through a small angle and released and that no slipping occurs, determine the frequency of small oscillations.

**19.89**  The 10 kg rod $AB$ is attached to two 4 kg disks as shown. Knowing that the disks roll without sliding, determine the frequency of small oscillations of the system.

**Fig. *P19.88***

**Fig. *P19.89***

**19.90** Three collars, each of mass $m$, are connected by pins to bars $AC$ and $BC$, each of length $l$ and negligible mass. Collars $A$ and $B$ can slide without friction on a horizontal rod and are connected by a spring of constant $k$. Collar $C$ can slide without friction on a vertical rod, and the system is in equilibrium in the position shown. Knowing that collar $C$ is given a small displacement and released, determine the frequency of the resulting motion of the system.

**Fig. P19.90**

**19.91** Two 2.7 kg uniform semicircular plates are attached to the 1.8 kg rod $AB$ as shown. Knowing that the plates roll without sliding, determine the period of small oscillations of the system.

**Fig. P19.91**

**19.92** A uniform 2.7 kg disk can roll without slipping on a cylindrical surface and is attached to a 1.8 kg uniform slender bar $AB$. The bar is attached to a spring of constant $k = 292$ N/m and can rotate freely in the vertical plane about point $A$. Knowing that end $B$ is given a small displacement and released, determine the period of the resulting vibration.

**Fig. P19.92**

**19.93** A slender rod of mass $m$ and length $l$ is suspended from two vertical springs, each of constant $k$ as shown. The rod is in equilibrium when it is given a small rotation about a horizontal axis through $G$ and released. Determine the frequency of small oscillations.

**Fig. P19.93**

**19.94**  A section of uniform pipe is suspended from two vertical cables attached at $A$ and $B$. Determine the frequency of oscillation when the pipe is given a small rotation about the centroidal axis $OO'$ and released.

**19.95**  A half section of pipe is placed on a horizontal surface, rotated through a small angle, and then released. Assuming that the pipe section rolls without sliding, determine the period of oscillation.

Fig. **P19.95**          Fig. **P19.96**

Fig. **P19.94**

*19.96  A thin plate of length $l$ rests on a half cylinder of radius $r$. Derive an expression for the period of small oscillations of the plate.

*19.97  As a submerged body moves through a fluid, the particles of the fluid flow around the body and thus acquire kinetic energy. In the case of a sphere moving in an ideal fluid, the total kinetic energy acquired by the fluid is $\frac{1}{4}(\gamma/g)Vv^2$, where $\gamma$ is the specific weight of the fluid, $V$ is the volume of the sphere, and $v$ is the velocity of the sphere. Consider a 0.45 kg hollow spherical shell of radius 10.2 cm which is held submerged in a tank of water by a spring of constant 584 N/m. (*a*) Neglecting fluid friction, determine the period of vibration of the shell when it is displaced vertically and then released. (*b*) Solve part *a*, assuming that the tank is accelerated upward at the constant rate of 7.3 m/s$^2$.

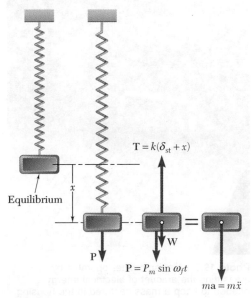

Fig. **P19.97**

## 19.7. FORCED VIBRATIONS

The most important vibrations from the point of view of engineering applications are the *forced vibrations* of a system. These vibrations occur when a system is subjected to a periodic force or when it is elastically connected to a support which has an alternating motion.

Consider first the case of a body of mass $m$ suspended from a spring and subjected to a periodic force **P** of magnitude $P = P_m \sin \omega_f t$, where $\omega_f$ is the circular frequency of **P** and is referred to as the *forced circular frequency* of the motion (Fig. 19.7). This force may be an actual external force applied to the body, or it may be a centrifugal force produced by the rotation of some unbalanced part of the body (see Sample Prob. 19.5). Denoting by $x$ the displacement of the body measured from its equilibrium position, we write the equation of motion,

$+\downarrow \Sigma F = ma$:    $P_m \sin \omega_f t + W - k(\delta_{st} + x) = m\ddot{x}$

Recalling that $W = k\delta_{st}$, we have

$$m\ddot{x} + kx = P_m \sin \omega_f t \qquad (19.30)$$

**Fig. 19.7**

**Fig. 19.8**

**Photo 19.1**  A seismometer operates by measuring the amount of electrical energy needed to keep a mass centered in the housing in the presence of strong ground shaking.

Next we consider the case of a body of mass $m$ suspended from a spring attached to a moving support whose displacement $\delta$ is equal to $\delta_m \sin \omega_f t$ (Fig. 19.8). Measuring the displacement $x$ of the body from the position of static equilibrium corresponding to $\omega_f t = 0$, we find that the total elongation of the spring at time $t$ is $\delta_{st} + x - \delta_m \sin \omega_f t$. The equation of motion is thus

$$+\downarrow \Sigma F = ma: \qquad W - k(\delta_{st} + x - \delta_m \sin \omega_f t) = m\ddot{x}$$

Recalling that $W = k\delta_{st}$, we have

$$m\ddot{x} + kx = k\delta_m \sin \omega_f t \qquad (19.31)$$

We note that Eqs. (19.30) and (19.31) are of the same form and that a solution of the first equation will satisfy the second if we set $P_m = k\delta_m$.

A differential equation such as (19.30) or (19.31), possessing a right-hand member different from zero, is said to be *nonhomogeneous*. Its general solution is obtained by adding a particular solution of the given equation to the general solution of the corresponding *homogeneous* equation (with right-hand member equal to zero). A *particular solution* of (19.30) or (19.31) can be obtained by trying a solution of the form

$$x_{part} = x_m \sin \omega_f t \qquad (19.32)$$

Substituting $x_{part}$ for $x$ into (19.30), we find

$$-m\omega_f^2 x_m \sin \omega_f t + kx_m \sin \omega_f t = P_m \sin \omega_f t$$

which can be solved for the amplitude,

$$x_m = \frac{P_m}{k - m\omega_f^2}$$

Recalling from (19.4) that $k/m = \omega_n^2$, where $\omega_n$ is the natural circular frequency of the system, we write

$$x_m = \frac{P_m/k}{1 - (\omega_f/\omega_n)^2} \qquad (19.33)$$

Substituting from (19.32) into (19.31), we obtain in a similar way

$$x_m = \frac{\delta_m}{1 - (\omega_f/\omega_n)^2} \qquad (19.33')$$

The homogeneous equation corresponding to (19.30) or (19.31) is Eq. (19.2), which defines the free vibration of the body. Its general solution, called the *complementary function,* was found in Sec. 19.2:

$$x_{comp} = C_1 \sin \omega_n t + C_2 \cos \omega_n t \qquad (19.34)$$

Adding the particular solution (19.32) to the complementary function (19.34), we obtain the *general solution* of Eqs. (19.30) and (19.31):

$$x = C_1 \sin \omega_n t + C_2 \cos \omega_n t + x_m \sin \omega_f t \qquad (19.35)$$

We note that the vibration obtained consists of two superposed vibrations. The first two terms in Eq. (19.35) represent a free vibration of the system. The frequency of this vibration is the *natural frequency* of the system, which depends only upon the constant $k$ of the spring and the mass $m$ of the body, and the constants $C_1$ and $C_2$ can be determined from the initial conditions. This free vibration is also called a *transient* vibration, since in actual practice it will soon be damped out by friction forces (Sec. 19.9).

The last term in (19.35) represents the *steady-state* vibration produced and maintained by the impressed force or impressed support movement. Its frequency is the *forced frequency* imposed by this force or movement, and its amplitude $x_m$, defined by (19.33) or (19.33'), depends upon the *frequency ratio* $\omega_f/\omega_n$. The ratio of the amplitude $x_m$ of the steady-state vibration to the static deflection $P_m/k$ caused by a force $P_m$, or to the amplitude $\delta_m$ of the support movement, is called the *magnification factor*. From (19.33) and (19.33'), we obtain

$$\text{Magnification factor} = \frac{x_m}{P_m/k} = \frac{x_m}{\delta_m} = \frac{1}{1 - (\omega_f/\omega_n)^2} \qquad (19.36)$$

The magnification factor has been plotted in Fig. 19.9 against the frequency ratio $\omega_f/\omega_n$. We note that when $\omega_f = \omega_n$, the amplitude of the forced vibration becomes infinite. The impressed force or impressed support movement is said to be in *resonance* with the given system. Actually, the amplitude of the vibration remains finite because of damping forces (Sec. 19.9); nevertheless, such a situation should be avoided, and the forced frequency should not be chosen too close to the natural frequency of the system. We also note that for $\omega_f < \omega_n$ the coefficient of $\sin \omega_f t$ in (19.35) is positive, while for $\omega_f > \omega_n$ this coefficient is negative. In the first case the forced vibration is *in phase* with the impressed force or impressed support movement, while in the second case it is 180° *out of phase*.

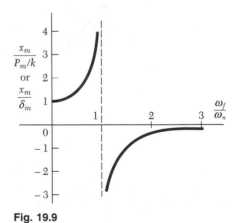

**Fig. 19.9**

Finally, let us observe that the velocity and the acceleration in the steady-state vibration can be obtained by differentiating twice with respect to $t$ the last term of Eq. (19.35). Their maximum values are given by expressions similar to those of Eqs. (19.15) of Sec. 19.2, except that these expressions now involve the amplitude and the circular frequency of the forced vibration:

$$v_m = x_m \omega_f \qquad a_m = x_m \omega_f^2 \qquad (19.37)$$

# SAMPLE PROBLEM 19.5

A motor weighing 158.7 kg is supported by four springs, each having a constant of 131.3 kN/m. The unbalance of the rotor is equivalent to a weight of 0.03 kg located 15.2 cm from the axis of rotation. Knowing that the motor is constrained to move vertically, determine (a) the speed in rpm at which resonance will occur, (b) the amplitude of the vibration of the motor at a speed of 1200 rpm.

# SOLUTION

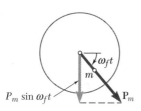

*a. Resonance Speed.* The resonance speed is equal to the natural circular frequency $\omega_n$ (in rpm) of the free vibration of the motor. The mass of the motor and the equivalent constant of the supporting springs are

$$m = 158.7 \text{ kg}$$

$$k = 4(131.3 \text{ kN/m}) = 525.2 \text{ kN/m}$$

$$\omega_n = \sqrt{\frac{k}{m}} = \sqrt{\frac{525.2}{158.7}} = 1.82 \text{ rad/s} = 17 \text{ rpm}$$

Resonance speed = 17 rpm ◄

*b. Amplitude of Vibration at 1200 rpm.* The angular velocity of the motor and the mass of the equivalent 0.03 kg weight are

$$\omega = 1200 \text{ rpm} = 125.7 \text{ rad/s}$$

$$m = 0.03 \text{ kg}$$

The magnitude of the centrifugal force due to the unbalance of the rotor is

$$P_m = ma_n = mr\omega^2 = (0.03 \text{ kg})(15.2 \text{ cm})(125.7 \text{ rad/s})^2 = 72 \text{ N}$$

The static deflection that would be caused by a constant load $P_m$ is

$$\frac{P_m}{k} = \frac{72 \text{ N}}{525.2 \times 10^3 \text{ N/m}} = 0.00014 \text{ m}$$

The forced circular frequency $\omega_f$ of the motion is the angular velocity of the motor,

$$\omega_f = \omega = 125.7 \text{ rad/s}$$

Substituting the values of $P_m/k$, $\omega_f$, and $\omega_n$ into Eq. (19.33), we obtain

$$x_m = \frac{P_m/k}{1 - (\omega_f/\omega_n)^2} = \frac{0.00014 \text{ m}}{1 - (125.7/1.82)^2} = -2.9 \times 10^{-8} \text{ m.}$$

$$x_m = 2.9 \times 10^{-8} \text{ m (out of phase)} \blacktriangleleft$$

*Note.* Since $\omega_f > \omega_n$, the vibration is 180° out of phase with the centrifugal force due to the unbalance of the rotor. For example, when the unbalanced mass is directly below the axis of rotation, the position of the motor is $x_m = 2.9 \times 10^{-8}$ m above the position of equilibrium.

# SOLVING PROBLEMS
## ON YOUR OWN

This lesson was devoted to the analysis of the *forced vibrations* of a mechanical system. These vibrations occur either when the system is subjected to a periodic force **P** (Fig. 19.7), or when it is elastically connected to a support which has an alternating motion (Fig. 19.8). In the first case, the motion of the system is defined by the differential equation

$$m\ddot{x} + kx = P_m \sin \omega_f t \tag{19.30}$$

where the right-hand member represents the magnitude of the force **P** at a given instant. In the second case, the motion is defined by the differential equation

$$m\ddot{x} + kx = k\delta_m \sin \omega_f t \tag{19.31}$$

where the right-hand member is the product of the spring constant $k$ and the displacement of the support at a given instant. You will be concerned only with the *steady-state* motion of the system, which is defined by a *particular solution* of these equations, of the form

$$x_{\text{part}} = x_m \sin \omega_f t \tag{19.32}$$

**1. If the forced vibration is caused by a periodic force P,** of amplitude $P_m$ and circular frequency $\omega_f$, the amplitude of the vibration is

$$x_m = \frac{P_m/k}{1 - (\omega_f/\omega_n)^2} \tag{19.33}$$

where $\omega_n$ is the *natural circular frequency* of the system $\omega_n = \sqrt{k/m}$, and $k$ is the spring constant. Note that the circular frequency of the vibration is $\omega_f$ and that the amplitude $x_m$ does not depend upon the initial conditions. For $\omega_f = \omega_n$, the denominator in Eq. (19.33) is zero and $x_m$ is infinite (Fig. 19.9); the impressed force **P** is said to be in *resonance* with the system. Also, for $\omega_f < \omega_n$, $x_m$ is positive and the vibration is *in phase* with **P**, while, for $\omega_f > \omega_n$, $x_m$ is negative and the vibration is *out of phase*.

*a. In the problems which follow, you may be asked to determine one of the parameters* in Eq. (19.33) when the others are known. We suggest that you keep Fig. 19.9 in front of you when solving these problems. For example, if you are asked to find the frequency at which the amplitude of a forced vibration has a given value, but you do not know whether the vibration is in or out of phase with respect to the impressed force, you should note from Fig. 19.9 that there can be two frequencies satisfying this requirement, one corresponding to a positive value of $x_m$ and to a vibration in phase with the impressed force, and the other corresponding to a negative value of $x_m$ and to a vibration out of phase with the impressed force.

(*continued*)

*b. Once you have obtained the amplitude $x_m$ of the motion of a compo-*nent of the system from Eq. (19.33), you can use Eqs. (19.37) to determine the maximum values of the velocity and acceleration of that component:

$$v_m = x_m \omega_f \qquad a_m = x_m \omega_f^2 \qquad (19.37)$$

*c. When the impressed force P is due to the unbalance of the rotor of a motor,* its maximum value is $P_m = mr\omega_f^2$, where $m$ is the mass of the rotor, $r$ is the distance between its mass center and the axis of rotation, and $\omega_f$ is equal to the angular velocity $\omega$ of the rotor expressed in rad/s [Sample Prob. 19.5].

*2. If the forced vibration is caused by the simple harmonic motion of a support,* of amplitude $\delta_m$ and circular frequency $\omega_f$, the amplitude of the vibration is

$$x_m = \frac{\delta_m}{1 - (\omega_f/\omega_n)^2} \qquad (19.33')$$

where $\omega_n$ is the *natural circular frequency* of the system, $\omega_n = \sqrt{k/m}$. Again, note that the circular frequency of the vibration is $\omega_f$ and that the amplitude $x_m$ does not depend upon the initial conditions.

*a. Be sure to read our comments in paragraphs 1, 1a, and 1b,* since they apply equally well to a vibration caused by the motion of a support.

*b. If the maximum acceleration $a_m$ of the support is specified,* rather than its maximum displacement $\delta_m$, remember that, since the motion of the support is a simple harmonic motion, you can use the relation $a_m = \delta_m \omega_f^2$ to determine $\delta_m$; the value obtained is then substituted into Eq. (19.33').

# Problems

**19.98** A 4-kg collar can slide on a frictionless horizontal rod and is attached to a spring of constant 450 N/m. It is acted upon by a periodic force of magnitude $P = P_m \sin \omega_f t$, where $P_m = 13$ N. Determine the amplitude of the motion of the collar if (a) $\omega_f = 5$ rad/s, (b) $\omega_f = 10$ rad/s.

**19.99** A 4-kg collar can slide on a frictionless horizontal rod and is attached to a spring of constant $k$. It is acted upon by a periodic force of magnitude $P = P_m \sin \omega_f t$, where $P_m = 9$ N and $\omega_f = 5$ rad/s. Determine the value of the spring constant $k$ knowing that the motion of the collar has an amplitude of 150 mm and is (a) in phase with the applied force, (b) out of phase with the applied force.

**19.100** A collar of mass $m$ which slides on a frictionless horizontal rod is attached to a spring of constant $k$ and is acted upon by a periodic force of magnitude $P = P_m \sin \omega_f t$. Determine the range of values of $\omega_f$ for which the amplitude of the vibration exceeds three times the static deflection caused by a constant force of magnitude $P_m$.

$\mathbf{P} = P_m \sin \omega_f t$

**Fig. P19.98, P19.99, and *P19.100***

**19.101** A small 18 kg block $A$ is attached to the rod $BC$ of negligible mass which is supported at $B$ by a pin and bracket and at $C$ by a spring of constant $k = 2043$ N/m. The system can move in a vertical plane and is in equilibrium when the rod is horizontal. The rod is acted upon at $C$ by a periodic force $\mathbf{P}$ of magnitude $P = P_m \sin \omega_f t$, where $P_m = 6.2$ N. Knowing that $b = 20.3$ cm, determine the range of values of $\omega_f$ for which the amplitude of vibration of block $A$ exceeds 0.36 cm.

**19.102** A small 18 kg block $A$ is attached to the rod $BC$ of negligible mass which is supported at $B$ by a pin and bracket and at $C$ by a spring of constant $k = 2043$ N/m. The system can move in a vertical plane and is in equilibrium when the rod is horizontal. The rod is acted upon at $C$ by a periodic force $\mathbf{P}$ of magnitude $P = P_m \sin \omega_f t$, where $P_m = 6.2$ N. Knowing that $\omega_f = 15$ rad/s, determine the range of values of $b$ for which the amplitude of vibration of block $A$ exceeds 0.36 cm.

**19.103** The 1.2-kg bob of a simple pendulum of length $l = 600$ mm is suspended from a 1.4-kg collar $C$. Knowing that the collar is acted upon by a periodic force $\mathbf{P} = P_m \sin \omega_f t$, where $P_m = 0.5$ N and $\omega_f = 3$ rad/s, determine the amplitude and phase of the motion of the bob.

81.3 cm

$b$

$A$

$k$

$C$

$B$

$P = P_m \sin \omega_f t$

**Fig. P19.101 and P19.102**

$\mathbf{P} = P_m \sin \omega_f t$

$x_C$

$C$

$l$

$\theta$

$B$

$x_B$

**Fig. *P19.103***

$\delta = \delta_m \sin \omega_f t$

**Fig. P19.104**

**Fig. P19.105 and P19.106**

**Fig. P19.108**

$\delta = \delta_m \sin \omega_f t$

**Fig. P19.109**

**19.104** A cantilever beam $AB$ supports a block which causes a static deflection of 40 mm at $B$. Assuming that the support at $A$ undergoes a vertical periodic displacement $\delta = \delta_m \sin \omega_f t$, where $\delta_m = 10$ mm, determine the range of values of $\omega_f$ for which the amplitude of the motion of the block will be less than 20 mm. Neglect the mass of the beam and assume that the block does not leave the beam.

**19.105** A 2-kg block $A$ slides in a vertical frictionless slot and is connected to a moving support $B$ by means of a spring $AB$ of constant $k = 117$ N/m. Knowing that the displacement of the support is $\delta = \delta_m \sin \omega_f t$, where $\delta_m = 100$ mm and $\omega_f = 5$ rad/s, determine (a) the amplitude of the motion of the block, (b) the amplitude of the fluctuating force exerted by the spring on the block.

**19.106** A 7.3-kg block $A$ slides in a vertical frictionless slot and is connected to a moving support $B$ by means of a spring $AB$ of constant $k = 1897$ N/m. Knowing that the displacement of the support is $\delta = \delta_m \sin \omega_f t$, where $\delta_m = 15.2$ cm, determine the range of values of $\omega_f$ for which the amplitude of the fluctuating force exerted by the spring on the block is less than 13.7 kg.

**19.107** A 2.3 kg block $A$ is attached to a spring of constant $k = 58.4$ N/m and a bar $BCD$ of negligible weight. The bar is connected at $D$ to a moving support $E$ by means of an identical spring. Knowing that the support $E$ undergoes a displacement $\delta = \delta_m \sin \omega_f t$, where $\delta_m = 0.5$ cm and $\omega_f = 10$ rad/s, determine (a) the magnitude of the maximum acceleration of block $A$, (b) the magnitude of the maximum force transmitted to the support at $C$.

**Fig. P19.107**

**19.108** A small 2 kg sphere $B$ is attached to the bar $AB$ of negligible mass which is supported at $A$ by a pin and bracket and connected at $C$ to a moving support $D$ by means of a spring of constant $k = 3.6$ kN/m. Knowing that support $D$ undergoes a vertical displacement $\delta = \delta_m \sin \omega_f t$, where $\delta_m = 3$ mm and $\omega_f = 15$ rad/s, determine (a) the magnitude of the maximum angular velocity of bar $AB$, (b) the magnitude of the maximum acceleration of sphere $B$.

**19.109** A beam $ABC$ is supported by a pin connection at $A$ and by rollers at $B$. A 120 kg block placed on the end of the beam causes a static deflection of 15 mm at $C$. Assuming that the support at $A$ undergoes a vertical periodic displacement $\delta = \delta_m \sin \omega_f t$, where $\delta_m = 10$ mm and $\omega_f = 18$ rad/s, and the support at $B$ does not move, determine the maximum acceleration of the block at $C$. Neglect the weight of the beam and assume that the block does not leave the beam.

**19.110** A simple pendulum of length $l$ is suspended from a collar $C$ which is forced to move horizontally according to the relation $x_c = \delta_m \sin \omega_f t$. Determine the range of values of $\omega_f$ for which the amplitude of the motion of the bob is less than $\delta_m$. (Assume that $\delta_m$ is small compared with the length $l$ of the pendulum.)

**19.111** In Prob. 19.110, determine the range of values of $\omega_f$ for which the amplitude of the motion of the bob exceeds $2\delta_m$.

**19.112** A 200-kg motor is supported by springs having a total constant of 215 kN/m. The unbalance of the rotor is equivalent to a 30-g mass located 200 mm from the axis of rotation. Determine the range of allowable values of the motor speed if the amplitude of the vibration is not to exceed 1.5 mm.

**19.113** A motor of mass 18 kg is supported by four springs, each of constant 40 kN/m. The motor is constrained to move vertically, and the amplitude of its motion is observed to be 1.5 mm at a speed of 1200 rpm. Knowing that the mass of the rotor is 4 kg, determine the distance between the mass center of the rotor and the axis of the shaft.

**Fig. P19.110**

**Fig. P19.113**

**19.114** A 163.3 kg motor is bolted to a light horizontal beam. The unbalance of its rotor is equivalent to a 0.03 kg weight located 19 cm from the axis of rotation, and the static deflection of the beam due to the weight of the motor is 1.5 cm. The amplitude of the vibration due to the unbalance can be decreased by adding a plate to the base of the motor. If the amplitude of vibration is to be less than $5.5 \times 10^{-5}$ m for motor speeds above 300 rpm, determine the required weight of the plate.

**Fig. P19.114**

**19.115** A motor of mass $M$ is supported by springs with an equivalent spring constant $k$. The unbalance of its rotor is equivalent to a mass $m$ located at a distance $r$ from the axis of rotation. Show that when the angular velocity of the rotor is $\omega_f$, the amplitude $x_m$ of the motion of the motor is

$$x_m = \frac{r(m/M)(\omega_f/\omega_n)^2}{1 - (\omega_f/\omega_n)^2}$$

where $\omega_n = \sqrt{k/M}$.

**19.116** Rod $AB$ is rigidly attached to the frame of a motor running at a constant speed. When a collar of mass $m$ is placed on the spring, it is observed to vibrate with an amplitude of 15 mm. When two collars, each of mass $m$, are placed on the spring, the amplitude is observed to be 18 mm. What amplitude of vibration should be expected when three collars, each of mass $m$, are placed on the spring? (Obtain two answers.)

(a)      (b)      (c)

**Fig. P19.116**

**Fig. P19.119**

**Fig. P19.121**

**Fig. P19.122**

**19.117** Solve Prob. 19.116, assuming that the speed of the motor is changed and that one collar has an amplitude of 9 mm and two collars have an amplitude of 3 mm.

**19.118** The unbalance of the rotor of a 181 kg motor is equivalent to a 0.024 kg weight located 15.2 cm from the axis of rotation. Knowing that the motor is supported by four springs, each of constant 73 kN/m, determine the range of allowable values of the motor speed if the maximum vertical acceleration of the motor is to be not to exceed 0.12 m/s$^2$.

**19.119** A counter-rotating eccentric mass exciter consisting of two rotating 0.1 kg weights describing circles of radius $r$ at the same speed but in opposite senses is placed on a machine element to induce a steady-state vibration of the element. The total weight of the system is 272 kg, the constant of each spring is $k = 595$ kN/m, and the rotational speed of the exciter is 1200 rpm. Knowing that the amplitude of the total fluctuating force exerted on the foundation is 16.4 kg, determine the radius $r$.

**19.120** Figures (1) and (2) show how springs can be used to support a block in two different situations. In Fig. (1) they help decrease the amplitude of the fluctuating force transmitted by the block to the foundation. In Fig. (2) they help decrease the amplitude of the fluctuating displacement transmitted by the foundation to the block. The ratio of the transmitted force to the impressed force or the ratio of the transmitted displacement to the impressed displacement is called the *transmissibility*. Derive an equation for the transmissibility for each situation. Give your answer in terms of the ratio $\omega_f/\omega_n$ of the frequency $\omega_f$ of the impressed force or impressed displacement to the natural frequency $\omega_n$ of the spring-mass system. Show that in order to cause any reduction in transmissibility, the ratio $\omega_f/\omega_n$ must be greater than $\sqrt{2}$.

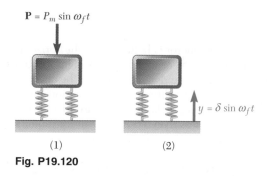

$$\mathbf{P} = P_m \sin \omega_f t$$

$$y = \delta \sin \omega_f t$$

(1)        (2)

**Fig. P19.120**

**19.121** A 27 kg disk is attached with an eccentricity $e = 150$ μm to the midpoint of a vertical shaft $AB$ which revolves at a constant angular velocity $\omega_f$. Knowing that the spring constant $k$ for horizontal movement of the disk is 580 kN/m, determine (a) the angular velocity $\omega_f$ at which resonance will occur, (b) the deflection $r$ of the shaft when $\omega_f = 1200$ rpm.

**19.122** A small trailer and its load have a total weight of 227 kg. The trailer is supported by two springs, each of constant 5108 N/m, and is pulled over a road, the surface of which can be approximated by a sine curve with an amplitude of 5 cm and a wavelength of 4.6 m (that is, the distance between successive crests is 4.6 m and the vertical distance from crest to trough is 10.2 cm). Determine (a) the speed at which resonance will occur, (b) the amplitude of the vibration of the trailer at a speed of 48 km/h.

**19.123**  Block $A$ can move without friction in the slot as shown and is acted upon by a vertical periodic force of magnitude $P = P_m \sin \omega_f t$, where $\omega_f = 2$ rad/s and $P_m = 22.2$ N. A spring of constant $k$ is attached to the bottom of block $A$ and to a 20 kg block $B$. Determine (a) the value of the constant $k$ which will prevent a steady-state vibration of block $A$, (b) the corresponding amplitude of the vibration of block $B$.

**19.124**  A vibrometer used to measure the amplitude of vibrations consists of a box containing a mass-spring system with a known natural frequency of 150 Hz. The box is rigidly attached to a surface which is moving according to the equation $y = \delta_m \sin \omega_f t$. If the amplitude $z_m$ of the motion of the mass relative to the box is used as a measure of the amplitude $\delta_m$ of the vibration of the surface, determine (a) the percent error when the frequency of the vibration is 750 Hz, (b) the frequency at which the error is zero.

**19.125**  A certain accelerometer consists essentially of a box containing a mass-spring system with a known natural frequency of 1760 Hz. The box is rigidly attached to a surface which is moving according to the equation $y = \delta_m \sin \omega_f t$. If the amplitude $z_m$ of the motion of the mass relative to the box times a scale factor $\omega_n^2$ is used as a measure of the maximum acceleration $a_m = \delta_m \omega_f^2$ of the vibrating surface, determine the percent error when the frequency of the vibration is 480 Hz.

**Fig. P19.123**

**Fig. _P19.124_ and _P19.125_**

# DAMPED VIBRATIONS

## *19.8. DAMPED FREE VIBRATIONS

The vibrating systems considered in the first part of this chapter were assumed free of damping. Actually all vibrations are damped to some degree by friction forces. These forces can be caused by *dry friction,* or *Coulomb friction,* between rigid bodies, by *fluid friction* when a rigid body moves in a fluid, or by *internal friction* between the molecules of a seemingly elastic body.

A type of damping of special interest is the *viscous damping* caused by fluid friction at low and moderate speeds. Viscous damping is characterized by the fact that the friction force is *directly proportional and opposite to the velocity* of the moving body. As an example, let us again consider a body of mass $m$ suspended from a spring of constant $k$, assuming that the body is attached to the plunger of a dashpot (Fig. 19.10). The magnitude of the friction force exerted on the plunger by the surrounding fluid is equal to $c\dot{x}$, where the constant $c$, expressed in N · s/m or lb · s/ft and known as the *coefficient of viscous damping,* depends upon the physical properties of the fluid and the construction of the dashpot. The equation of motion is

$+\downarrow \Sigma F = ma:$ $\qquad W - k(\delta_{st} + x) - c\dot{x} = m\ddot{x}$

**Fig. 19.10**

Recalling that $W = k\delta_{st}$, we write

$$m\ddot{x} + c\dot{x} + kx = 0 \tag{19.38}$$

Substituting $x = e^{\lambda t}$ into (19.38) and dividing through by $e^{\lambda t}$ we write the *characteristic equation*

$$m\lambda^2 + c\lambda + k = 0 \tag{19.39}$$

and obtain the roots

$$\lambda = -\frac{c}{2m} \pm \sqrt{\left(\frac{c}{2m}\right)^2 - \frac{k}{m}} \tag{19.40}$$

Defining the *critical damping coefficient* $c_c$ as the value of $c$ which makes the radical in Eq. (19.40) equal to zero, we write

$$\left(\frac{c_c}{2m}\right)^2 - \frac{k}{m} = 0 \qquad c_c = 2m\sqrt{\frac{k}{m}} = 2m\omega_n \tag{19.41}$$

where $\omega_n$ is the natural circular frequency of the system in the absence of damping. We can distinguish three different cases of damping, depending upon the value of the coefficient $c$.

1. *Heavy damping: $c > c_c$.* The roots $\lambda_1$ and $\lambda_2$ of the characteristic equation (19.39) are real and distinct, and the general solution of the differential equation (19.38) is

$$x = C_1 e^{\lambda_1 t} + C_2 e^{\lambda_2 t} \tag{19.42}$$

This solution corresponds to a nonvibratory motion. Since $\lambda_1$ and $\lambda_2$ are both negative, $x$ approaches zero as $t$ increases indefinitely. However, the system actually regains its equilibrium position after a finite time.

2. *Critical damping: $c = c_c$.* The characteristic equation has a double root $\lambda = -c_c/2m = -\omega_n$, and the general solution of (19.38) is

$$x = (C_1 + C_2 t)e^{-\omega_n t} \tag{19.43}$$

The motion obtained is again nonvibratory. Critically damped systems are of special interest in engineering applications since they regain their equilibrium position in the shortest possible time without oscillation.

3. *Light damping: $c < c_c$.* The roots of Eq. (19.39) are complex and conjugate, and the general solution of (19.38) is of the form

$$x = e^{-(c/2m)t}(C_1 \sin \omega_d t + C_2 \cos \omega_d t) \tag{19.44}$$

where $\omega_d$ is defined by the relation

$$\omega_d^2 = \frac{k}{m} - \left(\frac{c}{2m}\right)^2$$

Substituting $k/m = \omega_n^2$ and recalling (19.41), we write

$$\omega_d = \omega_n \sqrt{1 - \left(\frac{c}{c_c}\right)^2} \qquad (19.45)$$

where the constant $c/c_c$ is known as the *damping factor*. Even though the motion does not actually repeat itself, the constant $\omega_d$ is commonly referred to as the *circular frequency* of the damped vibration. A substitution similar to the one used in Sec. 19.2 enables us to write the general solution of Eq. (19.38) in the form

$$x = x_0 e^{-(c/2m)t} \sin(\omega_d t + \phi) \qquad (19.46)$$

The motion defined by Eq. (19.46) is vibratory with diminishing amplitude (Fig. 19.11), and the time interval $\tau_d = 2\pi/\omega_d$ separating two successive points where the curve defined by Eq. (19.46) touches one of the limiting curves shown in Fig. 19.11 is commonly referred to as the *period of the damped vibration*. Recalling Eq. (19.45), we observe that $\omega_d < \omega_n$ and, thus, that $\tau_d$ is larger than the period of vibration $\tau_n$ of the corresponding undamped system.

**Fig. 19.11**

**Photo 19.2** The automobile suspension shown consists essentially of a spring and a shock absorber, which will cause the body of the car to undergo *damped forced vibrations* when the car is driven over an uneven road.

## *19.9. DAMPED FORCED VIBRATIONS

If the system considered in the preceding section is subjected to a periodic force **P** of magnitude $P = P_m \sin \omega_f t$, the equation of motion becomes

$$m\ddot{x} + c\dot{x} + kx = P_m \sin \omega_f t \tag{19.47}$$

The general solution of (19.47) is obtained by adding a particular solution of (19.47) to the complementary function or general solution of the homogeneous Eq. (19.38). The complementary function is given by (19.42), (19.43), or (19.44), depending upon the type of damping considered. It represents a *transient* motion which is eventually damped out.

Our interest in this section is centered on the steady-state vibration represented by a particular solution of (19.47) of the form

$$x_{\text{part}} = x_m \sin (\omega_f t - \varphi) \tag{19.48}$$

Substituting $x_{\text{part}}$ for $x$ into (19.47), we obtain

$$-m\omega_f^2 x_m \sin (\omega_f t - \varphi) + c\omega_f x_m \cos (\omega_f t - \varphi) + kx_m \sin (\omega_f t - \varphi) = P_m \sin \omega_f t$$

Making $\omega_f t - \varphi$ successively equal to 0 and to $\pi/2$, we write

$$c\omega_f x_m = P_m \sin \varphi \tag{19.49}$$

$$(k - m\omega_f^2) x_m = P_m \cos \varphi \tag{19.50}$$

Squaring both members of (19.49) and (19.50) and adding, we have

$$[(k - m\omega_f^2)^2 + (c\omega_f)^2] x_m^2 = P_m^2 \tag{19.51}$$

Solving (19.51) for $x_m$ and dividing (19.49) and (19.50) member by member, we obtain, respectively,

$$x_m = \frac{P_m}{\sqrt{(k - m\omega_f^2)^2 + (c\omega_f)^2}} \qquad \tan \varphi = \frac{c\omega_f}{k - m\omega_f^2} \tag{19.52}$$

Recalling from (19.4) that $k/m = \omega_n^2$, where $\omega_n$ is the circular frequency of the undamped free vibration, and from (19.41) that $2m\omega_n = c_c$, where $c_c$ is the critical damping coefficient of the system, we write

$$\frac{x_m}{P_m/k} = \frac{x_m}{\delta_m} = \frac{1}{\sqrt{[1 - (\omega_f/\omega_n)^2]^2 + [2(c/c_c)(\omega_f/\omega_n)]^2}} \tag{19.53}$$

$$\tan \varphi = \frac{2(c/c_c)(\omega_f/\omega_n)}{1 - (\omega_f/\omega_n)^2} \tag{19.54}$$

Formula (19.53) expresses the magnification factor in terms of the frequency ratio $\omega_f/\omega_n$ and damping factor $c/c_c$. It can be used to determine the amplitude of the steady-state vibration produced by an impressed force of magnitude $P = P_m \sin \omega_f t$ or by an impressed support movement $\delta = \delta_m \sin \omega_f t$. Formula (19.54) defines in terms of the same parameters the *phase difference* $\varphi$ between the impressed force or impressed support movement and the resulting steady-state vibration of the damped system. The magnification factor has been plotted against the frequency ratio in Fig. 19.12 for various values of the damping factor. We observe that the amplitude of a forced vibration can be kept small by choosing a large coefficient of viscous damping $c$ or by keeping the natural and forced frequencies far apart.

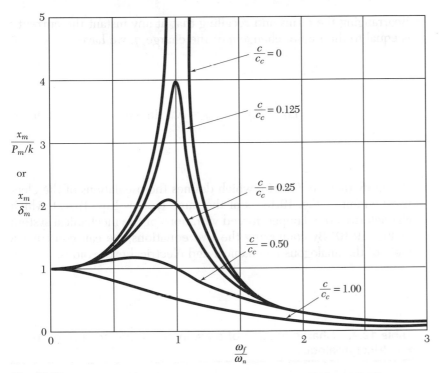

**Fig. 19.12**

# *19.10. ELECTRICAL ANALOGUES

Oscillating electrical circuits are characterized by differential equations of the same type as those obtained in the preceding sections. Their analysis is therefore similar to that of a mechanical system, and the results obtained for a given vibrating system can be readily extended to the equivalent circuit. Conversely, any result obtained for an electrical circuit will also apply to the corresponding mechanical system.

$E = E_m \sin \omega_f t$

**Fig. 19.13**

Consider an electrical circuit consisting of an inductor of inductance $L$, a resistor of resistance $R$, and a capacitor of capacitance $C$, connected in series with a source of alternating voltage $E = E_m \sin \omega_f t$ (Fig. 19.13). It is recalled from elementary circuit theory† that if $i$ denotes the current in the circuit and $q$ denotes the electric charge on the capacitor, the drop in potential is $L(di/dt)$ across the inductor, $Ri$ across the resistor, and $q/C$ across the capacitor. Expressing that the algebraic sum of the applied voltage and of the drops in potential around the circuit loop is zero, we write

$$E_m \sin \omega_f t - L\frac{di}{dt} - Ri - \frac{q}{C} = 0 \qquad (19.55)$$

Rearranging the terms and recalling that at any instant the current $i$ is equal to the rate of change $\dot{q}$ of the charge $q$, we have

$$L\ddot{q} + R\dot{q} + \frac{1}{C}q = E_m \sin \omega_f t \qquad (19.56)$$

We verify that Eq. (19.56), which defines the oscillations of the electrical circuit of Fig. 19.13, is of the same type as Eq. (19.47), which characterizes the damped forced vibrations of the mechanical system of Fig. 19.10. By comparing the two equations, we can construct a table of the analogous mechanical and electrical expressions.

**Table 19.2.  Characteristics of a Mechanical System and of Its Electrical Analogue**

| Mechanical System | | Electrical Circuit | |
|---|---|---|---|
| $m$ | Mass | $L$ | Inductance |
| $c$ | Coefficient of viscous damping | $R$ | Resistance |
| $k$ | Spring constant | $1/C$ | Reciprocal of capacitance |
| $x$ | Displacement | $q$ | Charge |
| $v$ | Velocity | $i$ | Current |
| $P$ | Applied force | $E$ | Applied voltage |

†See C. R. Paul, S. A. Nasar, and L. E. Unnewehr, *Introduction to Electrical Engineering*, 2nd ed., McGraw-Hill, New York, 1992.

Table 19.2 can be used to extend the results obtained in the preceding sections for various mechanical systems to their electrical analogues. For instance, the amplitude $i_m$ of the current in the circuit of Fig. 19.13 can be obtained by noting that it corresponds to the maximum value $v_m$ of the velocity in the analogous mechanical system. Recalling from the first of Eqs. (19.37) that $v_m = x_m \omega_f$, substituting for $x_m$ from Eq. (19.52), and replacing the constants of the mechanical system by the corresponding electrical expressions, we have

$$i_m = \frac{\omega_f E_m}{\sqrt{\left(\dfrac{1}{C} - L\omega_f^2\right)^2 + (R\omega_f)^2}}$$

$$i_m = \frac{E_m}{\sqrt{R^2 + \left(L\omega_f - \dfrac{1}{C\omega_f}\right)^2}} \qquad (19.57)$$

The radical in the expression obtained is known as the *impedance* of the electrical circuit.

The analogy between mechanical systems and electrical circuits holds for transient as well as steady-state oscillations. The oscillations of the circuit shown in Fig. 19.14, for instance, are analogous to the damped free vibrations of the system of Fig. 19.10. As far as the initial conditions are concerned, we should note that closing the switch $S$ when the charge on the capacitor is $q = q_0$ is equivalent to releasing the mass of the mechanical system with no initial velocity from the position $x = x_0$. We should also observe that if a battery of constant voltage $E$ is introduced in the electrical circuit of Fig. 19.14, closing the switch $S$ will be equivalent to suddenly applying a force of constant magnitude $P$ to the mass of the mechanical system of Fig. 19.10.

The discussion above would be of questionable value if its only result were to make it possible for mechanics students to analyze electrical circuits without learning the elements of circuit theory. It is hoped that this discussion will instead encourage students to apply to the solution of problems in mechanical vibrations the mathematical techniques they may learn in later courses in circuit theory. The chief value of the concept of electrical analogue, however, resides in its application to *experimental methods* for the determination of the characteristics of a given mechanical system. Indeed, an electrical circuit is much more easily constructed than is a mechanical model, and the fact that its characteristics can be modified by varying the inductance, resistance, or capacitance of its various components makes the use of the electrical analogue particularly convenient.

**Fig. 19.14**

$$\mathbf{P} = P_m \sin \omega_f t$$

**Fig. 19.15**

$$E = E_m \sin \omega_f t$$

**Fig. 19.16**

To determine the electrical analogue of a given mechanical system, we focus our attention on each moving mass in the system and observe which springs, dashpots, or external forces are applied directly to it. An equivalent electrical loop can then be constructed to match each of the mechanical units thus defined; the various loops obtained in that way will together form the desired circuit. Consider, for instance, the mechanical system of Fig. 19.15. We observe that the mass $m_1$ is acted upon by two springs of constants $k_1$ and $k_2$ and by two dashpots characterized by the coefficients of viscous damping $c_1$ and $c_2$. The electrical circuit should therefore include a loop consisting of an inductor of inductance $L_1$ proportional to $m_1$, of two capacitors of capacitance $C_1$ and $C_2$ inversely proportional to $k_1$ and $k_2$, respectively, and of two resistors of resistance $R_1$ and $R_2$, proportional to $c_1$ and $c_2$, respectively. Since the mass $m_2$ is acted upon by the spring $k_2$ and the dashpot $c_2$, as well as the force $P = P_m \sin \omega_f t$, the circuit should also include a loop containing the capacitor $C_2$, the resistor $R_2$, the new inductor $L_2$, and the voltage source $E = E_m \sin \omega_f t$ (Fig. 19.16).

To check that the mechanical system of Fig. 19.15 and the electrical circuit of Fig. 19.16 actually satisfy the same differential equations, the equations of motion for $m_1$ and $m_2$ will first be derived. Denoting, respectively, by $x_1$ and $x_2$ the displacements of $m_1$ and $m_2$ from their equilibrium positions, we observe that the elongation of the spring $k_1$ (measured from the equilibrium position) is equal to $x_1$, while the elongation of the spring $k_2$ is equal to the relative displacement $x_2 - x_1$ of $m_2$ with respect to $m_1$. The equations of motion for $m_1$ and $m_2$ are therefore

$$m_1 \ddot{x}_1 + c_1 \dot{x}_1 + c_2(\dot{x}_1 - \dot{x}_2) + k_1 x_1 + k_2(x_1 - x_2) = 0 \qquad (19.58)$$

$$m_2 \ddot{x}_2 + c_2(\dot{x}_2 - \dot{x}_1) + k_2(x_2 - x_1) = P_m \sin \omega_f t \quad (19.59)$$

Consider now the electrical circuit of Fig. 19.16; we denote, respectively, by $i_1$ and $i_2$ the current in the first and second loops, and by $q_1$ and $q_2$ the integrals $\int i_2 \, dt$ and $\int i_2 \, dt$. Noting that the charge on the capacitor $C_1$ is $q_1$, while the charge on $C_2$ is $q_1 - q_2$, we express that the sum of the potential differences in each loop is zero and obtain the following equations

$$L_1 \ddot{q}_1 + R_1 \dot{q}_1 + R_2(\dot{q}_1 - \dot{q}_2) + \frac{q_1}{C_1} + \frac{q_1 - q_2}{C_2} = 0 \qquad (19.60)$$

$$L_2 \ddot{q}_2 + R_2(\dot{q}_2 - \dot{q}_1) + \frac{q_1 - q_2}{C_2} = E_m \sin \omega_f t \quad (19.61)$$

We easily check that Eqs. (19.60) and (19.61) reduce to (19.58) and (19.59), respectively, when the substitutions indicated in Table 19.2 are performed.

# SOLVING PROBLEMS
## ON YOUR OWN

In this lesson a more realistic model of a vibrating system was developed by including the effect of the *viscous damping* caused by fluid friction. Viscous damping was represented in Fig. 19.10 by the force exerted on the moving body by a plunger moving in a dashpot. This force is equal in magnitude to $c\dot{x}$, where the constant $c$, expressed in N · s/m or lb · s/ft, is known as the *coefficient of viscous damping*. Keep in mind that the same sign convention should be used for $x$, $\dot{x}$, and $\ddot{x}$.

1. **Damped free vibrations.** The differential equation defining this motion was found to be

$$m\ddot{x} + c\dot{x} + kx = 0 \qquad (19.38)$$

To obtain the solution of this equation, calculate the *critical damping coefficient* $c_c$, using the formula

$$c_c = 2m\sqrt{k/m} = 2m\omega_n \qquad (19.41)$$

where $\omega_n$ is the natural circular frequency of the *undamped* system.

    *a.* **If $c > c_c$ (heavy damping),** the solution of Eq. (19.38) is

$$x = C_1 e^{\lambda_1 t} + C_2 e^{\lambda_2 t} \qquad (19.42)$$

where

$$\lambda_{1,2} = -\frac{c}{2m} \pm \sqrt{\left(\frac{c}{2m}\right)^2 - \frac{k}{m}} \qquad (19.40)$$

and where the constants $C_1$ and $C_2$ can be determined from the initial conditions $x(0)$ and $\dot{x}(0)$. This solution corresponds to a nonvibratory motion.

    *b.* **If $c = c_c$ (critical damping),** the solution of Eq. (19.38) is

$$x = (C_1 + C_2 t)e^{-\omega_n t} \qquad (19.43)$$

which also corresponds to a nonvibratory motion.

    *c.* **If $c < c_c$ (light damping),** the solution of Eq. (19.38) is

$$x = x_0 e^{-(c/2m)t} \sin(\omega_d t + \phi) \qquad (19.46)$$

where

$$\omega_d = \omega_n \sqrt{1 - \left(\frac{c}{c_c}\right)^2} \qquad (19.45)$$

and where $x_0$ and $\phi$ can be determined from the initial conditions $x(0)$ *and* $\dot{x}(0)$. This solution corresponds to oscillations of decreasing amplitude and of period $\tau_d = 2\pi/\omega_d$ (Fig. 19.11).

*(continued)*

**2. *Damped forced vibrations.*** These vibrations occur when a system with viscous damping is subjected to a periodic force **P** of magnitude $P = P_m \sin \omega_f t$ or when it is elastically connected to a support with an alternating motion $\delta = \delta_m \sin \omega_f t$. In the first case the motion is defined by the differential equation

$$m\ddot{x} + c\dot{x} + kx = P_m \sin \omega_f t \qquad (19.47)$$

and in the second case by a similar equation obtained by replacing $P_m$ with $k\delta_m$. You will be concerned only with the *steady-state* motion of the system, which is defined by a *particular solution* of these equations, of the form

$$x_{\text{part}} = x_m \sin (\omega_f t - \varphi) \qquad (19.48)$$

where

$$\frac{x_m}{P_m/k} = \frac{x_m}{\delta_m} = \frac{1}{\sqrt{[1 - (\omega_f/\omega_n)^2]^2 + [2(c/c_c)(\omega_f/\omega_n)]^2}} \qquad (19.53)$$

and

$$\tan \varphi = \frac{2(c/c_c)(\omega_f/\omega_n)}{1 - (\omega_f/\omega_n)^2} \qquad (19.54)$$

The expression given in Eq. (19.53) is referred to as the *magnification factor* and has been plotted against the frequency ratio $\omega_f/\omega_n$ in Fig. 19.12 for various values of the damping factor $c/c_c$. In the problems which follow you may be asked to determine one of the parameters in Eqs. (19.53) and (19.54) when the others are known.

# Problems

**19.126** Show that in the case of heavy damping $(c > c_c)$, a body never passes through its position of equilibrium $O$ (a) if it is released with no initial velocity from an arbitrary position, or (b) if it is started from $O$ with an arbitrary initial velocity.

**19.127** Show that in the case of heavy damping $(c > c_c)$, a body released from an arbitrary position with an arbitrary initial velocity cannot pass more than once through its equilibrium position.

**19.128** In the case of light damping, the displacements $x_1, x_2, x_3$, shown in Fig. 19.11 may be assumed equal to the maximum displacements. Show that the ratio of any two successive maximum displacements $x_n$ and $x_{n+1}$ is constant and that the natural logarithm of this ratio, called the *logarithmic decrement*, is

$$\ln \frac{x_n}{x_{n+1}} = \frac{2\pi(c/c_c)}{\sqrt{1 - (c/c_c)^2}}$$

**19.129** In practice, it is often difficult to determine the logarithmic decrement of a system with light damping defined in Prob. 19.128 by measuring two successive maximum displacements. Show that the logarithmic decrement can also be expressed as $(1/k) \ln (x_n/x_{n+k})$, where $k$ is the number of cycles between readings of the maximum displacement.

**19.130** In a system with light damping $(c < c_c)$, the period of vibration is commonly defined as the time interval $\tau_d = 2\pi/\omega_d$ corresponding to two successive points where the displacement-time curve touches one of the limiting curves shown in Fig. 19.11. Show that the interval of time (a) between a maximum positive displacement and the following negative displacement is $\frac{1}{2}\tau_d$, (b) between two successive zero displacements is $\frac{1}{2}\tau_d$, (c) between a maximum positive displacement and the following zero displacement is greater than $\frac{1}{4}\tau_d$.

**19.131** Successive maximum displacements of a spring-mass-dashpot system similar to that shown in Fig. 19.10 are 3.2, 1.9, and 1.1 cm. Knowing that $W = 16.3$ kg and $k = 2554$ N/m, determine (a) the damping factor $c/c_c$, (b) the value of the coefficient of viscous damping $c$. (*Hint:* See Probs. 19.128 and 19.129.)

**19.132** A 2-kg block is supported by a spring of constant $k = 128$ N/m and a dashpot with a coefficient of viscous damping $c = 0.6$ N · s/m. The block is in equilibrium when it is struck from below by a hammer which imparts to the block an upward velocity of 0.4 m/s. Determine (a) the logarithmic decrement, (b) the maximum upward displacement of the block from equilibrium after two cycles. (*Hint:* See Probs. 19.128 and 19.129.)

$k = 128$ N/m    $c = 0.6$ N·s/m

2 kg

**Fig. P19.132**

**19.133** The barrel of a field gun weighs 816 kg and is returned into firing position after recoil by a recuperator of constant $c = 19.3$ kN · s/m. Determine (a) the constant $k$ which should be used for the recuperator to return the barrel into firing position in the shortest possible time without any oscillation, (b) the time needed for the barrel to move back two-thirds of the way from its maximum-recoil position to its firing position.

**19.134** A 0.9 kg block $B$ is connected by a cord to a 2.4 kg block $A$ which is suspended as shown from two springs, each of constant $k = 180$ N/m, and a dashpot of damping coefficient $c = 7.5$ N · s/m. Knowing that the system is at rest when the cord connecting $A$ and $B$ is cut, determine the minimum tension that will occur in each spring during the resulting motion.

$k$    $c$    $k$

$A$

$B$

**Fig. P19.134 and P19.135**

**19.135** A 0.9 kg block $B$ is connected by a cord to a 2.4-kg block $A$ which is suspended as shown from two springs, each of constant $k = 180$ N/m, and a dashpot of damping coefficient $c = 60$ N · s/m. Knowing that the system is at rest when the cord connecting $A$ and $B$ is cut, determine the velocity of block $A$ after 0.1 s.

**19.136** A 1.8 kg uniform rod is supported by a pin at $O$ and a spring at $A$ and is connected to a dashpot at $B$. Determine (a) the differential equation of motion for small oscillations, (b) the angle that the rod will form with the horizontal 2.5 s after end $B$ has been pushed 23 mm down and released.

**Fig. P19.136**

**19.137** A 2.7 kg slender rod $AB$ is bolted to a 4.5 kg uniform disk. A dashpot of damping coefficient $c = 8.8$ N · s/m is attached to the disk as shown. Determine (a) the differential equation of motion for small oscillations, (b) the damping factor $c/c_c$.

**Fig. P19.137**

**19.138** A platform of weight 90.7 kg, supported by two springs each of constant $k = 44$ kN/m, is subjected to a periodic force of maximum magnitude equal to 556 N. Knowing that the coefficient of damping is 1576 N · s/m, determine (a) the natural frequency in rpm of the platform if there were no damping, (b) the frequency in rpm of the periodic force corresponding to the maximum value of the magnification factor, assuming damping, (c) the amplitude of the actual motion of the platform for each of the frequencies found in parts a and b.

**Fig. P19.138**

**19.139** Solve Prob. 19.138, assuming that the coefficient of damping is increased to 2102 N · s/m.

**19.140** In the case of the forced vibration of a system, determine the range of values of the damping factor $c/c_c$ for which the magnification factor will always decrease as the frequency ratio $\omega_f/\omega_n$ increases.

**19.141** Show that for a small value of the damping factor $c/c_c$, the maximum amplitude of a forced vibration occurs when $\omega_f \approx \omega_n$ and that the corresponding value of the magnification factor is $\dfrac{1}{2}(c_c/c)$.

**Fig. *P19.142* and P19.143**

***19.142*** A 16.3 kg motor is bolted to a light horizontal beam which has a static deflection of 0.19 cm due to the weight of the motor. Knowing that the unbalance of the rotor is equivalent to a weight of 0.02 kg located 15.8 cm. from the axis of rotation, determine the amplitude of the vibration of the motor at a speed of 900 rpm, assuming (*a*) that no damping is present, (*b*) that the damping factor $c/c_c$ is equal to 0.055.

**19.143** A 45 kg motor is bolted to a light horizontal beam which has a static deflection of 6 mm due to the weight of the motor. The unbalance of the motor is equivalent to a mass of 110 g located 75 mm from the axis of rotation. Knowing that the amplitude of the vibration of the motor is 0.25 mm at a speed of 300 rpm, determine (*a*) the damping factor $c/c_c$, (*b*) the coefficient of damping *c*.

**19.144** Solve Prob. 19.113, assuming that a dashpot having a coefficient of damping $c = 350$ N · s/m has been connected to the motor and to the ground.

**19.145** The unbalance of the rotor of a 180 kg motor is equivalent to a mass of 85 g located 150 mm from the axis of rotation. The pad which is placed between the motor and the foundation is equivalent to a spring of constant $k = 7.5$ kN/m in parallel with a dashpot of constant *c*. Knowing that the magnitude of the maximum acceleration of the motor is 9 mm/s² at a speed of 100 rpm, determine the damping factor $c/c_c$.

**Fig. *P19.146***

**Fig. P19.145**

**19.146** A 90.7 kg motor is supported by two springs, each of constant 219 kN/m, and is connected to the ground by a dashpot having a coefficient of damping $c = 7151$ N · s/m. The motor is constrained to move vertically, and the amplitude of its motion is observed to be 0.25 cm at a speed of 1200 rpm. Knowing that the weight of the rotor is 13.6 kg, determine the distance between the mass center of the rotor and the axis of the shaft.

**19.147** A machine element is supported by springs and is connected to a dashpot as shown. Show that if a periodic force of magnitude $P = P_m \sin \omega_f t$ is applied to the element, the amplitude of the fluctuating force transmitted to the foundation is

$$F_m = P_m \sqrt{\frac{1 + [2(c/c_c)(\omega_f/\omega_n)]^2}{[1 - (\omega_f/\omega_n)^2]^2 + [2(c/c_c)(\omega_f/\omega_n)]^2}}$$

**19.148** A 91 kg machine element supported by four springs, each of constant 175 N/m, is subjected to a periodic force of frequency 0.8 Hz and amplitude 89 N. Determine the amplitude of the fluctuating force transmitted to the foundation if (*a*) a dashpot with a coefficient of damping $c = 365$ N · s/m is connected to the machine and to the ground, (*b*) the dashpot is removed.

$\mathbf{P} = P_m \sin \omega_f t$

**Fig. P19.147 and *P19.148***

***19.149*** For a steady-state vibration with damping under a harmonic force, show that the mechanical energy dissipated per cycle by the dashpot is $E = \pi c x_m^2 \omega_f$, where $c$ is the coefficient of damping, $x_m$ is the amplitude of the motion, and $\omega_f$ is the circular frequency of the harmonic force.

***19.150*** The suspension of an automobile can be approximated by the simplified spring-dashpot system shown. (*a*) Write the differential equation defining the vertical displacement of the mass $m$ when the system moves at a speed $v$ over a road with a sinusoidal cross section of amplitude $\delta_m$ and wavelength $L$. (*b*) Derive an expression for the amplitude of the vertical displacement of the mass $m$.

**Fig. P19.150**

***19.151*** Two blocks $A$ and $B$, each of mass $m$, are supported as shown by three springs of the same constant $k$. Blocks $A$ and $B$ are connected by a dashpot and block $B$ is connected to the ground by two dashpots, each dashpot having the same coefficient of damping $c$. Block $A$ is subjected to a force of magnitude $P = P_m \sin \omega_f t$. Write the differential equations defining the displacements $x_A$ and $x_B$ of the two blocks from their equilibrium positions.

**Fig. P19.151**

**19.152** Express in terms of $L$, $C$, and $E$ the range of values of the resistance $R$ for which oscillations will take place in the circuit shown when switch $S$ is closed.

**Fig. P19.152**

**19.153** Consider the circuit of Prob. 19.152 when the capacitor $C$ is removed. If switch $S$ is closed at time $t = 0$, determine (a) the final value of the current in the circuit, (b) the time $t$ at which the current will have reached $(1 - 1/e)$ times its final value. (The desired value of $t$ is known as the time constant of the circuit.)

**19.154 and 19.155** Draw the electrical analogue of the mechanical system shown. (*Hint:* Draw the loops corresponding to the free bodies $m$ and $A$.)

**Fig. P19.154 and P19.156**          **Fig. P19.155 and P19.157**

**19.156 and 19.157** Write the differential equations defining (a) the displacements of the mass $m$ and of the point $A$, (b) the charges on the capacitors of the electrical analogue.

This chapter was devoted to the study of *mechanical vibrations,* that is, to the analysis of the motion of particles and rigid bodies oscillating about a position of equilibrium. In the first part of the chapter [Secs. 19.2 through 19.7], we considered *vibrations without damping,* while the second part was devoted to *damped vibrations* [Secs. 19.8 through 19.10].

In Sec. 19.2, we considered the *free vibrations of a particle,* that is, the motion of a particle $P$ subjected to a restoring force proportional to the displacement of the particle—such as the force exerted by a spring. If the displacement $x$ of the particle $P$ is measured from its equilibrium position $O$ (Fig. 19.17), the resultant $\mathbf{F}$ of the forces acting on $P$ (including its weight) has a magnitude $kx$ and is directed toward $O$. Applying Newton's second law $F = ma$ and recalling that $a = \ddot{x}$, we wrote the differential equation

$$m\ddot{x} + kx = 0 \qquad (19.2)$$

or, setting $\omega_n^2 = k/m$,

$$\ddot{x} + \omega_n^2 x = 0 \qquad (19.6)$$

Free vibrations of a particle

**Fig. 19.17**

The motion defined by this equation is called a *simple harmonic motion.*

The solution of Eq. (19.6), which represents the displacement of the particle $P$, was expressed as

$$x = x_m \sin(\omega_n t + \phi) \qquad (19.10)$$

where $x_m$ = amplitude of the vibration

$\omega_n = \sqrt{k/m}$ = natural circular frequency

$\phi$ = phase angle

The *period of the vibration* (that is, the time required for a full cycle) and its *natural frequency* (that is, the number of cycles per second) were expressed as

$$\text{Period} = \tau_n = \frac{2\pi}{\omega_n} \qquad (19.13)$$

$$\text{Natural frequency} = f_n = \frac{1}{\tau_n} = \frac{\omega_n}{2\pi} \qquad (19.14)$$

The velocity and acceleration of the particle were obtained by differentiating Eq. (19.10), and their maximum values were found to be

$$v_m = x_m \omega_n \qquad a_m = x_m \omega_n^2 \qquad (19.15)$$

Since all the above parameters depend directly upon the natural circular frequency $\omega_n$ and thus upon the ratio $k/m$, it is essential in any given problem to calculate the value of the constant $k$; this can be done by determining the relation between the restoring force and the corresponding displacement of the particle [Sample Prob. 19.1].

It was also shown that the oscillatory motion of the particle $P$ can be represented by the projection on the $x$ axis of the motion of a point $Q$ describing an auxiliary circle of radius $x_m$ with the constant angular velocity $\omega_n$ (Fig. 19.18). The instantaneous values of the velocity and acceleration of $P$ can then be obtained by projecting on the $x$ axis the vectors $\mathbf{v}_m$ and $\mathbf{a}_m$ representing, respectively, the velocity and acceleration of $Q$.

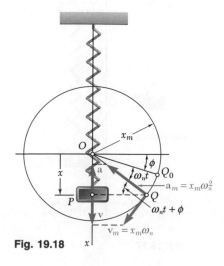

**Fig. 19.18**

While the motion of a *simple pendulum* is not truly a simple harmonic motion, the formulas given above can be used with $\omega_n^2 = g/l$ to calculate the period and natural frequency of the *small oscillations* of a simple pendulum [Sec. 19.3]. Large-amplitude oscillations of a simple pendulum were discussed in Sec. 19.4.

Simple pendulum

The *free vibrations of a rigid body* can be analyzed by choosing an appropriate variable, such as a distance $x$ or an angle $\theta$, to define the position of the body, drawing a free-body-diagram equation to express the equivalence of the external and effective forces, and writing an equation relating the selected variable and its second derivative [Sec. 19.5]. If the equation obtained is of the form

Free vibrations of a rigid body

$$\ddot{x} + \omega_n^2 x = 0 \qquad \text{or} \qquad \ddot{\theta} + \omega_n^2 \theta = 0 \qquad (19.21)$$

the vibration considered is a simple harmonic motion and its period and natural frequency can be obtained *by identifying* $\omega_n$ and substituting its value into Eqs. (19.13) and (19.14) [Sample Probs. 19.2 and 19.3].

The *principle of conservation of energy* can be used as an alternative method for the determination of the period and natural frequency of the simple harmonic motion of a particle or rigid body [Sec. 19.6]. Choosing again an appropriate variable, such as $\theta$, to define the position of the system, we express that the total energy of the system is conserved, $T_1 + V_1 = T_2 + V_2$, between the position of maximum displacement ($\theta_1 = \theta_m$) and the position of maximum velocity ($\dot{\theta}_2 = \dot{\theta}_m$). If the motion considered is simple harmonic, the two members of the equation obtained consist of homogeneous quadratic expressions in $\theta_m$ and $\dot{\theta}_m$, respectively.† Substituting $\dot{\theta}_m = \theta_m \omega_n$ in this equation, we can factor out $\theta_m^2$ and solve for the circular frequency $\omega_n$ [Sample Prob. 19.4].

Using the principle of conservation of energy

In Sec. 19.7, we considered the *forced vibrations* of a mechanical system. These vibrations occur when the system is subjected to a periodic force (Fig. 19.19) or when it is elastically connected to a support which has an alternating motion (Fig. 19.20). Denoting by $\omega_f$ the forced circular frequency, we found that in the first case, the motion of the system was defined by the differential equation

Forced vibrations

$$m\ddot{x} + kx = P_m \sin \omega_f t \qquad (19.30)$$

and that in the second case it was defined by the differential equation

$$m\ddot{x} + kx = k\delta_m \sin \omega_f t \qquad (19.31)$$

The general solution of these equations is obtained by adding a particular solution of the form

$$x_{\text{part}} = x_m \sin \omega_f t \qquad (19.32)$$

---

†If the motion considered can only be *approximated* by a simple harmonic motion, such as for the small oscillations of a body under gravity, the potential energy must be approximated by a quadratic expression in $\theta_m$.

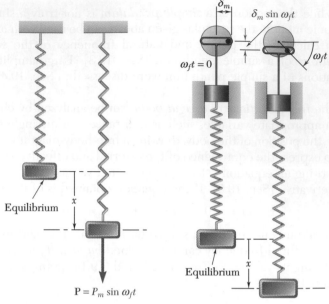

**Fig. 19.19**  **Fig. 19.20**

to the general solution of the corresponding homogeneous equation. The particular solution (19.32) represents a *steady-state vibration* of the system, while the solution of the homogeneous equation represents a *transient free vibration* which can generally be neglected.

Dividing the amplitude $x_m$ of the steady-state vibration by $P_m/k$ in the case of a periodic force, or by $\delta_m$ in the case of an oscillating support, we defined the *magnification factor* of the vibration and found that

$$\text{Magnification factor} = \frac{x_m}{P_m/k} = \frac{x_m}{\delta_m} = \frac{1}{1 - (\omega_f/\omega_n)^2} \quad (19.36)$$

According to Eq. (19.36), the amplitude $x_m$ of the forced vibration *becomes infinite when $\omega_f = \omega_n$*, that is, *when the forced frequency is equal to the natural frequency of the system*. The impressed force or impressed support movement is then said to be in *resonance* with the system [Sample Prob. 19.5]. Actually the amplitude of the vibration remains finite, due to damping forces.

Damped free vibrations

In the last part of the chapter, we considered the *damped vibrations* of a mechanical system. First, we analyzed the *damped free vibrations* of a system with *viscous damping* [Sec. 19.8]. We found that the motion of such a system was defined by the differential equation

$$m\ddot{x} + c\dot{x} + kx = 0 \quad (19.38)$$

where $c$ is a constant called the *coefficient of viscous damping*. Defining the *critical damping coefficient* $c_c$ as

$$c_c = 2m \sqrt{\frac{k}{m}} = 2m\omega_n \qquad (19.41)$$

where $\omega_n$ is the natural circular frequency of the system in the absence of damping, we distinguished three different cases of damping, namely, (1) *heavy damping*, when $c > c_c$; (2) *critical damping*, when $c = c_c$; and (3) *light damping*, when $c < c_c$. In the first two cases, the system when disturbed tends to regain its equilibrium position without any oscillation. In the third case, the motion is vibratory with diminishing amplitude.

In Sec. 19.9, we considered the *damped forced vibrations* of a mechanical system. These vibrations occur when a system with viscous damping is subjected to a periodic force $\mathbf{P}$ of magnitude $P = P_m \sin \omega_f t$ or when it is elastically connected to a support with an alternating motion $\delta = \delta_m \sin \omega_f t$. In the first case, the motion of the system was defined by the differential equation

Damped forced vibrations

$$m\ddot{x} + c\dot{x} + kx = P_m \sin \omega_f t \qquad (19.47)$$

and in the second case by a similar equation obtained by replacing $P_m$ by $k\delta_m$ in (19.47).

The *steady-state vibration* of the system is represented by a particular solution of Eq. (19.47) of the form

$$x_{\text{part}} = x_m \sin (\omega_f t - \varphi) \qquad (19.48)$$

Dividing the amplitude $x_m$ of the steady-state vibration by $P_m/k$ in the case of a periodic force, or by $\delta_m$ in the case of an oscillating support, we obtained the following expression for the magnification factor:

$$\frac{x_m}{P_m/k} = \frac{x_m}{\delta_m} = \frac{1}{\sqrt{[1 - (\omega_f/\omega_n)^2]^2 + [2(c/c_c)(\omega_f/\omega_n)]^2}} \qquad (19.53)$$

where $\omega_n = \sqrt{k/m}$ = natural circular frequency
of undamped system
$c_c = 2m\omega_n$ = critical damping coefficient
$c/c_c$ = damping factor

We also found that the *phase difference* $\varphi$ between the impressed force or support movement and the resulting steady-state vibration of the damped system was defined by the relation

$$\tan \varphi = \frac{2(c/c_c)(\omega_f/\omega_n)}{1 - (\omega_f/\omega_n)^2} \qquad (19.54)$$

The chapter ended with a discussion of *electrical analogues* [Sec. 19.10], in which it was shown that the vibrations of mechanical systems and the oscillations of electrical circuits are defined by the same differential equations. Electrical analogues of mechanical systems can therefore be used to study or predict the behavior of these systems.

Electrical analogues

**19.158**  The bob of a simple pendulum of length $l = 102$ cm is released from rest when $\theta = +5°$. Assuming simple harmonic motion, determine 1.6 s after release (a) the angle $\theta$, (b) the magnitudes of the velocity and acceleration of the bob.

**Fig. P19.158**

**19.159**  A 50 kg block is supported by the spring arrangement shown. The block is moved vertically downward from its equilibrium position and released. Knowing that the amplitude of the resulting motion is 60 mm, determine (a) the period and frequency of the motion, (b) the maximum velocity and maximum acceleration of the block.

**19.160**  A rod of mass $m$ and length $L$ rests on two pulleys $A$ and $B$ which rotate in opposite directions as shown. Denoting by $\mu_k$ the coefficient of kinetic friction between the rod and the pulleys, determine the frequency of vibration if the rod is given a small displacement to the right and released.

24 kN/m

24 kN/m

50 kg

**Fig. *P19.159***

**Fig. P19.160**

600 mm

**Fig. P19.161**

**19.161**  A 270 kg flywheel has a diameter of 1.2 m and a radius of gyration of 0.5 m. A belt is placed over the rim and attached to two springs, each of constant $k = 13$ kN/m. The initial tension in the belt is sufficient to prevent slipping. If the end $C$ of the belt is pulled 25 mm to the right and released, determine (a) the period of vibration, (b) the maximum angular velocity of the flywheel.

**19.162** Two small weights $w$ are attached at $A$ and $B$ to the rim of a uniform disk of radius $r$ and weight $W$. Denoting by $\tau_0$ the period of small oscillations when $\beta = 0$, determine the angle $\beta$ for which the period of small oscillations is $2\tau_0$.

**19.163** Two uniform rods, each of weight $W = 11$ kg and length $L = 102$ cm, are welded together to form the assembly shown. Knowing that the constant of each spring is $k = 730$ N/m and that end $A$ is given a small displacement and released, determine the frequency of the resulting motion.

**Fig. P19.162**

**Fig. *P19.163***

**19.164** A 2.7 kg slender rod $AB$ is bolted to a 4.5 kg uniform disk. A spring of constant 365 N/m is attached to the disk and is unstretched in the position shown. If end $B$ of the rod is given a small displacement and released, determine the period of vibration of the system.

**Fig. P19.164**

**19.165** Two uniform rods $AB$ and $CD$, each of length $l$ and mass $m$, are attached to gears as shown. Knowing that the mass of gear $C$ is $m$ and that the mass of gear $A$ is $4m$, determine the period of small oscillations of the system.

**Fig. P19.165**

**19.166** An inverted pendulum consisting of a sphere of weight $W$ and a bar $ABC$ of length $l$ and negligible weight is supported by a pin and bracket at $C$. A spring of constant $k$ is attached to the bar at $B$ and is undeformed when the bar is in the vertical position shown. Determine (a) the frequency of small oscillations, (b) the smallest value of $a$ for which the oscillations will occur.

**Fig. P19.166**

**19.167** A 29 kg block is attached to a spring of constant $k = 14.6$ kN/m and can move without friction in a vertical slot as shown. It is acted upon by a periodic force of magnitude $P = P_m \sin \omega_f t$, where $\omega_f = 10$ rad/s. Knowing that the amplitude of the motion is 1.9 cm, determine $P_m$.

**Fig. P19.167**

**19.168** A 360-g ball is connected to a paddle by means of an elastic cord $AB$ of constant $k = 70$ N/m. Knowing that the paddle is moved vertically according to the relation $\delta = \delta_m \sin \omega_f t$, with an amplitude $\delta_m = 200$ mm and a frequency $f_f = 0.5$ Hz, determine the steady-state amplitude of the motion of the ball.

**Fig. P19.168**

**19.169** A uniform rod of mass $m$ is supported by a pin at $A$ and a spring of constant $k$ at $B$ and is connected at $D$ to a dashpot of damping coefficient $c$. Determine in terms of $m$, $k$, and $c$, for small oscillations, (a) the differential equation of motion, (b) the critical damping coefficient $c_c$.

**Fig. P19.169**

# Computer Problems

**19.C1** The force-deflection equation for a class of nonlinear springs fixed at one end is $F = 0.5x^{1/n}$ where $F$ is the magnitude, expressed in pounds, of the force applied at the other end of the spring and $x$ is the deflection expressed in meter. Knowing that a block of weight $W$ is suspended from the spring and is given a small downward displacement from its equilibrium position, use computational software to calculate and plot the frequency of vibration of the block for values of $W$ equal to 0.23, 0.34 and .45 kg and values of $n$ from 1 to 2. Assume that the slope of the force deflection curve at the point corresponding to $F = W$ can be used as an equivalent spring constant.

**19.C2**   The slender rod $AB$ of weight $W_{AB}$ is attached to two identical collars, each of weight $W$. The system lies in a horizontal plane and is in equilibrium in the position shown when the collar $A$ is given a small displacement and released. Calculate and plot the period of vibration for values of the angle $\beta$ from 10° to 80° for weight ratios $W/W_{AB}$ of 0, 2 and 4.

**Fig. P19.C2**

**19.C3**   A 13.6 kg motor is supported by four springs, each of constant 61.3 kN/m. The unbalance of the motor is equivalent to a weight of 0.02 kg located 12.7 cm from the axis of rotation. Knowing that the motor is constrained to move vertically, use computational software to calculate and plot the amplitude of the vibration and the maximum acceleration of the motor for motor speeds from 0 to 1250 rpm and from 1300 to 2500 rpm.

**19.C4**   Solve Prob. 19.C3, assuming that a dashpot having a coefficient of damping $c = 2335$ N · s/m has been connected to the motor base and to the ground. Consider motor speeds from 0 to 2500 rpm.

**19.C5**   A machine element supported by springs and connected to a dashpot is subjected to a periodic force of magnitude $P = P_m \sin w_f t$. The *transmissibility* $T_m$ of the system is defined as the ratio $F_m/P_m$ of the maximum value $F_m$ of the fluctuating periodic force transmitted to the foundation to the maximum value $P_m$ of the periodic force applied to the machine element. Use computational software to calculate and plot the value of $T_m$ for frequency ratio $w_f/w_n$ from 0 to 5 and for damping factors $c/c_c$ equal to 0.1, 0.4, 0.7, and 1. (*Hint:* Use the formula given in Prob. 19.147.)

$\mathbf{P} = P_m \sin \omega_f t$

**Fig. P19.C5**

**19.C6**  The suspension of an automobile can be approximated by the simplified spring-and-dashpot system shown. Write the differential equation defining the vertical motion of the mass relative to the road when the system moves at a speed $v$ over a road with a sinusoidal cross section of amplitude $\delta_m$ and wave length $L$. Determine and plot the amplitude of the displacement of the mass relative to the road as a function of the speed $v$ for damping factors $c/c_c = 0.1$, 0.5, and 0.9.

**Fig. P19.C6**

**19.C7**  A block of mass $m = 90$ kg, supported by two springs of constants $k_1 = k_2 = 85$ kN/m, is subjected to a periodic force of maximum magnitude $P_m = 500$ N. The coefficient of damping is $c = 2500$ N · s/m. (*a*) Calculate and plot the amplitude of the steady-state vibration of the block for values of $w_f$ from 0 to 100 rad/s. (*b*) Determine the maximum amplitude and the corresponding frequency.

**Fig. P19.C7**

# A P P E N D I X

# Some Useful Definitions and Properties of Vector Algebra

The following definitions and properties of vector algebra were discussed fully in Chaps. 2 and 3 of *Vector Mechanics for Engineers: Statics*. They are summarized here for the convenience of the reader, with references to the appropriate sections of the *Statics* volume. Equation and illustration numbers are those used in the original presentation.

## A.1. ADDITION OF VECTORS (SECS. 2.3 AND 2.4)

Vectors are defined as *mathematical expressions possessing magnitude and direction, which add according to the parallelogram law*. Thus the sum of two vectors **P** and **Q** is obtained by attaching the two vectors to the same point $A$ and constructing a parallelogram, using **P** and **Q** as two sides of the parallelogram (Fig A.2). The diagonal that passes through $A$ represents the sum of the vectors **P** and **Q**, and this sum is denoted by **P** + **Q**. Vector addition is *associative* and *commutative*.

**Fig. A.1**

**Fig. A.2**

The *negative vector* of a given vector **P** is defined as a vector having the same magnitude $P$ and a direction opposite to that of **P** (Fig. A.1); the negative of the vector **P** is denoted by −**P**. Clearly, we have

$$\mathbf{P} + (-\mathbf{P}) = 0$$

**1289**

## A.2. PRODUCT OF A SCALAR AND A VECTOR (SEC. 2.4)

The product $k\mathbf{P}$ of a scalar $k$ and a vector $\mathbf{P}$ is defined as a vector having the same direction as $\mathbf{P}$ (if $k$ is positive), or a direction opposite to that of $\mathbf{P}$ (if $k$ is negative), and a magnitude equal to the product of the magnitude $P$ and the absolute value of $k$ (Fig. A.3).

**Fig. A.3**

## A.3. UNIT VECTORS. RESOLUTION OF A VECTOR INTO RECTANGULAR COMPONENTS (SECS. 2.7 AND 2.12)

The vectors $\mathbf{i}$, $\mathbf{j}$, and $\mathbf{k}$, called *unit vectors*, are defined as vectors of magnitude 1, directed, respectively, along the positive $x$, $y$, and $z$ axes (Fig. A.4).

**Fig. A.4**

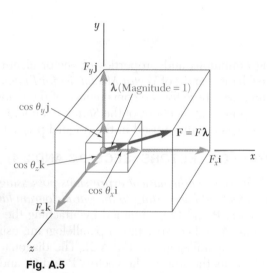

**Fig. A.5**

Denoting by $F_x$, $F_y$, and $F_z$, the scalar components of a vector $\mathbf{F}$, we have (Fig. A.5)

$$\mathbf{F} = F_x\mathbf{i} + F_y\mathbf{j} + F_z\mathbf{k} \tag{2.20}$$

In the particular case of a unit vector $\boldsymbol{\lambda}$ directed along a line forming angles $\theta_x$, $\theta_y$, and $\theta_z$ with the coordinate axes, we have

$$\boldsymbol{\lambda} = \cos\theta_x\mathbf{i} + \cos\theta_y\mathbf{j} + \cos\theta_z\mathbf{k} \tag{2.22}$$

## A.4. VECTOR PRODUCT OF TWO VECTORS (SECS. 3.4 AND 3.5)

The vector product, or *cross product*, of two vectors $\mathbf{P}$ and $\mathbf{Q}$ is defined as the vector

$$\mathbf{V} = \mathbf{P} \times \mathbf{Q}$$

which satisfies the following conditions:

1.  The line of action of **V** is perpendicular to the plane containing **P** and **Q** (Fig. A.6).
2.  The magnitude of **V** is the product of the magnitudes of **P** and **Q** and of the sine of the angle $\theta$ formed by **P** and **Q** (the measure of which will always be 180° or less); we thus have

$$V = PQ \sin \theta \qquad (3.1)$$

3.  The direction of **V** is obtained from the *right-hand rule*. Close your right hand and hold it so that your fingers are curled in the same sense as the rotation through $\theta$ which brings the vector **P** in line with the vector **Q**; your thumb will then indicate the direction of the vector **V** (Fig. A.6b). Note that if **P** and **Q** do not have a common point of application, they should first be redrawn from the same point. The three vectors **P**, **Q**, and **V**—taken in that order—are said to form a *right-handed triad*.

Vector products are *distributive* but *not commutative*. We have

$$\mathbf{Q} \times \mathbf{P} = -(\mathbf{P} \times \mathbf{Q}) \qquad (3.4)$$

**Vector Products of Unit Vectors.** It follows from the definition of the vector product of two vectors that

$$
\begin{array}{lll}
\mathbf{i} \times \mathbf{i} = 0 & \mathbf{j} \times \mathbf{i} = -\mathbf{k} & \mathbf{k} \times \mathbf{i} = \mathbf{j} \\
\mathbf{i} \times \mathbf{j} = \mathbf{k} & \mathbf{j} \times \mathbf{j} = 0 & \mathbf{k} \times \mathbf{j} = -\mathbf{i} \quad (3.7) \\
\mathbf{i} \times \mathbf{k} = -\mathbf{j} & \mathbf{j} \times \mathbf{k} = \mathbf{i} & \mathbf{k} \times \mathbf{k} = 0
\end{array}
$$

**Rectangular Components of Vector Product.** Resolving the vectors **P** and **Q** into rectangular components, we obtain the following expressions for the components of their vector product **V**:

$$
\begin{aligned}
V_x &= P_y Q_z - P_z Q_y \\
V_y &= P_z Q_x - P_x Q_z \\
V_z &= P_x Q_y - P_y Q_x
\end{aligned} \qquad (3.9)
$$

In determinant form, we have

$$\mathbf{V} = \mathbf{P} \times \mathbf{Q} = \begin{vmatrix} \mathbf{i} & \mathbf{j} & \mathbf{k} \\ P_x & P_y & P_z \\ Q_x & Q_y & Q_z \end{vmatrix} \qquad (3.10)$$

## A.5. MOMENT OF A FORCE ABOUT A POINT (SECS. 3.6 AND 3.8)

The moment of a force **F** (or, more generally, of a vector **F**) about a point $O$ is defined as the vector product

$$\mathbf{M}_O = \mathbf{r} \times \mathbf{F} \qquad (3.11)$$

where **r** denotes the *position vector* of the point of application $A$ of **F** (Fig. A.7a).

According to the definition of the vector product of two vectors given in Sec. A.4, the moment $\mathbf{M}_O$ must be perpendicular to the plane containing $O$ and the force **F**. Its magnitude is equal to

$$M_O = rF \sin \theta = Fd \qquad (3.12)$$

(a)

(b)

**Fig. A.6**

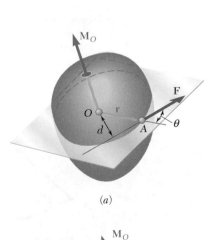

(a)

(b)

**Fig. A.7**

where $d$ is the perpendicular distance from $O$ to the line of action of $\mathbf{F}$, and its sense is defined by the sense of the rotation which would bring the vector $\mathbf{r}$ in line with the vector $\mathbf{F}$; this rotation should be viewed as *counterclockwise* by an observer located at the tip of $\mathbf{M}_O$. Another way of defining the sense of $\mathbf{M}_O$ is furnished by a variation of the *right-hand rule:* Close your right hand and hold it so that your fingers are curled in the sense of the rotation that $\mathbf{F}$ would impart to the rigid body about a fixed axis directed along the line of action of $\mathbf{M}_O$; your thumb will indicate the sense of the moment $\mathbf{M}_O$ (Fig. A.7*b*).

**Rectangular Components of the Moment of a Force.** Denoting by $x$, $y$, and $z$ the coordinates of the point of application $A$ of $\mathbf{F}$, we obtain the following expressions for the components of the moment $\mathbf{M}_O$ of $\mathbf{F}$:

$$
\begin{aligned}
M_x &= yF_z - zF_y \\
M_y &= zF_x - xF_z \\
M_z &= xF_y - yF_x
\end{aligned}
\tag{3.18}
$$

In determinant form, we have

$$
\mathbf{M}_O = \mathbf{r} \times \mathbf{F} = \begin{vmatrix} \mathbf{i} & \mathbf{j} & \mathbf{k} \\ x & y & z \\ F_x & F_y & F_z \end{vmatrix}
\tag{3.19}
$$

To compute the moment $\mathbf{M}_B$ about an arbitrary point $B$ of a force $\mathbf{F}$ applied at $A$, we must use the vector $\mathbf{r}_{A/B} = \mathbf{r}_A - \mathbf{r}_B$ drawn from $B$ to $A$ instead of the vector $\mathbf{r}$. We write

$$
\mathbf{M}_B = \mathbf{r}_{A/B} \times \mathbf{F} = (\mathbf{r}_A - \mathbf{r}_B) \times \mathbf{F}
\tag{3.20}
$$

or, using the determinant form,

$$
\mathbf{M}_B = \begin{vmatrix} \mathbf{i} & \mathbf{j} & \mathbf{k} \\ x_{A/B} & y_{A/B} & z_{A/B} \\ F_x & F_y & F_z \end{vmatrix}
\tag{3.21}
$$

where $x_{A/B}$, $y_{A/B}$, $z_{A/B}$ are the components of the vector $\mathbf{r}_{A/B}$:

$$
x_{A/B} = x_A - x_B \qquad y_{A/B} = y_A - y_B \qquad z_{A/B} = z_A - z_B
$$

## A.6. SCALAR PRODUCT OF TWO VECTORS (SEC. 3.9)

The scalar product, or *dot product*, of two vectors $\mathbf{P}$ and $\mathbf{Q}$ is defined as the product of the magnitudes of $\mathbf{P}$ and $\mathbf{Q}$ and of the cosine of the angle $\theta$ formed by $\mathbf{P}$ and $\mathbf{Q}$ (Fig. A.8). The scalar product of $\mathbf{P}$ and $\mathbf{Q}$ is denoted by $\mathbf{P} \cdot \mathbf{Q}$. We write

$$
\mathbf{P} \cdot \mathbf{Q} = PQ \cos \theta
\tag{3.24}
$$

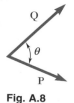

**Fig. A.8**

Scalar products are *commutative* and *distributive*.

**Scalar Products of Unit Vectors.** It follows from the definition of the scalar product of two vectors that

$$
\begin{aligned}
\mathbf{i} \cdot \mathbf{i} &= 1 & \mathbf{j} \cdot \mathbf{j} &= 1 & \mathbf{k} \cdot \mathbf{k} &= 1 \\
\mathbf{i} \cdot \mathbf{j} &= 0 & \mathbf{j} \cdot \mathbf{k} &= 0 & \mathbf{k} \cdot \mathbf{i} &= 0
\end{aligned}
\tag{3.29}
$$

**Scalar Product Expressed in Terms of Rectangular Components.** Resolving the vectors $\mathbf{P}$ and $\mathbf{Q}$ into rectangular components, we obtain

$$\mathbf{P} \cdot \mathbf{Q} = P_x Q_x + P_y Q_y + P_z Q_z \tag{3.30}$$

**Angle Formed by Two Vectors.** It follows from (3.24) and (3.29) that

$$\cos \theta = \frac{\mathbf{P} \cdot \mathbf{Q}}{PQ} = \frac{P_x Q_x + P_y Q_y + P_z Q_z}{PQ} \tag{3.32}$$

**Projection of a Vector on a Given Axis.** The projection of a vector $\mathbf{P}$ on the axis $OL$ defined by the unit vector $\boldsymbol{\lambda}$ (Fig. A.9) is

$$P_{OL} = OA = \mathbf{P} \cdot \boldsymbol{\lambda} \tag{3.36}$$

**Fig. A.9**

## A.7. MIXED TRIPLE PRODUCT OF THREE VECTORS (SEC. 3.10)

The mixed triple product of the three vectors $\mathbf{S}, \mathbf{P}$, and $\mathbf{Q}$ is defined as the scalar expression

$$\mathbf{S} \cdot (\mathbf{P} \times \mathbf{Q}) \tag{3.38}$$

obtained by forming the scalar product of $\mathbf{S}$ with the vector product of $\mathbf{P}$ and $\mathbf{Q}$. Mixed triple products are invariant under *circular permutations* but change sign under any other permutation:

$$\mathbf{S} \cdot (\mathbf{P} \times \mathbf{Q}) = \mathbf{P} \cdot (\mathbf{Q} \times \mathbf{S}) = \mathbf{Q} \cdot (\mathbf{S} \times \mathbf{P})$$
$$= -\mathbf{S} \cdot (\mathbf{Q} \times \mathbf{P}) = -\mathbf{P} \cdot (\mathbf{S} \times \mathbf{Q}) = -\mathbf{Q} \cdot (\mathbf{P} \times \mathbf{S}) \tag{3.39}$$

**Mixed Triple Product Expressed in Terms of Rectangular Components.** The mixed triple product of $\mathbf{S}, \mathbf{P}$, and $\mathbf{Q}$ may be expressed in the form of a determinant

$$\mathbf{S} \cdot (\mathbf{P} \times \mathbf{Q}) = \begin{vmatrix} S_x & S_y & S_z \\ P_x & P_y & P_z \\ Q_x & Q_y & Q_z \end{vmatrix} \tag{3.41}$$

The mixed triple product $\mathbf{S} \cdot (\mathbf{P} \times \mathbf{Q})$ measures the volume of the parallelepiped having the vectors $\mathbf{S}, \mathbf{P}$, and $\mathbf{Q}$ for sides (Fig. A.10)

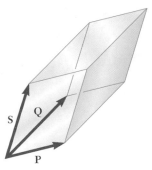

**Fig. A.10**

## A.8. MOMENT OF A FORCE ABOUT A GIVEN AXIS (SEC. 3.11)

The moment $M_{OL}$ of a force $\mathbf{F}$ (or, more generally, of a vector $\mathbf{F}$) about an axis $OL$ is defined as the projection $OC$ on the axis $OL$ of the moment $\mathbf{M}_O$ of $\mathbf{F}$ about $O$ (Fig. A.11). Denoting by $\boldsymbol{\lambda}$ the unit vector along $OL$, we have

$$M_{OL} = \boldsymbol{\lambda} \cdot \mathbf{M}_O = \boldsymbol{\lambda} \cdot (\mathbf{r} \times \mathbf{F}) \tag{3.42}$$

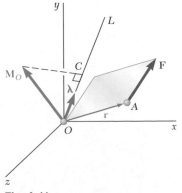

**Fig. A.11**

or, in determinant form,

$$M_{OL} = \begin{vmatrix} \lambda_x & \lambda_y & \lambda_z \\ x & y & z \\ F_x & F_y & F_z \end{vmatrix} \tag{3.43}$$

where $\lambda_x, \lambda_y, \lambda_z$ = direction cosines of axis $OL$
$\quad\quad x, y, z$ = coordinates of point of application of **F**
$\quad F_x, F_y, F_z$ = components of force **F**

The moments of the force **F** about the three coordinate axes are given by the expressions (3.18) obtained earlier for the rectangular components of the moment $\mathbf{M}_O$ of **F** about $O$:

$$\begin{aligned} M_x &= yF_z - zF_y \\ M_y &= zF_x - xF_z \\ M_z &= xF_y - yF_x \end{aligned} \tag{3.18}$$

**Fig. A.12**

More generally, the moment of a force **F** applied at $A$ about an axis which does not pass through the origin is obtained by choosing an arbitrary point $B$ on the axis (Fig. A.12) and determining the projection on the axis $BL$ of the moment $\mathbf{M}_B$ of **F** about $B$. We write

$$M_{BL} = \boldsymbol{\lambda} \cdot \mathbf{M}_B = \boldsymbol{\lambda} \cdot (\mathbf{r}_{A/B} \times \mathbf{F}) \tag{3.45}$$

where $\mathbf{r}_{A/B} = \mathbf{r}_A - \mathbf{r}_B$ represents the vector drawn from $B$ to $A$. Expressing $M_{BL}$ in the form of a determinant, we have

$$M_{BL} = \begin{vmatrix} \lambda_x & \lambda_y & \lambda_z \\ x_{A/B} & y_{A/B} & z_{A/B} \\ F_x & F_y & F_z \end{vmatrix} \tag{3.46}$$

where $\lambda_x, \lambda_y, \lambda_z$ = direction cosines of axis $BL$

$$x_{A/B} = x_A - x_B, y_{A/B} = y_A - y_B, z_{A/B} = z_A - z_B$$
$$F_x, F_y, F_z = \text{components of force } \mathbf{F}$$

It should be noted that the result obtained is independent of the choice of the point $B$ on the given axis; the same result would have been obtained if point $C$ had been chosen instead of $B$.

# B

# Moments of Inertia of Masses

# MOMENTS OF INERTIA OF MASSES

## B.1. MOMENT OF INERTIA OF A MASS

Consider a small mass $\Delta m$ mounted on a rod of negligible mass which can rotate freely about an axis $AA'$ (Fig. B.1$a$). If a couple is applied to the system, the rod and mass, assumed to be initially at rest, will start rotating about $AA'$. The details of this motion will be studied later in dynamics. At present, we wish only to indicate that the time required for the system to reach a given speed of rotation is proportional to the mass $\Delta m$ and to the square of the distance $r$. The product $r^2 \Delta m$ provides, therefore, a measure of the *inertia* of the system, that is, a measure of the resistance the system offers when we try to set it in motion. For this reason, the product $r^2 \Delta m$ is called the *moment of inertia* of the mass $\Delta m$ with respect to the axis $AA'$.

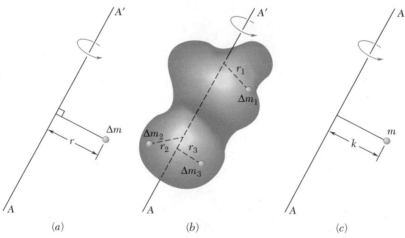

**Fig. B.1**

Consider now a body of mass $m$ which is to be rotated about an axis $AA'$ (Fig. B.1$b$). Dividing the body into elements of mass $\Delta m_1$, $\Delta m_2$, etc., we find that the body's resistance to being rotated is measured by the sum $r_1^2 \Delta m_1 + r_2^2 \Delta m_2 + \cdots$. This sum defines, therefore, the moment of inertia of the body with respect to the axis $AA'$. Increasing the number of elements, we find that the moment of inertia is equal, in the limit, to the integral

$$I = \int r^2 \, dm \tag{B.1}$$

The *radius of gyration k* of the body with respect to the axis $AA'$ is defined by the relation

$$I = k^2 m \quad \text{or} \quad k = \sqrt{\frac{I}{m}} \quad \text{(B.2)}$$

The radius of gyration $k$ represents, therefore, the distance at which the entire mass of the body should be concentrated if its moment of inertia with respect to $AA'$ is to remain unchanged (Fig. B.1c). Whether it is kept in its original shape (Fig. B.1b) or whether it is concentrated as shown in Fig. B.1c, the mass $m$ will react in the same way to a rotation, or *gyration,* about $AA'$.

If SI units are used, the radius of gyration $k$ is expressed in meters and the mass $m$ in kilograms, and thus the unit used for the moment of inertia of a mass is $kg \cdot m^2$. If U.S. customary units are used, the radius of gyration is expressed in feet and the mass in slugs (that is, in $lb \cdot s^2/ft$), and thus the derived unit used for the moment of inertia of a mass is $lb \cdot ft \cdot s^2$.†

The moment of inertia of a body with respect to a coordinate axis can easily be expressed in terms of the coordinates $x$, $y$, and $z$ of the element of mass $dm$ (Fig. B.2). Noting, for example, that the square of the distance $r$ from the element $dm$ to the $y$ axis is $z^2 + x^2$, we express the moment of inertia of the body with respect to the $y$ axis as

$$I_y = \int r^2 \, dm = \int (z^2 + x^2) \, dm$$

Similar expressions can be obtained for the moments of inertia with respect to the $x$ and $z$ axes. We write

$$I_x = \int (y^2 + z^2) \, dm$$

$$I_y = \int (z^2 + x^2) \, dm \quad \text{(B.3)}$$

$$I_z = \int (x^2 + y^2) \, dm$$

**Fig. B.2**

**Photo B.1** As you will discuss in your dynamics course, the rotational behavior of the camshaft shown is dependent upon the mass moment of inertia of the camshaft with respect to its axis of rotation.

---

†It should be kept in mind when converting the moment of inertia of a mass from U.S. customary units to SI units that the base unit *pound* used in the derived unit $lb \cdot ft \cdot s^2$ is a unit of force (*not* of mass) and should therefore be converted into newtons. We have

$$1 \, lb \cdot ft \cdot s^2 = (4.45 \, N)(0.3048 \, m)(1 \, s)^2 = 1.356 \, N \cdot m \cdot s^2$$

or, since $1 \, N = 1 \, kg \cdot m/s^2$,

$$1 \, lb \cdot ft \cdot s^2 = 1.356 \, kg \cdot m^2$$

**Fig. B.3**

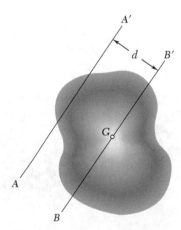

**Fig. B.4**

## B.2. PARALLEL-AXIS THEOREM

Consider a body of mass $m$. Let $Oxyz$ be a system of rectangular co-ordinates whose origin is at the arbitrary point $O$, and $Gx'y'z'$ a system of parallel *centroidal axes,* that is, a system whose origin is at the center of gravity $G$ of the body† and whose axes $x'$, $y'$, and $z'$ are parallel to the $x$, $y$, and $z$ axes, respectively (Fig. B.3). Denoting by $\bar{x}$, $\bar{y}$, and $\bar{z}$ the coordinates of $G$ with respect to $Oxyz$, we write the following relations between the coordinates $x$, $y$, and $z$ of the element $dm$ with respect to $Oxyz$ and its coordinates $x'$, $y'$, and $z'$ with respect to the centroidal axes $Gx'y'z'$:

$$x = x' + \bar{x} \qquad y = y' + \bar{y} \qquad z = z' + \bar{z} \tag{B.4}$$

Referring to Eqs. (B.3), we can express the moment of inertia of the body with respect to the $x$ axis as follows:

$$I_x = \int (y^2 + z^2)\, dm = \int [(y' + \bar{y})^2 + (z' + \bar{z})^2]\, dm$$
$$= \int (y'^2 + z'^2)\, dm + 2\bar{y} \int y'\, dm + 2\bar{z} \int z'\, dm + (\bar{y}^2 + \bar{z}^2) \int dm$$

The first integral in this expression represents the moment of inertia $I_{x'}$ of the body with respect to the centroidal axis $x'$; the second and third integrals represent the first moment of the body with respect to the $z'x'$ and $x'y'$ planes, respectively, and, since both planes contain $G$, the two integrals are *zero;* the last integral is equal to the total mass $m$ of the body. We write, therefore,

$$I_x = \bar{I}_{x'} + m(\bar{y}^2 + \bar{z}^2) \tag{B.5}$$

and, similarly,

$$I_y = \bar{I}_{y'} + m(\bar{z}^2 + \bar{x}^2) \qquad I_z = \bar{I}_{z'} + m(\bar{x}^2 + \bar{y}^2) \tag{B.5'}$$

We easily verify from Fig. B.3 that the sum $\bar{z}^2 + \bar{x}^2$ represents the square of the distance $OB$, between the $y$ and $y'$ axes. Similarly, $\bar{y}^2 + \bar{z}^2$ and $\bar{x}^2 + \bar{y}^2$ represent the squares of the distance between the $x$ and $x'$ axes and the $z$ and $z'$ axes, respectively. Denoting by $d$ the distance between an arbitrary axis $AA'$ and the parallel centroidal axis $BB'$ (Fig. B.4), we can, therefore, write the following general relation between the moment of inertia $I$ of the body with respect to $AA'$ and its moment of inertia $\bar{I}$ with respect to $BB'$:

$$I = \bar{I} + md^2 \tag{B.6}$$

Expressing the moments of inertia in terms of the corresponding radii of gyration, we can also write

$$k^2 = \bar{k}^2 + d^2 \tag{B.7}$$

where $k$ and $\bar{k}$ represent the radii of gyration of the body about $AA'$ and $BB'$, respectively.

---

†Note that the term *centroidal* is used here to define an axis passing through the center of gravity $G$ of the body, whether or not $G$ coincides with the centroid of the volume of the body.

Consider a thin plate of uniform thickness $t$, which is made of a homogeneous material of density $\rho$ (density = mass per unit volume). The mass moment of inertia of the plate with respect to an axis $AA'$ *contained in the plane* of the plate (Fig. B.5$a$) is

$$I_{AA',\,\text{mass}} = \int r^2\, dm$$

Since $dm = \rho t\, dA$, we write

$$I_{AA',\,\text{mass}} = \rho t \int r^2\, dA$$

But $r$ represents the distance of the element of area $dA$ to the axis

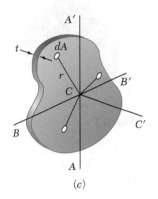

$(a)$ $\qquad\qquad$ $(b)$ $\qquad\qquad$ $(c)$

**Fig. B.5**

$AA'$; the integral is therefore equal to the moment of inertia of the area of the plate with respect to $AA'$. We have

$$I_{AA',\,\text{mass}} = \rho t I_{AA',\,\text{area}} \tag{B.8}$$

Similarly, for an axis $BB'$ which is contained in the plane of the plate and is perpendicular to $AA'$ (Fig. B.5$b$), we have

$$I_{BB',\,\text{mass}} = \rho t I_{BB',\,\text{area}} \tag{B.9}$$

Considering now the axis $CC'$ which is *perpendicular* to the plate and passes through the point of intersection $C$ of $AA'$ and $BB'$ (Fig. B.5$c$), we write

$$I_{CC',\,\text{mass}} = \rho t J_{C,\,\text{area}} \tag{B.10}$$

where $J_C$ is the *polar* moment of inertia of the area of the plate with respect to point $C$.

Recalling the relation $J_C = I_{AA'} + I_{BB'}$ which exists between polar and rectangular moments of inertia of an area, we write the following relation between the mass moments of inertia of a thin plate:

$$I_{CC'} = I_{AA'} + I_{BB'} \tag{B.11}$$

**Fig. B.6**

**Fig. B.7**

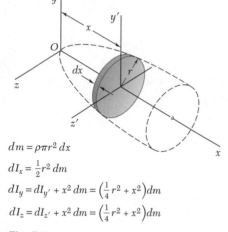

$dm = \rho\pi r^2\, dx$

$dI_x = \frac{1}{2}r^2\, dm$

$dI_y = dI_{y'} + x^2\, dm = \left(\frac{1}{4}r^2 + x^2\right)dm$

$dI_z = dI_{z'} + x^2\, dm = \left(\frac{1}{4}r^2 + x^2\right)dm$

**Fig. B.8** Determination of the moment of inertia of a body of revolution.

**Rectangular Plate.** In the case of a rectangular plate of sides $a$ and $b$ (Fig. B.6), we obtain the following mass moments of inertia with respect to axes through the center of gravity of the plate:

$$I_{AA',\,\text{mass}} = \rho t I_{AA',\,\text{area}} = \rho t(\tfrac{1}{12}a^3 b)$$
$$I_{BB',\,\text{mass}} = \rho t I_{BB',\,\text{area}} = \rho t(\tfrac{1}{12}ab^3)$$

Observing that the product $\rho abt$ is equal to the mass $m$ of the plate, we write the mass moments of inertia of a thin rectangular plate as follows:

$$I_{AA'} = \tfrac{1}{12}ma^2 \qquad I_{BB'} = \tfrac{1}{12}mb^2 \tag{B.12}$$
$$I_{CC'} = I_{AA'} + I_{BB'} = \tfrac{1}{12}m(a^2 + b^2) \tag{B.13}$$

**Circular Plate.** In the case of a circular plate, or disk, of radius $r$ (Fig. B.7), we write

$$I_{AA',\,\text{mass}} = \rho t I_{AA',\,\text{area}} = \rho t(\tfrac{1}{4}\pi r^4)$$

Observing that the product $\rho\pi r^2 t$ is equal to the mass $m$ of the plate and that $I_{AA'} = I_{BB'}$, we write the mass moments of inertia of a circular plate as follows:

$$I_{AA'} = I_{BB'} = \tfrac{1}{4}mr^2 \tag{B.14}$$
$$I_{CC'} = I_{AA'} + I_{BB'} = \tfrac{1}{2}mr^2 \tag{B.15}$$

## B.4. DETERMINATION OF THE MOMENT OF INERTIA OF A THREE-DIMENSIONAL BODY BY INTEGRATION

The moment of inertia of a three-dimensional body is obtained by evaluating the integral $I = \int r^2\, dm$. If the body is made of a homogeneous material of density $\rho$, the element of mass $dm$ is equal to $\rho dV$ and we can write $I = \rho \int r^2\, dV$. This integral depends only upon the shape of the body. Thus, in order to compute the moment of inertia of a three-dimensional body, it will generally be necessary to perform a triple, or at least a double, integration.

However, if the body possesses two planes of symmetry, it is usually possible to determine the body's moment of inertia with a single integration by choosing as the element of mass $dm$ a thin slab which is perpendicular to the planes of symmetry. In the case of bodies of revolution, for example, the element of mass would be a thin disk (Fig. B.8). Using formula (B.15), the moment of inertia of the disk with respect to the axis of revolution can be expressed as indicated in Fig. B.8. Its moment of inertia with respect to each of the other two coordinate axes is obtained by using formula (B.14) and the parallel-axis theorem. Integration of the expression obtained yields the desired moment of inertia of the body.

## B.5. MOMENTS OF INERTIA OF COMPOSITE BODIES

The moments of inertia of a few common shapes are shown in Fig. B.9. For a body consisting of several of these simple shapes, the moment of inertia of the body with respect to a given axis can be obtained by first computing the moments of inertia of its component parts about the desired axis and then adding them together. As was the case for areas, the radius of gyration of a composite body *cannot* be obtained by adding the radii of gyration of its component parts.

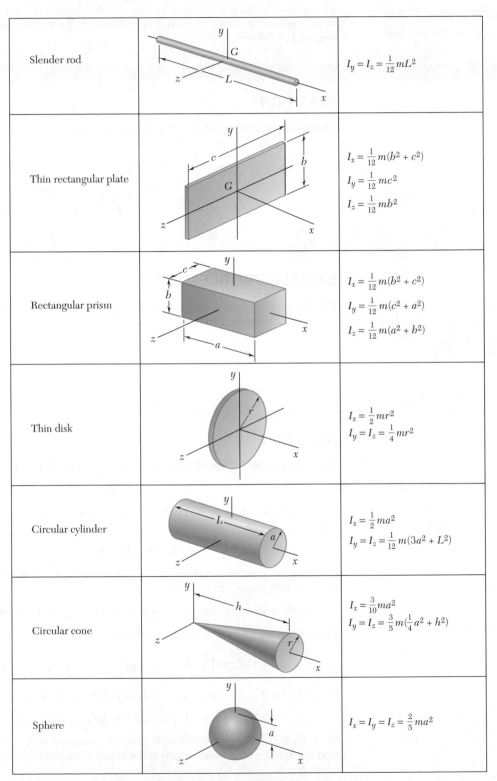

| Slender rod | | $I_y = I_z = \frac{1}{12} mL^2$ |
|---|---|---|
| Thin rectangular plate | | $I_x = \frac{1}{12} m(b^2 + c^2)$<br>$I_y = \frac{1}{12} mc^2$<br>$I_z = \frac{1}{12} mb^2$ |
| Rectangular prism | | $I_x = \frac{1}{12} m(b^2 + c^2)$<br>$I_y = \frac{1}{12} m(c^2 + a^2)$<br>$I_z = \frac{1}{12} m(a^2 + b^2)$ |
| Thin disk | | $I_x = \frac{1}{2} mr^2$<br>$I_y = I_z = \frac{1}{4} mr^2$ |
| Circular cylinder | | $I_x = \frac{1}{2} ma^2$<br>$I_y = I_z = \frac{1}{12} m(3a^2 + L^2)$ |
| Circular cone | | $I_x = \frac{3}{10} ma^2$<br>$I_y = I_z = \frac{3}{5} m(\frac{1}{4} a^2 + h^2)$ |
| Sphere | | $I_x = I_y = I_z = \frac{2}{5} ma^2$ |

**Fig. B.9**  Mass moments of inertia of common geometric shapes.

## SAMPLE PROBLEM B.1

Determine the mass moment of inertia of a slender rod of length $L$ and mass $m$ with respect to an axis which is perpendicular to the rod and passes through one end of the rod.

## SOLUTION

Choosing the differential element of mass shown, we write

$$dm = \frac{m}{L}\, dx$$

$$I_y = \int x^2\, dm = \int_0^L x^2\, \frac{m}{L}\, dx = \left[\frac{m}{L}\frac{x^3}{3}\right]_0^L \qquad I_y = \tfrac{1}{3}mL^2 \quad \blacktriangleleft$$

## SAMPLE PROBLEM B.2

For the homogeneous rectangular prism shown, determine the mass moment of inertia with respect to the $z$ axis.

## SOLUTION

We choose as the differential element of mass the thin slab shown; thus

$$dm = \rho bc\, dx$$

Referring to Sec. B.3, we find that the mass moment of inertia of the element with respect to the $z'$ axis is

$$dI_{z'} = \tfrac{1}{12}b^2\, dm$$

Applying the parallel-axis theorem, we obtain the mass moment of inertia of the slab with respect to the $z$ axis.

$$dI_z = dI_{z'} + x^2\, dm = \tfrac{1}{12}b^2\, dm + x^2\, dm = (\tfrac{1}{12}b^2 + x^2)\, \rho bc\, dx$$

Integrating from $x = 0$ to $x = a$, we obtain

$$I_z = \int dI_z = \int_0^a (\tfrac{1}{12}b^2 + x^2)\, \rho bc\, dx = \rho abc(\tfrac{1}{12}b^2 + \tfrac{1}{3}a^2)$$

Since the total mass of the prism is $m = \rho abc$, we can write

$$I_z = m(\tfrac{1}{12}b^2 + \tfrac{1}{3}a^2) \qquad I_z = \tfrac{1}{12}m(4a^2 + b^2) \quad \blacktriangleleft$$

We note that if the prism is thin, $b$ is small compared to $a$, and the expression for $I_z$ reduces to $\tfrac{1}{3}ma^2$, which is the result obtained in Sample Prob. B.1 when $L = a$.

## SAMPLE PROBLEM B.3

Determine the mass moment of inertia of a right circular cone with respect to (a) its longitudinal axis, (b) an axis through the apex of the cone and perpendicular to its longitudinal axis, (c) an axis through the centroid of the cone and perpendicular to its longitudinal axis.

## SOLUTION

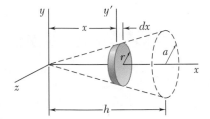

We choose the differential element of mass shown.

$$r = a\frac{x}{h} \qquad dm = \rho\pi r^2\, dx = \rho\pi\frac{a^2}{h^2}x^2\, dx$$

**a. Mass Moment of Inertia $I_x$.** Using the expression derived in Sec. B.3 for a thin disk, we compute the mass moment of inertia of the differential element with respect to the $x$ axis.

$$dI_x = \tfrac{1}{2}r^2\, dm = \tfrac{1}{2}\left(a\frac{x}{h}\right)^2\left(\rho\pi\frac{a^2}{h^2}x^2\, dx\right) = \tfrac{1}{2}\rho\pi\frac{a^4}{h^4}x^4\, dx$$

Integrating from $x = 0$ to $x = h$, we obtain

$$I_x = \int dI_x = \int_0^h \tfrac{1}{2}\rho\pi\frac{a^4}{h^4}x^4\, dx = \tfrac{1}{2}\rho\pi\frac{a^4}{h^4}\frac{h^5}{5} = \tfrac{1}{10}\rho\pi a^4 h$$

Since the total mass of the cone is $m = \tfrac{1}{3}\rho\pi a^2 h$, we can write

$$I_x = \tfrac{1}{10}\rho\pi a^4 h = \tfrac{3}{10}a^2(\tfrac{1}{3}\rho\pi a^2 h) = \tfrac{3}{10}ma^2 \qquad I_x = \tfrac{3}{10}ma^2 \quad \blacktriangleleft$$

**b. Mass Moment of Inertia $I_y$.** The same differential element is used. Applying the parallel-axis theorem and using the expression derived in Sec. B.3 for a thin disk, we write

$$dI_y = dI_{y'} + x^2\, dm = \tfrac{1}{4}r^2\, dm + x^2\, dm = (\tfrac{1}{4}r^2 + x^2)\, dm$$

Substituting the expressions for $r$ and $dm$ into the equation, we obtain

$$dI_y = \left(\frac{1}{4}\frac{a^2}{h^2}x^2 + x^2\right)\left(\rho\pi\frac{a^2}{h^2}x^2\, dx\right) = \rho\pi\frac{a^2}{h^2}\left(\frac{a^2}{4h^2} + 1\right)x^4\, dx$$

$$I_y = \int dI_y = \int_0^h \rho\pi\frac{a^2}{h^2}\left(\frac{a^2}{4h^2} + 1\right)x^4\, dx = \rho\pi\frac{a^2}{h^2}\left(\frac{a^2}{4h^2} + 1\right)\frac{h^5}{5}$$

Introducing the total mass of the cone $m$, we rewrite $I_y$ as follows:

$$I_y = \tfrac{3}{5}(\tfrac{1}{4}a^2 + h^2)\tfrac{1}{3}\rho\pi a^2 h \qquad I_y = \tfrac{3}{5}m(\tfrac{1}{4}a^2 + h^2) \quad \blacktriangleleft$$

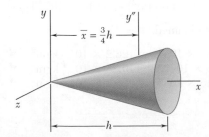

**c. Mass Moment of Inertia $\bar{I}_{y''}$.** We apply the parallel-axis theorem and write

$$I_y = \bar{I}_{y''} + m\bar{x}^2$$

Solving for $\bar{I}_{y''}$ and recalling that $\bar{x} = \tfrac{3}{4}h$, we have

$$\bar{I}_{y''} = I_y - m\bar{x}^2 = \tfrac{3}{5}m(\tfrac{1}{4}a^2 + h^2) - m(\tfrac{3}{4}h)^2$$

$$\bar{I}_{y''} = \tfrac{3}{20}m(a^2 + \tfrac{1}{4}h^2) \quad \blacktriangleleft$$

## SAMPLE PROBLEM B.4

A steel forging consists of a 15 × 5 × 5 cm rectangular prism and two cylinders of diameter 5 cm and length 7.6 cm as shown. Determine the mass moments of inertia of the forging with respect to the coordinate axes knowing that the specific weight of steel is 7849 kg/m³.

## SOLUTION

### Computation of Masses

**Prism**

$$V = (5 \text{ cm})(5 \text{ cm})(15 \text{ cm}) = 375 \text{ cm}^3 = .000375 \text{ m}^3$$

$$m = (375 \text{ cm}^3)(7849 \text{ kg/m}^3) = 2.9 \text{ kg}$$

**Each Cylinder**

$$V = \pi(2.5 \text{ cm})^2(7.6 \text{ cm}) = 149.2 \text{ cm}^3 = 1.492 \times 10^{-4} \text{ m}^3$$

$$m = (1.492 \times 10^{-4})(7849 \text{ kg/m}^3) = 1.17 \text{ kg}$$

**Mass Moments of Inertia.**   The mass moments of inertia of each component are computed from Fig. B.9, using the parallel-axis theorem when necessary. Note that all lengths are expressed in feet.

**Prism**

$$I_x = I_z = \tfrac{1}{12}(2.9 \text{ kg})[(.15 \text{ m})^2 + (.05 \text{ m})^2] = 6.04 \times 10^{-3} \text{ kgm}^2$$
$$I_y = \tfrac{1}{12}(2.9 \text{ kg})(.05 \text{ m})^2(.05 \text{ m})^2] = 1.208 \times 10^{-3} \text{ kgm}^2$$

**Each Cylinder**

$$I_x = \tfrac{1}{2}ma^2 + m\overline{y}^2 = \tfrac{1}{2}(1.17 \text{ kg})(.025 \text{ m})^2 + (1.17 \text{ kg})(.05 \text{ m})^2$$
$$= 32.9 \times 10^{-4} \text{ kgm}^2$$
$$I_y = \tfrac{1}{12}m(3a^2 + L^2) + m\overline{x}^2 = \tfrac{1}{12}(1.17 \text{ kg})[3(.025 \text{ m})^2 + (.076 \text{ m})^2]$$
$$+ (1.17 \text{ kg})(.063 \text{ m})^2 = 53.9 \times 10^{-4} \text{ kgm}^2$$
$$I_z = \tfrac{1}{12}m(3a^2 + L^2) + m(\overline{x}^2 + \overline{y}^2) = \tfrac{1}{12}(1.17 \text{ kg})[3(.025 \text{ m})^2 + (.076 \text{ m})^2]$$
$$+ (1.17 \text{ kg})[(.063 \text{ m})^2 + (.05 \text{ m})^2] = 83.15 \times 10^{-4} \text{ kgm}^2$$

**Entire Body.**   Adding the values obtained,

$$I_x = 60.4 \times 10^{-4} + 2(32.9 \times 10^{-4}) = 126.2 \times 10^{-4} \text{ kgm}^2 \quad \blacktriangleleft$$
$$I_y = 12.08 \times 10^{-4} + 2(53.9 \times 10^{-4}) = 119.88 \times 10^{-4} \text{ kgm}^2 \quad \blacktriangleleft$$
$$I_z = 60.4 \times 10^{-4} + 2(83.15 \times 10^{-4}) = 226.7 \times 10^{-4} \text{ kgm}^2 \quad \blacktriangleleft$$

Dimensions in mm

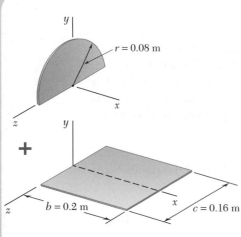

A thin steel plate which is 4 mm thick is cut and bent to form the machine part shown. Knowing that the density of steel is 7850 kg/m³, determine the mass moments of inertia of the machine part with respect to the coordinate axes.

## SOLUTION

We observe that the machine part consists of a semicircular plate and a rectangular plate from which a circular plate has been removed.

**Computation of Masses.** *Semicircular Plate*

$$V_1 = \tfrac{1}{2}\pi r^2 t = \tfrac{1}{2}\pi(0.08 \text{ m})^2(0.004 \text{ m}) = 40.21 \times 10^{-6} \text{ m}^3$$
$$m_1 = \rho V_1 = (7.85 \times 10^3 \text{ kg/m}^3)(40.21 \times 10^{-6} \text{ m}^3) = 0.3156 \text{ kg}$$

*Rectangular Plate*

$$V_2 = (0.200 \text{ m})(0.160 \text{ m})(0.004 \text{ m}) = 128 \times 10^{-6} \text{ m}^3$$
$$m_2 = \rho V_2 = (7.85 \times 10^3 \text{ kg/m}^3)(128 \times 10^{-6} \text{ m}^3) = 1.005 \text{ kg}$$

*Circular Plate*

$$V_3 = \pi a^2 t = \pi(0.050 \text{ m})^2(0.004 \text{ m}) = 31.42 \times 10^{-6} \text{ m}^3$$
$$m_3 = \rho V_3 = (7.85 \times 10^3 \text{ kg/m}^3)(31.42 \times 10^{-6} \text{ m}^3) = 0.2466 \text{ kg}$$

**Mass Moments of Inertia.** Using the method presented in Sec. B.3, we compute the mass moments of inertia of each component.

*Semicircular Plate.* From Fig. B.9, we observe that for a circular plate of mass $m$ and radius $r$

$$I_x = \tfrac{1}{2}mr^2 \qquad I_y = I_z = \tfrac{1}{4}mr^2$$

Because of symmetry, we note that for a semicircular plate

$$I_x = \tfrac{1}{2}(\tfrac{1}{2}mr^2) \qquad I_y = I_z = \tfrac{1}{2}(\tfrac{1}{4}mr^2)$$

Since the mass of the semicircular plate is $m_1 = \tfrac{1}{2}m$, we have

$$I_x = \tfrac{1}{2}m_1 r^2 = \tfrac{1}{2}(0.3156 \text{ kg})(0.08 \text{ m})^2 = 1.010 \times 10^{-3} \text{ kg} \cdot \text{m}^2$$
$$I_y = I_z = \tfrac{1}{4}(\tfrac{1}{2}mr^2) = \tfrac{1}{4}m_1 r^2 = \tfrac{1}{4}(0.3156 \text{ kg})(0.08 \text{ m})^2 = 0.505 \times 10^{-3} \text{ kg} \cdot \text{m}^2$$

*Rectangular Plate*

$$I_x = \tfrac{1}{12}m_2 c^2 = \tfrac{1}{12}(1.005 \text{ kg})(0.16 \text{ m})^2 = 2.144 \times 10^{-3} \text{ kg} \cdot \text{m}^2$$
$$I_z = \tfrac{1}{3}m_2 b^2 = \tfrac{1}{3}(1.005 \text{ kg})(0.2 \text{ m})^2 = 13.400 \times 10^{-3} \text{ kg} \cdot \text{m}^2$$
$$I_y = I_x + I_z = (2.144 + 13.400)(10^{-3}) = 15.544 \times 10^{-3} \text{ kg} \cdot \text{m}^2$$

*Circular Plate*

$$I_x = \tfrac{1}{4}m_3 a^2 = \tfrac{1}{4}(0.2466 \text{ kg})(0.05 \text{ m})^2 = 0.154 \times 10^{-3} \text{ kg} \cdot \text{m}^2$$
$$I_y = \tfrac{1}{2}m_3 a^2 + m_3 d^2$$
$$= \tfrac{1}{2}(0.2466 \text{ kg})(0.05 \text{ m})^2 + (0.2466 \text{ kg})(0.1 \text{ m})^2 = 2.774 \times 10^{-3} \text{ kg} \cdot \text{m}^2$$
$$I_z = \tfrac{1}{4}m_3 a^2 + m_3 d^2 = \tfrac{1}{4}(0.2466 \text{ kg})(0.05 \text{ m})^2 + (0.2466 \text{ kg})(0.1 \text{ m})^2$$
$$= 2.620 \times 10^{-3} \text{ kg} \cdot \text{m}^2$$

*Entire Machine Part*

$$I_x = (1.010 + 2.144 - 0.154)(10^{-3}) \text{ kg} \cdot \text{m}^2 \qquad I_x = 3.00 \times 10^{-3} \text{ kg} \cdot \text{m}^2 \blacktriangleleft$$
$$I_y = (0.505 + 15.544 - 2.774)(10^{-3}) \text{ kg} \cdot \text{m}^2 \qquad I_y = 13.28 \times 10^{-3} \text{ kg} \cdot \text{m}^2 \blacktriangleleft$$
$$I_z = (0.505 + 13.400 - 2.620)(10^{-3}) \text{ kg} \cdot \text{m}^2 \qquad I_z = 11.29 \times 10^{-3} \text{ kg} \cdot \text{m}^2 \blacktriangleleft$$

# SOLVING PROBLEMS
# ON YOUR OWN

In this lesson we introduced the *mass moment of inertia* and the *radius of gyration* of a three-dimensional body with respect to a given axis [Eqs. (B.1) and (B.2)]. We also derived a *parallel-axis theorem* for use with mass moments of inertia and discussed the computation of the mass moments of inertia of thin plates and three-dimensional bodies.

**1. *Computing mass moments of inertia.*** The mass moment of inertia $I$ of a body with respect to a given axis can be calculated directly from the definition given in Eq. (B.1) for simple shapes [Sample Prob. B.1]. In most cases, however, it is necessary to divide the body into thin slabs, compute the mass moment of inertia of a typical slab with respect to the given axis—using the parallel-axis theorem if necessary—and integrate the expression obtained.

**2. *Applying the parallel-axis theorem.*** In Sec. B.2 we derived the parallel-axis theorem for mass moments of inertia

$$I = \bar{I} + md^2 \tag{B.6}$$

which states that the mass moment of inertia $I$ of a body of mass $m$ with respect to a given axis is equal to the sum of the mass moment of inertia $\bar{I}$ of that body with respect to the *parallel centroidal axis* and the product $md^2$, where $d$ is the distance between the two axes. When the mass moment of inertia of a three-dimensional body is calculated with respect to one of the coordinate axes, $d^2$ can be replaced by the sum of the squares of distances measured along the other two coordinate axes [Eqs. (B.5) and (B.5')].

**3. *Avoiding unit-related errors.*** To avoid errors, it is essential that you be consistent in your use of units. Thus, all lengths should be expressed in meters or feet, as appropriate, and for problems using U.S. customary units, masses should be given in $lb \cdot s^2/ft$. In addition, we strongly recommend that you include units as you perform your calculations [Sample Probs. B.4 and B.5].

**4. *Calculating the mass moment of inertia of thin plates.*** We showed in Sec. B.3 that the mass moment of inertia of a thin plate with respect to a given axis can be obtained by multiplying the corresponding moment of inertia of the area of the plate by the density $\rho$ and the thickness $t$ of the plate [Eqs. (B.8) through (B.10)]. Note that since the axis $CC'$ in Fig. B.5c is *perpendicular to the plate*, $I_{CC', \text{mass}}$ is associated with the *polar* moment of inertia $J_{C, \text{area}}$.

Instead of calculating directly the mass moment of inertia of a thin plate with respect to a specified axis, you may sometimes find it convenient to first compute its mass moment of inertia with respect to an axis parallel to the specified axis and then apply the parallel-axis theorem. Further, to determine the mass moment of inertia of a thin plate with respect to an axis perpendicular to the plate, you may wish to first determine its mass moments of inertia with respect to two perpendicular in-plane axes and then use Eq. (B.11). Finally, remember that the mass of a plate of area $A$, thickness $t$, and density $\rho$ is $m = \rho tA$.

**5. Determining the mass moment of inertia of a body by direct single integration.** We discussed in Sec. B.4 and illustrated in Sample Probs. B.2 and B.3 how single integration can be used to compute the mass moment of inertia of a body that can be divided into a series of thin, parallel slabs. For such cases, you will often need to express the mass of the body in terms of the body's density and dimensions. Assuming that the body has been divided, as in the sample problems, into thin slabs perpendicular to the $x$ axis, you will need to express the dimensions of each slab as functions of the variable $x$.

 *a. In the special case of a body of revolution,* the elemental slab is a thin disk, and the equations given in Fig. B.8 should be used to determine the mass moments of inertia of the body [Sample Prob. B.3].

 *b. In the general case, when the body is not of revolution,* the differential element is not a disk, but a thin slab of a different shape, and the equations of Fig. B.8 cannot be used. See, for example, Sample Prob. B.2, where the element was a thin, rectangular slab. For more complex configurations, you may want to use one or more of the following equations, which are based on Eqs. (B.5) and (B.5′) of Sec. B.2.

$$dI_x = dI_{x'} + (\bar{y}_{el}^2 + \bar{z}_{el}^2)\, dm$$
$$dI_y = dI_{y'} + (\bar{z}_{el}^2 + \bar{x}_{el}^2)\, dm$$
$$dI_z = dI_{z'} + (\bar{x}_{el}^2 + \bar{y}_{el}^2)\, dm$$

where the primes denote the centroidal axes of each elemental slab, and where $\bar{x}_{el}$, $\bar{y}_{el}$, and $\bar{z}_{el}$ represent the coordinates of its centroid. The centroidal mass moments of inertia of the slab are determined in the manner described earlier for a thin plate: Referring to Fig. 9.12 on page 485, calculate the corresponding moments of inertia of the area of the slab and multiply the result by the density $\rho$ and the thickness $t$ of the slab. Also, assuming that the body has been divided into thin slabs perpendicular to the $x$ axis, remember that you can obtain $dI_{x'}$ by adding $dI_{y'}$ and $dI_{z'}$ instead of computing it directly. Finally, using the geometry of the body, express the result obtained in terms of the single variable $x$ and integrate in $x$.

**6. Computing the mass moment of inertia of a composite body.** As stated in Sec. B.5, the mass moment of inertia of a composite body with respect to a specified axis is equal to the sum of the moments of its components with respect to that axis. Sample Probs. B.4 and B.5 illustrate the appropriate method of solution. You must also remember that the mass moment of inertia of a component will be negative only if the component is *removed* (as in the case of a hole).

Although the composite-body problems in this lesson are relatively straightforward, you will have to work carefully to avoid computational errors. In addition, if some of the mass moments of inertia that you need are not given in Fig. B.9, you will have to derive your own formulas using the techniques of this lesson.

**B.1** The quarter ring shown has a mass $m$ and was cut from a thin, uniform plate. Knowing that $r_1 = \frac{1}{2}r_2$, determine the mass moment of inertia of the quarter ring with respect to (*a*) axis $AA'$, (*b*) the centroidal axis $CC'$ that is perpendicular to the plane of the quarter ring.

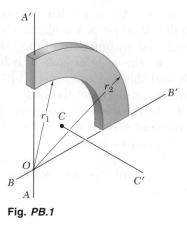

**Fig. PB.1**

**B.2** The spacer shown was cut from a thin, uniform plate. Denoting the mass of the component by $m$, determine its mass moment of inertia with respect to (*a*) the axis $AA'$, (*b*) the centroidal axis $CC'$ that is perpendicular to the plane of the component.

**Fig. PB.2**

**B.3** A thin plate of mass $m$ is cut in the shape of an isosceles triangle of base $b$ and height $h$. Determine the mass moments of inertia of the plate with respect to (*a*) the centroidal axes $AA'$ and $BB'$ in the plane of the plate, (*b*) the centroidal axis $CC'$ perpendicular to the plate.

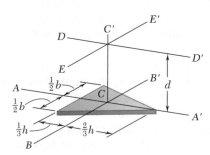

**Fig. PB.3 and PB.4**

**B.4** A thin plate of mass $m$ is cut in the shape of an isosceles triangle of base $b$ and height $h$. Determine the mass moments of inertia of the plate with respect to axes $DD'$ and $EE'$ parallel to the centroidal axes $AA'$ and $BB'$, respectively, and located at a distance $d$ from the plane of the plate.

**B.5**   A thin plate of mass $m$ has the trapezoidal shape shown. Deter-mine the mass moment of inertia of the plate with respect to (a) the $x$ axis, (b) the $y$ axis.

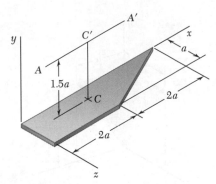

**Fig. PB.5 and PB.6**

**B.6**   A thin plate of mass $m$ has the trapezoidal shape shown. Deter-mine the mass moment of inertia of the plate with respect to (a) the cen-troidal axis $CC'$ that is perpendicular to the plate, (b) the axis $AA'$ which is parallel to the $x$ axis and is located at a distance $1.5a$ from the plate.

**B.7**   The parabolic spandrel shown is revolved about the $x$ axis to form a homogeneous solid of revolution of mass $m$. Using direct integration, ex-press the mass moment of inertia of the solid with respect to the $x$ axis in terms of $m$ and $b$.

**Fig. PB.7**

**B.8**   Determine by direct integration the mass moment of inertia with respect to the $z$ axis of the truncated right circular cone shown knowing that the radius of the base $r_1 = 2r_2$ and assuming that the cone has a uniform density and a mass $m$.

**Fig. PB.8**

**B.9**   The area shown is revolved about the $x$ axis to form a homoge-neous solid of revolution of mass $m$. Determine by direct integration the mass moment of inertia of the solid with respect to (a) the $x$ axis, (b) the $y$ axis. Express your answers in terms of $m$ and $a$.

**Fig. PB.9**

**B.10** Determine by direct integration the mass moment of inertia with respect to the $x$ axis of the pyramid shown assuming that it has a uniform density and a mass $m$.

**Fig. PB.10 and PB.11**

**B.11** Determine by direct integration the mass moment of inertia with respect to the $y$ axis of the pyramid shown assuming that it has a uniform density and a mass $m$.

**B.12** Determine by direct integration the mass moment of inertia and the radius of gyration with respect to the $y$ axis of the paraboloid shown assuming that it has a uniform density and a mass $m$.

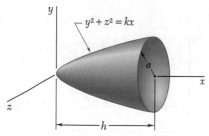

**Fig. PB.12**

**\*B.13** A thin steel wire is bent into the shape shown. Denoting the mass per unit length of the wire by $m'$, determine by direct integration the mass moment of inertia of the wire with respect to each of the coordinate axes.

**Fig. PB.13**

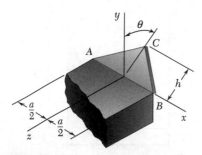

**Fig. PB.14**

**B.14** A thin triangular plate of mass $m$ is welded along its base $AB$ to a block as shown. Knowing that the plate forms an angle $\theta$ with the $y$ axis, determine by direct integration the mass moment of inertia of the plate with respect to (a) the $x$ axis, (b) the $y$ axis, (c) the $z$ axis.

**B.15** Shown is the cross section of the wheel of a caster. Determine its mass moment of inertia and radius of gyration with respect to axis $AA'$. (The specific weight of bronze is 8581 kg/m³; of steel, 7861 kg/m³; and of hard rubber, 1190 kg/m³.)

**Fig. PB.15**

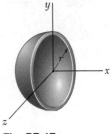

**Fig. *PB.16***

**B.16** Shown is the cross section of an idler roller. Determine its mass moment of inertia and its radius of gyration with respect to the axis $AA'$. (The density of bronze is 8580 kg/m³; of aluminum, 2770 kg/m³; and of neoprene, 1250 kg/m³.)

**B.17** Knowing that the thin hemispherical shell shown is of mass $m$ and thickness $t$, determine the mass moment of inertia and the radius of gyration of the shell with respect to the $x$ axis. (*Hint:* Consider the shell as formed by removing a hemisphere of radius $r$ from a hemisphere of radius $r + t$; then neglect the terms containing $t^2$ and $t^3$ and keep those terms containing $t$.)

**Fig. PB.17**

**B.18** For the homogenous ring of density $\rho$ shown, determine (*a*) the mass moment of inertia with respect to the axis $BB'$, (*b*) the value of $a_1$ for which, given $a_2$ and $h$, $I_{BB'}$ is maximum, (*c*) the corresponding value of $I_{BB'}$.

**Fig. PB.18**

**Fig. PB.19**

**Fig. PB.21**

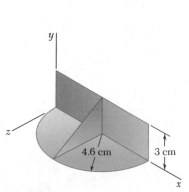

**Fig. PB.23**

**B.19** The steel machine component shown is formed by machining a hemisphere into the base of a truncated cone. Knowing that the density of steel is 7850 kg/m$^3$, determine the mass moment of inertia of the component with respect to the $y$ axis.

**B.20** After a period of use, one of the blades of a shredder has been worn to the shape shown and weighs .18 kg. Knowing that the mass moments of inertia of the blade with respect to the $AA'$ and $BB'$ axes are $8.3 \times 10^{-5}$ kgm$^2$ and $1.74 \times 10^{-4}$ kgm$^2$, respectively, determine ($a$) the location of the centroidal axis $GG'$, ($b$) the radius of gyration with respect to axis $GG'$.

**Fig. PB.20**

**B.21** The cups and the arms of an anemometer are fabricated from a material of density $\rho$. Knowing that the mass moment of inertia of a thin, hemispherical shell of mass $m$ and thickness $t$ with respect to its centroidal axis $GG'$ is $5ma^2/12$, determine ($a$) the mass moment of inertia of the anemometer with respect to the axis $AA'$, ($b$) the ratio of $a$ to $l$ for which the centroidal mass moment of inertia of the cups is equal to 1 percent of the mass moment of inertia of the cups with respect to the axis $AA'$.

**B.22** A square hole is centered in and extends through the aluminum machine component shown. Determine ($a$) the value of $a$ for which the mass moment of inertia of the component with respect to the axis $AA'$, which bisects the top surface of the hole, is maximum, ($b$) the corresponding values of the mass moment of inertia and the radius of gyration with respect to the axis $AA'$. (The density of aluminum is 2800 kg/m$^3$.)

**Fig. PB.22**

**B.23** The machine component shown is fabricated from 0.2 cm thick sheet steel. Knowing that the specific weight of steel is 7849 kg/m$^3$, determine the mass moment of inertia of the component with respect to each of the coordinate axes.

**B.24** A 3 mm thick piece of sheet metal is cut and bent into the machine component shown. Knowing that the density of steel is 7850 kg/m³, determine the mass moment of inertia of the component with respect to each of the coordinate axes.

Fig. *PB.24*

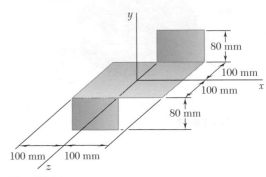

Fig. *PB.25*

**B.25** A 2 mm thick piece of sheet steel is cut and bent into the machine component shown. Knowing that the density of steel is 7850 kg/m³, determine the mass moment of inertia of the component with respect to each of the coordinate axes.

**B.26** A framing anchor is formed from 2-mm-thick galvanized steel. Determine the mass moment of inertia of the anchor with respect to each of the coordinate axes. (The density of galvanized steel is 7530 kg/m³.)

**B.27** A .25 cm thick piece of sheet steel is cut and bent into the machine component shown. Knowing that the specific weight of steel is 7849 kg/m³, determine the mass moment of inertia of the component with respect to each of the coordinate axes.

Fig. *PB.26*

Fig. PB.27

*B.28** The piece of roof flashing shown is formed from sheet copper that is 0.8 mm thick. Knowing that the density of copper is 8940 kg/m³, determine the mass moment of inertia of the flashing with respect to each of the coordinate axes.

Fig. PB.28

**Fig. PB.29**

**B.29** The machine element shown is fabricated from steel. Determine the mass moment of inertia of the assembly with respect to (a) the x axis, (b) the y axis, (c) the z axis. (The specific weight of steel is 7861 kg/m³.)

**B.30** Determine the mass moment of inertia of the steel machine element shown with respect to the y axis. (The density of steel is 7850 kg/m³.)

**Fig. PB.30 and PB.31**

**B.31** Determine the mass moment of inertia of the steel machine element shown with respect to the z axis. (The density of steel is 7850 kg/m³.)

**B.32** An aluminum casting has the shape shown. Knowing that the specific weight of aluminum is 2768 kg/m³, determine the mass moment of inertia of the casting with respect to the z axis.

**Fig. PB.32**

**Fig. *PB.33***

**B.33** Determine the mass moment of inertia of the steel machine element shown with respect to (a) the x axis, (b) the y axis, (c) the z axis. (The specific weight of steel is 7849 kg/m³.)

**B.34** Aluminum wire with a mass per unit length of 0.049 kg/m is used to form the circle and the straight members of the figure shown. Determine the mass moment of inertia of the assembly with respect to each of the co-ordinate axes.

**Fig. PB.34**

**B.35** The figure shown is formed from 3-mm-diameter steel wire. Knowing that the density of the steel is 7850 kg/m³, determine the mass moment of inertia of the wire with respect to each of the coordinate axes.

**Fig. PB.35**

**Fig. PB.36**

**B.36** A homogeneous wire with a weight per unit length of $6.1 \times 10^{-2}$ kg/m is used to form the figure shown. Determine the mass moment of inertia of the wire with respect to each of the coordinate axes.

---

## *B.6. MOMENT OF INERTIA OF A BODY WITH RESPECT TO AN ARBITRARY AXIS THROUGH O. MASS PRODUCTS OF INERTIA

In this section you will see how the moment of inertia of a body can be determined with respect to an arbitrary axis $OL$ through the origin (Fig. B.10) if its moments of inertia with respect to the three coordinate axes, as well as certain other quantities to be defined below, have already been determined.

The moment of inertia $I_{OL}$ of the body with respect to $OL$ is equal to $\int p^2 \, dm$, where $p$ denotes the perpendicular distance from the element of mass $dm$ to the axis $OL$. If we denote by $\boldsymbol{\lambda}$ the unit vector along $OL$ and by $\mathbf{r}$ the position vector of the element $dm$, we observe that the perpendicular distance $p$ is equal to $r \sin \theta$, which is the magnitude of the vector product $\boldsymbol{\lambda} \times \mathbf{r}$. We therefore write

$$I_{OL} = \int p^2 \, dm = \int |\boldsymbol{\lambda} \times \mathbf{r}|^2 \, dm \qquad (B.16)$$

Expressing $|\boldsymbol{\lambda} \times \mathbf{r}|^2$ in terms of the rectangular components of the vector product, we have

$$I_{OL} = \int [(\lambda_x y - \lambda_y x)^2 + (\lambda_y z - \lambda_z y)^2 + (\lambda_z x - \lambda_x z)^2] \, dm$$

where the components $\lambda_x$, $\lambda_y$, and $\lambda_z$ of the unit vector $\boldsymbol{\lambda}$ represent the direction cosines of the axis $OL$ and the components $x$, $y$, and $z$ of $\mathbf{r}$ represent the coordinates of the element of mass $dm$. Expanding the squares and rearranging the terms, we write

$$I_{OL} = \lambda_x^2 \int (y^2 + z^2) \, dm + \lambda_y^2 \int (z^2 + x^2) \, dm + \lambda_z^2 \int (x^2 + y^2) \, dm$$

$$- 2\lambda_x \lambda_y \int xy \, dm - 2\lambda_y \lambda_z \int yz \, dm - 2\lambda_z \lambda_x \int zx \, dm \qquad (B.17)$$

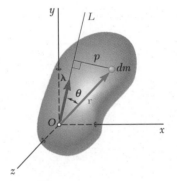

**Fig. B.10**

Referring to Eqs. (B.3), we note that the first three integrals in (B.17) represent, respectively, the moments of inertia $I_x$, $I_y$, and $I_z$ of the body with respect to the coordinate axes. The last three integrals in (B.17), which involve products of coordinates, are called the *products of inertia* of the body with respect to the $x$ and $y$ axes, the $y$ and $z$ axes, and the $z$ and $x$ axes, respectively. We write

$$I_{xy} = \int xy \, dm \qquad I_{yz} = \int yz \, dm \qquad I_{zx} = \int zx \, dm \qquad \text{(B.18)}$$

Rewriting Eq. (B.17) in terms of the integrals defined in Eqs. (B.3) and (B.18), we have

$$I_{OL} = I_x\lambda_x^2 + I_y\lambda_y^2 + I_z\lambda_z^2 - 2I_{xy}\lambda_x\lambda_y - 2I_{yz}\lambda_y\lambda_z - 2I_{zx}\lambda_z\lambda_x \quad \text{(B.19)}$$

We note that the definition of the products of inertia of a mass given in Eqs. (B.18) is an extension of the definition of the product of inertia of an area (Sec. 9.8). Mass products of inertia reduce to zero under the same conditions of symmetry as do products of inertia of areas, and the parallel-axis theorem for mass products of inertia is expressed by relations similar to the formula derived for the product of inertia of an area. Substituting the expressions for $x$, $y$, and $z$ given in Eqs. (B.4) into Eqs. (B.18), we find that

$$\begin{aligned} I_{xy} &= \bar{I}_{x'y'} + m\bar{x}\bar{y} \\ I_{yz} &= \bar{I}_{y'z'} + m\bar{y}\bar{z} \\ I_{zx} &= \bar{I}_{z'x'} + m\bar{z}\bar{x} \end{aligned} \qquad \text{(B.20)}$$

where $\bar{x}$, $\bar{y}$, and $\bar{z}$ are the coordinates of the center of gravity $G$ of the body and $\bar{I}_{x'y'}$, $\bar{I}_{y'z'}$, and $\bar{I}_{z'x'}$ denote the products of inertia of the body with respect to the centroidal axes $x'$, $y'$, and $z'$ (Fig. B.3).

## *B.7. ELLIPSOID OF INERTIA. PRINCIPAL AXES OF INERTIA

Let us assume that the moment of inertia of the body considered in the preceding section has been determined with respect to a large number of axes $OL$ through the fixed point $O$ and that a point $Q$ has been plotted on each axis $OL$ at a distance $OQ = 1/\sqrt{I_{OL}}$ from $O$. The locus of the points $Q$ thus obtained forms a surface (Fig. B.11). The equation of that surface can be obtained by substituting $1/(OQ)^2$ for $I_{OL}$ in Eq. (B.19) and then multiplying both sides of the equation by $(OQ)^2$. Observing that

$$(OQ)\lambda_x = x \qquad (OQ)\lambda_y = y \qquad (OQ)\lambda_z = z$$

where $x$, $y$, and $z$ denote the rectangular coordinates of $Q$, we write

$$I_x x^2 + I_y y^2 + I_z z^2 - 2I_{xy}xy - 2I_{yz}yz - 2I_{zx}zx = 1 \quad \text{(B.21)}$$

The equation obtained is the equation of a *quadric surface*. Since the moment of inertia $I_{OL}$ is different from zero for every axis $OL$, no point $Q$ can be at an infinite distance from $O$. Thus, the quadric

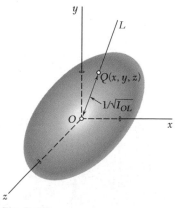

**Fig. B.11**

surface obtained is an *ellipsoid*. This ellipsoid, which defines the moment of inertia of the body with respect to any axis through $O$, is known as the *ellipsoid of inertia* of the body at $O$.

We observe that if the axes in Fig. B.11 are rotated, the coefficients of the equation defining the ellipsoid change, since they are equal to the moments and products of inertia of the body with respect to the rotated coordinate axes. However, the *ellipsoid itself remains unaffected,* since its shape depends only upon the distribution of mass in the given body. Suppose that we choose as coordinate axes the principal axes $x'$, $y'$, and $z'$ of the ellipsoid of inertia (Fig. B.12). The equation of the ellipsoid with respect to these coordinate axes is known to be of the form

$$I_{x'}x'^2 + I_{y'}y'^2 + I_{z'}z'^2 = 1 \qquad (B.22)$$

which does not contain any products of the coordinates. Comparing Eqs. (B.21) and (B.22), we observe that the products of inertia of the body with respect to the $x'$, $y'$, and $z'$ axes must be zero. The $x'$, $y'$, and $z'$ axes are known as the *principal axes of inertia* of the body at $O$, and the coefficients $I_{x'}$, $I_{y'}$, and $I_{z'}$ are referred to as the *principal moments of inertia* of the body at $O$. Note that, given a body of arbitrary shape and a point $O$, it is always possible to find axes which are the principal axes of inertia of the body at $O$, that is, axes with respect to which the products of inertia of the body are zero. Indeed, whatever the shape of the body, the moments and products of inertia of the body with respect to $x$, $y$, and $z$ axes through $O$ will define an ellipsoid, and this ellipsoid will have principal axes which, by definition, are the principal axes of inertia of the body at $O$.

If the principal axes of inertia $x'$, $y'$, and $z'$ are used as coordinate axes, the expression obtained in Eq. (B.19) for the moment of inertia of a body with respect to an arbitrary axis through $O$ reduces to

$$I_{OL} = I_{x'}\lambda_{x'}^2 + I_{y'}\lambda_{y'}^2 + I_{z'}\lambda_{z'}^2 \qquad (B.23)$$

The determination of the principal axes of inertia of a body of arbitrary shape is somewhat involved and will be discussed in the next section. There are many cases, however, where these axes can be spotted immediately. Consider, for instance, the homogeneous cone of elliptical base shown in Fig. B.13; this cone possesses two mutually perpendicular planes of symmetry $OAA'$ and $OBB'$. From the definition (B.18), we observe that if the $x'y'$ and $y'z'$ planes are chosen to coincide with the two planes of symmetry, all of the products of inertia are zero. The $x'$, $y'$, and $z'$ axes thus selected are therefore the principal axes of inertia of the cone at $O$. In the case of the homogeneous regular tetrahedron $OABC$ shown in Fig. B.14, the line joining the corner $O$ to the center $D$ of the opposite face is a principal axis of inertia at $O$, and any line through $O$ perpendicular to $OD$ is also a principal axis of inertia at $O$. This property is apparent if we observe that rotating the tetrahedron through $120°$ about $OD$ leaves its shape and mass distribution unchanged. It follows that the ellipsoid of inertia at $O$ also remains unchanged under this rotation. The ellipsoid, therefore, is a body of revolution whose axis of revolution is $OD$, and the line $OD$, as well as any perpendicular line through $O$, must be a principal axis of the ellipsoid.

**Fig. B.12**

**Fig. B.13**

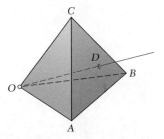

**Fig. B.14**

## *B.8. DETERMINATION OF THE PRINCIPAL AXES AND PRINCIPAL MOMENTS OF INERTIA OF A BODY OF ARBITRARY SHAPE

The method of analysis described in this section should be used when the body under consideration has no obvious property of symmetry.

Consider the ellipsoid of inertia of the body at a given point $O$ (Fig. B.15); let $\mathbf{r}$ be the radius vector of a point $P$ on the surface of the ellipsoid and let $\mathbf{n}$ be the unit vector along the normal to that surface at $P$. We observe that the only points where $\mathbf{r}$ and $\mathbf{n}$ are collinear are the points $P_1$, $P_2$, and $P_3$, where the principal axes intersect the visible portion of the surface of the ellipsoid, and the corresponding points on the other side of the ellipsoid.

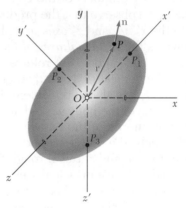

**Fig. B.15**

We now recall from calculus that the direction of the normal to a surface of equation $f(x, y, z) = 0$ at a point $P(x, y, z)$ is defined by the gradient $\nabla f$ of the function $f$ at that point. To obtain the points where the principal axes intersect the surface of the ellipsoid of inertia, we must therefore write that $\mathbf{r}$ and $\nabla f$ are collinear,

$$\nabla f = (2K)\mathbf{r} \tag{B.24}$$

where $K$ is a constant, $\mathbf{r} = x\mathbf{i} + y\mathbf{j} + z\mathbf{k}$, and

$$\nabla f = \frac{\partial f}{\partial x}\mathbf{i} + \frac{\partial f}{\partial y}\mathbf{j} + \frac{\partial f}{\partial x}\mathbf{k}$$

Recalling Eq. (B.21), we note that the function $f(x, y, z)$ corresponding to the ellipsoid of inertia is

$$f(x, y, z) = I_x x^2 + I_y y^2 + I_z z^2 - 2I_{xy}xy - 2I_{yz}yz - 2I_{zx}zx - 1$$

Substituting for $\mathbf{r}$ and $\nabla f$ into Eq. (B.24) and equating the coefficients of the unit vectors, we write

$$\begin{aligned} I_x x &- I_{xy}\,y - I_{zx}\,z = Kx \\ -I_{xy}\,x &+ I_y\,y - I_{yz}\,z = Ky \\ -I_{zx}\,x &- I_{yz}\,y + I_z\,z = Kz \end{aligned} \tag{B.25}$$

Dividing each term by the distance $r$ from $O$ to $P$, we obtain similar equations involving the direction cosines $\lambda_x$, $\lambda_y$, and $\lambda_z$:

$$I_x\lambda_x - I_{xy}\lambda_y - I_{zx}\lambda_z = K\lambda_x$$
$$-I_{xy}\lambda_x + I_y\lambda_y - I_{yz}\lambda_z = K\lambda_y \qquad \text{(B.26)}$$
$$-I_{zx}\lambda_x - I_{yz}\lambda_y + I_z\lambda_z = K\lambda_z$$

Transposing the right-hand members leads to the following homogeneous linear equations:

$$(I_x - K)\lambda_x - I_{xy}\lambda_y - I_{zx}\lambda_z = 0$$
$$-I_{xy}\lambda_x + (I_y - K)\lambda_y - I_{yz}\lambda_z = 0 \qquad \text{(B.27)}$$
$$-I_{zx}\lambda_x - I_{yz}\lambda_y + (I_z - K)\lambda_z = 0$$

For this system of equations to have a solution different from $\lambda_x = \lambda_y = \lambda_z = 0$, its discriminant must be zero:

$$\begin{vmatrix} I_x - K & -I_{xy} & -I_{zx} \\ -I_{xy} & I_y - K & -I_{yz} \\ -I_{zx} & -I_{yz} & I_z - K \end{vmatrix} = 0 \qquad \text{(B.28)}$$

Expanding this determinant and changing signs, we write

$$K^3 - (I_x + I_y + I_z)K^2 + (I_xI_y + I_yI_z + I_zI_x - I_{xy}^2 - I_{yz}^2 - I_{zx}^2)K$$
$$- (I_xI_yI_z - I_xI_{yz}^2 - I_yI_{zx}^2 - I_zI_{xy}^2 - 2I_{xy}I_{yz}I_{zx}) = 0 \qquad \text{(B.29)}$$

This is a cubic equation in $K$, which yields three real, positive roots $K_1$, $K_2$, and $K_3$.

To obtain the direction cosines of the principal axis corresponding to the root $K_1$, we substitute $K_1$ for $K$ in Eqs. (B.27). Since these equations are now linearly dependent, only two of them may be used to determine $\lambda_x$, $\lambda_y$, and $\lambda_z$. An additional equation may be obtained, however, by recalling from Sec. 2.12 that the direction cosines must satisfy the relation

$$\lambda_x^2 + \lambda_y^2 + \lambda_z^2 = 1 \qquad \text{(B.30)}$$

Repeating this procedure with $K_2$ and $K_3$, we obtain the direction cosines of the other two principal axes.

We will now show that *the roots $K_1$, $K_2$, and $K_3$ of Eq. (B.29) are the principal moments of inertia of the given body*. Let us substitute for $K$ in Eqs. (B.26) the root $K_1$, and for $\lambda_x$, $\lambda_y$, and $\lambda_z$ the corresponding values $(\lambda_x)_1$, $(\lambda_y)_1$, and $(\lambda_z)_1$ of the direction cosines; the three equations will be satisfied. We now multiply by $(\lambda_x)_1$, $(\lambda_y)_1$, and $(\lambda_z)_1$, respectively, each term in the first, second, and third equation and add the equations obtained in this way. We write

$$I_x^2(\lambda_x)_1^2 + I_y^2(\lambda_y)_1^2 + I_z^2(\lambda_z)_1^2 - 2I_{xy}(\lambda_x)_1(\lambda_y)_1$$
$$- 2I_{yz}(\lambda_y)_1(\lambda_z)_1 - 2I_{zx}(\lambda_z)_1(\lambda_x)_1 = K_1[(\lambda_x)_1^2 + (\lambda_y)_1^2 + (\lambda_z)_1^2]$$

Recalling Eq. (B.19), we observe that the left-hand member of this equation represents the moment of inertia of the body with respect to the principal axis corresponding to $K_1$; it is thus the principal moment of inertia corresponding to that root. On the other hand, recalling Eq. (B.30), we note that the right-hand member reduces to $K_1$. Thus $K_1$ itself is the principal moment of inertia. We can show in the same fashion that $K_2$ and $K_3$ are the other two principal moments of inertia of the body.

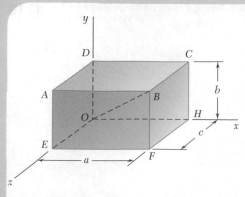

## SAMPLE PROBLEM B.6

Consider a rectangular prism of mass $m$ and sides $a$, $b$, and $c$. Determine (a) the mass moments and products of inertia of the prism with respect to the coordinate axes shown, (b) its mass moment of inertia with respect to the diagonal $OB$.

## SOLUTION

**a. Mass Moments and Products of Inertia with Respect to the Coordinate Axes.** *Mass Moments of Inertia.* Introducing the centroidal axes $x'$, $y'$, and $z'$, with respect to which the mass moments of inertia are given in Fig. B.9, we apply the parallel-axis theorem:

$$I_x = \bar{I}_{x'} + m(\bar{y}^2 + \bar{z}^2) = \tfrac{1}{12}m(b^2 + c^2) + m(\tfrac{1}{4}b^2 + \tfrac{1}{4}c^2)$$
$$I_x = \tfrac{1}{3}m(b^2 + c^2) \blacktriangleleft$$

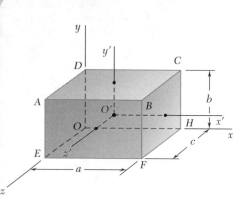

Similarly, $\qquad I_y = \tfrac{1}{3}m(c^2 + a^2) \qquad I_z = \tfrac{1}{3}m(a^2 + b^2) \blacktriangleleft$

*Mass Products of Inertia.* Because of symmetry, the mass products of inertia with respect to the centroidal axes $x'$, $y'$, and $z'$ are zero, and these axes are principal axes of inertia. Using the parallel-axis theorem, we have

$$I_{xy} = \bar{I}_{x'y'} + m\bar{x}\bar{y} = 0 + m(\tfrac{1}{2}a)(\tfrac{1}{2}b) \qquad I_{xy} = \tfrac{1}{4}mab \blacktriangleleft$$

Similarly, $\qquad\qquad I_{yz} = \tfrac{1}{4}mbc \qquad I_{zx} = \tfrac{1}{4}mca \blacktriangleleft$

**b. Mass Moment of Inertia with Respect to OB.** We recall Eq. (B.19):

$$I_{OB} = I_x\lambda_x^2 + I_y\lambda_y^2 + I_z\lambda_z^2 - 2I_{xy}\lambda_x\lambda_y - 2I_{yz}\lambda_y\lambda_z - 2I_{zx}\lambda_z\lambda_x$$

where the direction cosines of $OB$ are

$$\lambda_x = \cos\theta_x = \frac{OH}{OB} = \frac{a}{(a^2 + b^2 + c^2)^{1/2}}$$

$$\lambda_y = \frac{b}{(a^2 + b^2 + c^2)^{1/2}} \qquad \lambda_z = \frac{c}{(a^2 + b^2 + c^2)^{1/2}}$$

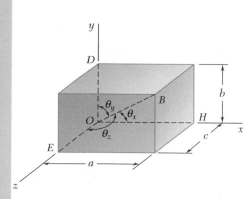

Substituting the values obtained for the mass moments and products of inertia and for the direction cosines into the equation for $I_{OB}$, we have

$$I_{OB} = \frac{1}{a^2 + b^2 + c^2}[\tfrac{1}{3}m(b^2 + c^2)a^2 + \tfrac{1}{3}m(c^2 + a^2)b^2 + \tfrac{1}{3}m(a^2 + b^2)c^2$$
$$- \tfrac{1}{2}ma^2b^2 - \tfrac{1}{2}mb^2c^2 - \tfrac{1}{2}mc^2a^2]$$
$$I_{OB} = \frac{m}{6}\frac{a^2b^2 + b^2c^2 + c^2a^2}{a^2 + b^2 + c^2} \blacktriangleleft$$

*Alternative Solution.* The mass moment of inertia $I_{OB}$ can be obtained directly from the principal mass moments of inertia $\bar{I}_{x'}$, $\bar{I}_{y'}$, and $\bar{I}_{z'}$, since the line $OB$ passes through the centroid $O'$. Since the $x'$, $y'$, and $z'$ axes are principal axes of inertia, we use Eq. (B.23) to write

$$I_{OB} = \bar{I}_{x'}\lambda_x^2 + \bar{I}_{y'}\lambda_y^2 + \bar{I}_{z'}\lambda_z^2$$
$$= \frac{1}{a^2 + b^2 + c^2}\left[\frac{m}{12}(b^2 + c^2)a^2 + \frac{m}{12}(c^2 + a^2)b^2 + \frac{m}{12}(a^2 + b^2)c^2\right]$$
$$I_{OB} = \frac{m}{6}\frac{a^2b^2 + b^2c^2 + c^2a^2}{a^2 + b^2 + c^2} \blacktriangleleft$$

## SAMPLE PROBLEM B.7

If $a = 3c$ and $b = 2c$ for the rectangular prism of Sample Prob. B.6, determine (*a*) the principal mass moments of inertia at the origin $O$, (*b*) the principal axes of inertia at $O$.

## SOLUTION

*a. Principal Mass Moments of Inertia at the Origin $O$.*  Substituting $a = 3c$ and $b = 2c$ into the solution to Sample Prob. B.6, we have

$$I_x = \tfrac{5}{3}mc^2 \qquad I_y = \tfrac{10}{3}mc^2 \qquad I_z = \tfrac{13}{3}mc^2$$
$$I_{xy} = \tfrac{3}{2}mc^2 \qquad I_{yz} = \tfrac{1}{2}mc^2 \qquad I_{zx} = \tfrac{3}{4}mc^2$$

Substituting the values of the mass moments and products of inertia into Eq. (B.29) and collecting terms yields

$$K^3 - (\tfrac{28}{3}mc^2)K^2 + (\tfrac{3479}{144}m^2c^4)K - \tfrac{589}{54}m^3c^6 = 0$$

We then solve for the roots of this equation; from the discussion in Sec. B.18, it follows that these roots are the principal mass moments of inertia of the body at the origin.

$$K_1 = 0.568867mc^2 \qquad K_2 = 4.20885mc^2 \qquad K_3 = 4.55562mc^2$$
$$K_1 = 0.569mc^2 \qquad K_2 = 4.21mc^2 \qquad K_3 = 4.56mc^2 \qquad \blacktriangleleft$$

*b. Principal Axes of Inertia at $O$.*  To determine the direction of a principal axis of inertia, we first substitute the corresponding value of $K$ into two of the equations (B.27); the resulting equations together with Eq. (B.30) constitute a system of three equations from which the direction cosines of the corresponding principal axis can be determined. Thus, we have for the first principal mass moment of inertia $K_1$:

$$(\tfrac{5}{3} - 0.568867)\,mc^2(\lambda_x)_1 - \tfrac{3}{2}mc^2(\lambda_y)_1 - \tfrac{3}{4}mc^2(\lambda_z)_1 = 0$$
$$-\tfrac{3}{2}mc^2(\lambda_x)_1 + (\tfrac{10}{3} - 0.568867)\,mc^2(\lambda_y)_1 - \tfrac{1}{2}mc^2(\lambda_z)_1 = 0$$
$$(\lambda_x)_1^2 + (\lambda_y)_1^2 + (\lambda_z)_1^2 = 1$$

Solving yields

$$(\lambda_x)_1 = 0.836600 \qquad (\lambda_y)_1 = 0.496001 \qquad (\lambda_z)_1 = 0.232557$$

The angles that the first principal axis of inertia forms with the coordinate axes are then

$$(\theta_x)_1 = 33.2° \qquad (\theta_y)_1 = 60.3° \qquad (\theta_z)_1 = 76.6° \qquad \blacktriangleleft$$

Using the same set of equations successively with $K_2$ and $K_3$, we find that the angles associated with the second and third principal mass moments of inertia at the origin are, respectively,

$$(\theta_x)_2 = 57.8° \qquad (\theta_y)_2 = 146.6° \qquad (\theta_z)_2 = 98.0° \qquad \blacktriangleleft$$

and

$$(\theta_x)_3 = 82.8° \qquad (\theta_y)_3 = 76.1° \qquad (\theta_z)_3 = 164.3° \qquad \blacktriangleleft$$

In this lesson we defined the *mass products of inertia* $I_{xy}$, $I_{yz}$, and $I_{zx}$ of a body and showed you how to determine the mass moments of inertia of that body with respect to an arbitrary axis passing through the origin $O$. You also learned how to determine at the origin $O$ the *principal axes of inertia* of a body and the corresponding *principal mass moments of inertia*.

**1. Determining the mass products of inertia of a composite body.** The mass products of inertia of a composite body with respect to the coordinate axes can be expressed as the sums of the mass products of inertia of its component parts with respect to those axes. For each component part, we can use the parallel-axis theorem and write Eqs. (B.20)

$$I_{xy} = \bar{I}_{x'y'} + m\bar{x}\bar{y} \qquad I_{yz} = \bar{I}_{y'z'} + m\bar{y}\bar{z} \qquad I_{zx} = \bar{I}_{z'x'} + m\bar{z}\bar{x}$$

where the primes denote the centroidal axes of each component part and where $\bar{x}$, $\bar{y}$, and $\bar{z}$ represent the coordinates of its center of gravity. Keep in mind that the mass products of inertia can be positive, negative, or zero, and be sure to take into account the signs of $\bar{x}$, $\bar{y}$, and $\bar{z}$.

  *a. From the properties of symmetry of a component part,* you can deduce that two or all three of its centroidal mass products of inertia are zero. For instance, you can verify that for a thin plate parallel to the $xy$ plane; a wire lying in a plane parallel to the $xy$ plane; a body with a plane of symmetry parallel to the $xy$ plane; and a body with an axis of symmetry parallel to the $z$ axis, *the mass products of inertia* $\bar{I}_{y'z'}$ *and* $\bar{I}_{z'x'}$ *are zero.*

  For rectangular, circular, or semicircular plates with axes of symmetry parallel to the coordinate axes; straight wires parallel to a coordinate axis; circular and semicircular wires with axes of symmetry parallel to the coordinate axes; and rectangular prisms with axes of symmetry parallel to the coordinate axes, *the mass products of inertia* $\bar{I}_{x'y'}$, $\bar{I}_{y'z'}$, *and* $\bar{I}_{z'x'}$ *are all zero.*

  *b. Mass products of inertia which are different from zero* can be computed from Eqs. (B.18). Although, in general, a triple integration is required to determine a mass product of inertia, a single integration can be used if the given body can be divided into a series of thin, parallel slabs. The computations are then similar to those discussed in the previous lesson for mass moments of inertia.

**2. Computing the mass moment of inertia of a body with respect to an arbitrary axis OL.** An expression for the mass moment of inertia $I_{OL}$ was derived in Sec. B.6 and is given in Eq. (B.19). Before computing $I_{OL}$, you must first determine the mass moments and products of inertia of the body with respect to the given coordinate axes as well as the direction cosines of the unit vector $\boldsymbol{\lambda}$ along $OL$.

**3. Calculating the principal mass moments of inertia of a body and determining its principal axes of inertia.** You saw in Sec. B.7 that it is always possible to find an orientation of the coordinate axes for which the mass products of inertia are zero. These axes are referred to as the *principal axes of inertia* and the corresponding mass moments of inertia are known as the *principal mass moments of inertia* of the body. In many cases, the principal axes of inertia of a body can be determined from its properties of symmetry. The procedure required to determine the principal mass moments and principal axes of inertia of a body with no obvious property of symmetry was discussed in Sec. B.8 and was illustrated in Sample Prob. B.7. It consists of the following steps.

*a. Expand the determinant in Eq. (B.28) and solve the resulting cubic equation.* The solution can be obtained by trial and error or, preferably, with an advanced scientific calculator or with the appropriate computer software. The roots $K_1$, $K_2$, and $K_3$ of this equation are the principal mass moments of inertia of the body.

*b. To determine the direction of the principal axis corresponding to $K_1$,* substitute this value for $K$ in two of the equations (B.27) and solve these equations together with Eq. (B.30) for the direction cosines of the principal axis corresponding to $K_1$.

*c. Repeat this procedure with $K_2$ and $K_3$* to determine the directions of the other two principal axes. As a check of your computations, you may wish to verify that the scalar product of any two of the unit vectors along the three axes you have obtained is zero and, thus, that these axes are perpendicular to each other.

*d. When a principal mass moment of inertia is approximately equal to a mass moment of inertia with respect to a coordinate axis,* the calculated values of the corresponding direction cosines will be very sensitive to the number of significant figures used in your computations. Thus, for this case we suggest you express your intermediate answers in terms of six or seven significant figures to avoid possible errors.

# Problems

**B.37**  Determine the mass products of inertia $I_{xy}$, $I_{yz}$, and $I_{zx}$ of the steel machine element shown. (The specific weight of steel is 7849 kg/m³.)

**Fig. PB.37**

**Fig. PB.38**

**B.38**  Determine the mass products of inertia $I_{xy}$, $I_{yz}$, and $I_{zx}$ of the aluminum casting shown. (The specific weight of aluminum is 2768 kg/m³.)

**B.39** and **B.40**  Determine the mass products of inertia $I_{xy}$, $I_{yz}$, and $I_{zx}$ of the cast aluminum machine component shown. (The density of aluminum is 2700 kg/m³.)

**Fig. PB.39**

**Fig. PB.40**

**B.41 through B.43** A section of sheet steel 3 mm thick is cut and bent into the machine component shown. Knowing that the density of the steel is 7860 kg/m³, determine the mass products of inertia $I_{xy}$, $I_{yz}$, and $I_{zx}$ of the component.

Fig. PB.41

Fig. *PB.42*

Fig. **PB.43**

Fig. *PB.44*

**B.44** A section of sheet steel .2 cm thick is cut and bent into the machine component shown. Knowing that the specific weight of steel is 7849 kg/m³, determine the mass products of inertia $I_{xy}$, $I_{yz}$, and $I_{zx}$ of the component.

**B.45 and B.46** Brass wire with a weight per unit length $w$ is used to form the figure shown. Determine the mass products of inertia $I_{xy}$, $I_{yz}$, and $I_{zx}$ of the wire figure.

Fig. **PB.45**

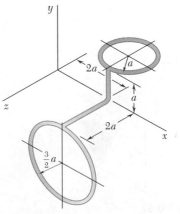

Fig. *PB.46*

**B.47**   The figure shown is formed from 0.2 cm.-diameter aluminum wire. Knowing that the specific weight of aluminum is 2768 kg/m$^3$, determine the mass products of inertia $I_{xy}$, $I_{yz}$, and $I_{zx}$ of the wire figure.

**Fig. PB.47**

**Fig. PB.48**

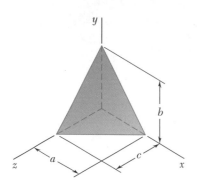

**Fig. PB.50**

**B.48**   A homogeneous wire with a mass per unit length of 1.8 kg/m is used to form the figure shown. Determine the mass products of inertia $I_{xy}$, $I_{yz}$, and $I_{zx}$ of the wire figure.

**B.49**   Complete the derivation of Eqs. (B.20), which express the parallel-axis theorem for mass products of inertia.

**B.50**   For the homogeneous tetrahedron of mass $m$ shown, (a) determine by direct integration the mass product of inertia $I_{zx}$, (b) deduce $I_{yz}$ and $I_{xy}$ from the results obtained in part $a$.

**Fig. PB.51**

**B.51**   The homogeneous circular cylinder shown has a mass $m$. Determine the mass moment of inertia of the cylinder with respect to the line joining the origin $O$ and point $A$ which is located on the perimeter of the top surface of the cylinder.

**B.52**   The homogeneous circular cone shown has a mass $m$. Determine the mass moment of inertia of the cone with respect to the line joining the origin $O$ and point $A$.

**Fig. PB.52**

**B.53**  Shown is the machine element of Prob. B.29. Determine its mass moment of inertia with respect to the line joining the origin $O$ and point $A$.

**Fig. PB.53**

**B.54**  Determine the mass moment of inertia of the steel machine element of Probs. B.33 and B.37 with respect to the axis through the origin which forms equal angles with the $x$, $y$, and $z$ axes.

**B.55**  The thin bent plate shown is of uniform density and weight $W$. Determine its mass moment of inertia with respect to the line joining the origin $O$ and point $A$.

**Fig. PB.55**

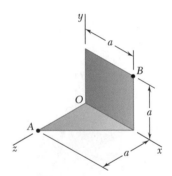

**Fig. PB.56**

**B.56**  A piece of sheet metal of thickness $t$ and density $\rho$ is cut and bent into the shape shown. Determine its mass moment of inertia with respect to a line joining points $A$ and $B$.

**B.57**  Determine the mass moment of inertia of the machine component of Probs. B.24 and B.43 with respect to the axis through the origin defined by the unit vector $\boldsymbol{\lambda} = (-4\mathbf{i} + 8\mathbf{j} + \mathbf{k})/9$.

**B.58 through B.60**  For the wire figure of the problem indicated, determine the mass moment of inertia of the figure with respect to the axis through the origin defined by the unit vector $\boldsymbol{\lambda} = (-3\mathbf{i} - 6\mathbf{j} + 2\mathbf{k})/7$.
 **B.58**  Prob. B.36
 **B.59**  Prob. B.35
 **B.60**  Prob. B.34

**B.61**  For the rectangular prism shown, determine the values of the ratios $b/a$ and $c/a$ so that the ellipsoid of inertia of the prism is a sphere when computed ($a$) at point $A$, ($b$) at point $B$.

**Fig. PB.61**

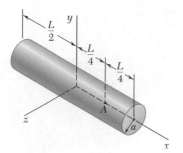

**Fig. PB.63**

**B.62** For the right circular cone of Sample Prob. B.3, determine the value of the ratio $a/h$ for which the ellipsoid of inertia of the cone is a sphere when computed (a) at the apex of the cone, (b) at the center of the base of the cone.

**B.63** For the homogeneous circular cylinder of radius $a$ and length $L$ shown, determine the value of the ratio $a/L$ for which the ellipsoid of inertia of the cylinder is a sphere when computed (a) at the centroid of the cylinder, (b) at point A.

**B.64** Given an arbitrary body and three rectangular axes $x$, $y$, and $z$, prove that the mass moment of inertia of the body with respect to any one of the three axes cannot be larger than the sum of the mass moments of inertia of the body with respect to the other two axes. That is, prove that the inequality $I_x \leq I_y + I_z$ and the two similar inequalities are satisfied. Further, prove that $I_y \geq \frac{1}{2}I_x$ if the body is a homogeneous solid of revolution, where $x$ is the axis of revolution and $y$ is a transverse axis.

**B.65** Consider a cube of mass $m$ and side $a$. (a) Show that the ellipsoid of inertia at the center of the cube is a sphere, and use this property to determine the mass moment of inertia of the cube with respect to one of its diagonals. (b) Show that the ellipsoid of inertia at one of the corners of the cube is an ellipsoid of revolution, and determine the principal mass moments of inertia of the cube at that point.

**B.66** Given a homogeneous body of mass $m$ and arbitrary shape and three rectangular axes $x$, $y$, and $z$ with origin at $O$, prove that the sum $I_x + I_y + I_z$ of the mass moments of inertia of the body cannot be smaller than the similar sum computed for a sphere of the same mass and the same material centered at $O$. Further, using the results of Prob. B.64, prove that if the body is a solid of revolution, where $x$ is the axis of revolution, its mass moment of inertia $I_y$ about a transverse axis $y$ cannot be smaller than $3ma^2/10$, where $a$ is the radius of the sphere of the same mass and the same material.

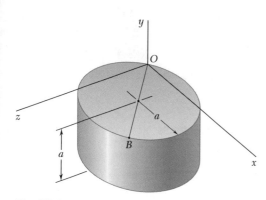

**Fig. PB.67**

**\*B.67** The homogeneous circular cylinder shown has a mass $m$, and the diameter $OB$ of its top surface forms 45° angles with the $x$ and $z$ axes. (a) Determine the principal mass moments of inertia of the cylinder at the origin $O$. (b) Compute the angles that the principal axes of inertia at $O$ form with the coordinate axes. (c) Sketch the cylinder, and show the orientation of the principal axes of inertia relative to the $x$, $y$, and $z$ axes.

**B.68 through B.72** For the component described in the problem indicated, determine (a) the principal mass moments of inertia at the origin, (b) the principal axes of inertia at the origin. Sketch the body and show the orientation of the principal axes of inertia relative to the $x$, $y$, and $z$ axes.

    \*B.68 Prob. B.53
    \*B.69 Probs. B.33 and B.37
    \*B.70 Prob. B.55
    \*B.71 Prob. B.56
    \*B.72 Probs. B.36 and B.58

The second half of the chapter was devoted to the determination of *moments of inertia of masses,* which are encountered in dynamics in problems involving the rotation of a rigid body about an axis. The mass moment of inertia of a body with respect to an axis $AA'$ (Fig. B.16) was defined as

Moments of inertia of masses

$$I = \int r^2\, dm \qquad \text{(B.1)}$$

where $r$ is the distance from $AA'$ to the element of mass [Sec. B.1]. The *radius of gyration* of the body was defined as

$$k = \sqrt{\frac{I}{m}} \qquad \text{(B.2)}$$

The mass moments of inertia of a body with respect to the coordinate axes were expressed as

$$I_x = \int (y^2 + z^2)\, dm$$

$$I_y = \int (z^2 + x^2)\, dm \qquad \text{(B.3)}$$

$$I_z = \int (x^2 + y^2)\, dm$$

We saw that the *parallel-axis theorem* also applies to mass moments of inertia [Sec. B.2]. Thus, the mass moment of inertia $I$ of a body with respect to an arbitrary axis $AA'$ (Fig. B.17) can be expressed as

Parallel-axis theorem

$$I = \bar{I} + md^2 \qquad \text{(B.6)}$$

**Fig. B.16**

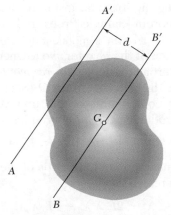

**Fig. B.17**

where $\bar{I}$ is the mass moment of inertia of the body with respect to the centroidal axis $BB'$ which is parallel to the axis $AA'$, $m$ is the mass of the body, and $d$ is the distance between the two axes.

Mass Moments of inertia of thin plates

The mass moments of inertia of *thin plates* can be readily obtained from the moments of inertia of their areas [Sec. B.3]. We found that for a *rectangular plate* the mass moments of inertia with respect to the axes shown (Fig. B.18) are

$$I_{AA'} = \tfrac{1}{12} ma^2 \qquad I_{BB'} = \tfrac{1}{12} mb^2 \qquad \text{(B.12)}$$

$$I_{CC'} = I_{AA'} + I_{BB'} = \tfrac{1}{12} m(a^2 + b^2) \qquad \text{(B.13)}$$

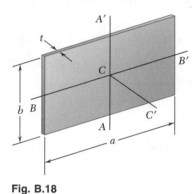

**Fig. B.18**                    **Fig. B.19**

while for a *circular plate* (Fig. B.19) they are

$$I_{AA'} = I_{BB'} = \tfrac{1}{4} mr^2 \qquad \text{(B.14)}$$

$$I_{CC'} = I_{AA'} + I_{BB'} = \tfrac{1}{2} mr^2 \qquad \text{(B.15)}$$

Composite bodies

When a body possesses *two planes of symmetry*, it is usually possible to use a single integration to determine its mass moment of inertia with respect to a given axis by selecting the element of mass $dm$ to be a thin plate [Sample Probs. B.2 and B.3]. On the other hand, when a body consists of *several common geometric shapes*, its mass moment of inertia with respect to a given axis can be obtained by using the formulas given in Fig. B.9 together with the parallel-axis theorem [Sample Probs. B.4 and B.5].

Mass Moment of inertia with respect to an arbitrary axis

In the last portion of the chapter, we learned to determine the mass moment of inertia of a body *with respect to an arbitrary axis* $OL$ which is drawn through the origin $O$ [Sec. B.6]. Denoting by $\lambda_x$, $\lambda_y$, and $\lambda_z$ the components of the unit vector $\boldsymbol{\lambda}$ along $OL$ (Fig. B.20) and introducing the *mass products of inertia*

$$I_{xy} = \int xy\, dm \qquad I_{yz} = \int yz\, dm \qquad I_{zx} = \int zx\, dm \qquad \text{(B.18)}$$

we found that the mass moment of inertia of the body with respect to $OL$ could be expressed as

$$I_{OL} = I_x\lambda_x^2 + I_y\lambda_y^2 + I_z\lambda_z^2 - 2I_{xy}\lambda_x\lambda_y - 2I_{yz}\lambda_y\lambda_z - 2I_{zx}\lambda_z\lambda_x \qquad \text{(B.19)}$$

**Fig. B.20**

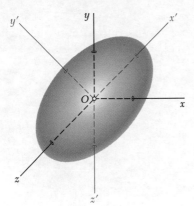

**Fig. B.21**

By plotting a point $Q$ along each axis $OL$ at a distance $OQ = 1/\sqrt{I_{OL}}$ from $O$ [Sec. B.7], we obtained the surface of an ellipsoid, known as the *ellipsoid of inertia* of the body at point $O$. The principal axes $x'$, $y'$, and $z'$ of this ellipsoid (Fig. B.21) are the *principal axes of inertia* of the body; that is, the products of inertia $I_{x'y'}$, $I_{y'z'}$, and $I_{z'x'}$ of the body with respect to these axes are all zero. There are many situations when the principal axes of inertia of a body can be deduced from properties of symmetry of the body. Choosing these axes to be the coordinate axes, we can then express $I_{OL}$ as

$$I_{OL} = I_{x'}\lambda_{x'}^2 + I_{y'}\lambda_{y'}^2 + I_{z'}\lambda_{z'}^2 \qquad \text{(B.23)}$$

where $I_{x'}$, $I_{y'}$, and $I_{z'}$ are the *principal mass moments of inertia* of the body at $O$.

When the principal axes of inertia cannot be obtained by observation [Sec. B.7], it is necessary to solve the cubic equation

$$K^3 - (I_x + I_y + I_z)K^2 + (I_xI_y + I_yI_z + I_zI_x - I_{xy}^2 - I_{yz}^2 - I_{zx}^2)K$$
$$- (I_xI_yI_z - I_xI_{yz}^2 - I_yI_{zx}^2 - I_zI_{xy}^2 - 2I_{xy}I_{yz}I_{zx}) = 0 \qquad \text{(B.29)}$$

We found [Sec. B.8] that the roots $K_1$, $K_2$, and $K_3$ of this equation are the principal mass moments of inertia of the given body. The direction cosines $(\lambda_x)_1$, $(\lambda_y)_1$, and $(\lambda_z)_1$ of the principal axis corresponding to the principal mass moment of inertia $K_1$ are then determined by substituting $K_1$ into Eqs. (B.27) and solving two of these equations and Eq. (B.30) simultaneously. The same procedure is then repeated using $K_2$ and $K_3$ to determine the direction cosines of the other two principal axes [Sample Prob. B.7].

Ellipsoid of inertia

Principal axes of inertia
Principal mass moments of inertia

# C

## Fundamentals of Engineering Examination

Engineers are required to be licensed when their work directly affects the public health, safety, and welfare. The intent is to ensure that engineers have met minimum qualifications involving competence, ability, experience, and character. The licensing process involves an initial exam, called the *Fundamentals of Engineering Examination*, professional experience, and a second exam, called the *Principles and Practice of Engineering*. Those who successfully complete these requirements are licensed as a *Professional Engineer*. The exams are developed under the auspices of the *National Council of Examiners for Engineering and Surveying*.

The first exam, the *Fundamentals of Engineering Examination*, can be taken just before or after graduation from a four-year accredited engineering program. The exam stresses subject material in a typical undergraduate engineering program, including dynamics. The topics included in the exam cover much of the material in this book. The following is a list of the main topic areas, with references to the appropriate sections in this book. Also included are problems that can be solved to review this material.

**Kinematics (11.1–11.6; 11.9–11.14; 15.2–15.8)**
Problems: 11.4, 11.5, 11.38, 11.64, 11.69, 11.102, 15.4, 15.40, 15.59, 15.64, 15.79, 15.112, 15.140

**Force, Mass, and Acceleration (12.1–12.6; 16.2–16.8)**
Problems: 12.6, 12.5, 12.30, 12.32, 12.37, 12.48, 12.53, 12.61, 16.2, 16.7, 16.13, 16.22, 16.31, 16.43, 16.56, 16.58, 16.73, 16.79, 16.131

**Work and Energy (13.1–13.6; 13.8, 17.1–17.7)**
Problems: 13.6, 13.8, 13.15, 13.18, 13.45, 13.46, 13.47, 13.56, 13.61, 13.66, 17.2, 17.4, 17.20, 17.23

**Impulse and Momentum (13.10–13.15; 17.8–17.12)**
    Problems: 131.123, 13.130, 13.140, 13.146, 13.157, 13.160, 13.173, 17.45, 17.49, 17.60, 17.63, 17.89, 17.95, 17.98

**Vibrations (19.1–19.3; 19.5–19.7)**
    Problems: 19.1, 19.3, 19.12, 19.19, 19.24, 19.28, 19.50, 19.57, 19.63, 19.77, 19.84, 19.86, 19.99, 19.105, 19.113

**Friction** (Problems involving friction occur in each of the above subjects.)

# PHOTO CREDITS

## CHAPTER 11

**Opener:** © Royalty-Free/CORBIS; **p. 605:** © Jonathan Nouroek/ PhotoEdit; **p. 646:** © Steven E. Sutton/CORBIS; **p. 647:** © U.S Navy; **p. 664:** © Phyllis Picardi/Index Stock Imagery.

## CHAPTER 12

**Opener:** © George Tiedemann/NewSport/CORBIS; **p. 698:** © All-sport Concepts/Getty Images: **p. 718:** © Kevin R. Morris/CORBIS; **p. 720:** © J. Von Rummelhoff, Mediaworks, University of California, Davis; **p. 732:** © NASA.

## CHAPTER 13

**Opener:** © Tim Pannell/CORBIS; **p. 808:** © BANUT, Sweden; **p. 823:** © Andrew Davidhazy/RIT; **p. 824:** © Tom McCarthy/Index Stock; **p. 826:** © Sabina Dowell.

## CHAPTER 14

**Opener:** © AP Photos/ St. Joseph News-Press, Ival Lawhon Jr.; **p. 864:** © NASA; **p. 875:** © Peter M. Fisher/CORBIS; **p. 888:** © The Boeing Company; **p. 890:** © NASA.

## CHAPTER 15

**Opener:** © Ford Motor Company; **p. 920:** © CORBIS; **p. 922:** © Joseph Nettis/Stock Boston Inc./PictureQuest; **p. 958:** © Royalty-Free/CORBIS; **p. 986:** © Northrop Grumman/Index Stock Imagery/ PictureQuest; **p. 1000:** © Stocktrek/CORBIS.

## CHAPTER 16

**Opener:** © dpa/HO/Landov; **p. 1027:** © Neil Rabinowitz/CORBIS; **p. 1030:** © Larry Bray/Getty Images; **p. 1032:** © Tony Arruza/ CORBIS; **p. 1054:** © Robert E. Daemmrich.

## CHAPTER 17

**Opener:** © ADREES LATIF/Reuters/Landov; **p. 1087:** © Patrick Giardino/CORBIS; **p. 1107 (both):** © Photography by Leah; **p. 1122:** Courtesy MTS Systems Corporation.

## CHAPTER 18

**Opener:** © age fotostock/SuperStock; **p. 1150:** © SuperStock; **p. 1152:** © Matthias Kulka/CORBIS; **p. 1168:** © Roger Ressmeyer/CORBIS; **p. 1169:** Courtesy of Caterpillar Engine Division.

## CHAPTER 19

**Opener:** © Ford Motor Company; **p. 1252:** © Tony Freeman/Index Stock; **p. 1264:** © Sabina Dowell.

# Index

Absolute acceleration, 956–958
Absolute motion of a particle, 647
Absolute system of units, 695
Absolute velocity, 934–935
Acceleration, 602, 632, 642
  absolute, 956–958
  angular, 920, 985
  components of, normal, 663–665
    radial, 666–667
    rectangular, 645–646
    tangential, 663–665
    transverse, 666–667
  Coriolis: in plane motion, 973–975
    in three-dimensional motion, 998–999
  in general rigid-body motion, 987
  of gravity, 618, 722
  in motion of rigid body about a fixed point, 985
  in plane motion, 956–958
  relative: of a particle: in general motion, 987
    in plane motion, 956–958
    with respect to a frame in general motion, 999–1000
    with respect to a frame in translation, 647
    with respect to a rotating frame, 973–975, 998–999
  in rotation, 920–921, 998–1000
  in three-dimensional motion, 998–999
  of two particles, 648–649
Acceleration-time curve, 604–605, 630
Amplitude, 1214, 1216
Analogue, electrical, 1265–1268
Angle
  phase, 1216
Angular acceleration, 920–921, 985
Angular coordinate, 919–920
Angular impulse, 875–876
Angular momentum
  conservation of, 720–721, 864, 1107
  of a particle, 718–721
  rate of change of, 718–719, 860, 863, 1027, 1165–1166
  of a rigid body, 1027–1028, 1105–1107, 1147–1151
  of a system of particles, 859–860, 862–864

Angular velocity, 920–921, 985–986
Apogee, 734
Areal velocity, 721
Auxiliary circle, 1216
Axioms of mechanics, 1030–1031
Axis
  of precession, 1186–1187
  of rotation, 916
    instantaneous, 945, 998
  of spin, 1185
Axisymmetrical body under no force, motion of, 1186–1187

Balancing of rotating shafts, 1169
Ballistic missiles, 734
Banking of curves, 698
Binormal, 665
Body centrode, 947
Body cone, 985, 1187
Borda's mouthpiece, 900

Cardan's suspension, 1183
Cathode-ray tube, 717
Center. *See also* Mass center
  of oscillation, 1229
  of percussion, 1062, 1114
  of rotation, instantaneous, 945–946
Central force, 720–721, 786
Central impact, 820–826
Centrifugal force, 699, 1053
Centrode, 947
Centroidal frame of reference, 863, 864, 1027, 1147, 1152–1153
Centroidal rotation
  effective forces in, 1030
  kinetic energy in, 1084–1085
  momentum in, 1106, 1107
Circular frequency, 1216
Circular orbit, 734
Coefficient
  of critical damping, 1262

Coefficient—*Cont.*
   of damping, 1261
   of restitution, 821–823, 1120, 1122
   of viscous damping, 1261
Complementary acceleration, 974
Complementary function, 1264
Components
   of acceleration (*see* Acceleration, components of)
   of angular momentum: of a particle, 720–721
       of a rigid body, 1149–1151
   of derivative of a vector function, 644
   of velocity (*see* Velocity)
Composite bodies
   moments of inertia of, 1300
Compound pendulum, 1237
Conic section, 732
Conservation
   of angular momentum, 720–721, 864, 1107
   of energy: for particles, 784–785
       for rigid bodies, 1186–1187
       for a system of particles, 864, 875
       in vibrations, 1240–1241
   of linear momentum, 695, 805–807, 864
   of momentum, 864–865
Conservative force, 783–785, 786
Constrained plane motion, 1051–1054
Coordinate
   angular, 919–920
   position, 603, 604
       relative, 618
Coriolis acceleration
   in plane motion, 973–975
   in three-dimensional motion, 998–999
Couples
   gyroscopic, 1185
   inertia, 1031
   momentum, 1105, 1106–1107, 1151–1152
Critical damping, coefficient of, 1262
Cross product. *See* Vector product
Curvature, radius of, 664–665
Curvilinear motion of a particle
   kinematics of, 641–679
   kinetics of, 697–744
Curvilinear translation, 916
Customary units, U.S., 696–697
Cylindrical coordinates, 667

D'Alembert, Jean, 1029
D'Alembert's principle, 1029, 1167

Damped vibrations. *See* Vibrations, damped
Damping
   coefficient of, 1261
   critical, 1262
   heavy, 1262
   light, 1262
   viscous, 1261
Damping coefficient, 1261
Damping factor, 1263
Deceleration, 605
Decrement, logarithmic, 1271
Deformation, period of, 820, 821, 1120
Density, 1299
Dependent motion, 619
Derivative
   of product of scalar and vector functions, 641
   of scalar product, 641
   of vector function, 643–645
   of vector product, 644
Determinant form
   for angular momentum, 718
Diagram
   acceleration-time, 604–605, 630
   displacement-time, 603
   velocity-time, 603, 632
Differential elements
   for moments of inertia: of areas
       of masses, 1300
Direct central impact, 820–823
Direct precession, 1187
Distance, 603
Dynamic balancing, 1169
Dynamic equilibrium
   of a particle, 699
   of a rigid body: in noncentroidal rotation, 1052
       in plane motion, 1031
Dynamics, definition of, 602

Earth satellites, 722, 732, 786
Eccentric impact, 1120–1122
Eccentricity, 732, 1260
Effective forces, 856
   for a rigid body: in plane motion, 1029, 1030
       in three dimensions, 1167
Efficiency, 764
Elastic impact, 822–823
Electrical analogue, 1265–1268
Ellipsoid
   of inertia, 1316–1317
   Poinsot, 1198

Elliptic orbit, 733, 734
Energy
    conservation of (*see* Conservation, of energy)
    kinetic: of a particle, 760–762
        of a rigid body: in plane motion, 1084–1085
            in rotation, 1085–1086
            in three dimensions, 1152–1153
        of a system of particles, 873–874
    potential, 781–782, 785, 1086
    total mechanical, 784–785
Equations
    of motion: for an axisymmetrical body, 1186–1187
        for a gyroscope, 1183–1184
        for a particle, 697–698, 719–721
        for a system of particles, 860–862, 863, 868–869
        for a three-dimensional body
            about a fixed point, 1167–1168
            about its mass center, 1165–1166
            in general motion, 1146, 1165
            in rotation about a fixed axis, 1168–1169
        for a two-dimensional body: in noncentroidal
            rotation, 1052–1053
            in plane motion, 1029
Equilibrium
    dynamic (*see* Dynamic equilibrium)
Equinoxes, precession of, 1193
Equipollence of external forces and effective forces,
        856, 1027, 1146
Equivalence of external forces and effective forces for
        a rigid body, 1029, 1166–1167
Escape velocity, 616, 734
Euler, Leonhard, 1166
Eulerian angles, 1183–1184
Euler's equations of motion, 1166
Euler's theorem, 984
External forces, 856


Fan, 889
Fluid flow, 886–889
Fluid friction, 1261
Focus of conic section, 732
Foot, 696, 697
Force
    central, 720, 721, 786
    centrifugal, 699, 1053
    conservative, 783–784, 785, 786, 827
    effective (*see* Effective forces)
    external, 856
    gravitational, 721–722, 759–760, 781–782, 786

Force—*Cont.*
    impulsive, 808
    inertia, 699
    internal, 856–857
    nonconservative, 785
    nonimpulsive, 808
    reversed effect (*see* Inertia vector)
Forced frequency, 1253
Forced vibrations
    damped, 1264–1265
    undamped, 1251–1253
Frame of reference, 698
    centroidal, 863, 864, 1147–1148, 1152–1153
    in general motion, 999–1000
    newtonian, 693–694
    rotating, 971–975, 998–999
    in translation, 644, 646, 647
Free precession, 1187, 1188
Free vibrations
    damped, 1261–1263
    undamped, 1214–1241
Free-body-diagram equation, 1031
Freely falling body, 618
Frequency, 1214
    circular, 1215
    forced, 1251, 1253
    natural, 1217
Frequency ratio, 1253
Friction
    fluid, 1261


Gears
    analysis of, 935, 959
    planetary, 941
General motion of a rigid body, 987
Geneva mechanism, 976, 977
Geosynchronous orbit, 728
Gradient, 784–785
Gram, 696
Graphical methods for solution of rectilinear-motion
        problems, 630, 631
Gravitation
    constant of, 722
    Newton's law, 721–722
Gravitational forces, 721–722, 759–760, 781–782, 786
Gravitational potential energy, 781–782
Gravitational system of units, 696
Gravity
    acceleration of, 616, 722

Gun, recoil, 1272
Gyration, radius of, 1297
Gyroscope, 1183–1186
Gyroscopic couple, 1185

Harmonic motion, simple, 1214–1218
Helicopter, 899, 1118
Hertz (unit), 1217
Hodograph, 642
Horsepower, 764
Hydraulic jump, 901
Hyperbolic trajectory, 733

Impact, 820–826
    central: direct, 820–823
        oblique, 823–826
    eccentric, 1120–1122
    elastic, 822–823
    line of, 820
    plastic, 822
Impulse
    angular, 875, 876
    linear, 805–806
    and momentum, principle of (*see* Principle of
        impulse and momentum)
Impulsive force, 807–808
Impulsive motion, 807–808, 1120
Inclined axes, moments of inertia, 1315, 1317
Inertia
    ellipsoid of, 1316–1317
    moments of (*see* Moments of inertia)
    principal axes of: for an area
        for a mass, 1316–1319
    products of: of areas
        of masses, 1315–1316
        parallel-axis theorem for, 1316
Inertia couple, 1031
Inertia force, 699
Inertia tensor, 1149
Inertia vector
    for a particle, 699
    for a rigid body in plane motion, 1031
Inertial system, 694*n.*
Input power, 764
Instantaneous axis of rotation, 945–946, 998
Instantaneous center of rotation, 945–947
Internal forces, 856–857
International system of units, 695–696
Invariable plane, 1198

Jet engine, 888

Kepler, Johann, 735
Kepler's laws, 735
Kilogram, 695
Kilometer, 696
Kilonewton, 696
Kilowatt, 763
Kinematics
    of particles, 602–683
        in curvilinear motion, 641–683
        in rectilinear motion, 603–634
        in relative motion, 618–619, 646, 647, 973–975,
            998–1000
    of rigid bodies: in general motion, 987
        in motion about a fixed point, 984–985, 986
        in plane motion, 932–958
        in rotation about a fixed axis, 919, 921
        in translation, 918–919
Kinetic energy. *See* Energy, kinetic
Kinetics, 602
    of particles, 692–843
    of rigid bodies, 1026–1201
    of systems of particles, 856–905

Laws
    Kepler's, 735
    Newton's (*see* Newton's law)
Linear impulse, 806
Linear momentum
    conservation of, 694–695, 806, 864
    of a particle, 694–695
    rate of change of, 859–860
    of a rigid body, 694–695, 1105–1106, 1151–1152
    of a system of particles, 859–860
Logarithmic decrement, 1271

Magnification factor, 1253, 1265
Mass, 693
Mass center
    of a rigid body, 1029
    of a system of particles, 860–886, 1104–1107
Mass moments of inertia, 1296–1317
Mass products of inertia, 1315–1316
Mechanical efficiency, 764
Mechanical energy, 785
Mechanics
    principles of, 1030–1031

Megagram, 696
Meter, 695
Metric ton, 696*n.*
Metric units, 695–696
Millimeter, 696
Moment
    of momentum (*see* Angular momentum)
Moment-area method, 631
Moments of inertia, 1296–1317
    of common geometric shapes, 1301
    of composite bodies, 1300
    determination of, by integration, 1300
    inclined axes, 1315, 1317
    of masses, 1296–1317
        parallel-axis theorem for, 1298
        principal, 1316–1319
        rectangular, 1297
        of thin plates, 1299–1300
Momentum
    angular (*see* Angular momentum)
    conservation of, 864–865
    linear (*see* Linear momentum)
    of a particle, 694–695
    of a rigid body, 1104–1107, 1147–1151
Momentum couple, 1105, 1106, 1150, 1151
Momentum vector, 1105, 1106, 1150, 1151
Motion
    about a fixed point, 984–986, 1167–1168
    about mass center, 862–864, 1035, 1165–1167
    absolute, 647
    of axisymmetrical body under no force, 1186–1188
    under a central force, 731–732
    curvilinear, 692–747
    equations of (*see* Equations, of motion)
    general, 987
    harmonic, simple, 1214–1218
    impulsive, 808, 1120
    of mass center, 860–862, 1035, 1167
    Newton's laws of (*see* Newton's laws of motion)
    of a particle, 602–843
    plane (*see* Plane motion)
    rectilinear, 603–640
    relative (*see* Relative motion)
    of a rigid body, 916–1202
    rolling, 1053–1054
    of a system of particles, 856–903
Motion curves, 606, 630–631

Natural Frequency, 1217
Newtonian frame of reference, 693–694

Newton's law
    of gravitation, 721–722
    of motion: first
        second, 693
Noncentroidal rotation
    dynamic equilibrium in, 1052
    effective forces in, 1053–1054
    equations of motion in, 1051, 1052
    kinetic energy in, 1086
    momentum in, 1106
Nonconservative force, 785
Nonimpulsive force, 808
Normal components of acceleration, 663–664
Nutation, 1183

Oblique central impact, 823–826
Orbit, 733–734
    geosynchronous, 728
Oscillation, center of, 1229
Oscillations
    of a rigid body, 1186–1187, 1240–1241
    of a simple pendulum, 1229
Osculating plane, 665
Output power, 764

Parabolic trajectory, 646, 733–734
Parallel-axis theorem
    for moments of inertia: of areas
        of masses, 1298
    for products of inertia: of areas
        of masses, 1316
Particles
    kinematics of (*see* Kinematics, of particles)
    kinetics of, 692–843
    relative motion of, 618–619, 646–647,
        973–977, 987
    systems of (*see* Systems, of particles)
    vibrations of (*see* Vibrations)
Pendulum
    compound, 1237
    simple, 1218–1219
Percussion, center of, 1062, 1114
Perigee, 734
Period
    of deformation, 820, 821, 1120
    of restitution, 820, 821, 1120
    of vibration, 1214, 1216–1217
        damped, 1263
        undamped, 1216–1217

Periodic time, 732–735
Phase angle, 1216–1217
Phase difference, 1265
Plane motion, 917
    constrained, 1051–1054
    Coriolis acceleration in, 973–975
    dynamic equilibrium in, 1031
    effective forces in, 1029, 1030
    equations of motion in, 1029
    kinematics of, 932–958
    kinetic energy in, 1084–1085
    momentum in, 1104–1106
Planetary gears, 941
Planetary motion, 735
Plastic impact, 822
Poinsot ellipsoid, 1198
Polar coordinates, 666
Polhode, 1198
Position coordinate, 602, 603
    relative, 618
Position vector, 641
    relative, 647
Potential energy, 781–785, 1086–1087
Potential function, 783
Pound force, 696–697
Pound mass, 697
Power, 763–764, 1087
Precession, 1183, 1185–1188
    direct, 1187
    of the equinoxes, 1193
    free, 1187, 1188, 1193
    retrograde, 1187
    steady, 1185–1188
Principal axes of inertia
    of masses, 1316–1317
Principal moments of inertia
    of masses, 1316
Principal normal, 665
Principle
    of impulse and momentum: for a particle, 805–807
        for a rigid body, 1104–1106, 1151–1152
        for a system of particles, 875–876
    of work and energy: for a particle, 760–763
        for a rigid body, 1083–1084, 1152–1153
        for a system of particles, 875
Principles of mechanics, 1030–1031
Product
    of inertia: of an area
        of mass, 1315–1316
    vector triple, 920

Projectile, 646, 698
Propeller, 889

Radial component
    of acceleration, 666–667
    of velocity, 666–667
Radius
    of curvature, 664–665
    of gyration, 1297
Rate of change
    of angular momentum, 718–719, 836, 1027, 1165–1166
    of linear momentum, 859–860
    of a vector, 644–645
        with respect to a rotating frame, 971–973
Rated speed, 703
Rectangular components
    of acceleration, 645–646
    of angular momentum: of a particle, 718–719
        of a rigid body, 1148–1151
    of derivative of a vector function, 644
    of velocity, 645–646
Rectilinear motion of a particle, 603–640
    uniform, 616
    uniformly accelerated, 617–618
Rectilinear translation, 916
Rectilinear-motion problems, solution of
    analytical, 613–616
    graphical, 630–631
Relative acceleration. *See* Acceleration, relative
Relative motion
    of a particle: with respect to a frame in general motion, 999–1000
        with respect to a frame in translation, 646–647
        with respect to a rotating frame, 973–975, 998–999
    of two particles, 618–619, 646–647
Relative position, 618, 647
Relative velocity
    of a particle: in general motion, 987
        in plane motion, 934–935
        with respect to a frame in general motion, 999–1000
        with respect to a frame in translation, 647
        with respect to a rotating frame, 973–975, 998
    of two particles, 618, 646–647
Resonance, 1253
Restitution
    coefficient of, 821–823, 1120, 1122
    period of, 820, 821, 1120

Retrograde precession, 1187
Reversed effective force. *See* Inertia vector
Revolution
    body of, 1300
Rigid body
    kinematics of (*see* Kinematics, of rigid bodies)
    kinetics of, 1026–1201
    vibrations of, 1228–1229, 1241
Rocket, 890
Rolling motion, 1053–1054
Rotating frame of reference, 971–975, 998–999
Rotating shafts, 1169
Rotation, 916–918
    centroidal (*see* Centroidal rotation)
    dynamic equilibrium in, 1052
    effective forces in, 1030, 1052
    equations of motion in, 1052–1054
    instantaneous axis of, 945–946, 998
    instantaneous center of, 945–947
    kinematics of, 919–921
    kinetic energy in, 1084–1085
    momentum in, 1105, 1106
    noncentroidal (*see* Noncentroidal rotation)
    uniform, 922
    uniformly accelerated, 922


Satellites, 722, 732, 734, 736, 790
Scalar product
    derivative of, 641
Second, 695, 696
Semimajor axis, 735
Semiminor axis, 735
Shafts, rotating, 1169
SI units, 695–696
Simple harmonic motion, 1214–1218
Simple pendulum, 1218–1219
Slipstream, 889
Slug, 696
Space centrode, 947
Space cone, 985, 1187
Space mechanics, 732–735, 786, 790
Space shuttle, 902
Space vehicles, 732, 734, 741–743, 799
Speed, 604, 641–642. *See also* Velocity
    rated, 703
Spin, 1180
Spring
    force exerted by, 759
    potential energy, 782–783

Spring constant, 759
Steady precession
    of gyroscope, 1185–1186
    of top, 1191
Steady-state vibrations, 1253, 1264, 1265
Stream of particles, 886–889
Systems
    of particles: angular momentum of, 859–860,
        862–864
        equations of motion for, 857, 858, 860–864
        impulse-momentum principle for, 875–876
        kinetic energy of, 873–874
        mass center of, 860–862
        variable, 886
        work-energy principle for, 875
    of units, 695–697


Tangential component of acceleration, 663–664
Tensor, inertia, 1149
Thrust, 889, 890
Top
    steady precession of, 1191
Torsional vibrations, 1231
Trajectory
    of projectile, 646, 734
    of space vehicle, 733–734
Transfer formula. *See* Parallel-axis theorem
Transient vibrations, 1264
Translation, 916
    curvilinear, 916
    effective forces in, 1030
    kinematics of, 918–919
    kinetic energy in, 1086
    momentum in, 1105
    rectilinear, 916
Transmissibility, principle of, 1030
Transverse component
    of acceleration, 668–669
    of velocity, 668–669


Unbalanced disk, 1054
Uniform rectilinear motion, 616
Uniform rotation, 992
Uniformly accelerated rectilinear motion, 617–618
Uniformly accelerated rotation, 922
Units, 695–697. *See also specific systems of units*
Universal joints, 996
U.S. customary units, 696–697

Variable systems of particles, 886
Vector function, 641, 644
    derivative of, 643–645
Vector product
    derivative of, 641
Vector triple product, 920
Vectors
    inertia, 699, 1027
    momentum, 1105, 1106, 1150, 1151
    position, 641, 647
    rate of change of, 971–973
Velocity, 603, 641–642
    absolute, 934–935, 945–947
    angular, 920–921, 985, 986
    areal, 721
    components of: radial, 666– 667
        rectangular, 645–646
        transverse, 666–667
    escape, 734
    in general rigid-body motion, 987, 988
    in motion of rigid body about a fixed point, 985
    in plane motion, 934–935, 945–947
    relative (*see* Relative velocity)
    in rotation, 919–920
Velocity-displacement curve, 631
Velocity-time curve, 630, 631

Vibrations, 1214–1281
    damped: forced, 1264–1265
        free, 1261–1263
    forced, 1251–1253
    free, 1214–1251, 1261–1263
    frequency of, 1217–1218
    period of (*see* Period, of vibration)
    of rigid bodies, 1228–1229, 1240–1241
    steady-state, 1253, 1264, 1267
    torsional, 1231
    transient, 1253, 1264, 1267
    undamped: forced, 1251–1253
        free, 1214–1241
Viscous damping, 1261

Watt, 764
Weight, 695–696, 722
Wheels, 1053–1054
Work
    and energy, principle of (*see* Principle, of work and
        energy)
    of a force, 756–760
    of force exerted by spring, 759
    of forces on a rigid body, 1083–1084
    of gravitational force, 759–760
    of a weight, 758–759

# Answers to Problems

Answers to problems with a number set in straight type are given on this and the following pages. Answers to problems set in italic are not listed. Answers to computer problems are given at www.mhhe.com/beerjohnston8.

## CHAPTER 11

| | |
|---|---|
| **11.1** | (a) $t = 3.33$ s. (b) $x = 11.07$ m, $v = 6.33$ m/s. |
| **11.2** | (a) $t = 0.333$ s. (b) $x = -2.74$ m, $v = 3.67$ m/s. |
| **11.3** | $x = 52$ m, $v = 115$ m/s, $a = 192$ m/s$^2$. |
| **11.4** | $x = 562$ m, $v = 770$ m/s, $a = 764$ m/s$^2$. |
| **11.5** | $x = 240$ mm, $v = 4.39$ m/s, $a = -24.0$ m/s$^2$. |
| **11.6** | $x = 24.0$ m, $a = -43.9$ mm/s$^2$. |
| **11.9** | $v = 1.427$ m/s, $x = 0.363$ m. |
| **11.10** | $v = 0.1273$ m/s, $x = 0.598$ m. |
| **11.11** | $v = -1.360$ m/s, $x = 0.393$ m. |
| **11.12** | $v = -1070$ mm/s, $x = -520$ mm, 2580 mm. |
| **11.13** | $v = 0.3$ m/s, $x = -10.38$ m, 0.975 m. |
| **11.16** | (a) 1.921 m/s. (b) 0.0981 m, 0.402 m. |
| **11.17** | $x = 0.271$ m, $v = 0.455$ m/s. |
| **11.18** | 3.84 m$^{-2}$. |
| **11.19** | $v = 34.3$ m/s, $x = 0.779$ m. |
| **11.22** | 67.5 m. |
| **11.23** | (a) 187.5 mm. (b) 11.51 s. |
| **11.24** | (a) 56.8 m. (b) ∞. |
| **11.25** | (a) 5.74 m/s. (b) 1.079 s. |
| **11.26** | (a) 22.5 m. (b) 38.4 m/s. |
| **11.29** | (a) $x = 29.8$ m, $a = 6.65$ m/s$^2$. (b) $x = 122.5$ m, $a = 6.30$ m/s$^2$. |
| **11.30** | (a) 7.15 km. (b) $-52.1 \times 10^{-6}$ m/s$^2$. (c) 49.9 min. |
| **11.34** | (a) $x = 2.36v_0 T$, $a = \pi v_0 / T$. (b) $0.363v_0$. |
| **11.35** | (a) 3.05 m/s$^2$. (b) 8.77 m/s$^2$. |
| **11.36** | (a) 125.3 m. (b) 3.17 s. |
| **11.37** | (a) 1.8 m/s$^2$. (b) 54 m/s. |
| **11.38** | (a) 6 s. (b) 55.8 m. |
| **11.41** | 11.60 s, 50.4 m. |
| **11.42** | (a) 9.63 s, 68.0 m. (b) $v_A = 35.8$ km/h, $v_B = 9.13$ km/h. |
| **11.43** | (a) 340 m. (b) 2.8 m. |
| **11.44** | (a) 42.67 m/s ↑. (b) 39.24 m/s ↑. |
| **11.45** | (a) 2.08 m/s$^2$. (b) 3.27 m/s$^2$. |
| **11.46** | (a) 3.59 m/s$^2$. (b) 7.08 s. (c) 278 m. |
| **11.49** | (a) 4 m/s ↑. (b) 1 m/s ↓. (c) 3 m/s ↓. |
| **11.50** | (a) $\mathbf{a}_A = 0.04$ m/s$^2$ ↑; $\mathbf{a}_C = 0.16$ m/s$^2$ ↓. (b) $\mathbf{v}_B = 0.16$ m/s ↑; 0.16 m. |
| **11.51** | (a) 2.4 m/s ↑. (b) 1.2 m/s ↑. |
| **11.52** | (a) 0.24 m/s$^2$ ↓. (b) 0.96 m/s$^2$ ↑. |
| **11.53** | (a) 600 mm/s →. (b) 1200 mm/s ←. (c) 900 mm/s ←. |
| **11.54** | (a) $\mathbf{a}_A = 50.8$ mm/s$^2$ →; $\mathbf{a}_B = 25.4$ mm/s$^2$ ←. (b) $\mathbf{v}_B = 152.5$ mm/s ←; 458 mm ←. |
| **11.57** | (a) 50 mm/s →. (b) 0. (c) $\mathbf{a}_A = 40$ mm/s$^2$ ←; $\mathbf{a}_C = 30$ mm/s$^2$ ←. |
| **11.58** | (a) $\mathbf{a}_B = 105$ mm/s$^2$ ←; $\mathbf{a}_C = 150$ mm/s$^2$ →. (b) $\mathbf{v}_A = 120$ mm/s ←; $\mathbf{v}_B = 180$ mm/s →. (c) 360 mm/s ←. |
| **11.59** | (a) 5.07 s. (b) 0.036 cm. ↓. |
| **11.60** | (a) 3 s. (b) 19 cm. ↓. |
| **11.63** | (b) $x = -4$ m, $v = 4$ m/s, 44 m. |
| **11.64** | (b) 5.83 s. |
| **11.65** | (a) 49.1 m. (b) 18 s, 30 s. |
| **11.66** | (a) 80.2 m. (b) 8 s, 36 s. |
| **11.69** | (a) 300 s. (b) 1.778 m/s. (c) 0.0474 m/s$^2$. |
| **11.70** | 1.64 m/s$^2$. |
| **11.71** | (a) 0.609 s. (b) $v = 0.38$ m/s, $x = 2.93$ m. |
| **11.72** | (a) $t_A = 50$ s, $t_B = 50.2$ s. (b) 3 m. |
| **11.75** | 622 m. |
| **11.76** | 3.97 s after the elevator starts. |
| **11.77** | 11.23 s. |
| **11.78** | 10.06 s, 112.0 km/h. |
| **11.79** | (a) 1.973 s. (b) 0.365 m/s, 0.1825 m/s. |
| **11.80** | (a) 9.58 min. (b) 31.3 km/h. |
| **11.83** | (a) 1.913 m/s. (b) 0.836 m. |
| **11.84** | (a) 97.7 m/s. (b) 1980 m. |
| **11.85** | (a) 3.19 s. (b) 62.6 m. |
| **11.86** | (a) 5.6 m/s$^2$. (b) 3.33 m/s$^2$. |
| **11.88** | $-8$ m. |
| **11.91** | (a) $\mathbf{v} = 8.25$ m/s ⬊ 76.0°; $\mathbf{a} = 24.1$ m/s$^2$ ∡ 85.2°. (b) $\mathbf{v} = 3.82$ m/s ⬊ 38.3°; $\mathbf{a} = 5.15$ m/s$^2$ ∡ 67.2°. |
| **11.92** | (a) $\mathbf{v} = 6$ m/s ⬊ 36.9°; $\mathbf{a} = 6$ m/s$^2$ ⬊ 36.9°. (b) $\mathbf{v} = 9$ m/s ⬊ 36.9°; $\mathbf{a} = 0$. (c) $\mathbf{v} = 6$ m/s ⬊ 36.9°; $\mathbf{a} = 6$ m/s$^2$ ⬈ 36.9°. |
| **11.93** | (a) 9.42 m/s ↑. (b) 7.26 m/s ←. (c) 3.14 m/s ↓. |
| **11.96** | (a) $\mathbf{r} = 0.5$ m ↑; $\mathbf{v} = 1.09$ m/s ⬊ 46.3°; $\mathbf{a} = 2.38$ m/s$^2$ ⬀ 85.4°. (b) $\mathbf{r} = 0.5$ m ⬊ 6.0°; $\mathbf{v} = 0.14$ m/s ∡ 31.8°; $\mathbf{a} = 0.2$ m/s$^2$ ⬈ 86.9°. |
| **11.97** | $v = \sqrt{R^2(1 + \omega_n^2 t^2) + c^2}$, $a = \omega_n R \sqrt{4 + \omega_n^2 t^2}$. |
| **11.98** | (a) $v = 3$ m/s, $a = 3.61$ m/s$^2$. (b) 3.82 s. |
| **11.99** | (a) 2.94 s. (b) 84.9 m. (c) 10.62 m. |
| **11.100** | 0.625 m $< h <$ 2.31 m. |
| **11.101** | (a) 5.0 m/s. (b) 4.8 m/s $< v_0 <$ 6.1 m/s. |
| **11.102** | (a) 264.4 m. (b) 161.3 m. |
| **11.105** | (a) Yes. (b) 0.937 m. |
| **11.106** | (a) 2.38 m. (b) 6.46 m. |
| **11.107** | (a) 6.4 m/s. (b) 1.33 m. |
| **11.108** | (a) $v_0 = 36.1$ m/s, $d = 18.3$ m. (b) $v_0 = 38.7$ m/s, $d = 26.6$ m. |
| **11.109** | 6.98 m/s. |
| **11.110** | (a) 9.34 m/s. (b) 9.27 m/s. |
| **11.113** | (a) 3.96°. (b) 375 m. (c) 17.51 s. |
| **11.114** | (a) 14.9°. (b) 0.1052 s. |
| **11.117** | (a) 2.55 m. (b) 61.8°. |
| **11.118** | (a) 91 m. (b) 30°. (c) 22.8 m. |
| **11.119** | 48.7° west of south. |
| **11.120** | 22.4 km/h, 51.4° west of north. |
| **11.121** | 76.3 km/h ∡ 57.0°. |
| **11.122** | (a) 1.44 m/s ⬀ 62.8°. (b) 1.48 m/s ⬊ 60°. |
| **11.125** | (a) 0.4 m/s$^2$ ↑. (b) 0.89 m/s ∡ 63.4°. |
| **11.126** | (a) 17.4 cm/s ⬋ 66.6°. (b) 14.3 cm/s ⬊ 68.6°. |

**11.127** (a) 52.5 mm/s² ⬹ 31.4°. (b) 157.5 mm/s ⬹ 31.4°.

**11.128** (a) 0.979 m. (b) 12.55 m/s ⬹ 86.5°.

**11.129** 6 m/s ⬹ 85.1°.

**11.130** 3.43 m/s ⬷ 21.0°.

**11.131** 14.30 km/h ⦞ 36.4°.

**11.132** (a) 405 km/h ⦞ 30.6°. (b) 74.7 km/h ⬷ 26.6°.

**11.135** (a) 0.1235 m/s². (b) 0.01008 m/s². (c) 0.001796 m/s².

**11.136** (a) 25 m/s. (b) 24 m/s.

**11.139** (a) 1.694 m/s². (b) 1.406 m/s².

**11.140** 2.18 m/s².

**11.141** 8.56 s.

**11.142** (a) 24.1 m/s. (b) 3.27 m/s².

**11.145** (a) 7.96 m. (b) 4.38 m.

**11.146** (a) 14.48 m/s. (b) 21.3 m.

**11.147** (a) 273 m. (b) 52.4 m.

**11.148** (a) 0.75 m. (b) 1.155 m. (c) 0.75 m.

**11.149** $\mathbf{a} = 1.5$ m/s² ↑, $\rho = 26.3$ m.

**11.150** $v_B^3/gv_A$.

**11.153** $(R^2 + c^2)/2\omega_n R$.

**11.154** 2.50 m.

**11.155** 149.8 Gm.

**11.156** 1425 Gm.

**11.157** $7.2 \times 10^3$ m/s.

**11.158** $3.42 \times 10^3$ m/s.

**11.161** 3670 km/h.

**11.162** 50.7 h.

**11.164** (a) $(0.204$ mm/s$)\mathbf{e}_r - (5.71$ mm/s$)\mathbf{e}_\theta$.
(b) $-(22.8$ mm/s²$)\mathbf{e}_r + (1.633$ mm/s²$)\mathbf{e}_\theta$.
(c) $(0.0583$ mm/s²$)\mathbf{e}_r$.

**11.165** (a) $v = A$, $a = A^2/B$. (b) Path is circle of radius $\rho = B$.

**11.166** (a) $\mathbf{v} = 3\pi b\ \mathbf{e}_\theta$; $\mathbf{a} = -4\pi^2 b\ \mathbf{e}_r$.
(b) $\theta = 2N\pi$, N = 0, 1, 2. . . .

**11.167** (a) $\mathbf{v} = (6$ m/s$)\mathbf{e}_r$; $\mathbf{a} = (24$ m/s²$)\mathbf{e}_\theta$.
(b) $\mathbf{v} = (12.73$ m/s$)\mathbf{e}_r + (4.24$ m/s$)\mathbf{e}_\theta$;
$\mathbf{a} = (16.97$ m/s²$)\mathbf{e}_r + (16.97$ m/s²$)\mathbf{e}_\theta$.

**11.168** (a) $v = 6\sqrt{2}\,t$ m/s, $a = 6\sqrt{2}$ m/s². (b) $\rho = \infty$, Path is a straight line.

**11.171** $\dfrac{\dot{\theta} d \sin \beta}{\sin^2(\beta - \theta)}$.

**11.172** (a) $\dot{r} = -\omega d/2$, $\dot{\theta} = \omega/2$. (b) $\ddot{r} = -\sqrt{3\omega^2 d/4}$, $\ddot{\theta} = 0$.

**11.173** 306 m/s.

**11.174** $v = 152.7$ m/s, $\alpha = 26°$.

**11.177** $b\theta\omega^2 e^{\frac{1}{2}\theta^2}\sqrt{4 + \theta^2}$.

**11.178** $\dfrac{b\omega^2}{\theta^4}\sqrt{36 + 4\theta^2 + \theta^4}$.

**11.182** $\tan^{-1}\left[R(2 + \omega_n^2 t^2)/c\sqrt{4 + \omega_n^2 t^2}\right]$.

**11.183** (a) $\theta_x = 90°$, $\theta_y = 123.7°$, $\theta_z = 33.7°$.
(b) $\theta = 103.4°$, $\theta_y = 134.3°$, $\theta_z = 47.4°$.

**11.184** $x = 11 + 2.15t + 0.002t^4$ m, $v = 2.15 + 0.00829t^3$ m/s.

**11.185** (a) 0.1328 s/m. (b) 434 m. (c) 7.53 m/s.

**11.187** (a) $\mathbf{a}_A = 38$ mm/s² ↓; $\mathbf{a}_B = 25.3$ mm/s² ↑;
$\mathbf{a}_C = 12.67$ mm/s² ↓.
(b) $\mathbf{v}_B = 203$ mm/s ↑; 811 mm ↑.

**11.188** (b) 1383 m. (c) 9 s, 49.5 s.

**11.190** (a) 115.3 km/h $\leq v_0 \leq$ 148.0 km/h.
(b) 6.66°, 4.05°.

**11.191** 2.67 m/s.

**11.193** (a) 2 cm/s². (b) 2.67 cm/s².

**11.194** (a) 0.634 m. (b) 9.07 m.

## CHAPTER 12

**12.1** 75 N.

**12.2** $m = 2.000$ kg at all latitudes. (a) 19.56 N. (b) 19.61 N.
(c) 19.64 N.

**12.5** 1.5 m/s

**12.6** (a) 5 s. (b) 0.612.

**12.7** 63.9 m.

**12.8** (a) 0.955. (b) 64.8 m.

**12.9** (a) 4.53 m (b) 4.5 s.

**12.10** 612 N.

**12.11** (a) 51.7 m. (b) 61.9 m.

**12.13** (a) $\mathbf{a}_A = 1.327$ m/s² ⬷ 30°; $\mathbf{a}_B = 0.442$ m/s² ⦞ 30°.
(b) 18.79 N.

**12.16** (a) 0.99 m/s² ⬳ 25°. (b) 46.8 N.

**12.17** (a) 2.55 m/s² ⬳ 25°. (b) 58.13 N.

**12.19** (a) $\mathbf{a}_A = 2.3$ m/s² ↓; $\mathbf{a}_B = 0.75$ m/s² ↑;
$\mathbf{a}_C = 0.75$ m/s² ↓.
(b) $T_A = 67.76$ N, $T_C = 81.25$ N.

**12.20** *System 1:* (a) 3.22 m/s² ↓. (b) 3.1 m/s ↓. (c) 0.932 s.
*System 2:* (a) 4.85 m/s² ↓. (b) 3.8 m/s ↓. (c) 0.618 s.
*System 3:* (a) 0.23 m/s² ↓. (b) 0.83 m/s ↓. (c) 13.04 s.

**12.21** (a) $\mathbf{a} = 8.3$ m/s² →; $T = 14.6$ kN. (b) 2.45 m/s².

**12.22** 0.357

**12.23** (a) 11.52 m/s² ←. (b) 0.787 m →.

**12.26** $0.347 m_0 v_0^2/F_0$.

**12.29** $\sqrt{k/m}\,(\sqrt{l^2 + x_0^2} - l)$.

**12.30** (a) 0.409 m ↑. (b) 65.4 N.

**12.31** (a) 2.78 m/s² ⬷ 30°. (b) 0.637 m/s ⬷ 30°.

**12.32** (a) $\mathbf{a}_A = 8.9$ m/s² →; $\mathbf{a}_B = 11.9$ m/s² →;
$\mathbf{a}_C = 12.6$ m/s² →.
(b) 27 N.

**12.33** (a) $\mathbf{a}_A = 6.14$ m/s² →; $\mathbf{a}_B = 8.8$ m/s² →;
$\mathbf{a}_C = 9$ m/s² →.
(b) 24.5 N. (c) 197.4 N →.

**12.36** (a) 34.2°. (b) 23.7 N.

**12.37** 2.49 m/s.

**12.40** (a) 3.13 m/s. (b) 13.87 N.

**12.41** (a) 2.24 N. (b) 1.421 N ⬹ 53.8°

**12.42** 0.732 m/s $\leq v \leq$ 4.34 m/s.

**12.44** (a) 0.742W. (b) 0.940W.

**12.45** $t = 11.32$ s.

**12.46** 331 N.

**12.47** (a) 607 N ⬳ 74.6°. (b) 581.2 N ⬳ 80°.

**12.48** (a) 201 m. (b) 515 N ↑.

**12.49** 0.534 m/s ↓.

**12.52** $\sqrt{gr\tan(\theta - \phi_s)} \leq v \leq \sqrt{gr\tan(\theta + \phi_s)}$.

**12.53** (a) 51.9°. (b) 0.449. (c) 115.5 km/h.

**12.54** (a) 0.247W. (b) 13.3°.

**12.55** 6.90°.

**12.56** 0°, 69.6°, 180°.

**12.57** (a) Does not slide; 0.611 N ⦞ 75°. (b) Slides up;
0.957 N ⬷ 40°.

**12.60** (a) 14.8 m/s. (b) $2587 \times 10^{-6}$ N.

**12.61** 0.324.

**12.64** $elVL/mdv_0^2$.

**12.65** $\dfrac{d}{l} > \dfrac{1.085}{v_0}\sqrt{\dfrac{eV}{mv_0^2}}$

**12.66** (a) 1.226 m/s ⬳ 60°. (b) 10.90 m/s² ⬹ 60°.

**12.67** (a) 7.47 N ⦞ 45°. (b) 6.94 m/s² ⬹ 45°.

**12.68** (a) $F_r = 10.8$ N, $F_\theta = 0$. (b) $F_r = -57.6$ N, $F_\theta = 57.6$ N.

**12.69** (a) $F_r = -319.7$ N, $F_\theta = 0$. (b) $F_r = -106.6$ N, $F_\theta = 213.2$ N.

**12.72** (a) 335 mm. (b) 1.304 m/s.

**12.73** $a_r = -84.2$ m/s$^2$, $a_\theta = 90.1$ m/s$^2$.

**12.74** (a) 114.26 N. (b) 5.5 m/s$^2$ →. (c) 4.76 m/s$^2$ ↓.

**12.75** (a) 127.3 N. (b) 6.1 m/s$^2$ →. (c) 4.2 m/s$^2$ ↓.

**12.76** $v_r = v_0 \sin 2\theta / \sqrt{\cos 2\theta}$, $v_\theta = v_0 \sqrt{\cos 2\theta}$.

**12.81** (a) 35800 km. (b) 3.07 km/h.

**12.82** 0.0247 m/s$^2$.

**12.83** (a) $2.1 \times 10^{30}$ kg. (b) 273 m/s$^2$.

**12.84** (a) $1.071 \times 10^6$ km. (b) 8.21 km/s.

**12.85** (a) 1685 km. (b) 25.2 min.

**12.88** $7.35 \times 10^3$ m/s $\measuredangle$ 45°.

**12.89** $2.73 \times 10^3$ m/s

**12.91** 15.94 m/s.

**12.92** (a) $T = 21.6$ N, $\mathbf{a}_{A/OE} = 8.5$ m/s$^2$ →. (b) 1.2 m/s.

**12.93** (a) 1.05 m/s. (b) $T = 10.4$ N, $\mathbf{a}_{A/OE} = 0.99$ m/s$^2$ ←.

**12.102** 105.5.

**12.103** 3760 m/s, 1426 m/s.

**12.105** (a) $1.636 \times 10^3$ m/s. (b) $0.73 \times 10^3$ m/s. (c) 0.333.

**12.108** (a) 159.0 m/s. (b) 70.1 m/s. (c) $2.37 \times 10^3$ m/s.

**12.109** (a) $499 \times 10^3$ km. (b) $\Delta v_B = 43.2$ m/s, $\Delta v_C = 2.42 \times 10^3$ m/s.

**12.110** 123.6 h.

**12.111** $5.27 \times 10^9$ km.

**12.114** 1.036 h.

**12.115** 95.1 min.

**12.116** $\cos^{-1}[(1 - n\beta^2)/(1 - \beta^2)]$.

**12.117** 80 m/s.

**12.121** (a) $(r_1 - r_0)/(r_1 + r_0)$. (b) 6065 $14.4 \times 10^9$ m.

**12.124** (a) 64.06 m. (b) 51.5 m/s.

**12.125** (a) 60.4 m. (b) 19.16 kN.

**12.127** (a) 693.25 N. (b) 1017.8 N.

**12.128** (a) 33.0 km/h. (b) 51.8 km/h.

**12.130** (a) $F_r = -31.6$ N, $F_\theta = -6.96$ N. (b) $F_r = -2.41$ N, $F_\theta = 4.93$ N.

**12.131** (a) $F_r = 2.4465 \tan^2 \theta \sec \theta$ N, $F_\theta = 2.4465 \tan \theta \sec \theta$ N. (b) $\mathbf{P} = 2.4465 \tan \theta \sec^3 \theta$ N $\measuredangle \theta$, $\mathbf{Q} = 2.4465 \tan^2 \theta \sec^2 \theta$ N →.

**12.133** (a) 1.819 kN. (b) 2510 km. (c) 1.617 m/s$^2$.

**12.134** (a) $(a_B)_r = (a_B)_\theta = a_\theta = 0$. (b) 18.4 m/s$^2$. (c) 0.89 m/s.


## CHAPTER 13

**13.1** 8.8 MN · m

**13.2** 2.36 GJ.

**13.5** 6.06 m.

**13.6** (a) 15.7 m. (b) 7.86 m/s $\measuredangle$ 20°. (c) 987 N · m

**13.7** (a) 17.54 m/s. (b) 0.893.

**13.8** −56.0 kJ.

**13.11** 2.86 m/s $\measuredangle$ 15°.

**13.12** −4.94 J.

**13.15** (a) 47.1 m. (b) $F_{AB} = 83.6$ kN $T$; $F_{BC} = 36.6$ kN $T$.

**13.16** (a) 107.7 m. (b) $F_{AB} = 83.6$ kN $C$; $F_{BC} = 36.6$ kN $C$.

**13.17** 8.12 m/s.

**13.18** (a) 9.5 m. (b) 5999 N.

**13.19** (a) 1.7 m/s $\measuredangle$ 30°. (b) 59.7 N · m

**13.20** (a) 1.09 m/s $\measuredangle$ 30°. (b) 17.14 N.

**13.23** 1.683 m/s.

**13.24** (a) 3.96 m/s. (b) 5.60 m/s.

**13.27** (a) 0.41 m/s ↑. (b) 0.046 m.

**13.28** (a) 0.41 m/s ↑. (b) 0.045 m.

**13.29** (a) 0.1590. (b) 1.8 m/s.

**13.32** (a) 3.72 m. (b) 120 m/s$^2$.

**13.33** $0.759 \sqrt{\dfrac{paA}{m}}$.

**13.34** $g_h = \dfrac{g_0}{[(h/R) + 1]^2}$. (a) 0.0314%. (b) 25.3%.

**13.35** $1/[1 - (v_0^2 - v^2)/2g_m R_m]$

**13.38** (a) 111.1 km. (b) 118.7 km.

**13.39** (a) 0.635 m/s. (b) 0.648 m/s.

**13.40** (a) 0.782 m/s. (b) 0.827 m/s.

**13.41** (a) $v_0 = \sqrt{3gl}$. (b) $v_0 = \sqrt{2gl}$.

**13.42** 12.3°

**13.45** 731 N.

**13.46** $N_{\min} = 731$ N at $B$; $N_{\max} = 5520$ N at $D$.

**13.47** 962.8 N · m/s

**13.50** (a) $0.278 \times 10^{-6}$ $mgb/\eta$. (b) $1.010 \times 10^{-3}$ $Wb/\eta$.

**13.51** (a) $t = 18.75$ s, $x = 238$ m. (b) $t = 26.2$ s, $x = 462$ m.

**13.52** (a) 86.9 W. (b) 16.45 m.

**13.55** (a) $k_e = k_1 k_2/(k_1 + k_2)$. (b) $k_e = k_1 + k_2$.

**13.56** (a) 0.526 m, $v_m = 4$ m/s. (b) 0.526 m, $v_m = 2.53$ m/s.

**13.57** 21.9 m/s.

**13.58** 23.1 m/s.

**13.61** (a) 1.12 m. (b) 4.67 m/s.

**13.62** (a) 0.01 m. (b) 0.02 m.

**13.63** (a) 0.292 m. (b) 2.39 m/s.

**13.64** (a) 43.1°. (b) 2.43 m/s.

**13.67** (a) 1.044 m/s. (b) 1.075 m/s. (c) 392 mm.

**13.68** (a) 1.004 m/s. (b) 0.576 m/s.

**13.69** (a) 2.39 m/s. (b) 5.78 N ↑.

**13.70** (a) 2.92 m/s. (b) −(33.9 N)$\mathbf{i}$ + (33.3 N)$\mathbf{j}$.

**13.71** (a) $v = 1.16$ m/s, $N = 16.32$ N →. (b) $v = 1$ m/s, $N = 33$ N ↓.

**13.72** (a) 300 N/m. (b) 1.132 m/s.

**13.75** $a = \dfrac{3}{5}l$.

**13.76** (a) 1089 N ↑. (b) 284 N ↑. (c) Yes.

**13.78** (b) $V = -\ln xyz$.

**13.79** (b) $V = \dfrac{1}{\sqrt{x^2 + y^2 + z^2}}$.

**13.80** (a) Not conservative, $U_{ABCA} - (k - 1)a^2/2$. (b) Conservative, $U_{ABCA} = 0$.

**13.85** $v_B = 25.1$ Mm/h.

**13.86** (a) 2810 kJ/kg. (b) 1345 kJ/kg.

**13.87** $5842.5 \times 10^3$ m.

**13.88** (a) $6.92 \times 10^9$ N · m (b) $168.1 \times 10^9$ N · m

**13.89** (a) 90.5 GJ. (b) 208 GJ.

**13.90** (a) $V = mgR[1 - (R/r)]$. (b) $T = \dfrac{1}{2} mgR^2/r$. (c) $E = mgR[1 - (R/2r)]$.

**13.93** $V_{2r} = 6.44$ m/s, $V_{\theta 2} = 2.4$ m/s.

**13.94** (a) 0.760 m. (b) 1.580 m/s.

**13.95** (a) 1.06 m. (b) 3.6 rad/s. (c) 144.1 N · m

**13.96** (a) 0.91 m. (b) 2.27 m. (c) 5 rad/s.

**13.100** (a) 1704 m/s. (b) 29.6 m/s.

**13.101** (a) 2.44 km/s. (b) 1.470 km/s.

**13.102** (a) 2.9 km/s. (b) 2.6 km/s.

**13.103** (a) 5613.3 km. (b) $16.98 \times 10^{23}$ kg.

**13.106** 3450 m/s.

**13.107** 1478 km.

**13.108** (a) 7.34 km/s. (b) 44.5°.

**13.109** (a) 10.66 km/s. (b) $15.43 \times 10^3$ km.

**13.112** (a) 3.45 km/s. (b) 4.17 km/s.
**13.113** 9.39 km/s; 58.9°.
**13.118** (b) $v_{esc}\sqrt{\alpha/(1 + \alpha)} \le v_0 \le v_{esc}\sqrt{(1 + \alpha)/(2 + \alpha)}$.
**13.119** 4.87 m/s ⬈ 76.4°
**13.121** 73.3 N.
**13.122** (a) 3.06 s. (b) 0.963 s.
**13.123** (a) 11.1 kN. (b) 3 s.
**13.124** (a) 39.9 m/s. (b) 10.88 kN.
**13.125** (a) 88.1 km/h. (b) 33.6 s.
**13.126** 206 N
**13.129** (a) 5.28 s. (b) 17.05 kN $C$.
**13.130** (a) 16.32 s. (b) 3.06 kN $T$.
**13.131** (a) 32.6 s. (b) 11.03 kN $T$.
**13.132** (a) 1.64 m/s ⬈ 30°. (b) 14.63 N.
**13.133** (a) $v_A$ = 2.5 m/s →; $v_B$ = 4.95 m/s →; $v_C$ = 4.35 m/s →.
(b) 0.114 N.
**13.136** (a) 5.19 m/s ↑. (b) 15.38 m/s ↑.
**13.137** (a) 15.38 m/s ↑. (b) 4.57 s.
**13.140** 180.8 N ⬉ 51.5°.
**13.141** 1.86$W$.
**13.142** 111.2 kN ⬉ 69.0°
**13.143** 3.86 N · s.
**13.146** (a) 1.71 m/s →. (b) 11000 N · s.
**13.147** (a) 432.2 m/s. (b) 12.88 N · s.
**13.148** (a) 6.30 N · s ⬉ 34.7°. (b) 11.28 J.
**13.149** (a) 1.154 m/s ←. (b) 1.292 m/s ←. (c) 1.280 m/s ←.
**13.150** (a) 1.85 m/s ↓. (b) 0.085 N · m.
**13.151** (a) 187.9 N · s, 589 J. (b) 150.3 N · s, 471 J.
**13.152** (a) 4.5 N · s, 13.5 J. (b) 3.79 N · s, 11.37 J.
**13.157** (a) $v'_A$ = 2.3 m/s ←; $v'_B$ = 2.2 m/s →. (b) 2.84 J.
**13.158** 0.875
**13.159** (a) 3.2 m/s →. (b) 6.96 m/s ←.
**13.160** (a) 2.07 m/s ←. (b) 1.874 m/s →.
**13.163** (a) 0.82 m/s →. (b) 0.59 m/s →.
**13.164** $v'_A$ = 1.773 m/s ⬈ 15.00°; $v'_B$ = 2.19 m/s ⬊ 50°.
**13.165** $v'_A$ = 0.711 $v_0$ ⬈ 39.3°; $v'_B$ = 0.636 $v_0$ ⬊ 45°.
**13.166** $v'_A$ = 1.56 m/s →; $v'_B$ = 2.76 m/s ⬈ 22.6°.
**13.167** $v'_A$ = 1.53 m/s →; $v'_B$ = 2.97 m/s → .
**13.171** (a) 0.941 m/s ⬈ 89.0°. (b) 0.01090.
**13.172** 156.1 mm.
**13.173** (a) 0.6. (b) 0.9.
**13.174** 3.96 m.
**13.176** (a) 0.2 m. (b) 0.269 m.
**13.177** (a) 2.65 m/s →. (b) 75.7 N · m.
**13.178** (a) 31.2 km/h. (b) 0.241.
**13.179** (a) $e_{AB}$ = 0.506, $e_{BC}$ = 0.333. (b) 44.0 mm.
**13.180** (a) $v'_C$ = 0, $v'_A$ = 1.372 m/s ↓. (b) 6.86 N · s ↑.
**13.181** (a) $v_C$ = 2.74 m/s ↓, $v_A$ = 1.372 m/s ↓.
(b) $v_C$ = 1.372 m/s ↓, $v_A$ = 2.47 m/s ↓.
**13.184** 96.0 mm.
**13.185** (a) 3.70 m/s ⬈ 62.2°. (b) 1175 N.
**13.186** 0.309
**13.187** 0.13 m ≤ $h$ ≤ 0.52 m.
**13.190** 265 km/h.
**13.191** (a) 106.3 mm. (b) 2.36 m/s.
**13.193** 2.06 m/s.
**13.194** 7.59 km/s.
**13.196** 2.86 kN.
**13.198** $\mathbf{v}_A$ = 0.9 m/s ←; $\mathbf{v}_B$ = 0.3 m/s ←; $\mathbf{v}_C$ = 0.137 m/s ←.
**13.199** $\mathbf{v}_A$ = 3.52 m/s ⬈ 72.2°; $\mathbf{v}_B$ = 0.035 m/s ⬈ 40°.
**13.201** (a) 62.7°. (b) 0.14 $v_0^2$ N · m (for $v_0$ in m/s).

**1348**

## CHAPTER 14

**14.1** $\mathbf{v}_A$ = 10.67 km/h ←; $\mathbf{v}_B$ = $\mathbf{v}_C$ = 4.27 km/h ←.
**14.2** (a) 1.706 m/s. (b) 0.576 m from left of $B$.
**14.3** 0.219 m/s →.
**14.4** 221 kN.
**14.5** (a) 5.2 km/h →. (b) 4 km/h →.
**14.6** (a) 43.4 g. (b) 293 m/s.
**14.9** (a) (1.333 m/s)$\mathbf{j}$; (0.333 m/s)$\mathbf{i}$; (1 m/s)$\mathbf{k}$.
(b) (1.8 kg · m²/s)$\mathbf{i}$ + (0.9 kg · m²/s)$\mathbf{j}$ − (3.6 kg · m²/s)$\mathbf{k}$.
**14.10** (a) (3.33 m/s)$\mathbf{j}$; (1.667 m/s)$\mathbf{i}$; (3.33 m/s)$\mathbf{k}$.
(b) (9 kg · m²/s)$\mathbf{i}$ + (4.5 kg · m²/s)$\mathbf{j}$ − (9 kg · m²/s)$\mathbf{k}$.
**14.11** zero
**14.12** (a) (0.8 m)$\mathbf{i}$ + (0.4 m)$\mathbf{j}$ + (0.8 m)$\mathbf{k}$.
(b) (114.3 N · s)$\mathbf{j}$ − (76 N · s)$\mathbf{k}$.
(c) (121.8 N · m · s)$\mathbf{i}$ − (61 N · m · s)$\mathbf{j}$ −
(91.4 N · m · s)$\mathbf{k}$.
**14.15** (120 m)$\mathbf{i}$ − (26.0 m)$\mathbf{j}$ + (13.33 m)$\mathbf{k}$.
**14.16** (3300 m)$\mathbf{i}$ + (750 m)$\mathbf{j}$ + (900 m)$\mathbf{k}$.
**14.17** 83.5 m/s, 7.8 s
**14.18** $v_B$ = 19 m/s; $v_C$ = 46 m/s.
**14.21** (a) 2.13 m/s →. (b) $\theta$ = 36.6°, $v_C$ = 3.46 m/s,
$v_D$ = 2.91 m/s.
**14.22** (a) 2.01 m/s. (b) 2.27 m/s.
**14.23** 798.3 m/s east, 215.8 m/s south, 1190 m/s down.
**14.24** (15.75 m)$\mathbf{i}$ + (2.79 m)$\mathbf{k}$.
**14.25** $v_A$ = 919 m/s; $v_B$ = 717 m/s; $v_C$ = 619 m/s.
**14.31** (a) 3550 J. (b) 1830 J.
**14.32** 2670 J.
**14.33** 4339.8 N · m
**14.34** 391348.8 N · m.
**14.37** (a) $\mathbf{v}_A$ = 0.2 $v_0$ ←; $\mathbf{v}_B$ = 0.693 $v_0$ ⬈ 30°;
$\mathbf{v}_C$ = 0.693 $v_0$ ⬊ 30°.
(b) $\mathbf{v}_A$ = 0.25 $v_0$ ⬈ 60°; $\mathbf{v}_B$ = 0.866 $v_0$ ⬈ 30°;
$\mathbf{v}_C$ = 0.433 $v_0$ ⬊ 30°.
**14.38** (a) $\dfrac{m_A}{m_A + m_B} v_0$ →. (b) $\dfrac{m_A}{m_A + m_B} \dfrac{v_0^2}{2g}$.
**14.39** $v_A$ = 2.29 m/s; $v_B$ = 1.97 m/s; $v_C$ = 3.43 m/s.
**14.40** $v_A$ = 3.23 m/s; $v_B$ = 2.79 m/s; $v_C$ = 1.6 m/s.
**14.41** $\mathbf{v}_A$ = 0.1714 $v_0$ ⬊ 19.5°; $\mathbf{v}_B$ = 0.985 $v_0$ ⬊ 80.1°;
$\mathbf{v}_C$ = 0.0286 $v_0$ ↑.
**14.42** (a) $\mathbf{v}_A$ = 0.1667 $v_0$ ⬊ 19.5°; $\mathbf{v}_B$ = 0.957 $v_0$ ⬈ 80.5°;
$\mathbf{v}_C$ = 0. (b) 0.0556.
**14.45** (a) $v_C$ = 3.50 m/s; $v_D$ = 1.750 m/s. (b) 0.786.
**14.46** (a) $v_C$ = 4.25 m/s; $v_D$ = 2.13 m/s. (b) 0.727.
**14.47** (a) $\mathbf{L} = mv_0 \mathbf{i}$; $\mathbf{H}_G = -\dfrac{2}{3} lmv_0 \mathbf{k}$. (b) $\mathbf{v}_A = -\dfrac{1}{3} v_0 \mathbf{i}$;
$\mathbf{v}_B = \dfrac{2}{3} v_0 \mathbf{i}$.
**14.48** (60.9 m/s)$\mathbf{i}$ + (955 m/s)$\mathbf{j}$ + (5033.8 m/s)$\mathbf{k}$.
**14.49** (a) 0.5 $v_0$. (b) 0.75 $v_0$. (c) 0.1875.
**14.50** (a) −(0.25$l$)$\mathbf{i}$ − (0.25$v_0$ $t$)$\mathbf{j}$. (b) (1.75$mlv_0$)$\mathbf{k}$. (c) $mv_0^2$
**14.53** (a) $\mathbf{v}_A$ = 2.075 m/s ↑; $\mathbf{v}_B$ = 5.98 m/s ⬊ 28°. (b) 3.15 m.
**14.54** (a) 1.4 m/s ⬈ 38.7°. (b) 0.54 m. (c) 10.00 rad/s ↰.
**14.55** (a) $\mathbf{v}_B$ = 2.4 m/s ⬈ 53.1°; $\mathbf{v}_C$ = 2.56 m/s →.
(b) $c$ = 1.059 m.
**14.56** (a) $\mathbf{v}_A$ = 2.4 m/s ↓; $\mathbf{v}_B$ = 3 m/s ⬈ 53.1°. (b) $a$ = 1.864 m.
**14.59** 312 N.
**14.60** 4.18 m/s.
**14.61** 117.2 N →; 56.8 N ↑.
**14.62** 146 N →; 71 N ↑.
**14.65** 4040 N ↑.

**14.67** $\mathbf{A} = 274$ N $\searchans 20.0°$; $M_A = 46.0$ N $\cdot$ m $\downarrow$.

**14.68** $C_x = 0, C_y = 130.6$ N $\downarrow$; $D_x = 215.2$ N $\rightarrow$, $D_y = 242.18$ N $\downarrow$.

**14.71** 100 kg/s.

**14.72** 25.6 m/s $\leftarrow$.

**14.73** $\mathbf{A} = 1245$ N $\uparrow$; $\mathbf{B} = 25582.7$ N $\searchans 84.3°$.

**14.74** (a) 42.0 kN, 1.143 m below $B$. (b) 30.0 kN, 3.20 m below $B$.

**14.75** 29.3 kN.

**14.76** 233.4 m/s.

**14.79** (a) $\frac{1}{2} v_A \rightarrow$. (b) $\frac{1}{4} \rho A v_A^3 (1 - \cos \theta)$.

$\quad\quad$ (c) $2\left(1 - \dfrac{V}{v_A}\right)\dfrac{V}{v_A}(1 - \cos \theta)$.

**14.80** (a) $11.51 \times 10^6$ W. (b) $20.97 \times 10^6$ W. (c) 0.55.

**14.83** $b\sqrt{\dfrac{1}{2} g d_1 d_2 (d_1 + d_2)}$.

**14.84** 15.09 m³/s.

**14.87** (a) $m_0 e^{qL/m_0 v_0}$. (b) $v_0 e^{-qL/m_0 v_0}$.

**14.88** $\mathbf{C} = 1.712$ kN $\uparrow$; $\mathbf{D} = 2.29$ kN $\uparrow$.

**14.89** (a) $\dfrac{m}{l}\left(gy + v^2\right)$. (b) $mg\left(1 - \dfrac{y}{l}\right)\uparrow$.

**14.90** (a) $mgy/l$. (b) $(m/l)\left[g(l - y) + v^2\right]\uparrow$.

**14.93** 3888 kN.

**14.94** 4550 N/s.

**14.95** 5500 m/s.

**14.96** (a) 116.0 kg/s. (b) 1810 kg.

**14.97** 2200 kg.

**14.98** 1498 m/s.

**14.101** 4150 m.

**14.102** 87.2 km.

**14.103** (a) 153.4 km. (b) 5670 m/s $\uparrow$.

**14.104** 56.4 km.

**14.107** (a) 1.3 m/s $\rightarrow$. (b) 1.3 m/s $\rightarrow$.

**14.109** $-(4.8$ kg $\cdot$ m²/s)$\mathbf{j} + (9.6$ kg $\cdot$ m²/s)$\mathbf{k}$.

**14.110** (a) $(1.867$ m)$\mathbf{i} + (1.533$ m)$\mathbf{j} + (0.667$ m)$\mathbf{k}$.

$\quad\quad$ (b) $(12$ kg $\cdot$ m/s)$\mathbf{i} + (28$ kg $\cdot$ m/s)$\mathbf{j} + (14$ kg $\cdot$ m/s)$\mathbf{k}$.

$\quad\quad$ (c) $-(2.8$ kg $\cdot$ m²/s)$\mathbf{i} + (13.33$ kg $\cdot$ m²/s)$\mathbf{j} -$

$\quad\quad$ $(24.3$ kg $\cdot$ m²/s)$\mathbf{k}$.

**14.111** $-1001.3$ m 657.7 m.

**14.113** $\mathbf{v}_A = 9.67$ m/s $\nearrow 60.3°$; $\mathbf{v}_B = 12.87$ m/s $\searchans 17.8°$.

**14.115** (a) 41.8°. (b) 200 N $\downarrow$.

**14.116** $\mathbf{C} = 134.1$ N $\downarrow$; D $= 178.6$ N $\uparrow$.

**14.117** (a) 30.8 m/s. (b) 96.7 m³/s. (c) 55.4 kW.


## CHAPTER 15

**15.1** (a) $t = 0.25$ s. (b) $\theta = -18.25$ rad, $\omega = 22.5$ rad/s.

**15.2** (a) 0.5 rad; $-4.71$ rad/s; $-34.5$ rad/s². (b) 0; $-1.934$ rad/s; 36.5 rad/s².

**15.3** (a) 0.253 rad; $-0.927$ rad/s; $-36.6$ rad/s². (b) 0; 0; 0.

**15.4** (a) 36 s. (b) 10.54 s.

**15.5** (a) 80 rev. (b) 800 rev.

**15.6** $\omega = 14.27$ rad/s, $\theta = 3.63$ rad.

**15.9** $\mathbf{v}_B = -(0.95$ m/s)$\mathbf{i} + (0.305$ m/s)$\mathbf{j} - (0.396$ m/s)$\mathbf{k}$.

$\quad\quad$ $\mathbf{a}_B = -(3.2$ m/s)$\mathbf{i} - (1.88$ m/s²)$\mathbf{j} + (6.23$ m/s²)$\mathbf{k}$.

**15.10** $\mathbf{v}_B = -(0.475$ m/s)$\mathbf{i} + (0.15$ m/s)$\mathbf{j} - (0.198$ m/s)$\mathbf{k}$.

$\quad\quad$ $\mathbf{a}_B = -(0.8$ m/s²)$\mathbf{i} - (0.7$ m/s²)$\mathbf{j} + (1.85$ m/s²)$\mathbf{k}$.

**15.11** $\mathbf{v}_E = (1.08$ m/s)$\mathbf{i} + (2.4$ m/s)$\mathbf{j}$; $\mathbf{a}_E = -(11.52$ m/s²)$\mathbf{i} + (5.18$ m/s²)$\mathbf{j} + (23.1$ m/s²)$\mathbf{k}$.

**15.12** $\mathbf{v}_C = -(2.4$ m/s)$\mathbf{j} - (1.8$ m/s)$\mathbf{k}$; $\mathbf{a}_C = (18$ m/s²)$\mathbf{i} + (19.2$ m/s²)$\mathbf{j} - (15.6$ m/s²)$\mathbf{k}$.

**15.15** $2.967 \times 10^4$ m/s, $5.9 \times 10^{-3}$ m/s²

**15.17** (a) $\omega = 2.5$ rad/s $\curvearrowleft$; $\alpha = 1.5$ rad/s² $\downarrow$.

$\quad\quad$ (b) 771 mm/s² $\searchans 76.5°$.

**15.18** 12 rad/s², $\curvearrowleft$ or $\downarrow$.

**15.19** (a) $\omega_A = 18$ rad/s $\downarrow$; $\alpha_A = 144$ rad/s² $\downarrow$; $\omega_C = 28.8$ rad/s $\downarrow$; $\alpha_C = 230$ rad/s² $\downarrow$. (b) 109.5 m/s² $\nearrow 15.52°$.

**15.20** (a) 24 m/s². (b) 244.6 m/s² $\nearrow 5.64°$.

**15.23** (a) $r_1 \omega_A / r_2$. (b) $\mathbf{a}_A = r_1 \omega_A^2 \downarrow$; $\mathbf{a}_B = r_1^2 \omega_A^2 / r_2 \downarrow$.

**15.24** (a) 5 rad/s. (b) $\mathbf{a}_A = 7.94$ m/s² $\downarrow$; $\mathbf{a}_B = 1.58$ m/s² $\downarrow$.

$\quad\quad$ (c) 2.225 m/s².

**15.27** (a) 1.6 m/s² $\rightarrow$. (b) 150 mm $\leftarrow$.

**15.28** (a) 1.934 m/s² $\rightarrow$. (b) 183.0 mm $\leftarrow$.

**15.29** (a) 2.75 rev. (b) $\mathbf{v}_B = 1.425$ m/s $\downarrow$; $\Delta y_B = 2.58$ m/s $\downarrow$.

$\quad\quad$ (c) 0.707 m/s² $\measuredangle 32.0°$.

**15.30** (a) $\mathbf{v}_A = .96$ m/s $\uparrow$; $\Delta y_A = 1.92$ m $\uparrow$.

$\quad\quad$ (b) $\mathbf{v}_B = 1.44$ m/s $\downarrow$; $\Delta y_B = 2.88$ m $\downarrow$.

**15.31** (a) 3.14 rad/s². (b) 8 s.

**15.32** (a) $\alpha_A = 4.34$ rad/s²; $\alpha_B = 1.832$ rad/s² $\downarrow$. (b) 3.70 s.

**15.35** $a = b\omega_0^2 / 2\pi \rightarrow$.

**15.36** $\alpha = bv^2 / 2\pi r^3 \downarrow$.

**15.37** (a) 0.378 rad/s $\downarrow$. (b) 160.4 mm/s $\uparrow$.

**15.38** (a) 0.615 rad/s $\curvearrowleft$. (b) 276 mm/s $\searchans 15°$.

**15.39** (a) 2.7 rad/s $\downarrow$. (b) 1.18 m/s $\nearrow 67.4°$.

**15.40** (a) 1.98 m/s $\rightarrow$. (b) 1.85 m/s $\measuredangle 67.4°$.

**15.43** (a) $-(2$ rad/s)$\mathbf{k}$. (b) $(100$ mm/s)$\mathbf{i} + (175$ mm/s)$\mathbf{j}$.

**15.45** (a) 1.8 rad/s $\curvearrowleft$. (b) 0.63 m/s $\measuredangle 69.4°$.

**15.46** (a) 2.81 rad/s $\curvearrowleft$. (b) 0.44 m/s $\measuredangle 35.3°$.

**15.47** (a) 8.33 rad/s $\downarrow$. (b) 2.78 rad/s $\downarrow$.

**15.48** (a) 30 rad/s $\curvearrowleft$. (b) 13.33 rad/s $\curvearrowleft$.

**15.49** (a) 6.03 m/s. (b) 9000 rpm. (c) 1508.

**15.50** (a) 3 rad/s $\downarrow$. (b) 1.5 rad/s $\downarrow$.

**15.51** (a) 105 rpm $\downarrow$. (b) 127.5 rpm $\downarrow$.

**15.52** (a) 1.5. (b) $\omega_A / 3 \curvearrowleft$.

**15.55** 497 mm/s $\leftarrow$.

**15.56** 791 mm/s $\leftarrow$.

**15.57** (a) $\omega_{BD} = 4.38$ rad/s $\downarrow$; $\mathbf{v}_D = 0.31$ m/s $\uparrow$.

$\quad\quad$ (b) $\omega_{BD} = 0$; $\mathbf{v}_D = 1.065$ m/s $\downarrow$.

$\quad\quad$ (c) $\omega_{BD} = 4.38$ rad/s $\curvearrowleft$; $\mathbf{v}_D = 0.31$ m/s $\downarrow$.

**15.58** (b) For $\theta = 22.9°$: $\omega_{BD} = 5.60$ rad/s $\downarrow$; For $\theta = 192.6°$: $\omega_{BD} = 5.60$ rad/s $\curvearrowleft$.

**15.59** (a) 4.27 rad/s $\downarrow$. (b) 1.330 m/s $\downarrow$. (c) 1.557 m/s $\measuredangle 34.7°$.

**15.60** (a) 5.13 rad/s $\curvearrowleft$. (b) 0.924 m/s $\leftarrow$. (c) 1.873 m/s $\nearrow 34.7°$.

**15.63** $\omega_{BD} = 2.5$ rad/s $\curvearrowleft$; $\omega_{DE} = 6.67$ rad/s $\downarrow$.

**15.64** $\omega_{AB} = \omega_{BD} = 16.04$ rad/s $\curvearrowleft$; $\omega_{DE} = 16.04$ rad/s $\downarrow$.

**15.65** (a) 14.43 rad/s $\downarrow$. (b) 4.4 m/s $\measuredangle 55.4°$.

**15.66** (a) 35 rad/s $\curvearrowleft$. (b) 0.203 m above $D$.

**15.69** $\mathbf{v}_B = 44.4$ m/s $\rightarrow$; $\mathbf{v}_C = 0$; $\mathbf{v}_D = 42.9$ m/s $\measuredangle 15°$;

$\quad\quad$ $\mathbf{v}_E = 31.4$ m/s $\searchans 45°$.

**15.70** (a) $\mathbf{v}_B = 0.43$ m/s $\leftarrow$; $\omega_{AB} = 0$. (b) $\mathbf{v}_B = 0.896$ m/s $\leftarrow$;

$\quad\quad$ $\omega_{AB} = 2.37$ rad/s $\downarrow$.

**15.71** Vertical line intersecting $zx$ plane at $x = 0$, $z = 2.65$ m.

**15.72** (a) 0.5 m to the right of $A$. (b) 0.1667 m/s $\uparrow$.

**15.73** (a) *Gear A*: 0.375 m left of A;

$\quad\quad$ *Gear C*: 0.75 m left of C.

$\quad\quad$ (b) $\omega_A = 4$ rad/s $\downarrow$; $\omega_C = 2$ rad/s $\curvearrowleft$.

**15.74** (a) 6 rad/s $\curvearrowleft$. (b) 0.76 m/s $\leftarrow$. (c) 0.61 m/s, wound.

**15.75** (a) 42.9 mm below $A$. (b) 22.5 m/s $\rightarrow$.

$\quad\quad$ (c) 15.95 m/s $\measuredangle 41.2°$.

**15.76** (a) 33.3 mm above $A$. (b) 25.5 m/s $\rightarrow$.

$\quad\quad$ (c) 18.06 m/s $\searchans 48.4°$.

**15.79** (a) 4.27 rad/s $\downarrow$. (b) 1.330 m/s $\downarrow$. (c) 1.557 m/s $\measuredangle 34.7°$.

**15.80** (a) 12 rad/s $\curvearrowleft$. (b) 3.9 m/s $\nearrow 67.4°$

**15.81** (a) 31.5 rad/s $\downarrow$. (b) 18 rad/s $\uparrow$. (c) 4.57 m/s $\nearrow$ 30°.
**15.82** (a) 16.1 rad/s $\uparrow$. (b) 9.36 rad/s $\downarrow$. (c) 9.36 rad/s $\downarrow$.
**15.85** (a) 0.9 m. (b) 10.8 m/s $\downarrow$.
**15.86** (a) 3 rad/s $\uparrow$. (b) 1.559 m/s $\uparrow$.
**15.87** (a) 2.55 m/s $\measuredangle$ 45°. (b) 2.57 rad/s $\downarrow$. (c) 2 m/s $\measuredangle$ 20°.
**15.88** (a) 8 rad/s $\downarrow$. (b) 41.6 rad/s $\uparrow$. (c) 6.48 m/s $\searrow$ 19.10°.
**15.91** (a) 1.5 m/s $\uparrow$. (b) 1.25 rad/s $\uparrow$.
**15.92** (a) $\omega_{AB} = 1.023$ rad/s $\uparrow$; $\omega_{BD} = 0.341$ rad/s $\downarrow$.
  (b) 0.086 m/s $\rightarrow$.
**15.93** (a) 2.49 rad/s $\uparrow$. (b) 3.73 rad/s $\downarrow$. (c) 0.835 m/s $\searrow$ 53.6°.
**15.94** (a) 4.8 rad/s $\downarrow$. (b) 8 rad/s $\uparrow$. (c) 1.5 m/s $\nearrow$ 53.1°.
**15.97** *Space centrode:* Lower rack. *Body centrode:*
  Circumference of gear.
**15.98** *Space centrode:* Quarter circle of 300-mm. radius centered
  at intersection of tracks. *Body centrode:* Semicircle of
  150-mm radius with center on rod $AB$ equidistant from $A$
  and $B$.
**15.103** (a) 3.66 m/s$^2$ $\leftarrow$. (b) 4.58 m/s$^2$ $\leftarrow$.
**15.104** (a) 3.66 m/s$^2$ $\rightarrow$. (b) 10.08 m/s$^2$ $\rightarrow$.
**15.105** (a) 16 rad/s$^2$ $\uparrow$. (b) 7.2 m/s$^2$ $\searrow$ 60°. (c) 12.47 m/s$^2$ $\downarrow$.
**15.106** (a) 5.4 m/s$^2$ $\leftarrow$. (b) 5.4 m/s$^2$ $\searrow$ 60°; (c) 9.35 m/s$^2$ $\uparrow$.
**15.109** (a) 2 rad/s$^2$ $\downarrow$. (b) 0.224 m/s$^2$ $\searrow$ 63.4°.
**15.110** $x = -0.1464$ m, $y = -0.1098$ m.
**15.111** (a) 1.37 m/s$^2$ $\nearrow$ 33.7°. (b) 3.27 m/s$^2$ $\searrow$ 35.5°.
  (c) 3.04 m/s$^2$ $\uparrow$.
**15.112** (a) 267 mm/s$^2$ $\nearrow$ 61.0°. (b) 252 mm/s$^2$ $\measuredangle$ 64.0°.
**15.115** $\mathbf{a}_C = 0.57$ m/s$^2$ $\searrow$ 58.0°; $\mathbf{a}_D = 0.6$ m/s$^2$ $\measuredangle$ 66.0°.
**15.116** 1.02 m/s$^2$ $\searrow$ 33.6°.
**15.117** (a) 28.8 rad/s$^2$ $\uparrow$. (b) 17.28 rad/s$^2$ $\downarrow$.
**15.118** (a) 3 rad/s$^2$ $\uparrow$. (b) 3 rad/s$^2$ $\downarrow$.
**15.119** 151.4 m/s$^2$ $\downarrow$.
**15.120** 301.56 m/s$^2$ $\uparrow$.
**15.121** (a) 24.4 m/s$^2$ $\leftarrow$. (b) 19.86 m/s$^2$ $\leftarrow$.
**15.122** (a) 4.83 m/s$^2$ $\leftarrow$. (b) 8.05 m/s$^2$ $\nearrow$ 72.5°.
**15.125** (a) 57.7 rad/s$^2$ $\uparrow$. (b) 28.9 rad/s$^2$ $\downarrow$.
**15.126** (a) 24.8 rad/s$^2$ $\uparrow$. (b) 30.6 rad/s$^2$ $\downarrow$.
**15.127** (a) 227 rad/s$^2$ $\uparrow$. (b) 100 rad/s$^2$ $\downarrow$.
**15.128** (a) 42 m/s$^2$ $\searrow$ 78.6°. (b) 62 m/s$^2$ $\measuredangle$ 19.5°.
**15.129** (a) 4 rad/s $\uparrow$. (b) 24 rad/s$^2$ $\uparrow$. (c) 10 m/s$^2$ $\uparrow$.
**15.131** (a) 11.10 rad/s$^2$ $\downarrow$. (b) 4.53 rad/s$^2$ $\downarrow$.
**15.137** $b\omega \left\{ \sin\theta - \dfrac{b(1 + \sin\theta)\cos\theta}{\sqrt{l^2 - b^2(1 + \sin\theta)^2}} \right\} \downarrow$.
**15.138** $\omega^2 \left\{ \dfrac{b^3 \cos^2\theta(1 + \sin\theta)}{[l^2 - b^2(1 + \sin\theta)^2]^{3/2}} - \dfrac{b\sin\theta}{[l^2 - b^2(1 + \sin\theta)^2]^{1/2}} \right\} \downarrow$.
**15.140** $\omega = \dfrac{bv_A}{b^2 + x_A^2}$ $\uparrow$; $\alpha = \dfrac{2bx_A v_A^2}{(b^2 + x_A^2)^2}$ $\uparrow$.
**15.141** $(v_B)_x = v_A - \dfrac{lb^2 v_A}{(b^2 + x_A^2)^{3/2}}$ $\rightarrow$; $(v_B)_y = \dfrac{lbx_A v_A}{(b^2 + x_A^2)^{3/2}}$ $\uparrow$.
**15.144** $\omega_{BD} = \dfrac{\omega b(b + l\cos\theta)}{l^2 + b^2 + 2bl\cos\theta}$ $\downarrow$;
  $v_E = \dfrac{\omega bl\sin\theta}{\sqrt{l^2 + b^2 + 2bl\cos\theta}}$ $\searrow \tan^{-1}\left( \dfrac{b\sin\theta}{l + b\cos\theta} \right)$.
**15.145** $\dfrac{\omega^2 bl(l^2 - b^2)\sin\theta}{(l^2 + b^2 + 2bl\cos\theta)^2}$ $\uparrow$.
**15.146** $\mathbf{v}_P = r\omega\left[ \cos\dfrac{r\omega t}{R - r} - \cos\omega t \right]\mathbf{i} + r\omega\left[ \sin\dfrac{r\omega t}{R - r} + \sin\omega t \right]\mathbf{j}$.
**15.147** $\mathbf{v} = (R\omega \sin\omega t)\mathbf{j}$; $\mathbf{a} = (R\omega^2 \cos\omega t)\mathbf{j}$.

**15.148** (a) 5.16 rad/s $\downarrow$. (b) 1.340 m/s $\searrow$ 60°.
**15.149** 1167 mm/s $\nearrow$ 51.8°.
**15.150** (a) $\omega_{BD} = 3.81$ rad/s $\downarrow$; $\mathbf{v}_{P/BD} = 8.3$ m/s $\measuredangle$ 16.26°.
  (b) $\omega_{BD} = 3.00$ rad/s $\downarrow$; $\mathbf{v}_{P/BD} = 5.08$ m/s $\rightarrow$.
**15.151** $\omega_{BD} = 3.82$ rad/s $\downarrow$; $\mathbf{v}_{P/BD} = 7.7$ m/s $\measuredangle$ 10.49°.
**15.152** (a) 11.25 rad/s $\uparrow$. (b) 1.875 m/s $\rightarrow$.
**15.153** (a) 13.33 rad/s $\uparrow$. (b) 0.625 m/s $\rightarrow$.
**15.158** $\mathbf{a}_1 = r\omega^2\,\mathbf{i} - 2u\omega\,\mathbf{j}$; $\mathbf{a}_2 = 2u\omega\,\mathbf{i} - r\omega^2\,\mathbf{j}$;
  $\mathbf{a}_3 = -(r\omega^2 + u^2/r + 2u\omega)\mathbf{i}$; $\mathbf{a}_4 = (r\omega^2 + 2u\omega)\mathbf{j}$.
**15.159** $\mathbf{a}_1 = r\omega^2\mathbf{i} + 2u\omega\mathbf{j}$; $\mathbf{a}_2 = -2u\omega\mathbf{i} - r\omega^2\mathbf{j}$;
  $\mathbf{a}_3 = (-r\omega^2 - u^2/r + 2u\omega)\mathbf{i}$; $\mathbf{a}_4 = (r\omega^2 - 2u\omega)\mathbf{j}$.
**15.160** (a) 54 rad/s$^2$ $\downarrow$. (b) 10 m/s$^2$ $\measuredangle$ 45°.
**15.161** (u) 38.8 rad/s$^2$ $\downarrow$. (b) 12 m/s$^2$ $\uparrow$.
**15.162** 0.025 m/s$^2$ to left of sled.
**15.164** (a) $\mathbf{v}_P = 4.42$ m/s $\searrow$ 35.8°; $\mathbf{a}_P = 27$ m/s$^2$ $\nearrow$ 72.0°.
  (b) $\mathbf{v}_D = 3.57$ m/s $\leftarrow$; $\mathbf{a}_D = 52.46$ m/s$^2$ $\nearrow$ 36.8°.
**15.165** (a) $\omega_{PD} = 21.1$ rad/s $\uparrow$; $\omega_{DE} = 2.24$ rad/s $\uparrow$.
  (b) $\alpha_{PD} = 77.8$ rad/s$^2$ $\downarrow$; $\alpha_{DE} = 278$ rad/s$^2$ $\uparrow$.
**15.166** (a) 1.702 m/s$^2$ $\searrow$ 21.5°. (b) 2.55 m/s$^2$ $\searrow$ 3.24°.
**15.167** (a) 2.38 m/s$^2$ $\nearrow$ 48.3°. (b) 1.438 m/s$^2$ $\nearrow$ 64.3°.
**15.170** (a) 2.32 m/s$^2$ $\uparrow$. (b) 11.94 m/s$^2$ $\searrow$ 37.1°. (c) 16.72 m/s$^2$ $\downarrow$.
**15.172** (a) 0.436 rad/s $\uparrow$. (b) 0.271 rad/s $\uparrow$.
**15.173** (a) 0.354 rad/s $\uparrow$. (b) 0.125 rad/s$^2$ $\uparrow$.
**15.174** (a) 8.3 rad/s $\downarrow$. (b) 10.67 m/s$^2$ $\nearrow$ 16.26°.
**15.175** (a) 25.8 rad/s$^2$ $\uparrow$. (b) 60.58 m/s$^2$ $\leftarrow$.
**15.178** $\omega_S = 0.38$ rad/s $\downarrow$; $\alpha_S = 4.5$ rad/s$^2$ $\downarrow$.
**15.179** $\omega_S = 1.487$ rad/s $\downarrow$; $\alpha_S = 59$ rad/s$^2$ $\downarrow$.
**15.180** 11.43 m/s$^2$ $\searrow$ 36.7°.
**15.181** 100 mm/s.
**15.182** (a) $(30.3$ rad/s$)\mathbf{i} - (40.4$ rad/s$)\mathbf{k}$.
  (b) $(4.4$ m/s$)\mathbf{i} + (3.3$ m/s$)\mathbf{k}$.
**15.183** (a) $(40.4$ rad/s$)\mathbf{i} - (30.3$ rad/s$)\mathbf{k}$.
  (b) $(3.3$ m/s$)\mathbf{i} + (4.4$ m/s$)\mathbf{k}$.
**15.184** $-(20$ rad/s$^2)\mathbf{i}$.
**15.185** $-(14.80$ rad/s$^2)\mathbf{k}$.
**15.188** (a) $-(20$ rad/s$^2)\mathbf{j}$. (b) $-(0.96$ m/s$^2)\mathbf{i} + (2.4$ m/s$^2)\mathbf{k}$.
  (c) $-(2.46$ m/s$^2)\mathbf{j}$.
**15.189** $-(0.831$ m/s$^2)\mathbf{i} - (1.230$ m/s$^2)\mathbf{j} + (2.08$ m/s$^2)\mathbf{k}$.
**15.190** (a) $-(0.1745$ rad/s$)\mathbf{i} - (0.524$ rad/s$)\mathbf{j}$.
  (b) $-(0.0914$ rad/s$)\mathbf{k}$.
  (c) $\mathbf{v}_P = -(1.675$ m/s$)\mathbf{i} + (0.5584$ m/s$)\mathbf{j} - (0.3228$ m/s$)\mathbf{k}$;
  $\mathbf{a}_P = (0.338$ m/s$^2)\mathbf{i} - (0.056$ m/s$^2)\mathbf{j} - (0.9743$ m/s$^2)\mathbf{k}$.
**15.191** (a) $(20$ rad/s$)\mathbf{i} - (7.5$ rad/s$)\mathbf{j}$. (b) $-(150$ rad/s$^2)\mathbf{k}$.
  (c) $-(5.7$ m/s$^2)\mathbf{i} - (60.9$ m/s$^2)\mathbf{j}$.
**15.192** $\omega_1^2 r \sin\theta \left( \dfrac{L}{r}\sin\theta - \cos\theta \right)\left[ \left( \dfrac{L}{r}\cos\theta + \sin\theta \right)\mathbf{i} + \left( \dfrac{L}{r}\sin\theta - \cos\theta \right)\mathbf{j} \right]$.
**15.193** (a) $\omega_1/\sin\beta$. (b) $-(\omega_1/\tan\beta)\mathbf{i}$. (c) $(\omega_1^2/\tan\beta)\mathbf{k}$.
**15.196** $(0.417$ m/s$)\mathbf{j}$.
**15.197** $-(1.2$ m/s$)\mathbf{j}$.
**15.198** $-(1.2$ m/s$)\mathbf{i} - (0.8$ m/s$)\mathbf{j}$.
**15.199** $(0.45$ m/s$)\mathbf{i} + (0.3$ m/s$)\mathbf{j}$.
**15.202** $-(0.333$ m/s$)\mathbf{j}$.
**15.203** $-(0.12$ m/s$)\mathbf{j}$.
**15.204** $\omega_1/\cos 20°$.
**15.205** $\omega_1 \cos 20°$.
**15.206** (a) $(1.45$ rad/s$)\mathbf{i} + (0.1563$ rad/s$)\mathbf{j} + (0.1249$ rad/s$)\mathbf{k}$.
  (b) $-(0.065$ m/s$)\mathbf{i}$.
**15.210** $(1.45$ m/s$^2)\mathbf{i} + (0.99$ m/s$^2)\mathbf{j}$.
**15.211** $(3.69$ m/s$^2)\mathbf{i} + (2.48$ m/s$^2)\mathbf{j}$.

15.212 $-(0.01 \text{ m/s}^2)\mathbf{j}$.

15.213 $-(0.011 \text{ m/s}^2)\mathbf{j}$.

15.214 (a) $(1.83 \text{ m/s})\mathbf{i} + (0.762 \text{ m/s})\mathbf{j} - (1.218 \text{ m/s})\mathbf{k}$.
(b) $-(7.3 \text{ m/s}^2)\mathbf{i} - (21.96 \text{ m/s}^2)\mathbf{k}$.

15.215 (a) $(0.685 \text{ m/s})\mathbf{i} + (1.186 \text{ m/s})\mathbf{j} - (2.2 \text{ m/s})\mathbf{k}$.
(b) $-(26.67 \text{ m/s}^2)\mathbf{i} + (2.7 \text{ m/s}^2)\mathbf{j} - (13.7 \text{ m/s}^2)\mathbf{k}$.

15.216 (a) $(0.48 \text{ m/s})\mathbf{i} + (1.5 \text{ m/s})\mathbf{j} + (2.64 \text{ m/s})\mathbf{k}$.
(b) $-(22.8 \text{ m/s}^2)\mathbf{j} + (15 \text{ m/s}^2)\mathbf{k}$.

15.217 (a) $-(1.125 \text{ m/s})\mathbf{i} + (0.915 \text{ m/s})\mathbf{j} - (0.78 \text{ m/s})\mathbf{k}$.
(b) $-(7.28 \text{ m/s}^2)\mathbf{i} - (6.75 \text{ m/s}^2)\mathbf{j}$.

15.220 (a) $(0.493 \text{ m/s})\mathbf{i} + (1.353 \text{ m/s})\mathbf{j} - (1.015 \text{ m/s})\mathbf{k}$.
(b) $-(8.46 \text{ m/s}^2)\mathbf{i} + (1.970 \text{ m/s}^2)\mathbf{j} - (2.96 \text{ m/s}^2)\mathbf{k}$.

15.221 (a) $(1.033 \text{ m/s})\mathbf{i} + (1.353 \text{ m/s})\mathbf{j} - (1.015 \text{ m/s})\mathbf{k}$.
(b) $-(8.46 \text{ m/s}^2)\mathbf{i} + (1.970 \text{ m/s}^2)\mathbf{j} - (4.58 \text{ m/s}^2)\mathbf{k}$.

15.222 (a) $(0.48 \text{ m/s})\mathbf{i} + (1.5 \text{ m/s})\mathbf{j} + (2.64 \text{ m/s})\mathbf{k}$.
(b) $(2.4 \text{ m/s}^2)\mathbf{i} - (25.1 \text{ m/s}^2)\mathbf{j} + (8.7 \text{ m/s}^2)\mathbf{k}$.

15.223 (a) $-(1.125 \text{ m/s})\mathbf{i} + (0.915 \text{ m/s})\mathbf{j} - (0.78 \text{ m/s})\mathbf{k}$.
(b) $-(7.58 \text{ m/s}^2)\mathbf{i} - (9.6 \text{ m/s}^2)\mathbf{j} + (5.2 \text{ m/s}^2)\mathbf{k}$.

15.224 (a) $-(1.54 \text{ m/s})\mathbf{i} - (1.368 \text{ m/s})\mathbf{j} + (2.05 \text{ m/s})\mathbf{k}$.
(b) $(32.3 \text{ m/s}^2)\mathbf{i} - (31.6 \text{ m/s}^2)\mathbf{j} - (26.66 \text{ m/s}^2)\mathbf{k}$.

15.225 (a) $\omega_1 \sin\theta\left[\left(\cos\theta - \dfrac{L}{r}\sin\theta\right)\mathbf{i} + \left[\left(\sin\theta + \dfrac{L}{r}\cos\theta\right)\mathbf{j}\right]\right.$.
(b) $\omega_1^2 \sin\theta\left(\dfrac{L}{r}\sin\theta - \cos\theta\right)\mathbf{k}$.

15.226 (a) $(\omega_1 - \omega_2)\left(\cos\theta - \dfrac{L}{r}\sin\theta\right)(\sin\theta\,\mathbf{i} - \cos\theta\,\mathbf{j}) + \omega_1\,\mathbf{j}$.
(b) $\omega_1(\omega_1 - \omega_2)\left(\dfrac{L}{r}\sin\theta - \cos\theta\right)\mathbf{k}$.

15.229 $\mathbf{v}_B = (1.299 \text{ m/s})\mathbf{i} - (1.853 \text{ m/s})\mathbf{j} + (1.590 \text{ m/s})\mathbf{k}$;
$\mathbf{a}_B = (0.795 \text{ m/s}^2)\mathbf{i} - (0.792 \text{ m/s}^2)\mathbf{j} - (0.976 \text{ m/s}^2)\mathbf{k}$.

15.230 $\mathbf{v}_A = (0.0905 \text{ m/s})\mathbf{i} + (3.55 \text{ m/s})\mathbf{j} - (3.55 \text{ m/s})\mathbf{k}$;
$\mathbf{a}_A = -(5.68 \text{ m/s}^2)\mathbf{i} - (111.7 \text{ m/s}^2)\mathbf{j} - (111.7 \text{ m/s}^2)\mathbf{k}$.

15.231 $\mathbf{v}_A = -(1.4 \text{ m/s})\mathbf{i} + (0.8 \text{ m/s})\mathbf{j} - (1.2 \text{ m/s})\mathbf{k}$;
$\mathbf{a}_A = -(21.08 \text{ m/s}^2)\mathbf{i} - (11.2 \text{ m/s}^2)\mathbf{j} + (33.6 \text{ m/s}^2)\mathbf{k}$.

15.232 $\mathbf{v}_A = -(1.4 \text{ m/s})\mathbf{i} + (0.8 \text{ m/s})\mathbf{j} - (1.2 \text{ m/s})\mathbf{k}$;
$\mathbf{a}_A = -(22.56 \text{ m/s}^2)\mathbf{i} - (10.18 \text{ m/s}^2)\mathbf{j} + (35.2 \text{ m/s}^2)\mathbf{k}$.

15.235 $\mathbf{v}_A = -(6.35 \text{ m/s})\mathbf{i} - (1.5 \text{ m/s})\mathbf{k}$;
$\mathbf{a}_A = -(12 \text{ m/s}^2)\mathbf{i} - (32.9 \text{ m/s}^2)\mathbf{j} + (72.8 \text{ m/s}^2)\mathbf{k}$.

15.236 $\mathbf{v}_B = -(0.85 \text{ m/s})\mathbf{i} - (1.5 \text{ m/s})\mathbf{k}$;
$\mathbf{a}_B = -(12 \text{ m/s}^2)\mathbf{i} + (32.9 \text{ m/s}^2)\mathbf{j} - (15.2 \text{ m/s}^2)\mathbf{k}$.

15.237 $\mathbf{v}_A = (60 \text{ mm/s})\mathbf{i} - (36 \text{ mm/s})\mathbf{j}$;
$\mathbf{a}_A = -(270 \text{ mm/s}^2)\mathbf{i} + (180 \text{ mm/s}^2)\mathbf{j} + (172.8 \text{ mm/s}^2)\mathbf{k}$.

15.238 $\mathbf{v}_A = (180 \text{ mm/s})\mathbf{i} - (156 \text{ mm/s})\mathbf{j} + (144 \text{ mm/s})\mathbf{k}$;
$\mathbf{a}_A = (180 \text{ mm/s}^2)\mathbf{i} - (443 \text{ mm/s}^2)\mathbf{j} - (115.2 \text{ mm/s}^2)\mathbf{k}$.

15.241 (a) $(544 \text{ mm/s}^2)\mathbf{i} - (135 \text{ mm/s}^2)\mathbf{j}$.
(b) $(256 \text{ mm/s}^2)\mathbf{i} - (153.6 \text{ mm/s}^2)\mathbf{k}$.

15.242 (a) $-(32 \text{ mm/s}^2)\mathbf{i} + (135 \text{ mm/s}^2)\mathbf{j}$.
(b) $(256 \text{ mm/s}^2)\mathbf{i} + (153.6 \text{ mm/s}^2)\mathbf{k}$.

15.243 $\mathbf{v}_C = -(0.6 \text{ m/s})\mathbf{i} - (1.6 \text{ m/s})\mathbf{j} + (2 \text{ m/s})\mathbf{k}$.
$\mathbf{a}_C = (9.6 \text{ m/s}^2)\mathbf{i} + (13.6 \text{ m/s}^2)\mathbf{j} + (8 \text{ m/s}^2)\mathbf{k}$.

15.244 (a) $v_C = 6.38 \text{ m/s}, a_C = 1602.5 \text{ m/s}^2$.
(b) $v_C = 0.638 \text{ m/s}, a_C = 16 \text{ m/s}^2$.

15.246 $\omega_{AB} = 42.4 \text{ rad/s} \downarrow$; $\omega_{BD} = 30 \text{ rad/s} \uparrow$;
$\omega_{DE} = 33.9 \text{ rad/s} \downarrow$.

15.247 (a) $3 \text{ rad/s} \downarrow$. (b) $180 \text{ mm/s} \rightarrow$. (c) unwound $300 \text{ mm/s}$.

15.249 $\mathbf{a}_A = 54 \text{ m/s}^2 \searcharrow 88°$; $\mathbf{a}_B = 53.1 \text{ m/s}^2 \nearrow 0.98°$;
$\mathbf{a}_C = 54 \text{ m/s}^2 \uparrow$.

15.251 $2.40 \text{ m/s} \searcharrow 73.9°$.

15.252 $\mathbf{a}_1 = 0.381 \text{ m/s}^2 \rightarrow$; $\mathbf{a}_2 = 0.213 \text{ m/s}^2 \nearrow 57.7°$.

15.254 (a) $(0.3 \text{ m/s})\mathbf{i} + (0.520 \text{ m/s})\mathbf{j} + (0.48 \text{ m/s})\mathbf{k}$.
(b) $-(2.60 \text{ m/s}^2)\mathbf{i} - (2.34 \text{ m/s}^2)\mathbf{j} + (8.31 \text{ m/s}^2)\mathbf{k}$.

## CHAPTER 16

16.1 (a) $\mathbf{A} = 295 \text{ N} \measuredangle 85.2°$; $\mathbf{B} = 145.0 \text{ N} \leftarrow$. (b) $0.0848$.

16.2 (a) $5.66 \text{ m/s}^2 \rightarrow$. (b) $0.577$.

16.3 (a) $4.9 \text{ m/s}^2 \searcharrow 30°$.
(b) $\mathbf{A} = 0.242 \text{ N} \nearrow 60°$; $\mathbf{C} = 7 \text{ N} \measuredangle 60°$.

16.4 (a) $18.43°$. (b) $3.1 \text{ m/s}^2 \searcharrow 18.43°$.

16.5 (a) $11.0 \text{ m}$ (b) $12.7 \text{ m}$.

16.6 (a) $7.85 \text{ m/s}^2$. (b) $3.74 \text{ m/s}^2$. (c) $4.06 \text{ m/s}^2$.

16.7 (a) $0.419g$. (b) $3.33$.

16.8 (a) $0.295g$. (b) $3.33$.

16.11 (a) $4.9 \text{ m/s}^2 \searcharrow 30°$. (b) $F_{AC} = 0$; $F_{BD} = 61.14 \text{ N}C$.

16.12 (a) $8.5 \text{ m/s}^2 \searcharrow 60°$. (b) $10 \text{ N} \uparrow$.

16.13 (a) $1.572 \text{ m/s}^2$. (b) $20.8 \text{ N}$.

16.14 (a) $1.629 \text{ N} \cdot \text{m}$. (b) $18.04 \text{ N}$.

16.17 *Just above B:* $|V| = 15.3 \text{ N}$, $|M| = 3.516 \text{ N} \cdot \text{m}$.

16.18 $V_B = -194.0 \text{ N}$; $|M|_{\max} = 36.4 \text{ N} \cdot \text{m}$.

16.21 (a) $12.77 \text{ N} \cdot \text{m}$. (b) $t = 658 \text{ s}$.

16.22 (a) $0.327 \text{ m}$ (b) $11{,}250 \text{ rev}$.

16.23 $127.9 \text{ N}$.

16.24 $74.5 \text{ s}$.

16.25 (a) $2.25 \text{ rad/s}^2 \uparrow$. (b) $77.2 \text{ N}\, C$.

16.26 (a) $13.68 \text{ N} \cdot \text{m}$. (b) $176.4 \text{ N} \searcharrow 44.0°$.

16.29 $27.57 \text{ N} \cdot \text{m} \cdot \text{s}^2$.

16.30 $\alpha_A = 22.5 \text{ rad/s}^2 \uparrow$; $\alpha_B = 15.00 \text{ rad/s}^2 \uparrow$.

16.31 (a) $11.87 \text{ rad/s}^2 \uparrow$. (b) $T_A = 38.74 \text{ N}, T_B = 27.7 \text{ N}$.

16.32 (a) $\alpha_A = 9.05 \text{ rad/s}^2 \downarrow$; $\alpha_B = 13.5 \text{ rad/s}^2 \uparrow$.
(b) $1.85 \text{ N} \downarrow$.

16.33 (a) $\alpha_A = 13.42 \text{ rad/s}^2 \uparrow$; $\alpha_B = 30.14 \text{ rad/s}^2 \downarrow$.
(b) $8.3 \text{ N} \downarrow$.

16.34 (a) $\alpha_A = 15.3 \text{ rad/s}^2 \downarrow$; $\alpha_B = 15.3 \text{ rad/s}^2 \uparrow$. (b) $0.4875$.

16.35 (a) $\alpha_A = 20.8 \text{ rad/s}^2 \downarrow$; $\alpha_B = 28.7 \text{ rad/s}^2 \downarrow$.

16.36 $\dfrac{2\mu_k g \sin\phi}{r(\sin 2\phi - \mu_k \cos 2\phi)} \uparrow$.

16.37 $(2g/r) \sin\phi \uparrow$.

16.40 (a) $\alpha_A = 13.5 \text{ rad/s}^2 \uparrow$; $\alpha_B = 3.2 \text{ rad/s}^2 \uparrow$.
(b) $\omega_A = 897 \text{ rpm} \uparrow$; $\omega_B = 538 \text{ rpm} \downarrow$.

16.43 (a) $(2.47 \text{ m/s}^2)$. (b) $0$.

16.45 (a) $2.67 \text{ m/s}^2 \rightarrow$. (b) $1.333 \text{ m/s}^2 \leftarrow$.

16.46 (a) $\alpha = -(1.200 \text{ rad/s}^2)\mathbf{j}$; $\mathbf{a}_G = 0$;
(b) $\alpha = -(0.900 \text{ rad/s}^2)\mathbf{j}$; $\mathbf{a}_G = -(0.1350 \text{ m/s}^2)\mathbf{i}$.

16.47 $0.8148 \text{ m/s}^2$. (b) $2.28 \text{ m/s}^2$.

16.48 (a) $(1.3 \text{ m/s}^2)\mathbf{i} - (1.085 \text{ m/s}^2)\mathbf{k}$
(b) $(1.3 \text{ m/s}^2)\mathbf{i} + (1.085 \text{ m/s}^2)\mathbf{k}$.

16.50 (a) $0.690 \text{ rad/s}^2 \uparrow$. (b) $0.136 \text{ m/s}^2 \uparrow$.

16.51 (a) $3108 \text{ N}$. (b) $0.32 \text{ rad/s}^2 \uparrow$.

16.52 $\mathbf{a}_A = 2.71 \text{ m/s}^2 \uparrow$; $\mathbf{a}_B = 1.496 \text{ m/s}^2 \uparrow$.

16.53 $170.9 \text{ mm}$.

16.56 (a) $11 \text{ rad/s}^2 \uparrow$. (b) $3.92 \text{ m/s}^2 \nearrow 32.6°$.

16.57 (a) $10 \text{ rad/s}^2 \uparrow$. (b) $3.50 \text{ m/s}^2 \nearrow 31.1°$.

16.58 (a) $3g/2L \downarrow$. (b) $g/4 \uparrow$. (c) $5g/4 \downarrow$.

16.59 (a) $2g/L \downarrow$. (b) $g/3 \uparrow$. (c) $5g/3 \downarrow$.

16.60 (a) $3g/L \downarrow$. (b) $1.323g \measuredangle 49.1°$. (c) $2.18g \searcharrow 66.6°$.

16.61 (a) $g/4 \uparrow$. (b) $5g/4 \downarrow$.

16.62 (a) $0$. (b) $g \downarrow$.

16.64 (a) $5v_0/2r$. (b) $v_0/\mu_k g$. (c) $v_0^2/2\mu_k g$.

16.65 (a) $1.718 \text{ s}$. (b) $3.31 \text{ m/s}$. (c) $7.14 \text{ m}$.

16.66 (a) $1.980 \text{ s}$. (b) $3.06 \text{ m/s}$. (c) $7.98 \text{ m}$.

16.67 (a) $2v_1/7\mu_k g$. (b) $\mathbf{v} = 2v_1/7 \rightarrow$; $\omega = 5v_1/7r \uparrow$.

16.71 (a) $11.36 \text{ rad/s}^2 \downarrow$. (b) $\mathbf{C}_x = 54.9 \text{ N} \leftarrow$; $\mathbf{C}_y = 44.5 \text{ N} \uparrow$.

16.72 (a) $0.152 \text{ m}$ (b) $96.53 \text{ rad/s}^2 \downarrow$.

16.73 (a) $11.67 \text{ rad/s}^2 \downarrow$. (b) $\mathbf{A}_x = 1.75 \text{ N} \leftarrow$; $\mathbf{A}_y = 9.81 \text{ N} \uparrow$.

**16.74** (a) 600 mm. (b) 7.78 rad/s² $\downarrow$.

**16.77** 505 N $\rightarrow$.

**16.78** (a) $\mathbf{M} = 0.0687$ N $\cdot$ m $\gamma$; $\mathbf{R} = 5.12$ N $\searrow$ 71.5°.
(b) $\mathbf{M} = 0$; $\mathbf{R} = 6.11$ N $\uparrow$.

**16.79** (a) $mg/4 \uparrow$. (b) $3g/2 \downarrow$.

**16.80** (a) $L/3$. (b) $g/2 \uparrow$; $3\ mg/4 \uparrow$.

**16.81** (a) $-(24g/19)\mathbf{j}$. (b) $-(4g/19)[(4/\pi)\mathbf{i} + 3\mathbf{j}]$.

**16.82** (a) $3g/4L$ $\gamma$. (b) $T_{AE} = mg/2$, $T_{BF} = 3\ mg/8$.

**16.84** (a) 4 rad/s² $\downarrow$. (b) 6.4 N.

**16.85** (a) 10.8 rad/s² $\downarrow$. (b) 5.56 N $\cdot$ m.

**16.89** 604 mm.

**16.90** $\tan \beta = \mu_s[1 + (r/\overline{k})^2]$.

**16.91** (a) 6.12 m. (b) 1.748 m.

**16.92** (a) 2.4 m/s² $\measuredangle$ 15°. (b) 0.1264.

**16.93** (a) 15.4 rad/s² $\downarrow$; 3.126 m/s² $\rightarrow$. (b) 0.18.

**16.94** (a) 23.2 rad/s² $\downarrow$; 4.7 m/s² $\rightarrow$. (b) 0.02.

**16.95** (a) 7.76 rad/s² $\downarrow$; 1.575 m/s² $\rightarrow$. (b) 0.338.

**16.96** (a) 7.73 rad/s² $\gamma$; 1.58 m/s² $\leftarrow$. (b) 0.32.

**16.97** (a) Does not slide. (b) 16 rad/s² $\downarrow$; 2.56 m/s² $\rightarrow$.

**16.98** (a) Does not slide. (b) 24 rad/s² $\downarrow$; 3.84 m/s² $\rightarrow$.

**16.101** (a) 7.5 m/s² $\nearrow$ 45°. (b) 1.87 N $\downarrow$.

**16.102** (a) 7.25 m/s² $\nearrow$ 45°. (b) 1.435 N $\nearrow$ 45°.

**16.103** (a) 72.4 rad/s² $\downarrow$. (b) 7.24 m/s² $\downarrow$.

**16.104** (a) 72.4 rad/s² $\gamma$. (b) 7.24 m/s² $\downarrow$.

**16.105** $\dfrac{25g \sin \theta}{26r}$ $\gamma$.

**16.106** $\dfrac{2(3\pi - 4)g \sin \theta}{(9\pi - 16)r}$ $\gamma$.

**16.107** (a) 0.688 $P/mr$ $\downarrow$. (b) 0.25 $P/(mg + P)$.

**16.108** (a) 0.1843 $P/mr$ $\downarrow$. (b) 0.933 $P/mg$.

**16.111** (a) 11.56 rad/s² $\downarrow$. (b) 25 N (c) 41.7 N $\uparrow$.

**16.112** (a) 6.28 rad/s² $\downarrow$. (b) 43 N.

**16.113** (a) 31.4 rad/s² $\downarrow$. (b) 5.71 N $\leftarrow$.

**16.114** 27.5 rad/s² $\downarrow$.

**16.117** 34.28 N $\rightarrow$.

**16.118** 22.7 N $\rightarrow$.

**16.119** 25.9 N $\searrow$ 60°.

**16.120** 15.01 N $\swarrow$ 60°.

**16.121** 57.0 N $\searrow$ 60°.

**16.124** 2.62 rad/s $\downarrow$.

**16.125** 1.164 m/s² $\rightarrow$.

**16.126** 35.2 N $\nearrow$ 30°.

**16.127** 13.57 N $\nearrow$ 30°.

**16.128** (a) 24.8 rad/s² $\downarrow$. (b) 1.32 N $\uparrow$.

**16.129** (a) 26.4 N $\cdot$ m. (b) 3.67 N $\uparrow$.

**16.130** $\mathbf{A} = 8.13$ N $\uparrow$; $\mathbf{B} = 8.61$ N $\uparrow$.

**16.131** $\mathbf{A}_x = 3.24$ N $\rightarrow$; $\mathbf{A}_y = 9.81$ N $\uparrow$; $\mathbf{B}_x = 1.08$ N $\leftarrow$;
$\mathbf{B}_y = 9.81$ N $\uparrow$.

**16.134** $\alpha_{AB} = 26.1$ rad/s² $\downarrow$; $\alpha_{BD} = 26.1$ rad/s² $\gamma$;
$\alpha_{DE} = 26.1$ rad/s² $\gamma$.

**16.135** (a) 9.34 rad/s² $\downarrow$. (b) 9.93 m/s² $\searrow$ 20°. (c) 27.3 N $\measuredangle$ 70°.

**16.136** (a) 50.2 N $\measuredangle$ 60.3°. (b) 0.273.

**16.137** (a) $\alpha_A = 2g/5r$ $\gamma$; $\alpha_B = 2g/5r$ $\downarrow$. (b) $mg/5$.
(c) $4g/5 \downarrow$.

**16.138** $\mathbf{a}_A = 2P/7m \rightarrow$; $\mathbf{a}_B = 22P/7m \leftarrow$.

**16.139** $\alpha_{AB} = 11.04$ rad/s² $\downarrow$; $\alpha_{BC} = 55.2$ rad/s² $\gamma$

**16.140** (a) 0.29 m (b) 7.03 rad/s² $\gamma$.

**16.143** Just to the right of $A$, $V = mg/4$;
$L/3$ to the right of $A$, $M = mgL/27$.

**16.144** (a) 29 N $\leftarrow$. (b) $F_{CF} = F_{BE} = 94$ N $T$.

**16.145** 125.2 rev.

**16.146** (a) $\alpha_A = 10.23$ rad/s² $\gamma$; $\alpha_B = 28.1$ rad/s² $\gamma$.
(b) $\omega_A = 320$ rpm $\downarrow$; $\omega_B = 440$ rpm $\gamma$.

**16.148** $T_A = 955.5$ N, $T_B = 932.8$ N.

**16.150** (a) $45g/26 \downarrow$. (b) $25mg/52 \uparrow$.

**16.151** $\dfrac{r^2g \sin \beta}{(r^2 + \overline{k}^2)}$.

**16.153** (a) 6.30 rad/s² $\gamma$. (b) 37.4 N $\searrow$ 30°.

**16.154** 98.7 N $\uparrow$.

## CHAPTER 17

**17.1** 8798 rev.

**17.2** 12.77 N $\cdot$ m.

**17.3** (a) 29.4 kg $\cdot$ m² . (b) 13.27 rev.

**17.4** (a) 0.3268 m (b) 11,204 rev.

**17.5** $\theta_A = \dfrac{3r\omega_0^2}{32\pi\mu_k g}$ rev $\downarrow$; $\theta_B = \dfrac{r\omega_0^2}{32\pi\mu_k g}$ rev $\gamma$.

**17.6** (a) $\omega_A = 25$ rad/s $\downarrow$; $\omega_B = 25$ rad/s $\gamma$. (b) 0.380.

**17.7** 0.24 m

**17.8** 1.260$L$.

**17.9** (a) 2.93 m/s $\downarrow$. (b) 2.36 m.

**17.10** 317 N $\downarrow$.

**17.11** 365 N $\downarrow$.

**17.14** (a) $l/\sqrt{12}$. (b) $\omega = 1.861\sqrt{g/l}$; $\mathbf{C} = 2mg \uparrow$.

**17.15** (a) 6.10 rad/s $\downarrow$. (b) 4 rad/s $\downarrow$.

**17.16** (a) 54.0° (b) 0.767$\sqrt{gr}$ $\searrow$ 54.0°.

**17.17** 6.8 rad/s.

**17.20** $\sqrt{4gs/3} \downarrow$.

**17.21** (a) 3.16 m/s $\rightarrow$. (b) 33.3 N $\leftarrow$.

**17.22** (a) $1.142\sqrt{g/r}$ $\downarrow$. (b) 1.553 $mg \uparrow$.

**17.23** $0.392\sqrt{g/r}$.

**17.26** (a) 8.9 rad/s $\downarrow$. (b) 21.7 rad/s $\gamma$.

**17.27** 1.497 N $\cdot$ m.

**17.29** $\mathbf{v}_A = 0.775\sqrt{gL}$ $\leftarrow$; $v_B = 0.775\sqrt{gL}$ $\nearrow$ 60°

**17.30** 4.65 m/s $\rightarrow$.

**17.31** 806 N/m.

**17.32** 1.35 rad/s $\downarrow$.

**17.33** 0.99 rad/s $\gamma$.

**17.36** 0.490 J.

**17.37** 3.87 m/s $\rightarrow$.

**17.38** 1.1 m/s $\leftarrow$.

**17.39** (a) 33.0 kW. (b) 88.0 kW.

**17.40** (a) 1.471 kW. (b) 2.02 kW.

**17.41** 1157 rpm.

**17.42** 258.2 N $\cdot$ m/s

**17.43** 2.1 N $\cdot$ m

**17.44** (a) 320 mm. (b) 11.2 min.

**17.45** 17 rad/s $\downarrow$.

**17.46** 33.5 N $\cdot$ m

**17.47** 7.72 s.

**17.48** $\dfrac{v}{2g\mu_k}$.

**17.49** 3.82 s.

**17.50** $\dfrac{2r\omega_0(1 + \mu_k^2)}{5g\mu_k(1 + \mu_k)}$

**17.51** $\left(\dfrac{M}{mr^2} - \dfrac{2\mu_k g}{r}\right)t$ $\gamma$.

**17.52** $\omega_A = 230.2$ rad/s $\downarrow$; $\omega_B = 96.2$ rad/s $\gamma$.

**17.55** (a) 22.9 N. (b) 8.75 N.

**17.56** $m\bar{v}\bar{k}^2\omega/\bar{v}$.

**17.60** 0.850 m.

**17.61** (a) 28.5 rad/s $\curvearrowright$. (b) 40.9 N $\cdot$ s $\nearrow$ 30°.

**17.62** (a) $\frac{1}{2}gt \downarrow$. (b) $\frac{2}{3}gt \downarrow$.

**17.63** (a) 4.57 m/s $\rightarrow$. (b) 0.152 m/s $\rightarrow$.

**17.66** $P_A = 5.25$ N, $P_B = 9$ N.

**17.67** (a) Pipe rolls without sliding.
(b) *Pipe:* 0.857 m/s $\rightarrow$,
*Plate:* 1.714 m/s $\rightarrow$.

**17.68** (a) $\dfrac{2r\omega_0}{7\mu_k g}$. (b) $\mathbf{v} = 2r\omega_0/7 \rightarrow$; $\omega = 2\omega_0/7 \downarrow$.

**17.69** (a) $5\bar{v}_0/2r$. (b) $\bar{v}_0/\mu_k g$.

**17.70** $5\omega_0/6$.

**17.71** (a) 2.54 rad/s. (b) 1.902 J.

**17.72** $\omega_A = \omega_B = 159.3$ rpm $\downarrow$; $\omega_P = 20.68$ rpm $\curvearrowright$.

**17.73** $\omega_{AB} = 71.23$ rpm $\curvearrowright$; $\omega_{DISK} = 288.77$ rpm $\downarrow$.

**17.74** (a) 5 rad/s. (b) 3.13 rad/s.

**17.75** $\omega_{BC} = 36.6$ rpm $\downarrow$; $\omega_A = 16.87$ rpm $\curvearrowright$.

**17.78** 2.51 m/s.

**17.79** (a) 15 rad/s. (b) 7.86 m/s.

**17.80** $\omega = 17.13$ rad/s, $I_{CD} = 0.0554$ N$\cdot$m$^2$s.

**17.81** (a) 0.6 m (b) 3.18 rad/s.

**17.84** $\mathbf{v} = 0.044$ m/s $\leftarrow$; $\omega = 0.770$ rad/s $\downarrow$.

**17.85** (a) 7.8 m/s $\leftarrow$. (b) 1.8 m/s $\rightarrow$.

**17.86** 4.9 rad/s $\downarrow$.

**17.87** 274 m.

**17.88** $\omega = 1.534$ rad/s $\curvearrowright$; $\mathbf{v}_A = 0.460$ m/s $\downarrow$.

**17.89** (a) (0.367 rad/s)$\mathbf{i}$. (b) $-(1144$ N)$\mathbf{k}$.

**17.90** (a) 0.5 m. (b) $-(0.1049$ m/s)$\mathbf{k}$.

**17.93** 2.4 rad/s $\downarrow$.

**17.94** (a) 19.47°. (b) 6.81 N $\cdot$ s $\measuredangle$ 75.6°.

**17.95** $\dfrac{L}{\sqrt{3}}$.

**17.96** $\omega = \frac{1}{7}(2 + 5\cos\beta)\omega_1 \curvearrowright$; $\bar{v} = \frac{1}{7}(2 + 5\cos\beta)\bar{v}_1 \leftarrow$.

**17.97** $\pi L/3$.

**17.98** (a) $\omega = 3\bar{v}_1/L \downarrow$; $\bar{v}_1/2 \downarrow$. (b) $\omega = 3\bar{v}_1/L \curvearrowright$; $\bar{v}_1/2 \uparrow$.
(c) $\omega = 0$; $\bar{v}_1 \uparrow$.

**17.101** $\omega = 3\sqrt{2}v_0/5a \curvearrowright$; $\bar{\mathbf{v}} - 0.825 v_0 \searrow 76.0°$.

**17.102** (a) $\sqrt{3g\sin\beta/L} \curvearrowright$. (b) $\mathbf{A}\Delta t = m\sqrt{gL\sin\beta/3} \uparrow$;
$\mathbf{B}\Delta t = 2m\sqrt{gL\sin\beta/3} \uparrow$.

**17.103** (a) 0.816.

**17.104** 399.4 m/s.

**17.105** 8.4°.

**17.108** (a) $\omega_0/4 \downarrow$. (b) 15/16. (c) 3.05°.

**17.109** $0.606\sqrt{gL} \rightarrow$.

**17.110** $\sqrt{3gL}/2$.

**17.111** (a) 2.975 rad/s $\downarrow$. (b) 0.727 m/s $\rightarrow$.

**17.112** (a) 5.961 rad/s $\downarrow$. (b) 1.456 m/s $\rightarrow$.

**17.115** 0.41 m.

**17.116** 105.4°.

**17.117** (a) $3v_0(1 + e)/2L \downarrow$. (b) $v_0(1 + e)\left(m_C + \dfrac{m_{AB}}{4}\right)\uparrow$.

**17.118** (a) $\bar{\mathbf{v}}_A = 0$, $\omega_A = \bar{v}_1/r \downarrow$; $\bar{\mathbf{v}}_B = \bar{v}_1 \rightarrow$, $\omega_B = 0$.
(b) $\bar{\mathbf{v}}_A = 2\bar{v}_1/7 \rightarrow$; $\bar{\mathbf{v}}_B = 5\bar{v}_1/7 \rightarrow$.

**17.119** $5v_0/4r$.

**17.120** (a) $\bar{\mathbf{v}}_A = \bar{v}_0\sin\theta\,\bar{\mathbf{j}}$, $\bar{\mathbf{v}}_B = \bar{v}_0\cos\theta\,\mathbf{i}$;
$\omega_A = (\bar{v}_0/r)(-\sin\theta\,\mathbf{i} + \cos\theta\,\mathbf{j})$, $\omega_B = 0$.
(b) $\frac{5}{7}(v_0\cos\theta)\mathbf{i}$.

**17.121** 10.25 rad/s $\curvearrowright$.

**17.122** 5.44 rad/s $\downarrow$.

**17.124** (a) $0.926\sqrt{gL} \leftarrow$. (b) $1.225\sqrt{gL} \leftarrow$.

**17.125** 5.30 s.

**17.127** 84.2 rpm.

**17.128** $-24.4$ rpm.

**17.130** (a) 1/3. (b) $\omega_1 L/4 \uparrow$. (c) $m\omega_1 L/4 \uparrow$.

**17.131** 60.6°

## CHAPTER 18

**18.1** $(1.296$ kg $\cdot$ m$^2$/s)$\mathbf{i} - (0.702$ kg $\cdot$ m$^2$/s)$\mathbf{k}$.

**18.2** $\dfrac{ma^2\omega}{12}[3\mathbf{j} + 2\mathbf{k}]$.

**18.3** $(2.0037$ kg $\cdot$ m$^2$/s)$\mathbf{i} + (5.505$ kg $\cdot$ m$^2$/s)$\mathbf{j}$.

**18.4** 9.7°.

**18.7** $(0.08$ kg $\cdot$ m$^2$/s)$\mathbf{j} + (0.48$ kg $\cdot$ m$^2$/s)$\mathbf{k}$.

**18.8** $(0.307$ kg $\cdot$ m$^2$/s)$\mathbf{i} + (1.229$ kg $\cdot$ m$^2$/s)$\mathbf{j}$.

**18.9** (a) 8.97 rad/s. (b) 0.016 rad/s.

**18.10** $(0.88$ kg $\cdot$ m$^2$/s)$\mathbf{i} - (0.0025$ kg $\cdot$ m$^2$/s)$\mathbf{j} - (14.5$ kg $\cdot$ m$^2$/s)$\mathbf{k}$.

**18.15** (a) $(5.65$ kg $\cdot$ m$^2$/s)$\mathbf{i} - (1.885$ kg $\cdot$ m$^2$/s)$\mathbf{j} +$
$(12.57$ kg $\cdot$ m$^2$/s)$\mathbf{k}$. (b) 25.4°.

**18.16** (a) $(5.65$ kg $\cdot$ m$^2$/s)$\mathbf{i} - (1.885$ kg $\cdot$ m$^2$/s)$\mathbf{j} +$
$(12.57$ kg $\cdot$ m$^2$/s)$\mathbf{k}$. (b) 154.6°.

**18.17** (a) $(1.45$ kg $\cdot$ m$^2$/s)$\mathbf{i} - (0.87096$ kg $\cdot$ m$^2$/s)$\mathbf{k}$. (b) 31.0°.

**18.18** (a) $(1.45$ kg $\cdot$ m$^2$/s)$\mathbf{i} - (0.87096$ kg $\cdot$ m$^2$/s)$\mathbf{k}$.
(b) $(1.45$ kg $\cdot$ m$^2$/s)$\mathbf{i} - (0.87096$ kg $\cdot$ m$^2$/s)$\mathbf{k}$.

**18.21** (a) $(1.439$ m/s)$\mathbf{j}$.
(b) $-(3.61$ rad/s)$\mathbf{i} - (3.20$ rad/s)$\mathbf{j} + (9.59$ rad/s)$\mathbf{k}$.

**18.22** (a) $-(1.439$ m/s)$\mathbf{i}$. (b) $-(2.47$ rad/s)$\mathbf{j} - (3.20$ rad/s)$\mathbf{k}$.

**18.23** (a) 0. (b) $(3F\Delta t/4md)(4\mathbf{i} - \mathbf{k})$.

**18.24** (a) $-(F\Delta t / 3m)\mathbf{k}$. (b) $-(3F\Delta t/4md)(4\mathbf{i} + \mathbf{j})$.

**18.25** (a) 0. (b) $\left(\dfrac{F\Delta t}{ma}\right)(2.50\mathbf{i} - 1.454\mathbf{j} + 2.19\mathbf{k})$.

**18.26** (a) $-(F\Delta t/m)\mathbf{k}$. (b) $\left(\dfrac{F\Delta t}{ma}\right)(3.75\mathbf{i} - 1.875\mathbf{j} + 1.250\mathbf{k})$.

**18.29** $(0.0125$ rad/s)$\mathbf{i} - (0.1396$ rad/s)$\mathbf{j} - (0.1814$ rad/s)$\mathbf{k}$.

**18.30** (a) $-0.726$ rad/s. (b) $-(729$ m/s)$\mathbf{i} - (2880$ m/s)$\mathbf{j} +$
$(558$ m/s)$\mathbf{k}$.

**18.31** (a) $(\omega_0/6)(-\mathbf{i} + \mathbf{j})$. (b) $(a\omega_0/6)\mathbf{k}$.

**18.32** (a) $(5ma\omega_0/24)\mathbf{k}$. (b) $-(ma\omega_0/24)\mathbf{k}$.

**18.33** (a) $\Delta t_A = 0.1290$ s; $\Delta t_B = 1.086$ s. (b) $-(50.6$ mm/s)$\mathbf{j}$.

**18.34** (a) 941 ms. (b) $(0.01693$ rad/s)$\mathbf{k}$. (c) $-(39.2$ mm/s)$\mathbf{j}$.

**18.39** 6.48 J

**18.40** $ma^2\omega^2/8$.

**18.41** 346 kg $\cdot$ m$^2$/s$^2$.

**18.42** 0.235 $mr^2\,\omega^2$.

**18.45** 9.59 J.

**18.46** 23.5 J.

**18.47** 237 J.

**18.48** 8.7 kg $\cdot$ m$^2$/s$^2$.

**18.49** 27 J.

**18.50** $\dfrac{7(F\Delta t)^2}{2m}$

**18.53** 5 $ma^2\omega_0^2/48$.

**18.54** $m\bar{v}_0^2/14$.

**18.55** $(7.02$ N $\cdot$ m)$\mathbf{j}$.

**18.56** $(ma^2\omega^2/6)\mathbf{i}$.

**18.57** $-(252$ kg $\cdot$ m$^2$/s$^2$.)$\mathbf{k}$.

**18.58** $(\frac{1}{8}mr^2\omega^2\sin 2\beta)\mathbf{k}$.

**18.59** $(1.92 \text{ N} \cdot \text{m})\mathbf{i}$.

**18.60** $(9.83 \text{ N} \cdot \text{m})\mathbf{k}$.

**18.63** $\mathbf{A} = -(4.74 \text{ N})\mathbf{j} - (3.95 \text{ N})\mathbf{k}$;
$\mathbf{B} = (4.74 \text{ N})\mathbf{j} + (3.95 \text{ N})\mathbf{k}$.

**18.64** $\mathbf{A} = -(10.66 \text{ N})\mathbf{j} - (8.88 \text{ N})\mathbf{k}$;
$\mathbf{B} = (10.66 \text{ N})\mathbf{j} + (8.88 \text{ N})\mathbf{k}$.

**18.65** $\mathbf{A} = -(1.836 \text{ N})\mathbf{j}$; $\mathbf{B} = (1.836 \text{ N})\mathbf{j}$.

**18.66** $\mathbf{A} = (2.417 \text{ N})\mathbf{k}$; $\mathbf{B} = -(2.417 \text{ N})\mathbf{k}$.

**18.69** (a) 60 rad/s². (b) $\mathbf{A} = -(15 \text{ N})\mathbf{k}$;
$\mathbf{B} = +(15 \text{ N})\mathbf{k}$.

**18.70** (a) $(24.8 \times 10^{-3} \text{ N} \cdot \text{m})\mathbf{k}$.
(b) $\mathbf{A} = (7.50 \times 10^{-3} \text{ N})\mathbf{i} + (15.00 \times 10^{-3} \text{ N})\mathbf{j}$;
$\mathbf{B} = -(7.50 \times 10^{-3} \text{ N})\mathbf{i} - (15.00 \times 10^{-3} \text{ N})\mathbf{j}$.

**18.71** (a) $(223.16 \text{ rad/s}^2)\mathbf{i}$.
(b) $\mathbf{A} = (4.1 \text{ N})\mathbf{k}$; $\mathbf{B} = -(4.1 \text{ N})\mathbf{k}$.

**18.72** (a) 0.434 kg · m²/s².
(b) $\mathbf{A} = (0.906 \text{ N})\mathbf{j}$; $\mathbf{B} = -(0.906 \text{ N})\mathbf{j}$.

**18.75** (a) $(0.9 \text{ N} \cdot \text{m})\mathbf{i}$. (b) $\mathbf{A} = -(0.938 \text{ N})\mathbf{j} + (1.125 \text{ N})\mathbf{k}$;
$\mathbf{B} = (0.938 \text{ N})\mathbf{j} - (1.125 \text{ N})\mathbf{k}$.

**18.76** (a) $(7.2 \text{ rad/s})\mathbf{k}$. (b) $\mathbf{A} = -(57.3 \text{ mN})\mathbf{i} + (47.4 \text{ mN})\mathbf{j}$;
$\mathbf{B} = (57.3 \text{ mN})\mathbf{i} - (47.4 \text{ mN})\mathbf{j}$.

**18.77** 34.3 N ↑.

**18.78** 1°; $A$ will move up.

**18.79** $\mathbf{A} = (1.527 \text{ N})\mathbf{j}$; $\mathbf{B} = -(1.527 \text{ N})\mathbf{j}$.

**18.80** 4290 N · m.

**18.83** (a) 27.0°. (b) 8.09 rad/s.

**18.84** (a) 7.53 rad/s. (b) 7.00 rad/s.

**18.85** (a) 71.2°. (b) 5.67 rad/s.

**18.86** (a) 5.51 rad/s. (b) 4.63 rad/s.

**18.89** 7.89 rad/s.

**18.90** 15.24 rad/s.

**18.91** $\mathbf{A} = -\dfrac{1}{4}ma\omega_1\omega_2\mathbf{k}$; $\mathbf{B} = \dfrac{1}{4}ma\omega_1\omega_2\mathbf{k}$.

**18.92** $\mathbf{A} = -(51.9 \text{ N})\mathbf{i}$; $\mathbf{M}_A = (3.8988 \text{ kg} \cdot \text{m}^2/\text{s}^2)\mathbf{i} + (7.8128 \text{ kg} \cdot \text{m}^2/\text{s}^2)\mathbf{k}$.

**18.93** $\mathbf{A} = (0.884 \text{ N})\mathbf{k}$; $\mathbf{B} = -(0.884 \text{ N})\mathbf{k}$.

**18.94** 6.79 rad/s.

**18.95** (a) $\mathbf{C} = -(592 \text{ N})\mathbf{j}$; $\mathbf{D} = (592 \text{ N})\mathbf{j}$.
(b) $\mathbf{C} = \mathbf{D} = 0$.

**18.96** 35.5 rpm.

**18.99** (a) $\dfrac{1}{4}ma^2\alpha_1\mathbf{i}$. (b) $\mathbf{A} = -\dfrac{1}{4}ma\omega_1\omega_2\mathbf{k}$;

$\mathbf{B} = \dfrac{1}{4}ma\omega_1\omega_2\mathbf{k}$.

**18.100** $\mathbf{A} = -(0.0075 \text{ N})\mathbf{j} + (0.135 \text{ N})\mathbf{k}$;
$\mathbf{B} = (0.0075 \text{ N})\mathbf{j} - (0.135 \text{ N})\mathbf{k}$.

**18.101** $\mathbf{R} = -(15.02 \text{ N})\mathbf{i} - (11.285 \text{ N})\mathbf{k}$; $\mathbf{M}_A^R = (94.185 \text{ N} \cdot \text{m})\mathbf{i}$.

**18.102** (a) $(7.9 \text{ N} \cdot \text{m})\mathbf{j}$.
(b) $\mathbf{R} = -(75.4 \text{ N})\mathbf{i} - (75 \text{ N})\mathbf{k}$; $\mathbf{M}_A^R = (223 \text{ N} \cdot \text{m})\mathbf{i}$.

**18.105** 31 rpm, 632.3 rpm.

**18.106** (c) −5.14%.

**18.109** 44.8°.

**18.110** 2340 rpm.

**18.111** (a) 4.51 rpm. (b) 4.57 rpm, 363 rpm

**18.112** $\cos\beta = \dfrac{2d^2\dot\psi}{(h^2 - d^2)\dot\phi}$.

**18.113** (a) 131.27 rad/s. (b) 0.055 m

**18.114** $2.163 \times 10^{21}$ kg · m

**18.122** (a) $\beta = 23.8°$. (b) Precession, 82.6 rpm; spin, 128.8 rpm.

**18.123** (a) 13.19°. (b) 1242 rpm (retrograde).

**18.124** Precession axis: $\theta_x = 125.0°$, $\theta_y = 36.6°$, $\theta_z = 80.7°$;
precession, 0.9426 rad/s (retrograde); spin, 0.1583 rad/s.

**18.125** Precession axis: $\theta_x = 90°$, $\theta_y = 26.0°$, $\theta_z = 64.0°$;
precession, 0.845 rad/s (retrograde); spin, 0.161 rad/s.

**18.128** Precession axis: $\theta_x = 99.04°$, $\theta_y = 90°$, $\theta_z = 9.04°$;
precession, 121.58 rpm (retrograde); spin, 60.3 rpm.

**18.129** (a) 8 rad/s. (b) 11.31 rad/s.

**18.130** (a) $-45° \le \theta \le 45°$. (b) 6 rad/s. (c) 4.90 rad/s.

**18.131** (a) 4 rad/s. (b) 5.66 rad/s.

**18.134** (a) $\sqrt{17g/11a}$. (b) $\sqrt{44g/17a}$.

**18.135** (a) 48.7°. (b) Spin, 54.5 rad/s; precession, −6.87 rad/s.

**18.136** (a) 41 rad/s. (b) Spin, 55.46 rad/s; precession,
−16.7 rad/s.

**18.145** 0.3347 kg · m²/s; $\theta_x = 48.7°$, $\theta_y = 41.4°$, $\theta_z = 90°$.

**18.146** 1.328 N · m

**18.148** (a) $B$ and $C$. (b) $\Delta t_B = 2.24$ s, $\Delta t_C = 3.82$ s. (c) 0.242 s.

**18.149** (a) $\dfrac{1}{8}\omega_0(-\mathbf{i} + \mathbf{j})$. (b) $0.0884\omega_0 a\mathbf{k}$.

**18.151** $-(1.767 \text{ N} \cdot \text{m})\mathbf{i} - (2.008 \text{ N} \cdot \text{m})\mathbf{j} + (3.0123 \text{ N} \cdot \text{m})\mathbf{k}$.

**18.152** (a) 0.00145 m above the axis,
$I_{xy} = 2.383 \times 10^{-3} \text{ N} \cdot \text{m s}^2$, $I_{zx} = 0$.
(b) 1.53 N at $A$; 0.137 N at $E$.

**18.154** $\mathbf{D} = -(34.4 \text{ N})\mathbf{j} + (21.6 \text{ N})\mathbf{k}$; $\mathbf{E} = -(8.8 \text{ N})\mathbf{j} + (21.6 \text{ N})\mathbf{k}$.

**18.155** (a) $(2.00 \text{ N} \cdot \text{m})\mathbf{i}$. (b) $\mathbf{D} = -(32.4 \text{ N})\mathbf{j} + (23.6 \text{ N})\mathbf{k}$; $\mathbf{E} = -(6.78 \text{ N})\mathbf{j} + (23.6 \text{ N})\mathbf{k}$.

## CHAPTER 19

**19.1** 7.312 m/s²; 1.571 s.

**19.2** 10 mm; 3.18 Hz.

**19.3** 0.032 m; 45.48 m/s².

**19.4** (a) 0.0056 m; 6.66 Hz. (b) 0.234 m/s; 9.8 m/s².

**19.5** (a) 0.0982 s; 10.183 Hz. (b) 0.047 m; 192.4 m/s².

**19.6** (a) 280 rpm. (b) 1.703 m/s.

**19.7** (a) 11.29°. (b) 1.933 m/s².

**19.10** (a) 0.2 s. (b) 30.8 mm. (c) 43.1°.

**19.11** (a) 2.48 mm. (b) 0.621 mm.

**19.12** (a) 0.0247 s. (b) 2.29 m/s ↑; 17.14 m/s² ↓.

**19.13** 70.3 mm ↑; 2.20 m/s ↓; 20.1 m/s² ↓.

**19.14** (a) 23.52 m (b) 161.78 m/s².

**19.15** (a) 0.892 s. (b) 3 m/s ↓.

**19.18** (a) 0.361 s; 2.77 Hz. (b) 0.765 m/s; 13.31 m/s².

**19.19** (a) 0.203 s; 4.93 Hz. (b) 1.859 m/s; 57.6 m/s².

**19.20** 2.634 s.

**19.21** 4 or 1/4.

**19.24** (a) 45.0 kg. (b) 14.22 kN/m.

**19.25** (a) 6.22 kN/m. (b) 2.27 kg.

**19.26** (a) 7451.55 N/m. (b) 30054.6 N/m.

**19.27** (a) 86.9 N. (b) 136.63 N/m.

**19.28** (a) 2.91 kN/m. (b) 763 N/m.

**19.29** (a) 427 kg. (b) 68.5 mm/s.

**19.33** (a) 60.1 mm. (b) 1.437 Hz.

**19.35** 23.1°.

**19.36** (a) 1.7943 s. (b) 1.825 s. (c) 2.12 s.

**19.37** 1.94 m.

**19.38** (a) 3.65 Hz. (b) 0.00763°.

**19.39** (a) 0.529 s. (b) 238 mm/s.

**19.40** (a) 0.858 s. (b) 0.093 m/s.

**19.41** (a) 7.53 Hz. (b) 97.3 rad/s².

**19.42** (a) 2.45 kg. (b) 457 mm/s.

**19.45** (a) 2.79 s. (b) 1.933 m.

**19.46** 0.776 s.

**19.47** (a) $6.82\sqrt{r/g}$. (b) $7.38\sqrt{r/g}$.

**19.48** (a) $6.33\sqrt{b/g}$. (b) $6.67\sqrt{b/g}$.

**19.49** (a) 0.5. (b) 0.707.

**19.50** 128.8 mm.

**19.55** $f_n = (1/2\pi)\sqrt{\dfrac{6k}{5m} + \dfrac{9g}{10l}}$ Hz.

**19.56** $f_n = (1/2\pi)\sqrt{\dfrac{2k}{3m} + \dfrac{4g}{3L}}$ Hz.

**19.57** (a) 3 Hz. (b) 81.57 N/m.

**19.58** 1.9 Hz.

**19.59** $1.407$ rad/s$^2$ ↓.

**19.60** 0.365 m/s$^2$.

**19.63** 11.17 mm.

**19.64** (a) 1.125 s. (b) 4.39 m/s.

**19.65** 5.2654 s.

**19.66** (a) 0.400 s. (b) 0.089 m/s.

**19.69** 0.351 m/s.

**19.70** 19.02 mm.

**19.71** $f_n = (1/2\pi)\sqrt{\dfrac{k}{5m}}$ Hz.

**19.72** (a) 0.368 s. (b) 0.574 m/s.

**19.73** $1.460$ kg · m$^2$.

**19.74** $\dfrac{l}{\sqrt{12}}$.

**19.77** 5.32 Hz.

**19.78** $f_n = (1/2\pi)\sqrt{\dfrac{2k}{3m} + \dfrac{4g}{3L}}$ Hz.

**19.79** (a) 0.720 s. (b) 0.0873 m/s.

**19.80** 2.05 Hz.

**19.83** 1.586 s.

**19.84** 1.627 s.

**19.85** 0.567 Hz.

**19.86** 1.073 s.

**19.89** 0.918 Hz.

**19.90** $f_n = (1/2\pi)\sqrt{\dfrac{12k}{7m} + \dfrac{8g}{7\sqrt{3}l}}$ Hz.

**19.91** 1.192 s.

**19.92** 1.305 s.

**19.93** $f_n = \dfrac{a}{2\pi l}\sqrt{\dfrac{6k}{m}}$.

**19.94** $f_n = (1/2\pi)\sqrt{\dfrac{3g}{l}}$.

**19.97** (a) 0.214 s. (b) 0.214 s.

**19.98** (a) 37.1 mm (in phase). (b) 260 mm (in phase).

**19.99** (a) 160 N/m. (b) 40 N/m.

**19.101** 37.67 rad/s $< \omega_f <$ 46.77 rad/s.

**19.102** 0.0929 m $< b <$ 3.55 m.

**19.104** $\omega_f <$ 11.07 rad/s; $\omega_f >$ 19.18 rad/s.

**19.105** (a) 174.6 mm. (b) 8.73 N.

**19.106** $\omega_f <$ 9.09 rad/s.

**19.107** (a) 0.94 m/s$^2$. (b) 2.42 N.

**19.108** (a) 0.450 rad/s. (b) 2.70 m/s$^2$.

**19.109** 3.21 m/s$^2$.

**19.112** $\omega_f \leq$ 310 rpm; $\omega_f \geq$ 316 rpm.

**19.113** 2.95 mm.

**19.114** 1557 N.

**19.116** $-5.63$ mm or 22.5 mm.

**19.117** 1.286 mm or 1.800 mm.

**19.120** Transmissibility $= \dfrac{1}{\left(1 - \omega_f^2/\omega_n^2\right)}$.

**19.121** (a) 1400 rpm. (b) 0.416 mm.

**19.122** (a) 4.9 m/s. (b) 0.084 m (out of phase).

**19.123** (a) 80 N/m. (b) 0.2775 m (out of phase).

**19.132** (a) 0.1178. (b) 38.4 mm.

**19.133** (a) 114.1 kN/m. (b) 0.1935 s.

**19.134** 8.82 N.

**19.135** 106.5 mm/s ↑.

**19.137** (a) $0.1534\dfrac{d^2\theta}{dt^2} + 0.2756\dfrac{d\theta}{dt} + 5.0325\,\theta = 0$   (b) 0.1565.

**19.138** (a) 297.5 rpm. (b) 273 rpm. (c) 0.02 m, 0.018 m

**19.139** (a) 297 rpm. (b) 252 rpm. (c) 0.085 m, 0.009 m

**19.140** $c/c_c \geq \sqrt{1/2}$.

**19.143** (a) 0.1269. (b) 462 N · s/m.

**19.144** 3.13 mm.

**19.145** 0.487.

**19.151** $m\dfrac{d^2x_A}{dt^2} + c\left(\dfrac{dx_A}{dt} - \dfrac{dx_B}{dt}\right) + 2k(x_A - x_B) = P_m \sin \omega_f t$.
$m\dfrac{d^2x_B}{dt^2} + 3c\dfrac{dx_B}{dt} - c\dfrac{dx_A}{dt} + 3kx_B - 2kx_A = 0$.

**19.152** $R < 2\sqrt{L/C}$.

**19.153** (a) $E/R$. (b) $L/R$.

**19.156** (a) $c\dfrac{d}{dt}(x_A - x_m) + kx_A = 0$.
$m\dfrac{d^2x_m}{dt^2} + c\dfrac{d}{dt}(x_m - x_A) = P_m \sin \omega_f t$.
(b) $R\dfrac{d}{dt}(q_A - q_m) + \dfrac{1}{C}q_A = 0$.
$L\dfrac{d^2q_m}{dt^2} + R\dfrac{d}{dt}(q_m - q_A) = E_m \sin \omega_f t$.

**19.157** (a) $c_1\dfrac{dx_A}{dt} + (k_1 + k_2)x_A - k_2x_m = 0$.
$m\dfrac{d^2x_m}{dt^2} + c_2\dfrac{dx_m}{dt} + k_2(x_m - x_A) = 0$.
(b) $R_1\dfrac{dq_A}{dt} + \left(\dfrac{1}{C_1} + \dfrac{1}{C_2}\right)q_A - \dfrac{1}{C_2}q_m = 0$.
$L\dfrac{d^2q_m}{dt^2} + R_2\dfrac{dq_m}{dt} + \dfrac{1}{C_2}(q_m - q_A) = 0$.

**19.158** (a) 1.22°. (b) 0.2675 m/s; 0.6776 m/s$^2$.

**19.160** $(1/2\pi)\sqrt{2\mu_k g/l}$.

**19.161** (a) 0.534 s. (b) 0.491 rad/s.

**19.162** 75.5°.

**19.164** 0.8289 s.

**19.165** $\tau_n = 2\pi\sqrt{\dfrac{60r^2 + 10l^2}{9gl}}$.

**19.167** 222.26 N.

**19.168** 211 mm.

## SI Prefixes

| Multiplication Factor | Prefix† | Symbol |
|---|---|---|
| $1\ 000\ 000\ 000\ 000 = 10^{12}$ | tera | T |
| $1\ 000\ 000\ 000 = 10^{9}$ | giga | G |
| $1\ 000\ 000 = 10^{6}$ | mega | M |
| $1\ 000 = 10^{3}$ | kilo | k |
| $100 = 10^{2}$ | hecto‡ | h |
| $10 = 10^{1}$ | deka‡ | da |
| $0.1 = 10^{-1}$ | deci‡ | d |
| $0.01 = 10^{-2}$ | centi‡ | c |
| $0.001 = 10^{-3}$ | milli | m |
| $0.000\ 001 = 10^{-6}$ | micro | $\mu$ |
| $0.\ 000\ 000\ 001 = 10^{-9}$ | nano | n |
| $0.000\ 000\ 000\ 001 = 10^{-12}$ | pico | p |
| $0.000\ 000\ 000\ 000\ 001 = 10^{-15}$ | femto | f |
| $0.000\ 000\ 000\ 000\ 000\ 001 = 10^{-18}$ | atto | a |

† The first syllable of every prefix is accented so that the prefix will retain its identity. Thus, the preferred pronunciation of kilometer places the accent on the first syllable, not the second.

‡ The use of these prefixes should be avoided, except for the measurement of areas and volumes and for the nontechnical use of centimeter, as for body and clothing measurements.

## Principal SI Units Used in Mechanics

| Quantity | Unit | Symbol | Formula |
|---|---|---|---|
| Acceleration | Meter per second squared | ... | $m/s^2$ |
| Angle | Radian | rad | † |
| Angular acceleration | Radian per second squared | ... | $rad/s^2$ |
| Angular velocity | Radian per second | ... | $rad/s$ |
| Area | Square meter | ... | $m^2$ |
| Density | Kilogram per cubic meter | ... | $kg/m^3$ |
| Energy | Joule | J | $N \cdot m$ |
| Force | Newton | N | $kg \cdot m/s^2$ |
| Frequency | Hertz | Hz | $s^{-1}$ |
| Impulse | Newton-second | ... | $kg \cdot m/s$ |
| Length | Meter | m | ‡ |
| Mass | Kilogram | kg | ‡ |
| Moment of a force | Newton-meter | ... | $N \cdot m$ |
| Power | Watt | W | $J/s$ |
| Pressure | Pascal | Pa | $N/m^2$ |
| Time | Second | s | ‡ |
| Velocity | Meter per second | ... | $m/s$ |
| Volume, solids | Cubic meter | ... | $m^3$ |
| Liquids | Liter | L | $10^{-3}\ m^3$ |
| Work | Joule | J | $N \cdot m$ |

† Supplementary unit (1 revolution $= 2\pi$ rad $= 360°$).
‡ Base unit.

## Centroids of Common Shapes of Areas and Lines

| Shape | | $\bar{x}$ | $\bar{v}$ | Area |
|-------|---|-----------|-----------|------|
| Triangular area | | | $\dfrac{h}{3}$ | $\dfrac{bh}{2}$ |
| Quarter-circular area | | $\dfrac{4r}{3\pi}$ | $\dfrac{4r}{3\pi}$ | $\dfrac{\pi r^2}{4}$ |
| Semicircular area | | $0$ | $\dfrac{4r}{3\pi}$ | $\dfrac{\pi r^2}{2}$ |
| Semiparabolic area | | $\dfrac{3a}{8}$ | $\dfrac{3h}{5}$ | $\dfrac{2ah}{3}$ |
| Parabolic area | | $0$ | $\dfrac{3h}{5}$ | $\dfrac{4ah}{3}$ |
| Parabolic spandrel | | $\dfrac{3a}{4}$ | $\dfrac{3h}{10}$ | $\dfrac{ah}{3}$ |
| Circular sector | | $\dfrac{2r\sin\alpha}{3\alpha}$ | $0$ | $\alpha r^2$ |
| Quarter-circular arc | | $\dfrac{2r}{\pi}$ | $\dfrac{2r}{\pi}$ | $\dfrac{\pi r}{2}$ |
| Semicircular arc | | $0$ | $\dfrac{2r}{\pi}$ | $\pi r$ |
| Arc of circle | | $\dfrac{r\sin\alpha}{\alpha}$ | $0$ | $2\alpha r$ |